BRASSEY'S MULTILINGUAL MILITARY DICTIONARY

BRASSEY'S: DICTIONNAIRE MULTILINGUE MILITAIRE

BRASSEY'S: DICCIONARIO POLÍGLOTO MILITAR

BRASSEY'S: MILITÄR-WÖRTERBUCH IN SECHS SPRACHEN

БРАССИ: МНОГОЯЗЫЧНЫЙ ВОЕННЫЙ СЛОВАРЬ

قاموس براسيس العسكري المتعدد اللغات

Also available from Brassey's

Brassey's Air Power: Aircraft, Weapons Systems and Technology series
Mason – Air Power: An Overview of Roles, Vol. 1
Walker – Air-to-Ground Operations, Vol. 2

Brassey's Sea Power: Naval Vessels, Weapons Systems and Technology series
Till – Modern Sea Power, Vol. 1
Gates & Lynn – Ships, Submarines and the Sea, Vol. 2
Gates – Surface Ships, Vol. 3
Ladd – Amphibious Warfare, Vol. 4

Brasseys Battlefield Weapons Systems and Technology series, 12 volume set

Alsudiary – Five War Zones: The Views of Local Military Leaders
Hilmes – Main Battle Tanks: Developments in Design since 1945
Laffin – War Annual
Perkins – Weapons and Warfare: Conventional Weapons and their Roles in Battle
RUSI/Brassey's Defence Yearbook
Simpkin – Race to the Swift: Thoughts on 21st Century Warfare
Simpkin – Deep Battle: The Brainchild of Marshal Tukhachevskii
Sloan – Mine Warfare on Land
West – Naval Forces and Western Security

International Journals
Defence Analysis
Defence Attaché

For full details of Brassey's titles please write to your local
Pergamon/Brassey's office.

BRASSEY'S MULTILINGUAL MILITARY DICTIONARY

BRASSEY'S: DICTIONNAIRE MULTILINGUE MILITAIRE

BRASSEY'S: DICCIONARIO POLÍGLOTO MILITAR

BRASSEY'S: MILITÄR-WÖRTERBUCH IN SECHS SPRACHEN

БРАССИ: МНОГОЯЗЫЧНЫЙ ВОЕННЫЙ СЛОВАРЬ

قاموس براسيس العسكري المتعدد اللغات

BRASSEY'S DEFENCE PUBLISHERS
(a member of the Pergamon Group)

LONDON · OXFORD · WASHINGTON · NEW YORK
BEIJING · FRANKFURT · SÃO PAULO · SYDNEY · TOKYO · TORONTO

U.K. (Editorial)	Brassey's Defence Publishers, 24 Gray's Inn Road, London WC1X 8HR
(Orders)	Brassey's Defence Publishers, Headington Hill Hall, Oxford OX3 0BW, England
U.S.A. (Editorial)	Pergamon-Brassey's International Defense Publishers, 1340 Old Chain Bridge Road, McLean, Virginia 22101, U.S.A.
(Orders)	Pergamon Press, Maxwell House, Fairview Park, Elmsford, New York 10523, U.S.A.
PEOPLE'S REPUBLIC OF CHINA	Pergamon Press, Room 4037, Qianmen Hotel, Beijing, People's Republic of China
FEDERAL REPUBLIC OF GERMANY	Pergamon Press, Hammerweg 6, D-6242 Kronberg, Federal Republic of Germany
BRAZIL	Pergamon Editora, Rua Eça de Queiros, 346, CEP 04011, Paraiso, São Paulo, Brazil
AUSTRALIA	Pergamon-Brassey's Defence Publishers, P.O. Box 544, Potts Point, N.S.W. 2011, Australia
JAPAN	Pergamon Press, 8th Floor, Matsuoka Central Building, 1–7–1 Nishishinjuku, Shinjuku-ku, Tokyo 160, Japan
CANADA	Pergamon Press Canada, Suite No. 271, 253 College Street, Toronto, Ontario, Canada M5T 1R5

Copyright © 1987 Brassey's Defence Publishers Ltd.

First edition 1987

Library of Congress Cataloging-in-Publication Data
Brassey's multilingual military dictionary=Brassey's,
dictionnaire multilingue militaire=Brassey's,
diccionario polígloto militar=Brassey's,
Militär-Wörterbuch in sechs Sprachen=[Brassi,
mnogoιazychnyï voennyï slovarʹ].
Includes index.
1. Military art and science—Dictionaries—
Polyglot.   2. Dictionaries, Polyglot.   I. Brassey's
Defence Publishers.   II. Title: Brassey's,
dictionnaire multilingue militaire.   III. Title:
Brassey's, diccionario polígloto militar.   IV. Title:
Brassey's, Militär-Wörterbuch in sechs Sprachen.
V. Title: Brassi, mnogoιazychnyï voennyï slovarʹ.
U25.B66 1987   355'.003   87–6965

British Library Cataloguing in Publication Data
Brassey's multilingual military dictionary.
1. Military art and science—Dictionaries
—Polyglot   2. Dictionaries, Polyglot
355'.003   U25
ISBN 0-08-027032-8

Printed in Great Britain by A. Wheaton & Co. Ltd., Exeter

TABLE OF CONTENTS

TABLE OF CONTENTS

TABLE DES MATIERES

INDICE DE MATERIAS

INHALTSVER-ZEICHNIS

ОГЛАВЛЕНИЕ (с)

قائمة المحتويات

ACKNOWLEDGEMENTS

Henry Stanhope, Diplomatic Correspondent at *The Times,* made the initial selection of terms. These were edited and translated into French, German and Spanish by Brigadier (Ret'd) Richard Simpkin with the assistance of a team of experts: Major Tony Valdes Scott, 14/20 H, Lieutenant-Colonel Jean Yves Binet, Infanterie de Marine, Oberstleutnant i.G. Elmar Dinter (Artillerie), Colonel Francisco M. Pariente, Infanteria y Estado Mayor, Dr Holger and Dr Dorothea Klein, and Daniel and Danielle Limon.

It was next decided to enlarge the scope of the Dictionary and increase the number of languages. Brigadier Simpkin was by then unable to undertake the added workload. The project was handed over to Lieutenant Colonel John MacFarlane, Royal Army Educational Corps, who added the Russian and Arabic and expanded the selection of terms in all languages in consultation with a team of his military linguist colleagues. The terms in this dictionary are held on computer. This means that any of the languages can appear as the key language to suit the needs of the user.

REMERCIEMENTS

Henry Stanhope, correspondant diplomatique au *Times,* a fait la première sélection de termes. Les termes choisis ont été, par la suite, revus puis traduits en français, en allemand et en espagnol par le Général de Brigade Richard Simpkin avec le concours d'une équipe d'experts composée du Major Tony Valdes-Scott, 14/20 H, du Lieutenant-Colonel Jean Yves Binet, Infanterie de Marine, de l'Oberstleutnant i.G. Elmar Dinter (Artillerie), du Colonel Francisco M. Pariente, Infanteria y Estado Mayor, du Docteur Holger et du Docteur Dorothea Klein et de Daniel et Danielle Limon.

Il a par la suite été décidé d'étendre le contenu du dictionnaire et d'y ajouter un nombre supplémentaire de langues. Le Général de Brigade Simpkin n'était pas à ce moment-là en mesure d'entreprendre cette nouvelle tâche et, par conséquent, l'entreprise a été confiée au Lieutenant-Colonel John MacFarlane du Royal Army Educational Corps qui a ajouté les traductions russes et arabes et a étendu la sélection des termes en toutes les langues en consultation avec une équipe de collègues linguistes militaires. Les expressions dans ce dictionnaire sont informatisées. De ce fait, chacune des langues du dictionnaire peut être employée comme langue-clef pour satisfaire aux besoins des utilisateurs.

AGRADECIMIENTOS

Al principio, Henry Stanhope, coresponsal diplomático de *The Times* londinense, escogió los términos. El General de Brigada (retirado) Richard Simpkin los preparó y los tradujo al francés, alemán y español con la ayuda de un grupo de peritos; Major Tony Valdes Scott, 14/20 H, Lieutenant-Colonel Jean Yves Binet, Infanterie de Marine, Oberstleutnant i.G. Elmar Dinter (Artillerie), Coronel Francisco M Pariente, Infantería y Estado Mayor, Dr Holger y Dr Dorothea Klein, Daniel y Danielle Limon.

Luego se decidió extender el alcance del diccionario y, al mismo tiempo, aumentar el número de idiomas. El general Simpkin no podía entonces emprender el trabajo adicional. Se le entregó el proyecto a Lieutenant Colonel John MacFarlane, Royal Army Educational Corps, el que agregó la versión rusa y árabe y extendió la selección de términos en todos los idiomas con la ayuda de su equipo de colegas políglotas militares. Los términos incluídos en este diccionario se guardan en computador. La consecuencia de esto es que no importa cuál idioma puede servir para el idioma llave, según las necesidades del usuario.

ANERKENNUNGEN

Henry Stanhope, diplomatischer Korrespondent bei der Londoner *Times,* hat die Anfangsauswahl der Begriffe getroffen. Diese wurden von Brigadegeneral (a.D.) Richard Simpkin bearbeitet und mit Hilfe einer Gruppe von Experten, d.h. Major Tony Valdes Scott, 14/20 Ħ; Oberstleutnant Jean Yves Binet, Infanterie de Marine; Oberstleutnant (i.G.) Elmar Dinter, Artillerie; Oberst Francisco M. Pariente, Infantería y Estado Mayor; sowie Dr. Holger und Dr. Dorothea Klein und Daniel und Danielle Limon ins Französische, Deutsche und Spanische übersetzt.

Man kam danach zu dem Entschluss den Anwendungsbereich des Wörterbuchs sowie die Anzahl der Sprachen zu vergrössern. Brigadegeneral Simpkin konnte zu diesem Zeitpunkt die damit verbundene Arbeitssteigerung nicht auf sich nehmen. Die Arbeit wurde Oberstleutnant John MacFarlane, Royal Army Educational Corps, übergeben, der mit Hilfe seines Lehrpersonals Russisch und Arabisch zufügte, sowie die Anzahl der Begriffe in allen Sprachen vergrösserte. Die Begriffe des Wörterbuchs sind im Computer gespeichert, d.h. jede Sprache kann als Schlüsselsprache dienen, dem Bedürfnis des Gebrauchers entsprechend.

Признательность

Завершил первоначальный выбор терминов дипломатический корреспондент лондонской газеты "Таймс" Генри Станхуп, подготовил к печати и перевел термины на французский, немецкий и испанский языки бригадный генерал в отставке Ричард Симпкин при содействии некоторых своих коллег-экспертов, именно: майора 14/20 гусаров Энтони Валдьеса Скотта, подполковника морской пехоты Жана Ива Бинэ, подполковника артиллерии в отставке Эльмара Дентера, полковника пехоты и генерального штаба Франциско М. Парьенте, доктора Хольгера и доктора Доротеи Клейн, Дениела и Дениел Лимон.

Было принято решение расширить охват словаря, увеличить количество языков. К тому времени бригадный генерал Симпкин оказался не в состоянии принять на себя дополнительное бремя работ, из-за чего проект был передан подполковнику британской образовательной службы Джону МакФерлану. Тот, при совещании со своими сотрудниками-лингвистами, добавил к этому словарю русский и арабский языки и расширил выбор терминов на всех представленных здесь языках.

Термины этого словаря хранятся на базе данных компьютера; это значит, что любой язык словаря может употребляться как ключевой язык, удовлетворяющий потребности всякого пользующегося им читателя.

الشكر

عمل « هنري استنحوب » وهو المراسل الديبلوماسي لصحيفة « تيمز » اختيار العبارات الإبتدائي . ان العميد « رتشارد سمكن » حرر هذه العبارات وترجمها إلى فرنسي والماني واسباني ساعده فريق من الخبراء : الرائد « توني فلدس سكوت » 14/20H المقدم « جين ايف بني » المشاة البحرية الفرنسية المقدم « الماردنتر » المدفية الالمانية . العقيد « فرنسكو بريني » المشاة الاسبانية . الدكتور « حولغر » والدكتور « دروتيا كلين » و « دنيل » و « دنيلا لمون » .

ثم قرر توسيع مدى القاموس بزيادة في عدد اللغات . وفي هذا الوقت لم يقدر العميد « سمكن » ان يقوم بالعمل الاضافي فسلم المشروع للمقدم « جون مكفارلن » ـ سلاح التعليم الملكي ـ الذين اضاف الروسية والعربية ووسع مجموعة العبارات المختارة بكل اللغات بالتشاور مع فريق يتألف من زملائه اللغويين والعسكرين حفظت العبارات في هذا القاموس في آلة حاسبة فمن الممكن أن تظهر أي لغة من اللغات كالمفتاح كما يناسب حاجات المستعمل .

INTRODUCTION

Modern technology has vastly increased our ability, both as nations and as individuals, to communicate information to one another and to come into closer personal and economic contact. This is particularly the case in the military sphere. Strategic and tactical co-operation among allies, out-of-area interventions, counter-terrorism operations, contacts between forces at points of tension, defense sales, training, international relief: all these demand a greater degree of communication and understanding.

The aim of this dictionary is to assist in this process. It is designed to meet the general needs of the Service user and to provide the basic military vocabulary for a wide variety of situations in which direct word-for-word translation is required in an army, navy or airforce context.

INTRODUCTION

La technologie moderne a augmenté en grande mesure notre capacité, en tant que nations et qu'individus, pour la communication d'informations entre nous et pour la formation de contacts de plus en plus directs sur le plan personnel et économique. Cette observation est particulièrement juste dans le domaine militaire. La coopération stratégique et tactique interalliée, les interventions rapides hors-limites, les opérations anti-terroristes, les contacts entre des forces militaires aux points sensibles, les ventes militaires, l'entrainement, les opérations de secours internationales: tout cela exige un niveau plus élevé de communication et de compréhension.

Ce dictionnaire vise à aider dans ce processus. Il est conçu pour répondre aux besoins de l'usage militaire et pour fournir un vocabulaire militaire de base dans un grand nombre de situations où une traduction textuelle est nécessaire que ce soit dans le contexte de l'Armée de Terre, de l'Armée de l'Air ou de la Marine.

INTRODUCCIÓN

La tecnología actual ha aumentado muchísimo nuestra capacidad, como países tanto como individuos, para comunicarnos y entrar en más estrecho contacto personal y económico. Esto se refiere sobre todo a la esfera militar. La cooperación estratégica y táctica entre fuerzas aliadas, las intervenciones militares lejnas, las operaciones contra-terroristas, los contactos entre fuerzas en sitios de tensión, las ventas de equipo de defensa, los ejercicios militares, el socorro internacional – todo esto necesita más comunicación y más comprensión mutua.

Este diccionario tiene por objeto ayudar este proceso. Está diseñado para satisfacer los requisitos generales del consultador militar y, con este fin, proporciona un vocabulario básico militar que se puede emplear en una gran variedad de situaciones en las que se precisa una traducción literal en un contexto pertinente al ejército, a la marina o a la fuerza aérea.

VORBEMERKUNG

Die moderne Technologie hat unsere Fähigkeit, Informationen zu übertragen und engere menschliche und ökonomische Verbindungen zu knüpfen, in grossem Masse gefördert, sowohl als Nation als auch als Einzelner. Dieses gilt besonders in dem militärischen Bereich. Die strategische und taktische Kooperation zwischen den Alliierten, z.B. gemeinsame Vermittlungen oder Eingriffe; Zusammenarbeit gegen den Terrorismus; Kontakte zwischen den Streitkräften an Spannungsorten; Waffenhandel; Ausbildung und internationale Hilfsaktionen: alle verlangen ein hohes Mass an Verständigung und klare Verbindungswege.

Dieses Wörterbuch versucht, diese Aufgabe zu erleichtern. Es soll dazu dienen, den allgemeinen Bedarf des Soldaten zu decken, und bietet ihm ein allgemeines Militärvokabular, das in verschiedenen Bereichen und Situationen verwendet werden kann, sei es im Heer, in der Marine oder Luftwaffe, in denen eine direkte, wörtliche Übersetzung verlangt wird.

ВВЕДЕНИЕ

Современная технология в значительной степени повысила наши прирожденные национальные и индивидуальные способности обмениваться информацией и устанавливать друг с другом более тесные личные и экономические контакты. Это особенно верно в военной области. Стратегическое и тактическое взаимодействие союзников, интервенция в отдаленных местах, антитеррористические операции, взаимосвязь между военными силами в районах высокой напряженности, продажа военной техники, подготовка, оказание международной помощи – все это требует повышенной степени сообщения и понимания.

Способствовать этому процессу – цель настоящего словаря, который предназначен для того, чтобы удовлетворять потребности военнослужащих и снабдить их лексическим минимумом военных терминов в разнообразных применениях, нуждающихся в непосредственном дословном переводе, как в военной, так и военно-морской или военно-воздушной областях.

المقدمة

لقد ساعدت التكنولوجيا الحديثة إلى أبعد الحدود على زيادة قدراتنا كشعوب وأفراد لتبادل المعلومات ولأن نقترب أكثر من بعضنا البعض في مجال الإتصالات الشخصية والإقتصادية .

وهذه هي القضية عملياً في مجال القاموس العسكري . فالتعاون الإستراتيجي والتكتيكي بين الحلفاء ، والتدخل خارج مناطق النفوذ ، والعمليات المضادة للإرهاب ، والإتصال بين القوات عند مناطق التوتر ، ومبيعات الأسلحة ، والتدريب ، والمساعدات الدولية ، كل هذه تتطلب درجة عالية من الإتصالات والفهم .

والغرض من هذا القاموس هو المساعدة في هذه المجالات ، ولقد صُمم لخدمة الإحتياجات العامة لرجال القوات المسلحة الذين يتعاملون في هذه التخصصات والذين بحاجة إلى الكلمات العسكرية الأساسية لخدمة مواقف عديدة ومتنوعة والتي يحتاج فيها إلى ترجمة مباشرة للكلمات المختلفة سواءً في الجيش أو البحرية أو الطيران .

HOW TO USE THE DICTIONARY

American-English* is the key language in this Edition.

To translate from French, German, Spanish, Russian and Arabic, go to the appropriate language index (see contents) and use the adjacent number to find the word or phrase in the main part of the dictionary. Phrases are arranged alphabetically by the first word in the phrase (excluding prepositions).

Subject fields are shown where necessary in the serial number column if they cover all languages, otherwise in the language column(s) concerned. Genders of all nouns are given, otherwise grammatical abbreviations are used only where needed for clarity.

You will find the key to subject field and grammatical abbreviations directly after this section. Useful key word groups are given in the appendices.

* British-English speakers may need occasionally to consult the British-English index.

COMMENT EMPLOYER CE DICTIONNAIRE

L'anglais-américain est la langue-clef de cette édition.

Pour traduire à partir du français, de l'allemand, de l'espagnol, du russe ou de l'arabe, cherchez dans l'index de la langue en question et employez le numéro à côté pour trouver le mot ou l'expression dans la partie principale du dictionnaire. Les expressions sont arrangées par ordre alphabétique selon le premier mot de l'expression (sans compter les prépositions).

Les sujets concernés sont indiqués, au besoin, dans la colonne de numérotage si les sujets sont valides pour toutes les langues. Sinon, les sujets sont indiqués dans la colonne vocabulaire. A part les genres des substantifs, qui sont tous donnés, les abréviations sont employées seulement dans les cas où le sens doit être éclairci.

La clef pour les sujets et les indicatifs grammaticaux vient à la suite de cette section. Certains groupes d'expressions utiles se trouvent dans les appendices du dictionnaire.

MODO DE USAR ESTE DICCIONARIO

El inglés-americano es el idioma clave de esta edición.

Para traducir del inglés, francés, alemán, ruso y árabe, consulte el índice correspondiente al idioma (véase la tabla de materias) y emplee el número contiguo para encontrar la palabra o frase en la sección principal del diccionario. Las frases se ordenan alfabéticamente, por la palabra inicial de la frase (excepto preposiciones).

Las esferas figuran cuando sea necesario en la columna de números de serie solamente cuando se refieran a todos los idiomas; si no, figuran en la(s) columna(s) correspondiente(s). Aparece el género de todos los sustantivos; en otros casos, se emplean abreviaturas gramáticas solamente cuando haga falta para más claridad.

La clave de las abreviaturas de esferas y gramáticas se halla inmediatamente detrás de esta sección. Los grupos de palabras claves útiles aparecen en los apéndices.

HINWEISE ZUM GEBRAUCH DES WÖRTERBUCHES

Die amerikanisch-englische Sprache ist die Schlüsselsprache in dieser Ausgabe.

Um aus der französischen, deutschen, spanischen, russischen oder arabischen Sprache zu übersetzen, ist es notwendig, in dem zugehörigen Wörterverzeichnis nachzuschlagen, und die danebenstehende Nummer dazu zu verwenden, das Wort oder den Ausdruck in dem Hauptteil des Wörterbuches zu finden. Die Ausdrücke sind alphabetisch nach dem ersten Wort eingeteilt worden (ausgeschlossen die Präpositionen).

Bestimmte Fachgebiete sind in der Spalte mit der laufenden Reihennummer angezeigt worden, wenn diese alle Sprachen einbeziehen, sonst sind sie in der Sprachenspalte angezeigt. Das Geschlecht der Hauptwörter ist angezeigt, sonst werden Abkürzungen nur verwendet, wenn sie der Klarheit dienen.

Der Schlüssel zu den Fachgebieten und grammatischen Abkürzungen ist unmittelbar nach diesem Absatz. Nützliche Schlüsselworte werden in dem Anhang angezeigt.

О ПОЛЬЗОВАНИИ СЛОВАРЕМ

Основным языком этого словаря является т.н. "американско-английский".

Чтобы перевести с французского, немецкого, испанского, русского, арабского языков, надо обратиться к соответствующему индексу языков (см. содержание), найти нужное слово (выражение) в главной части словаря. Выражения расположены в алфавитном порядке по первой букве выражения (за исключением предлогов). Когда касается всех языков, предметное поле показывается в колонке порядковых номеров; кроме этого, в колонке соответствующего языка. Дается род каждого имени существительного, а кроме того употребляется грамматическая аббревиатура, только в интересах ясности. Индекс предметных полей и грамматической аббревиатуры находится прямо после этого участка. Полезные группировки ключевых слов даются в приложениях.

كيفية إستخدام القاموس

تم إستخدام اللغة الأمريكية ـ الإنجليزية كمفتاح لهذه الطبعة .

وللترجمة من الفرنسية أو الالمانية أو الأسبانية أو الروسية أو العربية فيجب عليك إستخدام الفهرس الخاص بهذه اللغة (أنظر المحتويات) وأيضاً إستخدام الرقم المقارن لهذه الكلمة أو المصطلح وذلك للعثور على الكلمة في القسم الرئيسي للقاموس ، وقد تم ترتيب هذه المصطلحات أبجدياً تبعاً لأول كلمة في المصطلح (بغض النظر عن حروف الجر) .

أما فيما يتعلق بالمجالات الخاصة بالمواضيع فقد تم توضيحها حسب الحاجة إذا غطت كل اللغات وإلا فانه قد تم وضعها في عامود اللغة الخاص بها ، وكذلك تم أعطاء الجنس (مذكر ـ مؤنث) الخاص بكل إسم وأيضاً تم إستخدام إختصارات نحوية في حالة الحاجة إلى توضيح ما .

بإمكانك العثور على المفتاح الخاص بالمجالات الخاصة بالمواضيع وأيضاً بالإختصارات النحوية بعد هذا القسم مباشرة ، وهناك أيضاً مفتاح مفيد لمجموعات الكلمات في الجزء الخاص بالفهرس .

ABBREVIATIONS ABREVIATIONS ABREVIATURAS

1.	Subject indicators	Sujet concerné	Indicativos de esfera
N.B.	*Bold capitals in serial numbers column, sometimes in language column.*	*Majuscules en vedette dans la colonne de numérotage, parfois dans la colonne vocabulaire.*	*Mayúsculas negritas dentro columna de números de serie, a veces dentro columna principal.*
A	aviation	aviation	aviación
B	construction, civil/field engineering	bâtiment, génie civil/militaire	construcción, ingenería civil/militar
C	chemical (warfare)	chimique (guerre c.)	química (guerra qu.)
D	data processing	traitement de l'information	proceso de datos
E	equipment, engineering	matériel, technologie	material, technología
F	financial	financier	financiero
G	guided missiles	engins téléguidés	proyectiles/misiles teledirigidos
H	human organism, medicine	corps humain, médecine	cuerpo humano, medicina
I	intelligence	renseignements	información
L	land forces	armée de terre	fuerzas terrestres
M	maritime	maritime	marino
N	nuclear	nucléaire	nuclear
P	police, politics, security	police, politique, sécurité	polícia, política, seguridad
R	radar, telecommunications	radar, télécommunications	radar, telecomunicaciones
S	gunnery, weapons	tir, armes	tiro, armas
T	topography	topographie	topografía
U	universal, general	universel, général	universal, general
V	vehicles	véhicules	vehículos

ABKÜRZUNGEN

СОКРАЩЕНИЯ (с.мн.)

المختصرات

Bereichzeichen	**Предметные указатели**	مؤشرات الموضوع	**1.**
Das Bereichzeichen wird durch fette Grossbuchstaben in der Nummernspalte oder manchmal in der entsprechenden Hauptspalte angegeben.	*Показываются отчётливыми прописными буквами то в колонке порядковых номеров, то в колонке языка.*	تُعطىٰ أحيانا في عمود الأرقام المتسلسل ، وأحيانا في عمود اللغة .	*N.B.*
Flugwesen	авиация	مِلاحَة جَوّية / طَيَران	**A**
Hoch- und Tiefbau Ingenieurwesen, Pionierwesen	строительство, инженерно-строительное искусство, инженерно-саперное строительство	إنشاء ـ هندسة مدنية / حربية	**B**
Chemie (Kampfstoffe)	химическая война	حرب كيماوية	**C**
Datenverarbeitung	обработка данных	مُعالجة المُعطيات	**D**
Gerät, Technik	техника, строительство	مُهمات ـ هندسة	**E**
Wirtschaft/Finanzen	финансовый, деловой	مالي	**F**
Fernsteuerraketen	управляемые ракеты	صاروخ موجه	**G**
menschlicher Körper, Medizin	человеческий организм, медицина	أعضاء بشرية ـ دواء	**H**
Nachrichtenwesen	разведка	مُخابرات	**I**
Landstreitkräfte	сухопутные войска	قوات أرضية	**L**
Schiffahrt	морской, военно-морской	بحري	**M**
Kernwaffen	атомный, ядерный	نووي	**N**
Polizei, Politik, Sicherheit, Geheimhaltung	полицейский, политический, безопасность	شُرطة ـ سياسة ـ أمن	**P**
Radar, Fernmeldetechnik	радиолокационный, радиотехнический	رادار ـ وسائل الإتصال (سلكي ـ لاسلكي)	**R**
Waffen- und Schiesskunde	артиллерийский, оружие	مدفعي ـ أسلحة	**S**
Topographie	топография	طُبوغرافيا	**T**
Allgemeines	универсальный, общий	عالمي ـ عام	**U**
Kraftfahrzeugwesen	автомашины, машины	عربات	**V**

2. Grammatical indicators – indicatifs grammaticaux – indicativos gramaticales – grammatische Zeichen – грамматические показатели – مؤشرات نحوية

NB. Lower case italics in language column.
Minuscules italiques dans la colonne vocabulaire.
Minúsculas en bastardilla dentro columna principal.
Kursive Kleinbuchstaben in der Hauptspalte geben die grammatikalischen Begriffe.
указываются курсивом строчными буквами в столбце языка

الطباعة المائلة في عمود اللغة

adj	*adjective*	*adjectif*	*adjectivo*
adv	*adverb*	*adverbe*	*adverbio*
c	*compound*	*composé*	*compuesto*
f	*feminine*	*féminin*	*femenino*
ins	*inseparable*	*inséparable*	*inseparable*
inv	*invariable*	*invariable*	*invariable*
lat am	*Latin American*	*latino-américain*	*latinoamericano*
m	*masculine*	*masculin*	*masculino*
n	*noun*	*nom*	*sustantivo (s)*
neut	*neuter*	*neutre*	*neutro*
pl	*plural*	*pluriel*	*plural*
sep	*separable*	*séparable*	*separable*
sing	*singular*	*singulier*	*singular*
v	*verb*	*verbe*	*verbo*

Adjektiv	имя прилагательное (прилаг.)	صفة ـ حال	*adj*
Umstandswort	наречие (нареч.)		*adv*
Kompositum (komp.)	составное наречие (с) (сост. нар.)	كلمة مُركبة	*c*
Femininum	женский род (ж.)	مُؤنث	*f*
untrennbar (untr.)	неотделимый (неотдел.)	كلمة ثابتة	*ins*
invariabel	неизменяемый, неизменяющийся (неизмен.)		*inv*
latein-amerikanisch	латиноамериканский (лат. амер.)	أمـريكا اللاتينية	*lat am*
Maskulinum	мужской род (м.)	مُذَكَّر	*m*
Substantiv (s)	имя существительное (сущ.)	مُتعادل (إسم غير مذكر أو مؤنث)	*n*
Neutrum	средний род (ср.)	الإسـم	*neut*
Plural	множественное число (мн. ч.)	جَمع	*pl*
trennbar (tr.)	отделимый (отдел.)	قابل للفصل	*sep*
singular	единственное число (един. ч.)	ملازم لِ	*sing*
Verb	глагол (гл.)	فِعل	*v*

A

1
abaft
F (à l')arrière
E (a) popa
D achtern
R (на) корме
A خَلْف / في مُؤْخِرَةِ السَّفينة

2
abandon
F abandonner
E abandonar
D verlassen
R оставлять
A تَرَك ـ هَجَر

3
abandon ship stations
F postes d'abandon du navire *m. pl.*
E puestos de abandano (del buque) *m. pl.*
D Bergestationen *f. pl.*
R места по аварийному расписанию *ср. мн. ч.*
A مَوَاقِع تَرْك السَّفينَة

4
abeam
F (par le) travers
E (por el) través
D querab
R (на) траверзе
A مُقَابِلاً لمنتصف جانب السفينة

5
aboard
F (à) bord
E (a) bordo
D (an) Bord
R (на) борту
A على مَتْن (سَفينَة)

6 [A,G]
abort
F avorter
E abortar
D fehlschlagen [A] *tr.;* verlassen [G] *untr.*
R снять с задачи [A]; аварийное прекращение [G]
A أَحْبَط

7
absence without leave
F absence illégale *f.*
E ausencia sin permiso *f.*
D unerlaubte Abwesenheit *f.*
R самовольная отлучка *ж.*
A غِيَاب بغيرِ إِذْن

8
absentee
F manquant(e) *m.*
E ausente *m. f.*
D Abwesende(r) *m.*
R отсутствующий *м.*
A غَائِب *m*

9
academy
F académie militaire *f.*
E escuela (superior) *f.*
D Akademie *f.*
R академия *ж.*
A مَعْهَد *m*

10
accelerate
F accélérer
E acelerar(se)
D beschleunigen
R ускорять(ся)
A عَجَّل

11
accident
F accident *m.*
E accidente *m.*
D Unfall *m.*
R авария *ж.;* несчастный случай *м.*
A حَادِث *m*

12
accidental
F accidentel
E accidental
D Unfall- *komp.;* zufällig *adj.*
R случайный
A عَرْضِي

13
accommodate
F loger
E alojar
D unterbringen *tr.*
R размещать
A أَسْكَن

14
accommodation
F logement *m.*
E alojamiento *m.*
D Unterkunft *f.*
R размещение *ср.;* помещение *ср.*
A سَكَن *m*

15
accompany
F accompagner
E acompañar
D begleiten *untr.*
R сопровождать
A رَافَق

16 [F]
account
F compte *m.*
E cuenta *f.*
D Konto *neut.*
R счёт *м.*
A حِسَاب *m*

17

accountable
F responsable
E responsable
D verantwortlich
R подотчётный
A مُحَاسَب

18

account for v.
F justifier
E (dar) cuenta de
D Rechenschaft ablegen
   über tr.
R отчитываться
A حاسب

19

accounting unit
F unité autonome f.
E unidad autónoma f.
D Wirtschaftstruppenteil
   m.
R счётная часть ж.
A وَحْدَة مُحَاسَبَة f

20

accounts (section etc.)
F comptabilité f.
E teneduría de libros f.
D Buchhaltung f.
R отделение учёта и
   отчётности ср.
A قِسْم المُحَاسَبَة m

21 [U]
accuracy
F précision f.
E precisión f.
D Genauigkeit f.
R точность ж.
A دِقّة f

22 [G,S]
accuracy
F précision f.
E precisión f.
D Treffgenauigkeit f.
R точность стрельбы ж.
A دِقّة f

23

accurate
F précis
E preciso; certero
D genau
R точный; кучный
A دَقِيق

24

acknowledge
F donner l'aperçu
E acusar
D Empfang bestätigen m.
R подтверждать
   получение
A أَشْعَر بِتَسْلِيم

25

acknowledgement
F aperçu m.
E acuse m.
D Empfangsbestätigung f.
R подтверждение ср.
A إِشْعَار بِتَسَلّم

26

acoustic
F acoustique
E acústico
D Schall- komp.;
   akustisch adj.
R акустический;
   звуковой
A صَوْتِي / سَمْعِي

27

acoustics
F acoustique f.
E acústica f.
D Akustik f.
R акустика ж.
A عِلْم الصَوْت

28

action
F combat m.
E batalla f.
D Gefecht neut.
R действие ср.; бой м.
A قِتَال m

29

(out of) action
F (hors de) combat
E (fuera de) acción
D (ausser) Betrieb
R (из) строя
A مُعَطَّل

30

action stations
F postes de combat m. pl.
E puestos de combate m.
   pl.
D Gefechtsstationen f. pl.
R боевой пост м.
A مَوَاقِع القِتَال

31 [A,C,E,L,M]
activate
F activer
E activar
D aktivieren
R активировать
   [ACLM]; включать
   [E]
A نَشَّط

32 [A,C,E,L,M]
activation
F activation f.
E activación f.
D Aktivierung f.
R включение ср.
A تنشيط

33

activator
F déclencheur m.
E activador m.
D Aktivierungsmittel neut.
R активатор м.
A مُنَشِّط

34 [R]
active
F actif
E activo
D aktiv
R активный
A فَعَّال

35
actual
F effectif
E efectivo
D Ist- *komp.;* wirklich *adj.*
R действительный
A فِعْلِي

36 [E,U,B,S]
actuate
F actionner [U];
   déclencher [EBS]
E accionar
D betätigen [U]; scharf
   machen [EBS]
R приводить в действие
   [S]; приводить в
   движение [B];
   побуждать [U];
   возбуждать [E]
A شَغَّل

37 [A]
actuator
F déclencheur *m.*
E impulsor (del
   disparador) *m.*
D Betätigungsorgan *neut.*
R (силовой) привод *м.*
A مُشَغِّل *m*

38
acute
F aigu
E agudo
D spitz
R острый; резкий
A حَادّ/ ماهِر/ ذَكِي

39
acute angle
F angle aigu *m.*
E ángulo agudo *m.*
D spitzer Winkel *m.*
R острый угол *м.*
A زَاوِيَة حَادَّة *f*

40
adjacent unit
F unité voisine *f.*
E unidad contigua
D (benachbarte) Einheit *f.*
R (соседное)
   подразделение *ср.*
A وحده مجاورة *f*

41 [U,S]
adjust
F régler
E ajustar
D anpassen *tr.*
R регулировать
A ضَبَط

42 [E]
adjust
F régler
E ajustar
D einstellen *tr.;* regeln
R корректировать
A ضَبَط

43
adjutant
F officier-adjoint *m.*
E ayudante *m.*
D Adjutant *m.*
R адъютант *м.*
A مُسَاعِد *m*

44
administration
F administration *f.*
E administración *f.*
D Verwaltung *f.*
R управление *ср.;*
   администрация *ж.*
A إِدَارَة *f*

45
administrative
F administratif
E administrativo
D Verwaltungs- *komp.;*
   verwaltungsmässig
   adj.
R административный
A إِدَارِي

46
adrift
F (à la) dérive
E (a la) deriva; (al) garete
D treibend
R (по) течению; (по)
   воле волн;
   дрейфующий
A مَعَ ٱلتَّيَّار

47
advance *v.*
F avancer
E avanzar
D vorrücken (lassen) *v. tr.*
R наступать;
   продвигаться вперёд
A تَقَدَّم

48
advance *n.*
F avance *f.*
E avance *m.*
D Vormarsch *m.*
R наступление *ср.;*
   продвижение вперёд
   ср.; сближение с
   противником *ср.*
A تَقَدُّم *m*

49
advanced
F poussée
E avanzado
D fortgeschritten
R передовой
A مُتَقَدِّم

50
advance guard
F avant-garde *f.*
E vanguardia *f.*
D Vorhut *f.*
R авангард *м.*
A الْمُقَدَّمَة *f*

3

51

advice
F conseil *m.*
E consejo *m.*
D Rat(schlag) *m.*
R совет *м.*
A نَصِيحَة *f*

52

advise
F conseiller
E aconsejar
D beraten
R советовать
A نَصَح

53

adviser
F conseiller *m.*
E consejero *m.*
D Berater *m.*
R советник *м.*
A مُسْتَشَار *m*

54

advisory
F consultatif
E consultivo
D Beratungs- *komp.;*
   beratend *adj.*
R совещательный;
   консультативный
A إِسْتِشَارِي

55

aerial *adj.*
F aérien
E aéreo
D Luft- *komp.;* fliegerisch
   adj.
R воздушный
A هَوَائِي

56

aerodynamic
F aérodynamique
E aerodinámico
D aerodynamisch
R аэродинамический
A حَرَكِي جَوِّي

57

aerodynamics
F aérodynamique *f.*
E aerodinámica *f.*
D Aerodynamik *f.*
R аэродинамика *ж.*
A عِلْم الْحَرَكَة الْجَوَّيَة *m*

58

aeromedical
F aéromedical
E aeromédico
D Luftfahrtmedizins-
   komp.
R авиамедицинский
A طِبّي جَوِّي

59

aeromedicine
F médecine de l'air *f.*
E medicina de aviación *f.*
D Luftfahrtmedizin *f.*
R авиамедицина *ж.*
A الطِبّ الْجَوِّي *m*

60

aeronautical
F aéronautique
E aeronáutico
D Luftfahrt- *komp.;*
   aeronautisch *adj.*
R авиационный
A طَيْرَانِي

61

aeronautics
F aéronautique *f.*
E aeronáutica *f.*
D Flugwesen *neut.*
R аэронавтика *ж.*
A عِلْم الطَيْرَان *m*

62

aerospace *adj.*
F aérospatial
E espacial
D Raumfahrt- *komp.;*
   raumfahrttechnisch
   adj.
R воздушно-космический
A جَوِّي

63

affirm
F affirmer
E afirmar
D bestätigen
R подтверждать
A أَكَّد

64

affirmative
F affirmatif
E afirmativo
D Bejahungs- *komp.;*
   bejahend *adj.*
R утвердительный;
   понятно
A إِيجَابِي

65

afloat
F (à) flot
E (a) flote
D schwimmend
R наплаву; (на) воде
A طَافٍ / (طافي)

66 [M]

aft(er) *adj.*
F arrière
E (a) popa
D achter(n)
R (на) корме
A وَرَاء

67

afterburner
F dispositif de réchauffe
   m.
E cámara de
   poscombustión *f.*
D Nachbrenner *m.*
R форсажная камера *ж.*
A مُحْرَق لاَحِقٌ *m*

68

afterburning *n.*
F post-combustion *f.*
E poscombustión *f.*
D Nachbrennen *s. neut.*
R дожигание топлива *ср.*
A الإِحْتِرَاق اللاَحِق *m*

69
age group
F classe de recrutement *f.*
E clase *f.*
D Wehrpflichtsgruppe *f.*
R возрастная категория
 ж.; призывной
 возраст *м.*
A مجموعة عمر

70
agency
F organisme *m.*
E organismo *m.;* órgano
 m.
D Dienststelle *f.*
R агенство *ср.*; орган *м.*
A وَكَالَة *f*

71 [I]
agent
F agent *m.*
E agente *m.*
D Agent *m.*
R агент *м.*
A وَكِيل *m*

72 [B]
aggregate
F agrégat *m.*
E agregado *m.*
D Aggregat *neut.*
R агрегат *м.*
A مَجْمُوع *m*

73
agreement
F accord *m.*
E acuerdo *m.*
D Vereinbarung *f.*
R согласие *ср.*; договор
 м.
A إِتِّفَاقِيَة *f*

74
aground
F échoué
E encallado
D gestrandet
R (на) мели
A جَانِح

75 [M]
ahead
F (en) avant
E avante
D voraus
R впереди; вперёд
A في الأَمَام

76
ahoy!
F ohé!
E ¡ah del barco!
D ahoi!
R (на) корабле!
A يَا

77
aid *v.*
F aider
E ayudar
D helfen
R помогать
A ساعد

78
aid *n.*
F secours *m.*
E ayuda *f.*
D Hilfe *f.*
R помощь *ж.*
A مُسَاعَدَة *f*

79
aide
F aide-de-camp *m.*
E ayudante *m.;* edecán *m.*
D Adjutant *m.*
R адъютант *м.*
A مُرَافِق *m*

80
aileron
F aileron *m.*
E alerón *m.*
D Querruder *neut.*
R элерон *м.*
A جُنَيِّح *m*

81 [U]
aim *v.*
F viser
E aspirar
D richten; zielen
R стремиться
A سَدَّد صَوَّب

82 [S]
aim *v.*
F viser; pointer
E apuntar
D richten; zielen
R прицеливаться;
 наводить; целиться
A سَدَّد صَوَّب

83 [U]
aim *n.*
F but *m.*
E fin *m.;* objetivo *m.*
D Ziel *neut.*
R цель *ж.*
A هَدَف *m*

84
air *n. adj.*
F air *m.;* aérien *adj.*
E aire *m.;* aéreo *adj.*
D Luft *f.;* Luft- *komp.*
R воздух *м.*
A جَوّ

85
airbase
F terrain d'aviation *m.*
E base aérea *f.*
D Luftwaffenbasis *f.;*
 Luftstützpunkt *m.*
R авиабаза *ж.*
A قاعدة جوِّية *f*

86 [A]
airborne
F (en) vol
E (en) vuelo
D (in der) Luft
R (в) воздухе;
 воздушный
A مَحْمُول جوًّا

5

87 [E]
airborne
F (de) bord
E aerotransportado; (de, por) avión
D Bord- *komp.*
R воздушнодесантный
A مَحْمُول جَوًّا

88 [U]
airborne
F aéroporté
E aerotransportado
D Luftlande- *komp.*
R авиационный; аэромобильный
A مَحْمُول جَوًّا

89
airborne early warning and control
F station radar de bord d'alerte avancée et de contrôle *f.*
E alerta y control lejano por medios aerotransportados *f. m.*
D Frühwarnung und Überwachung aus der Luft *f. f.*
R система раннего воздушного оповещения и наведения *ж.*
A نِظَام محمولٍ جَوًّا لِلإِنْذَار المُبَكِّر والسَيْطَرَة

90
airburst *n.*
F explosion aérienne *f.*
E explosión en el aire *f.*
D Luftsprengpunkt *m.*
R воздушный взрыв *м.*
A إِنْفِجَار جَوِّي *m*

91
air corridor
F corridor aérien *m.*
E corredor aéreo *m.*
D Luftkorridor *m.*
R воздушный коридор *м.*
A مَمَر جَوِّي *m*

92
aircraft
F avion *m.*
E avión *m.*
D Flugzeug *neut.*
R самолёт(ы) *м. (мн. ч.)*
A طَائِرَة *f*

93
(communications) aircraft
F avion de liaison *m.*
E avión de enlace *m.*
D Verbindungsflugzeug *neut.*
R самолёт связи *м.*
A طائرة اتصالات

94
(ground attack) aircraft
F avion d'attaque au sol *m.*
E avión de ataque a tierra *m.*
D Erdkampfflugzeug *neut.*
R штурмовой самолёт *м.*; ударный самолёт *м.*
A طَائِره هجوم أرضي *f*

95
(lead) aircraft
F avion leader *m.*
E avión de cabeza *m.*
D Führungsflugzeug *neut.*
R ведущий самолёт *м.*
A طَائِرَة المُقَدَّمَة *f*

96
(pathfinder) aircraft
F avion marqueur *m.*
E avión de señalización *m.*
D Zielbeleuchter *m.*
R самолёт наведения *м.*
A طائرة كشافة

97
(rotary wing) aircraft
F appareil à voilure tournante *m.;* hélicoptère *m.*
E aparato de ala giratoria *m.;* helicóptero *m.*
D Hubschrauber *m.*
R вертолёт *м.*
A طَائِرَة ذات أَجْنِحَة دَوَّارَة

98
(stealth) aircraft
F avion furtif *m.*
E avión antidetectable *m.*
D „Stealth" -Flugzeug , (eng. Begriff) *neut.*
R скрытый самолёт *м.*
A طائرة مختفية

99
(strike) aircraft
F avion d'assaut *m.*
E avión de ataque *m.*
D Schlachtflugzeug *neut.*
R ударный самолёт *м.*
A طَائِرَة مُهَاجَمَة *f*

100
(swept-wing) aircraft
F avion à aile en flèche *m.*
E avión de alas en flecha *m.*
D Flugzeug mit pfeilförmigen Tragflächen
R самолёт с стреловидными крыльями *м.*
A طائرة ذات اجنحة سهمية

101
(transport) aircraft
F avion de transport *m.*
E avión de transporte *m.*
D Transportflugzeug *neut.*
R транспортный самолёт *м.*; авиатранспортёр *м.*
A طائرة نقل

102
(variable geometry) aircraft
F avion à géométrie variable *m.*
E avión de geometría variable (de ala variable) *m.*
D Schwenkflügelflugzeug *neut.*
R самолёт с изменяемой геометрией крыла
A طَائِره ذَات هَنْدَسَة مُتَغَيِّره

103
(vertical takeoff and landing) aircraft (VTOL)
F avion à décollage et atterrissage vertical *m.*
E avión de despegue y aterrizaje vertical *m.*
D Senkrechtstarter (meist eng. Abkürzung)
R самолёт с вертикальным взлётом и посадкой
A طَائِرة عَمُودِيَّة الإقْلَاع والهُبُوط

104 [A,M]
aircraft arresting system
F système d'arrêt *m.*
E dispositivo de detención *m.*
D Bremsvorrichtung (für Flugzeuge) *f.*
R (палубное) тормозное устройство *ср.*
A نِظَام تَوْقِيف الطَّائِرات *m*

105
aircraft carrier
F porte-avions *m. inv.*
E porta(a)viones *m. inv.*
D Flugzeugträger *m.*
R авианосец *м.*
A حَامِلة طَائِرَات *f*

106
aircrew
F équipage d'avion *m.*
E tripulación de avión *f.*
D Flugzeugbesatzung *f.*
R экипаж самолёта *м.*; лётный экипаж *м.*
A طَاقِم جوِّي *m*

107
air defense *adj.*
F antiaérien
E antiaéreo
D Flugabwehr- (Fla-) *komp.*
R противовоздушный
A دِفَاع جَوِّي

108
air defense *n.*
F défense aérienne *f.*
E defensa antiaérea *f.*
D Luftverteidigung *f.*
R противовоздушная оборона (ПВО) *ж.*
A الدِفَاع الجَوِّي *m*

109
airdrop
F largage *m.*
E lanzamiento *m.*
D Abwurf (aus der Luft) *m.*
R выброска с воздуха *ж.*
A هُبُوط / إنْزَال *m*

110
air evacuation
F évacuation par air *f.*
E evacuación aérea *f.*
D Evakuierung durch die Luft *f.*
R эвакуация по воздуху *ж.*
A إجَلاء جوِّي *m*

111
airfield
F terrain d'aviation *m.*
E campo de aviación *m.*
D Flugplatz *m.*
R аэродром *м. м.*
A مَطَار *m*

112
air force
F armée de l'air *f.*
E aviación militar *f.*
D Luftwaffe *f.*
R военновоздушные силы (ВВС) *ж. мн. ч.*
A سِلَاح الجَوّ *m*

113
airframe
F cellule d'avion *f.*
E armazón (del fuselaje) *f.*
D Flugzeugzelle *f.*
R корпус самолёта *м.*
A هَيْكَل الطَّائِرة *m*

114
air-ground
F air-sol
E aire-tierra *inv.*
D Bord-Boden
R воздух-земля
A جو ـ أرض

115
airhead
F tête de pont aérien *f.*
E cabeza de puente aéreo *f.*
D Luftlandebrückenkopf *m.*
R плацдарм десантирования *м.*
A رَأْس جِسْر جوِّي *m*

116

airlanded

F déposé (par
   avion/hélicoptère)
E transportado en avión
D (mit) Flugzeugen
   gelandet
R посадочно-десантный
A الهبوط على الأرض

117

airlift

F aérotransport *m.*
E puente aéreo *m.;*
   transporte aéreo *m.*
D Luftbrücke *f.*
R воздушные перевозки
   ж. мн. ч.
A جِسْر جوّي *m*

118

airline

F service de transports
   aériens *m.*
E línea aérea *f.*
D Fluglinie *f.*
R авиалиния *ж.*
A خَطّ جَوّي *m*

119 [M]

air-line

F tuyau d'alimentation
   d'oxygène *m.*
E tubo de aire *m.*
D Sauerstoffschlauch *m.*
R воздухопровод *м.*
A أنبوب هوائي *m*

120 [R]

airline

F fréquence prémarquée
   f.
E enlace radio *m.*
D Frequenzkennlinie *f.*
R канал радиосвязи *м.*
A مسار جوي *m*

121 [B,M]

air lock

F sas pneumatique *m.*
E esclusa de aire *f.*
D Gasschleuse *f.*
R тамбур газоубежища
   м.; воздушная
   пробка *ж.*
A غَلْق هوائِي *m*

122

airman

F aviateur *m.*
E aviador *m.*
D Flieger *m.*
R лётчик *м.*
A طَيّار *m*

123

airmobile

F aéromobile
E aeromóvil *m.*
D luftbeweglich
R аэромобильный
A مُتَحَرِّك جَوًّا *m*

124

airmobility

F aéromobilité (ALAT) *f.*
E mobilidad aérea *f.*
D Luftbeweglichkeit *f.*
R воздушная
   подвижность *ж.*
A النقل جوا *m*

125

air movement

F mouvement par voie
   aérienne *m.*
E movimiento aéreo *m.*
D Lufttransport *m.*
R авиатранспорт *м.;*
   воздушная перевозка
   ж.
A التَحَرُّك جوًّا *m*

126 [M]

air plot

F tracé de route aérienne
   m.
E gráfico aéreo *m.*
D Fliegerdarstellung *f.*
R планшет воздушной
   обстановки *м.*
A تَخْطِيط الطريق / تحديد
   الموقع *m*

127

air pocket

F trou d'air *m.*
E bache aéreo *m.*
D Luftloch *neut.*
R воздушная яма *ж.;*
   раковина *ж.;*
   газовый пузырь *м.*
A جَيْب هوائي *m*

128

air portable

F aérotransportable
E aerotransportable
D Luftlande- *komp.*
R приспособленный для
   воздушной
   перевозки
A قَابِل لِلحَمْل جَوًّا *m*

129

air pressure

F pression de l'air *f.*
E presión de aire *f.*
D Luftdruck *m.*
R давление воздуха *ср.*
A ضَغْط الهَوَاء *m*

130

air-sea rescue

F sauvetage aéro-
   maritime *m.*
E salvamento aeronaval
   m.
D Bergung *f.*
R авиационная
   спасательная служба
   ж.
A إنْقَاذ جَوّي بَحْري *m*

131

airspace

F espace aérien *m.*

E espacio aéreo *m.*

D Luftraum *m.*

R воздушное
пространство *cp.*

A *m* حَيِّز هَوَائِي / مجال جوي

132

airspeed

F vitesse relative *f.*

E velocidad relativa al
aire *f.*

D Fluggeschwindigkeit *f.*

R воздушная скорость
ж.; скорость
самолёта *ж.*

A *f* السُّرْعَة الهَوَائِية / الجوية

133

air strike

F attaque aérienne *f.*

E ataque aéreo *m.*

D Luftangriff *m.*

R удар авиации *м.*

A *f* ضربة جوية

134

airstrip

F terrain d'atterrissage *m.*

E pista de aterrizaje *f.*

D Feldflugplatz *m.*

R полоса *м.*; полевой
аэродром *м.*

A *m* مَدْرَج المَطَار

135

air superiority

F supériorité aérienne *f.*

E superioridad aérea *f.*

D Luftüberlegenheit *f.*

R воздушное
превосходство *cp.*

A *m* تَفَوُّق جَوِّي

136

air supremacy

F suprématie aérienne *f.*

E supremacía aérea *f.*

D Luftüberlegenheit *f.*

R воздушное господство
cp.

A *f* سِيَادَة جَوِّية

137

air surveillance

F surveillance aérienne *f.*

E vigilancia aérea *f.*

D Luftüberwachung *f.*

R воздушное
наблюдение *cp.*

A *m* إِشْرَاف جَوِّي

138

air-to-air

F air-air *inv.*

E aire-aire *inv.*

D Luftkampf- *komp.*

R воздух-воздух

A مِن الجَوِّ إلى الجَوِّ (جَوّ - جَوّ)

139

air-to-surface

F air-sol *inv.*

E aire-tierra *inv.*

D Bord-Boden- *komp.*

R воздух-земля

A من الجو إلى الارض جو ـ
أرض

140

air traffic

F circulation aérienne *f.*

E circulación aérea *f.*

D Luftverkehr *m.*

R воздушное движение
cp.

A *m* المُرُور الجَوِّي

141

air traffic control

F contrôle de la
circulation aérienne *f.*

E control tráfico aéreo *m.*

D Flugsicherung *f.*

R управление
воздушным
движением *cp.*

A *f* مُرَاقَبَة المُرُور الجَوِّي

142

air transportable

F aérotransportable

E transportable por avión

D (mit) Flugzeugen
transportfähig *adj.*

R аэротранспортабельный

A يُمكِن نقله جواً

143 [A]

**airworthiness
(certificate of)**

F navigabilité, certificat
de *m.*

E navegabilidad,
certificado de *m.*

D Lufttüchtigkeitszeugnis
neut.

R удостоверение
пригодности к
полёту *cp.*

A *f* شَهَادَة صَلاَحِية للطَّيَران

144 [A]

airworthy

F navigable

E (en) condiciones de
vuelo

D lufttüchtig

R годный к полёту

A صَالِحَة للطَّيَران

145

alarm

F alerte *f.*

E alarma *f.*

D Alarm *m.*

R тревога *ж.*

A *m* إنذار

146
alert *n.*
F alerte *f.*
E alerta *m.*
D Alarm *m.*
R сигнал тревоги *м.*
A إنذار *m*

147
alert *adj.*
F alerte
E alerta *inv.*
D alarmbereit
R бдительный; готовый
   (к боевым
   действиям)
A حَذِر

148
alert *v.*
F alerter
E alertar
D alarmieren
R поднимать тревогу
A أَنْذَر

149 [A,M]
alight *v.*
F amerrir
E amerizar
D landern [A]; wassern
   [M]
R приземляться [A];
   высаживаться из
   лодки [M]
A هَبَط

150
alighting area
F aire d'amerrissage *f.*
E zona de amerizaje *f.*
D Wasserungszone *f.*
R посадочная площадка
   ж.
A مِنْطَقَة هُبُوط *f*

151
align
F aligner
E alinear
D ausrichten (nach) *tr.*
R выстраивать в линию;
   наводить
A صَفّ . حَاذى

152
alignment
F alignement *m.*
E alineación *f.*
D Ausrichten *neut.*
R выравнивание *ср.*;
   наводка *ж.*
A صَفّ *m* . مُحَاذَاة *f*

153
all hands
F tout le monde
E todos
D alle Mann
R вся команда; все
   наверх!
A الجَمِيع

154
alliance
F alliance *f.*
E alianza *f.*
D Bündnis *neut.*
R союз *м.*; альянс *м.*
A حِلْف *m*

155
allocate
F allouer
E asignar
D zuweisen *tr.*
R распределять;
   назначать
A خَصَّص

156
allocation
F allocation *f.*
E asignación *f.*
D Zuteilung *f.*; Zuweisung
   f.
R распределение *ср.*;
   назначение *ср.*
A تَخْصِيص *m*

157
alloy
F alliage *m.*
E aleación *f.*
D Legierung *f.*
R сплав *м.*
A سَبِيكَة *f*

158
all-round
F (en) hérisson (défense)
E circular
D umfassend *adj.*
R круговой
A شَامِل

159
all weather *adj.*
F tous temps *adj. inv.*
E (para) todo tiempo *adj.*
   inv.
D Allwetter- *komp.*
R всепогодный
A فِي كُلّ أَحْوَال الجَوّ

160
ally *v.*
F (s')allier
E aliar(se)
D (sich) verbünden (mit)
   untr.
R соединять(ся) с
A حَالَف . تَحَالَف

161
ally *n.*
F allié *m.*
E aliado *m.*
D Alliierte(r) *(m.)*
R союзник *м.*
A حَلِيف *m*

162
alongside
F (le) long du (bord, etc...)
E (al) costado
D längsseits
R рядом; борт о борт
A بِمُحَاذاة

163
alter
F modifier
E cambiar
D ändern
R изменять(ся)
A بَدَّل . حَوَّل . غَيَّر

164
alter course
F changer le cap
E cambiar el rumbo
D (den) Kurs ändern
R менять курс
A بَدَّل الاتِّجَاه

165
alternate *adj.*
F tous les deux...
E cada dos ...
D abwechselnd
R переменный; чередующий
A بَدِيل . مُتَناوِب

166
alternate *v.*
F (faire) alterner
E alternar
D abwechseln (lassen) *tr.*
R чередовать(ся)
A تَنَاوب

167
alternate airfield
F terrain de dégagement *m.*
E campo eventual *m.*
D Ausweichflugplatz *m.*
R промежуточный аэродром *м.*
A مَطَار بَدِيل *m*

168
alternative *n.*
F alternative *f.*
E alternativa *f.*
D Alternative *f.*
R альтернатива *ж.*
A بَدِيل *m*

169
alternative *adj.*
F alternatif
E alternativo; optativo
D verschieden
R альтернативный
A بَدِيل . مُتَناوِب

170
alternative airfield
F terrain de déroutement *m.*
E campo de urgencia *m.*
D Notlandeplatz *m.*
R запасный аэродром *м.*
A مَطَار بَدِيل *m*

171
altimeter
F altimètre *m.*
E altímetro *m.*
D Höhenmesser *m.*
R альтиметр *м.*
A مِقْيَاس الإِرْتِفَاع *m*

172
altitude
F altitude *f.*
E altitud *f.*
D Höhe *f.*
R высота *ж.*
A إِرْتِفَاع *m*

173
aluminum
F aluminium *m.*
E aluminio *m.*
D Aluminium *neut.*
R алюминий *м.*
A ألُومِينِيُوم *m*

174
ambient
F ambiant
E ambiente
D umgebend *adj.*; Umgebungs- *komp.*
R окружающий со всех сторон
A مُحِيط

175
ambulance *n.*
F ambulance *f.*
E ambulancia *f.*
D Sanitätswagen *m.*
R санитарная машина *ж.*
A سَيَّارَة إِسْعَاف *f*

176
ambulance *adj.*
F sanitaire
E (de) ambulancia
D Sanitäts- *komp.*
R санитарный
A إِسْعَاف

177
ambush *n.*
F embuscade *f.*
E emboscada *f.*
D Hinterhalt *m.*; Überfall *m.*
R засада *ж.*
A كَمِين

178
ambush *v.*
F attirer dans une embuscade
E tender una emboscada
D (aus dem) Hinterhalt überfallen *untr.*
R устраивать засаду
A كَمَن

11

179

amendment
F modification *f.*
E corrección *f.*
D Richtigstellung *f.;*
　Ergänzung *f.*
R поправка *ж.*
A تعديل *m*

180

amidships
F (au) milieu
E (en) medio del navío
D mittschiffs
R посередине судна
A مُنتصف السفينة

181

ammeter
F ampèremètre *m.*
E amperímetro *m.*
D Amperemeter *m.;*
　Strommesser *m.*
R амперметр *м.*
A أَمْبِير متر (مِقْيَاس الأَمْبِير) *m*

182

ammunition
F munition *f.*
E municiones *f. pl.*
D Munition *f.*
R боеприпасы *м. мн. ч.*
A ذَخِيرة *f*

183

ammunition dump
F dépôt provisoire de
　munitions *m.*
E depósito de municiones
　m.
D Munitionslager *neut.*
R (полевой) склад
　боеприпасов *м.*
A كُدس الذَخِيرَة *m*

184

amphibian *n.*
F véhicule amphibie *m.*
E vehículo anfibio *m.*
D Schwimmfahrzeug *neut.*
R амфибия *ж.*
A بَرْمَائِي *m*

185

amphibious *adj.*
F amphibie
E anfibio
D Schwimm- *komp.;*
　amphibisch *adj.*
R плавающий
A بَرْمَائِي

186 [D]

analog computer
F ordinateur analogique
　m.
E computador analógico
　m.
D Analogrechner *m.*
R аналогичная
　электронная
　вычислительная
　машина *ж.;*
　аналогичный
　компъютер *м.*
A حَاسِبَة بالتَشابُه *f*

187

analyse
F analyser
E analizar
D analysieren
R анализировать
A حَلَّل

188

analysis
F analyse *f.*
E análisis *m.*
D Analyse *f.*
R анализ *м.*
A تَحْلِيل *m*

189

analyst
F analyste *m.*
E analizador *m.*
D Analytiker *m.*
R аналитик *м.*
A مُحَلِّل *m*

190

anchor *n.*
F ancre *m.*
E ancla *f.*
D Anker *m.*
R якорь *м.*
A مِرْسَاة *f*

191

anchor *v.*
F jeter l'ancre
E anclar
D ankern verankern *untr.;*
R бросать якорь;
　становиться на
　якорь
A أَرْسَى

192

anchorage
F mouillage *m.*
E ancladero *m.;*
　fondeadero *m.*
D Ankerplatz *m.*
R якорная стоянка *ж.*
A مَرْسَى *m*

193

anchor buoy
F bouée d'ancre *f.*
E boya de ancla *f.*
D Ankerboje *f.*
R буй якоря *м.*
A عَوَامَة الإرْسَاء *f*

194

anchor light
F feu de mouillage *m.*
E luz de fondeadero *f.*
D Ankerlicht *neut.*
R якорный огонь *м.*
A ضَوء الإرْسَاء *m*

195

anchor watch
F garde au mouillage *f.*
E guardia de ancla *f.;*
　vigilante de muelle *m.*
D Ankerwache *f.*
R вахта на якорной
　стоянке *ж.*
A مُرَاقَبة الإرساء *f*

196
anemometer
F anémomètre m.
E anemómetro m.
D Windstärkemesser m.
R анемометр м.
A مِقْيَاس سُرعَة الرِيح m

197
anesthetic n.
F anesthésiant m.
E anestésico m.
D Betäubungsmittel neut.
R обезболивающее
средство ср.
A مُخَدِّر m

198
angle n.
F angle m.
E ángulo m.
D Winkel m.
R угол м.
A زَاوِيَة f

199
angled deck
F pont oblique m.
E cubierta oblicua f.
D abgewinkeltes Flugdeck
neut.
R угловая палуба ж.
A سَطْح زَاوِي m

200 [A,M,S]
angle of depression
F angle de dépression m.
E ángulo de depresión m.
D Senkungswinkel m.;
negativer
Geländewinkel m.
R угол склонения м.
A زَاوِيَة الإِنْخِفَاض f

201 [A,M]
angle of drift
F angle de dérive m.
E ángulo de deriva m.
D Abtriebswinkel [M] m.;
Windwinkel [A] m.
R угол сноса [AM] м.;
угол дериваации [A]
м.
A زاوية الحَيْد f

202 [A,M,S]
angle of elevation
F angle d'élévation m.
E ángulo de elevación m.
D Erhöhungswinkel m.
R угол места [AM]; угол
возвышения [S]
A زاوية الإرْتِفَاع f

203 [M,S]
angle of incidence
F angle d'incidence m.
E ángulo de incidencia m.
D Einfallswinkel m.
R угол атаки [M] м.;
угол встречи [S] м.
A زاوية السُّقُوط f

204 [A,M,S]
angle of lead
F angle de dérive m.
E corrección objetivo f.
D Vorhaltewinkel m.
R угол упреждения м.
A زاوية السَّبْق f

205
angle of safety
F angle de sécurité m.
E ángulo de seguridad m.
D Sicherheitswinkel m.
R угол безопасности м.
A زاوية الأَمَان f

206 [S]
angle of site
F angle de site m.
E ángulo de situación (del
blanco) m.
D Geländewinkel m.
R угол места м.
A زاوية النَظْر f

207
angle of view
F angle de vue m.
E ángulo visual m.
D Blickfeld neut.
R угол м.; поле зрения
ср.
A زاوية المَنْظَر f

208
annex (document) v.
F joindre
E adjuntar
D beilegen tr.
R прилагать
A أَلْحَق

209
(document) annex n.
F annexe f.
E apéndice m.
D Anlage f.
R приложение ср.
A مُلْحَق m

210
(logistics) annex
F annexe logistique f.
E anexo logístico m.
D Anlage über Logistik in
Befehlen f.
R приложение по тылу
ср.
A مُلْحَق إداري m

211
annex (territory) v.
F annexer
E anexar
D besetzen
R аннексировать;
присоединять
A إِسْتَوْلَى على / ضَمَّ

13

212
annihilate
F anéantir
E aniquilar
D vernichten
R уничтожать
A أباد

213
annihilation
F anéantissement *m.*
E aniquilación *f.*
D Vernichtung *f.*
R уничтожение *cp.*
A إبَادَة *f*

214
annotate
F annoter
E anotar
D kommentieren
R аннотировать
A علّق / فسّر

215
annotation
F annotation *f.*
E nota *f.*
D Anmerkung *f.*
R примечание *cp.*;
   аннотация *ж.*
A تعليق / تفسير *m*

216
antarctic *adj.*
F antarctique
E antártico
D Antarktis- *komp.*;
   antarktisch *adj.*
R антарктический;
   южнополярный
A قُطبِي جُنوبي

217 [R]
antenna
F antenne *f.*
E antena *f.*
D Antenne *f.*
R антенна *ж.*
A هَوَائي *m*

218
antiaircraft
F anti-aérien
E anti-aéreo
D Fliegerabwehr- *komp.*;
   Flugabwehr- *komp.*
R зенитный *прилаг.*
A مُضاد للطائرات

219
**antiballistic missile
   system**
F défense antifusée *f.*
E defensa contra misiles
   f.; sistema de misiles
   antibalísticos *m.*
D Raketenabwehrsystem
   neut.
R противоракетная
   система *ж.*
A نِظَام مُضادٌ للصَوَارِيخ
   المَقْذُوفية *m*

220
antibiotic *n.*
F antibiotique *m.*
E antibiótico *m.*
D Antibiotikum *neut.*
R антибиотик *м.*
A مُضاد حيوي

221
antifreeze
F antigel *m.*
E solución anticongelante
   f.
D Frostschutzmittel *neut.*
R антифриз *м.*;
   незамерзающая
   смесь *ж.*
A مُقَاوِم التَجَمُّد *m*

222 [A]
Anti-G suit
F combinaison anti-G *f.*
E traje presurizado *m.*;
   traje anti-G *m.*
D Schwerkraftsanzug *m.*
R скафандр *м.*;
   пневмокостюм *м.*
A رِداء مُضادٌ للجاذبية

223
antipersonnel
F antipersonnel *inv.*
E contra personal *inv.*
D gegen Personal *komp.*;
   Splitter-
R противопехотный;
   осколочный
A مُضادٌ لِلاشْخَاص *m*

224
antirust
F antirouille *inv.*
E antioxidante
D Rostschutz- *komp.*
R противоржавый
A مُقَاوِم للصَدَأ *m*

225
antiseptic
F antiseptique
E antiséptico
D antiseptisch
R антисептический
   прилаг.;
   антисептическое
   средство *cp.*
A مُطَهِّر *m*

226
antiskid
F antidérapant
E antiderrapante
D Gleitschutz- *komp.*
R противозаносный
A مُضادٌ لِلإنْزِلاق *m*

227
antisubmarine
F anti-sous-marin
E antisubmarino
D U-Boot-Abwehr- *komp.*
R противолодочный
A مُضادٌ لِلغَوّاصَات *m*

228
antitank
F antichar
E antitanques *inv.;*
  contracarro *m.*
D Panzerabwehr-*komp.*
R противотанковый
A مُضَادٌ لِلدَّبَابَات *m*

229
aperture
F ouverture *f.*
E abertura *f.*
D Öffnung *f.*
R отверстие *cp.*
A فَتْحَة *f*

230
apogee
F apogée *m.*
E apogeo *m.*
D Apogäum *neut.;*
  Höhepunkt *m.*
R апогей *м.*
A نُقْطَة الأَوْج *f*

231
aport
F (à) babord
E (a) babor
D (nach) Backbord
R (к) бакпорту
A نحو الجانب الأَيسر من
  السفينة

232
approach *v.*
F (s')approcher (de)
E acercarse (a)
D (sich) nähern (an)
R приближаться (к)
A إقْتَرَب

233 [A,M,L]
approach *n.*
F approche *f.*
E acercamiento [AM] *m.;*
  aproche [L] *m.*
D Annäherung *f.*
R приближение [AM]
  cp.; подход [AL] *м.*
A إقْتِرَاب *m*

234 [M]
approach lane
F couloir d'approche *m.*
E canal de acceso *m.*
D Zufahrtsrinne *f.*
R подступ *м.;* путь
  подхода *м.*
A طَرِيق الاقْتِرَاب *m*

235
approach march
F marche d'approche *f.*
E marcha de
  aproximación *f.*
D Anmarsch *m.*
R подходный марш *м.*
A مَسِير الاقْتِرَاب *m*

236 [A]
approach path
F axe d'approche *m.*
E trayectoria de
  acercamiento *f.*
D Anflugschneise *f.*
R курс при подходе *м.*
A مَمَّر الاقْتِرَاب *m*

237 [A]
apron
F aire de trafic *f.*
E pista *f.*
D Vorfeld *neut.*
R рулежно-подходная
  полоса *ж.*
A مَريلة / صِدِرية *f*

238
arc
F arc *m.*
E ángulo *m.;* arco *m.*
D Bogen *m.*
R дуга *ж.*
A قَوْس *m*

239
arc of fire
F champ de tir *m.*
E ángulo de tiro *m.*
D Schussfeld *neut.*
R сектор стрельбы *м.*
A قَوْس النِيرَان *m*

240
arctic
F arctique
E ártico
D Polar- *komp.;* arktisch
  adj.
R арктический
A قُطْبِي شِمَالِي

241 [A,E,M,P]
area
F aire *f.*
E área *f.*
D Fläche *f.*
R район *м.;* площадь *ж.*
A مِنْطَقَة *f*

242 [T]
area
F zone *f.*
E zona *f.*
D Gebiet *neut.*
R округ *м.;* зона *ж.*
A مِنْطَقَة *f*

243
area bombing
F bombardement sur
  zone *m.*
E bombardeo de zona *m.*
D Bombenflächenwurf *m.*
R бомбометание по
  площади *cp.*
A قَصْف مِنْطَقَة *A*

244

area command

F commandement de zone
   m.
E jefe de zona *m.*
D Militärbereich *m.*
R районное
   командование *ср.*
A قِيَادَة مِنْطَقَة *f*

245

area forecast

F prévisions par zone *f.*
   pl.
E pronóstico de zona *m.*
D Gebietswettervorhersage
   f.
R прогноз погоды по
   округам *м.*
A تَنَبُّؤ جَوِّي لِلمِنْطَقَة *m*

246

area of interest

F zone d'intérêt *f.*
E zona de interés *f.*
D Interessengebiet *neut.*
R интересующая зона
   ж.
A مِنْطَقَة الإِهْتِمَام *f*

247

area of operations

F théâtre des opérations
   m.
E zona de operaciones *f.*
D Operationsgebiet *neut.*
R район боевых
   действий *м.*
A مِنْطَقَة العَمَلِيَات *f*

248

area of responsibility

F zone de responsabilité
   f.
E zona de responsibilidad
   f.
D Verantwortungsbereich
   m.
R район
   ответственности *м.*
A مِنْطَقَة المَسْؤُولِيَات *f*

16

249 [E,S,U]

arm *v.*

F armer
E armar
D bewaffnen [U]; scharf
   machen [ES]
R вооружать [SU];
   приводить
   в готовность [E];
   снаряжать[SU]
A سَلَّح

250

arm *n.*

F arme *f.*
E arma *f.*
D Waffengattung *f.*
R род войск *м.*
A سِلاح *m*

251

armament

F armement *m.*
E armamento *m.*
D Bewaffnung *f.*
R вооружение *ср.*
A سِلاح *m*

252

armaments *pl.*

F matériel de guerre *m.*
E material de guerra *m.*
D Ausrüstung *f.*
R оружие *ср.*
A أسلحة *f*

253 [E,S,U]

armed

F armé
E armado
D bewaffnet [U]; scharf
   [ES]
R вооружённый [SU];
   приведённый в
   готовность [E];
   снаряжённый[SU]
A مُسَلَّح

254

armed forces *pl.*

F forces armées *f. pl.*
E fuerzas armadas *f. pl.*
D Streitkräfte *f. pl.*
R вооружённые силы *ж.*
   мн. ч.
A القُوَات المُسَلَّحة *f*

255

armistice

F armistice *m.*
E armisticio *m.*
D Waffenstillstand *m.*
R перемирие *ср.*
A هُدْنَة *f*

256 [V]

armor

F blindage *m.*
E blindaje *m.*
D Panzerung *f.*
R броня *ж.*
A المُدَرَّعَات

257 [L]

armor

F forces blindées *f. pl.*
E fuerzas acorazadas *f.*
   pl.
D Panzertruppe(n) *f. (pl.)*
R бронетанковые войска
   ср. мн. ч.
A الدُرُوع

258

**armored personnel
   carrier (APC)**

F véhicule blindé de
   transport de troupes
   (VTT) *m.*
E transporte oruga
   acorazado (TOA) *m.*
D Transportpanzer *m.*
R бронетранспортёр
   (БТР) *м.*
A ناقِلة أشْخَاص مُدَرَّعة *f*

259

armored vest
F gilet pareballes *m.*
E chaleco blindado (antibalas) *m.*
D Panzerweste *f.*
R бронежилет *м.*
A صُدْرة مُقاوِمة لِلرَصاص *f*

260

armor piercing
F perforant
E perforante
D panzerbrechend
R бронебойный
A خَارِق لِلدُرُوع

261

armory
F dépôt d'armes *m.*
E armería *f.*
D Waffenkammer *f.*
R склад оружия *м.*
A مُسْتَوْدَع أَسْلِحَة *m*

262

arms *n. pl.*
F armes *f. pl.*
E armas *f. pl.*
D Waffen *f. pl.*
R оружие *ср.*
A أَسْلِحَة *f*

263

army (formation)
F armée *f.*
E ejército *m.*
D Armee *f.*
R армия *ж.*
A جَيْش المَيْدَان *m*

264

army (ground forces)
F armée de terre *f.*
E ejército *m.*
D Heer *neut.*
R сухопутные войска *ср.* мн. ч.
A جَيْش *m*

265 [P]

arrest *n.*
F arrêts *m. pl.*
E detención *f.*
D Arrest *m.*
R арест *м.*
A إعْتِقَال *m*

266

arrest *v.*
F arrêter
E detener
D verhaften *untr.*
R арестовывать; задерживать
A إعْتَقَل

267

arsenal
F arsenal *m.*
E arsenal *m.*
D Arsenal *neut.;* Zeughaus *neut.*
R арсенал *м.*
A مَصْنَع الأَسْلِحة . مَخْزَن الأَسْلِحة *m*

268

article (of treaty)
F article *m.*
E artículo *m.*
D Punkt *m.;* Ziffer *neut.*
R пункт *м.*
A شَرْط/ بند *m*

269

artificer
F artificier *m.*
E mecánico *m.*
D Mechaniker *m.*
R механик *м.;* техник *м.*
A صانع / ميكانيكي *m*

270

artificial
F artificiel
E artificial
D Kunst- *komp.*
R искусственный
A إصْطِنَاعي

271

artillery
F artillerie *f.*
E artillería *f.*
D Artillerie *f.*
R артиллерия *ж.*
A المِدْفَعية *f*

272

asbestos
F amiante *m.*
E amianto *m.*
D Asbest *m.*
R асбест *м.*
A أَسْبِسْتُوس *m*

273

ashore
F (à) terre
E (en la) orilla; (en) tierra
D (an) Land
R (на) берегу
A على الشَاطِىء . على الأَرْض

274

assault *n.*
F assaut *m.*
E asalto *m.*
D Angriff *m.*
R атака *ж.*
A هُجُوم . إقْتِحَام . إعتداء *m*

275

assault *v.*
F assaillir
E asaltar
D angreifen *tr.*
R атаковать; нападать
A هَجم على . إقْتَحَم . إعتدى

276

assault boat
F bateau d'assaut *m.*
E barco de asalto *m.*
D Sturmboot *neut.*
R десантный катер *м.*
A زورق إقتحام *m*

17

277

assault shipping
F bâtiments d'assaut *m.*
  pl.
E buques de asalto *m. pl.*
D Schiffe der
  Landungswelle *neut.*
  pl.
R суда первого эшелона
  десанта *ср. мн. ч.*
A سُفُن الإقْتِحَام *f*

278 [E]

assemble
F monter
E montar
D montieren
R собирать
A ركّب

279 [U]

assemble
F (r)assembler
E ajustar
D bereitstellen *tr.*
R собираться
A جَمع

280 [E]

assembly
F ensemble *m.*
E conjunto *m.*
D Baugruppe *m.*
R сборка *ж.;* агрегат *м.*
A تَرْكيب *m*

281

assembly area
F zone de déploiement
  d'attente *f.*
E zona de reunión *f.*
D Bereitstellungsraum *m.*
R район сосредоточения
  м.
A مِنْطَقَة التَجَمُّع *f*

18

282

assessment
F évaluation *f.*
E estudio *m.;* juicio *m.*
D Beurteilung *f.*
R оценка *ж.*
A تقدير *m*

283

assign
F affecter
E asignar
D unterstellen *untr.*
R назначать; поручать
A عَيَّن

284

assigning
F affectation *f.*
E asignación *f.*
D Unterstellung *f.*
R назначение *ср.;*
  поручение *ср.*
A تَعْيِين

285

assignment
F affectation *f.*
E cometido *m.*
D Auftrag *m.*
R задача *ж.;* назначение
  ср.
A مُهِمَّة *f*

286

astarboard
F (à) tribord
E (a) estribor
D (nach) Steuerbord
R (на) правом борту
A الجانب الأيسر للسفينة

287

astern
F (en) arrière
E atrás
D rückwärts
R позади; задним ходом
A في الوَرَاء

288

astro compass
F astrocompas *m.*
E brújula estelar *f.*
D Astrokompass *m.*
R астрокомпас *м.*
A بُوصَلة فلكية *f*

289

atmosphere
F atmosphère *f.*
E atmósfera *f.*
D Atmosphäre *f.*
R атмосфера *ж.*
A جَوّ *m*

290 [R]

atmospherics *pl.*
F parasites *m. pl.*
E parásitos *m. pl.*
D Störungen *f. pl.*
R атмосферные помехи
  ж. мн. ч.
A تَشْويشات جَوّية *f*

291

atom
F atome *m.*
E átomo *m.*
D Atom *neut.*
R атом *м.*
A ذَرَّة *f*

292

atomic
F atomique
E atómico
D Atom- *komp.*
R атомный
A ذَرَّي

293

attach *v.*
F détacher
E agregar
D zuteilen *tr.;* angliedern
  tr.
R прикомандировывать;
  прикреплять;
  придавать
A الحق

294
attached *adj.*
F détaché au près de
E agregado
D zugeteilt; angegliedert
R приданный
A مُلحق

295
attachment *n.*
F détachement *m.*
E agregación *f.*
D Angliederung *f.;*
   Zuteilung *f.*
R придание *cp.;*
   прикомандирование
   cp.
A الحاق *m*

296
attack *n.*
F attaque *f.*
E ataque *m.*
D Angriff *m.*
R атака *ж. cp.;*
   нападение
A هُجُوم *m*

297
attack *v.*
F attaquer
E atacar
D angreifen *tr.*
R атаковать; нападать
A هَجَم على

298
attenuate
F atténuer
E atenuar
D dämpfen
R затухать
A أوْهَن . أضْعَف

299
attenuation
F atténuation *f.*
E atenuación *f.*
D Dämpfung *f.*
R затухание *cp.*
A تَوْهين . إضعاف *m*

300 [U]
attitude
F attitude *f.*
E actitud *f.*
D Einstellung *f.*
R отношение *cp.*
A وَضْع ـ سِلوك *m*

301 [A]
attitude
F position *f.*
E posición (de vuelo) *f.*
D Fluglage *f.*
R положение *cp.;*
   ориентация *ж.*
A مَوْقف *m*

302
attrition
F usure *f.*
E desgaste *m.*
D Zermürbung *f.*
R истощение *cp.*
A إسْتِنْزاف *m*

303
audit *n.*
F verification des comptes
   f.
E intervención *f.*
D Rechnungsabnahme *f.;*
   Bücherrevision *f.*
R проверка *ж.;* ревизия
   ж.
A فَحْص حسابات

304
audit *v.*
F verifier
E intervenir
D abnehmen *tr.;*
   revidieren *untr.*
R проверять; ревизовать
A فَحَص حسابات

305
auditor
F verificateur *m.*
E interventor *m.*
D Bücherrevisor *m.;*
   Revisor *m.*
R ревизор *м.;* контролёр
   м.
A فاحص حسابات

306
authenticate
F authentifier
E autenticar
D authentisieren
R удостоверять
A صَدَّق على . وَثَّق . حَقَّقَ

307
authentication
F authentification *f.*
E autenticación *f.*
D Authentisierung *f.*
R удостоверение *cp.*
A تَصْديق . توثيق . تحقيق *m*

308
authenticator
F signe d'authentification
   m.
E autenticador *m.*
D Kennziffer(n) *f. (pl.)*
R удостоверение
   подлинности *cp.*
A إشارة التَعَرُّف *f*

309
authority
F autorité *f.*
E autoridad *f.*
D Autorität *f.*
R власть *ж.;*
   полномочие *cp.*
A سُلْطَة *f* . نُفُوذ *m*

310
authorization
F autorisation *f.*
E autorización *f.*
D Bevollmächtigung *f.;*
   Genehmigung *f.*
R разрешение *cp.*
A مَنْح السُّلْطة *m*

311
authorize
F autoriser
E autorizar
D (jemanden)
   bevollmächtigen;
   (etwas) genehmigen
R уполномочивать;
   санкционировать
A مَنْح السُّلْطة

312
automate
F automatiser
E automatizar
D automatisieren
R автоматизировать
A جَعَل تِلْقَائِياً

313
automatic
F automatique
E automático
D automatisch
R автоматический
A تِلْقائي

314
automatic pilot
F pilote automatique *m.*
E piloto automático *m.*
D Selbststeuergerät *neut.*
R автопилот *м.*
A طَيَّار آلي *m*

315
automation
F automatisation *f.*
E automatización *f.*
D Automation *f.*
R автоматизация *ж.*
A التَشْغِيل الذاتي *m*

20

316
automobile *n.*
F automobile *f.*
E automóvil *m.*
D Kraftfahrzeug *neut.*
R автомобиль *м.;*
   машина *ж.*
A سَيَّارة *f*

317
automotive
F automoteur(-trice)
E automotor
D Kraftfahrzeug- *komp.;*
   selbstbeweglich *adj.*
R автомобильный;
   самодвижущийся
A ذاتي الحركة

318
autumn
F automne *m.*
E otoño *m.*
D Herbst *m.*
R осень *ж.*
A الخَرِيف *m*

319
auxiliaries *pl.*
F auxiliaires *m. pl.*
E tropas auxiliares *f. pl.*
D Hilfstruppen *f. pl.*
R вспомогательные
   войска *cp. мн. ч.;*
   запасные войска *cp.*
   мн. ч.
A مُلْحَقَات *f*

320 [M]
auxiliary (motor)
F moteur auxiliaire *m.*
E motor auxiliar *m.*
D Hilfsmotor *m.*
R вспомогательный
   (мотор) *м.;*
   вспомогательный
   (двигатель) *м.*
A مُحَرِّك إضافي *m*

321 [M]
auxiliary (ship)
F navire auxiliaire *m.*
E buque auxiliar *m.*
D Hilfsschiff *neut.*
R вспомогательный
   (корабль)
A سَفِينة إضافية *f*

322
availability
F disponibilité *f.*
E disponibilidad *f.*
D Verfügbarkeit *f.*
R наличие *cp.*
A تَوَافُر *m*

323
available
F disponible
E disponible
D verfügbar
R наличный
A مُتَوافِر

324
average *v.*
F prendre la moyenne
E calcular el término
   medio
D (den) Durchschnitt
   nehmen
R выводить среднее
   число
A يَبْلُغ مُعَدَّله

325
average *n.*
F moyenne *f.*
E promedio *m.;* término
   medio *m.*
D Durchschnitt *m.*
R среднее (число) *cp.*
A مُعَدَّل *m*

326

average *adj.*

F moyen
E medio
D Durchschnitts- *komp.;*
   durchschnittlich *adj.*
R средний
A مُتَوَسِّط

327

avgas

F essence avion *f.*
E combustible para
   aviones *m.*
D Flugbenzin *neut.*
R авиационное горючее
   ср.
A بنزين طائرات *m*

328

aviation

F aviation *f.*
E aviación *f.*
D Flugwesen *neut.*
R авиация *ж.*
A طَيْران *m* . مِلاحة جوية *f*

329

avionics *pl.*

F avionique *f.*
E aviónica *f.*
D Luftfahrtelektronik *f.*
R авионика *ж.*
A إلكترونيات الطَيْران

330

awash

F (à) fleur d'eau
E (a) flor de agua
D überspült (z.B. Klippe);
   schwimmend
R (на) уровне воды;
   вровень с
   поверхностью воды
A على سَطْح الماء

331

ax

F hache *f.*
E hacha *f.*
D Axt *f.*
R топор *м.*
A فَأس *m* . بَلْطة *f*

332

axis

F axe *m.*
E eje *m.*
D Achse *f.*
R ось *ж.*
A مِحْوَر *m*

333

axle

F essieu *m.*
E eje *m.*
D Achse *f.*
R ось *ж.*
A مِحْوَر *m*

334

azimuth

F azimut *m.*
E azimut *m.*
D Azimut *m. neut.*
R азимут *м.*
A السَمْت *m*

B

335

back *n.*

F dos *m.*
E espalda *f.*
D Rücken *m.*
R спина *ж.;* тыльная
   сторона *ж.*
A ظَهْر *m* . خلف

336

back blast

F souffle arrière *m.*
E rebufo de culata *m.*
D Sog *m.*
R обратное пламя *ср.*
A عَصْف خَلْفي *m*

337

background

F arrière plan *m.*
E fondo *m.*
D Hintergrund *m.*
R фон *м.*
A الخَلْفية *f*

338

background radiation

F radioactivité naturelle *f.*
E radioactividad natural
   (ambiental) *f.*
D Hintergrundstrahlung *f.*
R природная
   (окружающая)
   радиация *ж.*
A إشْعاع خَلْفي *m*

339 [E]

backlash

F jeu *m.*
E juego *m.*
D Schlupf *m.*
R зазор *м.;* скольжение
   (винта) *ср.*
A حَرَكة إرْتجَاعية *f*

340

backloading

F évacuation (matériels) *f.*
E recuperación (de
   material) *f.*
D Abschub *m.*
R эвакуация *ж.*
A الإرْسال إلى الخَلْف

341

backscatter

F rétrodiffusion *f.*
E difusión retrograda *f.*
D Rückstreuung *f.*
R обратное рассеяние
   ср.
A تَشَتُّت خلفي *m*

342
backsight
F hausse *f.*
E alza *f.*
D Kimme *f.*
R прицел *м.*
A مُسَدِّدة خَلْفية *f*

343
backwards
F (en) arrière
E atrás
D rückwärts
R назад; обратно
A إلى الخَلْف

344
badge
F insigne *m.*
E distintivo *m.*
D Abzeichen *neut.*
R значок *м.*
A شارة *f.* . شِعَار *m*
علامة *f*

345
badge of rank
F insigne de grade *m.*
E insignias de grado *f. pl.*
D Dienstgradabzeichen
neut.
R знак различия *м.*
A شارة الرُّتْبة *f*

346
bag
F sac *m.*
E saco *m.*
D Beutel *m.*
R мешок *м.*; сумка *ж.*
A كِيس *m*

347
baggage
F bagages *m. pl.*
E bagaje *m.;* equipaje *m.*
D Gepäck *neut.*
R багаж *м.*; возимое
имущество *cp.*
A أَمْتِعَة *f*

348
baggagemaster
F responsable des bagages
m.
E jefe de carga *m.*
D Gepäckkommandant
m.
R заведующий багажом
м.
A المَسْؤول عن الأمْتِعة *m*

349
balance *v.*
F équilibrer
E equilibrar
D ausgleichen *tr.*
R уравновешивать
A وازِن

350
balance *n.*
F equilibre *m.*
E equilibrio *m.*
D Gleichgewicht *neut.*
R баланс *м.*; равновесие
cp.
A تَوازُن

351 [P]
balisage
F balisage *m.*
E balizaje de itinerario
m.; jalonamiento *m.*
D Streckenmarkierung *f.*
R балисаж *м.*; указание
пути *cp.*
A نِظام أنوار المِلاحة *f*

352
ballast
F lest *m.*
E lastre *m.*
D Ballast *m.*
R балласт *м.*
A صابُورة *f.* . ثُقل المِوازنة *m*

353
ball bearing
F roulement à billes *m.*
E cojinete de bolas *m.*
D Kugellager *neut.*
R шарикоподшипник *м.*
A مَحْمَل كُرَيَّات *m*

354
ballistic
F balistique
E balístico
D ballistisch
R баллистический
A مَقذوفي

355
ballistics *pl.*
F balistique *f.*
E balística *f.*
D Ballistik *f.*
R баллистика *ж.*
A عِلْم حَرَكة المَقْذوفات *m*

356
balloon
F ballon *m.*
E globo *m.*
D Ballon *m.*
R аэростат *м.*;
воздушный шар *м.*;
зонд *м.*
A مُنْطاد *m*

357
balsa wood
F balsa *m.*
E balsa *f.*
D Balsaholz *neut.*
R бальза *ж.*
A خَشَب البَلْسا *m*

358
ban *v.*
F interdire
E prohibir
D verbieten *untr.*
R запрещать
A حَرِّم

359
ban *n.*
F interdiction *f.*
E prohibición *f.*
D Verbot *neut.*
R запрет *м.*; запрещение *ср.*
A تَحْريم *m*

360 [R]
band
F bande de fréquences *f.*
E banda de ondas *f.*
D Frequenzband *neut.*
R полоса частот *ж.*
A نِطاق ذَبْذَبات *m*

361 [E]
band
F feuillard *m.*
E fleje *m.*
D Ring *m.*
R полоса *ж.*; лента *ж.*
A حِزام *m* . شَرِيط *m*

362
bandit
F bandit *m.*
E bandido *m.*
D Bandit *m.*
R бандит *м.*; разбойник *м.*
A لِصّ *m*

363
bandolier
F bandoulière *f.*
E bandolera *f.*
D Patronengurt *m.*
R патронташ *м.*
A نِجاد *m* . حَمالة *f*

364
bang
F détonation *f.*
E detonación *f.*
D Knall *m.*
R взрыв *м.*; удар *м.*
A ضَرْب

365 [T]
bank *n.*
F talus *m.*
E terraplén *m.*
D Böschung *f.*
R насыпь *ж.*;
A ضَفَّة *f*

366
(river) bank *n.*
F rive *f.*
E ribera *f.*
D Ufer *neut.*
R берег *м.* набережная *ж.*
A ضَفَّة *f*

367 [A]
bank *v.*
F virer
E cuivarse; ladearse
D (sich in die) Kurve legen
R крениться
A مال / يَميل

368
(sand) bank *n.*
F banc *m.*
E banco *m.*
D Bank (Bänke) *f.*
R банка *ж.*
A سَدّ [رَمْلي]

369 [P]
bar *v.*
F barrer
E obstruir
D sperren
R преграждать
A حَجَز

370 [M]
bar *n.*
F barre *f.*
E barra *f.*
D Barre *f.*
R отмель *ж.*; преграда *ж.*
A شَرِيط *m*

371 [E]
bar *n.*
F barre *f.*
E barra *f.*
D Stange *f.*
R штанга *ж.*; брус *м.*
A قَضِيب *m* . عَتَلة *f*

372
barbed wire entanglement
F réseau de barbelés *m.*
E alambrada *f.*
D Stacheldrahtverhau *m.*
R проволочное заграждение *ср.*
A شبكة أسلاك شائكة *f*

373
barge
F péniche *f.*
E barcaza *f.*
D Lastkahn *m.*
R баржа *ж.*
A صَنْدل *m*

374
barn
F grange *f.*
E granero *m.*
D Scheune *f.*
R сарай *м.*; амбар *м.*
A جُرْن *m*

375
barnacle
F bernache *f.*
E percebe *m.*
D Entenmuschel *m.*
R ракушка *ж.*; морская уточка *ж.*
A مَحارة . صَدَفة *f*

376
barograph
F barographe *m.*
E barógrafo *m.*
D Höhenschreiber *m.*
R барограф *м.*
A باروجراف *m*

377
barometer
F baromètre m.
E barómetro m.
D Barometer neut.
R барометр м.
A بارومتر(مِقْياس الضَّغط
الجَوّي) m

378
barometric
F barométrique
E barométrico
D barometrisch-
R барометрический
A بارومتري

379
barracks pl.
F caserne f.
E cuartel m.
D Kaserne f.
R казарма ж.
A ثُكَنات f

380 [S]
barrage
F tir de barrage m.
E cortina de fuego f.
D Sperrfeuer neut.
R заградительный огонь
м.; огневой вал м.
A سَدّ m

381
barrage balloon
F ballon de protection m.
E globo de barrera m.
D Sperrballon m.
R аэростат заграждения
м.
A مُنْطاد سَدّ m

382 [A]
barrel roll
F tonneau m.
E tonel abierto m.
D Walze f.
R бочка ж.
A تَقَلُّب m

383
barricade n.
F barricade f.
E barricada f.
D Barrikade f.
R баррикада ж.
A مِتْراس . حاجِز m

384
barricade v.
F barricader
E barrear
D sperren
R баррикадировать
A تَرَّس . حَجَز

385
barrier
F barrage m.
E barrera f.
D Sperre f.
R барьер м.; шлагбаум
м.
A حاجِز . مانِع m

386
base n.
F base f.
E base f.
D Stützpunkt m.;
Operationsbasis f.
R база ж.
A قاعدة f

387
base (on) v.
F baser (sur)
E basar (en)
D basieren (auf)
R основывать (на)
A أَسَّس (على)

388
base area
F base f.
E zona de base f.
D Stützpunkt m.
R тыловой район м.;
тыл м.
A مِنْطقة القاعِدة m

389
basic
F (de) base
E básico
D Grund- komp.;
grundlegend adj.
R основной
A أساسي . رئيسي

390 [T]
basin
F bassin m.
E cuenca f.
D Becken neut.
R бассейн м.; котловина
ж.
A حَوْض m

391 [M]
basin
F bassin m.
E dársena f.
D Hafenbecken neut.
R мелкая бухта ж.
A حَوْض m

392
basket
F corbeille f.; panier m.
E cesta f.
D Korb m.
R корзина ж.
A سَلّة f

393
bath unit
F unité de douches de
campagne f.
E unidad de duchas f.
D Bade- und
Wascheinheit f.
R банно-дезинфекционный
взвод м.
A وحدة حَمّام f

394 [P]
baton
F matraque f.
E porra f.
D Knüppel m.
R дубинка ж.
A هِراوة f

395
battalion
F bataillon *m.*
E batallón *m.*
D Bataillon *neut.*
R батальон *м.*
A كَتِيبة *f*

396
battalion task force
F groupement tactique *m.*
E grupo táctico *m.*
D Einsatz-kampfverband
   (auf Bataillons Ebene)
   m.
R батальонная группа
   специального
   назначения *ж.*
A كتيبة واجب القتال

397
batten *n.*
F latte *f.*
E listón *m.*
D Latte *f.*
R доска *ж.*; рейка *ж.*
A عارِضة *f*

398
batten (down) *v.*
F fixer des lattes
E asegurar con listones
D verschalken
R задраивать (люки)
A سَـدَّ

399 [S]
batter *v.*
F battre
E cañonear
D bombardieren
R колотить; пробивать
   огнём
A ضَرَب . دَكَّ

400 [S]
battery
F batterie *f.*
E batería *f.*
D Batterie *f.*
R батарея *ж.*
A سَرية مِدْفَعية *f*

401 [E]
battery
F pile *f.*
E pila *f.*
D Batterie *f.*
R батарея *ж.*;
   аккумулятор *м.*
A بطارية *f* . مَركَم *m*

402
battle *n.*
F bataille *f.*
E batalla *f.*
D Gefecht *neut.;* Schlacht
   f.
R бой *м.*; битва *ж.*
A مَعْرَكة *f*

403
battle *v.*
F lutter
E luchar
D kämpfen
R вести бой; сражаться
A قَاتَل

404
(order of) battle
F ordre de bataille *m.*
E orden de batalla *m.*
D Einsatzgliederung *f.*
R боевой порядок *м.*
A نظام المعركة *m*

405
battlefield
F champ de bataille *m.*
E campo de batalla *m.*
D Gefechtsfeld *neut.*
R поле боя *ср.*
A ساحَة المَعْرَكة *f*

406
battlefield illumination
F illumination du champ
   de bataille *f.*
E iluminación del campo
   de batalla *f.*
D Gefechtsfeldbeleuchtung
   f.
R освещение поля боя
   ср.
A إضاءة ساحة المَعْرَكة

407
battlefield mobility
F mobilité tous terrains
E movilidad a campo
   traviesa *f.*
D Geländegängigkeit *f.*
R тактическая
   мобильность *ж.*
A سُهُولة الحَرَكة في ساحة
   المَعْرَكة

408
battlefield surveillance
F surveillance du champ
   de bataille *f.*
E observación (continua)
   del campo de batalla
   f.
D Gefechtsfeld-
   überwachung *f.*
R наблюдение за полем
   боя *ср.*
A الإشْراف على ساحة المَعْرَكة

409 [M]
bay
F baie *f.*
E bahía *f.*
D Bucht *f.*
R бухта *ж.*
A شَرْم *m* . خَلِيج *m*

410
bayonet *v.*
F percer d'un coup de
   baïonette
E herir con la bayoneta
D (mit dem) Bajonett
   erstechen
R колоть штыком
A طَعَن (بالحَرْبة . بالسونكي)

25

411

bayonet *n.*
F baïonnette *f.*
E bayoneta *f.*
D Bajonett *neut.*
R штык *м.*
A سونكي . حَرْبة *f* *m*

412

beach *v.*
F (mettre au) sec
E varar
D stranden
R сесть на мель;
высадиться
A اَنْزَل (على الشاطِىء)

413

beach *n.*
F plage *f.*
E playa *f.*
D Strand *m.*
R берег *м.*; пляж *м.*
A شاطِىء *m*

414

beach group
F groupe de plage *m.*
E destacamento de playa
m.
D Strandkommando
neut.; Uferkommando
neut.
R береговая группа *ж.*;
высадочная группа
ж.
A مَجْموعة الشاطِىء *f*

415

beachhead
F tête de pont *f.*
E cabeza de playa *f.*
D Landekopf *m.*
R береговой плацдарм
м.; плацдарм
высадки *м.*
A رأس الشاطِىء *m*

416

beachmaster
F officier régulateur de
plage *m.*
E jefe de desembarco *m.*
D Uferkommandant *m.*
R комендант пункта
высадки *м.*
A قائد الشاطِىء *m*

417 [M]

beacon
F balise *f.*
E baliza *f.*
D Bake *f.*
R маяк *м.*
A مُرْشِد (التَّوْجيه والإنْذار) *m*

418

(fan marker) beacon
F radioborne en éventail
f.
E radiobaliza de abanico
f.
D Fächerfunkfeuer *neut.*
R веерный (маркерный)
радиомаяк *м.*
A مُرْشِد لاسِلْكي *m*

419

(homing) beacon
F balise d'autoguidage *f.*
E baliza automática *f.*
D Anfahrtfunkfeuer *neut.;*
Anflugfunkfeuer
R приводной радиомаяк
м.
A مُرْشِد لاصابة الهدف

420 [A,M]

(light) beacon
F fanal *m.*
E fanal [A] *m.;* aerofaro
[M] *m.*
D Signalfeuer *neut.*
R сигнальный огонь *м.*
A ضَوْء المَنارة *m*

421 [A]

(localizer) beacon
F radiophare
d'alignement de piste
m.
E radiofaro localizador
m.
D Funkleitstrahl *m.*
R посадочный
передатчик *м.*;
посадочный маяк *м.*;
курсовой радиомаяк
м.
A مُرْشِد مُحَدِّد المَوْقِع *m*

422

(pathfinder) beacon
F (radio)- balise
marqueuse *f.*
E (radio) baliza de
señalización *f.*
D Zielanflug(funk)feuer
neut.
R указатель курса *м.*
A مُرْشِد مُحَدِّد المَوْقِع

423

**(personal locator)
beacon**
F balise de repérage
personnelle *f.*
E baliza localizadora
personal *f.*
D persönliches
Platzfunkfeuer *neut.*
R личный радиомаяк *м.*
A جِهاز شَخْصِي لِتَعْيِين المَكَان

424

(radar) beacon
F balise radar *f.;* balise
répondeuse *f.*
E faro radar *m.;* baliza
respondente *f.*
D Radarbake *f.;*
Antwortbake *f.*
R радиолокационный
маяк *м.*
A مُرْشِد رادار *m*

425 [A,M,T]
(radio) beacon
F radio-balise *f.*
E radiofaro *m.*
D Funkfeuer *neut.*
R радиомаяк *м.*
A مُرْشِد لاسِلْكي *m*

426
(Z marker) beacon
F radioborne Z *f.*
E radiobaliza Z *f.*
D Z-Markierungsfeuer
   neut.
R радиомаяк-3 *м.*
A مِصْباح دَليل (زي)

427 [B]
beam *n.*
F poutre *f.*
E viga *f.*
D Balken *m.*
R балка *ж.*
A رافِدة *f* . عارضة *f*

428 [E,R]
beam *n.*
F faisceau *m.*
E haz *m.*
D Strahl *m.*
R луч *м.*
A شُعَاع *m* . حُزْمة *f*

429 [E,R]
beam *v.*
F émettre en faisceau
E emitir en haz
D (mit) Richtstrahler
   senden
R излучать [E];
   передавать [ER]
A شَعَ

430 [G]
beam riding
F guidé sur faisceau
E dirigido en haz; guiado
   por haz
D Leitstrahl- *komp.*
R наведение по лучу *ср.*
A مُوَجِّه بِالرادار

431 [A,M]
bear *v.*
F relever (l'objet) à ...
   degrés
E llevar ... grados
D (bei...Grad) liegen
R двигаться по
   направлению ...
   градусов
A إتِّجه

432
bear *v.*
F porter
E soportar; llevar (armas)
D tragen
R выдерживать; носить
   тяжесть
A تَحَمَّل

433
bearing *n.*
F roulement à billes *m.*
E cojinete *m.*
D Gleitlager *neut.*
R опора *ж.*
A كُرسي تَحميل

434 [A,M]
bearing
F relèvement *m.*
E marcación *f.*
D Peilung *f.;* Peilwinkel
   m.
R азимут *м.*
A إتِّجاه

435
(observer-target)
   bearing
F relèvement
   d'observation *m.*
E marcación observador-
   objetivo *f.*
D Beobachtungs-
   richtungswinkel *m.*
R угол
   наблюдатель-цель
   м.
A الاتِّجاه من الراصد
   إلى الهدف *m*

436
beat *v.*
F battre
E golpear; vencer (al
   enemigo)
D schlagen
R бить
A ضَرَب

437
bed
F lit *m.*
E cama *f.;* cauce *m.;*
   litera *f.*
D Bett *neut.*
R постель *ж.;* кровать
   ж.
A سَرير *m*

438
bedroll *n.*
F sac de couchage *m.*
E ropa de cama *f.*
D Schlafsack *m.*
R скатанная постель *ж.*
A كيس للنوم *m*

439 [M,L]
belay *v.*
F amarrer [M]; assurer [L]
E amarrar
D belegen [M]; fest
   machen [L]
R завёртывать;
   закреплять
A لَفّ الحَبْل . أوْثق بالحبل

440
bell
F cloche *f.*
E campana *f.*
D Glocke *f.*
R колокол *м.;* звонок *м.*
A جَرَس *m*

27

441

belligerent *adj.*

F belligérant
E beligerante
D kriegführend
R воюющий
A مُقَاتِل *m*

442 [A,H,M,V]

belly

F ventre *m.*
E barriga [H] *f.;* panza [AMV] *f.*
D Bauch *m.*
R живот [H] *м.;* трюм [AMV] *м.*
A بَطْن *f*

443 [M]

below

F (en) bas; sous
E abajo
D unten
R внизу
A أَسْفَل . تَحْت

444 [S]

belt

F bande-chargeur *f.*
E cinta *f.*
D Gurt *m.*
R пояс *м.*
A شَرِيط *m* ـ سِير *m*

445 [E]

belt

F courroie *f.*
E correa *f.*
D Riemen *m.*
R лента *ж.*
A نِطَاق *m*

446

(demolition) belt

F zone d'obstacles *f.*
E zona de destrucciones *f.;* zona de obstáculos *f.*
D Sperre *f.*
R зона разрушений *ж.*
A شَرِيط تَدْمِير *m*

447 [S]

(link) belt

F bande-chargeur articulée *f.*
E cinta metálica *f.*
D (Glieder)Gurt *m.*
R звенчатая (патронная) лента *ж.*
A شَرِيط مَفْصِلي *m*

448

(primary demolition) belt

F zone d'obstacles primaire *f.*
E zona de obstáculos principal *f.*
D Hauptsperrzone *f.*
R зона первичных разрушений *ж.*
A شَرِيط تَدْمِير ابْتِدَائِي *m*

449

(safety) belt

F ceinture de sécurité *f.*
E cinturón de seguridad *m.*
D Rettungsgurt [M] *m.;* Sicherheitsgurt [AV] *m.*
R спасательный пояс *м.*
A حِزَام الأمان

450

(seat) belt

F ceinture de sécurité *f.*
E cinturón *m.*
D Sicherheitsgurt *m.*
R предохранительный ремень *м.*
A حزام المقعد

451

(uniform) belt

F ceinturon *m.*
E cinturón *m.*
D Koppel *neut.*
R поясной ремень *м.*
A حِزَام *m*

452

benchmark

F repère *m.*
E cota *f.*
D Nivellierungszeichen *neut.*
R репер *м.;* отметка высоты *ж.*
A عَلامة مِساحة *f*

453

bend *v.*

F plier; virer
E encorvar; torcer
D (sich) biegen
R сгибаться
A عَطَف ـ ثَنى

454 [T]

bend *n.*

F virage *m.*
E curva *f.*
D Biegung *f.*
R сгиб *м.*
A مُنْعَطَف *m*

455 [M,H]

bends

F mal des caissons *m.*
E embolía gaseosa *f.*
D Luftdruckkrankheit *f.*
R кессонная болезнь *ж.*
A مَرَض إخْتِلاف الضَّغْط

456

berth *v.*

F accoster; mouiller
E atracar
D festmachen *tr.*
R ставить на якорь
A رَبَط بِالرَّصِيف

457

berth (for ship) *n.*

F poste *m.*
E amarradero *m.*
D Liegeplatz *m.*
R причал *м.;* якорная стоянка *ж.*
A مَرْسَى *m*

458

between decks *n.*

F entrepont *m.*

E entrepuente *m.*

D Zwischendeck *neut.*

R междупалубное
  пространство *ср.*

A بَيْن السُّطوح

459

**beyond economic
  repair**

F classé hors service

E reparación
  antieconómica *f.*

D aussonderungsreif

R (полностью)
  неисправный

A لا يُمْكِن إصلاحُه إقتصادياً

460

beyond local repair

F irréparable localement

E reparación en otro
  escalón *f.*

D reif für
  Depotinstandsetzung

R неремонтируемый на
  месте

A لا يمكن إصلاحه محلياً

461

bight (of rope)

F double *m.*

E gaza *f.*

D Taubucht *f.*

R шлаг *м.*

A عُرْوَة ـ عُقدة *f*

462

bilateral

F bilatéral

E bilateral

D bilateral; zweiseitig

R двусторонний

A ذو جانِبَيْن . ثُنائي

463

bilge

F cale *f.*

E pantoque *m.*

D Schiffsboden *m.*

R днище *ср.*

A قَعْر . قَاع *m*

464

billet *v.*

F cantonner

E alojar

D einquartieren *tr.*

R помещать; размешать

A أوى

465

billet *n.*

F logement *m.*

E alojamiento *m.*

D Quartier *neut.*

R помещение *ср.*;
  размещение *ср.*

A مَأْوى *m*

466 [M,H]

bill of health

F patente de santé *f.*

E patente de sanidad *m.*

D Gesundheitspass *m.*

R карантинное
  свидетельство *ср.*

A بَرَاءة الصِّحَّة
  كل شيء تمام

467

bill of lading

F connaissement *m.*

E conocimiento *m.*

D Konossement *neut.*

R коносамент *м.*;
  накладная *ж.*

A قَائِمة الشُّحْن *f*

468 [P,H]

bind *v.*

F lier [P]; panser [H]

E atar [P]; vendar [H]

D binden

R вязать; связывать;
  привязывать

A شَدّ . رَبَط

469

binding *n.*

F fixation *f.*

E fijación *f.*

D Bindung *f.*

R оковка *ж.*; связь *ж.*;
  сращивание *ср.*

A رَبْط *m*

470

binnacle

F habitacle *m.*

E bitácora *f.*

D Kompasshaus *neut.*

R нактоуз *м.*

A عُلْبة البوصَلة *f*

471

binocular *adj.*

F binoculaire

E binocular

D binokular (für beide
  Augen)

R бинокулярный

A ذو عَيْنَيْن

472

binoculars *pl.*

F jumelles *f. pl.*

E gemelos prismáticos *m.
  pl.*

D Doppelglas *neut.;*
  Fernglas *neut.*

R бинокль *м.*

A مِنْظار (ثنائي العَيْنية) *m*

473

biological

F biologique

E biológico

D biologisch

R биологический

A أَحْيَائي

29

474
biological agent
F agent biologique *m.*
E agente biológico *m.*
D biologischer Kampfstoff
   m.
R биологическое
   средство *ср.*
A عامِل أحْيَائي *m*

475
bird
F oiseau *m.*
E ave *f.*
D Vogel *m.*
R птица *ж.*
A طَيْر *m*

476 [A]
bird-strike
F dommages causés par
   un oiseau dans un
   réacteur d'avion
E avería causada por una
   ave dentro del reactor
   de un avión
D Vogelschlag *m.*
R столкновение с
   птицами в воздухе
   ср.
A إصْطِدام بِطَيْر

477 [D]
bit
F bit *m.*
E bit *m.;* bitio *m.*
D Bit *neut.;* Datenbit
   neut.
R бит *м.;* двоичная
   цифра *ж.*
A وحدة ذاكرة الكومبيوتر *f*

478
bivouac *v.*
F bivouaquer
E vivaquear
D biwakieren
R располагаться
   биваком; стоять
   биваком
A خَيَّم

479
bivouac *n.*
F bivouac *m.*
E vivaque *m.*
D Biwak *neut.*
R бивак *м.*
A مُخَيَّم *m* خيمة صغيرة *f*

480 [E]
blade
F ailette *f.*
E paleta *f.*
D Schaufel *f.*
R лопатка *ж.;* лопасть
   ж.
A نَصْل *m*

481 [A,M]
blade
F pale *f.*
E pala *f.*
D Flügel *m.*
R клинок [A] *м.;* руль
   [M] *м.*
A رِيشة *f*

482 [U]
blade
F lame *f.*
E hoja *f.*
D Blatt *neut.*
R лист *м.*
A شَفْرة *f*

483
blank cartridge
F cartouche à blanc *f.*
E cartucho de fogueo *m.*
D Platzpatrone *f.*
R холостой патрон *м.*
A طلقة فارغة *f*

484
blanket *n.*
F couverture *f.*
E manta *f.*
D Decke *f.*
R одеяло *ср.*
A بَطَّانية *f*

485 [S]
blanket *v.*
F dérober aux vues
E cegar; neutralizar
D abschirmen *tr.*
R покрывать
A حَجَب

486
blast *n.*
F souffle *m.*
E sacudida *f.*
D Luftdruck *m.*
R порыв *м.;* взрыв *м.*
A عَصْف *m*

487 [B]
blast *v.*
F (faire) sauter
E volar
D sprengen
R взрывать; дуть
A نَسَف

488 [N]
blast effect
F effet de souffle *m.*
E resultados de sacudida
   m. pl.
D Explosionswirkung *f.*
R действие взрывной
   волны *ср.*
A تأثير الإنفجار

489 [E]
bleed
F purger
E purgar
D entlüften
R спускать; продувать
A صرَف

490 [H]
bleed
F saigner
E sangrar
D bluten
R кровоточить
A نزَف

491 [U]
blind *adj.*
F aveugle
E ciego
D blind
R слепой
A أَعْمَى

492 [S]
blind *n. adj.*
F non-éclaté
E fallido
D Blindgänger *m.*
R неразорвавшийся
A قَذِيفة عَمْياء (غَيْر مُنْفَجِرة) *f*

493 [A]
blind
F (en) pilotage sans
   visibilité (en PSV) *m.*
E (a) ciegas
D blind *adj.;*
   Instrumenten- *komp.*
R слепой; (по) приборам
A أَعْمَى

494 [R]
blip
F blip *m.*
E cresta de eco *f.;*
   indicación visual *f.*
D Signal *neut.*
R отметка *ж.*; выброс
   м.
A نُقطة (أَوْ نَبْضة) على شاشة
   الرادار

495 [E,U]
block *n.*
F bloc *m.*
E bloque *m.*
D Block *m.*
R блок [U] *м.*; затор [E]
   м.
A كُتْلة *f*

496
block *v.*
F barrer
E cerrar
D sperren
R преграждать;
   задерживать
A سَدَّ . حَصَرَ . حَجَز

497
blockade *n.*
F blocus *m.*
E bloqueo *m.*
D Blockade *f.*
R блокада *ж.*
A حِصَار *m*

498
blockade *v.*
F bloquer; (faire le)
   blocus
E bloquear
D blockieren
R блокировать
A حاصَر

499
block and tackle
F palan *m.*
E aparejo *m.*
D Flaschenzug *m.*
R тали *м. мн. ч.*;
   такелаж *м.*
A بَكْرة وحبل

500
blood
F sang *m.*
E sangre *f.*
D Blut *neut.*
R кровь *ж.*
A دَم *m*

501 [H]
blood groups
F groupe sanguin *m.*
E grupos sanguíneos *m.*
   pl.
D Blutgruppen *f. pl.*
R группа крови *ж.*
A فصيلة الدم *f*

502 [H]
blood pressure
F pression artérielle *f.*
E presión sanguínea *f.*
D Blutdruck *m.*
R давление крови *ср.*
A ضغط الدم *m*

503
blow *v.*
F souffler
E soplar
D blasen
R дуть
A نَفَخ . عَصَف

504 [E]
blow (a fuze) *v.*
F (faire) fondre (un
   fusible)
E quemar(se) (un fusible)
D (eine Sicherung)
   durchbrennen *tr.*
R делать короткое
   замыкание;
   перетокать
A إحْتَرَق (صِمام)

505
blowtorch
F chalumeau *m.*
E soplete *m.*
D Lötlampe *f.*
R паяльная лампа *ж.*
A مَوْقِد لِحَام *m*

506
blow up
F (faire) sauter
E volar
D sprengen
R взрывать; подрывать
A نَسَف

507 [T]
bluff *n.*
F promontoire *m.*
E risco *m.*
D Steilufer *neut.*
R обрыв *м.*; отвесный
   берег *м.*
A جرف شديد الإنْحِدار *m*

31

508
bluff *v.*
F bluffer
E (hacer un) blof
D täuschen
R обманывать;
   запугивать
A خَدَع

509 [U]
bluff *n.*
F bluff *m.*
E blof *m.*
D Irreführung *f.*
R обман *м.*; запугивание
   ср.; блеф *м.*
A خُدْعَة *f*

510 [N]
board *v.*
F aborder
E abordar
D entern
R сесть на корабль;
   брать на абордаж
A رَكِب (سَفينة) . إقْتَحَم
   سفينة

511
boat
F embarcation *f.*
E barca *f.;* bote *m.*
D Boot *neut.*
R лодка *ж.*
A قارِب *m* - زَوْرَق *m*

512
boatload
F plein bateau (personnel)
   m.
E barcada *f.*
D Bootsladung *f.*
R перевозимая группа
   ж.
A قارِب مَلآن

513
body
F corps *m.*
E cuerpo *m.*
D Körper *m.*
R тело *ср.*
A جِسْم *m*

514
body armor
F gilet pareballes *m.*
E protección corporal
   personal *f.*
D Körperpanzer *m.*
R бронежилет *м.*
A دِرْع شَخْصي *m*

515 [P]
bodyguard
F garde du corps *m.*
E guardaespaldas *m. inv.;*
   guardia personal *f.*
D Leibwache *f.*
R телохранитель *м.*
A حارِس شخصي

516
boiler
F chaudière *f.*
E caldera *f.*
D Dampfkessel *m.*
R (паровой) котел *м.*
A مِرْجَل *m* . غلاية *f*

517
bollard
F pieu *m.*
E bolardo *m.*
D Poller *m.*
R пал *м.*; тумба *ж.*
A عَمُود الرَّبْط *m*

518 [V]
bolster
F sommier *m.*
E soporte *m.;* travesaño
   m.
D Querstrebe *f.*
R подкладка *ж.*; брус *м.*
A المِسْنَد

519 [E]
bolt *n.*
F boulon *m.*
E perno *m.*
D Bolzen *m.*
R болт *м.*
A تِرْبَاس *m*

520 [S]
bolt *n.*
F culasse *f.*
E cerrojo *m.*
D Schloss *neut.*
R затвор *м.*
A تِرْبَاس *m*

521
bomb *n.*
F bombe *f.*
E bomba *f.*
D Bombe *f.*
R бомба *ж.*
A قَنْبُلَة *f*

522
bombardier
F bombardier *m.*
E bombardero *m.*
D Bombenschütze *m.*
R бомбардир *м.*; капрал
   артиллерии *м.*
A رامٍ . مِدفَعي *m*

523
bombardment
F bombardement *m.*
E bombardeo *m.*
D Bombardement *neut.;*
   Beschiessung *f.*
R бомбардировка *ж.*
A قَصْف *m*

524
bomb bay
F soute à bombes *f.*
E almacén de bombas *m.*
D Bombenschacht *f.*
R бомбовый отсек *м.*
A مَخْزَن القَنَابِل *m*

525

bomb *v.;* **bombard**
F bombarder
E bombardear
D bombardieren
R бомбардировать
A قَصَف

526

bomb disposal
F déminage *m.*
E evacuación de bombas fallidas *f.*
D Bombenräumung *f.*
R обезвреживание неразорвавшихся боеприпасов *ср.*
A تأمين القنابل

527

bomb disposal squad
F équipe de déminage *f.*
E equipo de neutralización de bombas *m.*
D Bombensprengkommando *neut.*
R отделение обезвреживания *ср.;* отряд обезвреживания *м.*
A جَمَاعة تأمين القَنَابِل *f*

528 [A]

bomber
F bombardier *m.*
E bombardero *m.*
D Bomber *m.*
R бомбардировщик *м.*
A قاذِفة *f*

529

bombing angle
F angle de bombardement *m.*
E ángulo de bombardeo *m.*
D Bombenabwurfwinkel *m.*
R угол бомбометания *м.*
A زاوية إلْقَاء القَنَابِل *f*

530

bomblet
F petite bombe *f.*
E bomba pequeña *f.*
D Bomblet *neut.;* Kleinstbombe *f.;* Tochterbombe *f.*
R бомба малого калибра *ж.*
A قُنَيْبِلة ـ قُنْبُلة صغيرة *f*

531

bombproof
F blindé
E (a) prueba de bombas *inv.*
D bombensicher *adj.*
R непробиваемый бомбами
A صامِد للقنابِل

532

bombsight
F viseur *m.*
E visor de bombardeo *m.*
D Bombenzielgerät *neut.*
R бомбардировочный прицел *м.;* авиаприцел *м.*
A جِهَاز تَصْويب القَنابِل *m*

533 [M]

boom *n.*
F barrage *m.*
E barrera *f.*
D Sperre *f.*
R плавучий бон *м.*
A حاجِز عائم *m*

534 [A,S]

(sonic) boom *n.*
F détonation (sonique) *f.*
E estampido (sónico) *m.*
D (Überschall-)Knall *(komp.) m.*
R звуковой толчок *м.*
A دَوي إخْتِرَاق جِدار الصَوْت *m*

535 [G]

booster
F fusée de lancement *f.*
E cohete secundario *m.*
D Startrakete *f.*
R стартовая ракета *ж.;* ракетный ускоритель *м.*
A مُعَزِّز *m*

536 [H]

booster
F rappel *m.*
E inyección secundaria *f.*
D Wiederholungsimpfung *f.*
R усилительная доза *ж.*
A مُنَشِّط *m*

537 [R]

booster
F propulseur auxiliaire *m.*
E repetidor *m.*
D Hilfsmotor *m.;* Zusatzmotor *m.*
R ускоритель *м.*
A مُعَزِّز *m*

538

boot
F brodequin *m.*
E bota *f.*
D Stiefel *m.*
R ботинок *м.;* сапог *м.*
A حِذَاء ذورقبة *m*

539 [T]

border *n.*
F frontière *f.*
E frontera *f.*
D Grenze *f.*
R граница *ж.;* рубеж *м.*
A حُدود *f*

540 [B]

bore *v.*
F sonder
E sondar
D anbohren *v. tr.*
R сверлить
A ثَقَب

541
bore *n.*
F âme *f.*
E alma *f.*
D Seele *f.*
R канал ствола *м.*
A جَوْف *m* . قُطْر *m*

542
bosun
F maître d'équipage *m.*
E contramaestre *m.*
D (Ober-)Bootsmann *m.*
R боцман *м.*
A كبيرمَلاحي السفينة

543
bottle
F bouteille *f.*
E botella *f.*
D Flasche *f.*
R бутылька *ж.*
A قِنينة *f* . زُجاجة *f* .
قارورة *f*

544 [L]
bound
F bond *m.*
E salto *m.*
D Sprung *m.*
R бросок *м.;* переброска
ж.
A قفزة *f*

545 [L]
boundary
F limite *f.*
E límite *m.*
D Grenze *f.*
R рубеж *м.;* граница *ж.*
A حَدّ *m*

546
(out of) bounds
F accès interdit *m.*
E fuera de los límites
D Zutritt verboten
R запрещён(ный)
A خارج حدود المرور

547 [M]
bow
F avant *m.*
E proa *f.*
D Bug *m.*
R нос *м.*
A المقدَّمَة *f*

548
bowline
F nœud de chaise *m.*
E as de guía *m.*
D (einfacher) Palstek *m.*
R булинь *м.;*
беседочный узел *м.*
A حَبْل الشِّراع *m* . عُقْدَة
لِرَبْط الحِبال *f*

549
box *n.*
F boîte *f.*
E caja *f.*
D Kasten *m.*
R коробка *ж.;* ящик *м.*
A صَندوق *m*

550 [L]
box in *v.*
F cerner
E encerrar
D abriegeln *tr.*
R окружать
A يُحاصِر

551 [S]
bracket *v.*
F encadrer
E horquillar
D eingabeln *tr.*
R захватывать в вилку
A حَصَر (الهَدَف)

552 [E,U]
bracket *n.*
F support *m.*
E soporte *m.*
D Träger *m.*
R кронштейн [E] *м.;*
скоба [EU] *ж.*
A سَند *m* . حامِل *m*

553 [S]
bracket *n.*
F fourchette *f.*
E horquilla *f.*
D Gabel *f.*
R вилка *ж.*
A الحاصِرَة *f*

554
brain
F cerveau *m.*
E cerebro *m.*
D Gehirn *neut.*
R мозг *м.*
A دِماغ *m*

555
brain damage
F dommage cérébral *m.*
E herida cerebral *f.*
D Hirnverletzung *f.*
R повреждение мозга *ср.*
A ضَرَر الدِماغ *m*

556
brake *v.*
F freiner
E frenar
D bremsen
R тормозить
A كَبَح

557
brake *n.*
F frein *m.*
E freno *m.*
D Bremse *f.*
R тормоз *м.*
A كابِحة *f*

558 [V]
(steering) brake
F frein de direction *m.*
E freno de dirección *m.*
D Steuerbremse *f.*
R рулевой тормоз *м.*
A فَرامِل التوجيه

559
brake fluid
F liquide pour freins *m.*
E líquido para frenos *m.*
D Bremsflüssigkeit *f.*
R тормозная жидкость
  ж.
A زَيْت الكابِحة *m*

560
brake lining
F garniture (de frein) *f.*
E forro (del freno) *m.*
D Bremsbelag *m.*
R тормозная подкладка
  ж.
A بِطانة الكابِحة *f*

561
brake pedal
F pédale de frein *f.*
E pedal de freno *m.*
D Bremspedal *neut.*
R тормозная педаль *ж.*
A دَوّاسَة الكابِحة *f*

562 [A,L,M]
branch *n.*
F arme *f.*
E arma *f.*
D Waffengattung *f.*
R род войск *м.*; служба
  ж.
A سِلاح *m*

563 [T]
branch *v.*
F bifurquer
E separarse
D abzweigen *v. tr.*
R разветвляться
A تَفَرَّع

564 [U]
branch *n.*
F branche *f.*
E rama *f.*
D Zweig *m.*
R ветка *ж.*
A فَرْع *m*

565 [T,M]
branch *n.*
F bras (voie d'eau) *m.*
E brazo (vía fluvial) *m.*
D Arm (Gewässer) *m.*
R рукав (реки) *м.*
A فَرْع *m*

566
brass
F laiton *m.*
E latón *m.*
D Messing *neut.*
R латунь *ж.*; жёлтая
  медь *ж.*
A نُحاس أَصْفَر *m*

567
brassard
F brassard *m.*
E brazal que lleva
  insignia de grado *m.*
D Armbinde *f.;* Armband
  neut.
R нарукавная повязка
  ж.
A علامة تَمييز *f*

568
breach *n.*
F brèche *f.*
E brecha *f.*
D Bresche *f.*
R брешь *ж.*; пролом *м.*;
  отверстие *ср.*
A ثُغْرة *f*

569
breach *v.*
F ouvrir une brèche
E abrir brecha
D (eine) Bresche schlagen
  v.; überwinden
  (Minenfeld) *tr.*
R проламывать брешь;
  пробивать брешь
A فَتَح ثُغْرة

570
breadth
F largeur *f.*
E anchura *f.*
D Breite *f.*
R широта *ж.*
A عَرْض

571 [L,T]
break *n.*
F trouée *f.*
E abertura *f.*
D Unterbrechung *f.*
R прорыв *м.*
A إِنْقِطاع *m*

572
break *v.*
F rompre
E romper
D brechen
R ломать; разбивать
A كَسَر

573 [U]
break *n.*
F rupture *f.*
E ruptura *f.*
D Bruch *m.*
R прорыв *м.*; пролом *м.*
A كَسْر *m*

574 [M]
break bulk *v.*
F désarrimer
E desembalar
D (die) Last brechen
R начинать разгрузку
A فَكَّ الحُمَولة

575 [L]
break bulk *v.*
F subdiviser
E repartir
D anbrechen *tr.*
R подразделять;
  разделять
A فَكَّ الحُمَولة

35

576 [M,V]
breakdown *n.*
F panne *f.*
E avería *f.*
D Panne *f.*
R поломка *ж.*
A عُطْل *m*

577 [M,V]
break down *v.*
F (être en) panne
E averiarse
D (eine) Panne haben
R поломать
A تَعَطَّل

578 [E,F]
break down *v.*
F démonter [E]; détailler [F]
E analizar
D einteilen *tr.*
R разбивать (на категории); анализировать
A خَصَّص

579 [E,F]
breakdown *n.*
F démontage [E] *m.;* détail [F] *m.*
E análisis *m.*
D Einteilung *f.*
R разборка *ж.;* анализ *м.*
A تخصيص

580
breakers
F brisants *m. pl.*
E rompientes *m. pl.*
D Brandung *f.*
R буруны *м. мн. ч.*
A أَمْوَاج تَنْكَسِر على الشَّاطِىء . مُكسرات

581
break-in battle
F bataille de rupture *f.*
E batalla de ruptura *f.*
D Einbruchskampf *m.*
R вклинивающий бой *м.*
A مَعْرَكَة الاقتِحَام

582 [P,U]
break out *v.*
F (s')échapper
E estallar [U]; evadirse [P]
D ausbrechen *tr.*
R выламывать
A إنْفَجَر

583 [L]
breakout *n.*
F action de rupture *f.*
E rotura *f.*
D Ausbruch *m.*
R выход из окружения *м.*
A إخْتِراق الحصار

584 [L]
break through *v.*
F (faire une) percée
E efectuar una ruptura
D durchbrechen *tr.*
R прорывать
A إخْتَرَق

585 [L]
breakthrough *n.*
F percée *f.*
E ruptura *f.*
D Durchbruch *m.*
R прорыв *м.*
A إخْتِراق *m*

586
breakup *n.*
F bris *m.*
E desmembración *f.*
D Zerbrechen *neut.*
R развал *м.;* разруха *ж.*
A تَفَرُّق

587
breakwater
F brise-lames *m. inv.*
E rompeolas *m. inv.*
D Wellenbrecher *m.*
R волнолом *м.;* мол *м.*
A حَاجِز الأَمْوَاج

588
breathe
F respirer
E respirar
D atmen
R дышать
A تَنَفُّس

589
breathing apparatus
F appareil respiratoire *m.*
E aparato de respiración *m.*
D Atemgerät *neut.*
R дыхательный аппарат *м.*
A جِهَاز التَنَفُّس

590
breech(-block)
F culasse *f.*
E cierre *m.*
D Verschluss (-Stück) *m.* (*neut.*)
R казённая часть *ж.;* затвор *м.*
A مِغْلَاق *m* (كُتْلة المِغْلَاق *f*)

591
breeches buoy
F bouée-culotte *f.*
E boya pantalón *f.*
D Hosenboje *f.*
R спасательный круг со штанами *м.*
A طوق النَّجَاة *m*

592
breeze
F brise *f.*
E brisa *f.*
D Brise *f.*
R бриз *м.;* ветерок *м.*
A نَسِيم *m* . ريح *m*

593
brevity code
F code condensé *m.*
E código de brevedad *m.*
D Abkürzung *f.*
R код краткости *м.*
A رَمز وَجِيز *m*

594
bridge *v.*
F jeter un pont sur
E tender un puente sobre;
llenar (una brecha)
D überbrücken *untr.*
R соединять мостом
A جَسَّرَ . بَنَى الْجُسُور

595 [L,T]
bridge *n.*
F pont *m.*
E puente *m.*
D Brücke *f.*
R мост *м.*
A جِسْر *m*

596 [M]
bridge *n.*
F passerelle (de
commandement) *f.*
E puente de mando *m.*
D Kommandobrücke *f.*
R мостик *м.*
A عَبَّارة *f*

597
**(armored vehicle
launched) bridge
(AVLB)**
F char poseur de pont *m.*
E carro-puente *m.*
D Brückenlegepanzer *m.*
R (танковый)
бронированный
мостоукладчик
(БМТУ) *м.*
A جِسْر يُنزَل من مَرْكَبة مُدَرَّعة

598
bridgehead
F tête de pont *f.*
E cabeza de puente *f.*
D Brückenkopf *m.*
R плацдарм *м.*
A رَأْس جِسْر *m*

599
bridging (equipment)
F matériel de pontage *m.*
E material de puentes *m.*
D Übersetzungsmaterial
neut.
R переправочно-мостовые
средства *ср. мн. ч.*
A أَجْهزة التَّجْسير *f*

600
brief *v.*
F briefer
E (dar) instrucciones (a);
poner al corriente (de)
D einweisen *tr.*
R инструктировать
A أَوْجَز . أَعطى التَّعْليمات

601
briefcase
F serviette *f.*
E cartera *f.*
D Aktenmappe *f.*
R портфель *м.*
A شَنْطة *f*

602
briefing *n.*
F briefing *m.*
E (reunión para)
instrucciones *f.*
D Einweisung *f.*
R инструктаж *м.*;
брифинг *м.*
A إيجاز . إعْطاء التَّعْليمات *m*

603
broad
F large
E ancho
D breit
R широкий
A عَريض

604
broadcast *v.*
F radiodiffuser
E emitir
D (durch Radio)
übertragen *untr.*
R передавать по радио
A أذاع

605 [U]
broadcast *n.*
F émission
(d'informations) *f.*
E emisión *f.*
D Rundfunksendung *f.*
R радиопередача *ж.*
A إذاعة *f*

606 [M]
broadcast *n.*
F système amplificateur
m.
E sistema amplificador *m.*
D Lautsprecheranlage *f.*
R радиовещательная
сеть *ж.*
A إذاعة *f*

607 [E,U]
broken
F rompu
E averiado [E]; roto [U]
D kaputt
R разбитый [EU];
выведенный из строя
[EU];
неисправный[E]
A مَكْسُور

608 [T]
broken
F accidenté
E accidentado
D durchschnitten
R сломанный
A وَعْر

609 [M]
broken water
F clapotis *m.*
E hileros *m. pl.*
D Kabbelung *f.*
R толчея *ж.*
A مِياه وَعْرة *f*

610
bubble
F bulle (d'air) *f.*
E burbuja *f.*
D (Luft-)Blase *f.*
R пузырь *м.*
A فُقَّاعة *f*

611
bucket
F seau *m.*
E cubo *m.*
D Eimer *m.*
R ведро *ср.*
A سَطْل *m*

612 [U]
buckle *v.*
F boucler
E abrochar (con hebilla)
D (an- um- zu-)
  schnallen *tr.*
R застёгивать пряжку
A ثَبَّت (بالإبْزيم . بالمِشْبَك)

613 [T]
buckle *v.*
F (se) déformer
E torcer(se)
D (sich) verziehen
R сгибаться
A تَحَدَّب

614 [U]
buckle *n.*
F boucle *f.*
E hebilla *f.*
D Schnalle *f.*
R пряжка *ж.*
A إبْزيم ـ مِشْبَك *m*

615
budget
F budget *m.*
E presupuesto *m.*
D Haushalt *m.*
R бюджет *м.*
A مِيزانية *f*

616 [S]
buffer *n.*
F frein (de tir) *m.*
E freno (de retroceso) *m.*
D Rückstossdämpfer *m.*
R амортизатор *м.*
A مُخَمِّد . مُخَفِّف الصدمة *m*

617
buffer *n.*
F tampon *m.*
E tope *m.*
D Puffer *m.*
R буфер *м.*
A مِهْماد . مِصَدّ *m*

618
build *v.*
F construire
E construir
D bauen
R строить
A بَنَى

619
building
F bâtiment *m.*
E edificio *m.*
D Gebäude *neut.*
R здание *ср.*; постройка
  ж.
A بِناء *m*

620
buildup *n.*
F renforcement *m.*
E acumulación de fuerzas
  f.; concentración de
  esfuerzos *f.*
D Aufbau *m.*
R наращивание *ср.*
A تَقْوِية *f*

621
build up *v.*
F renforcer
E acrecentar
D aufbauen *tr.*
R наращивать
A قَوَّى

622 [T]
built up area
F zone urbanisée *f.*
E zona urbanizada *f.*
D geschlossene Ortschaft
  f.; bebautes Gelände
  neut.
R населённый пункт *м.*
A مساحة مُشَيَّدة *f*

623 [E]
bulb
F ampoule électrique *f.*
E bombilla *f.*
D Birne *f.*
R лампочка *ж.*
A بَصَلَة *f*

624
(in) bulk
F (en) vrac
E (a) granel
D lose
R навалом
A بالجُمْلة

625
bulkhead
F cloison *f.*
E mamparo *m.*
D Schott *neut.*
R переборка *ж.*
A حاجِز ـ مانِع *m*

626 [S]
bull's eye
F centre de la cible *m.*
E blanco *m.*
D Zentrum *neut.*;
  Kernschuss *m.*
R центр мишени *м.*;
  яблоко мишени *ср.*
A مَركَز الهدف *m*

627
bulldozer
F bulldozer *m.*
E motoniveladora *f.*
D Planierraupe *f.*
R бульдозер *м.*
A مِسْلَفَة كاشِطة _ *f*
(بُلْدُوزَر) *m*

628
bullet
F balle *f.*
E bala *f.*
D Kugel *f.*
R пуля *ж.*
A رَصاصة *f*

629
bulletin
F bulletin *m.*
E boletín *m.*
D Bericht *m.*
R бюллетень *м.*
A نَشْرَة *f* . بَيان *m*

630
bulletin board
F tableau d'affichage *m.*
E tablón de anuncios *m.*
D Anschlagtafel *f.;*
  schwarzes Brett *neut.*
R доска для объявлений
  ж.
A لَوْحة النَشَرات *f*

631
bulwark(s)
F pavois *m.*
E macarrón *m.*
D Verschanzung *f.*
R фальшборт *м.;* оплот
  м.
A أعْلى ظَهْر السَّفينة

632 [U]
bump *n.*
F choc *m.*
E choque *m.*
D Stoss *m.*
R столкновение *ср.*
A صَدْمَة *f*

633 [U]
bump *v.*
F (se) heurter (contre)
E chocar (contra)
D stossen (gegen)
R сталкиваться (с)
A صَدَم

634 [V]
bumper
F pare-chocs *m. inv.*
E parachoques *m. inv.*
D Stossstange *f.*
R буфер *м.*
A واقية مِن الصَّدْم *f*

635 [A]
bumpy
F cahoteux
E agitado
D böig
R ухабистый; тряский
A مُضْطَرِب

636 [L]
bunch *v.*
F (s')entasser
E juntarse
D (sich) zusammenziehen
  tr.
R скучиваться
A شكَّل مجموعة أثناء التعرض
  للنيران

637
bung *n.*
F bondon *m.*
E bitoque *m.*
D Spund *m.*
R пробка *ж.;* затычка
  ж.; втулка *ж.*
A مِصَدّ *m*

638 [M]
bunk *n.*
F couchette *f.*
E litera *f.*
D Koje *f.*
R койка *ж.*
A سَرير *m*

639 [L]
bunker
F blockhaus *m.*
E refugio *m.*
D Bunker *m.*
R бункер *м.;* бетонное
  пулемётное гнездо
  ср.
A دُشمة *f*

640
buoy
F bouée *f.*
E boya *f.*
D Boje *f.*
R буй *м.*
A طافية *f* . عَوَّامة *f*

641
buoyancy
F flottabilité *f.*
E flotabilidad *f.*
D Schwimmkraft *f.*
R плавучесть *ж.*
A قابِلية العَوْم

642
buoyant
F flottable
E boyante
D schwimmfähig
R плавучий
A طافٍ

643
bureau
F bureau *m.*
E oficina *f.*
D Büro *neut.*
R бюро *ср.*
A مَكْتَب _ *m* إدارة *f*

644
burn *v.*
F brûler
E quemar(se)
D (ver)brennen
R жечь; сжигать
A إحْتَرَق

39

645 [H]

burn *n.*

F brûlure *f.*
E quemadura *f.*
D Brandwunde *f.*
R ожог *м.*
A حُرْقَة *f* ـ حَرْق *m*

646 [G]

burnout

F fin de combustion *f.*
E fin de combustión *m.*
D Brennschluss *m.*
R окончание выгорания
   ср.
A إنْطِفاء

647 [S,G]

burst *n.*

F éclatement (obus) *m.;*
   rafale (mitrailleuse) *f.*
E explosión (granada) *f.;*
   ráfaga (ametralladora)
   f.
D Sprengpunkt (Granate)
   m.; Feuerstoss (MG)
   m.
R разрыв (граната) *м.;*
   очередь (пулемёта)
   ж.
A دَفْعَة (نيران) *f*

648

burst *v.*

F éclater
E reventar
D sprengen
R разрываться;
   взрываться
A انْفَجَر

649

bury *v.*

F enterrer
E enterrar
D vergraben
R погребать; хоронить
A دَفَن

650 [T]

bush

F buisson *m.*
E arbusto *m.*
D Gesträuch *neut.*
R куст *м.*
A دَغْل *m*

651 [E]

bush

F bague (de pallier) *f.*
E forro *m.*
D Buchse *f.*
R втулка *ж.;* вкладыш
   м.
A خاتِم *m* ـ حلقة *f*

652 [S]

butt (of rifle)

F crosse *f.*
E culata *f.*
D (Gewehr-) Kolben *m.*
R приклад (винтовки) *м.*
A أخْمَص *m*

653 [U]

button *v.*

F boutonner
E abotonar; abrochar
D (zu-)knöpfen
R застёгивать(ся) на
   пуговицы
A زَرَّر

654 [U]

button *n.*

F bouton *m.*
E botón *m.*
D Knopf *m.*
R кнопка *ж.;* пуговица
   (одежды) *ж.*
A زِرّ *m*

655 [V]

buttoned up

F (aux) volets fermés
E (con) escotillas cerradas
D (mit) geschlossenen
   Luken
R (с) закрытыми
   люками
A مَزْرُور ـ مُزَرَّر

656

butts

F butte de tir *f.*
E espaldón *m.*
D Scheibenstand *m.*
R мишенный вал *м.*
A جدار حجز الطلقات *m*

657 [S]

buttstock

F crosse *f.*
E culata *f.*
D Stossfuge *f.*
R приклад *м.*
A أخْمَص *m*

658 [L]

bypass *v.*

F contourner
E evitar
D umgehen *untr.*
R обходить; объезжать
A تَجَنَّب

659 [T]

bypass *n.*

F dérivation *f.*
E desvío de carretera *m.;*
   ruta de desvío *f.*
D Umgehungsstrasse *f.*
R обход
A مَمَرَّ فَرْعِي *m*

C

660 [V]
cab
F cabine *f.*
E cabina *f.*
D Führerhaus *neut.*
R кабина (водителя) *ж.*
A مَقْصُورة السائِق *f*

661 [A]
cabin
F soute *f.*
E cabina *f.*
D Kabine *f.*
R кабина *ж.*
A غُرْفة *f*

662 [M]
cabin
F cabine *f.*
E camarote *m.*
D Kajüte *f.*
R каюта *ж.*
A غُرْفة *f*

663 [E,M]
cable
F câble *m.*
E cable *m.*
D Kabel *neut.;*
   Kabellänge [M] *f.*
R кабель [E] *м.*; канат
   [EM] *м.*; трос [M] *м.*
A حَبْل أسْلاك *m*

664
cache
F cache *f.*
E alijo *m.*
D Lager *neut.;* Versteck
   neut.
R тайный склад оружия
   м.
A مُخبأ *m*

665
cadet
F élève officier *m.*
E cadete *m.*
D Kadett *m.;*
   Offiziersanwärter *m.*
R кадет *м.*
A طالب عسكري *m*

666
cadre
F cadre *m.*
E cuadro *m.*
D Kader *m.*
R кадры *мн. ч.*
A كادر *m*

667 [E,M]
caisson
F caisson *m.*
E cajón *m.*
D Senk- *komp.;*
   Schwimmkasten *m.*
R кессон *м.*
A قَيْسُون *m*

668
caliber
F calibre *m.*
E calibre *m.*
D Kaliber *neut.*
R калибр *м.*
A عِيَار *m*

669 [S]
calibrate
F calibrer
E calibrar
D kalibrieren
R состреливать
A عَايَر

670 [E]
calibrate
F étalonner
E calibrar
D eichen
R калибровать
A ضبط العيار

671
calipers *pl.*
F compas à calibrer
E calibrador *m.*
D Greifzirkel *m.*
R кронциркуль *м.*;
   нутромер *м.*
A أداة قِياس ذات فَكَّيْن *f* .
   فِرْجَار *m*

672 [R]
call *v.*
F appeler
E llamar
D anrufen *tr.*
R звать; вызывать
A نَادَى

673
call for
F demander
E pedir
D anfordern *tr.*
R требовать
A إسْتَدْعَى

674 [R]
call number
F indicatif d'appel *m.*
E prefijo de llamada *m.*
D Kennziffer *f.*
R позывной
A رقم استدعاء

675
call up *v.*
F appeler sous les
   drapeaux
E llamar al servicio
   militar
D aufrufen *tr.;* einberufen
   tr.
R призывать
A إستدعى

676
calm
F calme
E liso
D still
R тихий
A هادِئ

41

677

camera

F appareil photo *m.*
E cámara fotográfica *f.*
D Kamera *f.*
R фотоаппарат *м.*
A آلة تَصْوِير *f*

678

camouflage *v.*

F camoufler
E camuflar; mimetizar
D tarnen
R маскировать(ся)
A مَوَّه

679

camouflage *n.*

F camouflage *m.*
E camuflaje *m.*
D Tarnung *f.*
R маскировка *ж.*
A تَمْويه *m*

680

camp

F camp *m.*
E campo *m.*
D Lager *neut.*
R лагерь *м.*
A مُخَيَّم *m* . مُعَسْكَر *m*

681

campaign

F campagne *f.*
E campaña *f.*
D Feldzug *m.*
R кампания *ж.*; поход
A حَمْلَة *f*

682

camshaft

F arbre à cames *m.*
E árbol de levas *m.*
D Nockenwelle *f.*
R распределительный
(кулачковый) валик
м.
A عَمُود الكامة *m*

683

canal

F canal *m.*
E canal *m.*
D Kanal *m.*
R канал *м.*
A قَنَاة *f*

684 [L]

canalize

F canaliser
E canalizar
D begrenzen; kanalisieren
R канализировать
A تحديد حيِّز العمليات

685

cancel

F annuler; supprimer
E cancelar; suprimir
D annullieren; rückgängig
machen
R аннулировать;
отменять
A أَلْغَى

686

cannibalize

F cannibaliser
E reaprovechar
D ausschlachten *tr.*;
aufteilen *tr.*
R расчленять; разрывать
A أَصْــلِح بِنَقْل أَجْزَاء من آلة
أخرى

687

(automatic) cannon

F canon automatique *m.*
E cañón *m.*
D Maschinenkanone
(MK) *f.*
R автоматическая пушка
ж.
A مِدْفَع سَرِيع الطَلَقات *m*

688 [V,U]

canopy

F bâche *f.*
E toldo *m.*
D Plane [V] *f.;*
Überdachung [U] *f.*
R прикрытие [U] *ср.*;
брезент [V] *м.*
A حُجَيْرة واقية *f* . غطاء
مُنْزَلَق *m*

689 [A]

canopy

F voilure (de parachute)
f.
E campana (de
paracaídas) *f.*
D Fallschirmkappe *f.*
R купол (парашюта) *м.*
A حُجَيْرة واقية *f* . غطاء
مُنْزَلَق *m*

690

canteen

F bidon *m.*
E cantina *f.*
D Kantine *f.*
R лавка-столовая *ж.*;
фляга (для воды) *ж.*
A مَطْعَم *m*

691

canvas

F toile (à voiles, de tente)
f.
E lona *f.*
D Segeltuch *neut.*
R брезент *м.*; холст *м.*
A نَسِيج قِنَّب ـ خِيش *m*

692 [E,S]

cap *n.*

F amorce *f.*
E cápsula (fulminante) *f.*
D Sprengkapsel *f.*
R капсюль *м.*
A قَلَنْسُوَة *f*

42

693
cap *n.*
F calot *m.*
E gorra *f.*
D (Feld-)Mütze *f.*
R фуражка *ж.*; кепка *ж.*
A غِطاء *m*

694
capability
F capacité (de) *f.*
E capacidad *f.*
D Fähigkeit *f.*
R способность *ж.*
A مَقْدِرة *f*

695
capable (of)
F capable (de)
E capaz (de)
D fähig (zu)
R способный (на)
A قادِر (على)

696 [E]
capacity
F capacité *f.*
E capacidad *f.*
D Inhalt *m.;* Volumen *neut.*
R ёмкость *ж.*; объём *м.*
A سِعة *f*

697 [T]
cape
F cap *m.*
E cabo *m.*
D Kap *neut.*
R мыс *м.*
A رَأْس *m*

698 [U]
cape
F pèlerine *f.*
E capotillo *m.*
D Umhang *m.*
R накидка *ж.*
A مِعْطَف *m*

699 [M]
capsize *v.*
F (faire) chavirer
E (hacer) zozobrar
D kentern (lassen)
R опрокидывать(ся)
A إنْقَلَب

700
capstan
F cabestan *m.*
E cabrestante *m.*
D Gangspill *neut.*
R шпиль *м.*; кабестан *м.*
A رَحَويّة *f*

701
capsule
F capsule *f.*
E cápsula *f.*
D Kapsel *f.*
R капсула *м.*; капсюль *ж.*
A كَبْسُولة *f*

702 [E]
captive *adj.*
F prisonnier
E cautivo
D Fessel- *komp.*
R задержанный; привязной
A مُقَيَّد

703 [U]
captive *n.*
F prisonnier *m.*
E cautivo *m.*
D Gefangene(r) *m.*
R пленник *м.*; заключённый *м.*
A أسير *m*

704 [U]
capture *v.*
F capturer
E prender
D gefangennehmen *tr.*
R брать в плен
A أسَرَ . إسْتَوْلَى على

705
carbon fiber
F fibre de carbone *f.*
E fibra de carbono *f.*
D Kohlefaser *f.*
R углеродистое (углеродное) волокно *ср.*
A لِيفة كَرْبُونية *f*

706
carbureter
F carburateur *m.*
E carburador *m.*
D Vergaser *m.*
R карбюратор *м.*
A مُبَخِر السيارة ـ كاربوراتير *m*

707
cargo *n.*
F cargaison *f.*
E carga *f.*
D Fracht *f.*
R груз *м.*
A شَحْنة *f*

708 [A,M]
cargo(-carrying) *adj.*
F cargo *adj.*
E (de) carga
D Transport- [A] *komp.;* Fracht- [M] *komp.*
R грузовой
A شَحْنِ

709
cargo sling
F élingue de suspension *f.*
E eslinga de carga *f.*
D Gurt *m.*
R грузовая сетка *ж.*
A حَبْل لِرَفْع الشَّحْنة *m*

43

710

carpet bombing

F bombardement par
vagues *f.*

E bombardeo en oleadas
m.

D Flächenbombardierung
f.

R бомбометание по
площади *ср.*

A قَصْف مِنْطَقة

711 [A]

carrier air group

F groupe aérien
embarqué *m.*

E agrupación aérea
embarcada *f.*

D Trägerflugzeuggruppe *f.*

R авианосная
авиационная группа
ж.

A حامِلة مَجْمُوعة جَوِّية *f*

712 [M]

carrier group

F groupe porte-avions *m.*

E escuadra portaaviones
f.

D Flugzeugträgergruppe *f.*

R авианосная группа *ж.*

A مَجْمُوعة الحامِلات *f*

713

Cartesian coordinates
pl.

F coordonnées
cartésiennes *f. pl.*

E coordenadas cartesianas
f. pl.

D kartesische
Koordinaten *f. pl.*

R прямоугольная
система координат
ж.

A الإحْداثِيّات الديكارْتِيّة *f*

44

714

cartographic

F cartographique

E cartográfico

D kartographisch

R картографический

A خَرائِطي

715

cartridge

F cartouche *f.*

E cartucho *m.*

D Patrone *f.*

R патрон *м.*

A خَرْطُوشة *f*

716

cartridge case

F douille *f.*

E vaina *f.*

D Patronenhülse *f.*

R (патронная) гильза *ж.*

A ظَرْف الخَرْطُوشة *m*

717

casing *n.*

F enveloppe *f.*

E caja *f.*

D Gehäuse *neut.*

R оболочка *ж.*;
обшивка *ж.*

A غِلاف *m*

718 [U,R]

cassette

F cassette *f.*

E cargador *m.*

D Kassette *f.*

R кассета *ж.*

A مُمْسِك ـ كاسيت *m*

719

castaway *n.*

F naufragé *m.*

E náufrago *m.*

D Schiffbrüchige(r) *m.*

R потерпевший
кораблекрушение

A مَنْبُوذ *m*

720 [M]

cast off *v.*

F larguer les amarres

E desamarrar

D losmachen *tr.*

R отдавать швартовы

A أطْلَق حبل رَبْط السَّفِينة ـ
غادر الميناء

721

casualties *pl.*

F pertes *m. f.*

E perdidas *f. pl.*

D Verluste *m. pl.*

R потери *ж. мн. ч.*

A خَسائِر *f*

722

**casualty (wounded;
killed)**

F blessé *m.;* mort *m.*

E baja *f.*

D Verwundete(r) *m.;*
Tote(r) *m.*

R раненый *м.;* убитый
м.

A خَسارة *f*(جَريح . قَتيل)

723

catalyst

F catalyseur *m.*

E catalizador *m.*

D Katalysator *m.*

R катализатор *м.*

A حَفّاز ـ وسيط كيماوي *m*

724

catalyze

F catalyser

E catalisar

D katalysieren

R катализировать

A حَفَز

725 [A]
catapult *n.*
F catapulte (de
  lancement) *f.*
E catapulta *f.*
D Katapult *neut.;*
  Schleuderstarthilfe *f.*
R катапульта *ж.*
A مَنْجَنِيق *m*

726 [A]
catapult *v.*
F catapulter
E catapultar
D katapultieren
R катапультировать
A أَطْلَقَ بِالْمَنْجَنِيق

727
caution *n.*
F prudence *f.*
E prudencia *f.*
D Vorsicht *f.*
R осторожность *ж.;*
  предостережение *ср.*
A جَذَر *m*

728
cavalry
F cavalerie *f.*
E caballería *f.*
D Kavallerie *f.*
R кавалерия *ж.*
A سِلَاح الفُرسَان
  (الدبابات) *m*

729
ceasefire *n.*
F cessez-le-feu *m.*
E alto el fuego *m.*
D Feuereinstellung *f.*
R прекращение огня *ср.*
A وَقْف إطْلَاق النَّار *m*

730
cease firing *v.*
F cesser le feu
E cesar el fuego
D Feuer einstellen *tr.*
R прекращать огонь
A أوْقَف الرَّمَى

731 [A]
ceiling
F plafond *m.*
E techo *m.*
D Gipfelhöhe *f.*
R потолок *м.;*
  максимальная
  высота *ж.*
A أقْصَى إرْتِفَاع *m*

732
celestial
F céleste
E celestial
D Himmels- *komp.*
R небесный
A سَمَاوِيّ ـ فَلَكِي

733
celestial navigation
F navigation
  astronomique *f.*
E navegación astronómica
  f.
D Astronavigation *f.*
R австронавигация *ж.*
A الِمَلَاحة الفَلَكِيّة

734 [H,P,U]
cell
F cellule *f.*
E celda [P] *f.;* célula [HU]
  f.
D Zelle *f.*
R ячейка [HP] *ж.;*
  клетка [PU] *ж.*
A خَلِيّة [HU] *f* . وَحْدَة أوَّلِية
  في مُنَظَّمة [P] *f*

735
cellar
F cave *f.*
E sótano *m.*
D Keller *m.*
R подвал *м.*
A قَبْو *m*

736
cellular
F cellulaire
E celular
D Zell- *komp.;* zellular
  adj.
R клеточный; резервный
A على شَكْل خَلَايَا

737
cement
F ciment *m.*
E hormigón *m.*
D Zement *m.*
R цемент *м.;* бетон *м.*
A أسمنت *m*

738
cemetery
F cimitière *f.*
E cementario *m.*
D Friedhof *m.*
R кладбище *ср.*
A مَقبرة *f*

739
censor *v.*
F passer par le contrôle
E censurar
D prüfen
R просматривать;
  подвергать цензуре
A رَاقَب

740
censor *n.*
F censeur *m.*
E censor *m.*
D Zensor *m.*
R цензор *м.*
A مُرَاقِب *m*

741
center
F centre *m.*
E centro *m.*
D Mitte *f.;* Mittepunkt *m.*
R центр *м.;* середина *ж.*
A مُرْكَز *m* . مُنتَصَف *m*

742

centerline

F axe *m.;* ligne médiane *f.*
E eje *m.;* línea mediana *f.*
D Achse *f.;* Mittellinie *f.*
R ось *ж.;* центральная
  линия *ж.*
A خَطّ المُتَصَف

743

center of gravity

F centre de gravité *m.*
E centro de gravedad *m.*
D Schwerpunkt *m.*
R центр тяжести *м.*
A مَرْكَز الثُّقْل

744

centigrade

F centigrade
E centígrado
D Celsius- *komp.;* Grad
  Celsius *m.*
R стоградусный; (по)
  Цельсию
A سنتيجراد

745

central

F central
E central
D Mittel- *komp.;* zentral
  adj.
R центральный
A مَرْكَزِي . مُتَوَسِّط

746

centralize

F centraliser
E centralizar(se)
D zentralisieren
R централизировать
A رَكَّز

747

central nervous system

F système nerveux central
  m.
E sistema nervioso central
  m.
D Zentralnervensystem
  neut.
R главная нервная
  система *ж.*
A الجِهَاز العَصَبي المَرْكَزي

748

centrifugal force

F force centrifuge *f.*
E fuerza centrífuga *f.*
D Zentrifugalkraft *f.*
R центробежная сила *ж.*
A قُوَة طارِدة مَرْكَزية *f*

749

ceramic

F céramique
E cerámico
D Keramik- *komp.*
R гончарный;
  керамический
A خَزَفي

750

ceramics *pl.*

F céramique *f.*
E cerámica *f.*
D Keramik *f.*
R керамика *ж.*
A الخَزَفِيَّات *f*

751

ceremonial *adj.*

F cérémonial
E ceremonial
D feierlich; zeremoniell
R торжественный;
  церемониальный
A رَسمي

752

certificate *n.*

F certificat *m.*
E certificado *m.*
D Bescheinigung *f.;*
  Schein *m.*
R свидетельство *ср.;*
  удостоверение *ср.*
A شَهَادة *f*

753

certify

F certifier; constater
E certificar; declarar
D bescheinigen
R удостоверять;
  заверять
A أكَّد . شَهِد

754 [R]

chaff ('rope')

F plaquettes de brouillage
  f. pl.
E virutas *f. pl.*
D Düppel *m.;*
  Düppelstreifen *m. pl.*
R (дипольные)
  отражатели *м. мн.*
  ч.; 'мякина' *ж.*
A رقائق معدنية لتشويش
  الرادار *f*

755

chain *n.*

F chaîne *f.*
E cadena *f.*
D Kette *f.*
R цепь *ж.*
A سِلْسِلَة *f*

756

chain of command

F voie hiérarchique *f.*
E conducto regular *m.;*
  línea de mando *f.*
D Unterstellungsverhältnis
  neut.
R инстанции
  командования *ж.*
  мн. ч.
A سِلْسِلَة القِيَادة

757
chalk *n.*
F craie *f.*
E tiza *f.*
D Kreide *f.*
R мел *м.*
A طباشير *m*

758
chalk *v.*
F marquer à la craie
E marcar con tiza
D (mit) Kreide markieren
R писать мелом;
    рисовать мелом
A رَسَم بِالطَّبَاشِير

759
chalk commander
F chef de transport *m.*
E jefe de embarque *m.*
D Einsatzführer *m.*
R начальник десантной
    группы *м.*
A قَائِد الزُّمْرة *m*

760
chalk number
F numéro-repère *m.*
E número de serie *m.*
D Absatznummer *f.;*
    Verladenummer *f.*
R порядковый номер *м.*
A رَقْم الزُّمْرة *m*

761
chalk troops
F troupe numérotée *f.*
E tropa numerada *f.*
D Landungstruppen
    (meist eng. Begriff) *f.*
    pl.
R десантная группа *ж.*
A أفْراد الزُّمْرة *m*

762
challenge
F interpeller
E (dar el) alto; pedir el
    'santo y seña'
D anrufen *tr.*
R окликать
A تَحَدَّى

763
challenge *n.*
F procédé d'identification
    f.
E consigna *f.;*
    procedimiento de
    identificación *m.;*
    'santo y seña' *m.*
D Anruf *m.*
R оклик *м.*
A تَحَدٍّ *m*

764 [S]
chamber
F chambre *f.*
E recámara *f.*
D Laderaum *m.*
R патронник *м.*; камора
    ж.
A حُجْرة *f*

765 [M,R]
channel *n.*
F chenal [M] *m.;* chaîne
    [R] *f.*
E canal *m.*
D Fahrrinne [M] *f.;*
    Kanal [R] *m.*
R канал [M] *м.*; канал
    связи [R] *м.*
A مَمَرّ مائِي [M] *m* .
    شَرِيط التَرَدُّد [R] *m*

766
chaplain
F aumônier *m.*
E capellán *m.*
D Feldkaplan *m.*
R военный священник
    м.; капеллан (устар.)
    м.
A قَسّ ـ قِسِّيس

767 [P]
charge *n.*
F accusation *f.*
E acusación *f.*
D Anklage *f.*
R обвинительный акт *м.*
A مُقاضاه *f*

768 [B,E,L,M,S]
charge *n.*
F charge *f.*
E carga *f.*
D Sprengladung *f.*
R заряд *м.*
A حُشْوة

769 [A,B,E,R,V]
charge *v.*
F charger
E cargar
D (auf)laden *(tr.) untr.*
R заряжать
A شَحَن

770 [F]
charge *v.*
F charger
E cobrar
D berechnen; verlangen
R назначать;
    запрашивать (цену,
    плату)
A طَلَب أجراً

771 [L]
charge *n.*
F assaut *m.*
E carga *f.*
D Sturmangriff *m.*
R атака *ж.*; нападение
    ср.
A هجوم *m*

772
chart *n.*
F carte (marine) *f.*
E carta marítima *f.*
D Seekarte *f.*
R карта *ж.*; схема *ж.*;
  график *м.*
A خَرِيطة بَيَانية *f*

773
charter *v.*
F affréter
E alquilar; fletar
D chartern
R фрахтовать; нанимать
A إسْتَأْجَر

774
chase *v.*
F poursuivre
E perseguir
D verfolgen
R преследовать
A لاحَق

775 [A,E,R,S]
chassis
F châssis *m.*
E chasis *m.*
D Fahrgestell *neut.*
R шасси *м.*
A شاسية *m* ـ هيكل *m*

776
check *v.*
F contrôler
E controlar
D kontrollieren; prüfen;
  überprüfen
R проверять;
  контролировать
A فَحَص

777
checklist
F liste de contrôle *f.*
E lista de control *f.*
D Kontrolliste *f.*
R контрольный список
  м.; перечень *м.*
A قائِمة المُرَاجَعَة

778
checkpoint
F (point de) contrôle *m.*
E control *m.*
D Kontrollstelle *f.*
R контрольно-пропускной
  пункт (КПП) *м*
A نُقْطَة الفَحْص

779
chemical *n.*
F produit chimique *m.*
E sustancia química *f.*
D Chemikalie *f.*
R химикалии *мн. ч.*
A مَادَة كِيمِيائِّة *f*

780
chemical *adj.*
F chimique
E químico
D chemisch
R химический
A كِيمِيائِي

781
chemical agent
F agent chimique
E agente químico *m.*
D (chemischer)
  Kampfstoff *m.*
R (боевое) химическое
  вещество *ср.*
A كاشِف كِيمِيائي *m*

782
chemical defense
F défense NBC *f.*
E defensa antigás *f.*
D Gasabwehr *f.*
R противохимическая
  оборона *ж.*
A الدِفَاع الكِيمِيائي *m*

783 [U]
chest
F coffre *m.*
E cofre *m.*
D Kasten *m.*
R ящик *м.*
A صُندُوق *m*

784 [H]
chest
F poitrine *f.*
E pecho *m.*
D Brust *f.*
R грудь *ж.*
A صَدْر *m*

785
chief of staff
F chef d'état-major *m.*
E Jefe de Estado Mayor
  m.
D Stabschef *m.*
R начальник штаба
  (начштаба) *м.*
A رَئِيس الأرْكَان *m*

786
chin
F menton *m.*
E barbilla *f.*
D Kinn *neut.*
R подбородок *м.*
A ذَقْن *m*

787
chinstrap
F jugulaire *f.*
E barboquejo *m.*
D Kinnriemen *m.*
R подбородочный
  ремень *м.*
A زِنَاق *m*

788 [A,V]
chock *n.*
F cale *f.*
E calzo *m.*
D Bremsklotz *m.*
R клин [V] *м.*;
  тормозная колодка
  [A] *ж.*
A زَنَّاقَات *f*

789 [V]
choke n.
F starter m.
E estrangulador m.
D Starthilfe f.
R заслонка ж.; дроссель м.
A مُخْنَقة f ـ خانِق m

790 [H]
chronic
F chronique
E crónico
D chronisch
R хронический
A مُزْمِن

791 [N]
chronic dose
F dose chronique f.
E dosis crónica f.
D kritische Menge f.
R хроническая доза ж.
A مِقْدَار مُزْمِن m

792
chronometer
F chronomètre m.
E cronómetro m.
D Uhr f.
R хронометр м.
A مِقْياس الوَقْت m

793
cine camera
F caméra f.
E cámara (cinematográfica) f.
D Filmkamera f.
R киноаппарат м.
A آلة تَصْوير سِينَمائي f

794 [R]
cipher
F chiffre m.
E cifra f.
D Geheimschrift f.
R шифр м.
A رَمْز m . شَفْرة f

795 [A]
circle v.
F décrire des cercles
E (dar) vueltas; girar
D kreisen
R вращаться; окружать
A دار حَوْل

796
circle n.
F cercle m.
E círculo m.
D Kreis m.
R круг м.; кружок м.
A دائِرة f

797 [E]
circuit
F circuit m.
E circuito m.
D Stromkreis m.
R цепь ж.; схема ж.
A دائِرة f

798
circular
F rond
E circular
D rund
R круглый; циркулярный
A مُسْتَدِير . دائِرِي

799 [N,S]
circular error probable (CEP)
F écart circulaire probable m.
E desvío probable circular m.
D wahrscheinliche kreisförmige Streuung f.
R вероятная радиальная ошибка ж.
A خَطَأ دائِرِي مُحْتَمَل

800
circumference
F circonférence f.
E circunferencia f.
D Umfang m.
R окружность ж.; периферия ж.
A مُحِيط الدَائِرة m

801 [M]
citadel
F citadelle f.
E ciudadela f.
D Festung f.
R крепость ж.
A قَلْعة f

802
civil defense
F protection civile f.
E defensa pasiva f.
D ziviler Bevölkerungsschutz m.
R гражданская оборона ж.; организация противовоздушной обороны ж.
A الدفاع المَدَني m

803
civil disturbance
F agitation f.
E disturbio civil m.
D Unruhen f. pl.
R гражданские беспорядки м. мн. ч.
A شَغَب مَدَني m

804
civilian n.
F civil m.
E civil m.
D Zivilist m.; Zivilperson f.
R гражданин м.
A مَدَني m

805
(in) civilian clothes
F (en) civil
E (en) traje de paisano
D (in) Zivil
R (в) штатской одежде
A بِاللِّبَاسِ المَدَنِي

806
civil-military cooperation
F coopération civilo-militaire f.
E cooperación civilo-militar f.
D zivil-militärische Zusammenarbeit f.
R гражданско-военное сотрудничество ср.
A تعاونٌ عَسْكَرِي مَدَنِي

807
civil-military relations
F relations civilo-militaires f. pl.
E relaciones civilo-militares f. pl.
D zivil-militärische Beziehungen f. pl.
R военно-гражданские отношения ср. мн. ч.
A عَلاَقَاتٌ عَسْكَرِية مَدَنِية f

808
clamp v.
F bloquer
E sujetar
D befestigen
R зажимать; скреплять
A قَمَط

809 [E]
clamp n.
F bride de serrage f.
E abrazadera f.
D Klampe f.
R зажим м.
A قامِطة f

810
clandestine
F clandestin
E clandestino
D Geheim- komp.; heimlich adj.
R скрытный
A سِرِّي - خَفِي

811 [P]
clarification
F clarification f.
E aclaración f.
D Erklärung f.
R пояснение ср.; выяснение ср.
A تنقية f

812
clash
F conflit armé m.
E encuentro m.
D bewaffneter Zusammenstoss m.
R стычка ж.; схватка ж.
A إشتباك m

813
clasp knife
F couteau de poche m.
E navaja f.
D Klappmesser neut.
R складной нож м.
A مُدية f

814 [B,E]
classification
F classification f.
E capacidad (de puente) f.; clasificación (de vehículo) f.
D Lastenklasse (eines Fahrzeugs) f.; Tragfähigkeit (einer Brücke) f.
R определение ср.; классификация ж.
A تَصْنِيف m

815 [P]
classification
F classification f.
E clasificación f.
D Einstufung f.
R определение ср.
A حَظْر m

816 [P]
classified
F classifié
E reservado; secreto
D geheim
R секретный
A مَحْظُور

817
class of supplies
F catégorie d'approvisionnements f.
E clase de abastecimientos f.
D Nachschubklasse f.
R категория снабжения ж.
A مستويات الإمداد

818
clean adj.
F propre
E claro; limpio
D sauber
R чистый
A نَظِيف

819
clean v.
F nettoyer
E limpiar
D reinigen
R чистить; очищать
A نَظَّف

820 [P]

clear *v.*

F disculper
E absolver; aprobar
D freisprechen *tr.;*
  genehmigen *untr.*
R объявлять
  невиновным
A أَطْلَقَ ـ بَرَّأَ

821 [C,L]

clear *v.*

F dégager
E aclarar [C]; despejar [L]
D bahnen
R открывать [L];
  освобождать [CL]
A طَهَّرَ

822 [P]

clear (documents) *v.*

F chercher l'autorisation
  de
E aprobar
D genehmigen
R получать разрешение
  (на документы)
A صَدَّقَ

823 [L,M]

clear (mines) *v.*

F déminer [LM];
  désobuser [L]
E barrer [M]; limpiar (de
  minas) [L]
D räumen
R разминировать
A أَزَال

824 [A,L,M]

clear (route etc.) *adj.*

F libre (itinéraire etc.)
E despejado (camino etc.)
D frei (Strasse usw.)
R открытый [AML];
  свободный[AML];
  'зелёная улица'
  (разг.)[L]
A مَكْشُوف . سَالِك

825

clerk

F employé de bureau *m.*
E oficinista *m. f.*
D Schreiber *m.*
R писарь *м.*; конторский
  служащий *м.*
A كَاتِب *m*

826

climb *v.*

F escalader
E escalar; trepar
D steigen
R лазить; подниматься
A صَعِد

827

clockwise

F (dans le) sens des
  aiguilles d'une montre
E (en la) dirección de las
  agujas del reloj
D rechtsdrehend
R (по) часовой стрелке
A في إتجاه عقرب الساعة

828 [U]

close *v.*

F fermer
E cerrar
D (ver)schliessen
R закрывать
A أَغْلَق

829

close *adj.*

F serré; rapproché
  (combat)
E apretado; cercano;
  próximo
D dicht (an); nahe
R близкий; тесный
A قَرِيب

830 [L,M]

close (with) *v.*

F (s')approcher (de)
E acercarse; aproximarse
  (a)
D (sich) nähern (an);
  näherrücken (an) *tr.*
R сближаться;
  приближаться
A إقْتَرَب

831

close combat

F combat rapproché *m.*
E combate próximo *m.*
D Nahkampf *m.*
R ближний бой *м.*;
  рукопашный бой *м.*
A قِتَال مُتلاحِم *m*

832

closed area

F zone interdite *f.*
E zona prohibida *f.*
D Sperrgebiet *neut.*
R запретная зона *ж.*
A مِنْطَقَة مُغْلَقة *f*

833 [H]

clot *v.*

F (se) coaguler
E cuajar(se)
D gerinnen (lassen)
R запекаться
A تَجَلَّط

834

clothing

F vêtements *m. pl.*
E vestidos *m. pl.*
D Kleider *neut. pl.*
R одежда *ж.*
A أَلْبِسة *f*

835

cloud *n.*

F nuage *m.*
E nube *f.*
D Wolke *f.*
R облако *ср.*; туча *ж.*
A غَيْم . سَحَاب *m*

51

836 [S]
cluster *n.*
F grappe *f.*
E metralla *f.*
D Bündel *neut.*
R кисть *ж.;* пучок *м.*
A عُنقُود *m*

837
cluster bomb
F projectile à mitraille
  (d'habitude terme
  anglais) *m.*
E bomba rompadora de
  metralla *f.;* bomba
  beluga *f.*
D Mutterbombe (meist
  eng. Begriff) *f.*
R кассетная бомба *ж.*
A قُنْبُلة عُنقُودية *f*

838 [E]
clutch
F embrayage *m.*
E embrague *m.*
D Kupplung *f.*
R защёлка *ж.;*
  сцепление *ср.*
A الجِهَاز الفَاصِل *m*

839 [R]
clutter
F interférence *f.*
E interferencia *f.*
D Störzeichen *neut. pl.*
R местные помехи *ж.*
  мн. ч.
A تَشْويش *m*

840
coal
F charbon *m.*
E carbón *m.*
D Kohle *f.*
R уголь *м.*
A فَحْم حَجَرِي *m*

841
coast *n.*
F côte *f.*
E costa *f.*
D Küste *f.*
R побережье *ср.*
A سَاحِل *m*

842 [M]
coast *v.*
F longer la côte
E costear
D (an der) Küste entlang
  fahren
R плавать вдоль
  побережья
A أبْحر في مُحاذَاة سَاحِل

843
coast *v.*
F planer (moteur arrêté)
E (ir en) punto muerto
D Freilauf fahren
R двигаться по инерции;
  плыть по инерции
A تَحَرَّك بَطْنًا

844
coastal force
F force côtière *f.*
E fuerza costera *f.*
D Küstenverband *m.*
R морские части
  прибрежного
  действия *ж. мн. ч.*
A قُوة سَاحِلية *f*

845
coastguard
F garde-côte *m.*
E guardacostas *m. inv.*
D Küstenwacht(-mann) *f.*
  m.
R береговая охрана *ж.*
A خَفَر السَّواحِل *m*

846
coaxial
F coaxial
E montar
D koaxial
R коаксиальный;
  спаренный (пулемёт)
A مُتَّحِد المِحْوَر

847 [S]
cock *v.*
F armer
E amartillar
D spannen
R взводить (курок)
A نَصَب طَارِق (البُنْدَقِية)

848 [A]
cockpit
F habitacle *m.*
E cabina *f.;* carlinga *f.*
D Kanzel *f.*
R кабина *ж.*
A كابينة الطيّار *f*

849 [E]
cocoon *v.*
F (mettre en) cocon
E poner en capullo
D einwickeln *tr.*
R ставить на
  консервацию
A غَلَّف

850
code
F code *m.*
E cifra *f.;* clave *f.*
D Kode *m.*
R код *м.*
A رَمْز *m*

851
code name
F appellation
  conventionnelle *f.*
E nombre en clave *m.*
D Deckname *m.*
R кодированное
  название *ср.*
A إسْم رَمْزي *m*

852
cold *adj.*
F froid
E frío
D kalt
R холодный
A بارد

853 [G]
cold launch
F lancement à froid *m.*
E lanzamiento frío *m.*
D Kaltstart *m.*
R холодный пуск *м.*
A إِطْلَاق على البارِد

854
collate
F collationner
E cotejar
D vergleichen
R сличать; сопоставлять
A قَارَن

855 [N]
collateral damage
F degâts collatéraux *m. pl.*
E daños secundarios *m. pl.*
D Nebenschaden *m.*
R параллельные потери и ущерб *мн. ч.*
A خسائر إضافية

856
collect *v.*
F rassembler
E reunir(se)
D (sich) sammeln
R собирать(ся)
A جَمَع

857
collide (with)
F (se) heurter (contre); aborder
E chocar (con)
D zusammenstossen (mit) *tr.*
R сталкиваться (с)
A إِصْطَدَم بِ

858
collision
F collision *f.*
E choque *m.;* colisión *f.*
D Zusammenstoss *m.*
R столкновение *ср.*
A إِصْطِدَام *m*

859
color
F couleur *f.*
E color *m.*
D Farbe *f.*
R цвет *м.*
A لَوْن *m*

860
(camouflage) color-patch
F tache de couleur *f.*
E parche de color *m.*
D Farbfleck (Tarnung) *m.*
R маскировочный узор *м.*
A لَوْحَة مُلَوَّنة *f*

861
colors *n.*
F drapeau *m.*
E bandera *f.*
D Standarte *f.*
R флаг *м.;* знамя *ср.*
A عَلَم *m*

862 [L]
column
F colonne *f.*
E columna *f.*
D Kolonne *f.*
R колонна *ж.*
A رَتَل *m*

863
(steering) column
F colonne de direction *f.*
E columna de dirección *f.*
D Lenksäule *f.*
R рулевая колонка *ж.*
A عَمُود القِيَادة

864 [H]
coma
F coma *m.*
E coma *m.*
D Koma *s.*
R кома *ж.;* коматозное состояние *ср.*
A غيبوبة *f*

865
combat
F combat *m.*
E combate *m.*
D Kampf *m.*
R бой *м.*
A قِتَال *m*

866
combat clothing
F tenue de combat *f.*
E traje de campaña *m.*
D Kampfanzug *m.*
R боевая форма *ж.*
A لِباس القِتال *m*

867
combat effectiveness
F efficacité opérationnelle *f.*
E efectividad combativa *f.*
D Kampfwert *m.*
R боеспособность *ж.*
A فَعَّالية القِتَال *f*

868
combat engineering
F génie *m.*
E ingeniería *f.*
D Pionierwesen *neut.*
R сапёрные действия *ср. мн. ч.*
A هَنْدَسَة القِتَال *f*

869 [H]
combat fatigue
F épuisement au combat *m.*
E fatiga de combate *f.*
D Kampfermüdung *f.*
R психическая травма *ж.*
A إرهاق بسبب العمليات

53

870

combat information center

F centre d'information de combat *m.*

E centro de informaciones de combate *m.*

D Gefechtsnachrichtenstelle *f.*

R боевой информационный пост *м.*

A مَرْكَز مَعْلُومات القِتال *m*

871 [L]

combat maneuver forces

F forces de manoeuvre au combat *f. pl.*

E fuerzas de maniobra *f. pl.*

D Gefechtsgruppe *f.*

R силы боевого манёврирования *ж. мн. ч.*

A قوات المناورات القتالية

872

combat power

F puissance de combat *f.*

E potencia de combate *f.*

D vollkommene Kampfkraft *f.*

R боевая мощь *ж.*

A القدرة القتالية *f*

873

combat survival

F survie au combat *f.*

E supervivencia en combate *f.*

D Überleben *neut.*

R боевое выживание *ср.*

A البَقَاء في القِتَال

54

874

combat train

F train de combat *m.*

E tren de combate *m.;* tren de víveres y bagajes *m.*

D logistischer Bestandteil *m.*

R боевой обоз *м.*

A قافلة قتال *f*

875

combine *v.*

F combiner

E combinar

D (sich) zusammenstellen *tr.*

R объединять(ся)

A وَحَّد

876 [A]

combuster

F système de combustion *m.*

E comburente *m.*

D Verbrennungs-vorrichtung *f.*

R прибор прямоточного сгорания *м.*

A مُحْرق *m*

877

combustible

F combustible

E combustible

D brennbar

R горючий

A قَابِل لِلإحْتِرَاق

878

combustion

F combustion *f.*

E combustión *f.*

D Verbrennung *f.*

R горение *ср.;* сгорание *ср.*

A إحْتِراق *m*

879

command (give order) *v.*

F commander

E mandar

D befehlen; kommandieren

R приказывать

A أَمَر

880

command (to have) *v.*

F commander

E comandar; mandar

D (das) Kommando haben

R командовать

A قاد

881

command (act of) *n.*

F commandement *m.*

E orden *f.*

D Kommando *neut.;* Führung *f.*

R командование *ср.;* начальствование *ср.*

A قِيَادَة *f*

882

command (element commanded) *n.*

F commandement *m.*

E comandancia *f.*

D Kommando- *komp.;* Führung- *komp.*

R команда (под командой) *ж.;* командование (над) *ср.;* начальствование (над) *ср.;* управление *ср.*

A قِيَادَة *f*

883

command (at one's disposal) *v.*

F avoir à sa disposition

E disponer (de)

D (zur) Verfügung stehen

R (иметь в) распоряжении

A . . . تحت قِيَادَة

884
command (mastery) *n.*
F maîtrise *f.*
E dominio *m.*
D Herrschaft über *f.*
R господство *ср.*
A السَّيْطَرة *f*

885
command (order) *n.*
F ordre *m.*
E mando *m.*
D Befehl *m.*
R приказ *м.*
A أمْر

886
command (overlook)
v.
F dominer
E dominar
D beherrschen
R господствовать (над);
выходить (на)
A أشْرَف على

887
(take) command *v.*
F prendre le
commandement
E asumir el mando
D (Führung) übernehmen
R принимать
командование
A تقلد القيادة

888
commandant
F commandant *m.*
E comandante *m.;* jefe *m.*
D Kommandant *m.*
R комендант *м.*
A قائِد *m*

889
commandeer
F requisitioner
E requisar
D requirieren
R реквизировать
A صَادَر

890
commander
F officier commandant *m.*
E comandante *m.;* jefe *m.*
D Kommandeur *m.;* Chef
m.
R командир *м.;*
начальник *м.*
A قائِد *m*

891
command group
F groupe allégé de
commandement *m.*
E grupo de mando *m.*
D Kommandostab *m.*
R группа управления *ж.*
A مجموعة قيادة *f*

892
commando *n.*
F commando *m.*
E comando *m.*
D Sturmtruppe *f.*
R десантно-диверсионный
отряд *м.*
A فدائي *m*

893 [E]
commercial
F commercial
E comercial
D Handels- *komp.;*
handelsüblich *adj.*
R торговый
A تِجَاري

894
commissar
F commissaire (du
peuple) *m.*
E comisario *m.*
D Kommissar *m.*
R комиссар *м.*
A الضابط السياسي

895
commission *n.*
F grade d'officier *m.*
E despacho de oficial *m.*
D Offizierspatent *neut.*
R патент на офицерский
чин *м.;* офицерское
звание *ср.*
A شَهَادة ضابط

896
commission *v.*
F nommer officier
E nombrar oficial
D (zum) Offizier befördern
R присваивать
офицерское звание
A قَلَّد قِيَادَة لضابط

897 [M]
commission *v.*
F admettre au service
actif
E poner en servicio activo
D (in) Dienst stellen
R вводить в строй
A جَهَّز (سَفِينَة)

898
commitment
F engagement *m.*
E obligacíon *f.*
D Verpflichtung *f.*
R обязательство *ср.*
A تَعَهُّد *m*

899
committed force
F force engagée *f.*
E fuerza hipotecada *f.*
D (unter) Befehl stehende
Einheit *adj.*
R введённая в бой часть
ж.
A قوات مُخَصَّصة

900

commodity
F produit *m.*
E producto *m.*
D Waren *f. pl.*
R предмет *м.*; товар
A مَادّة ـ سلعة *f*

901

common *adj.*
F commun
E común
D gemeinsam
R общий
A مُشْتَرَك

902

commonality
F communauté *f.*
E comunidad *f.*
D Gemeinsamkeit *f.*
R общность *ж.*
A عَامَّة

903

common user items *pl.*
F produits communs *m. pl.*
E artículos de uso común *m. pl.*
D Artikel für gemeinsamen Gebrauch *m. pl.*
R предметы общего употребления *м. мн. ч.*
A مَوادّ للإِسْتِعْمَال المُشْتَرَك *f*

904 [U]

communicate (to)
F communiquer (à)
E comunicar (a)
D mitteilen *tr.*
R сообщать; передавать
A إرْتَبَط بِ

905 [R,T]

communicate (with)
F communiquer (avec)
E comunicarse (con)
D (in) Verbindung stehen (mit)
R поддерживать связь (с)
A إتّصَل بِ

906

communication center
F centre de transmissions *m.*
E centro de transmisiones *m.*
D Fernmeldezentrale *f.*
R узел связи *м.*
A مَرْكَز الإِشَارة *m*

907 [R]

communications *pl.*
F télécommunications *f. pl.*
E transmisiones *f. pl.*
D Fernmeldewesen *neut.*
R средства связи *ср. мн. ч.*
A إتّصَالات *f*

908 [T]

communications *pl.*
F moyens de transport *m. pl.*
E red de transporte *m.*
D Verkehrsnetz *neut.*
R пути сообщения *м. мн. ч.*
A مواصلات *f*

909

communiqué
F communiqué *m.*
E comunicado *m.*
D Verlautbarung *f.*
R коммюнике *ср.*
A بَلاغ رَسْمي

910

companionway
F escalier des cabines *m.*
E escalera de cámara *f.*
D Kajütentreppe *f.*
R сходной трап *м.*
A دَرَج ـ سُلّم *m*

911 [L]

company *n.*
F compagnie *f*
E compañía *f*
D Kompanie *f.*
R рота *ж.*
A سَرِية *f*

912 [M]

(ship's) company
F équipage *m.*
E tripulación *f.*
D Besatzung *f.*
R экипаж *м.*
A نُوتية سَفِينة / طاقِم سفينة *m*

913

company commander
F commandant de compagnie *m.*
E jefe de compañía *m.*
D Kompaniechef *m.*
R командир роты *м.*
A قائد سَرِية

914 [A,G,M,V]

compartment
F compartiment *m.*
E compartimiento *m.*
D Abteilung [G] *f.*; Kabine [AMV] *f.*
R отделение [MV] *ср.*; купе [VM] *ср.*; отсек [GM] *м.*
A مَقْصُورة *f*

915

compass
F boussole *f.*; compas *m.*
E brújula *f.*
D Kompass *m.*
R компас *м.*
A بُوصَلة *f*

916

(steering) compass
F compas de route *m.*
E brújula de derrota *f.*
D Steuerkompass *m.*
R путевой компас *м.*
A بُوصَلَة التَّوْجِيه

917

compass error
F erreur de compas *f.*
E error de brújula *m.*
D Kompassfehler *m.*
R девиация компаса *ж.*
A خَطَأ البُوصَلَة *m*

918

compatibility
F compatibilité *f.*
E compatibilidad *f.*
D Vereinbarkeit *f.;*
   Verträglichkeit *f.*
R совместимость *ж.*
A إِنْسِجَام *m*

919 [M]

complement
F effectif *m.*
E dotación *f.*
D Besatzung *f.*
R экипаж *м.;* команда
   ж.
A المَجْمُوعة الكامِلة *f*

920 [E]

component
F pièce *f.*
E pieza *f.*
D Bestandteil *m.;* Teil
   neut.
R деталь *ж.;* блок *м.*
A مُرَكَّب *m*

921

compound armor
F blindage composite *m.*
E blindaje compuesto *m.*
D Schottpanzerung *f.*
R многослойная броня
   ж.
A تَصْفِيح بِفُلَاذ مَخْلُوط *m*

922

comprehensive
F compréhensif
E extenso
D umfassend
R исчерпывающий
A شامِل

923 [E]

compress *v.*
F comprimer
E comprimir
D verdichten
R сжимать
A ضَغَط

924 [H]

compress *n.*
F compresse *f.*
E compresa *f.*
D Kompresse *f.*
R компресс *м.*
A رُباط ضاغط *m*

925

compression
F compression *f.*
E compresión *f.*
D Verdichtung *f.*
R сжатие *ср.*
A إِنْضِغَاط *m*

926

compression ratio
F taux de compression *m.*
E relación de compresión
   f.
D Verdichtungsverhältnis
   neut.
R отношение сжатия
   ср.; пропорция
   сжатия *ж.*
A نِسْبَة الإِنْضَغَاط *f*

927

compressor
F compresseur *m.*
E compresor *m.*
D Kompressor *m.*
R компрессор *м.*
A ضاغط *m*

928 [P]

compromise *v.*
F compromettre
E comprometer
D blosstellen *tr.*
R компрометировать
A إِتَّفَق إِتَّفَاقًا وَسَطًا

929

compute
F calculer
E calcular
D rechnen
R считать; вычислять
A حَسَب

930

computer
F ordinateur *m.*
E computador *m.*
D Elektronenrechner *m.*
R электрическая
   вычислительная
   машина (ЭВМ) *ж.*
A حاسِبَة *f*

931

computer programer
F programmeur *m.*
E programador *m.*
D Programmierer *m.*
R программист(ка)
   м.(ж.)
A مُدِير بَرْنَامِج الحاسِبة

932

conceal
F cacher
E ocultar
D verbergen
R скрывать
A أَخْفَى

933

concealment
F défilement *m.*
E encubrimiento *m.*
D Verbergen *neut.*
R скрытие *ср.*
A إِخْفَاء . إِسْتِتَار

57

934 [U]

concentrate *v.*

F (se) concentrer
E concentrar(se)
D (sich) konzentrieren;
   (sich) massieren
R сосредоточивать(ся);
   концентрировать(ся)
A رَكَّز

935 [S]

concentrate *v.*

F (faire) converger
E concentrar
D zusammenfassen *tr.*
R сосредоточивать
A تَجَمَّع

936 [L]

concentration

F concentration *f.*
E concentración *f.*
D Massierung *f.;*
   Schwerpunktbildung
   f.
R сосредоточение *cp.*
A حَشْد *m*

937 [U]

concentration

F concentration *f.*
E concentración *f.*
D Konzentration *f.*
R концентрация *ж.*
A تَجَمُّع *m*

938 [S]

concentration

F convergence de tir *f.*
E concentración del tiro *f.*
D Feuerzusammenfassung
   f.
R сосредоточение *cp.;*
   веер *м.*
A تَرْكِيز *m*

58

939

concentration area

F zone de concentration
   f.
E zona de concentración
   f.
D Aufmarschgebiet *neut.*
R район сосредоточения
   м.
A مِنْطَقَة حَشْد *f*

940

concept

F concept *m.*
E concepto *m.*
D Plan *m.*
R понятие *cp.;* замысел *м.*
A مَفْهُوم *m*

941

concept of operations

F conception de la
   manoeuvre *f.*
E idea de maniobra *f.*
D Operationsplan *m.*
R замысел боевых
   действий *м.*
A مَفْهُوم العَمَلِيات الحَرْبِية

942

conclusion

F conclusion *f.*
E conclusión *f.*
D Entschluss *m.;* Ergebnis
   neut.
R заключение *cp.*
A إستنتاج *m*

943

concrete *n.*

F béton *m.*
E hormigón *m.*
D Beton *m.*
R бетон *м.*
A خَرَسانة *f*

944

concurrent

F simultané
E concurrente
D gleichzeitig
R одновременный;
   совпадающий
A مُتلازِم

945

condensation

F condensation *f.*
E condensación *f.*
D Kondensation *f.*
R сгущение *cp.;*
   конденсация (воды)
   ж.
A تَكَاثُف *m*

946 [E]

condense

F (se) condenser
E condensar(se)
D kondensieren
R сгущать
A كَثَّف *m*

947

condenser

F condensateur *m.;*
   condenseur (de
   vapeur) *m.*
E condensador *m.*
D Kondensator *m.;*
   Kondensor (Linse) *m.*
   (f.)
R конденсатор *м.*
A مُكَثِّف *m*

948

condition

F condition *f.*
E estado *m.*
D Zustand *m.*
R состояние *cp.*
A حالة صحِّيّة *f*

949
condominium
F condominium *m.*
E condominio *m.*
D Kondominium *neut.*
R совместное
  господство *ср.*
A سِيَادة مُشْتَرَكة *f*

950 [E,R]
conduct *v.*
F (être) conducteur de
E conducir
D leiten
R проводить
A وصّل

951
conduct
F conduite *f.*
E comportamiento *m.*
D Benehmen *neut.*
R поведение *ср.*
A سُلوك *m*

952
cone
F cône *m.*
E cono *m.*
D Kegel *m.*
R конус *м.*
A مَخْروط *m*

953
conference
F conference *f.*
E conferencia *f.*
D Konferenz *f.*
R конференция *ж.*
A مؤتَمر *m*

954
confidential
F confidentiel
E confidencial
D vertraulich
R конфиденциальный
A سِرّيّ - خُصوصي

955
confirmation
F confirmation *f.*
E confirmación *f.*
D Bestätigung *f.*
R подтверждение *ср.*
A تَأكيد *m*

956
conic(al)
F conique
E cónico
D kegelförmig
R конический; конусный
A مَخْروطي

957
conning tower
F kiosque *m.*
E torreta *f.*
D Kommandoturm *m.*
R боевая рубка *ж.*
A بُرْج القِيَادة

958
conquer
F conquérir
E conquistar; vencer
D erobern
R завоёвывать
A فَتَح . إنْتَصَر

959
conscript *n.*
F appelé *m.*
E recluta *m.*
D Wehrdienstpflichtiger
  m.; Eingezogener *m.*
R призывник *м.*
A مُجنَّد

960 [A,R]
console *n.*
F console *f.*
E consola *f.*
D Instrumentenbank [A]
  f.; Sichtgerät [R] *neut.*
R пульт *м.;* панель *м.*
A نتوء *m* . لَوْحَة *f*

961 [L]
consolidation
F organisation *f.*
E consolidación *f.*
D Festigung *f.*
R закрепление *ср.*
A تَدْعيم *m*

962 [E]
constant *n.*
F constante *f.*
E constante *f.*
D Konstante *f.*
R константа *ж.;*
  постоянная величина
  ж.
A ثَابِت *m*

963
consumables *pl.*
F aliments *m. pl.*
E comestibles *m. pl.*
D Nahrungsmittel *m. pl.*
R продукты питания *м.*
  мн. ч.
A مَوادّ إسْتِهْلَاكية *f*

964
consumption rate
F taux de consommation
  m.
E ritmo de consumo *m.*
D Verbrauchs-
  geschwindigkeit *f.*
R темп потребления *м.;*
  скорость расхода *ж.*
A معدل الإستهلاك *m*

965
contact *n.*
F contact *m.*
E contacto *m.*
D Fühlung *f.;* Kontakt *m.*
R соприкосновение *ср.;*
  контакт *м.*
A إتِّصَال *m*

966

contact *v.*

F contacter
E poner(se) en contacto (con)
D (sich in) Verbindung setzen (mit)
R приходить в соприкосновение; соприкасаться; устанавливать связь
A إتَّصَل بِ

967 [L,U]

contain

F contenir
E contener
D abriegeln *tr.;* enthalten *untr.*
R содержать
A وَعَى ـ شَمَل

968 [V]

container

F conteneur *m.*
E container *m.*
D Transportbehälter *m.*
R контейнер *м.*
A وِعَاء . *m* صُنْدُوق *m*

969

contaminate

F contaminer
E contaminar
D verseuchen
R загрязнять
A لَوَّث

970

contamination

F contamination *f.*
E contaminación *f.*
D Verseuchung *f.*
R загрязнение *ср.*
A تَلْوِيث *m*

971

continent *n.*

F continent *m.*
E continente *m.*
D Festland *neut.*
R материк *м.;* континент *м.*
A قَارَّة *f*

972

continental shelf

F plateau continental *m.*
E plataforma continental *f.*
D Festlandssockel *m.*
R континентальный шельф *м.*
A الإفْرِيز القارّي *m*

973

contingency

F éventualité *f.*
E contingencia *f.*
D Möglichkeit *f.*
R случай *м.;* вариант обстановки *м.*
A حَادِثة *f* ـ طَارِىء *m*

974

contingency force

F force de deploiement rapide *f.*
E fuerza eventual *f.*
D Bereitschaftstruppe *f.*
R корпус быстрого развёртывания *м.*
A قوات الإنتشار السريع

975

contingent

F contingent *m.*
E contingente *m.*
D Kontingent *neut.*
R контингент *м.*
A فريق

976

contract

F contrat *m.*
E contrato *m.*
D Vertrag *m.*
R контракт *м.*
A عَقْد *m*

977 [A,M,S,V]

control *n.*

F organe de commande *m.*
E dispositivo de mando *m.*
D Steuerorgan *neut.*
R управление *ср.*
A مُرَاقبة *f*

978 [E,P]

control *n.*

F contrôle *m.*
E control *m.*
D Kontrolle *f.*
R контроль [P] *м.;* рычаг [E] *м.*
A إدَارَة *f*

979 [E,P]

control *v.*

F contrôler
E controlar
D kontrollieren
R контролировать
A أدار

980 [A,M,V]

control *v.*

F maîtriser
E dominar
D bedienen; führen
R управлять
A راقَب

981 [U]

control v.
F maîtriser; diriger
E dominar; dirigir
D beherrschen; leiten
R руководить;
   управлять;
   контролировать
A تَحَكُّم

982 [U]

control (of) n.
F direction f.
E manejo m.
D Leitung f.
R управление ср.;
   руководство ср.
A تَحَكُّم m

983 [U]

control (over) n.
F maîtrise f.
E control m.
D Herrschaft f.
R командование ср.
A سيطرة على

984

control column
F levier de commande m.
E palanca de comando f.
D Steuersäule f.
R рычаг управления м.
A عَمُود القِيَادة m

985

controlled airspace
F espace aérien contrôlé
   m.
E espacio aéreo
   controlado m.
D überwachter Luftraum
   m.
R регулированное
   воздушное
   пространство ср.
A حَيِّز جَوِّي مَحْكُوم m

986

controller
F contrôleur m.
E controlador m.
D Leiter m.
R контролёр м.
A مُدِير m

987

control measures
F directives pour la mise-
   en-place f. pl.
E medidas de control f.
   pl.
D Kontrollmassnahmen
   f. pl.; Überwachungs-
   massnahmen f. pl.
R управление
   операциями ср.
A تعليمات السيطرة

988

controls pl.
F commandes f. pl.
E aparatos de control m.
   pl.
D Steuerungen f. pl.
R рычаги м. мн. ч.;
   ручки ж. мн. ч.
A مَقَاوِد ـ قيادات f

989

conventional weapons
F armes conventionnelles
   f. pl.
E armas convencionales f.
   pl.
D konventionelle Waffen
   f. pl.
R обычные виды оружия
   м. мн. ч.
A أَسْلِحَة تَقْلِيدية

990

converge (on)
F (faire) converger (sur)
E converger (en)
D zusammenlaufen (bei)
   tr.
R сходиться (в, на);
   приближаться (к)
A تَقَارَب

991

convergence
F convergence f.
E convergencia f.
D Annäherung f.
R схождение ср.
A تَقَارُب m

992 [M,L]

convoy
F convoi m.
E convoy m.
D Geleitzug [ML] m.;
   Kolonne [L] f.
R автоколонна [L] ж.;
   конвой [M] м.
A قَافِلة f

993

convoy assembly
F formation de convoi f.
E formación de convoyes
   f.
D Geleitzugs- komp.;
   Zusammenstellung f.
R сбор конвоя м.
A إِجْتِمَاع القَافِلة

994

convoy escort
F escorte de convoi f.
E escolta (de convoy) f.
D Geleitschutz m.
R охранение конвоя ср.
A حَرَس القَافِلة m

995

convoy joiner
F navire ralliant un
   convoi m.
E buque que se une a un
   convoy m.
D stossendes Schiff neut.
R присоединяющийся к
   конвою м.
A (سَفِينة) مُلْتَحِقَة بالقَافِلة

996

convoy leaver
F navire se détachant
   d'un convoi *m.*
E buque separándose de
   un convoy *m.*
D ausscheidendes Schiff
   neut.
R отделяющийся от
   конвоя *м.*
A (سَفِينَة) تَتْرُك القافِلة

997

cook *v.*
F cuire
E cocinar
D kochen
R варить
A طَبَخ

998

cook *n.*
F cuisinier *m.*
E cocinero *m.*
D Koch *m.*
R повар *м.*; кок *м.*
A طَبَّاخ *m*

999

cooperation
F cooperation *f.*
E cooperación *f.*
D Kooperation *f.;*
   Zusammenarbeit *f.*
R сотрудничество *ср.;*
   взаимодействие *ср.*
A تعاون *m*

1000

coordinate *v.*
F coordonner
E coordinar
D koordinieren
R взаимодействовать;
   координировать
A نَسَّق

1001

coordinates *n. pl.*
F coordonneés *f. pl.*
E coordenadas *f. pl.*
D Koordinaten *f. pl.*
R координаты *м. мн. ч.*
A إحداثي *m* / إحداثيات *f*

1002

coordinating authority
F autorité de
   coordination *f.*
E autoridad de
   coordinación *f.*
D federführende
   Dienststelle *f.*
R координирующий
   орган *м.*
A هَيْئة التَّنْسِيق *f*

1003

cordite
F cordite *f.*
E cordita *f.*
D Kordit *neut.*
R кордит *м.*
A كُرْدِيت (نوع من البَارُود) *m*

1004

cordon *n.*
F cordon *m.*
E cordón *m.*
D Absperrkette *f.*
R кордон *м.;* охранение
   ср.
A نِطاق (الحِصار) *m*

1005

cordon (off) *v.*
F boucler
E acordonar
D absperren *tr.*
R оцеплять
A حاصَر

1006

corner *n.*
F coin *m.*
E esquina *f.;* rincón *m.*
D Ecke *f.*
R угол *м.*
A زَاوِية *f*

1007 [P]

corner *v.*
F acculer
E acorralar; cercar
D (in die) Ecke treiben
R загонять в угол;
   загонять в тупик
A حَصَر في زَاوِية

1008 [R]

corner reflector
F réflecteur polyédrique
   m.
E pantalla polyédrica *f.*
D Winkelreflektor *m.*
R уголковый
   отражатель *м.*
A عَاكِس زاوي *m*

1009

(army) corps
F corps (d'armée) *m.*
E cuerpo (de ejército) *m.*
D Korps *neut.;* Truppe *f.*
R корпус (армии) *м.*
A فَيْلَق *m*

1010

corpse
F cadavre *m.*
E cadáver *m.*
D Leiche *f.*
R труп *м.*
A جُثَّة *f*

1011

correct *adj.*
F juste
E correcto
D richtig
R правильный
A صَحِيح

1012

correction
F rectification *f.*
E rectificación *f.*
D Verbesserung *f.*
R исправление *ср.;*
   поправка *ж.*
A تَصْحِيح *m*

1013

correlate

F (mettre) en corrélation
(avec)

E correlacionar (a); tener
correlación (con)

D (in) Wechselbeziehung
bringen; stehen (zu)

R сопоставлять; сличать

A أقام عِلاقة

1014

correlation

F corrélation f.

E correlación f.

D Wechselbeziehung (zu)
f.

R соотношение ср.

A عَلاقة مُتَبَادَلة f

1015

corrode

F (se) corroder

E corroer(se)

D korrodieren

R разъедать

A تَأَكَّل

1016

corrosion

F corrosion f.

E corrosión f.

D Korrosion f.

R разъедание ср.;
ржавчина ж.

A تَـآكُـل m

1017

corrosive

F corrosif

E corrosivo

D zerfressend

R разъедающий

A آكِل

1018

corrugated

F ondulé

E ondulado

D Well- komp.

R гофрированный;
волнистое (железо)

A مُمَوَّج

1019

count v.

F compter

E contar

D zählen

R считать

A عَـدَّ

1020

countdown n.

F compte à rebours m.

E cuenta hacia atrás f.

D Vorbereitung zum
Start/Abschuss

R счёт времени
готовности м.

A العَدّ التَنازُلي m

1021

counter v.

F contrecarrer

E contestar

D abfangen tr.; entgegnen
untr.

R отбивать; отражать

A ضَادَّ

1022

counterattack

F contre-attaque f.

E contraataque m.

D Gegenangriff m.

R контратака ж.

A هُجُوم مُضَادّ m

1023

counterclockwise

F (en) sens invers des
aiguilles d'une montre

E (contra la) dirección de
las agujas del reloj
inv.

D linksdrehend adj.

R (против) часовой
стрелки

A في عكس إتجاه عقرب الساعة A

1024

counterespionage

F contre-espionnage m.

E contra-espionaje m.

D (Spionage-)Abwehr f.

R контршпионаж м.;
контрразведка ж.

A مُكَافَحَة التَجَسُّس f

1025 [S]

counterfire

F tir de destruction (de
neutralisation) m.

E contestación del fuego
f.

D Niederhalten neut.;
Zerschlagen neut.

R контратакующий
огонь м.

A الرَّمْي المُضَادّ m

1026

counterinsurgency adj.

F anti-insurrectionnel

E contra-subversivo

D Partisanenbekämpfungs-
komp.

R противопартизанский

A مُضَادّ لِلتَمَرُّد

1027

countermeasure

F contre-mesure f.

E contramedida f.

D Gegenmassnahme f.

R мера противодействия
ж.

A تَدْبِير مُضَادّ m

1028

countersign

F signe de reconnaissance *m.*

E contraseña *f.*

D Geheimzeichen *neut.*

R отзыв *м.*

A تصريح *m*

1029

country

F pays *m.*

E país *m.*

D Land *neut.*

R страна *ж.*

A بَلَد *f*

1030

country of origin

F pays de provenance *m.*

E país de origen *m.*

D Herkunftsland *neut.*

R родная страна *ж.*

A بلد أَصْله *m*

1031

courier

F estafette *f.*

E correo *m.*

D Eilbote *m.;* Kurier *m.*

R курьер *м.;* посыльный *м.*

A ساعٍ

1032 [A,M]

course

F cap *m.*

E rumbo *m.*

D Kurs *m.*

R курс *м.*

A إتِّجَاه *m*

1033

(off) course

F déviant (du cap fixé)

E desviado; fuera de ruta

D (vom Kurs) abweichend

R отклонённый от курса

A أَنْحِرَاف اتِّجَاهي

1034

(on) course

F suivant le cap fixé

E (en) ruta

D (den Kurs) einhaltend

R направленный на курсе

A في الطَّرِيق

1035 [A,G,M]

(pursuit) course

F cap de poursuite *m.*

E rumbo de caza *m.*

D Verfolgekurs *m.*

R кривая погони [G] *ж.;* курс преследования *м.*

A طريق المُلاحقة

1036

(training) course

F stage *m.*

E curso *m.*

D Kurs *m.*

R курс (обучения) *м.*

A دَوْرَة *f*

1037

course of action

F possibilité d'action *f.*

E línea de acción *f.*

D Einsatzplan *m.*

R образ действий *м.*

A إتِّجاه (العملية) *f*

1038

court martial

F cour martiale *f.*

E consejo de guerra *m.*

D Kriegsgericht *neut.*

R военно-полевой суд *м.*

A محكمة عسكرية *f*

1039

(air) cover *n.*

F couverture aérienne *f.*

E protección *f.*

D Deckung *f.*

R прикрытие с воздуха *ср.;* авиационное прикрытие *ср.*

A حماية جوية ستار جوي *m*

1040

cover (by fire) *v.*

F battre par le feu

E dominar

D decken

R держать под обстрелом

A سترب نار

1041

cover (by patrol) *v.*

F contrôler

E cubrir

D decken

R патрулировать; наблюдать; защищать

A سترب دَوْرية

1042

(cloud) cover *n.*

F couverture nuageuse *f.*

E cubierta *f.*

D (Wolken-) Decke *f.*

R облачность *ж.*

A غِطاء من السَّحَاب *m*

1043

(fighter) cover *n.*

F couverture de chasse *f.*

E cobertura de caza *f.*

D Jagdschutz *m.*

R истребительное прикрытие *ср.*

A حِماية المُقَاتِلات *f*

1044

cover (from attack) *v.*

F couvrir

E cubrir

D decken

R прикрывать

A سترعن الهجوم

1045
cover (from fire) *n.*
F abri *m.*
E protección *f.;* refugio
    m.
D Deckung *f.*
R укрытие *ср.*
A سِتْر عن النار

1046
**cover (from
    observation)** *n.*
F protection contre
    l'observation *f.*
E refugio *m.*
D Pflanzendecke *f.*
R покров *м.;* укрытие
    (от наблюдения) *ср.*
A استتار عن الأنْظَار

1047
(hatch) cover *n.*
F panneau *m.*
E cuartel *m.*
D Einstiegklappe *f.;*
    Lukendeckel *m.*
R крышка люка *ж.*
A غِطَاء كُوَّة *m*

1048
(natural) cover *n.*
F couverture naturelle *f.*
E abrigo natural *m.*
D natürliche Deckung *f.*
R естественное укрытие
    ср.
A سِتْر طَبِيعِي *m*

1049
(security) cover *n.*
F couverture *f.*
E pretexto *m.*
D Deckmantel *m.;*
    Schutzidentität *f.*
R личина *ср.;* маска *ср.*
A سِتَار أمن

1050
(take) cover *v.*
F (s')abriter; (se) mettre à
    couvert
E cubrirse
D (Deckung) nehmen
R укрываться
A إستتر

1051 [L]
(under) cover
F défilé; à couvert; à
    l'abri
E bajo abrigo
D gedeckt; geschützt
R (в) укрытии
A مَسْتُور

1052
cover (with a weapon)
    v.
F tenir en joue
E amenazar
D decken
R держать под
    прицелом
A صوّب سلاحًا إلى . . .

1053
covered approach
F route à l'abri des vues
    et des coups *f.*
E aproches protegidos *m.*
    pl.
D gedeckter Anmarschweg
    m.
R укрытый подступ *м.*
A خط إقتراب مأمون

1054
covering force
F force de couverture *f.;*
    force de sécurité *f.*
E fuerza de cobertura *f.*
D Sicherungstruppen *f. pl.*
R войска прикрытия *ср.*
    мн. ч.
A قُوة التَّغْطِية *f*

1055 [P]
covert
F (en) secret
E secreto
D verborgen *adj.*
R тайный; скрытный
A سِرِّي

1056
coxswain
F patron *m.*
E patrón *m.*
D Bootsführer *m.*
R старшина шлюпки *м.*
A مُوَجِّه الدَفَّة *m*

1057 [E]
crack *n.*
F fente *f.*
E grieta *f.*
D Riss *m.*
R трещина *ж.*
A صَدْع *m*

1058
crack unit
F unité d'élite *f.*
E unidad especial *f.;*
    unidad de elite *f.*
D Sondertrupp *m.*
R (замечательное)
    подразделение *ср.*
A وحدة النخبة

1059 [E]
cradle
F berceau *m.*
E cuna *f.*
D Gestell *neut.*
R люлька *ж.*
A سَرِير ـ مَهْد

1060 [M]
craft
F bâtiment *m.*
E embarcación *f.*
D Boot *neut.*
R судно (суда) *ср.(мн.ч.)*
A زَوْرَق *m* . زَوَارِق *f*

65

1061
crane
F grue *f.*
E grúa *f.*
D Kran *m.*
R подъёмный кран *м.*
A رَافِعة *f*

1062 [A]
crash *v.*
F (s')écraser
E estrellarse
D abstürzen *tr.*
R потерпеть аварию
A تَحَطَّم

1063 [A]
crash *n.*
F accident *m.*
E accidente *m.*
D Absturz *m.*
R авария *ж.*; крушение *ср.*
A تَحَطُّم *m*

1064 [M]
crash dive
F plongée en urgence *f.*
E inmersión de urgencia *f.*
D Schnelltauchen *neut.;* Nottauchen *neut.*
R срочное погружение *ср.*
A غَطْس سَريع

1065
crater *v.*
F interdire par des entonnoirs
E obstaculizar con cráteres
D (mit) Trichtern versehen
R усеивать воронками
A حَفَر

1066
crater *n.*
F entonnoir *m.*
E cráter *m.*
D (Granat- usw.) Trichter *m.*
R воронка *ж.*
A حُفْرَة *f*

1067 [T]
creek
F ruisseau *m.*
E arroyo *m.;* riachuelo *m.*
D Bach *m.*
R приток *м.*; ручей *м.*
A خَوْر - نَهْر *m*

1068 [M]
creek
F anse *f.*
E cala *f.;* ensenada *f.*
D Bucht *f.*
R бухта *ж.*; залив *м.*
A خَوْر - نَهْر *m*

1069
creeping barrage
F barrage roulant *m.*
E cortina de fuego móvil *f.*
D Feuerwalze *f.*
R ползущий огневой вал *м.*
A سَدّ زَاحِف *m*

1070 [S]
crest
F masque *m.*
E cresta *f.*
D toter Kamm *m.*
R угол прикрытия *м.*
A ذِرْوَة *f*

1071 [T]
crest
F crête *m.*
E cresta *f.*
D (Berg- usw.) Kamm *m.*
R гребень *м.*
A ذِرْوَة *f*

1072
crew
F équipage *m.*
E tripulación *f.;* equipo *m.*
D Besatzung *f.*
R экипаж *м.*
A طَاقِم *m*

1073
crisis
F crise *f.*
E crisis *f.*
D Krise *f.*
R кризис *м.*
A أزمة *f*

1074
critical
F critique
E crítico
D Grenz- *komp.;* kritisch *adj.*
R критический
A حَرِج

1075 [N]
critical mass
F masse critique *f.*
E masa crítica *f.*
D kritische Masse *f.*
R критическая масса *ж.*
A كُتْلَة حَرِجة *f*

1076
cross *v.*
F passer; traverser
E cruzar
D überqueren *untr.*
R переходить; переезжать; переправлять
A إجْتَاز - عَبَر

1077

cross-attachment
F détachement temporaire
   m.
E agregación f.
D Wechselangliederung f.
R обмен приданными
   подразделениями м.
A تبادُل الوحدات الثانوية أثناء
   القِتَال

1078

cross-country adj.
F tous terrains inv.
E (a) campo través
   (traviesa)
D geländegängig
R вездеходный
A إختِراق الضاحية

1079

crossfire
F (des) feux croisés m. pl.
E fuego cruzado m.
D Kreuzfeuer neut.
R перекрёстный огонь
   м.
A نيران مُتقاطِعة f

1080

crossroads pl.
F carrefour m.
E cruce m.
D Strassenkreuzung f.
R перекрёсток м.
A مُلْتَقَى الطُّرُق

1081

cross-servicing
F soutien logistique
   mutuel
E apoyo logístico común
   m.
D gemeinsame logistische
   Unterstützung f.
R взаимное
   материально-
   техническое
   обеспечение ср.
A إتِّفاقِية مُتَبادَلة لِلصِّيانة
   المعاونة الإدارية المتبادلة

1082

crowd
F foule f.
E muchedumbre f.; tropel
   m.
D Menschenmenge f.
R толпа ж.
A حَشْد m

1083

cruise v.
F (faire une) croisière
E cruzar
D kreuzen
R крейсировать
A طاف

1084

cruising altitude
F altitude de croisière f.
E altitud de crucero f.
D Reiseflughöhe f.
R крейсерская высота
   ж.
A إرْتِفَاع الطَّوَاف m

1085

cryptanalysis
F analyse
   cryptographique f.
E análisis criptográfico m.
D Geheimschriftanalyse f.
R дешифрование ср.
A تَحْلِيل الرُّمُوز m

1086

cryptography
F cryptographie f.
E criptografía f.
D Geheimschreibekunst f.
R шифрование ср.;
   криптография ж.
A الكِتَابة الرُّمْزية f

1087

cryptomaterial
F matériel
   cryptographique m.
E material criptográfico
   m.
D geheimschutzbedürftiges
   Material neut.
R шифровальный
   материал м.
A مَوَادّ الكِتَابة الرُّمْزية f

1088

cupola
F tourelleau m.
E cúpula f.
D Kuppel f.
R купол м.
A قُبَّة f

1089

curfew
F couvre-feu m. inv.
E toque de queda m.
D Ausgehverbot neut.
R комендантский час м.
A مَنْع التَجَوُّل m

1090

current n.
F courant m.
E corriente f.
D Strom(-stärke) m.(f.)
R ток м.; течение ср.
A تَيَّار m

1091

current adj.
F actuel
E corriente; actual
D gegenwärtig;
   augenblicklich
R текущий
A مؤخراً

1092

cut v.
F couper
E cortar
D schneiden
R резать
A قَطَع

1093 [E,L]
cut off *v.*
F couper
E cortar; copar
D abschneiden *tr.;*
   absprengen *tr.*
R прерывать; отрезать
A قَطَع

1094 [M]
cutter
F cotre *m.*
E cúter *m.*
D Kutter *m.*
R катер *м.*
A زَوْرَق *m*

1095
cyclone
F cyclone *m.*
E ciclón *m.*
D Zyklon *m.*
R циклон *м.*
A إعْصَار *m*

1096
cylinder
F cylindre *m.;* bouteille
   (de gaz) *f.*
E cilindro *m.*
D Zylinder *m.*
R цилиндр *м.;* валик *м.*
A أُسْطُوَانة *f*

1097
cylindrical
F cylindrique
E cilíndrico
D zylindrisch
R цилиндрический
A أُسْطُوَانِي

D

1098
dagger
F poignard *m.*
E puñal *m.*
D Dolch *m.*
R кинжал *м.*
A خَنْجَر *m*

1099 [E,M,U]
damage
F dommages *m. pl.*
E avería [E] *f.;* daño
   [MU] *m.*
D Schaden *m.*
R повреждение *ср.*
A عَطَب *m*

1100
damage assessment
F évaluation des
   dommages *f.*
E evaluación de averías *f.*
D Schadenfeststellung *f.*
R определение
   повреждения *ср.*
A تَقْدِير العَطَب *m*

1101
damage control
F organisation sécurité *f.*
E reparación de averías *f.*
D Schadenbegrenzung *f.*
R ремонтно-
   восстановительные
   работы *ж. мн. ч.;*
   служба живучести
   ж.
A حَصْر العَطَب *m*

1102
danger
F danger *m.*
E peligro *m.*
D Gefahr *f.*
R опасность *ж.*
A خَطَر *m*

1103
danger area
F zone dangereuse *f.*
E zona de peligro *f.*
D Gefahrenzone *f.*
R зона опасности *ж.*
A مِنْطَقة الخَطَر *f*

1104
dangerous
F dangereux
E peligroso
D gefährlich
R опасный
A خَطِر

1105
dark
F obscur
E oscuro
D dunkel
R тёмный
A مُظْلِم

1106
darken
F (s')assombrir
E oscurecer
D dunkel werden;
   verdunkeln
R темнеть; затемнять
A أظْلَم

1107
darkness
F obscurité *f.*
E oscuridad *f.*
D Dunkelheit *f.*
R темнота *ж.*
A ظَلاَم *m*

1108 [R]
dash
F trait *m.*
E raya *f.*
D Morsestrich *m.*
R тире *ср.*
A شَرْطَة *f*

1109
data *pl.*
F données *f. pl.*
E datos *m. pl.*
D Daten *neut. pl.*
R данные *ж. мн. ч.*
A مُعْطَيات *f*

1110 [D]
data base
F base de données *f.*
E base de datos *f.*
D Datenbasis *f.*
R база данных *ж.*
A قاعدة معلومات *f*

1111
date time group
F groupe (date/heure) *f.*
E hora de origen *f.*
D Zeit-
   Datumklassifikation *f.*
R временная группа *ж.*
A مجموعة التأريخ والزمن

1112
datum
F repère *m.*
E dato *m.*
D Bezugspunkt *m.*
R репер *м.*; исходный
   уровень *м.*
A مُعْطَى *m*

1113
datum level
F plan de comparaison *m.*
E nivel de comparación
   m.
D Bezugsebene *f.*
R нулевой уровень *м.*
A مُسْتَوَى المُعْطَى *m*

1114
davits *pl.*
F portemanteau *m.*
E pescantes *m. pl.*
D Davits *m. pl.*
R шлюпбалки *ж. мн. ч.*
A مَرافع - رَوافع *f*

1115
dawn
F aube *f.*
E alba *f.*; aurora *f.*
D (Morgen-)
   Dämmerung *f.*
R рассвет *м.*; (утренняя)
   заря *ж.*
A فَجْر *m*

1116
day
F jour *m.*
E día *m.*
D Tag *m.*
R день *м.*
A يَوْم *m*

1117
daylight
F plein jour *m.*
E luz (del día) *f.*
D Tageslicht *neut.*
R дневной свет *м.*
A ضَوْء النَّهار *m*

1118
dazzle *v.*
F éblouir
E deslumbrar
D blenden
R ослеплять;
   маскировать
A عَمَّى

1119
dazzle *n.*
F éblouissement *m.*
E brillante *m.*
D Blenden *neut.*
R ослепление *ср.*;
   маскировка краской
   ж.
A عَمْى *m*

1120
D-day
F (le) jour J *m.*
E día D *m.*
D Operationsbeginn *m.*
R день Д *м.*; день
   начала операций *м.*
A يَوْم « ي » (يَوْم بَدْء العَمَلِية)

1121
dead
F mort
E muerto
D tot
R погибший; мёртвый
A مَيِّت

1122
deadline
F date limite *f.*; heure
   limite *f.*
E fecha tope *f.*; hora tope
   f.
D (letzter) Termin *m.*
R крайний срок *м.*
A آخِر مَوْعِد (لإنْجَاز
   عَمَل ما) *m*

1123 [R,S]
dead space
F zone de silence [R] *f.*;
   zone en angle mort [S] *f.*
E espacio muerto [R] *m.*;
   zona en ángulo
   muerto [S] *f.*
D tote Zone *f.*
R мёртвая зона [S] *ж.*;
   зона молчания [R] *ж.*
A مَجَال مَيِّت *m*

1124 [G,R,S]
dead time
F temps mort *m.*
E tiempo muerto *m.*
D Totzeit *f.*; Verzugszeit
   f.
R время запаздывания
   ср.
A زَمَن العَطَالة *m*

1125
debarcation
F débarquement *m.*
E desembarque *m.*;
   desembarco *m.*
D Ausschiffung *f.*;
   Landung *f.*
R высадка *ж.*; выгрузка
   ж.
A نُزُول *m*

1126
debark
F débarquer
E desembarcar
D ausschiffen *tr.*;
   landen
R высаживать(ся);
   выгружать(ся)
A نَزَل

1127
debrief *v.*
F (faire un) compte-
rendu de mission
E interrogar
D befragen *untr.;*
ausfragen *tr.*
R опрашивать
A استجوب

1128 [N]
decay *n.*
F désintégration *f.*
E desintegración *f.*
D Zerfall *m.*
R распад *м.*
A تَناقُص *m*

1129
deceive
F tromper
E engañar
D täuschen
R обманывать
A خَدَع

1130
decentralize
F décentraliser
E descentralizar
D dezentralisieren
R децентрализировать
A أَبْطَل المَرْكَزِية

1131
deception
F déception *f.*
E engaño *m.*
D Täuschung *f.*
R обман *м.*;
дезориентирование
ср.
A خُدْعَة *f*

1132
decimal *adj.*
F décimal
E decimal
D Dezimal- *komp.*
R десятичный
A عُشْرِي

1133
decimal *n.*
F décimale *f.*
E fracción decimal *f.*
D Dezimalbruch *m.*
R десятичная дробь *ж.*
A كَسْر عُشْرِي *m*

1134
decimate
F décimer
E diezmar
D dezimieren; vernichten
R косить на десятую;
уничтожать
A أَهْلَك ـ أَباد

1135
decipher
F déchiffrer
E descifrar
D entziffern
R расшифровывать
A حَلْ الشُفْرة

1136
decision (on)
F décision (sur) *f.*
E decisión (por) *f.*
D Entscheidung (über) *f.*
R решение (по,о) *ср.*
A قَرَار (عَن) *m*

1137
deck
F pont *m.*
E cubierta *f.*
D Deck *neut.*
R палуба *ж.*
A سَطْح *m*

1138
declassify
F déclassifier
E desclasificar
D (die)
Geheimhaltungsstufe
aufheben
R рассекречивать
A غَيَّر من تَصْنِيف

1139
declination
F déclinaison *f.*
E declinación *f.*
D Deklination *f.*
R (магнитное) склонение
ср.
A مَيْل *m*

1140
decode
F décoder
E descifrar
D entziffern
R расшифровывать по
коду
A فكّ الرمز

1141 [A,M]
decompress
F décomprimer
E descomprimir
D (die) Verdichtung
vermindern
R уменьшать давление
[A]; производить
декомпрессию
[M]
A خَفَّف الضَغْط

1142 [M]
**decompression
chamber**
F chambre de
décompression *f.*
E cámara de
descompresión *f.*
D Unterdruckkammer *f.*
R декомпрессионная
камера *ж.*
A حُجْرة تَخْفِيف الضَغْط *f*

1143 [A,M,H]
**decompression
sickness**
F mal des caissons *m.*
E embolía gaseosa *f.*
D (Höhen- bzw.)
Tiefdruckkrankheit *f.*
R кессонная болезнь *ж.*
A مَرَض تخفيف الضَغْط *m*

1144 [C,H,N]

decontaminate

F décontaminer [CN]
E descontaminar
D entseuchen [CH];
   entstrahlen [N]
R дегазировать [C];
   дезактивировать
   [HN]
A طَهَّر

1145 [C,H,N]

decontamination

F désinfection f.;
   décontamination f.
E descontaminación f.
D Entseuchung f.;
   Entstrahlung f.
R дегазация ж.;
   дезактивизация ж.
A تَطْهِير m

1146

decoy v.

F leurrer
E atraer con señuelo
D verleiten
R приманивать
A خَدَع

1147

decoy n.

F leurre m.
E señuelo m.
D Schein- (z.B. Anlage)
   komp.
R макет м.; ложная цель
   ж.
A خَدِيعَة f

1148

decrypt

F décrypter
E descifrar
D entschlüsseln
R расшифровывать
A حَلَّ نَصًّا مَرْمُوزاً

1149 [S]

dedicated

F (de) mission réservée
E comprometido
D bestimmt adj.
R выделенный;
   назначенный
A مُكَرَّس

1150

deep adj.

F profond
E profundo
D tief
R глубокий
A عَمِيق

1151

deep sea diver

F scaphandrier m.
E buzo m.
D Tiefseetaucher m.
R глубоководный
   водолаз м.
A غَطَّاس بَحْرِي m

1152

defeat v.

F battre
E derrotar; vencer
D besiegen
R наносить поражение;
   побеждать
A هَزَم

1153

defeat n.

F défaite f.
E derrota f.
D Niederlage f.
R поражение ср.
A هَزِيمَة f

1154

defect n.

F défectuosité f.
E defecto m.
D Defekt m.
R недостаток м.;
   неисправность ж.
A عَيْب m

1155 [H]

defect v.

F déserter
E desertar
D überlaufen tr.;
   weglaufen tr.
R перебегать (на другую
   сторону)
A إِرْتَدَّ ـ مَرَق

1156

defection

F défection f.
E deserción f.
D Fahnenflucht f.
R дезертирство ср.;
   нарушение (долга)
   ср.; переход (на
   другую сторону) м.
A إِرْتَدَاد m

1157

defector

F transfuge m.
E desertor m.
D Fahnenflüchtige(r) m.
R перебежчик м.;
   отступник м.
A مُرْتَدّ m

1158

defend

F défendre
E defender
D verteidigen
R оборонять(ся);
   защищать(ся)
A دَافَع عَن

1159

defense

F défense f.
E defensa f.
D Verteidigung f.
R оборона ж.; защита
   ж.
A دِفَاع m

71

1160
(reverse slope) defense
F défense de contre-pente
  f.
E defensa de
  contrapendiente *f.*
D Hinterhangverteidigung
  f.
R оборона на обратном
  скате *ж.*
A دفاع السفح الخلفي

1161
defense in depth
F défense en profondeur
  f.
E defensa en profundidad
D tiefgegliederte
  Verteidigung *f.*
R эшелонированная
  оборона *ж.*
A دِفَاع في عُمْق *m*

1162
defense in place
F défense ferme *f.*
E defensa estática *f.*
D statische Verteidigung *f.*
R стабильная оборона
  ж.
A دِفَاع مَوْضِعِي *m*

1163
defense policy
F politique de défense *f.*
E política de defensa *f.*
D Verteidigungspolitik *f.*
R оборонительная
  политика *ж.*
A سِيَاسة دِفَاعية *f*

1164
defensive *adj.*
F défensif
E defensivo
D Verteidigungs- *komp.*
R оборонительный;
  защитный
A دِفَاعِي

72

1165
(on the) defensive *n.*
F (sur la) défensive *f.*
E (a la) defensiva *f.*
D (in der) Verteidigung *f.*
R (на) обороне
A إتَّخَذ وَضْع دِفَاعي

1166
deficiency
F manque *m.*
E falta *f.*
D Mangel *m.*
R недостаток *м.*
A نَقْص *m*

1167
defilade *v.*
F défiler
E desenfilar
D decken
R укрывать рельефом
A سَتَر

1168 [L]
(hull) defilade
F (à) défilement de caisse
E (a) desenfilada de casco
D halbgedeckt
R полузакрытым
  корпусом
A جِسْم مَسْتُور *m*

1169
(in) defilade *n.*
F (en) défilement *m.*
E (en) desenfilada *f.*
D Deckung (gegen Sicht)
  f.
R укрытие (в укрытии)
  ср.
A مُسْتَتَر

1170 [L]
(in turret) defilade
F (à) défilement de
  tourelle
E (a) desenfilada de torre
D (in) gedeckte(r) Stellung
  adj. s. f.
R укрытым корпусом
A البُرْج مَسْتُور

1171 [T]
defile *n.*
F défilé *m.*
E desfiladero *m.*
D Engpass *m.*
R теснина *ж.*; дефиле
  ср.
A مَمَرّ *m*

1172 [E]
definition
F netteté *f.*
E claridad *f.*
D Auflösungsvermögen
  neut.
R чёткость *ж.*; резкость
  ж.
A دِقَّة *f*

1173 [U]
definition
F définition *f.*
E definición *f.*
D Definition *f.*
R определение *ср.*
A تَعْرِيف *m*

1174
deflect
F défléchir
E desviar(se)
D ablenken *tr.;* abweichen
  tr.
R отклонять; отражать
A حَرَف

1175 [S]
deflection
F dérive *f.*
E desviación *f.*
D Ablenkung *f.;*
  Abweichung *f.*
R отклонение *ср.;*
  угломер *м.*
A أنْحِرَاف *m*

1176
defoliant *n.*
F agent défoliant *m.*
E agente *m.*
D Entlaubungsmittel *neut.*
R дефолиант *м.*
A مُجَرَّد من الوَرَق *m*

1177
degrade *v.*
F dégrader
E degradar
D degradieren *untr.;*
  herabsetzen *tr.*
R разжаловать
A عَزَل

1178
degree
F degré *m.*
E grado *m.*
D Grad *m.*
R градус *м.*
A دَرَجَة *f*

1179 [H]
dehydration
F déshydratation *f.*
E deshidratación *f.*
D Entwässerung *f.*
R обезвоживание *ср.*
A جفاف *m*

1180 [E]
deicer
F dégivreur *m.*
E deshelador *m.*
D Enteiser *m.*
R антиобледенитель *м.*
A إذابة الجليد

1181
delay *n.*
F retard *m.*
E dilación *f.;* retraso *m.*
D Verzögerung *f.*
R задержка *ж.*
A أَجَل ـ تأخِير *m*

1182 [U]
delay *v.*
F retarder; (s')attarder
E demorar(se)
D (sich) verzögern
R задерживать(ся)
A عَوَّق . أَجَّل

1183 [L]
delay *v.*
F retarder
E retardar
D hinhalten *tr.*
R задерживать;
  замедлять
A أَجَّل ـ أخّر

1184
delayed action
F (à) retardement
E (de) acción retardadora
D Verzögerungs- *komp.*
R замедленное действие
  ср.
A فِعْل مُعَوَّق *m*

1185
delaying tactics *pl.*
F manoeuvre retardatrice
  f.
E tácticas retardadoras *f.*
  pl.
D Verzögerungstaktik *f.*
R сдерживающие
  действия *ср. мн. ч.*
A تَعْبِئات إعَاقَة

1186
delegate *v.*
F déléguer
E delegar
D übertragen *untr.*
R уполномочивать;
  поручать
A أناب

1187 [A,M,L]
deliberate *adj.*
F préparé
E pausado
D planmässig
R планомерный
A مُدَبَّر

1188
deliberate crossing
F franchissement préparé
  m.
E paso deliberado *m.*
D vorbereiteter
  Flussübergang *m.*
R планомерное
  форсирование (реки)
  ср.
A عُبُور مُدَبَّر *m*

1189 [G,S]
deliver
F livrer
E lanzar
D abschiessen *tr.*
R наносить (удар)
A قَذَف

1190 [G,S]
delivery error
F dispersion *f.*
E dispersión *f.*
D Streuung *f.*
R ошибка доставки *ж.*
A خَطَأ الرَّمْي *m*

1191 [G,S]
delivery means
F moyens de lancement
  m. pl.
E medios de lanzamiento
  m. pl.
D Abschussgerät *neut.*
R средства доставки *ср.*
  мн. ч.
A وَسَائِل القَذْف *f*

1192
delouse
F épouiller
E despiojar
D entläusen *untr.*
R дезинсектировать;
   уничтожать
   паразитов
A قَتَل القَمَل

1193
demand *n.*
F demande *f.*
E demanda *f.*
D Anforderung *f.*
R запрос *м.*
A طلب مُعدات

1194
demilitarize
F démilitariser
E desmilitarizar
D entmilitarisieren
R демилитаризировать
A نَزَع السِّلاح *m*

1195
demobilization
F démobilisation *f.*
E desmovilización *f.*
D Entlassung *f.*
R демобилизация *ж.*
A تَسْريح *m*

1196
demolition
F destruction *f.*
E destrucción *f.*
D Zerstörung *f.*
R разрушение *ср.*;
   подрывание *ср.*
A تَدْمير *m*

1197 [P]
demonstrate
F manifester
E manifestarse
D demonstrieren
R демонстрировать
A تَظَاهَر

1198 [P]
demonstration
F manifestation *f.*
E manifestación *f.*
D Demonstration *f.*
R демонстрация *ж.*;
   манифестация *ж.*
A تَظَاهُر *m*

1199 [L]
demonstration
F demonstration *f.*
E exhibición *f.*
D Beweis *m.;* Darstellung
   f.
R ложная атака *ж.*;
   показ *м.*
A تحرُّكات خادِعة

1200 [P]
demonstrator
F manifestant *m.*
E manifestante *m. f.*
D Demonstrant *m.*
R демонстрант *м.*
A مُتَظَاهِر *m*

1201
demoralized *adj.*
F démoralisé
E desmoralizado
D entmutigt
R деморализованный
A روح معنوية مُنْخَفِضة

1202
denial measure
F mesure d'interdiction *f.*
E medida de interdicción
   f.
D Abriegelungsmassnahme
   f.
R мера по воспрещению
   ж.
A تَدْبير (رفض / إنكار)

1203
density
F densité *f.*
E densidad *f.*
D Dichte *f.*
R густота *ж.;* плотность
   ж.
A كَثَافة *f*

1204
dental *adj.*
F dentaire
E odontologio
D Zahn- *komp.;*
   zahnärztlich *adj.*
R зубоврачебный
A سِنّي

1205
dentist
F dentiste *m. f.*
E dentista *m. f.*
D Zahnarzt *m.*
R зубной врач *м.*
A طَبِيب الأَسْنَان *m*

1206 [L]
deny
F interdire
E cerrar (con fuego);
   impedir
D abriegeln *tr.*
R лишать; отрицать
A حَرَّم - مَنَع

1207
depart
F partir
E partir; salir
D abfahren *tr.*
R уходить (уезжать;
   улетать)
A غَادَر . ذَهَب

1208

departure

F départ *m.*

E partida *f.;* salida *f.*

D Abfahrt (Abflug usw) *f.*

R уход *м.;* отправление *ср.*

A مُغَادَرة *f* . ذَهَاب *m*

1209

dependent (on) *adj.*

F dépendant (de)

E dependiente (de)

D abhängig (von)

R зависимый (от)

A مُعْتَمِد على

1210

deploy

F (se) déployer

E desplegar(se)

D (sich) entwickeln

R развёртывать(ся);
распологать(ся)

A نَشَر - وزع

1211

deployment

F déploiement *m.*

E despliegue *m.*

D Entwicklung *f.*

R развёртывание *ср.;*
расположение *ср.*

A نَشَر - توزيع *m*

1212

depot

F dépôt *m.*

E depósito *m.*

D Nachschublager *neut.*

R склад *м.;* депо *ср.*

A مُسْتَوْدَع *m*

1213

depress

F déprimer

E deprimir

D niederdrücken *tr.;*
senken *untr.*

R подавлять; понижать

A خَفَض

1214 [T,U,S]

depression

F dépression *f.*

E depresión *f.*

D Senkung *f.*

R снижение [S] *ср.;*
впадина [T] *ж.;* спад
[U] *м.*

A إنْخِفَاض *m*

1215

depth

F profondeur *f.*

E profundidad *f.*

D Tiefe *f.*

R глубина *ж.*

A عُمْق *m*

1216

depth charge

F grenade sous-marine *f.*

E carga de profundidad *f.*

D Unterwasserbombe *f.*

R глубинная
(подлодочная)
бомба *ж.*

A عُبوة الأَعْمَاق *f*

1217

deputy *adj.*

F adjoint

E adjunto

D stellvertretend

R заместительный

A نَائِب *m*

1218

derelict

F abandonné

E abandonado; derrelicto

D herrenlos

R покинутый;
брошенный

A مَهجور - مَتروك

1219 [M]

derrick

F mât de charge *m.*

E pluma de carga *f.*

D Ladebaum *m.*

R подъёмная стрела *ж.;*
деррик-кран *м.*

A مِرْفَاع *m* ـ رافعة *f*

1220

derust

F dérouiller

E desherrumbrar

D entrosten

R очищать от ржавчины

A أزال الصَّدَأ

1221

desalt

F dessaler

E desalinar

D entsalzen

R опреснять

A أزال المِلْح

1222

desalting

F dessalement *m.*

E desalación *f.*

D Entsalzen *neut.*

R опреснение *ср.*

A إزَالة المِلْح *f*

1223

desert *v.*

F déserter

E desertar

D weglaufen *tr.;*
überlaufen *tr.*

R дезертировать

A فَرَّ

1224
desert n.
F désert m.
E desierto m.
D Wüste f.
R пустыня ж.
A صَحْرَاء f

1225
deserter
F déserteur m.
E desertor m.
D Fahnenflüchtiger m.
R дезертир м.
A فَارّ m

1226 [E]
design n.
F conception f.
E diseño m.
D Konstruktion f.;
  Entwurf m.
R проект м.;
  конструкция ж.
A تَصْمِيم m

1227
design v.
F concevoir
E diseñar
D konstruieren; entwerfen
R конструировать
A صَمَّم

1228
designate
F désigner
E nombrar
D ernennen
R назначать; определять
A عَيَّن

1229
destination
F destination f.
E destino m.
D Bestimmungsort m.
R место назначения ср.
A غَايَة f

1230
destroy
F détruire
E destruir
D zerstören
R разрушать;
  уничтожать
A دَمَّر

1231
destruction
F destruction f.
E destrucción f.
D Vernichtung f.
R разрушение ср.;
  уничтожение ср.
A تَدْمِير m

1232 [G]
destruct system
F système de destruction
  m.
E destructor m.
D Zerstörungsmechanismus
  m.
R система ликвидации
  ж.; система
  подрыва ж.
A جِهَاز إِتْلَاف الصَّارُوخ m

1233 [A,M,L]
detach
F détacher
E destacar
D abkommandieren tr.
R отделять
A أَفْرَز

1234 [A,M,L]
detachment
F détachement m.
E destacamento m.
D Kommando neut.
R отряд м.; команда ж.;
  расчёт м.
A مُفْرَزَة f

1235 [A,M,L]
detail n.
F détachement m.
E destacamento m.
D Kommando neut.
R наряд [AML] м.;
  команда [L] ж.
A زُمْرَة f

1236 [U]
detail n.
F détail m.
E detalle m.
D Einzelheit f.
R подробность ж.
A تَفْصِيل m

1237 [A,M,L]
detail (for) v.
F affecter (à)
E destacar (para)
D abstellen (zu) tr.
R назначать (в,на)
A عَيَّن

1238
detain v.
F detenir
E detener
D festhalten tr.
R задерживать;
  арестовывать
A حَجَز

1239 [E,R]
detect
F détecter
E detectar
D auffassen tr.
R обнаруживать
A كَشَفَ

1240 [E,R]
detection
F détection f.
E detección f.
D Entdeckung f.
R обнаружение ср.
A كَشْف m

1241
detective
F agent *m.*
E detective *m.*
D Detektiv *m.*
R агент-сыщик *м.*
A كَشْفِي - مُخْبِر *m*

1242
detector
F avertisseur *m.;*
  détecteur *m.*
E detector *m.*
D Anzeiger *m.;*
  Gleichrichter *m.*
R определитель *м.;*
  индикатор *м.*
A كَاشِف *m*

1243
deter
F dissuader
E disuadir
D abschrecken *tr.*
R удерживать;
  устрашать
A رَدَع

1244
deterrent *n.*
F arme de dissuasion *f.*
E arma de disuasión *f.*
D Abschreckungsmittel
  neut.
R средство устрашения
  ср.; средство
  сдерживания *ср.*
A رَادِع *m*

1245
detonate
F (faire) détoner; (faire)
  sauter (mine etc...)
E (hacer) detonar
D detonieren (lassen)
R детонировать;
  взрывать(ся)
A فَجَرَ

1246
detonation
F détonation *f.*
E detonación *f.*
D Explosion *f.*
R детонация *ж.;* взрыв
  м.
A تَفجِير *m*

1247
detonator
F détonateur *m.;* amorce
  f.
E detonador *m.*
D Zündkapsel *f.*
R капсюль *м.*
A مُفَجِّر *m*

1248
detour
F détour *m.*
E rodeo *m.*
D Umleitung *f.*
R обход *м.;* объезд *м.*
A دَوْرَة *f*

1249
detrain *v.*
F débarquer d'un train
E bajar del tren
D aussteigen *tr.*
R выгружать(ся);
  высаживать(ся) из
  поезда
A غادر القطار

1250
develop
F (se) développer
E desarrollar(se)
D (sich) entwickeln
R развивать(ся)
A طَوَّر

1251
development
F développement *m.;* mise
  au point *f.*
E desarrollo *m.*
D Entwicklung *f.*
R развитие *ср.*
A تَطْوِير *m*

1252 [S]
deviation
F écart *m.*
E desviación *f.*
D Streuung *f.*
R отклонение *ср.;*
  девиация (компаса)
  ж.
A إنْحِرَاف *m*

1253 [A,M]
deviation
F déviation *f.*
E desviación *f.*
D Missweisung *f.;*
  Versetzung *f.*
R отклонение *ср.*
A إنْحِرَاف *m* . مَحِيد *m*

1254
device
F dispositif *m.*
E aparato *m.*
D Vorrichtung *f.*
R устройство *ср.;*
  прибор *м.*
A مُعِدَّة *f*

1255
diagonal *adj.*
F diagonal
E diagonal
D schräg
R диагональный; косой
A مَائِل *m*

1256
diagram
F schéma
E diagrama *m.*
D Schema *neut.*
R схема *ж.;* график *м.*
A رَسْم تَخْطِيطِي *m*

1257
diameter
F diamètre *m.*
E diámetro *m.*
D Durchmesser *m.*
R диаметр *м.*
A قُطْر *m*

77

1258 [E]
diaphragm
F diaphragme *m.;*
 membrane *f.*
E diafragma *m.*
D Blende *f.;* Membran *f.*
R диафрагма *ж.;*
 мембрана *ж.*
A حِجَاب *m*

1259
diesel-electric
F diesel-électrique
E dieseléctrico
D dieselelektrisch
R дизельно-электрический
A ديزل ـ كَهْرُبَائِي

1260
differential *adj. n.*
F différentiel *adj. n. m.*
E diferencial *adj. s. m.*
D differential *adj. s. neut.*
R дифференциальный
 прилаг.;
 отличительный
 прилаг.;
 дифференциал *сущ.*
 м.
A جِهَاز تَفَاضُلِي *m*

1261
dig *v.*
F creuser
E excavar
D graben
R копать; рыть
A حَفَر

1262 [B,L]
digger
F excavatrice (pour
 tranchées) *f.*
E zanjadora (mecánica) *f.*
D Grabenbagger *m.*
R землеройная машина
 ж.; экскаватор *м.*
A حَفَّارة آلِية *f*

1263
dig in
F (s')enterrer
E atrincherarse
D (sich) eingraben *tr.*
R окапывать(ся)
A تَخَنْدَق

1264
digit
F chiffre *m.*
E dígito *m.*
D Ziffer *f.*
R цифра *ж.*
A رَقَم *m*

1265
digital
F digital; numérique
E digital
D digital
R цифровой
A رَقْمِي

1266
dimension
F cote *f.;* dimension *f.*
E dimensión *f.*
D Abmessung *f.;*
 Dimension *f.*
R измерение *ср.;* размах
 м.
A بُعْد *m*

1267 [A,M,R]
dip
F inclinaison *f.*
E inclinación *f.*
D Inklination *f.*
R наклонение *ср.*
A مَيْل *m*

1268
direct *adj.*
F direct
E directo
D direkt
R прямой
A مُبَاشِر

1269
direct *v.*
F diriger
E dirigir
D lenken
R управлять; наводить
A وَجَّه

1270
direction
F direction *f.*
E dirección *f.*
D Richtung *f.*
R направление *ср.*
A إتِّجَاه *m*

1271
directional
F directionnel
E direccional
D Richt(ungs)- *komp.*
R направленный
A إتِّجَاهِي

1272
directive
F directive *f.*
E instrucción *f.*
D Richtlinie *f.*
R распоряжение *ср.;*
 директива *ж.*
A تَوْجِيه *m*

1273 [S]
director
F goniomètre boussole *m.*
E director de tiro *m.*
D Richtkreis *m.*
R прибор управления
 огнём *м.;* буссоль
 ж.
A جِهَاز مُدِير *m*

1274
disarm
F désarmer
E desarmar
D entschärfen; entwaffnen
R разоружать(ся)
A نَزَع السِّلاح

1275
disarmament
F désarmement *m.*
E desarme *m.*
D Abrüstung *f.*
R разоружение *ср.*
A نَزْع السِّلاح *m*

1276
disband
F dissoudre (une unité)
E disolver; licenciar
D auflösen *tr.*
R распускать;
расформировать
A أنهى خدمة

1277 [S]
discharge
F (faire) feu
E disparar
D abschiessen *tr.*
R выстрелить
A رمى

1278 [A,L,M]
discharge
F rendre à la vie civile
E licenciar
D entlassen *untr.;*
ausmustern *tr.*
R увольнять в запас;
увольнять в отставку
A سَرَّح

1279 [A,L,M]
discharge
F décharger
E descargar
D ausladen *tr.*
R разгружать
A أفرغ

1280 [P]
discharge
F libérer
E poner en libertad
D entlassen *untr.;*
freilassen *tr.*
R освобождать
A أطلق سراح

1281 [E]
discharge
F émettre
E descargar
D freilassen *tr.;*
entladen
R выпускать
A أفرز

1282
discipline *v.*
F punir
E disciplinar
D disziplinieren
R дисциплинировать;
наказывать
A عَاقَب

1283
discipline *n.*
F discipline *f.*
E disciplina *f.*
D Disziplin *f.*
R дисциплина *ж.*
A إنضِباط *m*

1284
discrepancy
F divergence *f.*
E discrepancia *f.*
D Abweichung *f.*
R разногласие *ср.;*
расхождение *ср.*
A إخْتِلاف *m*

1285
disease
F maladie *f.*
E enfermedad *f.*
D Krankheit *f.*
R болезнь *ж.*
A مَرَض *m*

1286
disengage
F (se) dégager
E romper el contacto
D (sich) absetzen *tr.*
R выходить из боя
A خَلَّص

1287
disinfectant
F désinfectant *m.*
E desinfectante *m.*
D Desinfektionsmittel
neut.
R дезинфицирующее
средство *ср.*
A مُطَهِّر *m* مُبيد الجراثيم

1288
dismantle
F démanteler
E desmantelar
D abmontieren *tr.*
R разбирать;
демонтировать
A فَكَّك

1289
dismiss
F (faire) rompre les rangs
E despedir; licenciar
D entlassen
R отпускать; разойдись!
A صَرَف

1290 [L]
dismount
F débarquer; (mettre)
pied (à terre)
E desmontarse
D aussteigen *tr.*
R спешивать(ся)
A تَرَجَّل

1291 [U]
dispatch *v.*
F dépêcher
E enviar
D absenden *tr.*
R посылать
A أرْسَل

1292 [L]
dispatch *v.*
F (faire) partir
E despachar; enviar
D (in) Marsch setzen
R отправлять
A بَعَث

79

1293

dispatch n.
F dépêche f.
E despacho m.; parte m.
D Depesche f.
R отправка ж.; депеша ж.
A رِسَالة f

1294

dispatch rider
F estafette f.
E correo m.
D Meldefahrer m.
R мотоциклист связи м.
A سَاعٍ راكب m

1295 [A]

dispersal area
F zone de dispersion f.
E zona de dispersión f.
D Auflockerungsraum m.
R район рассредоточения м.
A مِنْطَقة التَّفَرُّق f

1296

disperse
F (se) disperser
E dispersar(se)
D auflockern tr.
R рассеиваться; разгонять; рассредоточивать
A إنْتَشَرَ

1297 [A,L,M,S]

dispersion
F dispersion f.
E dispersión f.; difusión [S] f.
D Auflockerung f.; Streuung [S] f.
R рассеивание ср.; разлёт м.
A انتشار ــ نشر القوات ــ تشتت الطلقات

1298 [U]

displace
F déplacer
E cambiar de sitio
D versetzen
R вытеснять
A حَلَّ مَحَلَّ

1299 [L]

displace
F (se) replier
E replegar(se); retirar(se)
D verlegen
R перемещать(ся); передвигать(ся)
A تَحَرَّك

1300 [M]

displace
F déplacer
E desplazar
D verdrängen
R (иметь) водоизмещение (в)
A أزَاح

1301 [M]

displacement
F déplacement m.
E desplazamiento m.
D Wasserverdrängung f.
R водоизмещение ср.
A إزَاحَة f

1302 [D,R]

display v.
F afficher
E desplegar
D wiedergeben untr.
R проявлять; показывать
A عَرَض

1303 [D,R]

display n.
F affichage m.
E despliegue m.
D Wiedergabe f.
R проявление ср.; дисплей м.
A عَرْض m

1304 [D,R]

display unit
F dispositif d'affichage m.
E grupo de despliegue m.
D Sichtgerät neut.
R дисплей м.; экран индикатора м.
A وَحْدَة عَرْض f

1305

(have at one's) disposal
F disposer (de)
E disponer (de)
D (über....) verfügen
R (иметь в) распоряжение
A تَحْت تَصَرُّفِه

1306

dispose (of)
F (se) défaire (de)
E deshacerse (de)
D abschaffen tr.; (an) ordnen
R располагать; ликвидировать
A تَخَلَّص من

1307 [A,L,M]

disposition
F dispositif m.
E disposición f.
D Aufstellung f.
R расположение ср.; дислокация ж.
A تَرْتِيب m

1308 [A,M,V]

disruptive pattern
F motif de camouflage m.
E desfiguración f.
D Tarnanstrich m.
R искажающий узор [AMV] м.; маскировочный узор [M] м.
A تَمْويه مُبَرْقَش m

1309 [I]
disseminate
F diffuser
E diseminar
D verbreiten
R распространять;
рассеивать
A نَشَر

1310 [I]
dissemination
F diffusion f.
E diseminación f.
D Verbreitung f.
R распространение ср.
A m نَشْر

1311
dissident adj. n.
F dissident adj. n. m.
E disidente adj. s. m.
D andersdenkend(-e(r))
adj. (m.)
R инакомыслящий м.;
диссидент(ный) м.
(прилаг.)
A m مُتَمَرِّد

1312
distance
F distance f.
E distancia f.
D Entfernung f.
R расстояние ср.;
дальность ж.
A مَسَافَة f

1313
(observer-target)
distance
F distance d'observation
f.
E distancia observador-
objetivo f.
D Entfernung f.
R дальность
наблюдатель-цель
ж.
A البُعد بين الراصد والهدف

1314
distilled water
F eau distillée f.
E agua destilada f.
D destilliertes Wasser
neut.
R дистиллированная
вода ж.;
опреснённая вода ж.
A ماء مُقَطِّر m

1315
distribute
F répartir
E repartir
D verteilen
R распределять;
раздавать
A وَزَّع

1316
distribution
F répartition f.
E reparto m.
D Verteilung f.
R распределение ср.
A m تَوزِيع

1317 [V]
distributor
F distributeur (de
l'allumage) m.
E distribuidor m.
D Verteiler m.
R диффузор м.;
распределитель
зажигания м.
A m مُوَزِّع

1318 [A]
ditch v.
F (faire un) amerrissage
forcé
E amerizar
D notwassern untr.
R (делать) вынужденную
посадку на воду
A هَبَط هُبُوطاً اِضْطِرَارِياً

1319
ditch n.
F fossé m.
E cuneta f.; foso m.
D Graben m.
R ров м.; канава ж.
A m خَنْدَق

1320 [A,M]
dive v.
F piquer [A]; plonger [M]
E picar [A]; submergirse
[M]
D stürzen [A]; tauchen
[M]
R пикировать [A];
погружаться [M]
A غَطَس

1321 [M]
diver
F scaphandrier m.
E buzo m.
D Taucher m.
R водолаз м.
A غَطَّاس

1322 [A,L,M]
diversion
F diversion f.
E diversión f.
D Ablenkungsmanöver
neut.
R отвлекающий манёвр
м.
A m تَحْوِيل

1323 [T]
diversion
F déviation f.
E desviación f.
D Umleitung f.
R обход м.; отвод м.
A m تَحْوِيل

1324
diversionary
F (de) diversion
E (de) diversión
D Ablenkungs- *komp.;*
  Täuschungs- *komp.*
R отвлекающий;
  ложный
A تَحْوِيلي

1325 [A,M,V]
divert
F dérouter
E desviar
D ablenken *tr.*
R отводить
A حَوَّل

1326
divide *v.*
F (se) diviser; (se)
  bifurquer
E dividir(se); bifurcarse
D (sich) teilen
R делить(ся);
  разделять(ся)
A فَرَّق

1327
divided highway
F route à double chaussée
  f.
E carretera de dos calles
  f.
D doppelte Fahrbahn *f.*
R двойная автострада
  ж.; двойное шоссе
  ср.
A طَرِيق مُفَرَّق *m*

1328
dividers *pl.*
F pointes sèches *f. pl.*
E compás de puntas *m.*
D Stechzirkel *m.*
R делительный циркуль
  м.
A فِرْجَار تَقْسِيم *m*

82

1329 [L]
division
F division *f.*
E división *f.*
D Division *f.*
R дивизия *ж.*
A فِرْقَة *f*

1330 [L]
divisional
F divisionnaire
E divisional
D Divisions- *komp.*
R дивизионный
A فِرْقي

1331
divisional slice
F tranche divisionnaire *f.*
E modulo logístico
  divisionario *m.*
D Versorgungspaket der
  Division *neut.*
R дивизионное
  довольствие *ср.*
A حِصَّة الفِرْقَة *f*

1332
divisional troops *pl.*
F éléments organiques
  divisionnaires *m. pl.*
E tropas divisionarias *f.*
  pl.
D Divisionstruppe *f.*
R дивизионные части *ж.*
  мн. ч.
A قوات الفِرْقَة *f*

1333 [M]
dock *v.*
F entrer au bassin
E atracar al muelle
D anlegen *tr.;* docken
  untr.
R ставить в док;
  входить в док
A أَدْخَل في الحَوْض ـ أَرْسى
  السفينة

1334
dock *n. sing.;* **docks**
  pl.
F bassin *m.;* docks *m. pl.*
E dársena *f.;* muelles *m.*
  pl.
D Dock *neut.;* Werft *f.*
R док *м.;* пристань *ж.;*
  верфь *ж.*
A حَوْض *m* أَحْوَاض *f/*
  السفينة

1335 [A]
docking
F arrimage *m.*
E acoplamiento *m.*
D Anlegen *neut.*
R докование *ср.;*
  швартовка *ж.*
A الإدْخَال في الحَظِيرَة ـ الرُسو

1336
dockyard
F arsenal maritime *m.;*
  chantier *m.*
E arsenal *m.;* astillero *m.*
D Werft *f.*
R судоремонтный завод
  м.;
  судостроительная
  верфь *ж.*
A دار صِنَاعة *m* ـ تَرسانة
  بحرية *f*

1337
doctor
F médecin *m.*
E médico *m.*
D Arzt *m.*
R врач *м.*
A طَبِيب *m*

1338
doctrine
F doctrine *f.*
E doctrina *f.*
D Doktrin *f.;* Lehre *f.*
R учение *ср.;* доктрина
  ж.
A عقيدة تعبوية *f*

1339

door

F porte *f.*
E puerta *f.*
D Tür *f.*
R дверь *ж.*
A بَاب *m*

1340

Doppler effect

F effet Doppler *m.*
E efecto Doppler *m.*
D Dopplereffekt *m.*
R доплеровский эффект
 м.
A ظَاهِرة دُوبْلِر *f*

1341

dose

F dose *f.*
E dosis *f.*
D Dosis *f.*
R доза *ж.*
A مِقْدَار *m*

1342

dosimeter

F dosimètre *m.*
E dosímetro *m.*
D Dosimeter *neut.*
R дозиметр *м.*
A مِقْيَاس الجُرْعَات *m*

1343 [I,P]

dossier

F dossier *m.*
E expediente *m.*
D Aktenbündel *neut. pl.;*
 Akten *f. pl.*
R досье *ср.;* дело *ср.*
A مِلَفّ *m*

1344

dot

F point *m.*
E punto *m.*
D Punkt *m.*
R точка *ж.*
A نُقْطَة *f*

1345

double *adj.*

F double
E doble
D Doppel- *komp.*
R двойной
A مُزْدَوِج *m*

1346

downgrade

F déclasser
E degradar
D degradieren *untr.;*
 herabsetzen *tr.*
R понижать
A نَزَّل رُتْبة *m*

1347

downstream

F (en, vers l') aval
E (aguas) abajo
D stromab(wärts)
R вниз; внизу; (по)
 течению
A مَع التَّيَار

1348

downwash

F déplacement d'air *m.*
E rebufo *m.*
D Abwind *m.*
R спускающаяся струя
 ж.
A الاجْتِرَاف السُفْلَى *m*

1349

downwind (of)

F (sous le) vent (de)
E sotavento (de)
D leewärts (von)
R подветренный
A بِاتِّجَاه الرِّيح

1350 [H]

draft *n.*

F conscription *f.*
E recluta *f.*
D Einberufung *f.*
R призыв *м.*
A جُرْعَة *f*

1351 [H]

draft *v.*

F appeler
E quintar
D einberufen *tr.*
R призывать
A رَسَم *m*

1352 [M]

draft *n.*

F tirant d'eau *m.*
E calado *m.*
D Tiefgang *m.*
R водоизмещение *ср.;*
 углубление *ср.*
A غَاطِس *m*

1353

draftee

F conscrit *m.*
E recluta *m.*
D Wehrpflichtige(r) *m.*
R призывник *м.*
A مُجَنَّد *m*

1354 [M,P]

drag *v.*

F draguer
E dragar
D schleppen
R тащить(ся) [MP];
 чистить драгой [M]
A سَحَب شَبَكَة صَيْد

1355

drag (anchor) *v.*

F chasser sur l'ancre
E garrar
D (den Anker) treiben
R дрейфовать при
 отданном якоре
A سَحَب المِرْسَاة

1356

drain *v.*

F (faire) écouler
E drenar
D entwässern
R дренировать; осушать
A صَرَف

1357

drain (channel) *n.*
F tranchée *f.*
E zanja de dranaje *f.*
D Abflussrinne *f.*
R дренажная канава *ж.*;
водосточная канава
ж.
A قَنَاة تَصْرِيف *f*

1358

drainpipe
F tuyau d'écoulement *m.*
E tubo de desogüe *m.*
D Abflussrohr *neut.*
R дренажная труба *ж.*
A أنْبُوب صَرْف *m*

1359

draw
F dessiner
E dibujar
D zeichnen
R рисовать
A رَسَم

1360

draw (rations)
F percevoir (des rations)
E cobrar
D beziehen
R получать
продовольствие
A سحب تعينات

1361 [S]

draw
F sortir
E sacar
D ziehen
R обнажать
A أظْهَر سلاحاً

1362 [M]

draw...
F caler...
E calar...
D Tiefgang haben
R (иметь) осадку;
(сидеть) в воде на ...
A غَاطِس السفينة *m*

1363

drawbar
F barre d'attelage *f.*
E barra de tracción *f.*
D Zugstange *f.*
R тяговый стержень *м.*
A قَضِيب جَرّ *m*

1364

drawbar pull
F traction (de barre)
d'attelage *f.*
E fuerza de tracción (al
gancho) *f.*
D Zugkraft *f.*
R тяговое усилие *ср.*
A جُهْد القَطْر *m*

1365

drawing *n.*
F dessin *m.;* plan *m.*
E dibujo *m.*
D Zeichnung *f.*
R рисунок *м.*
A رَسْم *m*

1366

dredge *v.*
F draguer
E dragar
D ausbaggern *tr.*
R углублять драгой;
драгировать
A كَسَح . حَفَر

1367

dredger
F drague *f.*
E draga *f.*
D Bagger *m.*
R землечерпалка *ж.*;
экскаватор *м.*
A حَفَّارة *f*

1368

dress uniform
F tenue de cérémonie *f.*
E uniforme de etiqueta *m.*
D Paradeuniform *f.*
R парадная форма
одежды *ж.*
A بزة رسمية

1369 [A,M]

drift *v.*
F dériver
E derivar; (ir a la) deriva
D treiben
R сноситься ветром [A];
плавать по течению
[M]; дрейфовать[M]
A إنْساق

1370 [S]

drift *n.*
F dérive *f.*
E derivación *f.*
D Seitenabweichung *f.*
R деривация *ж.*
A إنْحِرَاف *m*

1371 [A,M]

drift *n.*
F dérive *f.*
E deriva *f.*
D Versetzung *f.*
R дрейф [M] *м.*;
отклонение [AM] *ср.*
A إنْسِيَاق *m*

1372

drill *n.*
F exercice *m.*
E instrucción *f.*
D Exerzierausbildung *f.*
R строевая подготовка
ж.
A التعليم الأولي *m*

1373

drinking water
F eau potable *f.*
E agua potable *f.*
D Trinkwasser *neut.*
R питьевая вода *ж.*
A مَاء الشُّرْب *m*

1374 [E]
drive
F système d'entraînement
m.; système de
transmission m.
E accionamiento m.;
transmisión f.
D Antrieb m.
R привод м.; передача
ж.
A نَقْل الحَرَكَة m

1375 [E]
drive v.
F actionner; commander
E impulsar
D treiben
R приводить
A قَاد

1376 [V]
drive v.
F conduire
E conducir
D fahren
R управлять
A سَاق

1377
driver
F conducteur m.
E conductor m.
D Fahrer m.
R водитель м.; шофёр
м.
A سَائِق m

1378
driving band
F ceinture de transmission
f.
E aro guía m.
D Führungsband neut.
R ведущий поясок м.
A سَيْر نَقْل الحَرَكَة m

1379
driving wheel
F roue motrice f.
E rueda motriz m.;
volante
D Treibrad neut.
R ведущее колесо ср.
A عَجَلة قِيادة f

1380
drogue
F parachute de freinage
m.
E paracaídas de frenado
m. inv.
D Bremsschirm m.
R тормозной парашют
м.; буксируемая
мишень ж.
A هدف تقطره طائرة
للتدريب m

1381 [A]
drone
F avion téléguidé m.;
drone m.
E avión teledirigido m.
D Fernlenkflugzeug neut.
R радиоуправляемый
самолёт м.
A طَائِرة مُوَجَّهة f

1382 [A,E,H,L]
drop n.
F largage m.
E lanzamiento m.
D Absprung [ALH] m.;
Abwurf [E] m.
R сбрасывание [ALE]
ср.; выброска
[ALHE] ж.
A إسْقَاط m

1383 [A,L]
drop v.
F larguer
E lanzar
D abwerfen tr.
R сбрасывать
A أسْقَط

1384 [S]
drop v.
F tirer plus près
E reducir
D zurückfallen
R сползать; понижать
A إنقِصْ

1385
dropmaster
F chef largueur m.
E jefe de lanzamiento m.
D Absetzer m.
R начальник команды
по сбрасыванию м.
A مُرَاقِب الإسْقَاط m

1386
drown
F (se) noyer
E anegar(se); perecer
ahogado
D ertränken; ertrinken
R топить; тонуть
A غَرِق

1387 [S]
drum
F tambour m.
E depósito cilíndrico m.
D Trommel m.
R барабан м.; цилиндр
м.
A بَكَرَة f

1388
dry dock
F cale sèche f.
E dique seco m.
D Trockendock neut.
R сухой док м.
A حَوْض جَافّ m

1389
dry gap
F brèche sèche f.
E brecha seca f.
D Rinne f.; Schlucht f.
R сухой овраг м.;
суходол м.
A ثُغْرة جَافّة f

1390 [B]
duckboard
F caillebotis *m.*
E paseo entablado *m.*
D Laufbrett *neut.*
R решётчатый настил *м.*
A أرضية أَلواح خشب

1391 [B]
dugout
F abri enterré *m.*
E refugio subterráneo *m.*
D Unterstand *m.*
R блиндаж *ж.*; землянка
A مَلجَأ *m*

1392 [S]
dumdum
F balle dum-dum *f.*
E bala dum-dum *f.*
D Dumdumgeschoss *neut.*
R пуля 'дум-дум' *ж.*
A رصاص دَمدَم

1393
dump *n.*
F dépôt temporaire *m.*
E depósito *m.*
D Auslagerung *f.;* Lager *m.*
R склад *м.*
A مُستَودَع *m*

1394
dump *v.*
F (faire un) dépôt de ...
E almacenar; depositar
D lagern
R разгружать; хранить на складе
A فَرَّغ

1395
dumping program
F plan de stockage *m.*
E plan de almacenamiento *m.*
D Auslagerungsplan *m.*
R план по складыванию *м.*
A بَرنَامَج تَفرِيغ

86

1396
duplicate *n.*
F double *m.*
E duplicado *m.*
D Duplikat *neut.*
R дубликат *м.*
A نُسخَة *f*

1397
duplicate *v.*
F reproduire
E (hacer a) multicopista
D vervielfältigen
R снимать копию
A نَسَخ

1398 [E,R]
duplicate *v.*
F (faire le) double de ...
E duplicar; reproducir
D verdoppeln
R удваивать
A إستَنسَخ

1399
dusk
F crépuscule *m.*
E anochecer *m.*
D Abenddämmerung *f.*
R сумерки *мн. ч.*
A ظلام *m*

1400
duty *adj.*
F (de) jour; (de) service
E (de) día; (de) servicio
D diensthabend *adj.*
R дежурный
A مُنَاوِب *m*

1401
dynamic
F dynamique
E dinámico
D dynamisch
R динамический
A حَرَكي

E

1402 [H]
ear
F oreille *f.*
E oreja *f.*
D Ohr *neut.*
R ухо (уши) *ср.* (*мн. ч.*)
A أُذن *f*

1403
earmuffs *pl.*
F casque antibruit *m.*
E orejeras *f. pl.*
D Ohrenschützer *m. pl.*
R наушники *м. мн. ч.*
A حَوَافِظ الأُذن *f*

1404
earphones *pl.*
F casque *m.;* écouteurs *m. pl.*
E auriculares *m. pl.*
D Kopfhörer *m. pl.*
R наушники (радиоаппарата) *м. мн. ч.*
A سَمَّاعات *f*

1405 [T,E]
earth *n.*
F terre [T] *f.;* prise de terre [E] *f.*
E tierra [T] *f.;* toma de tierra [E] *f.*
D Erde *f.*
R земля *ж.*
A أَرض *f*

1406
earth *v.*
F (mettre à la) terre
E conectar a tierra
D erden
R заземлять
A وَصَل بِالأَرض

1407

earthworks *pl.*

F terrassement *m.*

E terraplén *m.*

D Stellungen *f. pl.*

R земляные укрепления
ср. мн. ч.

A تُحَصِينَات تُرَابِية *f*

1408

easting *n.*

F abscisse *f.*

E hacia el este

D Ostwert *m.*

R восточное отшествие
ср.; курс на ост *м.*

A الشَّرْقِيَات *f*

1409

ebb *v.*

F baisser

E bajar

D ebben *untr.;*
zurückfluten *tr.*

R отливать; отступать
(о море); убывать (о
воде)

A إنْحَسَـرالماء

1410

ebb (tide)

F marée descendante *f.;*
reflux *m.*

E marea menguante *f.;*
reflujo *m.*

D Ebbe *f.*

R отлив *м.*

A جَـزْر *m*

1411

echelon *n.*

F échelon *m.*

E escalón *m.*

D Staffel *f.*

R эшелон *м.*

A قَدَمَة *f* ـ تشكيـل طبـقي *m*

1412

(assault) echelon

F échelon d'assaut
(d'attaque) *m.*

E escalón de asalto *m.*

D Angriffswelle *f.*

R штурмовой эшелон *м.*

A تشكيل الإقْتِحَام *m*

1413

(followup) echelon

F échelon de
renforcement *m.*

E escalón de refuerzo *m.*

D nachfolgende Welle *f.*

R второй эшелон *м.*

A النَسَق الثَانِي *f*

1414

(forward) echelon

F échelon avant *m.*

E escalón avanzado *m.*

D vorgeschobene Staffel *f.*

R первый эшелон *м.*

A النَسَق الأمامي *m*

1415

(in) echelon *adj.*

F (en) échelon

E escalonado (en escalón)

D gestaffelt

R (в) эшелоне;
эшелоном

A مُنَسَّق *m*

1416

(rear) echelon

F échelon arrière *m.*

E escalón de retaguardia
m.

D zweite Staffel *f.*

R второй эшелон;
тыловой эшелон

A النسق الخلفي *m*

1417

(sea) echelon

F échelon maritime *m.*

E escalón marítimo *m.*

D Reservegruppe *f.*

R морской эшелон *м.*

A قَدَمَة بَحْرِية ـ نسق بحري *m*

1418 [S]

(support) echelon

F unité qui fournit feu
d'appui

E escalón de apoyo de
fuego *m.*

D Feuerunterstützungs-
staffel *m.*

R подразделение,
поддерживающее
взаимным огнём ср.

A قدمة السند الناري

1419

(support) echelon

F échelons administratifs
et logistiques *m. pl.*

E escalón de apoyo
logístico *m.*

D (logistischer)
Unterstützungsstaffel
m.

R звено системы
снабжения ср.

A قدمة السند

1420

echelonment

F échelonnement *m.*

E organización del
escalón *f.*

D Gliederung *f.*

R эшелонирование *ср.*

A تَنسيق *m*

1421

echo *v.*

F (faire) écho

E (hacer) eco

D widerhallen *untr.*

R отдаваться эхом

A أصْدَى

1422

echo *n.*

F écho *m.*

E eco *m.*

D Echo *neut.*

R эхо *ср.;* отголосок *м.*

A صدًى *m*

87

1423

echo sounder

F écho-sondeur *m.*
E sonda acústica *f.*
D Echolot *neut.*
R эхолот *м.*
A فاحِص بِالصَّدَى *m*

1424

echo sounding

F écho-sondage *m.*
E sondaje acústico *m.*
D Echolotung *f.*
R измерение эхолотом *ср.*
A فَحْص بِالصَّدَى

1425

economic(al)

F économique
E económico
D Spar- *komp.;* wirtschaftlich *adj.*
R экономический; экономный
A إقْتِصادي

1426

economize

F economiser
E economizar
D sparsam umgehen mit *tr.*
R экономить
A إقْتَصَد

1427

economy of force

F économie des forces *f.*
E economía de fuerzas *f.*
D ökonomischer Kräfteeinsatz *m.*
R экономия сил и средств *ж.*
A الإقْتِصاد في القُوَّة

88

1428

effect *n.*

F effet *m.*
E efecto *m.*
D Effekt *m.; Wirkung f.*
R эффект *м.;* воздействие *ср.*
A تَأْثِير *m*

1429 [A,M,L,U]

effective *adj.*

F valide [U]; efficace [ALM]
E efectivo [ALM]; eficaz [U]
D kampffähig [ALM]; wirksam [U]
R эффективный; действительный
A مُؤَثِّر

1430 [H]

effectives *pl.*

F effectifs *m. pl.*
E efectivos *m. pl.*
D Ist-Stärke *f.*
R наличный (боевой) состав *м.*
A صالِحون لِلخِدْمة *m*

1431

efficient

F efficace
E eficaz
D leistungsfähig *adj.*
R эффективный
A فعّال

1432 [S]

eject

F éjecter
E expulsar
D auswerfen *tr.*
R выбрасывать
A قَذَف

1433

ejection seat

F siège éjectable *m.*
E asiento proyectable *m.*
D Schleudersitz *m.*
R катапультируемое сиденье *ср.*
A مَقْعَد قاذِف *m*

1434 [S]

ejector

F éjecteur *m.*
E expulsor *m.*
D Auswerfer *m.*
R отражатель *м.*
A قَذّاف ـ قاذِف *m*

1435

electric

F électrique
E eléctrico
D elektrisch
R электрический
A كَهْرَبائي

1436

electricity

F électricité *f.*
E electricidad *f.*
D Elektrizität *f.*
R электричество *ср.*
A كَهْرَباء *m*

1437

electromagnetic pulse

F impulsion électromagnétique *f.*
E pulso electromagnético *m.*
D elektromagnetischer Impuls *m.*
R электромагнитный пульс *м.*
A نَبْضَة كَهرومغناطيسِيَّة *f*

1438

electron

F électron *m.*
E electrón *m.*
D Elektron *neut.*
R электрон *м.*
A إلِكْتُرون *m*

1439

electronic

F électronique
E electrónico
D Elektronen- *komp.;* elektronisch *adj.*
R электронный
A إلكْتَرُونِي

1440

electronic counter-countermeasures *pl.*

F contre-contremesures électroniques *f. pl.*
E contra-contramedidas electrónicas *f. pl.*
D elektronische Schutzmassnahmen *f. pl.*
R меры борьбы с радиопротиводействием *ж. мн. ч.*
A الإِجْرَاءَات الإِلكْترونية المُضادَّة للإجراءات المُضادَّة

1441

electronic countermeasures *pl.*

F contre-mesures électroniques *f. pl.*
E contramedidas electrónicas *f. pl.*
D elektronische Gegenmassnahmen *f. pl.*
R радиоразведка и радиопротиводействие *ж. ср.*
A الإِجْرَاءَات الإِلكْترونية المضادة *f*

1442

electronics

F électronique *f.*
E electrónica *f.*
D Elektronik *f.*
R электроника *ж.*
A عِلْم الإِلكْتَرُونات *m*

1443 [U]

element

F élément *m.*
E elemento *m.*
D Element *neut.*
R элемент *м.;* часть *ж.*
A عُنْصُر *m*

1444 [A,L,M]

element

F unité élémentaire *f.*
E elemento *m.*
D Truppenteil *m.*
R подразделение *ср.*
A عُنْصُر *m*

1445

elevate

F élever; pointer en hauteur
E elevar
D erhöhen
R поднимать; повышать
A رَفَع

1446 [S]

elevation (angle)

F (angle d')élevation *m.;* (angle de) hausse *m.*
E alza *f.;* (ángulo de) elevación *m.*
D Höhenwinkel *m.*
R (угол) возвышения *м.;* (угол) прицеливания *м.*
A زَاوِية الإِرْتِفَاع *f*

1447

embark

F (s')embarquer
E embarcar(se)
D (sich) einschiffen *tr.*
R садить(ся) на корабль; грузить(ся)
A أَرْكَب

1448

embarkation

F embarquement *m.*
E embarco *m.;* embarque *m.*
D Einschiffung *f.*
R посадка *ж.;* погрузка *ж.*
A إِرْكَاب *m*

1449

embassy

F ambassade *f.*
E embajada *f.*
D Botschaft *f.*
R посольство *ср.*
A سِفارة *f*

1450

emerge

F déboucher; émerger
E emerger; salir
D auftauchen *tr.*
R выходить; возникать
A بَرَز

1451

emergency *adj.*

F (d')exception; (de) secours; (d')urgence
E (de) auxilio; (de) emergencia; (de) urgencia
D Not- *komp.*
R запасный; вынужденный; неприкосновенный
A الطَوَارِىء *f*

1452

emergency *n.*

F cas d'urgence *m.*
E emergencia *f.*
D Notfall *m.*
R крайняя необходимость *ж.;* крайность *ж.*
A حَالة الطَوَارِىء *f*

1453

emergency airfield
F terrain de secours *m.*
E campo de emergencia *m.*
D Notflugplatz *m.*
R запасный аэродром *м.*
A أَرْض هُبُوط إِضْطِرَارِي *f*

1454

emergency anchorage
F mouillage auxiliaire *m.*
E ancladero *m.;*
fondeadero de auxilio *m.*
D Notankerplatz *m.*
R запасная стоянка *ж.*
A مَرْسَى إِضْطِرَارِي *m*

1455 [N]

emergency risk
F risque exceptionnel *m.*
E riesgo excepcional *m.*
D Notrisiko *neut.*
R риск первой степени *м.*
A مُخَاطِرَة إِضْطِرَارِية *f*

1456

emission
F émission *f.*
E emisión *f.*
D Ausstrahlung *f.;*
Emission *f.*
R излучение *ср.*; эмиссия *ж.*
A إِصْدَار *m*

1457

emission control
F contrôle d'émission *m.*
E control de emisión *m.*
D gelenkter Kreisverkehr *m.*
R управление излучениями *ср.*
A حَاكِم الإِصْدَار *m*

90

1458

emit
F émettre
E emitir
D ausstrahlen *tr.*
R излучать; издавать
A أَصْدَر

1459

emplacement
F emplacement *m.*
E emplazamiento *m.*
D Stand *m.*
R местоположение *ср.*
A مَوْضِع - مَوقِع *m*

1460

encircle
F cerner; encercler
E envolver
D einkesseln *tr.*
R окружать
A طَوَّق

1461

encirclement
F encerclement *m.*
E envolvimiento *m.*
D Einkesselung *f.*
R окружение *ср.*
A تَطْوِيق *m*

1462

encode
F enchiffrer
E cifrar
D verschlüsseln
R кодировать
A كَتَب بالرمز

1463

encrypt
F coder
E cifrar
D verschlüsseln
R зашифровывать
A كَتَب بالرُّمُوز

1464 [H,E,A]

endurance
F endurance [HE] *f.;*
durée [HE] *f.;*
autonomie [A] *f.*
E resistencia *f.*
D Ausdauer *f.*
R продолжительность *ж.*
A تَحَمُّل المَشَاقّ *m*

1465

endure
F durer; supporter
E resistir
D aushalten *tr.*
R терпеть; выносить
A تَحَمُّل المَشَاقّ

1466

enemy *adj. n.*
F ennemi *adj. n. m.*
E enemigo *adj. s. m.*
D Feind *m.;* feindlich *adj.*
R противник *м.*; враг *м.*; вражеский *прилаг.*
A عَدُوّ *m*

1467

energy
F énergie *f.*
E energía *f.*
D Energie *f.*
R энергия *ж.*; сила *ж.*
A طَاقَة *f*

1468

enfilade *v.*
F battre d'enfilade
E enfilar
D flankierend schiessen
R обстреливать продольным огнём
A أَطْلَق نِيرَان جانبية

1469

(in) enfilade

F (en) enfilade

E (en) enfilada

D (unter) flankierendem
Feuer

R (под) продольным
огнём

A جَانِبِي

1470

engage

F prendre à partie

E atacar; trabar combate

D bekämpfen

R бежать; избегать

A إِشْتَبَك

1471 [A,L,M]

engagement

F accrochage *m.*

E combate *m.*

D Gefecht *neut.*

R бой *м.*; стычка *ж.*

A إِشْتِبَاك *m*

1472

engine

F moteur *m.*

E motor *m.*

D Motor *m.*

R двигатель *м.*; мотор
м.

A مُحَرِّك *m*

1473 [E]

engineer

F ingénieur *m.*

E ingeniero *m.*

D Ingenieur *m.;* Techniker
m.

R инженер *м.*; механик
м.

A مِيكَانِيكِي *m*

1474 [L]

engineer(s) *pl.*

F sapeur(s) *m. (pl.)*

E ingeniero(s) *m. (pl.)*

D Pionier(e) *m. (pl.)*

R сапёры *м.* мн. ч.;
инженерно-сапёрные
войска *ср.* мн. ч.

A مُهَنْدِس *m*

1475 [E]

engineering

F technologie *f.*

E ingeniería *f.*

D Ingenieurwesen *neut.;*
Technik *f.*

R техника *ж.*;
инженерное
искусство *ср.*

A هَنْدَسَة *f*

1476 [L]

engineering

F génie *m.*

E ingeniería *f.*

D Pionierwesen *neut.*

R инженерно-сапёрное
искусство (дело) *ср.*

A هَنْدَسَة *f*

1477

enlarge

F agrandir

E ampliar

D vergrössern

R увеличивать(ся);
расширять(ся)

A وَسَّع ـ كَبَّر

1478

enlargement

F agrandissement *m.*

E ampliación *f.*

D Vergrösserung *f.*

R увеличение *ср.*;
расширение *ср.*

A تَوْسِيع ـ تَكْبِير *m*

1479

enlist

F (s')engager

E alistar(se)

D (sich) anwerben (lassen)

R поступать на военную
службу

A جَنَّد

1480

enlisted personnel

F sous officiers et
hommes du rang *m.*
pl. m. pl.

E clases de tropa *f. pl.;*
clases y soldados *m.*
pl.

D Unteroffiziere und
Mannschaften *m. pl.*
f. pl.

R военнослужащие *м.*
мн. ч.; личный
состав *м.*

A مُجَنَّدُون *m*

1481

enlistment

F engagement *m.*

E alistamiento *m.*

D Anwerbung *f.*

R поступление на
военную службу *ср.*

A تَجْنِيد *m*

1482

enriched uranium

F uranium enrichi *m.*

E uranio enriquecido *m.*

D angerreichertes Uran
neut.

R обогаченный уран *м.*

A يُورَانيوم (مُزَوَّد ـ مقوّى) *m*

1483 [M]

ensign

F pavillon *m.*

E bandera *f.*

D Flagge *f.*

R (кормовой) флаг *м.*

A بيرق بحري *m*

1484 [M,U,V]
enter
F entrer
E entrar
D einlaufen [M];
einsteigen [VU]
R входить; въезжать
A دَخَل

1485 [U,M,V]
entrance *n.*
F entrée *f.*
E entrada *f.*
D Eingang [U] *m.;*
Einfahrt [MV] *f.*
R вход [UM] *м.;* въезд
[UMV] *м.*
A مَدْخَل *m*

1486 [P]
entry
F inscription *f.*
E apunte *m.*
D Vermerk *m.*
R занесение *ср.*
A تَسْجِيل *m*

1487
envelop
F déborder; envelopper
E envolver
D umfassen *untr.*
R охватывать; окружать
A غَلَّف

1488
(sealed) envelope
F enveloppe (cachetée) *f.;*
(sous) pli cacheté *m.*
E sobre (cerrado) *m.*
D (versiegelter) Umschlag
m.
R (запечатанный)
конверт *м.*
A ظَرْف *m*

1489
enveloping movement
F manoeuvre
d'enveloppement *f.*
E movimiento envolvente
m.
D Umfassungsbewegung *f.*
R охватывающее
движение *ср.;*
охватывающий
манёвр *м.*
A حَرَكَة إلْتِفَاف *f*

1490
envelopment
F enveloppement *m.*
E envolvimiento *m.*
D Umfassung *f.*
R охват *м.;* окружение
ср.
A إلْتِفَاف *m*

1491
epaulette
F épaulette *f.*
E charretera *f.;* hombrera
f.
D Schulterstück *neut.*
R эполет *м.*
A كَتِفِيَّة ـ إسبليطة *f*

1492
equation
F équation *f.*
E ecuación *f.*
D Gleichung *f.*
R выравнивание *ср.*
A مُعَادَلَة ـ موازنة *f*

1493
equator
F équateur *m.*
E ecuador *m.*
D Äquator *m.*
R экватор *м.*
A خَطّ الإسْتِواء *m*

1494
equatorial
F équatorial
E ecuatorial
D äquatorial
R экваториальный
A إسْتِوَائي

1495
(in) equilibrium
F (en) équilibre *m.*
E (en) equilibrio *m.*
D (im) Gleichgewicht *m.*
R равновесие (в
равновесии) *ср.*
A مُتَوَازِن

1496
equinox
F équinoxe *m.*
E equinoccio *m.*
D Äquinoctium *neut.;*
Tag- und
Nachtgleiche *f.*
R равноденствие *ср.*
A إعْتِدَال *m*

1497
equip
F équiper (de); armer
E equipar (de)
D ausrüsten (mit) *tr.*
R оборудовать;
снабжать
A جَهَّز

1498
equipment
F équipement *m.;*
matériel *m.*
E equipo *m.;* material *m.*
D Ausrüstung *f.*
R оборудование *ср.;*
снабжение *ср.*
A جِهَاز *m*

1499

equivalence
F parité *f.*
E equivalencia *f.*
D Gleichwertigkeit *f.*
R эквивалентность *ж.*;
  равноценность *ж.*
A تَكَافُؤٌ *m*

1500

equivalent
F équivalent
E equivalente
D gleichwertig
R эквивалент *м.*
A مِكَافِىء

1501

escalate
F gravir un degré dans
  l'escalade
E escalar
D steigern
R переносить (на);
  расширять (на)
A صَعَّدَ

1502

escalation
F escalade *f.*
E escalamiento *m.*
D Steigerung *f.*
R эскалация *ж.*;
  расширение *ср.*;
  перенос *м.*
A تَصْعِيد *m*

1503 [P]

escape *n.*
F évasion *f.*
E escape *m.;* evasión *f.*
D Flucht *f.*
R побег *м.;* бегство *ср.*
A هُرُوب *m*

1504 [P]

escape *v.*
F (s')évader (de)
E escaparse (de)
D fliehen
R бежать; избегать
A هَرَب

1505

escapee
F évadé *m.*
E evadido *m.*
D Flüchtige(r) *m.*
R бежавший *м.*
A هَارِب *m*

1506

escape hatch
F panneau de sauvetage
  m.
E escotilla de escape *f.*
D Notausstieg *m.*
R спасательный люк *м.;*
  аварийный люк *м.*
A كُوَّة النِّجاة *f*

1507 [B,T]

escarpment
F escarpement *m.*
E escarpa *f.*
D Böschung [T] *f.;*
  Steilabhang [B] *m.*
R откос [T] *м.;* крутость
  [T] *ж.;* эскарп [B] *м.*
A إنْحِدَار *m*

1508

escort *v.*
F escorter
E escoltar
D begleiten
R конвоировать;
  сопровождать
A حَرَس

1509 [A,L,M,P]

escort *n.*
F escorte *f.*
E escolta *f.*
D Begleitschutz [LMP]
  m.; Bedeckung [A] *f.;*
  Geleit *neut.*
R эскорт [AM] *м.;*
  конвой [MP] *м.;*
  охрана [LP] *ж.*
A حَرَس *m* . مُفْرَزَة
  حِرَاسة *f*

1510

espionage
F espionnage *m.*
E espionaje *m.*
D Spionage *f.*
R шпионаж *м.*
A جَاسُوسِية *f*

1511

essential
F essentiel; indispensable;
  (de) première
  nécessité
E indispensable
D unentbehrlich
R существенный;
  необходимый
A ضَرُورِي

1512

establish
F établir
E establecer
D aufstellen *tr.*
R устанавливать
A أنْشَأ

1513

establishment
F établissement *m.;*
  effectif *m.*
E centro *m.;*
  establecimiento *m.;*
  organismo *m.*
D Dienststelle *f.*
R учреждение *ср.;* штат
  м.
A مُنْشَأة *f*

1514

estimate *v.*
F apprécier; estimer
E apreciar
D schätzen
R оценивать
A قَدَّر

1515

estimate *n.*
F appréciation *f.*
E estimación *f.*
D Schätzung *f.*
R оценка *ж.*
A تَقْدِير *m*

1516

estimated time of arrival (ETA)
F heure prévue d'arrivée *f.*
E hora estimada de llegada *f.*
D geschätzte Ankunftszeit *f.;* vorgesehene Ankunftszeit *f.*
R расчётное время прибытия *cp.*
A وَقْت الوُصُول المُقَدَّر *m*

1517

estimated time of departure (ETD)
F heure prévue de départ *f.*
E hora estimada de salida *f.*
D geschätzte Abflugzeit *f.;* vorgesehene Abflugzeit *f.*
R расчётное время убытия *cp.*
A وَقْت المُغَادَرَة المُقَدَّر *m*

1518

estimate of the situation
F appréciation de la situation *f.*
E apreciación de la situación *f.*
D Lagebeurteilung *f.*
R оценка обстановки *ж.*
A تَقْدِير المَوْقِف *m*

1519

estuary
F estuaire *m.*
E estuario *m.*
D Flussmündung *f.*
R устье *cp.;* дельта *ж.*
A مَصَبّ *m*

1520 [A,M,L,P]

evacuate
F évacuer
E evacuar
D evakuieren
R эвакуировать
A جَلَا . أَجْلَى

1521 [A,M,L,P]

evacuation
F évacuation *f.*
E evacuación *f.*
D Evakuierung *f.*
R эвакуация *ж.*
A إجْلَاء *m*

1522

evacuee
F évacué *m.*
E evacuado *m.*
D Evakuierte(r) *m.*
R эвакуированный *м.*
A مُجْلَى ـ مُخْلى

1523

evade
F éviter
E evadir
D ausweichen *tr.*
R избегать
A تَجَنَّب

1524

evaluate
F évaluer
E evaluar
D auswerten *tr.*
R оценивать
A قَيَّم ـ قَدَّر

1525

evaluation
F évaluation *f.*
E evaluación *f.*
D Auswertung *f.*
R оценка *ж.*
A تَقْوِيم ـ تَقْدِير

1526

evasion
F évitement *m.*
E evasión *f.*
D Ausweichen *neut.*
R бегство *cp.;* избежание *cp.;* уклонение *cp*
A تَجَنُّب

1527

evasion and escape
F évasion et récupération *f. f.*
E evasión y escape *f. m.*
D Fluchten und Ausweichen *f. neut.*
R избежание и побег *cp. м.*
A التَّجَنُّب والهَرَب

1528

evasive action
F manoeuvre d'évitement *f.*
E acción evasiva *f.*
D Ausweichmanöver *neut.*
R манёвр уклонения *м.*
A عَمَل التَّجَنُّب

1529

even keel, on an
F (à égal) tirant d'eau
E (en) iguales colados
D gleichlastig
R (на) ровный киль
A رَافِدة قَصّ سَوِيَّة

1530 [B]

excavator
F excavateur *m.*
E excavadora *f.*
D Bagger *m.*
R экскаватор *м.*
A حَفَّارة *f*

1531

exchange *v.*
F échanger (pour, entre)
E canjear (por, con)
D austauschen (mit, gegen) *tr.*
R обменивать; меняться
A تَبَادَل

1532 [U,P]

execute
F exécuter
E ejecutar
D ausführen [U] *tr.*; hinrichten [P] *tr.*
R выполнять [U]; казнить [P]
A نَفَّذ

1533 [N]

executing commander
F commandant utilisateur *m.*
E jefe usuario *m.*
D ausführender Truppenführer (meist eng. Begriff) *m.*
R ответственный начальник *м.*
A القائد المُنَفِّذ *m*

1534

exercise *n.*
F exercice *f.;* évolution *f.*
E ejercicio *m.*
D (Truppen-)Übung *f.*
R тренировка *ж.*; занятие *ср.*; учение *ср.*
A تَمْرين *m*

1535 [I]

exfiltration
F exfiltration *f.*
E exfiltración *f.*
D heimlich herausbringen *tr.*
R вывоз *м.*; удаление *ср.*
A تهريب شخص من أرض العدو

1536

exhaust *v.*
F épuiser
E agotar
D erschöpfen
R истощать (силу)
A خَرَج العادم

1537

exhaust *adj. n.*
F (d')échappement *(adj.) n. m.*
E (de) escape *(adj.) s. m.*
D Auspuff- *komp. m.*
R выхлопной *прилаг.*; выпускной *прилаг.*; выпуск *м.*
A خُروج العادم *m*

1538

exhaustion
F épuisement *m.*
E agotamiento *m.*
D Erschöpfung *f.*
R истощение *ср.*; изнурение *ср.*; изнеможение *ср.*
A وَهَن *m*

1539

exigencies *pl.*
F exigences *f. pl.*
E exigencias *f. pl.*
D Erfordernisse *neut. pl.*
R потребности *ж. мн. ч.*; нужды *ж. мн. ч.*
A ضرورة

1540

expedite
F activer; hâter
E acelerar
D beschleunigen
R ускорять; завершать
A عَجَّل

1541

expedition
F expédition *f.*
E expedición *f.*
D Feldzug *m.*
R экспедиция *ж.*; операция *ж.*
A حَمْلَة *f*

1542

expeditionary force
F corps expéditionnaire *m.*
E cuerpo expedicionario *m.*
D Expeditionskorps *neut.*
R экспедиционные войска *ср. мн. ч.*
A قُوَّة الحَمْلة *f*

1543 [S,U]

expend
F consommer; épuiser
E consumir; gastar; usar [S]
D verschiessen [S]; verbrauchen [U]
R тратить; расходовать
A أَنْفَق

1544

expendable
F sacrifiable
E prescindible
D entbehrlich *adj.*
R расходуемый; невозвратимый
A مستهلك

1545

expenditure
F consommation *f.*
E consumo *m.;* gasto *m.*
D Verbrauch *m.*
R расход *м.*
A إنْفَاق *m*

1546

experienced
F expérimenté
E experto; perito
D erfahren
R опытный
A ذو خبرة

1547

experiment
F essai m.; expérience f.
E ensayo m.; experimento m.
D Versuch m.
R опыт м.; испытание ср.; эксперимент м.
A تَجْرِبَة f

1548

experimental
F expérimental
E experimental
D Versuchs- komp.
R экспериментальный; испытательный
A تَجْرِيبِي

1549

expert n.
F expert m.
E especialista m. pl.
D Experte m.; Fachmann m.
R знаток м.; эксперт м.
A خبير

1550

explode
F (faire) éclater; (faire) sauter
E estellar; explotar
D explodieren; krepieren; sprengen
R взрывать(ся); подрывать(ся)
A فَجَّر

1551

exploit
F exploiter
E explotar (el éxito)
D ausnutzen tr.
R эксплуатировать; воспользоваться
A إِسْتَثْمَر

1552

exploitation
F exploitation f.
E explotación (del éxito) f.
D Ausnutzung f.
R эксплуатация ж.
A إِسْتِثْمَار m

1553

exploratory
F exploratoire
E exploratorio
D Untersuchungs- komp.
R исследовательский
A إِسْتِكْشَافِي

1554

explore
F explorer; sonder
E explorar; sondar
D erforschen
R исследовать; разведывать
A كَشَف

1555

explosion
F explosion f.
E explosión f.
D Explosion f.
R взрыв м.
A إِنْفِجَار m

1556

explosion-proof
F anti-déflagrant
E antiexplosivo
D beschusssicher; explosionsgeschützt
R взрывобезопасный
A مَنِيع على الإِنْفِجَار m

1557

explosive adj.
F explosible
E explosivo
D explosiv adj.; Spreng- komp.
R взрывчатый; взрывной
A مُتَفَجِّر m

1558

explosive n.
F explosif m.
E explosivo m.
D Sprengstoff m.
R взрывчатые вещества ср. мн. ч.
A مُتَفَجِّر m

1559 [U,N]

expose
F (s')exposer
E exponer(se)
D belichten untr.; aussetzen tr.; bestrahlen (lassen) [N]
R подвергать [U]; оставлять незащищенным [N]
A عَرَّض

1560 [H,N]

exposure
F exposition f.
E frío [H] m.; exposición [N] f.
D Erfrieren [H] neut.; Bestrahlung [N] f.
R подвергание (холоду) [H] ср.; облучение [N] ср.
A تَعْرِيض m

1561

exposure dose
F dose d'exposition f.
E dosis de exposición f.
D Strahlendosis f.
R доза облучения ж.
A مِقْدَار التَّعْرِيض m

1562

exposure time

F temps d'exposition *m.;*
temps de pose *m.*
E tiempo de exposición
m.
D Belichtungszeit *f.*
R время облучения *ср.*
A زَمَن التَّعْرِيض *m*

1563

extension

F prolongation *f.*
E prolongación *f.*
D Zeitverlängerung *neut.;*
Verlängerung *neut.*
R продление *ср.;*
удлинение *ср.*
A تمديد الوقت

1564

exterior *adj.*

F extérieur
E exterior
D Aussen- *komp.;*
äusserlich *adj.*
R внешний
A خَارِج

1565

external

F externe
E externo
D Aussen- *komp.;*
äusserlich *adj.*
R внешний
A خَارِجِي

1566 [S]

extractor

F extracteur *m.*
E extractor *m.*
D Auszieher *m.*
R выбрасыватель *м.*
A مُسْتَخْرِج *m*

1567

eye *n.*

F oeil (yeux) *m.(pl.)*
E ojo *m.*
D Auge *neut.*
R глаз *м.*
A عَيْن *f*

1568

eye injury

F blessure à l'oeil *f.*
E herida ocular *f.*
D Augenverletzung *f.*
R рана глаза *ж.;*
ранение глаза *ср.*
A جُرْح العَيْن *m*

1569

eyepiece

F oculaire *m.*
E ocular *m.*
D Okular *neut.*
R окуляр *м.;* линза *ж.*
A عَيْنِيَّة *f*

F

1570

face (the enemy)

F affronter (l'ennemi)
E enfrentarse con (el
enemigo)
D (dem Feind)
gegenüberstehen
R стоять лицом (к
противнику)
A واجه العدو

1571

facilities *pl.*

F installations *f. pl.;*
moyens *m. pl.*
E facilidades *f. pl.*
D Einrichtungen *f. pl.*
R средства *ср. мн. ч.;*
устройства *ср. мн. ч.*
A تَسْهِيلات *f*

1572

facings *pl.*

F parement *m.*
E paramentos *m. pl.;*
revestimientos *m. pl.*
D Aufschläge *m. pl.*
R отделка (мундира) *ж.*
A طبقة كاسية *f*

1573 [R]

facsimile *n.*

F fac-similé *m.*
E facsímil *m.*
D Bildübertragung(-
sgerät) *f. (neut.)*
R факсимиле *ср.*
A صُورَة ـ نُسْخَة *f*

1574

factor

F facteur *m.*
E factor *m.*
D Faktor *m.*
R фактор *м.*
A عَامِل

1575 [R]

fading

F fading *m.*
E desvanecimiento *m.*
D Schwund *m.;* Fading *m.*
R замирание *ср.*
A الخَبّو

1576 [E,U]

fail

F avoir une panne [E];
faillir [U]
E fallar [E]; fracasar [U]
D versagen [EU];
scheitern [U]
R потерпеть неудачу
A سَقَط . فَشَل

97

1577 [A,G,N,S]
fail-safe *adj.*
F (à) sûreté de mise de
    feu
E (de) seguro
D Sicherheits- *komp.*
R предохранительное
    устройство *cp.*
A أَمِن مِن التَّعَطُّل

1578 [D,E]
fail-safe *adj.*
F (à) sûreté intégrée
E (de) seguro
D Selbstzerlegungs- *komp.*
R предохранительный
    прилаг.
A أَمِن مِن التَّعَطُّل

1579 [U]
failure
F échec *m.;* défaut *m.*
E fracaso *m.*
D Fehlschlag *m.*
R неудача *ж.;* провал *м.*
A فَشَل *m*

1580 [E]
failure
F panne *f.*
E fallo *m.*
D Störung *f.*
R отказ *м.;* поломка *ж.*
A تَوَقُّف *m*

1581 [M]
fair-way
F chenal *m.*
E canalizo *m.*
D Fahrwasser *neut.*
R фарватер *м.*
A قَنَاة مِلاحِيَّة

1582
fall in *v.;* "Fall in!"
F former les rangs; "A
    vos rangs!"
E alinearse; "¡En filas!"
D antreten *tr.*
R становиться;
    "Становитесь!"
A تَجَمَّع

1583
fallout
F retombée (radioactive)
    f.
E polvillo radioactivo *m.*
D (radioaktiver)
    Niederschlag *m.*
R (радиоактивное)
    выпадение *cp.*
A غُبَار مُتَسَاقِط *m*

1584
fall out *v.;* "Fall out!";
    "Fall out!" (one
    person)
F rompre les rangs;
    "Rompez!"; "Sortez
    des rangs!"
E romper filas
D abtreten *tr.;*
    "Abtreten!"
R расходиться;
    "Разойдись!"
A تَفَرَّق

1585
false
F faux
E falso
D falsch
R ложный
A كَاذِب

1586
false alarm
F fausse alerte *f.*
E falsa alarma *f.*
D blinder Alarm *m.*
R ложная тревога *ж.*
A إِنْذَار زَائِف *m*

1587
fan
F ventilateur *m.*
E ventilador *m.*
D Ventilator *m.*
R вентилятор *м.*
A مِرْوَحَة *f*

1588
fan out (of troops)
F déployer en éventail
E avanzar separados
D ausschwärmen *tr.*
R распространяться
A إِنْتَشَر

1589
fascine
F fascine *f.*
E fajina *f.*
D Faschine *f.*
R фашина *ж.*
A حُزْمَة أَغْصَان *f*

1590 [U]
fast
F rapide
E rápido
D schnell
R быстрый
A سَرِيع

1591 [M]
(made) fast
F amarré
E amarrado
D fest (-gemacht)
R закреплённый
A رَبَط

1592 [M]
(make) fast
F (s')amarrer; saisir
E amarrar; echar los
    amarros
D festmachen *tr.*
R крепить; закреплять
A رَبَط

1593 [E,H]
fatigue
F fatigue *f.*
E fatiga *f.*
D Ermüdung *f.*
R усталость *ж.*
A تَعَب [H] *m* بَلَى [E] *m*

1594

fatigue duty
F service de corvée m.
E faena f.
D Arbeitsdienst m.
R хозяйственная работа
  ж.
A شُغْل m

1595

fatigues pl.
F tenue de corvée f.
E traje de faena m.
D Arbeitsanzug m.
R рабочая одежда ж.
A أشْغَال f

1596 [E,R]

fault
F court-circuit m.
E avería f.
D Störung f.
R неисправность ж.
A خَلَل m

1597 [T]

feature n.
F particularité du terrain
  f.
E rasgo topográfico m.
D Merkmal neut.
R местный предмет м.
A عارضة أرْضِية ـ تضاريس f

1598 [E,S]

feed
F alimentation f.
E alimentación f.
D Zuführung f.
R подача ж.
A غَذَّى

1599 [L,M]

feint
F diversion f.
E treta f.
D Täuschungsmanöver
  neut.; Scheinangriff
  m.
R ложный манёвр м.
A تَظَاهُر

1600

fence (in) v.
F clôturer
E encerrar
D einzaunen tr.
R ограждать
A سَيَّج ـ سَوَّر

1601

(wire) fence n.
F clôture (en fil
  métallique) f.
E cerca (de alambre) f.
D (Draht-) Zaun m.
R (проволочная)
  изгородь ж.
A سِيَاج ـ سور m

1602

fend (off)
F parer
E detener; repeler
D abwehren tr.
R отгонять
A دَفَع

1603 [M]

fender
F défense f.
E defensa f.
D Fender m.
R кранец м.
A مِصَدّ m

1604 [V]

fender
F pare-chocs m. inv.
E parachoques m. inv.
D Stossstange f.
R крыло ср.
A مِصَدّ m

1605

ferry n.
F bac m.
E barca f.; transbordador
  m.
D Fähre f.
R паром м.; перевоз м.
A مَعْبَر m عَبَّاره f

1606

ferry v.
F livrer; faire passer
E llevar; transportar
D überführen untr.;
  übersetzen tr.
R перевозить
A عَبَر

1607 [H]

fever
F fièvre f.
E fiebre f.
D Fieber neut.
R лихорадка ж.
A حُمَّى f

1608

fiber
F fibre f.
E fibra f.
D Faser f.
R волокно ср.
A لِيف m

1609

fiberglass
F fibre de verre f.
E fibra de vidrio f.
D Glasfaserkunststoff
  (GFK) m.
R стеклянное волокно
  ср.
A زُجَاج لِيفِى m

1610

fiber optics
F fibre optique f.
E óptica de fibra f.
D Glasfaseroptik f.
R волокнистая оптика
  ж.
A أجْهِزَة بصرية لِيفِيَّة f

99

1611

field *adj.*

F (en/de) campagne; roulant

E (de) campo; (de) campaña; ambulante

D Feld- *komp.*

R полевой

A مَيْداني

1612

fieldcraft

F utilisation du terrain *f.*

E técnicas de combate *f. pl.*

D Feldkampffertigkeit *f.*

R полевая тактика *ж.*

A المَهارة في المَيْدَان

1613

field dress

F tenue de combat *f.*

E uniforme de combate *m.*

D Feldanzug *m.*

R полевая одежда *ж.*

A لِبَاس المَيْدان *m*

1614 [H]

field dressing

F paquet de pansements *m.*

E vendaje de campaña *m.*

D Notverband *m.*

R полевая перевязка *ж.*

A ضِمَاد المَيْدان *m*

1615

field exercise

F exercice de combat *f.*

E ejercicio de combate *m.*

D Truppenübung *f.*

R полевое учение *ср.*

A تَمْرِين المَيْدان *m*

100

1616

field gun

F canon de campagne *m.*

E cañón de campaña *m.*

D Feldgeschütz *neut.*

R полевая пушка *ж.*; полевое орудие *ср.*

A مِدْفع المَيْدان *m*

1617

field kitchen

F "roulante" *f.*

E cocina de campaña *f.*

D Feldküche *f.*

R полевая кухня *ж.*

A مَطْبخ المَيْدان *m*

1618

field of fire

F secteur de tir *m.*

E campo de tiro *m.*

D Feuerbereich *m.*

R сектор обстрела *м.*

A مَجَال الرَّمْى *m*

1619

field of vision

F champ de vision *m.*

E campo visual *m.*

D Gesichtsfeld *neut.*

R поле зрения *ср.*

A مَجَال النَّظر *m*

1620

field ration

F rations de combat *f. pl.*

E víveres de campaña *m. pl.*

D Feldverpflegung *f.*

R полевой паёк *м.*

A أرْزَاق المَيْدان *f*

1621

field shop

F atelier de campagne *m.*

E taller de campaña *m.*

D Feldwerkstatt *f.*

R полевая мастерская *ж.*

A مَعْمَل مَيْداني *m*

1622

fight *v.*

F (se) battre (avec)

E batir (con)

D kämpfen (gegen)

R сражаться; воевать; вести бой

A قَاتَل

1623

fighter ground attack

F avion d'appui tactique *m.*

E caza de ataque a tierra *m.*

D Fliegerbodenangriff *m.*

R (штурмовой) истребитель *м.*

A طائرة هجوم أرْضي *f*

1624 [V]

fighting compartment

F compartiment de combat *m.*

E cámara de combate *f.*

D Kampfraum *m.*

R боевое отделение *ср.*

A حُجْرَة الرَّمْى *f*

1625

fighting through

F attaque dans la foulée *f.*

E asalto de fuego y maniobra *m.*

D (dem Ziel entgegen) kämpfen

R стремительный бой *м.*

A قِتَال مِن خِلال مَوْقع العَدو

1626

filament

F filament *m.*

E filamento *m.*

D (Glüh)-Faden *m.*

R нить *ж.*; волокно *ср.*

A فَتِيلَة *f*

1627 [L]
file *n.*
F file *f.*
E fila *f.*
D Rotte *f.*
R ряд *м.*
A رَتَل ـ صَفّ *m*

1628 [U]
file *n.*
F dossier *m.*
E carpeta *f.*
D Ordner *m.*
R регистратор *м.*; досье *ср.*
A مِلَفّ

1629 [U]
file *v.*
F joindre au dossier
E archivar
D ordnen
R регистрировать; хранить
A وَضَعَ في المِلافات

1630
(the) files
F archives *f. pl.*
E archivos *m. pl.*
D Akten *f. pl.;* Archiv *neut.*
R архив *м.*; дело *ср.*
A مِلَفّات *f*

1631
fill
F remplir (de)
E llenar (de)
D füllen
R наполнять(ся)
A مَلَأَ

1632 [A,V]
fill up
F faire le plein
E cargar
D tanken
R заправляться
A مَلَأَ

1633 [N]
film badge
F dosiphote *f.*
E fotoseriación *f.*
D Strahlendosimeter *m.*
R пленочный дозиметр *м.*
A شَارَة فِيلمِيَّة *f*

1634
filter *v.*
F filtrer
E filtrar
D filtern
R фильтровать
A رَشَّح

1635
filter *n.*
F filtre *m.*
E filtro *m.*
D Filter *m.*
R фильтр *м.*
A مُرَشِّح *m*

1636 [A,M]
fin
F empennage [A] *m.;* dérive [M] *f.*
E timón *m.*
D Flosse *f.*
R руль *м.*; киль *м.*
A زِعْنِفة *f*

1637 [G,S]
fin
F ailette *f.*
E aleta *f.*
D Flügel *m.*
R стабилизатор *м.*
A زِعْنِفة *f*

1638
final *adj.*
F final
E final
D End- *komp.*
R конечный
A نِهَائِي

1639 [A]
(automatic direction) finder
F radiocompas *m.*
E radiogoniómetro automático *m.*
D Orter *m.*
R автокомпас *м.*
A آلة تحديد الاتجاه

1640
(direction) finder
F goniomètre *m.*
E radiogoniómetro *m.*
D Peiler *m.*
R пеленгатор *м.*; радиопеленгатор *м.*
A آلة تَحْدِيد الإتِّجاه *f*

1641 [P,R]
(radio direction) finder
F radiogoniomètre *m.*
E radiogoniómetro *m.*
D Sucher *m.*
R радиопеленгатор *м.*
A راديو تَحْدِيد الاتجاه

1642
(range) finder
F télémètre *m.*
E telémetro *m.*
D Entfernungsmesser *m.*
R дальномер *м.*
A مُقَدِّرَة المَدَى *f*

1643
fin-stabilized
F stabilisé par ailettes
E estabilizado por aletas
D flügelstabilisiert
R (co) стабилизатором; стабилизированный оперением
A مُتَوَازِن بِزَعَانِف

1644 [S]
fire *v.*
F tirer
E hacer fuego; tirar
D schiessen
R стрелять; вести огонь
A رَمَى

1645 [S]
fire *n.*
F tir *m.*
E fuego *m.;* tiro *m.*
D Feuer *neut.*
R огонь *м.*
A نار *f*

1646 [S]
Fire!
F Feu!
E ¡Fuego!
D Feuer!
R Огонь!
A إرْم

1647
(accompanying) fire
F tir d'accompagnement *m.*
E fuegos de acompañamiento (inmediato) *m. pl.*
D Begleitfeuer *neut.*
R огонь *м.*
A رمي المرافقة

1648
(adjustment) fire
F tir de réglage *m.*
E disparo de corrección *m.*
D Justierungsfeuer *neut.*
R пристрелка *ж.*
A رمي الاحكام

1649
fire (a salute)
F rendre les honneurs militaires
E saludar
D (ein) Salut schiessen
R салютовать
A رمي التّحيّة *m*

1650
(barrage) fire
F tir de barrage *m.*
E fuego de barrera *m.*
D Sperrfeuer *neut.*
R (заградительный) огонь *м.*
A رمي السّدّ النّاري

1651
(bracket) fire
F tir à la fourchette *m.*
E disparo de horquilla *m.*
D Klammerfeuer *neut.*
R вилка *ж.*
A رمي الحَصْر *m*

1652
(calibration) fire
F tir de régimage *m.*
E disparo de ajuste *m.*
D Kalibrierungsfeuer *neut.*
R сострел орудий *м.*
A رمي المعايرة *m*

1653
(combing) fire
F tir de ratissage *m.*
E fuego de peinado *m.*
D Absuchfeuer *neut.*
R (прочёсывающий) огонь *м.*
A رمي التمشيط *m*

1654
(concentrated) fire
F tir de concentration *m.*
E fuego concentrado *m.*
D Feuerkonzentration *f.*
R (сосредоточенный) огонь *м.*
A رمي مركّز *m*

1655
(continuous) fire
F tir continu *m.*
E fuego continuo *m.*
D Dauerfeuer *neut.*
R (непрерывный) огонь *м.*
A رَمي مُسْتَمِرّ *m*

1656 [U]
(controlled) fire
F feu *m.*
E fuego *m.*
D Feuer *neut.*
R огонь *м.*
A رَمْيَ مَحْكُوم

1657
(counterbattery) fire
F tir de contre-batterie *m.*
E tiro de contrabatería *m.*
D Artilleriebekämpfung *f.*
R контрбатарейный огонь *м.*
A رمي مضادّ للمِدْفَعِية *m*

1658
(counterbattery) fire
F tir de contre-batterie *m.*
E tiro de contrabatería *m.*
D Artilleriebekämpfung *f.*
R контрбатарейный огонь *м.*
A رمي مضاد للمدفعية

1659
(counterpreparation) fire
F tir de contrepréparation *m.*
E fuego de contrapreparación *m.*
D Vorbereitungsgegenfeuer *neut.*
R контрподготовка *ж.*
A قصف مضادّ للتحضيرات *m*

1660
(covering) fire
F tir de protection *m.*
E fuego de cobertura *m.*
D Deckungsfeuer *neut.*
R (обеспечение) огнём *ср.*
A نيران سَاترة *f*

1661

(crisscross) fire

F feux croisés *m. pl.*
E fuego cruzado *m.*
D Kreuzfeuer *neut.*
R (перекрёстный) огонь
  м.
A رمي متقاطع *m*

1662

(demolition) fire

F tir de destruction *m.*
E fuego de destrucción *m.*
D Vernichtungsfeuer *neut.*
R огонь на разрушение
  м.
A رمي النَّسْف *m*

1663

(direct) fire

F tir à vue *m.*
E tiro directo *m.*
D direktes Feuer *neut.*
R огонь прямой
  наводкой *м.*
A رمي مباشر *m*

1664

(distributive) fire

F tir sur zone *m.*
E fuego disperso *m.*
D Streufeuer *neut.*
R (расходящий) огонь *м.*
A رمي موزّع *m*

1665

(enfilade) fire

F tir d'enfilade *m.*
E fuego de enfilada *m.*
D Längsfeuer *neut.*
R (продольный) огонь
  м.
A نيران جانبيّة *f*

1666

(final protective) fire

F tir d'arrêt *m.*
E fuegos de protección
  finales *m. pl.*
D Schlussdeckungsfeuer
  neut.
R последний
  заградительный
  огонь *м.*
A نار مَحْمِيَة نهائية

1667

(flank-protective) fire

F tir de flanquement *m.*
E fuego de protección de
  flanco *m.*
D Flankendeckungsfeuer
  neut.
R флангозащитный
  огонь *м.*
A نيران حماية الجناح

1668

(flat-trajectory) fire

F tir tendu *m.*
E tiro de trayectoria
  rasante *m.*
D Flachfeuer *neut.*
R (настильный) огонь *м.*
A رمي بِمَحْرَك مُنبَسِط *m*

1669

(grazing) fire

F tir rasant *m.*
E fuego rasante *m.*
D Streiffeuer *neut.*
R (настильный) огонь *м.*
A رمي بالتمَّاس *m*

1670

(harassing) fire

F tir de harcèlement *m.*
E fuego de hostigamiento
  m.
D Störfeuer *neut.*
R (беспокоящий) огонь
  м.
A نيران الازعاج *f*

1671

(high angle) fire

F tir vertical *m.*
E tiro de alta trayectoria
  m.
D Steilfeuer *neut.*
R огонь с большим
  углом возвышения
  м.
A رمي بزاوية عالية *m*

1672

(illuminating) fire

F tir éclairant *m.*
E tiro de iluminación *m.*
D Leuchtfeuer *neut.*
R (осветительный) огонь
  м.
A نيران مُضِيئة *f*

1673

(indirect) fire

F tir indirect *m.*
E tiro indirecto *m.*
D indirektes Feuer *neut.*
R стрельба непрямой
  наводкой *ж.*; огонь
  непрямой наводкой
  м.
A رَمْي غَيْرُ مُباشر *m*

1674

(interdiction) fire

F tir d'interdiction *m.*
E fuego de interdicción
  m.
D Sperrfeuer *neut.*
R (отсечный) огонь *м.*
A رمي المَنع *m*

1675

(interlocking) fire

F feux croisés *m. pl.*
E fuego superpuesto *m.*
D Schachtelfeuer *neut.*
R (согласованный) огонь
  м.
A نيران مُتَقَاطِعَة *f*

1676

(low angle) fire

F tir plongeant *m.*
E tiro de baja trayectoria *m.*
D Tieffeuer *neut.*
R огонь с низким углом возвышения *м.*
A رمي بِزَاوِية منخفضة *m*

1677

(marking) fire

F tir de balisage *m.*
E fuego de señalamiento *m.*
D Markierungsfeuer *neut.*
R (указательный) огонь *м.*
A نار تشير إلى الهدف *f*

1678

(mowing) fire

F tir de fauchage *m.*
E fuego de peinado *m.*
D Mähfeuer *neut.*
R (губительный) огонь *м.*
A نيران الحَصْد *f*

1679

(neutralization) fire

F tir de neutralisation *m.*
E fuego de neutralización *m.*
D Niederhaltungsfeuer *neut.*
R огонь на подавление *м.*
A نيران الإسْكات *f*

1680

(oblique) fire

F tir d'écharpe *m.*
E tiro oblicuo *m.*
D flankierendes Feuer *neut.*
R косоприцельный огонь *м.*
A رَمْي (مائل ـ مُنْحَرِف) *m*

1681

(observed) fire

F tir observé *m.*
E tiro observado *m.*
D beobachtetes Schiessen *neut.*
R стрельба с наблюдением *ж.*
A نَار مَرْصُودة *f*

1682

fire (on sight)

F tirer à vue
E tirar con blanco visible
D (sofort) schiessen; (auf) Anhieb schiessen
R вести огонь
A اطلاق النار فورا *m*

1683

(overhead) fire *n.*

F tir par-dessus les troupes *m.*
E fuego por encima de propias tropas *m.*
D Hochschuss *m.*
R огонь через голову *м.*
A رمي فوق الرُؤوس *m*

1684

(plunging) fire

F tir plongeant *m.*
E fuego desde una posición alta *m.*
D Sturzfeuer *neut.*
R (навесный) огонь *м.*
A رمي غاطِس *m*

1685

(point blank) fire

F tir à bout portant *m.*
E tiro a boca de jarro *m.*
D Kernschiessen *neut.*
R огонь в упор *м.*
A رمي قريب *m*

1686

(predicted) fire

F tir préparé *m.*
E tiro con predicción *m.*
D Planschiessen *neut.*
R стрельба по исчисленным данным *ж.*
A رمي حسب الخريطة

1687

(quick) fire

F tir rapide *m.*
E tiro rápido *m.*
D Schnellfeuer *neut.*
R беглый огонь *м.*
A رَمْي سَرِيع *m*

1688

(radar) fire

F tir au radar *m.*
E fuego controlado por radar *m.*
D Radarschiessen *neut.*
R (управляемый) огонь *м.*
A رمي الرادار *m*

1689

(raking) fire

F tir balayant *m.*
E fuego de rastrilleo *m.*
D Streichfeuer *neut.*
R (продольный) огонь *м.*
A رمي الجَرْف *m*

1690

(ranging) fire

F tir de réglage *m.*
E fuego de medición de distancia *m.*
D Einschiessen *neut.;* Reichschiessen *neut.*
R (пристрелочный) огонь *м.*
A رمي الإحكام *m*

1691

(registration) fire
F tir de réglage *m.*
E tiro de registro *m.*
D Erfassungsfeuer *neut.*
R пристрелка *ж.*
A رمي التَّسجيل *m*

1692

(defensive) fire(s) *pl.*
F tir défensif *m.*
E tiro defensivo *m.;* tiro de detención (protección) *m.*
D Abwehrfeuer *neut.*
R заградительный огонь *м.*
A نِيرَان دِفاعِية *f*

1693

(reinforcing) fire(s)
F tir(s) de renforcement *m. m.(f.)*
E tiro de apoyo *m.*
D Verstärkungsauftrag (Artilleriefeuer) *m.*
R подкрепление огнём *ср.*
A رمي التعزيز

1694

(searching) fire
F tir échelonné *m.*
E tiro progresivo *m.*
D Staffelfeuer *neut.*
R стрельба шкалой *ж.*
A رَمْي تَفْتِيش

1695

(supporting) fire
F feu d'appui *m.*
E tiro de apoyo *m.*
D Unterstützungsfeuer *neut.*
R (поддерживающий) огонь *м.*
A رمي المساندة *m*

1696

(suppressive) fire
F tir de neutralisation *m.*
E fuego de supresión *m.*
D Unterdrückungsfeuer *neut.*
R (подавляющий) огонь *м.*
A رمي الإسْكَات *m*

1697

(sweeping) fire
F tir fauchant *m.*
E fuego de barrido *m.*
D Kehrfeuer (mit Feuer bestreichen) *neut.*
R (прочёсывающий) огонь *м.*
A نيران كاسِحَة *f*

1698

(traversing) fire
F tir fauchant *m.*
E fuego de barrido lateral *m.*
D Querfeuer *neut.*
R (круговой) огонь *м.*
A رمي مُسْتَعْرِض *m*

1699

(trial) fire
F tir d'essai *m.*
E tiro de ensayo *m.*
D Probeschiessen *neut.*
R (учебный) огонь *м.*
A رمي التجريب *m*

1700

(uncontrolled) fire
F incendie *f.*
E incendio *m.*
D Brand *m.*
R пожар *м.*
A رمي بدون تحكم

1701

(under) fire
F (sous) feu
E (bajo el) fuego
D (unter) Beschuss
R (под) огнём
A تحت النار

1702

(unobserved) fire
F tir non contrôlé *m.*
E fuego sin observación *m.*
D unbeobachtes Schiessen *neut.*
R стрельба без наблюдения *ж.*
A رمي غَيرمَرْصُود *m*

1703

(withering) fire
F feu foudroyant *m.*
E tiro arrollador *m.*
D Vernichtungsfeuer *neut.*
R губительный огонь *м.*
A رمي صاعق

1704

fire alarm
F avertisseur d'incendie *m.*
E alarma de incendios *f.*
D Feuermelder *m.*
R пожарная тревога *ж.*
A إنْذَار الحَريق *m*

1705 [G,S]

Fire and forget
F missile de guidage automatique [G] *m.;* "Tir et n'y pense plus" [S]
E autónomo [G]; expendable [S]
D „Fire and forget" (eng. Begriff)
R самоуправляемый снаряд [G] *м.;* сбрасываемое оружие [S] *ср.*
A الاطلاق والنِّسي الرُّمي والنَّبْذ

1706 [L]
fire and maneuver
F feu et mouvement *m.*
  m.
E fuego y movimiento *m.*
  m.
D Feuer und Bewegung
  neut. f.
R огонь и манёвр *м. м.*
A الرَّمْى والْمُناوَرة

1707 [N]
fireball
F boule de feu *f.*
E bola de fuego *f.*
D Feuerball *m.*
R огневой шар *м.*
A وَهَّاج ـ نَيزَك *m*

1708
fire control
F conduite de tir *f.*
E conducción del tiro *m.*
D Feuerleitung *f.*
R управление огнём *ср.*
A السَّيْطرة على الرَّمْي

1709
fire control system
F système de commande
  de tir *m.*
E equipo de conducción
  del tiro *m.*
D Feuerleitanlage *f.*
R система управления
  огнём *ж.*
A جِهَاز مُراقَبَة النِّيَران *m*

1710 [S]
fire direction center
F poste central de tir *m.*
E central de tiro *m.*
D Feuerleitzentrale(-stelle)
  f.
R центр управления
  огнём *м.*
A مَرْكَز تَوْجِيه النِّيَران *m*

106

1711
fire drill
F exercice anti-incendie
  m.
E ejercicios de incendio
  m. pl.
D Feuerlöschübung *f.*
R противопожарные
  действия *ср. мн. ч.*
A اجراءات إطفاء الحَريق

1712
fire engine
F pompe à incendie *f.*
E bomba de incendios *f.*
D Feuerwehrfahrzeug
  neut.
R пожарная машина *ж.*
A سيارة إطْفَاء الحَريق *m*

1713
fire escape
F escalier de secours *m.*
E escalera de incendios *f.*
D Nottreppe *f.*
R пожарная лестница *ж.*
A سُلَّم النَّجَاة *m*

1714
fire extinguisher
F extincteur *m.*
E extintor *m.*
D Feuerlöscher *m.*
R огнетушитель *м.*
A مُطْفِئة حريق *f*

1715 [L]
firefight
F accrochage *m.*
E combate por el fuego
  m.
D Feuerkampf *m.*
R огневой бой *м.*
A قِتَال بالنَّار

1716 [U]
firefighting *adj.*
F contre l'incendie
E (de) extinción de
  incendios
D Brandbekämpfungs-
  komp.
R противопожарный
A مُكَافَحَة الحَريق

1717
fire for effect
F tir d'efficacité *m.*
E tiro de eficacia *m.*
D Wirkungsfeuer *neut.*
R огонь на поражение
  м.
A نار التَّأْثِير

1718
fire group
F equipe feu *f.*
E grupo de tiro *m.*
D Schiesstruppe *f.*
R стреляющая группа
  ж.
A مَجْمُوعَة رمي *f*

1719
fireman
F pompier *m.*
E bombero *m.*
D Feuerwehrmann *m.*
R пожарник *м.*
A رَجُل الحَريق *m*

1720
fire over open sights
F tirer à plein fouet
E hacer fuego con
  puntería directa
D (ohne) Visier schiessen
R вести огонь прямой
  наводкой
A رمي مُبَاشِرَاً

1721

firepower

F puissance de feu *f.*
E potencia de fuego *f.*
D Feuerkraft *f.*
R огневая мощь *ж.*
A قُوَّة النِّيرَان *f*

1722

fireproof *adj.*

F ignifugé
E incombustible
D feuerfest
R огнеупорный;
несгораемый
A مَنِيع على النار *m*

1723

firestorm

F tempête de feu *f.*
E descarga de fuego *f.*
D Feuersturm *m.*
R огнённый шторм *м.*
A عَاصِفة نارِيَّة *f*

1724

firing circuit

F circuit de mise à feu *m.*
E circuito de disparo *m.*
D Zündkreis *m.*
R подрывная сеть *ж.*
A دَائِرة إطْلاق النَّار *f*

1725

firing data

F éléments de tir *m. pl.*
E elementos de tiro *m. pl.*
D Schusswerte *m. pl.*
R данные для стрельбы
(огня) *ж. мн. ч.*
A مُعْطَيَات الرَّمْى *f*

1726

firing device

F allumeur *m.*
E dispositivo de dar fuego
m.
D Zündvorrichtung *f.*
R взрыватель *м.*
A جِهاز إطْلاق النَّار *m*

1727

firing squad

F peloton d'exécution *m.*
E pelotón de ejecución *m.*
D Exekutionskommando
neut.
R команда, наряженная
для расстрела *ж.*
A زُمْرة الرَّمي *f*

1728

firm *adj.*

F ferme
E firme
D fest
R твёрдый
A ثَابِت *m*

1729

first aid *n.*

F premiers secours *m. pl.*
E primera curación *f.*
D erste Hilfe *f.*
R первая помощь *ж.*;
скорая помощь *ж.*
A إسْعَاف أوَّلي *m*

1730

fish *v.*

F pêcher
E pescar
D fischen
R ловить рыбу
A إصطاد سَمَكًا

1731 [N]

fission

F fission *f.*
E fisión *f.*
D Spaltung *f.*
R расщепление *ср.*
A إنْشِطار *m*

1732 [N]

fissionable

F fissible
E fisionable
D spaltbar
R расщепляемый
A قابل للانشطار

1733

fission products *pl.*

F produits de fission *m.*
pl.
E productos de fisión *m.*
pl.
D Spaltprodukte *neut. pl.*
R продукты
расщепления *м. мн.*
ч.; произведения
расщепления *ср. мн.*
ч.
A نِتَاج الإنْشِطَار *m*

1734 [E]

fit *v.*

F monter
E montar
D montieren
R приспособлять;
оснащать
A ركَّب

1735

fit

F (en) pleine forme
E (en) buen estado físico
D fähig *adj.;* -tauglich
komp.
R здоровый; годный
A لائِق

1736

fitted for radio

F equipé radio
E adaptado para aparatos
de radio
D (für) Funkgeräte
ausgestattet
R снабжённый
средствами связи
A مُجَهَّز براديو *m*

1737 [L]

fitter

F mécanicien-ajusteur *m.*
E ajustador *m.*
D Monteur *m.;*
Mechaniker *m.*
R слесарь-монтажник *м.*
A برَّاد *m*

1738

(tie down) fitting

F point d'arrimage *m.*
E punto de anclaje *m.;*
   punto de amarre *m.*
D Befestigungspunkt *m.*
R прихватка *ж.*
A وَصْلَة *f*

1739 [A,M]

fix *v.*

F faire le point; repérer
E fijar (la posición)
D (den Standort) peilen
R определять место
A عَيَّن النُقطة

1740 [A,M]

fix *n.*

F point *m.;* relèvement *m.*
E posición *f.*
D Peilstandort [A] *m.;*
   Besteck [M] *neut.*
R засечка *ж.;*
   определение места
   ср.
A تَعْيِين النُقطة *m*

1741 [P,R]

(radio) fix

F localisation
   radiogoniométrique *f.*
E localizar por
   radiogoniometría
D Funkpeilung *f.*
R (радио)засечка *ж.*
A نقطة مُعَيَّنة *f*

1742

"Fix bayonets!"

F "Baïonette au canon!"
E "¡Calen bayoneta!"
D „Seitengewehr
   aufpflanzen!"
R "Примкнуть штык!"
A الحِراب ركِّب

108

1743

fixed ammunition

F munition encartouchée
   f.
E munición fija *f.*
D Patronenmunition *f.*
R унитарные снаряды *м.*
   мн. ч.; (патроны,
   боеприпасы)
A مَقْذُوف كامل *m*

1744 [A,M]

fixed light

F feu fixe *m.*
E luz fija *f.*
D Festfeuer *neut.*
R постоянный огонь *м.*
A نُور ثَابِت *m*

1745

flag *n.*

F drapeau *m.*
E bandera *f.*
D Flagge *f.*
R флаг *м.;* знамя *ср.*
A عَلَم *m*

1746

flag down *v.*

F arrêter
E detener
D anhalten *tr.*
R останавливать
A لَوَّح لِلسَّيَارة بالوُقُوف

1747

flag of truce

F drapeau blanc *m.*
E bandera de parlamento
   f.
D Parlamentärflagge *f.*
R парламентёрский флаг
   м.
A راية المفاوضة *f*

1748

flag rank

F grade d'officier
   supérieure de marine
   m.
E (de) alta graduación;
   contraalmirante y
   superior
D Flaggrang *m.*
R адмиральское звание
   ср.
A رتبة بحرية عالية

1749

flagstaff

F mât de drapeau *m.*
E asta de bandera *f.*
D Fahnenmast *m.*
R флагшток *м.*
A سَارِية العَلَم *f*

1750

flail

F fléau *m.*
E cadena *f.*
D Flegel *m.*
R трал *м.*
A درّاسة

1751

flak

F tir anti-aérien *m.*
E fuego antiaéreo *m.*
D Flak *f.*
R звенитная артиллерия
   ж.
A نيران م/ط

1752

flame *n.*

F flamme *f.*
E llama *f.*
D Flamme *f.*
R пламя *ср.*
A لَهَب *m*

1753
flameproof
F ignifugé
E ignifugo
D feuerfest
R пламеупорный;
   огнестойкий
A مَنِيع على اللَّهَب m

1754
flamethrower
F lance-flammes m. inv.
E lanzallamas m. inv.
D Flammenwerfer m.
R огнемёт м.
A قَاذِفة لَهَب f

1755
flammable
F inflammable
E inflamable
D brennbar
R пламеопасный;
   огнеопасный
A قَابِل لِلإلْتِهَاب ـ قَابِل لِلإشْتِعَال

1756
flank v.
F prendre de flanc
E flanquear
D flankieren
R обходить фланг;
   фланкировать
A هَاجَم جَنَاح العَدُوّ

1757
flank n.
F flanc m.
E flanco m.
D Flanke f.
R фланг м.
A جَانِب m جَنَاح m

1758
flank v.
F protéger les flancs
E flanquear
D flanken
R защищать фланг
A حمى الجناح

1759
flanker unit
F unité flanc-garde f.
E grupo del flanco m.
D Flankeinheit f.
R укрепление,
   прикрывающее
   фланг ср.
A وَحْدة الجناح f

1760
flank guard
F flanc-garde f.
E guarda flanco f.
D Flankendeckung f.
R боковое охранение ср.
A مَجْنَبَة f حَارِس الجَنب m

1761 [A,M]
flap
F volet [A] m.; ailette [M]
   f.
E aletazo m.
D Klappe f.
R закрылок м.
A قلَّابة f

1762
flare adj.
F éclairant
E (de) iluminación
D Leucht- komp.
R осветительный
A مشعل

1763
flare n.
F fusée éclairante f.
E bengala f.
D Leuchtsignal neut.
R осветительный патрон
   м.
A مِشْعَل m

1764 [A]
flare path
F piste balisée f.
E baliza f.
D Leuchtpfad m.
R освещаемая полоса
   ж.
A مَدْرَج مُنَوَّر m

1765
flare-up n.
F bagarre f.
E estallido m.; llamarada
   f.
D Aufbrausen neut.
R вспышка ж.
A إنْدِلاَع اللَهَب m

1766 [N,S,U]
flash
F éclair [NU] f.; lueur [S]
   f.
E destello [NU] m.;
   fogonazo [S] m.
D Mündungsfeuer [S]
   neut.; Strahl [NU] m.
R вспышка ж.; блеск м.
A وَمِيض m

1767
flash blindness
F aveuglement par éclair
   m.
E cegadura por destello f.
D Blindheit durch
   Blendung f.
R ослепление от
   вспышки ср.
A عَمَى الوَمِيض m

1768
flash burn
F brûlure par éclair f.
E quemadura por destello
   f.
D Verletzung durch
   Lichtblitz f.
R ожог от вспышки м.
A حَرْق الوميض m

1769 [A,M]
flashing light
F feu à éclats m.
E luz de destellos f.
D Blinkfeuer neut.
R проблесковый огонь
   м.; проблесковый
   свет м.
A ضَوْء وَمَّاض m

1770
flashlight
F torche électrique *f.*
E linterna eléctrica *f.*
D Taschenlampe *f.*
R переносный фонарь *м.*
A مِصْبَاح *m*

1771
flash ranging
F repérage par la lueur *m.*
E localización óptica *f.*
D Lichtmessung *f.*
R оптическая разведка *ж.*; светомерие *ср.*
A تَحْدِيد المَدَى بالوميض

1772 [T,U]
flat
F plat
E llano
D flach
R плоский
A مُسَطَّح

1773
flechette
F fléchette *f.*
E dardo *m.*
D Pfeilchen *neut.*
R подкалиберный снаряд *м.*
A سَهْم جوّي *m*

1774 [V]
fleet
F parc automobile *m.*
E escuadra *f.*
D Wagenpark *m.*
R парк *м.*
A قَافِلة *f*

1775 [M]
fleet
F flotte *f.*
E flota *f.*
D Flotte *f.*
R морской флот *м.*
A أُسْطُول *m*

1776 [A]
(air) fleet
F flotte aérienne *f.*
E flota aérea *f.*
D Luftflotte *f.*
R (воздушный) флот *м.*
A أُسْطُول (جَوّي) *m*

1777
(fishing) fleet
F flotille de pêche *f.*
E flota de pesca *f.;* pesquera *f.*
D Fischereiflotte *f.*
R рыболовственная флотилия (рыбфлот) *ж.*
A سُفُن صَيْد *f*

1778
flesh wound
F blessure superficielle *f.*
E herida superficial *f.*
D Fleischwunde *f.*
R лёгкая рана *ж.*
A جرح بسطح الجلد

1779
flex *n.*
F câble souple *m.*
E flexible *m.*
D Schnur *f.*
R (гибкий) шнур *м.*
A سِلك كَهْرَبَائِي معزول *m*

1780
flexible
F souple
E flexible
D geschmeidig
R гибкий
A مَرِن

1781
flight
F vol *m.*
E vuelo *m.*
D Flug *m.*
R полёт *м.*; рейс *м.*
A طَيْران *m*

1782
(contour) flight
F vol tactique en suivi de terrain *m.*
E vuelo a baja cota *m.*
D Konturfliegen *neut.*
R полёт по контурам *м.*
A الطيران على طول خط الارتفاعات المتساوية

1783
(hedge-hopping) flight
F vol rasant *m.*
E vuelo rasante *m.*
D Heckenhüpfen *neut.*
R (бреющий) полёт *м.*
A طيران مُسِفّ *m*

1784
(instrument) flight
F vol aux instruments *m.*
E instrumento *m.*
D Blindflug *m.*
R полёт по приборам *м.*; слепой полёт *м.*
A طيران بلا رؤية (بالآلات) *m*

1785
(low level) flight
F vol à basse altitude *m.*
E vuelo a bajo nivel *m.*
D Tieffliegen *neut.*
R полёт на небольшой высоте *м.*
A الطيران على مستوى منخفض *m*

1786
(maiden) flight
F vol inaugural *m.*
E vuelo primero *m.*
D Jungfernflug *m.*
R (первый) полёт *м.*
A الطيران الأوّلي *m*

1787
(tactical) flight
F vol tactique *m.*
E vuelo táctico *m.*
D taktisches Fliegen *neut.*
R (боевой) полёт *м.*
A طيران تعبوي *m*

1788

(visual) flight

F vol à vue *m.*
E vuelo visual *m.*
D Sichtflug
R визуальный полёт *м.*
A الطَّيْرَان بالنَّظَر

1789

flight deck

F pont d'envol *m.*
E cubierta de vuelo *f.*
D Tragdeck *neut.*
R взлётная палуба *ж.*
A سطح طيران *m*

1790

flight engineer

F mécanicien navigant *m.*
E mecánico de vuelo *m.*
D Bordmechaniker *m.*
R бортинженер *м.*
A مُهَنْدِس طَيْرَان *m*

1791

flight information

F renseignements de vol
  m. pl.
E informes de vuelo *m.*
  pl.
D Fluginformationen *f. pl.*
R сведения о полёте *ср.*
  мн. ч.
A مَعْلُومَات عن الطَّيْرَان *f*

1792 [A]

flight path

F axe de vol *m.*
E trayectoria de vuelo *f.*
D Flugweg *m.*
R траектория полёта *ж.*
A مَمَرّ جَوّي *m*

1793

flight profile

F profil de vol *m.*
E perfil de vuelo *m.*
D Flugprofil *neut.*
R профиль полёта *м.*
A صُورة جَانِبيَّة للطَّيْرَان *f*

1794

flight surgeon

F médecin militaire
  aéromédical *m.*
E médico aeronáutico *m.*
D Flugarzt *m.*
R военный
  аэромедицинский
  врач *м.*
A جرَّاح في الطيران *m*

1795

flippers

F palmes *f. pl.*
E aletas *f. pl.*
D Flossen *f. pl.*
R плавники *м. мн. ч.*
A زعانف *f*

1796

float *v.*

F flotter
E flotar
D schwimmen
R плавать
A طَاف . عَام

1797 [A,M]

float *n.*

F flotteur [A] *m.;* quai
  flottant [M] *m.*
E pontón *m.*
D Schwimmer [A] *m.;*
  Schwimmsteg [M] *m.*
R понтон [A] *м.;* плот
  [M] *м.;* паром [M] *м.*
A طَوْف *m* عوامة *f*

1798

(cause to) float

F faire flotter
E hacer flotar
D schwimmfähig machen
R спускать
A طَوّف

1799

float (downstream)

F descendre (la rivière)
E flotar
D schwimmen
  (stromabwärts)
R плавать (вниз по
  течению)
A عام

1800

float (upwards)

F flotter
E subir
D schwimmen (aufwärts)
R проноситься
A طاف

1801

floating dock

F dock flottant *m.*
E dique flotante *m.*
D Schwimmdock *m.*
R плавучий док *м.*
A حَوْض عائم *m*

1802 [U]

flood *v.*

F déborder; inonder;
  monter
E crecer; desbordar;
  inundar
D fluten; überlaufen *tr.;*
  überschwemmen *untr.*
R наводнять
A فَاض

1803 [M]

flood *n.*

F flux *m.*
E pleamar *f.*
D Flut *f.*
R затопление *ср.*
A فَيْض *m*

1804 [U]

flood *n.*

F inondation *f.*
E inundación *f.*
D Überschwemmung *f.*
R наводнение *ср.*
A فَيْضَان *m*

111

1805
flood (land)
F inonder
E inundar
D überschwemmen
R наводнять
A فاض

1806
flood (over)
F déborder
E desbordarse
D überschwemmen
R затоплять
A غَمَر

1807 [M]
flood (tide)
F flux m.
E flujo m.
D Flut f.
R подъём воды м.
A مَدّ

1808
floodlight
F projecteur m.
E foco m.
D Flutscheinwerferlicht
   neut.
R прожектор м.
A ضوء غَامِر

1809 [M]
flotation
F flottaison f.
E flotación f.
D Schwimmen neut.
R плавучесть ж.
A عَوْم . طَوْف m

1810 [V]
flotation
F flottabilité f.
E flotación f.
D Schwimmen neut.
R плавучесть ж.
A عَوْم m

1811 [E,G]
flotation collar
F flotteur m.
E cuello boyante m.
D Schwimmkragen m.
R пояс плавучести м.
A حِزام العَوْم

1812
flotilla
F flotille f.
E flotilla f.
D Flotille f.
R флотилия ж.
A أسْطُول صَغِير m

1813
flotsam
F épaves flottantes f. pl.
E restos flotantes m. pl.
D Treibgut neut.
R плавающие обломки
   м. мн. ч.
A حُطَام السفينة العائم m

1814
flow v.
F couler
E fluir
D fliessen
R течь
A سَال

1815
flow n.
F courant m.
E corriente m.
D Strom m.
R течение ср.
A سَيْل . تَيَّار m

1816
flow chart
F organigramme m.
E diagrama de flujo m.
D Leistungsdiagramm
   neut.; Tabelle f.
R график м.; схема ж.
A جدول بياني

1817
flow rate
F débit m.
E caudal m.
D Durchflussmenge f.
R темп м.; течение ср.
A سُرْعَة التَيَّار

1818
fluid adj.
F fluide
E flúido
D flüssig
R жидкий
A سَائِل

1819
fluid n.
F fluide m.
E flúido m.
D Flüssigkeit f.
R жидкость ж.
A سَائِل m

1820
fluorescence
F fluorescence f.
E fluorescencia f.
D Fluoreszenz f.
R свечение ср.;
   флуоресценция ж.
A ضوء فلورسنت f

1821
fluorescent
F fluorescent
E fluorescente
D fluoreszierend
R флуоресцентный
A ضوء فلورسنت

1822 [E]
flush (with) adj.
F (à) fleur (de)
E (a) raz (de)
D bündig (mit); versenkt
R вровень (с)
A على إسْتِوَاءِ (مع)

1823 [A]
fly v.
F voler
E volar
D fliegen
R летать
A طار

1824 [A]
fly
F transporter par avion
E transportar en avión
D (im) Flugzeug
   befördern; ausfliegen
R перевозить
A نَقَل بطائرة

1825 [L,M]
fly (flag) v.
F battre
E llevar
D führen
R нести (флаг)
A خَفَق

1826
**flying bridge
   (flybridge)**
F passerelle f.
E pasarela f.
D besondere
   Kontrollbrücke z.B.
   Nock (manchmal eng.
   Begriff) f.
R продольный мостик
   м.; перекидной
   (мостик) м.
A جِسْر وَقْتي m

1827 [E]
flywheel
F volant m.
E volante m.
D Schwungrad neut.
R маховик м.
A حَدّافة f

1828
foam
F écume f.; mousse f.
E espuma f.
D Schaum m.
R пена ж.
A زَبَد

1829
foam rubber
F caoutchouc mousse m.
E espuma de caucho f.
D Schaumgummi neut.
R резиновая пена ж.
A مَطّاط مُزْبَد ـ مَطّاط
   رَغوي m

1830
focal length
F distance focale f.
E distancia focal f.
D Brennweite f.
R фокусное расстояние
   ср.
A البُعْد البُؤْري

1831
focus n.
F foyer m.
E foco m.
D Brennpunkt m.
R фокус м.
A بُؤْرَة f

1832
focus (an instrument)
   v.
F concentrer; mettre au
   point
E enfocar
D bündeln; scharf
   einstellen tr.
R фокусировать
A ضبط

1833
focus (concentrate on)
   v.
F (se) concentrer (sur)
E concentrar(se) (en)
D (sich) konzentrieren
   (auf)
R сосредоточивать(ся)
A ركّز

1834
fog
F brouillard m.; brume f.
E niebla f.
D Nebel m.
R туман м.
A ضَبَاب m

1835
fogbound
F arrêté par le brouillard;
   pris dans la brume
E immovilizado por la
   niebla; ocultado en
   niebla
D (durch) Nebel
   behindert; (im) Nebel
   gehüllt
R задержанный туманом
A مُكْتَنَف بالضَّبَاب

1836
foggy
F brumeux; (de)
   brouillard
E brumoso; neblinoso
D neblig
R туманный
A ضَبَابي

1837
foghorn
F sirène de brume f.
E sirena (de niebla) f.
D Nebelhorn m.
R туманная сирена ж.
A بُوق الضَّبَاب m

113

1838

foglamp

F phare antibrouillard *m.*

E faro de niebla *m.*

D Nebellampe *f.*

R туманная лампа *ж.*

A مِصْبَاح الضَّبَاب *m*

1839 [E]

foil *n.*

F feuille *f.*

E hoja *f.*

D Folie *f.*

R фольга *ж.*; станиоль *м.*

A وَرَقَة فِلِزِّيَّة *f*

1840

fold (in the ground)

F repli de terrain *m.*

E pliegue *m.*

D Senkung *f.;* Bodenfalte *f.*

R складка местности *ж.*

A ثَنْيَة ارض *f*

1841

foliage

F feuillage *m.*

E follaje *m.*

D Laub *neut.*

R листва *ж.*

A ورق الشجر *m*

1842

follow

F suivre

E seguir

D (einem) folgen

R следовать (за)

A تَبِع

1843

follow and support force

F échelon de premier renfort *m.*

E fuerza de relevo *f.*

D „follow and support force" (eng. Begriff)

R войска второго эшелона *ср. мн. ч.;* подкрепляющие войска *ср. мн. ч.*

A قوّة سند لاحقة *f*

1844 [A,M,L]

followup *n.*

F poursuite *f.*

E persecución *f.;* seguimiento *m.*

D Verfolgung *f.*

R развитие наступления *ср.;* преследование *ср.*

A تَتَبُّع فَوْري *m*

1845 [A,M,L]

follow up *v.*

F poursuivre

E perseguir

D nachdrängen *tr.;* verfolgen *untr.*

R развивать наступление; преследовать

A تَتَبَّع

1846 [U]

follow up *v.*

F faire le suivi (de)

E perseguir

D nachfassen *tr.*

R следовать

A تَتَبَّع فوراً

1847 [U]

followup *n.*

F suivi *m.*

E continuación *f.;* seguimiento *m.*

D Nachuntersuchung *f.*

R последствия *ср. мн. ч.*

A تَتَبُّع *m*

1848

foolproof

F infaillible

E (a) prueba de mal trato; infalible

D narrensicher

R безопасный; надёжный

A لا يَتَعَطَّل حتّى مع سُوء إسْتِعْمَاله

1849

foot

F pied *m.*

E pie *m.*

D Fuss *m.*

R нога *ж.*; ступня *ж.*

A قَدَم *f*

1850

foothold

F position ferme *f.*

E posición establecida *f.*

D Halt *m.*

R опорная точка *ж.*

A قاعدة لتقدم اضافي *f*

1851 [B]

footing

F semelle de pont *f.*

E pie *m.*

D Grundlage *f.*

R опора *ж.*

A قاعدة أساس *f*

1852

footprint

F empreinte de pied *f.*

E huella *f.*

D Fußspur *f.*

R след *м.*

A أثر القدم *m*

1853 [E,P]

force *n.*

F force *f.*

E fuerza *f.*

D Kraft *f.;* Gewalt *f.*

R сила *ж.*

A قُوَّه *f*

114

1854
force *v.*
F forcer
E forzar
D zwingen
R заставлять
A أَجْبَر

1855
(in) force
F (en) vigueur
E vigente
D (in) Kraft (treten) *f.*
R (в) силе: (в) действии
A نَفاذ

1856
forced march
F marche forcée *f.*
E marcha forzada *f.*
D Eilmarsch *m.;*
Gewaltmarsch *m.*
R стремительный марш
м.
A سَيْر اضطراري *m*

1857 [A,M,L]
forces *pl.*
F armée(s) *f. (pl.)*
E fuerzas *f. pl.*
D Streitkräfte *f. pl.*
R вооружённые силы *ж.*
мн. ч.
A قُوَّات *f*

1858
ford *v.*
F (faire) passer à gué
E vadear
D durchwaten *untr.*
R переходить вброд
A خَاض

1859
ford
F gué *m.*
E vado *m.*
D Furt *f.*
R брод *м.*
A مخاضة *f*

1860
fordable
F guéable
E vadeable
D durchwatbar
R переправляемый
бродом
A قابل للخَوْض فيه

1861
**(deep) fording
capability**
F aptitude à franchir un
gué profond *f.*
E profundidad de vadeo
f.
D Watfähigkeit *f.*
R способность к
переправе *ж.*
A مَقْدِرة عُبُور المَخاضات
العَميقة

1862
**(shallow) fording
capability**
F aptitude à franchir un
gué peu profond *f.*
E capacidad de vadeo (sin
preparación) *f.*
D Watfähigkeit (ohne
Vorbereitung) *f.*
R способность к
переправе через
мелкие воды *ж.*
A قابلية الخَوْض (عميق)

1863
fore and aft *adj.*
F (de l') avant à l'arrière
E (de) popa a proa
D längschiff; vorn und
hinten
R продольный
A من الأمام إلى الوَراء

1864
forecastle
F gaillard *m.*
E castillo *m.*
D Back *f.;* Vorderdeck
neut.
R бак *м.;* полубак *м.*
A الطَّرَف الأمامي (في سَفينة)

1865
foreground
F premier plan *m.*
E primer plano *m.*
D Vordergrund *m.*
R передний план *м.*
A الأرض الأمامية *f*

1866
foreshore
F plage *f.*
E frente marítimo *m.;*
orilla *f.;* playa *f.*
D Uferland *neut.;* Watt
neut.
R береговая полоса
осушки *ж.*
A شاطى ء *m*

1867 [S]
foresight
F guidon *m.*
E guión de mira *m.*
D (Visier-)Korn *neut.*
R мушка *ж.*
A مُسَدَّدة أَمامِيَّة *f*

1868
forge
F forge *f.*
E herrería *f.*
D Schmiede *f.*
R кузница *ж.*
A مَصْنَع الحديد *m*

1869 [T]
fork
F embranchement *m.*
E bifurcación *f.*
D Abzweigung *f.*
R развилка *ж.*
A مَفْرَق *m*

115

1870
forklift (truck)
F chariot élévateur *m.*
E elevadora-
 transportadora de
 horquilla *f.*
D Gabelstapler *m.*
R подъёмная машина
 ж.
A رافعة شوكية *f*

1871
format
F format *m.*
E formato *m.*
D Format *neut.*
R формат *м.*
A شَكْل *m*

1872
formation (disposition)
F formation *f.*
E formación *f.*
D Aufbau *m.;* Gliederung
 f.
R строй *м.*; построение
A تَشْكِيل *m*

1873
forming up place
F position d'attaque *f.;*
 zone de démarrage *f.*
E zona de despliegue *f.*
D Sturmausgangsstellung
 f.
R исходное положение
A مَكَان التَشْكِيل *m*

1874
form up
F prendre une formation
E formar
D (sich) aufstellen *tr.*
R выстраивать(ся)
A جَمَع

116

1875
(field) fortifications
F fortifications de
 campagne *f. pl.*
E fortificaciones *f. pl.*
D Festungswerk *neut.*
R полевое
 оборонительное
 сооружение *ср.*
A تَحْصِينَات *f*

1876
fortify
F fortifier
E fortificar
D befestigen
R укреплять
A حصّن

1877 [A,M,L,U]
forward *adj.*
F (d', de l') avant
E avanzado [AML];
 delantero [U]
D Vorder- *komp.;* vorder
 adj.
R передовой; передний
A أمامي

1878
forward air controller
F contrôleur aérien
 avancé *m.*
E controlador avanzado
 de aviones *m.*
D vorgeschobener
 Luftüberwachungsleiter
 m.
R передовой
 авианаводчик *м.*
A مُسَيْطِر جوي أمامي

1879
forward defense
F défense avancée *f.*
E defensa avanzada *f.*
D Vorneverteidigung *f.*
R передовая оборона
 ж.; оборона
 переднего края *ж.*
A الدِفَاع الأمامي *m*

1880
**forward edge of the
 battle area**
F limite avancée de la
 zone de bataille *f.*
E borde anterior de la
 zona de combate *m.*
D vorderer Rand der
 Verteidigung *m.*
R передний край
 обороны *м.*
A الحَدّ الأمامي لِنْطَقَة
 المُعْرَكَة *m*

1881
**forward line of
 friendly forces**
F ligne avant des forces
 amies *f.*
E limite anterior de las
 fuerzas proprias *m.*
D vorderste Linie der
 eigenen Kräfte (meist
 eng. Begriff) *f.*
R передний край своих
 войск *м.*
A الحَدّ الأمامي للقُوَّات
 الصَّدِيقة *m*

1882
**forward maintenance
 area**
F zone d'entretien de
 l'avant *f.*
E zona avanzada de
 suministro *f.*
D vorgeschobener
 Versorgungsraum *m.*
R передовой район
 обеспечения *м.*
A مِنْطَقَة صِيَانة اماميَّة *f*

1883
forward observer
F observateur avancé *m.*
E observador avanzado
 m.
D vorgeschobener
 Beobachter *m.*
R передовой
 наблюдатель *м.*
A راصِد أمامي *m*

1884 [E,V]
forward repair group
F sections mobiles de
réparations *f. pl.*
E equipo avanzado de
reparaciones *m.*
D vorgeschobene
Instandsetzungstruppe
f.
R (передовая ремонтная)
команда *ж.*
A مجموعة تصليح أمامية

1885
fougasse
F fougasse *f.*
E fougasse (término
francés)
D „Napalm-Mine" *f.*
R фугас *м.*
A لُغْم صَغِير *m*

1886 [M]
foul *v.*
F aborder; entrer en
collision (avec un
autre navire)
E caer (sobre otro buque)
D zusammenstossen *tr.*
R запутывать(ся)
A إصْطَدَم

1887
foul anchor
F ancre engagée *f.*
E ancla enredada *f.*
D unklarer Anker *m.*
R нечистый якорь *м.*
A مِرْساة مُشْتَبِكة *f*

1888
founder
F sombrer
E hundirse
D sinken
R идти ко дну
A غَرِق

1889
foxhole
F trou individuel *m.*
E hoyo de protección *m.*
D Schützendeckungsloch
neut.
R одиночный окоп *м.*;
стрелковая ячейка
ж.
A حُفْرَة فَرْدِيَّة *f*

1890
**fractional orbital
bombardment
system**
F système de
bombardement à
orbite fractionnée
(terme anglais) *m.*
E FOBS (término inglés)
D Raketen mit erdnaher
Umlaufbahn (meist
eng. Begriff) *f. pl.*
R система обстрела с
частичной орбиты
ж.
A جِهَاز مَدَارِي جُزْئِي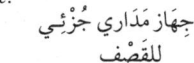
للقَصْف

1891
fragile
F fragile
E frágil
D zerbrechlich
R хрупкий; ломкий
A هَشّ

1892
frame *n.*
F châssis *m.;* cadre *m.*
E armazón *m.*
D Gestell *neut.*
R остов *м.;* корпус *м.;*
планёр *м.*
A إطار *m*

1893 [U]
frame(-work) *n.*
F cadre *m.*
E esqueleto *m.*
D Rahmen *m.*
R рама *ж.;* рамка *ж.*
A هَيْكَل *m*

1894 [B]
framework
F ossature *f.*
E armazón *f.*
D Gerüst *neut.;* Gerippe *f.*
R рама *ж.;* остов *м.*
A هَيْكل *m*

1895
fraternize
F fraterniser (avec)
E confraternizar
D fraternisieren
R брататься
A تصادق

1896
free
F libre
E libre
D frei
R свободный; вольный
A حُرّ

1897
free air overpressure
F surpression incidente *f.*
E sobrepresión de aire
libre *f.*
D Überdruck *m.*
R повышенное давление
(на открытом
пространстве) *ср.*
A ضَغْط زَائِد لِلْمُحِيط

1898
freeboard
F franc-bord *m.*
E francobordo *m.*
D Freibord *m.*
R надводный борт *м.*
A الجُزْء الطّافِي *m*

1899
freedom
F liberté *f.*
E libertad *f.*
D Freiheit *f.*
R свобода *ж.*
A حُرِّية *f*

1900
free drop
F largage en chute libre
   m.
E lanzamiento libre *m.*
D Abwurf (ohne
   Fallschirm) *m.*
R беспарашютное
   сбрасывание *ср.*
A سُقُوط ذاتِي *m*

1901
free fall
F saut en commandé *f.*
E caída libre *f.*
D Freifallspringen *neut.*
R свободное падение *ср.*;
   затяжной прыжок *м.*
A إسْقاط حُرّ *m*

1902
free fire area
F zone d'ouverture du feu
   f.
E zona de fuego libre *f.*
D offenes Beschussgebiet
   neut.
R участок
   неограниченного
   огня *м.*
A منطقة نار حرّة *f*

1903
free issue
F dotation gratuite *f.*
E entrega gratuita *f.;* sin
   cargo
D kostenlose Abgabe *f.*
R бесплатная выдача *ж.*
A تَسْليم بلا دَفْع ـ نُسْخَة مجانية

1904
freeze
F geler; congeler
E helar(se); congelar
D frieren; gefrieren;
   tiefkühlen
R мёрзнуть;
   замораживать
A جَمَّد

118

1905 [L]
freeze
F (se) figer sur place
E permanecer inmóvil
D stillstehen *tr.*
R приковываться
A تَجَمَّد

1906
freight *n.*
F fret *m.*
E flete *m.*
D Fracht *f.*
R груз *м.*; фрахт *м.*
A شَحْن *m*

1907
(rail) freight car
F wagon de marchandises
   m.
E vagón de mercancias *m.*
D Güterwagen *m.*
R товарный вагон *м.*;
   грузовой вагон *м.*
A عَرَبَة شَحْن *f*

1908
freight container
F conteneur *m.*
E contenedor *m.*
D Transportbehälter *m.*
R (товарный) контейнер
   м.
A مُسْتَوْدَع شَحْن *m*

1909 [A]
freighter
F avion de transport *m.*
E avión de carga *m.*
D Transportflugzeug *neut.*
R грузовой самолёт *м.*
A طائِرة شَحْن *f*

1910 [M]
freighter
F cargo *m.*
E buque de carga *m.*
D Frachtschiff *neut.*
R грузовое судно *ср.*
A سَفِينة شَحْن *f*

1911
frequency
F fréquence *f.*
E frecuencia *f.*
D Frequenz *f.*
R частота *ж.*
A تَرَدُّد *m*

1912
fresh
F frais
E recién hecho
D frisch
R свежий
A جَديد

1913
**"Friend!" (answer to
   challenge)**
F "Ami!"
E "¡Gente de paz!"
D „Freund!"
R "Свой!"
A صديق

1914 [P]
frisk
F fouiller
E registrar
D durchsuchen
R обыскивать
A فَتَّش

1915
frogman(-men)
F homme(s)-grenouille(s)
   m.
E hombre(s)-rana *m.*
D Kampfschwimmer *m.*
R лёгководолаз *м.*;
   ныряльщик с
   аквалангом *м.*
A ضُفْدَع بَشَرِي *m*

1916 [L]
front (also Soviet)
F front *m.*
E frente *m.*
D Front *f.*
R фронт *м.*
A جَبْهَة *f*

1917

frontage
F front *m.*
E extensión frontal *f.*
D Frontbreite *f.*
R ширина фронта *ж.*
A جَبْهَة *f*

1918 [V]

frontal arc
F débattement avant *f.*
E arco frontal *m.*
D frontaler
Panzershutzbereich *m.*
R лобовая часть *ж.*
A قَوْس جَبْهِي *m*

1919

frontal attack
F attaque frontale *f.*
E ataque frontal *m.*
D Frontalangriff *m.*
R фронтальный удар *м.;*
фронтальное
наступление *ср.*
A هُجُوم جَبْهِي *m*

1920

frontier
F frontière *f.*
E frontera *f.*
D Grenze *f.*
R граница *ж.*
A حُدود *f*

1921

frontsight
F guidon *m.*
E punto de mira *m.*
D Vordervisier *neut.;*
Korn *neut.*
R мушка *ж.*
A مُسَدَّدة أمامية

1922

frost *n.*
F gel *m.;* gelée *f.*
E helada *f.*
D Frost *m.*
R мороз *м.*
A صَقِيع *m*

1923

(degrees of) frost
F (degrés de) froid *m.*
E (grados) bajo cero
D Kältegrad *m.*
R (...) градусов мороза
мороза
A دَرَجات الصَّقِيع *f*

1924

frostbite
F gelure *f.*
E congelación *f.*
D Erfrierung *f.*
R обморожение *ср.*
A عَضَّة الصَّقِيع *f*

1925

frozen
F gelé
E helado
D (ein-, zu)gefroren
R замёрзший;
замороженный
A مُتَجَمِّد

1926

fuel
F carburant *m.*
E carburante *m.*
D Betriebstoff *m.;*
Treibstoff *m.*
R топливо *ср.;* горючее
ср.
A وُقُود *m*

1927

fuel can
F bidon *m.*
E bidón *m.*
D (Betriebstoff-) Kanister
m.
R топливная банка *ж.*
A صَفِيحة وُقُود *f*

1928

fuel drum
F fût *m.*
E bidón *m.*
D (Betriebstoff-) Fass *neut.*
R топливная бочка *ж.*
A بَرْمِيل وُقُود *m*

1929 [N]

fuel rod
F tringle combustible *m.*
E elemento combustible
m.
D Brennstab *m.*
R огневой стержень *м.*
A قَضِيب وُقُود *m*

1930

full moon
F pleine lune *f.*
E luna llena *f.*
D Vollmond *m.*
R полнолуние *ср.*
A القَمِر التامّ

1931

(at) full power
F (à) plein régime
E (a) pleno rendimiento
m.
D (bei) Höchstleistung *f.*
R полная сила *ж.;*
(полным ходом;
полной скоростью)
A بِكامِل القُوّة

1932 [M]

"Full speed ahead!"
F "(En) avant toute!"
E "¡Avante toda!"
D „Volle Kraft voraus!"
R полный ход вперёд!!
A الى الأمام بِأقْصَى سُرْعَة

1933 [M]

"Full speed astern!"
F "(En) arrière toute!"
E "¡Atras toda!"
D „Volle Kraft zurück!"
R "полный ход назад!"
A الى الوَرَاء بِأقْصَى سُرْعَة

119

1934
function *v.*
F fonctionner
E funcionar
D funktionieren
R действовать;
функционировать
A عَمِلَ

1935
function *n.*
F fonction *f.*
E función *f.;* cargo *m.*
D Aufgabe *f.;* Funktion *f.*
R назначение *ср.;*
функция *ж.*
A وَظِيفَة *f*

1936
functional
F fonctionnel
E funcional
D funktional
R функциональный;
действующий
A وَظِيفِي

1937 [M]
funnel *n.*
F cheminée *f.*
E chimenea *f.*
D Schornstein *m.*
R дымовая труба *ж.;*
дымоход *м.*
A مِدْخَنَة *f*

1938 [E,U]
funnel *n.*
F entonnoir *m.*
E embado *m.*
D Trichter *m.*
R воронка *ж.;* литник
м.
A قُمَع *m*

1939
furlough
F permission *f.*
E licencia *f.*
D Urlaub *m.*
R отпуск *м.*
A إجازة

1940
furnace
F four *m.*
E horno *m.*
D Ofen *m.*
R печь; горн
A فُرْن *m*

1941
furnace (of a boiler)
F foyer *m.;* brûleur (à
mazout) *m.*
E hogar *m.*
D Feuerung *f.*
R топка *ж.*
A فُرْن *m*

1942 [E,S]
fuse
F amorcer; armer
E armar
D (Zünder) anbringen an
tr.
R взводить взрыватель
A جهّز بصمامة

1943
fuze (electrical)
F fusible *m.*
E fusible *m.*
D Sicherung *f.*
R пробка *ж.;* плавкий
предохранитель *м.*
A صِمَامَة كَهْرَبَائِيَة *f* ـ فيوز *m*

1944
fuze (engineering)
F fusée *f.;* amorce *f.*
E plomo *m.*
D Sicherung *f.*
R запал *м.*
A مِصْهَر *m*

1945 [E]
fuze (of a charge)
F fusée *f.;* mèche *f.*
E espoleta *f.*
D Zündschnur *f.*
R взрыватель *м.*
A صِمَامَة *f* ـ فيوز *m*

1946 [E]
fuse (together)
F mettre en fusion
E fundir
D verschmelzen
R сплавлять
A صَهَر

1947 [N]
fusion
F fusion *f.*
E fusión *f.*
D Verschmelzung *f.*
R расплавление *ср.*
A صَهْر *m*

1948 [S]
fuze
F allumeur *m.;* fusée *f.*
E espoleta *f.*
D Zünder *m.*
R (ударная, снарядная)
трубка *ж.*
A صِمَامَة *f* ـ فيوز *m*

1949
(backup) fuze
F fusée de secours *f.*
E espoleta secundaria *f.*
D Reserve-Zweitzünder *m.*
R (запасной) взрыватель
м.
A صمامة ساندة

1950 [S]
(base) fuze
F fusée de culot *f.*
E espoleta de culote *f.*
D Grundzünder *m.*
R (донный) взрыватель
м.
A صمامة القاعدة

1951
(clockwork time) fuze
F fusée à mouvement
d'horlogerie *f.*
E espoleta de relojería *f.*
D Uhrwerkzünder *m.*
R (механический)
взрыватель *м.*
A صمامة زمنية ذات حركة
ساعية

1952
(controlled variable time) fuze
F fusée de proximité à réglage préalable f.
E espoleta de proximidad f.
D (zeitlich regelbarer) Zünder m.
R (дистанционно-неконтактный) взрыватель м.
A صمامة زمنية محكومة

1953 [E]
(detonation) fuze
F cordeau détonant m.
E espoleta de cápsula fulminante f.
D Sprengzünder m.
R (основной) взрыватель м.
A صمامة صاعقة

1954
(direction action) fuze
F fusée percutante f.
E espoleta a percusión f.
D Aufschlagzünder m.
R взрыватель мгновенного действия м.
A صِمَام ذو فِعْل مُباشِر m

1955
(impact action) fuze
F fusée percutante f.
E espoleta de contacto f.
D Aufschlagzünder m.
R взрыватель мгновенного действия м.
A صمامة مصادمة

1956
(influence) fuze
F fusée à influence f.
E espoleta por influencia f.
D Influenzzünder m.
R (неконтактный) взрыватель м.
A صِمَامة تَأْثير f

1957
(mechanical) fuze
F fusée à mouvement d'horlogerie f.
E espoleta mecánica f.
D (mechanischer) Zünder m.
R (механический) взрыватель м.
A صمامة آلية

1958
(percussion) fuze
F fusée percutante f.
E espoleta de percusión f.
D Aufschlagzünder m.
R взрыватель ударного действия м.
A صِمَامَة التفجِير f

1959
(point detonating) fuze
F fusée d'impact f.
E espoleta de impacto f.
D Punktaufschlagzünder m.
R (основной) взрыватель м.
A صمامة الرأس

1960
(pressure) fuze
F fusée à pression (dépression) f.
E espoleta de presión f.
D Druckzünder m.
R взрыватель нажимного действия м.
A صمامة ضغط

1961
(pressure release action) fuze
F allumeur à relâchement m.
E espoleta de alivio de presión f.
D Druckauslösezünder m.
R взрыватель нажимного действия м.
A صِمَامَة ضَغْطِيَّة f

1962
(proximity) fuze
F fusée de proximité f.
E espoleta de proximidad f.
D Annäherungszünder m.
R неконтактный взрыватель м.
A صِمَامَة تَقَارُبِيَّة f

1963
(proximity) fuze
F fusée de proximité f.
E espoleta de proximidad f.
D Abstandzünder m.
R (неконтактный) взрыватель м.
A صمامة تقاربية f

1964
(pull) fuze
F allumeur à traction m.
E espoleta de tracción f.
D Zugzünder m.
R взрыватель натяжного действия м.
A صمامة ذات سَحَّابَة

1965
(release) fuze
F allumeur à relâchement m.
E espoleta de suelta f.
D Auslösezünder m.
R взрыватель нажимного действия м.
A صمامة تَسْيِيب

1966 [E]
(safety) fuze
F mèche lente f.
E fusible m.
D Sicherheitszünder m.
R (огнепроводный) шнур м.
A صَهِيرة الأمن

1967

(self-destroying) fuze

F fusée autodestructrice *f.*

E espoleta
  autodestructora

D Selbstzerstörungszünder
  m.

R самоликвидатор *м.*;
  взрыватель с
  самоликвидатором
  м.

A صِمَامَة مُدَمَّرَة ذَاتِيَّة

1968

(time) fuze

F fusée à temps *f.*

E espoleta de tiempo *f.*

D Zeitzünder *m.*

R (дистанционный)
  взрыватель *м.*

A صمامة زمنية

1969

(trembler) fuze

F allumeur à bascule *m.*

E espoleta de influencia *f.*

D Bebzünder *m.*

R взрыватель
  дрожащего действия
  м.

A صمامة رجّاجة

1970

(variable delay) fuze

F fusée à retard réglable
  f.

E espoleta de retardo
  variable *f.*

D (regelbarer)
  Verzögerungszünder
  m.

R (непостоянный
  замедленный)
  взрыватель *м.*

A صمامة زمنية متغيرة

1971

(variable time) fuze

F fusée de proximité *f.*

E espoleta de
  radioproximidad *f.*

D Annäherungszünder *m.*

R радиовзрыватель *м.*

A صِمَامَة زَمَنِيَّة مُتَغَيِّرَة

1972

fuze assembly

F bouchon allumeur *m.*

E montaje de espoleta *m.*

D Sicherungsanlage *f.*

R комплект взрывателей
  м.

A مجموعة الصمامة

G

1973 [P]

gag (someone)

F baîllonner

E amordazar

D knebeln

R замыкать рот

A سدّ الفم

1974 [C]

G agent

F agent G *m.*

E agente G *m.*

D Nervengas (z.B. GB,
  GX) *neut.*

R отравляющие
  вещества (OB)
  нервно-
  паралитического
  действия *ср. мн. ч.*

A عَامِل مُثِيرٍ لِلْأَعْصَاب *m*

1975

gaiter

F guêtre *f.*

E polaina *f.*

D Gamasche *f.*

R гетры *м. мн. ч.*

A طُزلق

1976

gale (Force 8)

F coup de vent *m.*

E viento duro *m.*

D stürmischer Wind *m.*

R шторм *м.*; стормовой
  ветер *м.*

A رِيح شَدِيدَة *f*

1977

gallantry

F courage *m.*

E valor *m.*

D Tapferkeit *f.*

R храбрость *ж.*

A شجاعة

1978 [N]

gamma ray

F rayon gamma *m.*

E rayo gama *m.*

D Gammastrahl *m.*

R гамма-лучи *м. мн. ч.*

A شعاع جاما

1979

gangway

F passerelle
  (d'embarquement) *f.*

E plancha (de atraque) *f.*

D Laufsteg *m.*

R входной порт *м.*; вход
  с трапа *м.*

A مَمَّر (في سَفِينة) *m*

1980 [E]

gantry

F portique *f.*

E caballete *m.*

D Gerüst *neut.*

R помост *м.*; портал *м.*

A جِسْر دَارِج *m*

1981

(minefield) gap

F trouée *f.*

E brecha *f.*

D Minengasse *f.*

R проход (в минном
  поле) *м.*

A ثُغْرَة *f*

1982
garage *n.*
F garage *m.*
E garaje *m.*
D Garage *f.*
R гараж *м.*
A حَظِيرةِ سَيَارَات *f*

1983 [R]
garble
F erreur dans un message
   f.
E error *m.*
D Verstümmelung *f.*
R искажение *ср.*
A شَوّه

1984
garnishing
F garnissage (de
   camouflage) *m.*
E tira de guarnición *f.*
D Tarnmaterial *m.*
R маскировочная
   вплетённая материя
   ж.
A بَرْقَشَة *f*

1985
garrison *adj.*
F (de) garnison
E (de) cuartel; (de)
   guarnición
D Standort- *komp.*
R гарнизонный
A حَامِية *f*

1986
garrison *v.*
F placer une garnison
   (ville); garnir
   (position)
E guarnacer
D (mit einer) Garnison
   belegen
R ставить гарнизон
A وَضَعَ حَامِية

1987
garrison *n.*
F garnison *f.*
E guarnición *f.*
D Garnison *f.*
R гарнизон *м.*
A حَامِية *f*

1988
gas
F gaz *m.*
E gas *m.*
D Gas *neut.*
R газ *м.*
A غَاز *m*

1989
(blister) gas
F gaz vésicant *m.*
E gas vesicante *m.*
D Blasengas *neut.*
R кожно-нарывное
   отравляющее
   вещество (ОВ) *ср.*
A غَاز منفط

1990
(choking) gas
F gaz asphyxiant *m.*
E gas sofocante *m.*
D Stickgas *neut.*
R удушающее
   отравляющее
   вещество *ср.*
A غَاز خانق

1991
(incapacitating) gas
F gaz incapacitant *m.*
E gas nervioso *m.*
D Lähmungsgas *neut.*
R отравляющее
   вещество (ОВ) *ср.*
A غَاز معجز

1992
(inert) gas
F gaz inactif *m.*
E gas inerte *m.*
D Edelgas *neut.*
R инертный газ *м.*
A غَاز خَامِل *m*

1993
(irritant) gas
F gaz irritant *m.*
E gas irritante *m.*
D Reizgas *neut.*
R раздражающее
   отравляющее
   вещество *ср.*
A غاز مهيج

1994
(mustard) gas
F ypérite *f.*
E gas mostaza *m.*
D Senfgas *neut.*
R иприт *м.*; горчичный
   газ *м.*
A غاز الخردل

1995
(tear) gas
F gaz lacrymogène *m.*
E gas lacrimógeno *m.*
D Tränengas *neut.*
R слезоточивое
   отравляющее
   вещество *ср.*
A غاز مسيل للدموع

1996
gasket
F joint *m.*
E junta *f.*
D Dichtung *f.*
R прокладка *ж.*; сальник
   м.
A حَشِيَّة *f*

1997
gas mask
F masque à gaz *m.*
E máscara *f.*
D Gasmaske *f.*
R противогаз *м.*
A قناع الغاز

1998
gasoline
F essence *f.*
E gasolina *f.*
D Benzin *neut.*
R бензин *м.*
A بَنْزِين *m*

1999 [V]
gas pedal
F pédale d'accélération *f.*
E acelerador *m.*
D Gaspedal *neut.*
R ускоритель *м.;*
акселератор *м.*
A دَوَّاسَة السُّرعة *f*

2000
gas turbine
F turbine à gaz *f.*
E turbina a gas *f.*
D Gasturbine *f.*
R газовая турбина *ж.;*
газотурбинный
A تُربِينَة الغاز *f*

2001
gauge *v.*
F mesurer
E medir
D ausmessen *tr.*
R измерять;
калибровать
A قاس

2002 [E]
gauge *n.*
F calibre *m.*
E indicador *m.*
D Messinstrument *neut.*
R мера *ж.;* калибр *м.;*
измерительный
прибор *м.*
A مِقْيَاس *m*

2003 [V]
(loading) gauge *n.*
F gabarit (de chargement)
m.
E gálibo (de carga) *m.*
D Lademass *neut.*
R размер (груза) *м.*
A مِقْيَاس *m*

2004 [V]
gear
F vitesse *f.*
E marcha *f.*
D Gang *m.*
R передача *ж.*
A مُسَنَّن ـ تِرس *m*

2005 [A]
(landing) gear
F train d'atterrissage *m.*
E tren de aterrizaje *m.*
D Fahrgestell *neut.*
R (посадочное)
устройство *ср.*
A عجلات الهبوط *f*

2006 [E]
(lifting) gear
F appareil de levage *m.*
E equipos *m. pl.*
D Hebewinde *f.*
R (подъёмный)
механизм *м.*
A معدّات الرفع *f*

2007
(reverse) gear
F marche arrière *f.*
E marcha atrás *f.*
D Rückwärtsgang *m.*
R задний ход *м.;*
обратный ход *м.;*
задняя передача *ж.*
A مُسَنَّنَة السَّيرِالمَعْكُوس *m*

2008 [M,V]
(steering) gear
F direction [V] *f.;*
(appareil à)
gouvernail [M] *m.*
E (aparato de) gobierno
[M] *m.;* mecanismo
de dirección *m.*
D Lenkung [V] *f.;*
Steuereinrichtung [M]
f.
R рулевое устройство
ср.; рулевой
механизм *м.*
A جِهَاز التَّوْجِيه *m*

2009
gearbox
F boîte de vitesses *f.*
E caja de engranajes *f.;*
caja de cambios *f.*
D Getriebe *neut.*
R коробка передач *ж.*
A صَنْدوق (المَسَنَّنَات /
التروس) *m*

2010
gearing
F engrenage *m.*
E engranaje *m.*
D Triebwerk *neut.*
R зацепление *ср.;* привод
м.
A تَشْبِيك *m*

2011
gearshift
F (levier de) changement
de vitesse *m.*
E palanca de velocidades
f.
D Schalthebel *m.*
R рычаг переключения
передач *м.*
A تَبْدِيل السُّرعَة *m*

2012
geiger counter
F compteur Geiger *m.*
E contador Geiger *m.*
D Geigerzähler *m.*
R счётчик Гейгера *м.*
A عداد جيجر *m*

2013

gendarmerie
F gendarmerie *f.*
E Guardia Civil *f.;*
   Guardia Nacional
D Gendarm *m.*
R жандармерия *ж.*
A الدَّرك

2014 [L,U]

general *adj.*
F général
E general
D General- *komp.;*
   allgemein *adj.*
R общий; генеральный
A عَام

2015

general alert
F alerte générale *f.*
E alerta general *f.*
D allgemeiner Alarm
   (meist eng. Begriff) *m.*
R общая тревога; сигнал
   общей тревоги
A إنْذَار عَام *m*

2016

general orders
F consignes générales *f.*
   pl.
E reglamentos del cuartel
   general *m. pl.*
D Allgemeinbefehle *m. pl.*
R генеральный устав *м.*
A اوامر دائمة *f*

2017

general reserve
F réserve générale *f.*
E reserva estratégica *f.*
D strategische Reserve *f.*
R общий резерв *м.;*
   резерв общего
   назначения *м.*
A إحْتِيَاط عَام *m*

2018 [E]

generate
F générer
E generar
D erzeugen
R производить
A وَلَّد

2019

generator
F générateur *m.*
E generador *m.*
D Generator *m.*
R генератор *м.*
A مُوَلِّد *m*

2020

(smoke) generator
F fumigène *m.*
E productor de humo *m.*
D (Rauch-) Erzeuger *m.*
R дымообразователь *м.*
A مولّد دخّان *m*

2021

Geneva Convention
F Convention de Genève
   f.
E Convención de Ginebra
   f.
D Genfer Konvention *f.*
R Женевская конвенция
   ж.
A مِيثَاق جنيف *m*

2022

geographical
F géographique
E geográfico
D geographisch
R географический
A جُغْرَافِي

2023

georef
F géoref *m.*
E sistema georef *m.*
D Georef-System *neut.*
R система наведения по
   земным параметрам
   ж.
A مَعْلَم جُغْرَافِي *m*

2024

GI
F soldat américain *m.*
E soldado estadounidense
   m.
D amerikanischer Soldat
R американский солдат
   м.
A جندي امريكي

2025

gimbal
F cardan *m.*
E cardán *m.*
D Kardanring *m.*
R карданов подвес *м.*
A مِحْوَر

2026

girder
F poutre *f.*
E viga *f.*
D Tragbalken *m.*
R балка *ж.;* брус *м.*
A عَارِضة ـ كَمَرة *f*

2027 [M]

give way
F céder (à)
E ceder el paso (a)
D ausweichen *tr.*
R уступать; поддаваться
A تَرَاجَع

2028

glacier
F glacier *m.*
E glaciar *m.*
D Gletscher *m.*
R ледник *м.;* глетчер *м.*
A نَهْر ثَلْجِي *m*

2029

glare *n.*
F éblouissement *m.*
E deslumbramiento *m.*
D Blendung *f.*
R ослепление *ср.*; блеск
   м.
A نور سَاطِع *m*

2030

glass
F verre *m.*
E vidrio *m.*
D Glas *neut.*
R стекло *ср.*
A زُجاج *m*

2031

(plate) glass
F verre triple *m.*
E vidrio cilindrado *m.*
D Scheibenglass *neut.*
R зеркальное стекло *ср.*
A زجاج لوحي

2032

glass reinforced plastic (GRP)
F plastique renforcé de
   fibre de verre *m.*
E plástico reforzado con
   fibra de vidrio *m.*
D Glasfaserkunststoff
   (GFK) *m.*
R стеклянное волокно
   ср.
A زُجاج ليفي (بلاستيك) *m*

2033

glide *n.*
F vol plané *m.*
E planeo *m.*
D Gleitflug *m.*
R планирование *ср.*
A طَيْران شِرَاعي *m*

2034

glide *v.*
F planer
E planear
D gleiten
R планировать
A طار طَيْرَاناً شِرَاعياً

2035

glide mode
F (en) descente
   automatique *f.*
E descenso automático
   (término inglés) *m.*
D automatischer
   Landeanflug (meist
   eng. Begriff) *m.*
R автоматическая
   система
   приземления *ж.*
A طَريقَة الطَيْران الشِرَاعي *m*

2036

glide path
F axe de descente *m.*
E trayectoria de planeo *f.*
D Gleitweg *m.*
R траектория
   планирования *ж.*
A مَمَّر الإنْحِدَار *m*

2037

glider
F planeur *m.*
E planeador *m.*
D Segelflugzeug *neut.*
R планёр *м.*
A طَائِرة شِرَاعيّة *f*

2038

(immediate) goal
F objectif initial *m.*
E objetivo *m.*
D Ausgangsziel *neut.*
R (ближайшая) задача
   ж.
A الهدف الأوّلي

2039

goggles *pl.*
F lunettes (protectrices)
   m. pl.
E anteojos *m. pl.;* gafas
   (submarinas) *f. pl.*
D Schutzbrille *f.*
R защитные очки *м. мн.*
   ч.
A نَظَّارات واقية *f*

2040 [T]

gorge
F gorge *f.*
E cañón *m.*
D Schlucht *f.*
R ущелье *ср.*
A ممرّ ضيق

2041

government *adj.*
F gouvernemental *adj.*
E gubernamental
D Regierungs- *komp.*
R правительственный
   прилаг.
A حكومي

2042

government
F gouvernement *m.*
E gobierno *m.*
D Regierung *f.*
R правительство *ср.;*
   управление *ср.*
A حكومة *f*

2043

(military) governor
F gouverneur militaire *m.*
E gobernador (militar) *m.*
D Militärgouverneur *m.*
R (военный) губернатор
   м.
A حاكم عسكري

2044 [B]
grab
F benne preneuse *f.*
E cuchara de dos
   mandíbulas *f.*
D Greifer *m.*
R черпак *м.*
A كبّاش *m*

2045 [P,U]
grade *v.*
F classer
E clasificar
D einstufen *tr.*
R сортировать
A سَوّى

2046 [T]
grade *n.*
F rampe *f.*
E pendiente *f.*
D Steigung *f.*
R подъём *м.*; уклон *м.*
A مُنْحَدِر *m*

2047 [U]
grade *n.*
F grade *m.*
E grado *m.*
D Stufe *f.*
R степень *ж.*; ранг *м.*;
   класс *м.*
A دَرَجة *f*

2048 [B]
grade *v.*
F niveler
E nivelar
D planieren
R нивелировать
A دَرَّج

2049 [V]
grade (of fuel) *n.*
F qualité d'un carburant
   f.
E calidad *f.*
D Kraftstoffsorte *f.*
R качество *ср.*
A مَرْتَبة *f*

2050
grade crossing
F passage à niveau *m.*
E paso a nivel *m.*
D Bahnübergang *m.*
R переезд с шоссе *м.*
A مَمَرّ على مُسْتَوَى وَاحِد *m*

2051 [E]
grader
F niveleuse *f.*
E niveladora *f.*
D Planiermaschine *f.*
R нивелирующая
   машина *ж.*;
   нивелировщик *м.*
A آلة تَسْوِية *f*

2052 [P]
grading
F classification *f.*
E graduación *f.*
D Geheimhaltungsstufe *f.*
R сортирование *ср.*
A تَدْرِيج *m*

2053
gradual
F graduel
E gradual
D allmählich
R постепенный
A تَدْرِيجِي

2054 [E]
graduate
F graduer
E graduar
D graduieren
R градуировать;
   калибровать
A نال دَرَجة ـ خِرِّيج *m*

2055 [S]
graduation (sight)
F échelle de hausse *f.*
E elevación *f.*
D Visierskala *f.*
R деление *ср.*
A تدريج

2056
graph
F graphique
E gráfica *f.*
D Kurvenblatt *neut.*
R график *м.*
A رسم بياني

2057
grapnel
F grappin *m.*
E rezón *m.*
D Dregganker *m.*
R дрек *м.*; кошка *ж.*
A كُلّاب ـ مِخطاف *m*

2058
grating
F grille *f.*
E reja *f.*
D Gitter *neut.*
R решётка *ж.*
A غِطَاء مُصَبَّع *m* ـ مَشْرَبية
   حَدِيد *f*

2059
grave *n.*
F fosse *f.*
E sepultura *f.*
D Grab *neut.*
R могила *ж.*
A قَبْر *m*

2060
gravel
F gravier *m.*
E grava *f.*
D Kies *m.*
R гравий *м.*
A حصْباء

2061
graving dock
F cale sèche *f.*
E dique de carena *f.*
D Trockendock *neut.*
R (сухой) ремонтный
   док *м.*
A حَوْض جَافّ *m*

2062
gravity *adj.*
F (par la) pesanteur
E (por) efecto del peso
D Gravitations- *komp.*
R самотёком; напорный
A جَاذِبِيَّة *f*

2063
gravity *n.*
F pesanteur *f.*
E gravedad *f.*
D Schwere *f.*
R тяжесть *ж.*; тяготение *ср.*
A جَاذِبِيَّة *f*

2064 [S]
graze *v.*
F raser
E rozar
D bestreichen
R оцарапать
A كَشْط

2065
grease *n.*
F graisse *f.*
E grasa *f.*
D Fett *neut.*
R смазка *ж.*; жир *м.*
A شَحْم *m*

2066
grease *v.*
F graisser
E engrasar
D schmieren
R смазывать
A شَحَّم

2067
grease gun
F pistolet graisseur *m.*
E pistola engrasadora *f.*
D Schmierpresse *f.*
R тавотный шприц *м.*
A مسدس تَشْحِيم *m*

2068
great circle
F grand cercle *m.*
E círculo máximo *m.*
D Grosskreis *m.*
R большой круг *м.*
A دَائِرة عُظْمَى *f*

2069
Greenwich Mean Time
F heure de Greenwich *f.*
E tiempo medio de Greenwich (TMG) *m.*
D Greenwich-Zeit *f.*
R среднее время по гринвичскому меридиану *ср.*
A توقيت جرينتش المتوسط

2070
(drill) grenade
F grenade d'exercice *f.*
E granada de instrucción *f.*
D Drillgranate *f.*
R граната-болванка *ж.*
A قنبلة يدوية (للتعليم)

2071
(fragmentation) grenade
F grenade défensive *f.*
E granada rompedora *f.*
D Splittergranate *f.*
R (осколочная) граната *ж.*
A قنبلة يدوية (شظايا)

2072
(hand) grenade
F grenade à main *f.*
E granada de mano *f.*
D (Hand-)Granate *f.*
R (ручная) граната *ж.*
A قُنْبُلة يَدَوِيَّة *f*

2073
(smoke) grenade
F grenade fumigène *f.*
E granada fumígena *f.*
D Rauchgranate *f.*
R дымовая граната *ж.*
A قنبلة دخان يدوية

2074
(stick) grenade
F grenade à manche *f.*
E granada de mano *f.*
D Stielgranate *f.*
R ручная граната *ж.*
A قنبلة يدوية (لاصقة)

2075
(stun) grenade
F grenade neutralisante *f.*
E granada incapacitante *f.*
D Betäubungsgranate *f.*
R "оглушительная" граната *ж.*
A قنبلة يدوية مدوِّخة

2076 [T]
grid
F quadrillage *f.*
E cuadrícula *f.*
D Gitter *neut.*
R (координатная) сетка *ж.*; решётка *ж.*
A تَرْبِيع *m* ـ شَبَكَة خطوط *f*

2077
grid bearing
F gisement *m.*
E marcación de cuadriculado *f.*
D (Gitter-) Richtungswinkel *m.*
R дирекционный угол *м.*
A إنْجاه التَّرْبِيع *m*

2078
(map) gridding
F carroyage d'une carte *f.*
E cuadrículas *f. pl.*
D Gitternetz *neut.*
R решётка *ж.*
A التربيع *m*

2079

grid magnetic angle
F déclinaison magnétique
du quadrillage *m.*
E declinación magnética
de cuadriculado *f.*
D Gitterdeklination *f.;*
Nadelabweichung *f.*
R магнитное склонение
сетки *ср.*
A الزَّاوية المَغْنَطِيسِيَّة التَّرْبِيعِيَّة *f*

2080

grievance
F sujet de plainte *m.*
E injustia *f.*
D Beschwerde *f.*
R жалоба *ж.*
A ضَرَر *m*

2081

(pistol) grip
F poignée pistolet *f.*
E empuñadura *f.*
D Pistolengriff *m.*
R пистолетная рукоятка
ж.
A قبضت مسدس

2082

ground- *adj.*
F (de) terre
E (de) tierra
D Boden- *komp.*
R наземный
A أرْضِي

2083

ground *n.*
F sol *m.*
E suelo *m.*
D Boden *m.*
R грунт *м.*; земля *ж.*
A أرْض *f*

2084

ground (an aircraft) *v.*
F interdire à un avion de
voler
E permanecer en tierra
D stillegen (Flugwerkehr,
Flugzeug) *tr.*
R запрещать полёты
A منع طائرة من الطيران

2085

ground (arms) *v.*
F poser l'arme à terre
E rendir
D niederlegen (Waffen) *tr.*
R положить оружие
A وضع السلاح على الأرض

2086

(dead) ground
F angle mort *m.*
E terreno muerto *m.*
D toter Winkelbereich *m.*
R мёртвое пространство
ср.
A أرض ميتة *f*

2087

(killing) ground
F zone de destruction *f.*
E zona de destrucción *f.*
D Vernichtungsraum *m.*
R огневой мешок *м.*
A ارض القتل *f*

2088

(middle) ground
F banc médian *m.*
E terreno medio *m.*
D Mittelgrund *m.*
R (средняя) зона *ж.*;
(средняя) местность *ж.*
A الأرض الوُسْطى *f*

2089

(parade) ground
F terrain de manoeuvres
m.
E plaza de armas *f.*
D Paradeplatz *m.;*
Exerzierplatz *m.*
R плац-парад *м.;*
учебный плац *м.*
A ساحة العرض

2090 [E,S]

(proving) ground
F terrain d'essai *m.*
E zona de pruebas *f.*
D Versuchsgelände *neut.*
R испытательный
полигон *м.*
A ميدان الاختبار *m*

2091

(rough) ground
F terrain accidenté *m.*
E terreno escabroso *m.*
D (unebenes) Gelände
neut.
R (пересечённая)
местность *ж.*
A أرْض وَعْرَة *f*

2092

(soft) ground
F terre molle *f.*
E terreno blando *m.*
D (weicher) Boden *neut.*
R (слабый) грунт *м.*
A أرْض رَخْوة

2093

(trafficable) ground
F terrain practicable *m.*
E terreno transitable por
vehículos *m.*
D befahrbares Gelände
neut.
R (проходимая)
местность *ж.*
A أرض صالحة للعربات

129

2094

ground alert
F alerte au sol *f.*
E alerta en tierra *f.*
D (Alarm-)Bereitschaft am Boden *f.*
R наземная тревога *ж.*
A إنْذَار أَرْضِي *m*

2095

ground burst
F explosion au sol *f.*
E explosión en superficie *f.*
D Bodensprengpunkt *m.*
R наземный взрыв *м.*
A إنْفِجَار أَرْضِي *m*

2096

ground clearance
F garde au sol *f.*
E altura sobre el suelo *f.*
D Bodenfreiheit *m.*
R просвет *м.*; клиренс *м.*
A الفاصِل بَيْن المَرْكَبَة والأَرْض

2097

ground control
F contrôle du sol *m.*
E control desde tierra *m.*
D Flugüberwachung (vom Boden) *f.*
R наземное управление *ср.*
A مُرَاقَبَة أَرْضِيَّة *f*

2098

ground-controlled approach (GCA)
F approche contrôlée du sol *f.*
E aproximación controlada desde tierra *f.*
D bodenseitig geleiteter Anflug *m.*
R наземное управление посадкой *ср.*
A الإقْتِرَاب المُتَحَكَّم فيه مِن الأَرْض

2099

ground-controlled interception
F interception contrôlée du sol *f.*
E interceptación dirigida desde tierra *f.*
D bodengeleitetes Abfangen *neut.*
R наземное наведение истребителей-перехватчиков *ср.*
A الإعْتِرَاض المُتَحَكَّم فيه مِن الأَرْض

2100

ground crew
F personnel au sol *m.*
E personal de tierra *m.*
D Bodenpersonal *neut.*
R (наземный) обслуживающий экипаж *м.*
A طَاقِم أرضي

2101

ground effect machine (GEM)
F aéroglisseur *m.*; appareil à effet de sol *m.*
E aerodeslizador *m.*
D Luftkissenfahrzeug *neut.*
R машина на воздушной подушке *ж.*
A مَكَنَة تَسْتَعْمِل تَأْثِير الأَرْض

2102

(nominal) ground pressure
F pression au sol *f.*
E presión en tierra *f.*
D Bodendruck *m.*
R (номинальное) наземное давление *ср.*
A الضَّغْط الأرضي

2103

ground run
F faire tourner au sol
E (hacer) pruebas de tierra *f. pl.*
D Bodenprobe machen
R проверять на земле
A تَجْرِبَة على الأَرْض *f*

2104

ground to air
F sol-air
E tierra-aire *inv.*
D Boden-Luft- *komp.*
R земля-воздух
A أرض - جو

2105

ground to ground
F sol-sol
E tierra-tierra *inv.*
D Boden-Boden- *komp.*
R земля-земля
A أرض - أرض

2106

group *n.*
F groupe *m.*; groupement *m.*
E grupo *m.*
D Gruppe *f.*
R группа *ж.*
A جَمَاعَة *m*

2107

group *v.*
F (se) grouper
E agrupar(se)
D gruppieren
R группировать; собираться
A جَمَعَ

2108

group formation
F groupement *m.*
E grupo *m.*
D Verband *m.*
R соединение *ср.*
A تَشْكِيل جَمَاعِي *m*

2109 [M]
growler
F banquise *f.*
E iceberg que amenaza
   los buques *m.*
D „Growler" (eng.
   Begriff) *neut.*
R небольшой айсберг *м.*
A جبل ثلج عائم *m*

2110
G suit
F combinaison anti G *f.*
E traje anti G *m.*
D Anti-G-Anzug *m.;*
   Pilotendruckanzug *m.*
R противоперегрузочный
   комбинезон *м.;*
   костюм *м.*
A لِبَاس وَاقٍ لِلطَّيَّارِين *m*

2111
guard *n.*
F garde *f.*
E guardia *f. m.*
D Wache *f.*
R караул *м.*
A حَرَس *m*

2112
guard *v.*
F garder
E guardar
D überwachen *untr.*
R стоять на часах;
   сторожить
A حَرَس

2113
(demolition) guard
F détachement de
   protection d'un
   dispositif de
   destruction *m.*
E guardia de demolición
   f.
D Sprengkommandowache
   f.
R часовой отряд
   подрывных зарядов
   м.
A حرس التدمير *m*

2114
guard duty
F (être de) faction
E guardia *f.*
D Wachdienst *m.*
R караульная служба *ж.*
A خدمة الحراسة *f*

2115
guard house
F corps de garde *m.*
E cuartel de la guardia *m.*
D Wachstube *f.*
R караульное
   помещение *ср.*
A مخفر الحراس *m*

2116
guard of honor
F garde d'honneur *f.*
E guardia de honor *m.*
D Ehrenwache *f.*
R почётный караул *м.*
A حرس الشرف *m*

2117
guardship
F bâtiment de protection
   et d'escorte *m.*
E buque de guardia *m.;*
   navio de ronda *m.*
D Wachschiff *neut.*
R стационер *м.;*
   сторожевой
   (корабль) *м.*
A سَفِينَة حِرَاسَة *f*

2118
guerrilla
F partisan *m.*
E guerrillero *m.*
D Partisan *m.*
R партизан *м.*
A عِصابي *m*

2119
guesstimate
F calcul au pifomètre *m.*
E estimación aproximada
   f.
D Schätzung *f.*
R догадка *ж.*
A التقدير بلا احصاءات

2120
guidance
F guidage *m.*
E dirección *f.*
D Fernlenkung *f.*
R управление *ср.*
A تَوْجِيه *m*

2121 [S]
(command) guidance
F guidage *m.*
E dirección *f.*
D Lenkung *f.*
R управление *ср.;*
   наведение *ср.*
A نظام التوجيه

2122
guide *v.*
F guider
E dirigir
D führen
R вести; направлять;
   наводить
A دَلَّ

2123
guide *v.*
F guider
E guiar
D führen; lenken
R управлять; наводить
A وَجَّه

131

2124
guide *n.*
F guide *m.*
E guía *f.*
D Führer *m.*
R проводник *м.*; гид *м.*
A دَلِيل *m*

2125
guidon
F guidon *m.*
E banderola *f.*
D Standarte *f.*
R (остроконечный) флажок *м.*
A راية صغيرة

2126
gulf
F golfe *m.*
E golfo *m.*
D Golf *m.*
R залив *м.*
A خَلِيج *m*

2127 [T]
gully
F goulet *m.*
E barranco *m.*
D Rinne *f.*
R овраг *м.*; лощина *ж.*
A وهَدة

2128
gun
F canon *m.*
E cañón *m.*
D Kanone *f.*
R орудие *ср.*
A مِدْفَع *m*

2129
gunboat
F canonnière *f.*
E cañonero *m.;* lancha cañonera *f.*
D Kanonenboot *m.*
R канонерская лодка *ж.*
A زَوْرَق مُسَلَّح *m*

2130
gun carriage
F affût de canon *m.*
E cureña *f.*
D Lafette *f.*
R лафет *м.*; орудийный станок *м.*
A حَاضِن المِدْفَع *m*

2131
gunfire
F fusillade *f.;* feu d'artillerie *m.*
E cañoneo *m.*
D Geschützfeuer *neut.*
R орудийный огонь *m.*
A رَمْي المِدْفَع *m*

2132
gunner
F artilleur *m.;* canonnier *m.;* mitrailleur *m.*
E ametrallador *m.;* apuntador *m.;* cañonero *m.*
D Bordschütze *m.;* Richtschütze *m.*
R артиллерист *м.*; пулемётчик *м.*
A مِدْفَعِي *m*

2133
gunnery
F tir au canon *m.*
E tiro *m.*
D Schiesswesen *neut.*
R артиллерийское дело *ср.*
A عِلم المِدْفَعِية *m*

2134
(at) gun point
F (à la) force des armes
E (a) punta de pistola; forzado
D (mit) vorgehaltener Pistole
R силой оружия
A شهّر مسدس على

2135
gunrunner
F trafiquant d'armes *m.*
E contrabandista de armas *m.*
D Waffenschmuggler *m.*
R контрабандист по оружию *м.*
A مُهرِّب أسلحة

2136
gunship
F hélicoptère armé *m.*
E helicóptero armado (de ametralladoras y misiles) *m.*
D „Gunship" (eng. Begriff) *m.;* bewaffneter Hubschrauber *m.*
R боевой вертолёт *м.*
A طائرة عمودية مسلحة

2137
gunshot
F portée de fusil *f.*
E alcance *m.*
D Schussweite *f.*
R дальность выстрела *ж.*
A مدى المدفع

2138
gymnasium
F gymnase *m.*
E gimnasio *m.*
D Sporthalle *f.*
R спортзал *м.*
A ملعب الرياضة

2139
gym shoe
F chaussures de tennis *f. pl.*
E zapatillas de goma *f. pl.*
D Sport-Turnschuhe *m. pl.*
R спортивная обувь *ж.*
A حِذاء للرياضة

2140
gym suit
F survêtement m.
E traje deportivo m.
D Trainingsanzug m.
R спортивный костюм
м.
A لباس للرياضة

2141
gyrocompass
F gyro-compas m.
E girocompás m.
D Kreiselkompass m.
R гирокомпас м.
A بُوصَلة كَهْرَبَائِيَّة f

2142
gyromagnetic compass
F compas gyromagnétique
m.
E compás giromagnético
m.
D Kreiselmagnetkompass
m.
R гиромагнитный
компас м.
A بُوصَلة مَغْنَطِيسِية
جِيرُوسْكُوبية f

2143
gyro preset
F orientation initiale du
gyro f.
E ajuste del giro m.
D Kreiselgrundeinstellung
f.
R первое
ориентирование
гироскопа ср.
A سبق ضبط البوصلة
الكهربائية

2144
gyroscope
F gyroscope m.
E giróscopo m.
D Kreisel m.
R гироскоп м.
A جِيرُوسْكُوب m

2145
gyroscopic
F gyroscopique
E giroscópico
D Kreisel- komp.
R гироскопический,
A جِيرُوسْكُوبي

2146
gyrostabilizer
F gyro-stabilisateur m.
E establizador giroscópico
m.
D Kreiselstabilisierung f.
R гиростабилизатор м.
A مُقِرّ جِيرُوسْكُوبِي m

H

2147
habitable
F habitable
E habitable
D bewohnbar
R обитаемый
A صالح للسكن

2148
habitual association
F association habituelle f.
E agregación habitual f.
D ständige
Zusammenarbeit f.
R обычная близость ж.
A زمالة معتادة

2149
hachuring
F hachures f. pl.
E área surcada f.
D Schraffierung f.
R штриховка ж.
A تظليل

2150 [S]
half-loaded
F demi-approvisionné
E semicargada
D teilgeladen
R полузаряженный,
A نِصْف مُحَمَّل

2151
(at) half mast adj.
F (en) berne
E (a) media asta
D (auf) Halbmast
R приспущенный
A مُنَكَّس

2152 [M]
(at) half speed adj.
F (à) demi-vitesse
E (a) media máquina
D (mit) halbe(r) Kraft adj.
R (на) полскорости
A نِصْف السُّرْعَة

2153 [N]
**half thickness; half
value layer
(HVL)**
F demi-épaisseur f.
E semigrosor m.
D Halbwertschicht (HWS)
f.
R толщина защитного
слоя ж.
A نِصْف تَخَانَة

2154
half tide
F mi-marée f.
E marea media f.
D Gezeitenmitte f.
R половина прилива ж.
A نِصْف المَدّ

2155 [V]
half track adj.
F semi-chenillé
E semioruga
D Halbketten- komp.
R полугусеничный
A نِصْف شَاحِنة

2156
halt v.
F (s')arrêter
E parar(se)
D (an-)halten (lassen)
R останавливать(ся)
A تَوَقَّف

133

2157
"Halt!"
F "Halte!
E "¡Alto!"
D „Halt!"
R "стой!"
A قِف

2158 [L,V]
halts
F stationnements *m. pl.*
E altos *m. pl.*
D Etappen *f.*
R привалы *м. мн. ч.*
A توقفات *f*

2159
hand *n.*
F main *f.*
E mano *f.*
D Hand *f.*
R рука *ж.*
A يَد *f*

2160 [P]
handcuff *v.*
F (passer les) menottes (à) *f. pl.*
E maniatar
D Handschellen anlegen
R надевать наручники
A أصفاد *f*

2161 [S]
hand-hold
F (à) main
E (de) mano
D Hand- *komp.*
R ручной; переносный
A يَدَوِي . بالأيْدِي

2162 [A,M,V]
handle *v.*
F conduire; manœuvrer
E conducir; gobernar
D führen
R управлять; вести
A إسْتَعْمَل

2163 [U]
handle *v.*
F manier; manutentionner
E manejar
D behandeln; verladen *untr.*
R держать в руках; перекладывать
A عَالِج

2164
handle *n.*
F poignée *f.*
E asa *f.;* mango *m.;* palanca *f.*
D Griff *m.*
R ручка *ж.*
A مِقْبَض *m*

2165
(cocking) handle
F levier d'armement *m.*
E palanca (de montar) *f.*
D Gewehr-Spannhahn *m.*
R курок *м.*
A مقبض

2166 [U]
handle (to manage)
F (s')occuper de
E encargarse (de)
D (Buch)führen; leiten
R регулировать
A ادار

2167 [A,M,V]
handle (to manipulate)
F manœuvrer
E manejar
D umgehen mit *tr.*
R управлять; руководить
A قاد

2168 [U]
handle (to perform)
F manœuvrer
E comportarse
D Leistungen nachkommen *tr.*
R вести себя; действовать
A عمل

2169
handlebar
F guidon *m.*
E manillar *m.*
D Lenkstange *f.*
R руль *м.*
A مِقْوَد *m*

2170
(weapon) handling
F maniement d'armes *m.*
E manejo *m.*
D Handhabung einer Waffe *f.*
R обращение *ср.*
A (مهارة) إستعمال الأسلحة

2171 [L]
hand-off
F transmission de responsabilité *f.*
E entrega *f.*
D Unterteilung der Verantwortung *f.*
R передача задач *ж.*
A رفع المسؤولية

2172
hand over command to
F remettre le commandement à
E entregar el mando a
D (den) Befehl übergeben an *untr.*
R передавать командование
A تسليم القيادة

2173

hand-over-hand *adj.*

F main sur main
E mano sobre mano
D Hand über Hand
R рука за рукой; с руки
на руку
A تَسَلُّق الحِبَال

2174 [R]

handset

F combiné radio *m.*
E aparato *m.*
D Hörer *m.*
R ручной телефон *м.*
A جهاز يدوي *m*

2175 [S]

handspike

F levier de pointage *m.*
E palanca (para mover un obús) *f.*
D Hebestange *f.*
R правило *cp.*
A عتلة التسديد *f*

2176

hand to hand fight

F combat corps à corps *m.*
E combate cuerpo a cuerpo *m.*
D Nahkampf *m.*
R рукопашный бой *м.*
A القتال يداً بيد

2177 [S]

(elevating) handwheel

F volant de pointage en hauteur *m.*
E volante de puntería (en elevación) *m.*
D Höhenstellrad *neut.*
R маховик подъёмного механизма *м.*
A مدوّرة يدوية للأرتفاع *f*

2178 [S]

(traverse) handwheel

F volant de pointage en direction *m.*
E volante de puntería (en dirección) *m.*
D Schwenkstellrad *neut.*
R маховик поворотного механизма *м.*
A مدورة يدوية للتسديد *f*

2179

hangar

F hangar *m.*
E hangar *m.*
D Flugzeughalle *f.*
R ангар *м.*
A حَظِيرة *f*

2180

hangfire

F long feu *m.*
E (tiro a) inflamación retardada *f.*
D Nachbrenner *m.;* Nachzündung *f.*
R затяжной выстрел *м.*; осечка *ж.*
A تَأْخِير الرَّمْي *m*

2181

harass

F harceler
E hostilizar
D stören
R беспокоить; изводить
A أَزْعَج

2182 [C]

harassing agent

F toxique irritant *m.*
E agresivo de hostigamiento *m.*
D Reizwirkstoff *neut.*
R раздражающее ОВ *cp.*
A عامل مهيج *m*

2183

(electronic) harassment

F brouillage intensif *m.*
E interferencia *f.*
D Störung *f.*
R изматывающие помехи *ж. мн. ч.*
A تشويش شديد *m*

2184

harbor *n.*

F port *m.*
E puerto *m.*
D Hafen *m.*
R гавань *ж.*; порт *м.*
A مَرْفَأ *m*

2185

harbor master

F capitaine de port *m.*
E capitán de puerto *m.*
D Hafenmeister *m.*
R начальник порта *м.*
A قَائِد المَرْفَأ *m*

2186 [M]

hard

F cale d'embarquement *f.*
E muelle rígido *m.*
D Landestelle *f.*
R брод *м.*; причал *м.*
A رَصِيف الإِبْحَار *m*

2187

hardstand

F aire de stationnement/de stockage *f.*
E firme de estaciamento *m.*
D (befestigter) Stellplatz *m.*
R твёрдопокрытая площадка *ж.*
A رَصِيف *m*

135

2188

**(computer etc.)
hardware**
F matériel *m.*
E equipos *m. pl.*
D Werkstoff *m.*
R оборудование
  компьютера *ср.*
A الأدوات المعدنية

2189

harness *v.*
F exploiter
E utilizar
D nutzbar machen
R использовать
A ربط الأحزمة

2190 [E,R]

harness *n.*
F unité collective (UC) *f.*
E atalaje *m.*
D Sprechgeschirr *neut.*
R такелаж *м.*
A أحْزِمة *f*

2191 [A,V]

harness *n.*
F harnais *m.*
E atalaje *m.*
D Gurtzeug *neut.*
R снаряжение *ср.*
A عُدّة الوَصْل *f*

2192

harpoon *n.*
F harpon *m.*
E arpón *m.*
D Harpune *f.*
R гарпун *м.*
A رُمح للصَّيْد *m*

2193 [L]

hasty attack
F attaque accélérée *f.*
E ataque precipitado *m.*
D Schnellangriff *m.*
R стремительное
  нападение *ср.*
A هجوم عاجل

136

2194

hasty breaching
F déminage rapide *m.*
E levantamiento
  improvisado (de
  minas) *m.*
D Überwinden eines
  Minenfelds ohne
  grössere
  Vorbereitungen *neut.*
R проход с ходу *м.;*
  поспешный проход
A فَتْح ثُغْرَة بِسُرعة

2195

hasty crossing
F franchissement dans la
  foulée *m.*
E paso improvisado *m.*
D Übergang ohne
  grössere
  Vorbereitungen *m.*
R форсирование с ходу
  ср.
A عُبُور سَرِيع

2196

hasty defense
F défense improvisée *f.*
E defensa improvisada *f.*
D Verteidigung ohne
  grössere
  Vorbereitungen *f.*
R поспешная оборона *ж.*
A دِفاع عَاجِل

2197

hasty demolition
F démolition rapide *f.*
E destrucción
  improvisada *f.*
D Schnellsprengung *f.*
R поспешнопроведённое
  разрушение *ср.*
A تَدْمِير عَاجِل

2198 [M,V]

hatch
F écoutille *f.*
E escotilla *f.*
D Luke *f.*
R люк *м.*
A كُوّة *f*

2199

haul *v.*
F haler
E jalar
D holen (ein-, auf-, ab-)
R тащить; тянуть
A جَـرَّ

2200 [A]

(short) haul
F (avion) de court rayon
  d'action
E (de) autonomía
  limitada *f.*
D Kurzstreckenflugzeug
  neut.
R (самолёт) ближнего
  действия
A (طائرة) قصيرة المدى

2201

haulage
F transport routier *m.*
E transporte *m.*
D Beförderung *f.;*
  Transport *m.*
R перевозка *ж.*
A شحنة منقولة

2202

haversack
F sac de fantassin *m.*
E mochila *f.*
D Provianttasche *f.*
R заплечный ранец *м.*
A مِزْودة *f*

2203
hawser
F haussière *f.*
E guindaleza *f.*
D Trosse *f.*
R трос *м.*
A قَلْس *m* حبل أو سلك
سميك *m*

2204
hazard *n.*
F danger *m.*
E riesgo *m.*
D Gefahr *f.*
R риск *м.*; опасность *ж.*
A خَطَر *m*

2205
haze
F brume *f.*
E calina *f.*
D Dunst *m.*
R лёгкий туман *м.*; мгла *ж.*
A ضَبَاب خَفِيف *m*

2206
H Bomb
F bombe-H *f.*
E bomba-H *f.*
D Wasserstoffbombe (H-Bombe) *f.*
R водородная бомба *ж.*
A قنبلة نَوَوِيّة *f*

2207 [G,S]
head
F tête *f.*
E ojiva *f.*
D Kopf *m.;* Spitze *f.*
R боеголовка *ж.*
A رَأْس *m*

2208 [U]
head *n.*
F chef *m.*
E jefe *m.*
D Chef *m.*
R начальник *м.*; глава *ж.*
A رَأْس *m*

2209 [H]
head *n.*
F tête *f.*
E cabeza *m.*
D Kopf *m.*
R голова *ж.*
A رأس

2210 [M]
head for
F mettre le cap (sur)
E arrumbar (a)
D ansteuern (nach) *tr.*
R направлять(ся) (к); держать курс (на)
A إتَّجَه إلى

2211
heading
F cap *m.*
E rumbo *m.*
D Steuerkurs *m.*
R направление *ср.*; курс *м.*
A إتِّجَاه *m*

2212
headland
F promontoire *m.*
E promontorio *m.*
D Kap *neut.*
R мыс *м.*; нос *м.*
A رَأْس *m*

2213
headlight
F phare *m.*
E faro *m.*
D Scheinwerfer *m.*
R фара *ж.*; головной фонарь *м.*
A مِصْبَاح أَمَامِي *m*

2214
headquarters
F quartier général *m.*
E cuartel general *m.*
D Hauptquartier *neut.*
R штаб *м.*
A مَقَرّ *m* ـ قِيادات *f*

2215 [M]
heads
F chiotte *f.*
E lávabos *m. pl.*
D Klosett *neut.;* Klo
R уборная *ж.*
A مرحاض

2216
headset
F casque (à écouteurs) *m.*
E auriculares *m. pl.*
D Kopfhörer *m.*
R головной телефон *м.*
A سَمَّاعَات الرَّأْس *f*

2217
head-up display
F dispositif tête haute *m.*
E información que se presenta delante del piloto (término inglés) *f.*
D (in) Frontscheibe eingespiegelte Informationen (meist eng. Begriff)
R прямое визуальное изображение *ср.*
A لَوْحَة عَرْض في مُسْتَوَى رَأْس الطَّيَار *f*

2218 [T]
headwater
F sources *f. pl.*
E cabecera *f.*
D Oberlauf *m.*
R верхний горизонт воды *м.*
A منابع النهر

2219
headway
F marche (en avant) *f.*
E marcha (avante) *f.*
D Fahrt (voraus) *f.*
R поступательное движение *ср.*
A تَقَدُّم *m*

2220

heat *n.*
F chaleur *f.*
E calor *m.*
D Wärme *f.*
R жара *ж.*; жар *м.*
A حَرَارة *f*

2221

heat (up) *v.*
F chauffer
E calentar
D erhitzen
R нагревать(ся);
   разогревать
A سَخَّن

2222

heat exhaustion
F épuisement dû à la
   chaleur *m.*
E golpe de calor *m.*
D Hitzerschöpfung *f.*
R тепловое истощение
   ср.
A وهن بسبب الحرارة

2223 [G]

heat-seeking
F thermoguidé
E infrarrojo
D wärmesuchend
R теплосамоуправляемый
A موجه بحرارة الهدف

2224 [M]

heave *v.*
F lancer
E jalar
D hieven
R тянуть; поднимать
A جَرَّ

2225 [M]

heave to
F (se) mettre en panne
E ponerse al pairo
D beidrehen
R лечь в дрейф
A توقف

2226

heavy
F lourd
E pesado
D schwer
R тяжёлый
A ثَقيل

2227 [L]

heavy (armor)
F groupement blindé *m.*
E preponderante (en
   carros)
D panzerstark
R имеющий перевес
   брони
A ثقيل (الدروع)

2228

heavy artillery
F artillerie lourde *f.*
E artillería pesada *f.*
D schwere Artillerie *f.*
R тяжёлая артиллерия
   ж.
A المِدْفَعِيَّة الثَّقيلة *f*

2229

heavy duty *adj.*
F (à) haute résistance *f.*
E (de) servicio pesado
D Hochleistungs- *komp.*
R тяжёлого типа
A عَمَل شَاقّ

2230 [L]

heavy (infantry)
F groupement mécanisé
   m.
E preponderante (en
   infantería)
D infanteriestark
R имеющий перевес
   пехоты
A ثقيل (المشاة)

2231

heavy sea
F grosse mer *f.*
E ola grande *f.*
D schwerer Seegang *m.*
R бурное море *ср.*
A بَحْر شَديد الإضْطِرَاب

2232

heavy seas *pl.*
F mer houleuse *f.*
E oleada *f.*
D schwere See *f.*
R сильное волнение *ср.*
A بِحَار شَديدة الإضْطِرَاب

2233

heavy weather
F gros temps *m.*
E tiempo pesado *m.*
D schweres Wetter *neut.*
R бурная погода *ж.*
A جَوّ عَاصِف

2234 [T]

hedge
F haie *f.*
E seto vivo *m.*
D Hecke *f.*
R изгородь *ж.*
A سياج

2235

hedge-hop
F voler à rase-mottes
E volar a ras de tierra
D heckenhüpfen
R проводить бреющий
   полёт
A طار على وجه الأرض

2236

height
F hauteur *f.*
E altura *f.*
D Höhe *f.*
R высота *ж.*
A إرْتِفَاع *m*

2237 [N]

(cloud top) height

F altitude maximale de
   nuage f.
E altura máxima de nube
   nuclear f.
D Wolkendeckehöhe f.
R максимальная высота
   облачности ж.
A ارتفاع الغيوم الأقصى

2238 [H,E]

(drop) height

F hauteur de largage f.
E altura de lanzamiento f.
D Absprunghöhe [H] f.;
   Abwurfhöhe [E] f.
R высота сбрасывания
   ж.
A اِرْتِفَاع الاسْقَاط m

2239

**(parachute
deployment)
height**

F hauteur d'ouverture de
   parachute f.
E altura de abertura de
   paracaídas f.
D Fallschirmentfaltungs-
   höhe f.
R высота раскрытия
   парашюта ж.
A اِرْتِفَاع إنْفِتَاح المَظَلَّة m

2240 [N]

(safeburst) height

F hauteur de sécurité de
   retombée f.
E altura de explosión
   libre f.
D Sicherheitshöhe f.
R безопасная высота
   взрыва ж.
A علو الانفجار المأمون

2241

heighten

F (se) rehausser
E elevar(se)
D (sich) erhöhen
R повышать(ся);
   усиливать(ся)
A رَفَعَ

2242 [N]

**(optimum) height of
burst**

F hauteur optimum
   d'explosion f.
E altura óptima de
   explosión f.
D optimale
   Explosionshöhe f.
R оптическая высота
   взрыва ж.
A علو الانفجار الأفضل

2243 [N]

**(scaled) height of
burst**

F hauteur d'explosion
   convertie f.
E altura de explosión f.
D eingestufte
   Explosionshöhe f.
R высота взрыва по
   шкале ж.
A علو الانفجار المرسوم

2244

heliborne

F héliporté
E transportado en
   helicóptero
D (per) Hubschrauber
   befördert
R вертолётно-десантный
A منقول في طائرة عمودية

2245

helicopter

F hélicoptère m.
E helicóptero m.
D Hubschrauber m.
R вертолёт м.
A طَائِرَة عَمُودِيَّة f

2246

helilift

F hélitransport m.
E transportado en
   helicóptero
D Hubschrauberaufgabe f.
R перевозка вертолётом
   ж.
A انتقال بطائرة عمودية

2247

helipad

F plateforme de poser
   (d'hélicoptères) f.
E helipuerto m.
D Hubschrauberlandeplatz
   m.
R вертолётная площадка
   ж.
A مَهْبَط طَائِرَات عَمُودِيَّة m

2248

helitransport

F hélitransport m.
E transporte en
   helicóptero
D Hubschraubertransport
   m.
R вертолётный
   транспорт м.
A النقل بطائرات عمودية

2249

helium

F hélium m.
E helio m.
D Helium neut.
R гелий м.
A هِلْيُوم m

2250

helm n.

F barre f.
E timón m.
D Steuer neut.
R румпель м.
A سَاعِد سُكَّان ـ مِقْوَد الدَّفَة m

2251

helmet

F casque *m.*
E casco *m.*
D Helm *m.*
R шлем *м.*
A خُوذَة *f*

2252

helmet-liner

F sous-casque *m.*
E (a) harnés del casco *m.*
D Helmeinsatzstück *neut.*
R подкладка шлема *f.*
A بطانة الخوذة

2253

helmsman

F homme de barre *m.*
E timonel *m.*
D Steuermann *m.*
R штурман *м.*
A مُدِير السُّكَّان ـ مُوَجِّه الدَّفَّة *m*

2254

hermetically

F hermétiquement
E herméticamente
D hermetisch; luftdicht
R герметически-
A باحكام

2255

H-hour

F heure H *f.*
E hora H *f.*
D Angriffsbeginn *m.;* Stunde H *f.*
R час-Ч *м.*
A سَاعَة الصِّفْر *f*

2256 [P]

hide *n.*

F cache *f.*
E cuero *m.*
D Versteck *neut.*
R укрытие *ср.*
A مُخْبَأ *m*

2257

hide *v.*

F (se) cacher
E esconder(se)
D (sich) verbergen
R прятать(ся); скрывать(ся)
A أخْفَى

2258 [L]

hide *n.*

F position camouflée *f.*
E posición camuflada *f.*
D Tarnstellung *f.*
R место укрытия *ср.*
A مُخْبَأ

2259

high

F haut
E alto
D hoch *adj.;* Hoch- *komp*
R высокий
A عَالٍ

2260

high and dry

F (à) sec
E (en) seco
D gestrandet
R (на) суше; (на) берегу
A جَافّ

2261

high command

F haut commandement *m.*
E alto mando *m.*
D Oberkommando *neut.*
R верховное командование *ср.*
A القيادة العليا

2262

high explosive *adj.*

F explosif
E explosivo
D Spreng- *komp.*
R бризантный
A شَدِيد الإنْفِجَار

2263

high explosive *n.*

F explosif puissant *m.*
E alto explosivo *m.*
D hochexplosiver Sprengstoff
R бризантное взрывчатое вещество (ВВ) *ср.*
A مُتَفَجِّرَات شَدِيدَة الإنْفِجَار *f*

2264

high explosive antitank (HEAT) *adj.*

F explosif anti-chars
E (a) carga hueca
D Hohlladungs- *komp.*
R кумулятивный противотанковый
A شَدِيد الإنْفِجَار ضِدّ الدَّبَّابَات

2265

high explosive plastic (HEP)

F plastic *m.*
E plástico explosivo (término inglés) *m.*
D Quetschkopf- *komp.*
R пластичное бризантное ВВ *ср.*
A بَلَاسْتِيك شَدِيد الإنْفِجَار

2266

high level

F (à) très haut niveau
E (de) alto nivel
D (auf) hohem Niveau
R высокостепенный
A مستوى عال

2267 [A]

high octane

F (d')octane élevé
E (de) alto octanaje
D oktanreich *adj.*
R высокооктановый
A أوكتين عال

2268

high seas *adj. n.*
F (de) haute mer *(adj.) n. f.*
E (de) alta mar *(adj.) s. f.*
D Hochsee- *komp.;* hohe
   See *f.*
R открытое море *ср.*
A عُرْض البَحْر *m*

2269

high speed *adj.*
F (à) grande vitesse
E (de) alta velocidad
D Schnell- *komp.*
R быстроходный
A عَالِي السُّرْعَة

2270

high tension
F (à) haute tension
E (de) alta tensión
D Hochspannung *f.*
R высоковольтный
A ضغط عال

2271

high water
F marée haute *f.*
E pleamar *f.*
D Hochwasser *neut.*
R полная вода *ж.*
A أقْصَى المَدّ

2272

highway
F grande route *f.*
E carretera *f.*
D Hauptstrasse *f.*
R шоссе *ср.;* автострада
   ж.
A طَرِيق عَامّ *m*

2273

hijack
F détourner (par la force)
E secuestrar
D entführen
R угонять
A اختطف

2274

hijack
F détournement *m.*
E secuestro *m.*
D Entführung *f.*
R угон *м.*
A اختطاف

2275

hill
F colline *f.*
E colina *f.*
D Hügel *m.*
R холм *м.;* высота *ж.*
A تَلّ *m*

2276

hilly country
F terrain accidenté *m.*
E terreno accidentado *m.*
D hügeliges Gelände *neut.*
R холмистая местность
   ж.
A أرْض جَبَلِيَّة *f*

2277

hinterland
F hinterland *m.*
E traspaís *m.*
D Hinterland *m.*
R глубокий тыл *м.*
A خَلْف النهر والبحر ـ ظَهِير *m*

2278

hit *v.*
F frapper
E alcanzar
D treffen
R ударять; попадать
A أصاب *m*

2279

(direct) hit
F coup au but *m.*
E impacto directo *m.*
D Volltreffer *m.*
R прямое попадание *ср.*
A إصابة مباشرة *f*

2280 [S]

(over) hit *n.*
F coup long *m.*
E impacto largo *m.*
D Hochschuss *m.*
R перелёт *м.*
A إصابة مجاوزة *f*

2281 [S]

(short) hit *n.*
F coup court *m.*
E impacto corto *m.*
D Tiefschuss *m.*
R недолёт *м.*
A إصابة قاصرة *f*

2282

hit and run tactics
F tactique de harcèlement
   f.
E guerrilla *f.*
D Strippangriffstaktik *f.*
R внезапная тактика *ж.*
A إسلوب قتال المراوغة

2283 [B,E,M]

hitch *n.*
F nœud *m.*
E cote *m.*
D Stich *m.*
R узел *м.*
A عُقدَة *f*

2284 [M]

hoist *v.*
F hisser
E izar
D hissen
R поднимать; тянуть
A رَفَع

2285 [B,E]

hoist *v.*
F lever
E alzar
D hochziehen *tr.*
R поднимать
A رَفَع

141

2286
hoist *n.*
F appareil de levage *m.*
E montacargas *m. inv.*
D Hebewerk *m.*
R подъём *м.*; лебедка
ж.
A مِرْفَاع *m*

2287 [M]
hold *n.*
F cale *f.*
E bodega *f.*
D Laderaum *m.*
R трюм *м.*
A مُسْتَوْدَع *m*

2288 [A]
hold *v.*
F maintenir en attente
E detener
D halten
R держать
A إِحْتَفَظ بِمَوْضِع

2289
(weapons) hold
F tir prescrit *m.*
E prohibición del uso de
armas contra aviones
f.
D Benutzungsart *f.*
R запрет *м.*
A قبض على

2290
hold fire
F tir interdit!
E alto el fuego
D Feuer einstellen;
stopfen
R прекращать огонь;
воздерживаться от
огня
A كَفّ عن الرَّمْي

142

2291
hold firm
F tenir ferme
E mantenerse firme
D festhalten *tr.*
R стойко держаться;
обороняться
A قَاوَم

2292
hold ground
F occuper le terrain
E conservar (una
posición)
D (eine Stellung) halten
R владеть позицией;
занимать место;
занимать позицию
A تَشَبَّث بِالأَرْض

2293
holding anchorage
F mouillage d'attente *m.*
E fondeo de espera *m.*
D (Versammlungs-)Reede *f.*
R выжидательная
стоянка *ж.*
A مَرْسَى تَجَمُّع *m*

2294
holding area
F zone d'attente *f.*
E zona de espera *f.*
D Bereitschaftsgebiet *neut.*
R район ожидания *м.*
A منطقة الحشد *f*

2295
holding attack
F action de fixation *f.*
E ataque de detención *m.*
D Fesselungsangriff *m.*
R сдерживающая атака
ж.
A هُجُوم لِتَثْبِيت العَدُو *m*

2296
hold off
F tenir à distance
E contener; rechazar
D abwehren *tr.*
R отражать (атаку)
A صَدّ

2297
hold out (against)
F tenir bon (devant)
E resistir
D aushalten (gegen) *tr.*
R выстаивать;
выдерживать
A قَاوَم

2298
hole
F trou *m.*
E agujero *m.*
D Loch *neut.*
R дыра *ж.*; отверстие
ср.
A ثُقْب *m*

2299 [T]
hollow
F dépression *f.*
E hoyo *m.*
D Vertiefung *f.*
R лощина *ж.*
A جَوْف

2300 [S]
hollow base
F culot creux *m.*
E base hueca *f.*
D Hohllage *f.*
R пустотелое дно *ср.*
A قاعدة جوفاء

2301
hollow charge
F charge creuse *f.*
E carga hueca *f.*
D Hohlladung *f.*
R кумулятивный заряд
м.
A حشوة مفرغة

2302
holster
F étui (de revolver) *m.*
E pistolera *f.*
D Pistolentasche *f.*
R кобура *ж.*
A قِرَاب ـ جِرَاب *m*

2303
home port
F port d'attache *m.*
E puerto de matrícula *m.*
D Heimatshafen *m.*
R порт приписки *м.*
A المَرْفَأ الأَساسي

2304
(radio) homing
F radio ralliement *m.*
E radiogoniómetro *m.*
D Senderanflug *m.*
R самонаведение *ср.*
A تَوْجِيه لاَسِلْكي

2305
homing guidance
F autoguidage *m.*
E guía de encuentro *f.*
D Zielpeilung *f.*
R самонаведение *ср.*
A جهاز التوجيه

2306
homogeneous
F homogène
E homogéneo
D homogen; gleichartig
R однородный;
 сплошной
A مُتَجَانِس

2307
(to render) honors
F rendre les honneurs
E dar honores a (la
 bandera)
D Ehren erweisen
R (оказывать) почести
A أَدَّى المَرَاسِم

2308 [H]
hood *n.*
F capuchon *m.*
E capucha *f.*
D Kapuze *f.*
R капюшон *м.*; капор *м.*
A كُمَّة

2309 [V]
(engine) hood
F capot *m.*
E capó *m.*
D Motorhaube *f.*
R капот (двигателя) *м.*
A غِطَاء المَحَرِّك

2310 [M,V]
(folding) hood
F capote *f.*
E capota *f.*
D Verdeck *neut.*
R (складной) капот *м.*
A غِطَاء يُطْوَى

2311
hook *v.*
F crocher
E enganchar
D anhängen *tr.*
R зацеплять; прицеплять
A عَلِقَ

2312
hook *n.*
F crochet *m.*
E gancho *m.*
D Haken *m.*
R гак *м.*; крюк *м.*
A كُلَّاب

2313
(aircraft arrester)
hook
F crosse d'appontage *f.*
E gancho de frenaje *m.*
D Hafthaken *m.*
R тормозной крюк *м.*
A كُلَّاب الإِيقَاف للطائرة

2314 [A]
(tail) hook
F crochet arrière *m.;*
 crosse d'appontage *f.*
E gancho de cola *m.*
D Heckhaken *m.*
R тормозной гак *м.;*
 тыловой гак *м.*
A خُطَاف الذَّيْل

2315
(towing) hook
F crochet d'attelage *m.*
E gancho de remolque *m.*
D Schlepphaken *m.*
R буксирный крюк *м.*
A كلاب القطر

2316 [V]
hook up *v.*
F accrocher
E enganchar
D anhaken; anhängen
R зацеплять
A عَلِقَ

2317 [R]
hookup *n.*
F conjugaison *f.*
E transmisión en circuito
 f.
D Zusammenschaltung *f.*
R установление связи *ср.*
A إِتِّصَال

2318
hopper
F trémie *f.*
E tolva *f.*
D Trichter *m.*
R воронка *ж.*; бункер *м.*
A خَزَّان ذو قاع مصرِّف

2319
horizon
F horizon *m.*
E horizonte *m.*
D Horizont *m.*
R горизонт *м.*
A أُفْق

143

2320

(apparent) horizon

F horizon apparent *m.*
E horizonte aparante *m.*
D (scheinbarer) Horizont *m.*
R гирогоризонт *м.*
A أُفْق ظاهِر

2321

(true) horizon

F horizon vrai *m.*
E horizonte auténtico *m.*
D wahrer Horizont *m.*
R горизонт *м.*
A الأفق الصَّحيح

2322

(visible) horizon

F horizon visible *m.*
E horizonte visual *m.*
D Kimm *m.*
R видимый горизонт *м.*
A أُفْق مَرْئي

2323

horizontal

F horizontal
E horizontal
D waagerecht
R горизонтальный
A أُفْقِي

2324 [S]

horizontal error

F écart probable horizontal *m.*
E error horizontal *m.*
D Breitenstreuung *f.*
R горизонтальное отклонение *ср.*
A خَطَأ أُفْقِي

2325 [V]

horn

F avertisseur *m.*
E bocina *f.*
D Hupe *f.*
R гудок *м.*
A بُوق

2326

horse

F cheval
E caballo *m.*
D Pferd *neut.*
R лощадь *ж.*; конь *м.*
A حِصَان

2327

horsepower (HP)

F cheval (-vapeur) (CV) *m.*
E caballo (de fuerza) *m.*
D Pferdestärke (PS) *f.*
R лощадиная сила *ж.*
A قُدْرة حِصَانية

2328

hose *n.*

F tuyau flexible *m.*
E manga *f.*
D Schlauch *m.*
R шланг *м.*
A خُرْطُوم *m*

2329

hose (down) *v.*

F laver (à grande eau)
E regar (con manga)
D (mit einem Schlauch) bespritzen
R поливать из шланга
A إِسْتَعْمَل خُرْطُوم

2330

(base) hospital

F hôpital de l'arrière *m.*
E hospital principal *m.*
D Standlazarett *neut.*
R госпиталь *м.*
A مستشفى القاعدة

2331

(combat support) hospital

F ambulance divisionnaire
E hospital de apoyo de campaña *m.*
D Feldlazarett *neut.*
R полевой госпиталь *м.*
A مُسْتَشْفى سَنَد القِتال *m*

2332

(field) hospital

F hôpital de campagne *m.*
E ambulancia *f.*
D Feldlazarett *neut.*
R полевой госпиталь *м.*
A مُسْتَشْفى الْمَيْدان *m*

2333

(forward) hospital

F hôpital de l'avant *m.*
E hospital de campaña *m.* *m.*
D Feldlazarett *neut.*
R (эвакуационный) госпиталь *м.*
A مستشفى امامي

2334 [A,M,L,U]

(general) hospital

F hôpital militaire médico chirurgical *m.*
E hospital (militar) general *m.*
D Lazarett [AML] *neut.;* Krankenhaus [U] *neut.*
R общий стационарный госпиталь *м.*
A مُسْتَشْفى عَام *m*

2335 [V]

hospitalize

F hospitaliser
E hospitalizar
D (ins) Lazarett einliefern
R госпитализировать
A اقام في مستشفى

2336

hostage

F otage *m.*
E rehén *m.*
D Geisel *f.*
R заложник *м.*
A رَهْن *m* ـ رهينة *f*

2337

hostile

F ennemi; hostile
E enemigo; hostil
D feindlich
R неприятельский;
  вражеский
A عُدْوَاني

2338

hostile criteria

F critères d'identification
  ennemie m.pl.
E reglamentos por
  identificar buque o
  avión enemigo m. pl.
D feindliche Merkmale
  neut. pl.
R оценка возможных
  враждебных
  действий f.
A فيصل عدائي

2339 [R]

hostile track

F piste hostile f.
E trayectoria hostil f.
D Feindsignal neut.
R отметка цели ж.
A تَتَبُّع طَائِرة مُعَادِيَّة

2340

hostilities

F hostilités f. pl.
E hostilidades f. pl.
D Feindseligkeiten f. pl.
R боевые действия ср.
  мн. ч.; военные
  действия ср. мн. ч.
A أعْمَال عُدْوَانِيَّة f

2341 [R]

hotline

F téléphone rouge m.
E enlace directo m.
D heisser Draht m.
R горячая линия ж.
A نظام اتصال سريع

2342

hot pursuit

F poursuite de près f.
E seguimiento inmediato
  m.
D dichte Verfolgung f.
R неотступное
  преследование ср.
A مُلاَحَقة شَدِيدة f

2343 [E]

housing

F carter m.; enveloppe f.
E caja f.
D Gehäuse neut.
R корпус м.; кожух м.
A غِلاَف m حَوْض m

2344

hover v.

F faire du vol stationnaire
E cernerse; revolotear
D Schwebeflug m.
R парить; быть в
  режиме висения
A حَام

2345

hover n.

F vol stationnaire m.
E vuelo estacionario m.
D Schwebeflug m.
R висение ср.
A حَوَمَان m

2346

hover ceiling

F plafond de vol
  stationnaire m.
E techo de vuelo
  estacionario m.
D Schwebgipfelhöhe f.
R потолок висения м.
A أقْصَى إرْتِفَاع الحَوَمَان

2347

hovercraft

F aéroglisseur m.
E aerodeslizador m.
D Luftkissenfahrzeug
  neut.
R машина на воздушной
  подушке ж.
A حوّامة

2348

howitzer

F obusier m.
E obús m.
D Haubitze f.
R гаубица ж.
A مِدْفَع قَوْس m

2349

(pack) howitzer

F obusier compact léger
  m.
E obús desmontable m.
D Packhaubitze f.
R горно-вьючная
  гаубица ж.
A قذّاف محمول

2350

hub

F moyeu m.
E cubo m.
D Nabe f.
R ступица ж.
A سُرَّة f

2351 [M]

hulk

F épave f.
E barco viejo m.; casco
  m.
D Hulk m.
R корпус старого судна
  м.
A هيكل سفينة

145

2352 [A,M,V]
hull
F caisse (chars) *f.;* coque
f.
E casco *m.*
D Rumpf *m.;*
(Panzer-)Wanne *f.*
R корпус *м.;* остов *м.*
A بَدَن *m* قَوْقَعَة *f* هَيْكَل *m*

2353 [M]
hull down
F coque noyée
E más abajo del horizonte
D (hinter der) Kimm
R скрытым корпусом
A غَاطِس البَدَن *m*

2354
humid
F humide
E húmedo
D feucht
R влажный; сырой
A رَطْب *m*

2355
humidity
F humidité *f.*
E humedad *f.*
D Feuchtigkeit *f.*
R влажность *ж.;*
сырость *ж.*
A رُطُوبة *f*

2356
hunt *v.*
F chasser
E cazar
D jagen
R охотиться
A صَاد

2357
hunt *n.*
F chasse *f.*
E caza *f.*
D Jagd *f.*
R охота *ж.;* поиски *м.*
мн. ч.
A صَيْد *m*

2358 [M]
hunter killer group
F groupe chasseur de
sous-marin *m.*
E grupo cazasubmarinos
m.
D Verfolgungs- und
Vernichtungs-
geschwader *neut.*
R флотилия,
предназначенная для
обнаружения и
истребления
подлодок *ж.*
A مجموعة بحرية لقنص
الغواصات

2359 [L,M]
(mine) hunting
F chasse aux mines *f.*
E caza(minas) *m.*
D Minensuch- *komp.*
R автономная мина *ж.*
A كسح الالغام

2360
hurricane (Force 12)
F ouragan *m.*
E huracán *m.*
D Orkan *m.*
R ураган *м.*
A إعْصَار *m*

2361
hurricane lamp
F lampe-tempête *f.*
E lámpara a prueba de
viento *f.*
D Sturmlaterne *f.*
R фонарь молния *м.*
A مِصْبَاح إعْصَارِي *m*

2362
hut
F baraquement *m.*
E barraca *f.*
D Baracke *f.*
R барак *м.*
A كوخ

2363
hydrant
F bouche d'incendie *f.;*
prise d'eau *f.*
E boca de riego *f.;* boca
de incendias *f.*
D Hydrant *m.*
R гидрант *м.;*
водонапорный кран
м.
A صُنْبُور رَئِيسِي *m*

2364
hydraulic
F hydraulique
E hidráulico
D hydraulisch *adj.;*
Öldruck- *komp.;*
Wasserdruck- *komp.*
R гидравлический;
водяной
A هِدْرُولي ـ هيدْرُوليكي *m*

2365
hydrofoil
F hydrofoil *m.*
E aerodeslizador *m.;*
hidrofoil *m.*
D Tragfläche *f.*
R подводное крыло *ср.;*
лодка на подводных
крыльях *ж.*
A زَوْرَق زَلاق *m*

2366
hydrogen
F hydrogène *m.*
E hidrógeno *m.*
D Wasserstoff *m.*
R водород *м.*
A هِدْرُوجِين *m*

2367
hydrogen bomb
F bombe à hydrogène *f.*
E bomba de hidrógeno *f.*
D Wasserstoffbombe *f.*
R водородная бомба *ж.*
A قنبلة ـ هيدروجينية *f*

2368

hydrographic
F hydrographique
E hidrográfico
D hydrographisch
R гидрографический
A مُتَعَلِّق بِعِلْم وَصْف المِياه

2369

hydrography
F hydrographie *f.*
E hidrografia *f.*
D Gewässerkunde *f.;*
   Hydrographie *f.*
R гидрография *ж.*
A عِلْم وَصْف المِياه

2370

hydrophone
F hydrophone *m.*
E hidrófono *m.*
D Unterwasserhorchgerät
   neut.
R гидрофон *м.*
A مِسْمَاع مَائِي *m*

2371 [M]

hydroplane
F hydroglisseur *m.*
E hidroplano *m.*
D Gleitboot *neut.*
R глиссер *м.;* гидроплан
   м.
A زَلَّاقَة مَائِيَّة *f*

2372 [M]

hydroplanes
F barres de plongée *f. pl.*
E timones de profundidad
   m. pl.
D Tiefenruder *neut. pl.*
R подводные
   (управляющие) рули
   м. мн. ч.
A زَلَّاقَات مَائِيَّة *f*

2373

hydrostatic
F hydrostatique
E hidrostático
D hydrostatisch *adj.;*
   Wasser- *komp.*
R гидростатический
A هِذْرُوسْتاتي . إستاتيكا
   المَوائع *m*

2374

hyperbolic navigation
F navigation hyperbolique
   f.
E navegación hiperbólica
   f.
D Hyperbelverfahren *neut.*
R гиперболическая
   система навигации
   ж.
A المِلَاحة بطريقة زَائِدِيَّة المَقْطَع

2375

hypergolic fuel
F carburant hypergolique
   m.
E carburante hipergólico
   m.
D hypergolischer
   Treibstoff *m.*
R гиперголическое
   топливо *ср.;*
   самовоспламен-
   яющееся топливо *ср.*
A وُقُود تِلْقَائِي الإِشْتِعَال *m*

2376

hypersonic
F hypersonique
E hipersónico
D Überschall- *komp.*
R гиперзвуковой
A فَرْط صَوْتي *m*

2377

hypodermic syringe
F seringue hypodermique
   f.
E jeringa hipodérmica *f.*
D Injektionsspritze *f.*
R подкожный шприц *м.*
A مِحْقَنَة *f*

2378

hypsometry
F hypsométrie *f.*
E hipsometría *f.*
D Höhenmessung *f.*
R отметка высоты *ж.*
A إِبْسُومِتْر *m*

I

2379

ice
F glace *f.;* verglas *m.*
E hielo *m.*
D Eis *neut.*
R лёд *м.*
A جَلِيد *m*

2380

(pack) ice
F banquise *f.*
E témpanos flotantes *m.*
   f.
D Packeis *m.*
R плавучий лёд *м.;*
   паковый лёд *м.*
A كتلة جليد

2381

iceberg
F iceberg *m.*
E iceberg *m.*
D Eisberg *m.*
R айсберг *м.*
A جَبَل ثَلْج عَائِم *m*

2382

icebound
F retenu (navire); bloqué
   (port)
E preso entre (buque);
   bloqueado (puerto)
D eingefroren
R затертый льдами
   (корабль);
   скованная льдом
   (вода)
A مُغَطَّى بِالجَلِيد

147

2383
icebreaker
F brise-glace *m. inv.*
E rompehielos *m. inv.*
D Eisbrecher *m.*
R ледокол *м.*
A كَاسِحَة الجَلِيد *f*

2384
ice field
F champ de glace *m.*
E campo de hielo *m.*
D Eisfeld *neut.*
R ледяное поле *ср.*
A حَقْل جَلِيد *m*

2385
ice floe
F banquise *f.*
E témpano de hielo *m.*
D Treibeisscholle *f.*
R плавучая льдина *ж.*
A طَوْف جَلِيدي *m*

2386
ice free
F libre de glace
E libre de hielo
D eisfrei
R свободный от льда;
    незамерзающий
A خَالٍ من الجَلِيد

2387
ice up
F (se) givrer
E helarse
D vereisen
R обледенеть
A جَمَّد

2388
icing (up)
F givrage *m.*
E formación de hielo *f.*
D Vereisung *f.*
R обледенение *ср.*
A تَجْمِيد

148

2389
icy
F verglacé
E helado
D vereist
R ледяной; покрытый
    льдом
A جَلِيدي

2390
identification
F identification *f.*
E identificación *f.*
D Bestimmung *f.;*
    Feststellung *f.*
R опознание *ср.;*
    выяснение *ср.*
A إِثْبَات الشَّخْصِية

2391
**identification friend or
foe (IFF)**
F identification amie ou
    ennemie *f.*
E identifación amigo o
    enemigo *f.*
D Freund-Feind-Kennung
    (FFK) *f.*
R радиолокационное
    опознавание *ср.*
A تَمْيِيز الصَّدِيق والعَدُو

2392
identify (oneself)
F (s')identifier
E identificar(se)
D identifizieren (sich)
    untr.; ausweisen *tr.*
R удостоверять (свою)
    личность *ж.*
A حَقَّق الهُوَّية

2393
identity
F identité *f.*
E identidad *f.*
D Gleichheit *f.;* Identität
    f.
R личность *ж.*
A هُوَّية *f*

2394
identity card
F carte d'identité *f.*
E cédula personal *f.*
D Ausweis *m.;* Kennkarte
    f.
R удостоверение
    личности *ср.*
A بِطَاقَة هُوَّية *f*

2395
igniter
F allumeur *m.*
E espoleta iniciadora *f.*
D Zünder *m.*
R воспламенитель *м.*
A مُشْعِل

2396 [V]
ignition
F allumage *m.*
E encendido *m.*
D Zündung *f.*
R воспламенение *ср.*
A إِشْعال

2397
illegal
F illégal
E ilegal
D rechtswidrig
R нелегальный;
    незаконный
A غَيْر قَانُوني

2398
illuminate
F éclairer
E iluminar
D beleuchten
R освещать
A أَضَاء

2399
**illuminating
(munitions)**
F éclairant
E (de) iluminación
D Leucht- *komp.*
R осветительные
(боеприпасы) *м. мн.
ч.*
A تَنْوِير

2400
illumination
F éclairage *m.*
E iluminación *f.*
D Beleuchtung *f.*
R освещение *cp.;*
облучение *cp.*
A إِضَاءَة *f*

2401
image
F image *f.*
E imagen *f.*
D Abbildung *f.;* Bild *neut.*
R изображение *cp.;*
отражение *cp.*
A صُورَة *f*

2402
image intensification
F intensification d'image
f.
E intensificación de
imagen *f.*
D Restlichtverstärkung *f.*
R усиление изображения
cp.
A إِزَادَة شِدَّة الصُّورَة *f*

2403
image intensifier
F intensificateur d'image
m.
E intensificador de
imagen *m.*
D Restlichtverstärker *m.*
R усилитель
изображения *м.*
A مُزِيد شِدَّة الصُّورَة *m*

2404
immediate
F immédiat
E inmediato
D sofortig
R немедленный;
срочный
A فَوْرِي

2405 [R]
**immediate
(operational)**
F urgent-opérations
E urgente
D unverzüglich
R срочно
A فوري

2406 [N]
**immediate permanent
incapacitation
dose**
F dose d'irradiation
immédiatement fatale
f.
E dosis de radiacíon que
produce incapacidad
física inmediata
y permanente *f.*
D unmittelbare
Kampfunfähigkeitsdosis
f.
R немедленная
постоянная доза *ж.*
A مقدار الجرعة الإشعاعية
للإعجاز الفوري

2407 [N]
**immediate transient
incapacitation
dose (IT)**
F dose d'irradiation fatale
à la longue *f.*
E dosis de radiacíon que
produce incapacidad
física inmediata
y temporáneo *f.*
D unmittelbare
vorübergehende
Kampfunfähig-
keitsdosis *f.*
R немедленная
временная доза *ж.*
A مقدار الجرعة الإشعاعية
للإعجاز البطيء

2408
immerse
F (se) plonger
E sumergir(se)
D (sich) eintauchen *tr.*
R погружать(ся);
потоплять(ся)
A غَطَس

2409
immersed
F noyé
E sumergido
D versenkt
R погруженный;
потопленный
A غَاطِس *m*

2410
immersion
F immersion *f.*
E inmersión *f.*
D Untertauchen *neut.;*
Versenkung *f.*
R погружение *cp.*
A تَغْطِيس *m*

2411

immersion-proof
F étanche
E (a) prueba de la inmersión
D tauchfähig; wasserdicht
R водонепроницаемый
A مُقَاوِم لِلتَّغْطِيس

2412

immunisation
F immunisation *f.*
E imunización *f.*
D Immunisierung *f.*
R иммунизация *ж.*
A تَمْنِيع *m*

2413 [S,U]

impact *n.*
F impact [S] *m.;* choc [U] *m.*
E impacto *m.*
D Aufschlag [S] *m.;* Stoss [U] *m.*
R удар *м.;* столкновение *ср.*
A صَدْمَة *f*

2414

impact area
F zone d'impact *f.*
E zona de impactos *f.*
D Zielgebiet *neut.*
R поражаемый район *м.;* сектор стрельбы *м.*
A مِنْطَقَة الإِصَابَة *f*

2415

impassable
F infranchissable (rivière); impracticable (route)
E infranqueable; invadeable
D unpassierbar
R непроходимый
A لا يُعبر

150

2416

implode
F imploser
E implosionar
D implodieren
R взрывать(ся)
A إِنْفَجَر الى الدَّاخِل

2417

implosion
F implosion *f.*
E implosión *f.*
D Implosion *f.*
R взрыв, направленный внутрь *м.;* имплозия *ж.*
A إِنْفِجَار الى الدَّاخِل *m*

2418

imprint *n.*
F référence de publication *f.*
E pie de imprenta *m.*
D Impressum *neut.*
R отпечаток *м.*
A أَثَر مَطْبُوع *m*

2419

imprison
F emprisonner
E encarcelar
D einsperren *tr.*
R заключать (в тюрьму)
A حَبَس

2420

improvise
F improviser
E improvisar
D improvisieren
R импровизировать
A إِنْتَكَر

2421

improvised
F improvisé; (de) fortune
E improvisado; (de) fortuna
D Behelfs- *komp.*
R импровизированный
A مُبْتَكَر

2422

impulse charge
F charge d'impulsion *f.*
E carga de impulsión *f.*
D Initialzündung *f.*
R пороховой заряд *м.*
A شُحْنَة دافِعة *f*

2423

inactivate
F désactiver
E inactivar
D inaktivieren; unwirksam machen
R переводить в резерв
A أخمد/سكن

2424

inboard *adj.*
F intérieur
E interior
D Innenbord- *komp.*
R расположенный внутри
A فِي السَّفِينَة

2425 [A]

incendiary
F incendiaire
E incendiario *m.*
D Brand- *komp.;* Feuer- *komp.*
R зажигательный
A محرق

2426

incident
F incident *m.*
E incidente *m.*
D Zwischenfall *m.*
R инцидент *м.;* случай *м.*
A حَادِثَة *f*

2427

incline *v.*
F incliner
E inclinar(se)
D (sich) neigen
R наклонять(ся); склонять(ся)
A مَال

2428

incline n.
F pente f.
E cuesta f.
D Neigung f.
R наклон м.; склон м.
A مَيْل m

2429

inclined
F incliné
E inclinado
D schief; schräg
R наклонный; склонный
A مَائِل

2430

incoming unit
F unité relevante f.
E unidad entrante f.
D (einziehende) Einheit f.
R (приезжее)
  подразделение ср.
A وَحْدَة وَارِدة f

2431

increase (the range)
F allonger le tir
E aumentar (el alcance)
D vergrössern
R увеличивать
  (дальность)
A طَوّل المدى

2432

incursion
F incursion f.
E incursión f.
D Einfall m.
R вторжение ср.;
  нашествие ср.
A غزوة

2433

indefensible
F indéfendable
E indefendible
D unhaltbar; nicht zu
  verteidigen
R незащищаемый
A لا يمكن الدفاع عنه

2434

indefinite
F illimité; indéfini
E ilimitado; indefinido
D unbegrenzt; unbestimmt
R неопределённый;
  бессрочный
A غَيْر مُحَدَّد

2435

independent
F autonome; indépendant
E autónomo;
  independiente
D selbständig; unabhängig
R самостоятельный;
  независимый
A مُسْتَقِل

2436

indicated airspeed
F vitesse indiquée f.
E velocidad indicada f.
D angezeigte
  Eigengeschwindigkeit
  f.
R техническая скорость
  самолёта ж.
A السُّرْعَة النِّسْبِيَّة المُبَيَّنة f

2437

(target) indication
F désignation d'un
  objectif f.
E disignación del blanco
  f.
D Schussangabe f.;
  Zielangabe f.
R указание ср.;
  целеуказание ср.
A تعيين الهدف

2438

indicator
F indicateur m.
E indicador m.
D Zeiger m.
R индикатор м.;
  указатель м.
A مُبَيِّن m . دَلِيل m

2439

(ground position)
  indicator (GPI)
F indicateur de position-
  sol m.
E indicador de posición
  en el suelo m.
D Ortungsgerät neut.
R указатель (наземного)
  местоположения м.
A مُبَيِّن المَوْقِع الأَرْضِي

2440 [R]

(moving target)
  indicator
F éliminateur d'echos
  fixes m.
E supresor de ecos fijos
  m.
D Festzeichenunterdrücker
  m.
R индикатор
  движущейся цели м.
A مُبَيِّن الغَرَض المُتَحَرِّك m

2441 [R]

(plan position)
  indicator (PPI)
F radar oscilloscope m.
E indicador de posición
  en planta m.
D Rundumsichtradarschirm
  m.
R индикатор кругового
  обзора (ИКО) м.
A مُبَيِّن مَوَاقِع إسْقَاطِي m

2442

indirect
F indirect
E indirecto
D indirekt
R непрямой; косвенный
A غَيْر مُبَاشِر

2443

indoctrination
F endoctrinement m.
E adoctrinamiento m.
D Schulung f.
R индоктринация ж.
A رسّخ عقيدة في الذهن

151

2444 [E]
induce
F induire
E inducir
D induzieren
R индуктировать
A إِسْتَحَثّ

2445
induced radiation
F radiation induite f.
E radiación inducida f.
D induzierte
  Radioaktivität f.
R наведённая
  радиоактивность ж.
A فَاعِلِيَّة إِشْعَاعِيَّة مُسْتَحَثَّة f

2446 [E]
induction
F induction f.
E inducción f.
D Induktion f.
R индукция ж.
A حَثّ m

2447
inert (munition)
F inerte
E inerte
D gesichert; unwirksam
R неснаряженный;
  инертный;
  снаряженный
  инертными
  веществами
A بَاطِل ـ فَاسِد

2448
inertia
F inertie f.
E inercia f.
D Trägheit f.
R инерция ж.
A القُصُور الذَّاتِي m

2449
inertial guidance
F guidage par inertie m.
E dirección por inercia f.
D Trägheitslenkung f.
R инерциальное
  наведение ср.
A التَّوْجِيه بِالقُصُور الذَّاتِي

2450
inertial navigation
  system
F système de navigation à
  inertie m.
E sistema de navegación a
  inercia m.
D Trägheitsnavigations-
  system neut.
R инерциальная система
  навигации ж.
A جِهَاز المِلَاحَة بِالقُصُور
  الذَّاتِي

2451
infantry
F infanterie f.
E infantería f.
D Infanterie f.
R пехота ж.;
  (мотострелковые
  войска)(ср. мн. ч.
A مُشَاة f

2452
infantryman
F fantassin m.
E infante m.; soldado de
  infantería
D Infanterist m.
R пехотинец м.
A جُنْدِي مُشَاة m

2453 [L,P]
infiltrate
F (s')infiltrer; noyauter
E infiltrarse (en)
D einsickern tr.
R просачиваться;
  проникать
A تَسَلَّل

2454 [L,P]
infiltration
F infiltration f.;
  noyautage m.
E infiltración f.
D Unterwanderung f.
R просачивание ср.
A تَسَلُّل m

2455
inflammable
F inflammable
E inflamable
D brennbar
R огнеопасный
A قَابِل لِلالتِهاب

2456
inflatable adj.
F gonflable
E inflable
D aufblasbar
R надувной
A قَابِل لِلنَّفْخ

2457
inflatable n.
F canot pneumatique m.
E bote inflable m.
D Schlauchboot neut.
R надувная лодка ж.
A زَوْرَق مَطَّاط m

2458
inflight refuelling
F ravitaillement en vol m.
E reabastecimiento en
  vuelo m.
D Nachtanken in der Luft
  (meist eng. Begriff)
  neut.
R дозаправка в воздухе
  ж.
A تَمْوِين بِالوُقُود أَثْنَاء الطَّيْرَان

2459 [U]
inform
F avertir
E avisar
D mitteilen tr.
R сообщать; уведомлять
A أَبْلَغ

2460 [P]
inform
F dénoncer
E delatar (a)
D anzeigen *tr.*
R доносить
A وَشَى

2461
information
F renseignements *m. pl.*
E informes *m. pl.*
D Informationen *f. pl.*
R данные *ж. мн. ч.;*
сведения *ср. мн. ч.*
A مَعْلُومَات f

2462
informer
F dénonciateur *m.*
E denunciante *m. f.*
D Informant *m.;* Spitzel
m.
R доносчик *м.*
A وَاشٍ ـ مُخْبِر m

2463
infrared (IR) *adj.*
F infrarouge
E infrarrojo
D Infrarot- *komp.*
R инфракрасный
A تَحْت الحَمْرَاء

2464
infrared ray
F radiations infrarouges *f.
pl.*
E rayo infrarrojo *m.*
D Infrarotstrahlen *m. pl.*
R (инфракрасный) луч
м.
A أَشِعَّة تَحْت الحمراء f

2465
infrastructure
F infrastructure *f.*
E infraestructura *f.*
D Infrastruktur *f.*
R основание *ср.;*
инфраструктура *ж.*
A أَساسٌ m

2466 [A]
initial approach
F approche initiale *f.*
E acercamiento inicial *m.*
D Anflugbeginn *m.*
R первоначальное
сближение *ср.;*
предварительное
наведение *ср.*
A إِقْتِرَاب أَوَّلِي m

2467
initiate
F amorcer; prendre
l'initiative
E iniciar
D (in) Gang setzen
R начинать; приступать
(к)
A بَدَأ

2468 [N,S]
initiation
F amorçage *m.*
E iniciación *f.*
D Zündung *f.*
R инициирование *ср.;*
воспламенение *ср.
м.;* разрыв
A بدء العمل

2469 [N,S]
initiation
F amorçage *m.*
E comienzo *m.*
D Zündsatz *m.*
R зажжение *ср.;*
введение *ср.*
A بَدْء

2470
initiative
F initiative *f.*
E iniciativa *f.*
D Initiative *m.;*
Unternehmungsgeist
R инициатива *ж.*
A مُبَادَرَة f

2471
initiator
F amorce détonateur *f.*
E encendedor *m.*
D Anreger *m.*
R инициатор *м.*
A بَادِىء m

2472
inject
F injecter
E inyectar
D einspritzen *tr.*
R впрыскивать
A حَقَن

2473
injection
F injection *f.*
E inyección *f.*
D Injektion *f.;* Spritze *f.*
R впрыскивание *ср.*
A حَقْن m

2474
injure
F blesser
E herir
D verletzen
R вредить; ранить
A جَرَح

2475
injury
F blessure *f.*
E herida *f.*
D Verletzung *f.*
R повреждение *ср.;* рана
ж.
A إِصَابَة f

2476 [T]
inlet
F crique *f.*
E ensenada *f.*
D Meeresarm *m.*
R впуск *м.*; вход *м.*
A شَرْم *m*

2477
inner *adj.*
F intérieur
E interior
D inner
R внутренний
A دَاخِلي

2478 [D]
input(s) *n.*
F (données) d'entrée *(f. pl.) f.*
E (datos) de entrada *(m. pl.) f.*
D Eingabe(-daten) *f. (neut. pl.)*
R входные данные *ж. мн. ч.*; ввод *м.*
A مَعْلُومَات تُزَوَّد بِها آلَة حَاسِبة

2479
inquiry
F enquête *f.*
E investigación *f.*
D Untersuchung *f.*
R расследование *ср.*
A تَحْقِيق *m*

2480
insecure
F non-protégé
E inseguro
D ungesichert
R открытый; ненадёжный
A غير آمن

2481
insert *v.*
F introduire
E introducir
D einfügen *tr.*
R вставлять; вводить
A أَدْخَل

2482
insertion
F insertion *f.*
E introducción *f.*
D einsetzen *tr.*
R введение *ср.*
A إدخال

2483 [T]
inset
F carton intérieur *m.*
E tarjeta interior *f.*
D Nebenkarte *f.*
R вкладка *ж.*; вклейка *ж.*
A دَسّ *m*

2484
inshore *adj.*
F côtier
E costero; litoral
D Küsten- *komp.*
R прибрежный
A قَرِيب من البَرّ

2485 [A,L,M]
inspect
F faire l'inspection
E inspeccionar
D inspizieren
R производить осмотр
A فَتَش

2486 [E]
inspect
F inspecter
E inspeccionar
D kontrollieren
R проверять
A فَحَص

2487 [A,M,L]
inspection
F revue *f.*
E revista *f.*
D Besichtigung *f.*
R осмотр *м.*; инспекция *ж.*
A تَفْتِيش *m*

2488 [E]
inspection
F inspection *f.*
E inspección *f.*
D Inspektion *f.*; Überprüfung *f.*
R осмотр *м.*; проверка *ж.*
A فَحْص *m*

2489
"(for) inspection port arms!"
F "inspection des armes!" *f.*
E "¡(en) revista armas!"
D „(zur) Inspektion!"
R "(на) грудь!"
A عاليا إحمل !

2490
inspector
F inspecteur *m.*
E inspector *m.*
D Inspektor *m.*; Prüfer *m.*
R инспектор *м.*; контролёр *м.*
A مُفَتِّش *m*

2491
installation
F installation *f.*
E instalación *f.*
D Einrichtung *f.*
R объект *м.*
A مُنْشآت

154

2492

instruct

F instruire; charger
E instruir; mandar
D ausbilden; beauftragen
R обучать; учить
A دَرَّب

2493

instruction

F instruction *f.;* consigne
   f.
E instrucción *f.;* orden *f.*
D Ausbildung *f.;* Auftrag
   m.
R обучение *ср.*
A تَدْرِيب *m*

2494

instructional

F (d')application
E (de) instrucción
D Ausbildungs- *komp.*
R учебный
A تَدْرِيبِي

2495

instructor

F instructeur *m.*
E instructor *m.*
D Ausbilder *m.*
R инструктор м.
A مُدَرِّب *m*

2496

instrument

F instrument *m.*
E instrumento *m.*
D Instrument *neut.*
R прибор *м.;*
   инструмент *м.*
A آلَة *f*

2497

instrumentation

F instrumentation *f.*
E instrumentación *f.*
D Instrumentierung *f.*
R приборное
   оборудование *ср.*
A إسْتِخْدام ألآلَات

2498

**instrument landing
   system (ILS)**

F système d'atterrissage
   aux instruments (ou
   système ILS) *m.*
E sistema de aterrizaje
   por instrumentos *m.*
D Instrumentenlandesystem
   neut.
R система посадки по
   приборам *ж.*
A نِظَام الهُبوط الآلي

2499

insurgency

F état insurrectionnel *m.;*
   insurrection *f.*
E insurgencia *f.;*
   sublevación *f.*
D Aufruhr *m.;* Aufstand
   m.
R восстание *ср.;* мятеж
   м.
A تَمَرُّد *m*

2500

insurgent *n.*

F insurgé
E insurgente *m. f.*
D Aufrührer *m.*
R повстанец *м.;*
   мятежник *м.*
A مُتَمَرِّد *m*

2501 [E]

integral (with) *adj.*

F solidaire (de)
E integral (con)
D einteilig (mit);
   zugehörig (zu)
R входящий в состав;
   интегральный;
   составной
A مُتَكَامِل (مع)

2502

integrate (with)

F intégrer (dans, à)
E integrar (en)
D eingliedern (in)
R включать (в);
   объединять
A تَكَامَل (مع)

2503

integrated circuit

F circuit intégré *m.*
E circuito integrado *m.*
D integrierte Schaltung *f.*
R объединённая сеть *ж.*
A دَائِرَة مُتَكَامِلة *f*

2504

integration

F intégration *f.*
E integración *f.*
D Integration *f.*
R объединение *ср.*
A تَكَامُل *m*

2505

intelligence *n.*

F renseignement(s) *m.*
   (pl.)
E información(es) *f. (pl.)*
D (Feind-)Nachrichten
   f. pl.
R разведка *ж.*
A إسْتِخْبارَات *f*

2506

(acoustic) intelligence

F renseignements obtenus
   par écoute *m. pl.*
E información acústica *f.*
D Hörnachrichtenmaterial
   neut.
R (акустическая)
   разведка *ж.*
A الاستخبارات الصوتية

2507

(basic) intelligence

F documentation générale *f.*

E informaciones básicas *f. pl.*

D grundlegende Informationen *f. pl.*

R основные данные *ж. мн. ч.*

A إسْتِخْبَارات (أساسية رئيسية) *f*

2508

(biographical) intelligence

F renseignement(s) biographique(s) *m.(pl.)*

E información biográfica *f.*

D biographische Informationen *f. pl.*

R биографические данные *ж. мн. ч.*

A الإسْتِخْبَارات الفردية *f*

2509

(combat) intelligence

F renseignement tactique *m.*

E información de combate *f.*

D Kampfnachrichten- material *neut.*

R (боевая) разведка *ж.*

A الاستخبارات القتالية

2510

(communications) intelligence

F renseignements- télécommunications *m. pl.*

E información de transmisiones *f.*

D Meldenachrichten- material *neut.*

R радиоразведка *ж.*

A استخبارات الاتصالات

2511

(counter)intelligence

F défense contre le service de renseignement ennemi *f.*

E contrainteligencia *f.*

D Nachrichtenabwehr *f.*

R контрразведка *ж.*

A تجَسُّس مُضَاد *m*

2512

(electronic) intelligence

F renseignements par moyens électroniques *m. pl.*

E información electrónica *f.*

D (elektronisches) Nachrichtenmaterial *neut.*

R радиоразведка *ж.*

A الاستخبارات الالكترونية

2513

(imagery) intelligence

F renseignements par moyens électroniques et photographiques *m. pl.*

E información de fotografía aérea *f.*

D bildliches Nachrichtenmaterial *neut.*

R сведения из изображений *ср. мн. ч.*

A الإستخبارات الصورية *f*

2514

(military) intelligence

F service de renseignement *m.*

E información militar *f.*

D militärisches Nachrichtenwesen *neut.*

R военная разведка *ж.*

A الإسْتِخْبَارَات العَسْكَرِيَّة *f*

2515

(photographic) intelligence

F interprétation photo *f.*

E información fotográfica *f.*

D (photographisches) Nachrichtenmaterial *neut.*

R (фотографическая) разведка *ж.*

A تصوير الاستخبارات

2516

(signals) intelligence

F service d'écoutes *m.*

E información de transmisiones *f.*

D Funknachrichtenmaterial *neut.*

R радиоразведка *ж.*

A استخبارات الاشارة

2517

(tactical counter) intelligence

F contre-espionnage tactique *m.*

E contrainteligencia táctica *f.*

D taktische Abschirmung *f.*

R тактическая контрразведка *ж.*

A عمليات تعبوية لمكافحة استخبارات العدو

2518

(target) intelligence

F renseignement sur l'objectif *m.*

E información de objetivo *f.*

D Zielbeschreibung

R разведка цели *ж.*

A إسْتِخْبَارَات الهدف

2519

(technical) intelligence

F renseignement technique *m.*
E información técnica *f.*
D technische Aufklärung *f.*
R техническая разведка *ж.*
A اسْتِخْبَارَات فَنِّيَّة *f*

2520

intelligence cycle

F cycle du renseignement *m.*
E ciclo de información *m.*
D Nachrichtenverarbeitung *f.*
R процесс разведывательных данных *м.*
A دَوْرَة الإِسْتِخْبَارَات *f*

2521

intelligence estimate

F évaluation de l'ennemi *f.*; synthèse du renseignement *f.*
E apreciación de información *m.*
D Feindlagebeurteilung *f.*
R оценка разведывательных данных *ж.*
A تَقْدِير الإِسْتِخْبَارَات *m*

2522

intensifier

F intensificateur (d'images) *m.*
E aumentador *m.*
D Verstärker *m.*
R усилитель изображений *м.*
A مُزِّيد الشدة

2523 [I,R]

intercept

F renseignements obtenus par le service des écoutes *m. pl.*
E informaciones interceptadas *f. pl.*
D abgefangene Informationen *f. pl.*
R перехват; *м.* подслушивание *ср.*
A برقيات معترضة

2524 [A]

intercept *v.*

F intercepter
E interceptar
D abschneiden *tr.*; abfangen
R перехватывать
A إعْتَرَض

2525 [A]

interceptor

F intercepteur *m.*
E interceptor *m.*
D Abfangjäger *m.*
R истребитель-перехватчик *м.*
A طَائِرَة إعْتِرَاض *f*

2526

intercept receiver

F détecteur d'interception radioélectrique *m.*
E estación de captación *f.*
D Abfangdetektor *m.*
R разведывательный радиоприёмник *м.*
A مُسْتَقْبِل رَسَائِل مُلْتَقَطَة *m*

2527

interchange *v.*

F échanger
E intercambiar
D austauschen *tr.*
R обмениваться; переменять(ся)
A تَبَادَل

2528

interchangeability

F interchangeabilité *f.*
E intercambiabilidad *f.*
D Austauschbarkeit *f.*
R взаимозаменяемость *ж.*
A قَابِلِيَّة التَّبَادُل *m*

2529

interchangeable

F interchangeable
E intercambiable
D austauschbar
R взаимозаменяемый
A قَابِل لِلتَّبَادُل

2530

intercom

F intercom *m.*
E sistema de intercomunicación *m.*
D Bordsprechanlage *f.*; Gegensprechanlage *f.*
R селекторная связь *ж.*; переговорочное устройство *ср.*
A نِظَام إتّصَال داخِلِي *m*

2531

interdict *v.*

F interdire
E interdecir
D abriegeln *tr.*
R воспрещать; запрещать; изолировать
A حَرَّم

2532

interdiction

F interdiction *f.*
E interdicción *f.*
D Abriegelung *f.*
R воспрещение *ср.*; изоляция *ж.*
A تَحْرِيم *m*

2533

interface

F dialogue *m.;* interface *f.*
E entrecara *f.*
D Grenze *f.;* Grenzfläche *f.*
R стык *м.;* интерфейс *м.;* диалоговый режим *м.*
A السَّطْح الداخِلي *m*

2534 [R]

interference

F parasites électriques *m. pl.*
E interferencia *f.*
D Empfangsstörung *f.*
R помехи *ж. мн. ч.*
A تشويش *m*

2535

interim *adj.*

F provisoire
E provisional
D vorläufig
R временный; промежуточный
A مُؤَقَّت

2536

interim measure

F action provisoire *f.*
E medida provisional *f.*
D Zwischenmassnahme *f.*
R временная мера *ж.*
A إجْراء مُؤَقَّت *m*

2537

interior

F intérieur
E interior
D Innen- *komp.;* inner *adj.*
R внутренний
A داخِل

2538

intermediate

F intermédiaire
E intermedio
D Zwischen- *komp.*
R промежуточный; средний
A مُتَوَسِّط

2539

intermittent

F intermittent
E intermitiente
D zeitweilig *adj.;* aussetzend [E] *adj.;* Stör- [S] *komp.*
R прерывистый
A مُتَقَطِّع

2540

intern *v.*

F interner
E internar
D internieren
R интернировать
A إعْتَقَل

2541

internal

F interne
E interno
D Innen- *komp.;* inner *adj.*
R внутренний
A داخِلي

2542

international

F international
E internacional
D international
R международный
A دَوْلي

2543

international dateline

F ligne internationale de changement de date *f.*
E línea internacional de cambio de fecha *f.*
D internationale Datumsgrenze *f.*
R международная линия суточного времени *ж.*
A خَطّ التَّأْريخ الدَّوْلي *m*

2544

international law

F droit international *m.*
E derecho internacional *m.*
D Völkerrecht *neut.*
R международное право *ср.*
A القَانُون الدَّوْلي *m*

2545

internee

F interné *m.*
E internado *m.*
D Internierte(r) *m.*
R интернированный *м.*
A مُعْتَقَل *m*

2546

internment

F internement *m.*
E internamiento *m.*
D Internierung *f.*
R интернирование *ср.*
A إعْتَقال *m*

2547

interocular distance

F écart oculaire *m.*
E distancia interocular *f.*
D Augenabstand *m.*
R окулярная дальность *ж.*
A البُعْد بَيْن العَيْنَيْن *m*

2548

interoperability
F interopérabilité *f.*
E interoperabilidad *f.*
D Austauschbarkeit (meist
   eng. Begriff) *f.*
R взаимодействуемость
   ж.
A تَبَادُل الفِعْل

2549

interoperable
F interopérable
E interoperable
D austauschbar (meist
   eng. Begriff)
R взаимодействуемый
A مُتَبَادِل الفِعْل

2550

interpret
F interpréter
E interpretar
D dolmetschen; übersetzen
R переводить
A تَرْجَم

2551

interpretation of
   intelligence
F interprétation du
   renseignement *f.*
E interpretación de
   información *f.*
D Nachrichtenauswertung
   f.
R оценка разведки *ж.*
A تفسير الاستخبارات

2552

interpreter
F interprète *m. f.*
E intérprete *m. f.*
D Dolmetscher *m. ;*
   Übersetzer *m.*
R переводчик *м.*
A مُتَرْجِم *m*

2553

(consecutive)
   interpreter
F interprète capable
   d'interprétation
   consécutive *m.*
E intérprete (de forma
   alternativa) *m. f.*
D konsekutiv-Dolmetscher
   m.
R переводчик *м.*
A مترجم متتابع

2554

(simultaneous)
   interpreter
F interprète capable
   d'interprétation
   simultanée *m.*
E intérprete (de forma
   simultánea) *m. f.*
D Simultandolmetscher *m.*
R переводчик *м.*
A مترجم متواقت

2555

interrogate
F interroger
E interrogar
D verhören
R допрашивать
A إسْتَجْوَب

2556

interrogation
F interrogation *f.*
E interrogación *f.*
D Verhör *m.*
R допрос *м.*
A إسْتِجْوَاب *m*

2557

interrupt
F interrompre
E interrumpir
D unterbrechen *untr.*
R прерывать
A قَطَع

2558

interruption
F interruption *f.*
E interrupción *f.*
D Unterbrechung *f.*
R перерыв *м.;*
   прерывание *ср.*
A قَطْع *m*

2559

interval
F intervalle *m.*
E intervalo *m.*
D (Zeit-)Abstand *m.*
R промежуток *м.;*
   интервал *м.*
A فَاصِلة *f*

2560

(contour) interval
F équidistance des
   courbes *f.*
E equidistancia entre
   curvas *f.*
D Höhenlinienabstand *m.*
R высота сечения *ж.*
A فاصلة ما بين المنحنيات

2561

(grid) interval
F intervalle du
   quadrillage *m.*
E intervalo de
   cuadriculado *m.*
D Gitterabstand *m.*
R интервал сетки *м.*
A فَاصِلة التَّرْبِيع *f*

2562

(vertical) interval
F équidistance réelle *f.*
E intervalo vertical *m.*
D Höhenabstand *m.*
R высота сечения *ж.*
A فاصل رأسي *m*

2563

intervene
F intervenir
E intervenir
D eingreifen *tr.*
R вмешиваться
A تَدَخَّل

2564

intervention
F intervention *f.*
E intervención *f.*
D Eingreifen *neut.*
R вмешивание *ср.*;
   интервенция *ж.*
A تَدَخُّل *m*

2565

intervisibility
F intervisibilité *f.*
E visibilidad mutua *f.*
D Zwischensichtbarkeit *f.*
R взаимная видимость
   ж.
A تبادل الرؤية

2566

intravenous drip
F perfusion intraveineuse
   f.
E gotero intravenoso *m.*
D Intravenöstropfen *m.*
   pl.
R внутривенный шприт
   м.
A التغذية في الوريد

2567

intrenching tool
F outil individuel *m.*
E útil de mango corto *m.*
D Verschanzungsgerät
   neut.
R лопата *ж.*
A أداة الحفر

2568 [A]

intrude
F pénétrer
E penetrar
D eindringen *tr.*
R вторгаться
A إقْتَحَم

2569 [A]

intruder
F chasseur de pénétration
   m.; intruder *m.*
E caza de penetración *m.;*
   intruso *m.*
D Störflugzeug *neut.;*
   Eindringling *m.*
R самолёт вторжения *м.*
A مُعْتَدٍ *m*

2570

intruder alarm system
F dispositif d'alerte anti-
   intrusion *m.*
E sistema de alerta de
   intrusos *m.*
D Sabotage-Alarm-System
   neut.
R система обнаружения
   проникновения *ж.*
A نِظَام الإنْذَار ضِدّ السَّطْو *m*

2571

invade
F envahir
E invadir
D einfallen *tr.*
R вторгаться
A غزا

2572

invasion
F invasion *f.*
E invasión *f.*
D Invasion *f.*
R вторжение *ср.*
A غزوة *f*

2573

invent
F inventer
E inventar
D erfinden
R изобретать
A إخْتَرَع

2574

invention
F invention *f.*
E invención *f.*
D Erfindung *f.*
R изобретение *ср.*
A إخْتِرَاع *m*

2575

inventor
F inventeur *m.*
E inventor *m.*
D Erfinder *m.*
R изобретатель *м.*
A مُخْتَرِع *m*

2576

inventory
F inventaire *m.;* état *m.*
E inventario *m.*
D Inventar *neut.;*
   Lagerbestand *m.*
R опись *ж.;* инвентарь
   м.
A جَرْدَ *m*

2577

inventory control
F contrôle d'inventaire *m.*
E manejo de inventario
   m.
D materielle Versorgung *f.*
R управление учётом
   материальных
   средств *ср.*
A مُرَاقَبَة الجَرْد *m*

2578

invert
F intervertir; renverser
E invertir
D umkehren *tr.*
R перевёртывать
A عَكَس . قَلَب

2579
inverted
F interverti; renversé
E invertido
D umgekehrt
R перевернутый
A مَعْكوس . مَقْلُوب

2580 [L]
invest
F investir
E sitiar
D einschliessen *tr.*
R осаждать; обкладывать
A حَاصَر

2581 [P]
investigate
F enquêter (sur)
E investigar
D ermitteln; untersuchen
R расследовать
A حَقَّق

2582 [P]
investigation
F enquête *f.*
E investigación *f.*
D Untersuchung *f.;*
  Ermittlungen *f. pl.*
R расследование *ср.*
A تَحْقيق *m*

2583
ion
F ion *m.*
E ion *m.*
D Ion *neut.*
R ион *м.*
A أَيُون *m*

2584
ionization
F ionisation *f.*
E ionización *f.*
D Ionisierung *f.*
R ионизация *ж.*
A تَأَيُّن *m*

2585
ionosphere
F ionosphère *f.*
E ionosfera *f.*
D Ionosphäre *f.*
R ионосфера *ж.*
A الأَيُونوسفير *m*

2586
iron
F fer *m.*
E hierro *m.*
D Eisen *neut.*
R железо *ср.*
A حَديد *m*

2587
irradiate
F irradier
E irradiar
D bestrahlen
R облучать
A أَشَعَّ

2588 [L]
irregular
F soldat irrégulier *m.*
E guerrillero *m.*
D Freischärler *m.*
R иррегулярный солдат
  м.
A غَيْر نظامي

2589
island
F île *f.*
E isla *f.*
D Insel *f.*
R остров *м.*
A جَزيرة

2590
isobar
F isobare *f.*
E isobara *f.*
D Isobare *f.*
R изобара *ж.*
A إِيزُوبار

2591
isometric
F isométrique
E isométrico
D isometrisch
R изометрический
A إِيسُومتري

2592
isotherm(al) *n. (adj.)*
F isotherme *adj. n. f.*
E isoterma *f.;* isotérmico
  adj.
D isotherm *adj. s. f.*
R изотерма *ж.;*
  изотермический
  прилаг.
A مُتَساوي الحَرارة

2593
isotope
F isotope *m.*
E isótopo *m.*
D Isotop *neut.*
R изотоп *м.*
A نَظير *m*

2594
issue *n.*
F dotation *f.*
E distribución *f.*
D Ausgabe *f.;* Verteilung
  f.
R выдача *ж.;* отпуск *м.*
A تَوْزيع *m*

2595
issue *v.*
F distribuer
E dar; distribuir
D ausgeben; verteilen
R выдавать; отпускать
A وَزَّع

2596
(to be) issued with
F (être) doté de
E dado
D (damit) zugeteilt
R оснащённый
A مُزَوَّد ب . . .

161

2597
isthmus
F isthme *m.*
E istmo *m.*
D Isthmus *m.*
R перешеек *м.*
A بَرْزَخ

2598
item (on a list)
F point *m.;* article *m.*
E artículo *m.;* partida *f.*
D Posten *m.*
R предмет *м.*
A بَنْد *m*

2599
item (thing)
F article *m.*
E artículo *m.*
D Artikel *m.*
R вещь *ж.;* изделие *ср.*
A مَادَّة

2600
itemize
F détailler
E detallar
D (einzeln) aufführen *tr.*
R перечислять;
   классифицировать
A فَصَّل المُفْرَدَات

J

2601
jack
F verin *m.*
E gato *m.*
D Fahrzeugheber *m.*
R подъёмник *м.*
A رافعة سيارة *f*

2602 [E]
jacket *n.*
F chemise *f.*
E chaqueta *f.;* cubierta *f.*
D Mantel *m.*
R чехол *м.;* кожух *м.*
A غِلَاف *m*

162

2603 [H]
jacket *n.*
F veste *f.*
E guerrera *f.*
D Jacke *f.*
R куртка *ж.*
A سُتْرَة *f*

2604
(life) jacket
F gilet (de sauvetage) *m.*
E chaleco salvavidas *m.*
D Schwimmweste *f.*
R (спасательный) жилет
   м.
A سُترة النجاة

2605 [V]
jackknife *v.*
F (se) replier
E doblarse
D zusammenklappen *tr.*
R вилять прицепом
A إِنْطَوَى مِثْل مُدْية جَيْب

2606
jackknife *n.*
F couteau de poche *m.*
E navaja *f.*
D Klappmesser *neut.*
R складной нож *м.*
A مُدْيَة جَيْب *f*

2607 [S]
jam *v.*
F (s')enrayer
E encasquillarse
D Ladehemmung haben
R заклиниваться
A تَعَرْقَل

2608 [R]
jam *v.*
F brouiller
E interferir
   (intencionalmente)
D (absichtlich) stören
R заглушать
A شَوَّش

2609
(automatic search)
jammer
F brouilleur (à poursuite
   automatique) *m.*
E perturbador de
   búsqueda automática
   m.
D (automatischer)
   Störsender *m.*
R передатчик
   автоматического
   поиска *м.*
A جهاز تشويش تلقائي البحث
   m

2610
(repeater) jammer
F brouilleur (répondeur)
   m.
E perturbador de
   repetición *m.*
D Repetierstörsender *m.*
R передатчик
   непрерывного типа
   м.
A جهاز تشويش مكرر *m*

2611
(search) jammer
F brouilleur (chercheur)
   m.
E perturbador de
   exploración *m.*
D Suchstörsender *m.*
R передатчик
   прицельного поиска
   м.
A جهاز تشويش باحث *m*

2612
jamming
F brouillage *m.*
E interferencia intencional
   f.
D Störung *f.*
R активные помехи *ж.*
   мн. ч.
A تَشْوِيش *m*

2613
(barrage) jamming
F brouillage (en barrage) *m.*
E perturbador de barreamiento *m.*
D Sperrstörung *f.*
R (заградительная) помеха *ж.*
A سدّ تشويش

2614
(spot) jamming
F brouillage sélectif *m.*
E jamming en una frecuencia *m.*
D Punktstörung *f.*
R прицельная помеха *ж.*
A تشويش نقطي

2615
jargon
F jargon *m.*
E jerga *f.*
D Jargon *m.*
R жаргон *м.*
A رطانة

2616
jeep
F jeep *f.*
E jeep *m.*
D Jeep (Am. Begriff)
R ГАЗик *м.*
A عربة جيب

2617
jerrycan
F jerrycan *m.*
E bidón *m.*
D Benzinkanister *m.*
R бензиновая бочка *ж.*
A صفيحة لحمل البنزين

2618
jersey
F tricot *m.*
E jersé *m.*
D Strickjacke *f.*
R фуфайка *ж.*
A قماش صوفي

2619 [A]
jet *adj.*
F (à) réaction
E (a) reacción
D Düsen- *komp.*
R реактивный *прилаг.*
A نفّاث

2620 [E]
jet
F brûleur *m.*
E mechero *m.*
D Düse *f.*
R форсунка *ж.*
A نضّاحة *f*

2621 [U]
jet
F jet *m.*
E chorro *m.*
D Strahl *m.*
R жиклёр *м.;* форсунка *ж.*
A حَنَفِيّة ـ نافورة *f*

2622 [V]
jet
F gicleur *m.*
E brotador *m.*
D Düse *f.*
R жиклёр *м.*
A نضّاحة *f*

2623
jet engine
F moteur à réaction *m.;* réacteur *m.*
E motor a reacción *m.;* reactor *m.*
D Düsenmotor *m.*
R реактивный двигатель *м.;* реактивный мотор *м.*
A مُحَرِّك نفّاث

2624
jetsam
F épaves (rejetées) *f. pl.*
E desechado *m.;* echazón *f.*
D geworfene Ladung *f.;* Strandgut *m.*
R выброшенный груз *м.*
A مَطْرُوحَات

2625
jet stream
F courant-jet *m.;* jet-stream *m.*
E corriente de chorro *m.*
D Jetstream *m.*
R поток высотного ветра *м.*
A تيّار مُتَدَفِّق

2626 [A]
jettison
F jeter
E lanzar
D abwerfen *tr.*
R выбрасывать
A تَخَفَّف من . . .

2627 [M]
jettison
F larguer
E echar
D über Bord werfen
R сбрасывать за борт
A تَخَفَّف مِن . . .

2628
jettisonable
F largable
E prescindible
D abwerfbar
R сбрасываемый
A قابل للتفريغ

2629
jetty
F jetée *f.*
E muelle *m.*
D Hafendamm *m.;* Mole *f.;* Pier *m.*
R пристань *ж.;* мол *м.*
A رَصيف *m*

163

2630
jib (of crane)
F bras de levage *m.*
E puntal *m.*
D Ausleger *m.*
R стрела *ж.*
A ذراع الرافعة *m*

2631
join *v.*
F (s')unir (à)
E unir(se) (a)
D (sich) anschliessen (an)
    tr.
R соединять(ся)
A إلْتَحَق بـ

2632
joint *adj.*
F combiné *adj.;*
    interarmes *inv.*
E conjunto *adj.;* (de)
    conjunto
D gemeinsam *adj.*
R объединённый;
    совместный
A مُشْتَرِك

2633
joint operations center
F poste de
    commandement
    interarmes *m.*
E centro de operaciones
    conjuntas *m.*
D gemeinsamer
    Gefechtsstand *m.*
R центр совместных
    операций *м.*
A مَرْكَز العَمَلِيَّات المُشْتَرَكَة *m*

2634
joystick
F manche à balai *f.*
E palanca de control *f.*
D Steuerknüppel *m.*
R рычаг управления *м.*
A عصا القيادة

164

2635
judge advocate
F assesseur *m.*
E auditor de guerra *m.*
D Rechtsoffizier *m.*
R военный прокурор *м.*
A نائب الاحكام العام

2636 [S]
jump *n.*
F angle de cabrage *m.*
E ángulo de vibración *m.*
D Abgangsfehlerwinkel *m.*
R скачок *м.*
A إهْتِزَاز *m*

2637 [A]
jump *n.*
F saut *m.*
E salto *m.*
D Absprung *m.*
R прыжок *м.;* выброска
    ж.
A قَفْز *m*

2638 [A]
jump *v.*
F sauter en parachute
E saltar
D abspringen *tr.*
R прыгать (с
    парашютом)
A قَفَز

2639
jump altitude
F altitude de largage *f.*
E altura de lanzamiento *f.*
D Absetzhöhe *f.*
R высота выброски *ж.*
A إرْتِفاع القَفز *m*

2640
jump attitude
F position de largage *f.*
E posición de lanzamiento
    f.
D Absetzfluglage *f.*
R положение выброски
    ср.
A وَضْع القَفز *m*

2641
jumpmaster
F chef largueur *m.*
E jefe de salto *m.*
D Absetzer *m.*
R начальник выброски
    м.
A قائد القَفْز *m*

2642 [T]
junction
F bifurcation *f.;* jonction
    f.
E cruce *m.;* empalme *m.*
D Knotenpunkt *m.;*
    Kreuzung *f.*
R перекрёсток *м.;* стык
    дорог *м.*
A مَفْرَق *m*

2643
jungle
F jungle *f.*
E selva *f.*
D Dschungel *m.*
R джунгли *мн. ч.*
A دغل *m*

2644
jungle *adj.*
F (de la) jungle
E (de) selva
D Dschungel- *komp.*
R связанный с
    джунглями
A دغلي

2645
junior
F subalterne
E (de) menor grado;
    subalterno
D Rangniedriger
    m. rangniedrig *adj.*
R младший (офицер,
    сержант) *м.*
A صَغِير

2646
junta
F junte *f.*
E junta *f.*
D Junta *f.*
R хунта *ж.*
A مجلس سياسي

2647
jurisdiction
F compétence *f.;*
  jurisdiction *f.*
E competencia *f.;*
  jurisdicción *f.*
D Rechtsprechung *f.;*
  Zuständigkeit
  (-sbereich) *f. (m.)*
R юрисдикция *ж.;*
  подсудность *ж.*
A القضاء *m*

K

2648
kapok
F capoc *m.*
E capoc *m.*
D Kapok (Pflanzenfaser)
  m.
R капок *м.*
A ‹كابوك *m*

2649
kayak
F kayac *m.*
E kayac *m.*
D Faltboot *neut.;* Kayak
  m. neut.
R каяк *м.*
A كياك (زَوْرق جِلْدِي) *m*

2650
K-day(i.e. convoys)
F jour K *m.*
E día K *m.*
D Tag K *m.*
R день выдвижения
  конвоя *м.;* день-К *м.*
A يوم الإنْطِلَاق *m*

2651
keel
F quille *f.*
E quilla *f.*
D Kiel *m.*
R киль *м.*
A صَالِب *m* ـ رافِدة القَصّ *f*

2652
keel over
F (faire) chavirer
E (hacer) zozobrar
D kentern (lassen)
R опрокидываться
A إنْقَلَب

2653
keep watch (over)
F monter la garde (sur);
  surveiller
E estar de guardia;
  montar la guardia;
  velar (por)
D aufpassen (auf);
  überwachen
R наблюдать (за)
A تَرَقَّب

2654 [P,R]
key *n.*
F clef *f.*
E clave *f.*
D Schlüssel [P] *m.;* Kode
  [R] *m.*
R ключ [P] *м.;* код [R] *м.*
A مِفْتاح *m*

2655 [U]
key *n.*
F clef *f.;* clé *f.*
E llave *f.*
D Schlüssel *m.*
R ключ *м.*
A مِفْتَاح *m*

2656 [E]
key *n.*
F clavette *f.*
E chaveta *f.*
D Keil *m.*
R клин *м.;* шпонка *ж.*
A مِفْتَاح *m*

2657 [T]
key *n.*
F légende *f.*
E leyenda *f.*
D Zeichenerklärung *f.*
R условные знаки *м. мн.*
  ч.
A مِفْتاح *m*

2658
key *adj.*
F (-)clé; (-)clef (suffixe)
E clave (sigue la palabra)
D Haupt- *komp.;*
  Schlüssel- *komp.*
R ключевой; главный
A رَئِيسي

2659 [R]
(telegraphy) key *n.*
F manipulateur *m.*
E manipulador *m.*
D Taste *f.*
R ключ *м.;* кнопка *ж.*
A مِبْراق *m*

2660
keyboard
F clavier de touches *m.*
E teclado *m.*
D Tastatur *f.*
R коммутатор *м.;*
  ключевая доска *ж.*
A لَوْحَة مَفَاتِيح *f*

2661 [S]
kick
F recul *m.*
E retroceso *m.*
D Rückstoss *m.*
R отдача *ж.;* откат *м.*
A تراجع البندقية *m*

2662 [H]
kill *v.*
F tuer
E matar
D töten
R убивать
A قَتَل

165

2663

kill *v.*

F abattre; détruire
E derribar; detruir
D abschiessen
R уничтожать
A قَتَل

2664

kill *n.*

F destruction *f.*
E derribo *m.;* destrucción *f.*
D Abschuss *m.;* Vernichtung *f.*
R уничтожение *ср.*
A قَتْل

2665

killed in action

F tué au combat
E muerto en combate
D (im Felde) gefallen
R павший (в бою) *м.*
A قُتِلَ في العَمَلِيّات

2666

killer satellite

F satellite d'interception *m.*
E satélite de caza *m.*
D Jagdsatellit (meist eng. Begriff) *m.*
R боевой спутник *м.;* боевой космический корабль *м.*
A قمر صِنَاعِي قتّال *m*

2667

kill probability

F probabilité de destruction *f.*
E probabilidad de destrucción *f.*
D Vernichtungs-wahrscheinlichkeit *f.*
R вероятность поражения *ж.*
A إحْتِمَالات القَتْل *f*

166

2668

kill rate

F taux de destruction *m.*
E rasa de destrucción *f.*
D Verlustrate *f.*
R поражающая способность *ж.*
A مُعَدَّل القَتْل *m*

2669

kilometers per hour (kph)

F kilomètres par heure (km/h)
E kilómetros por hora (km/h)
D Stundenkilometer (km/h) *m. pl.*
R (...) километров в час
A كِيلُومِتْر (في ساعة)

2670

kilowatt

F kilowatt *m.*
E kilovatio *m.*
D Kilowatt *neut.*
R киловатт *м.*
A كِيلوواط

2671 [S]

kinetic energy *adj. n.*

F (à) énergie cinétique *(adj.) n. f.*
E energía cinética *adj. s. f.*
D kinetische Energie *f.;* Wucht- *komp.*
R кинетическая энергия *ж.*
A طاقة حَرَكيّة *f*

2672

(cleaning) kit

F trousse de nettoyage *f.*
E equipo de limpia *m.*
D Reinigungssatz *m.*
R комплект инструментов *м.*
A لوازم التنظيف

2673

(combat) kit

F tenue de combat *f.*
E equipo de combate *m.*
D Felduniform *m.*
R боевая одежда *ж.*
A بِزّه القتال

2674

(first aid) kit

F trousse de premier secours *f.*
E botiquín (de primeros auxilios) *m.*
D Selbsthilfesatz *m.*
R санитарная сумка *ж.*
A حَقيبة الإسْعَاف *f*

2675

(mess) kit

F gamelle (individuelle) *f.*
E cubierto de campaña *m.*
D Essgeschirr *neut.*
R парадная обеденная форма (англ.); *ж.* личные столовые принадлежности (амер.) *ж. мн. ч.*
A قَصْعَة فَرديّة *f*

2676

(modification) kit

F nécessaire de modification *m.*
E equipo de modificación *m.;* conjunto de modificación *m.*
D Abänderungssatz *m.*
R модификационный комплект *м.;* ремонтный комплект *м.*
A لَوَازِم التَّعْدِيل *f*

2677

(survival) kit

F trousse de survie *f.*
E equipo de emergencia *m.*
D Überlebenssatz *m.*
R спасательные имущества *ср. мн. ч.*
A اللوازم للبقاء على قيد الحياة *A*

2678
(tool) kit
F jeu d'outils *m.*
E juego de herramientas
   m.
D Werkzeugsatz *m.*
R комплект приборов
   м.; агрегат приборов
   м.
A حَقِيبَة أَدَوَات

2679
kitbag
F musette *f.*
E saco *m.*
D Kleidersack *m.*
R вещевая сумка *ж.*
A كِيس الْعُدَّة

2680
knapsack
F havresac *m.*
E machila *f.*
D Tornister *m.*
R ранец *м.*
A كِيس

2681 [M]
knife
F couteau *m.*
E cuchillo *m.*
D Messer *neut.*
R нож *м.*
A سِكِّين *m*

2682
knife-edge
F couteau de balance *m.*
E filo (de cuchillo) *m.*
D Messerschneide *f.*
R остриё ножа *ср.*
A حَدّ السِّكِّين *m*

2683 [T]
knife-edge
F arête *f.*
E cresta con laderas
   pronunciadas *f.*
D Kammlinie *f.*
R остриё гребня *ср.*
A حَرْف الْجِبَال *m*

2684 [T]
knoll
F monticule *m.;* mamelon
   m.
E otero *m.*
D Bergkuppe *f.*
R бугор *м.*
A تل *m*

2685 [A,M,U]
knot *n.*
F nœud *m.*
E nudo *m.*
D Knoten *m.*
R узел *м.*
A عُقْدَة *f*

2686
knot *v.*
F nouer
E anudar
D verknoten
R завязывать узлом
A عَقَد

L

2687
label *n.*
F étiquette *f.*
E etiqueta *f.;* rótulo *m.*
D Ettikett *neut.*
R ярлык *м.;* этикетка *ж.*
A بِطَاقَة *f*

2688
label *v.*
F étiqueter
E etiquetar; rotular
D etikettieren
R прикреплять ярлык;
   маркировать
A وَضَع بِطَاقَة عَلَى

2689
ladder
F échelle *f.*
E escalera *f.*
D Leiter *f.*
R лестница *ж.;* трап [M]
   м.
A سُلَّم *m*

2690
laden (with)
F chargé (de)
E cargado (con)
D beladen (mit)
R нагруженный
A مُحَمَّل (ب)

2691
lagoon
F lagune *f.*
E laguna *f.*
D Lagune *f.*
R лагуна *ж.*
A بُحَيْرَة مُتَّصِلة بالْبَحر *f*

2692
lake
F lac *m.*
E lago *m.*
D See *m.;* Teich *m.*
R озеро *ср.*
A بُحَيْرَة *f*

2693
laminated
F (en) feuilles; feuilleté;
   (à) lames
E laminado
D Lamellen- *komp.;*
   lamelliert *adj.*
R листовой; слоистый
A رَقَائِقِي

2694
land *n.*
F terre *f.*
E tierra *f.*
D Land *neut.*
R земля *ж.;* суша *ж.*
A بَرّ *m*

2695 [A,L,M]
land *v.*
F atterrir [A];
  apponter[AM];
  débarquer[LM]
E aterrizar [A]; amerizar
  [M]; desembarcar
  [LM]
D landen
R приземляться;
  садиться; высадить
A نَزَل إلى البَرّ

2696
(no man's) land
F zone neutre *f.*
E tierra de nadie *f.*
D Niemandsland *neut.*
R ничья земля *ж.*
A الارض الحرام

2697
(air) landed
F posé (par avion/hélico)
E puesto en tierra (por
  avión/helicóptero)
D (per Luft) gelandet
R посадочно-десантный
A أُنزِل من الجو

2698 [A,M]
landing
F appontage *m.*;
  atterrissage[A] *m.*;
  descente[M] *f.*
E aterrizaje [A] *m.*;
  desembarco [M] *m.*
D Landung *f.*
R десант *м.*; посадка *ж.*
A هُبُوط *m*

2699
(airborne) landing
F atterrissage d'assaut *m.*
E aterrizaje de asalto *m.*
D Luftlandeoperation *f.*
R воздушный десант *м.*
A الانزال من الجو

168

2700 [L,M]
(assault) landing
F débarquement de vive
  force *m.*
E desembarco de asalto
  m.
D Sturmangriff *m.*;
  Angriffslandung
  (amphibisch)
R высадка десанта *ж.*
A الاقتحام من البحر

2701 [A,L]
(assault) landing
F poser d'assaut
E aterrizaje de asalto *m.*
D Sturmangriff *m.*;
  Angriffslandung (aus
  der Luft)
R воздушный десант *м.*
A الاقتحام من الجو

2702
(belly) landing
F atterrissage sur le
  ventre *m.*
E aterrizaje de panza *m.*
D Bauchlandung *f.*
R посадка на брюхо *ж.*;
  посадка с убранным
  шасси *ж.*
A هُبوط بِلا عَجَلات

2703
(instrument) landing
F atterrissage aux
  instruments *m.*;
  atterrissage sans
  visibilité *m.*
E aterrizaje con
  instrumentos *m.*
D Blindlandung *f.*
R слепое приземление
  ср.
A الهبوط بالعدادات

2704
landing aid
F aide à l'atterrissage *m.*
E ayuda al aterrizaje *f.*
D Landehilfsmittel *s.*
R средство обеспечения
  приземления *ср.*
A مُسَاعِد هُبُوط *m*

2705 [L,M]
landing area
F zone de mise à terre *f.*
E zona de aterrizaje [L]
  f.; zona de
  desembarco [M] *f.*
D Absprungraum [L];
  Landeraum [M]
R район высадки *м.*
A مِنْطَقَة الهُبُوط *f*

2706
landing craft
F engin de débarquement
  m.
E barcaza de desembarco
  f.; embarcación de
  desembarco *f.*
D Landungsboot *neut.*
R десантное средство *ср.*
A زَوْرق إنْزَال *m*

2707
landing force
F force de débarquement
  f.
E fuerza de desembarco *f.*
D Landungstruppe *f.*
R (морско-)десантные
  войска *ср. мн. ч.*
A قُوَّة إنْزَال *f*

2708
landing lights *pl.*
F feux d'atterrissage *m.*
  pl.
E faros de aterrizaje *m.*
  pl.
D Landefeuer *neut. pl.*
R посадочные фары *ж.*
  мн. ч.; огни *м. мн. ч.*
A أنْوار الهُبُوط *f*

2709

landing mat

F tapis de débarquement *m.*

E emparillado de aterrizaje (de desembarco) *m.*

D Landematte

R десантный настил *м.*

A حَصِير هُبُوط *m*

2710

landlocked

F entouré par les terres; sans accès à la mer

E cercado de tierra

D vom Meer abgeschnitten

R окруженный сушей

A مُحَاط بِالأَرْض

2711

landmark

F repère *m.*

E referencia terrestre *f.*

D Grenzstein *m.;* Wahrzeichen *neut.*

R ориентир *м.;* местный предмет *м.*

A مَعْلَم أرْضي *m*

2712 [L]

lane

F cheminement *m.*

E trocha *f.*

D (Minen) Gasse *f.*

R проход *м.*

A مَمَرّ *m*

2713 [A]

lane

F voie (aérienne) *f.*

E via (aérea) *f.*

D (Ab-, An-, Ein-) Flugschneise *f.*

R трасса *ж.;* эшелон *м.*

A طَرِيق مِلاَحَة *m*

2714 [M]

lane

F route (de navigation) *f.*

E ruta (de navegación) *f.*

D Fahrrinne *f.*

R путь *м.;* проход *м.*

A طَرِيق مِلاَحَة *m*

2715 [V]

lane

F voie *f.*

E senda *f.*

D Fahrbahn *f.*

R ряд *м.;* трасса *ж.*

A مَمَرّ *m*

2716

language

F langue *f.*

E idioma *m.;* lengua *f.*

D Sprache *f.*

R язык *м.*

A لُغَة *f*

2717 [S]

lanyard

F tire-feu *m. inv.*

E tirafrictor *m.*

D Abzugsleine *f.*

R шнур *м.*

A حَبْل الرَّمْي *m*

2718 [M]

lanyard

F aiguillette *f.*

E acollador *m.*

D Schnur *f.*

R (тросовый) талреп *м.;* стропка *ж.*

A قِيطَان *m*

2719

laser

F laser *m.*

E laser *m.*

D Laser *m.*

R лазерное средство *ср.;* 'лазер' *м.*

A لِيزَر *m*

2720

laser beam

F faisceau de laser *m.*

E rayo laser(ico) *m.*

D Laserstrahl *m.*

R лазерный луч *м.*

A حُزْمَة لِيزَر *f*

2721

laser designator

F marqueur laser *m.*

E marcador laser *m.*

D Laser- Zielbeleuchtungsgerät *neut.*

R лазерное средство облучения цели *ср.*

A مُعَيِّن لِيزَر *m*

2722

laser guidance

F guidage par laser *m.*

E guía por laser *f.*

D Laser- (Zielflug-) Lenkung *f.*

R лазерное наведение *ср.*

A تَوْجِيه بِاللِّيزَر *m*

2723

laser illuminator

F illuminateur laser *m.*

E iluminador laser *m.*

D Laser- Beleuchtungsgerät *neut.*

R лазерное средство подсветки/ освещения цели *ср.*

A مُضِيء لِيزَر *m*

2724

lash *v.*

F arrimer; lier

E atar; sujetar

D befestigen *v.;* festbinden *tr.*

R привязывать; крепить

A رَبَط

2725

lashing

F aiguillette *f.;* amarre *f.*
E amarre *m.;* atadura *f.*
D Lasche *f.;* Schnur *f.*
R найтов *м.;* привязка
   ж.; закрепление *ср.*
A رَبْط *m*

2726

last post

F (sonnerie de)
   l'extinction des feux
   f.; sonnerie des morts
   f.
E toque de oración *m.*
D Zapfenstreich *m.*
R вечерняя заря *ж.*
A نهاية *f*

2727 [N]

latent lethality dose

F dose létale latente
E dosis mortal latente *f.*
D verborgene
   Strahlenbelastungsdosis
   f.
R скрытая смертельная
   доза *ж.*
A جرعة قاتلة كامنة *f*

2728

lateral

F latéral
E lateral
D Seiten *komp.;* seitlich
   adj.
R боковой
A جَانِبِي

2729

latitude

F latitude *f.*
E latitud *f.*
D Breite *f.*
R широта *ж.*
A خَطّ العَرْض *m*

170

2730

latrine

F latrine *f.*
E letrina *f.*
D Latrine *f.*
R уборная *ж.*
A مِرْحَاض *m*

2731 [U]

launch *v.*

F déclencher
E emprender
D laufen lassen
R наносить; бросать
A أطْلَق

2732 [A]

launch *v.*

F lancer
E lanzar
D katapultieren *untr.*
R катапультировать
A اطلق

2733 [L,U]

launch *v.*

F déclencher
E emprender
D starten *untr.;* ansetzen
   tr.
R подниматься в
   операцию
A شنّ

2734 [M]

launch *v.*

F mettre à l'eau
E botar; echar al agua
D (Schiff vom) Stapel
   laufenlassen
R спускать на воду
A انزل إلى البحر

2735 [G,S]

launch *v.*

F lancer
E lanzar
D abschiessen *tr.*
R запускать [G];
   стрелять [S]
A قَذَف

2736 [M]

launch *n.*

F vedette *f.*
E lancha *f.*
D Barkasse *f.*
R катер *м.*
A زورق

2737

launcher

F rampe de lancement *f.;*
   lance- *c.*
E rampa de lanzamiento
   [G] *f.;* lanza- [S] *c.*
D Abschussrampe [G] *f.;*
   Werfer [S] *m.*
R спусковая установка
   ж.; стартовая
   установка *ж.;*
   гранатомёт *м.*
A قَاذِفَة *f*

2738 [B]

launching nose

F bec de lancement *m.*
E pico de lanzamiento *m.*
D Schnabel *m.*
R пусковой бугель *м.*
A رَأْس الإطْلاق

2739

laws of war

F lois de la guerre *f. pl.*
E leyes de la guerra *f. pl.*
D Kriegsvölkerrecht *neut.*
R закон войны *м.*
A قوانين الحرب *f*

2740 [L]
lay
F poser
E sembrar; tender
D legen
R ставить;
　прокладывать
A وَضَع

2741 [S]
lay
F pointer
E apuntar
D richten
R наводить;
　прицеливать
A وَجَّه

2742 [L,M]
lay (cables or mines)
v.
F poser
E colocar; tender [L];
　sembrar [M]
D legen (Kabel/Minen)
R прокладывать кабель;
　минировать
A مدّ حبلاً ـ زرع الغاماً

2743 [S]
lay down a barrage *v.*
F établir un barrage
E establecer una cortina
　de fuego
D Sperrfeuer auslösen *tr.*
R ставить
　заградительный
　огонь
A اقام سداً

2744
laydown bombing
F bombardement en vol
　rasant *m.*
E bombardeo a ras de
　tierra *m.*
D Bombenabwurf mit
　Verzögerungszünder
　im Tiefflug (meist eng.
　Begriff) *m.*
R бомбометание с
　малых высот *ср.*
A قصف أفْقِي من عُلُوِّ الصِفْر

2745
layer
F couche *f.*
E capa *f.*
D Schicht *f.*
R слой *м.*; наводчик [S]
　м.
A طَبَقَة *f*

2746 [L]
(bridge) layer
F lanceur de pont *m.*
E tiendepuentes *m. inv.*
D Brückenlegepanzer *m.*
R мостоукладчик *м.*
A آلة لنصب جسراً

2747 [L]
(mine) layer
F poseur de mines *m.*
E sembrador de minas *m.*
D Minenleger *m.*
R минный раскладчик
　м.
A زراعة الالغام

2748 [S]
(indirect) laying
F pointage indirect *m.*
E puntería indirecta *f.*
D indirektes Richten *neut.*
R непрямая наводка *ж.*
A تَسْدِيد غَيْر مُبَاشِر *m*

2749 [L]
(pattern) laying
F pose de mines suivant
　schéma *f.*
E sembrar minas según
　un plan
D Reihenlegen *neut.*
R планированное
　минирование *ср.*
A زرع الالغام بحسب خطة

2750 [L,M]
(random) laying
F pose de mines au
　hasard *f.*
E sembrar minas
D (planloses) Minenlegen
　neut.
R непланированное
　минирование *ср.*
A وضع الالغام بلا خطة

2751 [L]
(scattered) laying
F pose dispersée de mines
　f.
E sembrar minas (de
　forma dispersa)
D verstreutes Minenlegen
　neut.
R рассеянное
　минирование *ср.*
A وضع الالغام بشكل منتشر

2752 [T]
lay of the land
F configuration du terrain
　f.
E configuración del
　terreno *f.*
D Verlauf des Gelände *m.*
R рельеф местности *м.*
A تشكلات الارض

2753
layout of forces
F déploiement des forces
　m.
E despliegue de fuerzas
　m.
D Truppendislozierung *f.*
R расположение войск
　ср.
A ترتيب القوات

2754 [A,L,M]
lead *v.*
F ouvrir la marche
E llevar la delantera
D führen
R идти первым; водить
A تقدم *m*

171

2755
lead (conduct) *v.*
F mener
E conducir
D führen
R водить; приводить
A قَاد

2756
lead (direct) *v.*
F diriger; mener
E dirigir
D führen
R водить; руководить
A اتجه إلى

2757 [E]
lead *n.*
F câble *m.*
E cable *m.*
D Leitung *f.*
R провод *м.*
A حُزْمَة أسْلاَك

2758 [S]
lead *n.*
F dérive *f.*
E deriva *f.*
D Vorhalt *m.*
R упреждение *ср.*; угол
    упреждения *м.*
A التَّصْوِيب أمام الهَدَف *m*

2759
leader
F chef *m.;* commandant
    m.
E jefe
D Führer *m.*
R руководитель *м.*
A قائد

2760
leadership
F direction *f.;*
    commandement *m.*
E dotes de mando *f. pl.*
D Führen *neut.;* Führung
    f.
R руководство *ср.;*
    водительство *ср.*
A قيادة *f*

172

2761
lead time
F délai *m.*
E plazo de anticipación
    m.
D Anlaufzeit *f.*
R осуществительное
    время *ср.;*
    упредительное время
    [S] *ср.*
A الوَقْت المَطْلُوب لِبِنَاء الطَائِرَة

2762
leakage
F fuite *f.*
E escape *m.*
D Lecken *neut.*
R течь *ж.*
A تسرب *m*

2763
leapfrog *v.*
F (se) déplacer en
    perroquet ou par
    bonds
E pasar de escalón
D überschlagend vorgehen
    tr.
R двигаться перекатами
A تَقَدَّم بِوُثْبَات

2764
learn
F apprendre
E aprender; saber
D erfahren; lernen
R учиться
A تَعَلَّم

2765
leave *n.*
F congé *m.;* permission *f.*
E permiso *m.*
D Urlaub *m.*
R отпуск *м.*
A إجَازَة *f*

2766
leaver
F navire détaché d'un
    convoi *m.*
E buque destacado de un
    convoy *m.*
D alleinfahrendes Schiff
    (z.B. aus den
    Geleitzug) *neut.*
R корабль,
    отделяющийся от
    конвоя *м.*
A سفينة تترك القافلة

2767
(to) leeward
F sous le vent
E (a) sotavento
D leewärts
R (под) ветер
A تَحْت الرِّيح

2768
leeway
F dérive *f.*
E deriva *f.*
D Abtrift *f.*
R снос *м.;* дрейф *м.*
A إنْجِرَاف (مَعَ الرِّيح)

2769
left *adj.*
F gauche
E izquierdo
D link
R левый
A يَسَار

2770
leg
F portion d'itinéraire *f.;*
    étape *f.*
E etapa *f.;* tramo *m.*
D Etappe *f.*
R отрезок маршрута *м.*
A مرحلة *f*

2771
legal
F légal
E legal
D gesetzlich *adj.;* Rechts-
    komp.
R легальный; законный
A قانوني

2772
(map) legend
F légende *f.*
E leyenda *f.*
D Zeichenerklärung *f.*
R условное обозначение
    ср.
A مفتاح الخريطة

2773
length
F longueur *f.*
E largo *m.*
D Länge *f.*
R длина *ж.*
A طُول *m*

2774
lens
F lentille *f.;* objectif *m.*
E lente *f.;* objetivo *m.*
D Linse *f.;* Objektiv *neut.*
R линза *ж.*
A عَدَسَة *f*

2775 [C,N,S]
lethal
F meurtrier
E mortífero
D tödlich
R смертоносный
A مُميت

2776 [H,V]
lethal
F létal
E letal
D tödlich
R смертельный
A مُميِّت

2777
lethality
F létalité *f.;* caractère
    meurtrier *m.*
E letalidad *f.*
D Tödlichkeit [CHNU] *f.;*
    Zielwirkung [S] *f.*
R поражаемость *ж.*
A إماتة *f*

2778
level *n.*
F niveau *m.;* plan *m.*
E nivel *m.*
D Ebene *f.;*
R уровень *м.*
A مُستوى *m*

2779
level *adj.*
F (de) niveau; (en) palier
E (a) nivel; nivelado
D eben
R ровный
A مُسْتَوٍ

2780 [B]
level *v.*
F niveler
E nivelar
D planieren
R выравнивать;
    нивелировать
A سَوَّى

2781 [B]
level *n.*
F niveau à bulle *m.*
E nivel *m.*
D Nivelliergerät *neut.*
R уровень *м.*
A مِسواه

2782 [T]
level *n.*
F partie platte *f.*
E llano *m.*
D Ebene *f.*
R равнина *ж.*
A سهل

2783
level of supply
F niveau des
    approvisionnements
    m.
E nivel de abastacimiento
    m.
D Bevorratungshöhe *f.*
R норма снабжения *ж.*
A مستوى التموين *m*

2784
lever *n.*
F levier *m.*
E palanca *f.*
D Hebel *m.*
R рычаг *м.;* ручка *ж.*
A عَتَلة ـ رَافِعَة *f*

2785
lever (up) *v.*
F soulever (à l'aide d'un
    barre ...)
E alzar (con palanca)
D (mit einer Stange usw.)
    heben
R поднимать (рычагом)
A رَفَع بِعَتَلة

2786
liaise (with)
F faire la liaison (entre)
E enlazar; establecer
    enlace (con)
D Verbindung aufnehmen
    (mit) *v. tr.*
R поддерживать связь
    (с)
A إرْتَبَط (ب)

2787
liaison
F liaison *f.*
E enlace *m.*
D Verbindung *f.*
R связь взаимодействия
    ж.
A إرْتِباط *m*

2788
liberate
F libérer
E libertar; poner en
libertad
D befreien
R освобождать
A حَرَّر

2789
liberty
F liberté f.
E libertad f.
D Freiheit f.
R свобода ж.
A حُرِّيَّة f

2790 [M]
liberty boat
F vedette des
permissionnaires f.
E lancha de servicio f.; de
marineria
D Landgängerboot neut.
R шлюпка с
увольняемыми на
берег ж.
A زَوْرَق الإِجازة m

2791 [P]
lie detector
F détecteur de mensonges
m.
E detector de mentiras m.
D Lügendetektor m.
R полиграф м.
A آلة لكشف الكذب

2792 [M]
lie off
F rester au large
E mantenerse a la altura
D vom Lande abhalten v.
tr.
R находиться на
некоротком
расстоянии (от
берега)
A بَقِيَ بعيداً بَعْض الشيْءِ عن
الشاطِىء

2793 [T]
lie of the land
F configuration du terrain
f.
E configuración del
terreno f.
D Verlauf des Gelände m.
R рельеф местности м.
A تشكلات الأرض

2794 [N]
(half) life
F période radioactive f.
E período de
radioactividad m.
D Halbwertzeit f.
R полураспад м.
A نصف العمر

2795 [E]
(service) life
F durée économique (de
vie) f.
E tiempo de vida m.
D Lebensdauer f.
R эксплуатационный
срок службы м.
A عُمْر الخِدْمَة

2796
(shelf) life
F durée de conservation f.
E vida f.
D Haltbarkeitsdauer f.
R сохранность ж.
A العمر الاقصى للخزن

2797
lifebelt
F ceinture de sauvetage f.
E cinturón salvavidas m.
D Rettungsgürtel m.
R спасательный пояс м.
A حِزَام النَّجاة m

2798
lifeboat
F canot de sauvetage m.
E lancha de socorro f.;
bote salvavidas (de
buque) m.
D Rettungsboot neut.
R спасательная шлюпка
ж.
A قَارِب النَّجاة m

2799
lifeline
F ligne de sauvetage f.;
garde-corps (d'un
bâtiment) m.
E andarivel de
salvamento (de
buque) m.; cuerda
salvavidas f.
D Rettungslinie f.
R спасательная верёвка
ж.
A حَبْل النَّجاة

2800
lift v.
F lever
E alzar
D heben
R поднимать
A رَفَع

2801 [G]
lift off v.
F décoller
E despegar
D starten
R стартоваться;
запускаться
A إِنْطَلَق

2802 [G]
liftoff n.
F décollage m.
E despegue m.
D Start m.
R запуск м.; отрыв м.
A إِنْطِلاق m

2803 [A,M,U,V]
light *n.*
F feux [AMV] *m. pl.;*
  lumière *f.*
E luz *f.*
D Licht *neut.;* Feuer
  [AMV] *neut.*
R свет *м.;* огонь *м.;*
  огонёк [AMV] *м.*
A نُور *m*

2804
light *v.*
F allumer; éclairer
E encender; alumbrar
D anzünden *tr.;*
  beleuchten *v.;*
  befeuern [AM] *v.*
R зажигать; светить
A اشعل

2805 [M,U]
light *adj.*
F (en) lest
E ligero; (en) lastre [M]
D leicht; unbeladen [M]
R лёгкий
A خَفِيف

2806
light (illuminate) *v.*
F éclairer
E alumbrar; iluminar
D beleuchten *untr.*
R освещать
A أضاء

2807
light (switch on) *v.*
F allumer
E encender
D Licht anmachen *tr.*
R включать
A فتح الأنوار

2808
(set) light (to) *v.*
F allumer
E encender
D Feuer anzünden *tr.*
R зажигать
A أشْعَل

2809
light alloy *adj. n.*
F (en) alliage léger
  (adj.) n. m.
E (de) aleación ligera
  (adj.) s. f.
D Leichtmetall *neut.*
R лёгкий сплав *м.;*
  лёгкосплавной
  прилаг.
A سَبِيكة خَفِيفَة *f*

2810
light buoy
F bouée lumineuse *f.*
E boya luminosa *f.*
D Leuchttonne *f.*
R освещаемый буй *м.*
A طَافَية مُضِيئَة *f*

2811 [M]
lighter
F péniche *f.;* chaland *m.*
E barcaza *f.;* gabarra *f.;*
  lanchón *m.*
D Leichter *m.*
R лихтер *м.*
A صَنْدَل *m*

2812
lighthouse
F phare *m.*
E faro *m.*
D Leuchtturm *m.*
R маяк *м.*
A مَنَارَة *f*

2813
light machinegun
F mitrailleuse légère *f.*
E ametralladora ligera *f.*
D leichtes
  Maschinengewehr
R лёгкий пулемёт *м.*
A رَشَاش خَفِيف *m*

2814
lightning
F éclairs *m. pl.;* foudre *f.*
E relámpago *m.;* rayo *m.*
D Blitz *m.*
R молния *ж.*
A بَرْق *m*

2815 [R]
lightning arrester
F parafoudre *m.*
E pararrayos *m. inv.*
D Blitzschutz *m.*
R молниеотвод *м.*
A واقية صواعق *f*

2816
lightning rod
F paratonnerre *m.*
E pararrayos *m. inv.*
D Blitzableiter *m.*
R молниеуводитель *м.;*
  молниеотвод *м.*
A مانعة صواعق *f*

2817
lightship
F bateau-feu *m.;* bateau-
  phare *m.*
E buque-faro *m.*
D Feuerschiff *neut.*
R плавучий маяк *м.*
A مَرْكَب مَنار *m*

2818
lights out
F extinction des feux *f.*
E hora de apagar las
  luces *f.*
D Lichter löschen
  (Tagesende)
R вечерняя заря *ж.*
A إطفاء الأنوار

2819
limb
F membre *m.*
E miembro *m.*
D Glied *neut.*
R член *м.*
A عُضْو *m*

2820
limber *n.*
F avant-train *m.*
E armón (de artillería) *m.*
D Protze *f.*
R передок *м.*
A عَرَبَة المِدْفَع *f*

2821
limit *adj.*
F limite
E limitado; límite (sigue la palabra)
D Grenz- *komp.*
R предел *м.*
A حَدّ *m*

2822
limit *n.*
F limite *f.*
E límite *m.*
D Grenze *f.*
R предельный
A حَدّ *m*

2823
limit (to) *v.*
F limiter (à)
E limitar (a)
D begrenzen einschränken (auf) *tr.*
R ограничивать
A حَدَّد *m*

2824
limit of advance
F limite d'action *f.*
E límite de acción *m.*
D Angriffsgrenze *f.*
R предельный рубеж *м.*; предел наступления *м.*
A حد التقدم

2825
limit of exploitation
F limite d'action *f.*
E límite de acción *m.*
D Verfolgungsgrenze *f.*
R конечный рубеж *м.*
A حد الاستغلال

2826
limit of fire
F limite de tir *f.*
E límite (de seguridad) de tiro *m.*
D Sicherheitsgrenze *f.*
R граница обстрела *ж.*
A حَدّ الرَّمْي *m*

2827 [S]
line
F direction *f.*
E dirección *f.*
D Schusslinie *f.*
R направление *ср.*
A خط التسديد

2828 [A,L]
(anchor) line
F câble de parachutage *m.*
E cable de paracaídas *m.*
D Haftseil *neut.*
R вытяжная верёвка *ж.*
A حبل التثبيت *m*

2829 [S]
(artillery control) line
F ligne de sécurité *f.*
E línea de seguridad *f.*
D Feuerleitlinie *f.*
R основное направление *ср.*
A خط ضبط المدفعية

2830
(battle) line
F ligne de bataille *f.*
E línea de batalla *f.*
D Kampflinie *f.*
R боевой порядок *м.*
A خط المعركة

2831 [G,S]
(bomb release) line
F point de lancement *m.*
E punto de lanzamiento *m.*
D Abschusspunkt *m.*
R точка сброса *ж.*; точка запуска *ж.*
A خَطّ إطْلاق القَنَابِل

2832
(contour) line
F courbe de niveau *f.*
E curva de nivel *f.*
D Höhenlinie *f.*
R горизонталь *м.*
A خَطّ المَنَاسِيب *m*

2833
(final coordination) line
F ligne de coordination des feux *f.*
E línea de coordinación final *f.*
D Endkoordinationslinie *f.*
R последний рубеж огневого вала *м.*
A خَطّ التَّنْسِيق النهائِي

2834
(final protective) line
F ligne d'arrêt *f.*
E línea de protección final *f.*
D letzte Deckungslinie *f.*
R предельный защитный рубеж *м.*
A خَطّ الحِمَايَة النهائي

2835
(firing) line
F ligne de tir *f.*
E línea de fuego *f.*
D Feuerstellung *f.*
R боевая линия *ж.*
A خَطّ الرَّمْي *m*

2836
(gun-target) line
F ligne tireur-but *f.*
E línea pieza-objetivo *f.*
D Schussrichtung *f.*; Ziellinie *f.*
R линия 'орудие-цель' *ж.*
A الخَطّ بَيْن المِدْفَع والهَدَف *m*

2837
(holding) line
F ligne de sûreté *f.*
E línea límite de fuegos *f.*
D Bereitschaftslinie *f.;*
  Wartelinie *f.*
R рубеж сопротивления
  м.
A خط القوات الساترة

2838 [T]
(interrupted) line
F ligne discontinue *f.*
E línea quebrada *f.*
D Strichlinie (bzw.
  Punktlinie) *f.*
R пунктирная линия *ж.*
A خطّ مَقْطُوع *m*

2839 [N]
(isodose rate) line
F ligne d'égale intensité
  (radioactive) *f.*
E línea de igual
  intensidad (radiáctiva)
  f.
D Isodosenkurve *f.*
R линия равных уровней
  радиации *ж.*
A خطّ جُرْعَات إشْعَاعِية
  مُتَسَاوية

2840
(isogonic) line
F ligne isogone *f.*
E línea isogónica *f.*
D Isogone *f.*
R изогоническая линия
  ж.
A خطّ التَحارُف

2841
(no-fire) line
F ligne de sécurité *f.*
E línea de seguridad *f.*
D Sicherheitsgrenze *f.*
R рубеж безопасности
  ведения огня *м.*
A خطّ عَدَم الرَّمْي *m*

2842
(nuclear safety) line
F ligne de sécurité
  nucléaire *f.*
E línea de seguridad
  nuclear *f.*
D Atom-Sicherheitsgrenze
  f.
R рубеж ядерной
  безопасности *м.*
A خَطّ الأمان النَّوَوِي *m*

2843
(observer-target) line
F ligne d'observation *f.*
E línea observador-
  objetivo *f.*
D Sehstreifen *m.*
R линия
  наблюдатель-цель
  ж.
A الخَطّ بين الرَّاصِد والهَدَف

2844 [R]
(off) line
F (hors) ligne
E (fuera de) línea
D Off-Line Betrieb *m.*
R автономный
A معالجه المعطيات بشكل
  غير مباشر

2845 [R]
(on) line
F (en) ligne
E (en) línea
D On-Line Betrieb *m.*
R управляемый
A معالجه المعطيات بشكل
  مباشر

2846
(orienting) line
F direction repère *f.*
E línea de orientación *f.*
D Richtungslinie *f.*
R ориентирная линия
  ж.; нулевая линия
  ж.
A خطّ التَّوْجِيه *m*

2847 [L]
(phase) line
F limite de bond *f.*
E límite de salto *m.*
D Durchlauflinie *f.*
R промежуточный
  уравнительный
  рубеж *м.*
A خط التبليغ

2848 [A,M]
(position) line
F ligne de position *f.*
E línea de posición *f.*
D Standlinie *f.*
R позиционная линия
  ж.
A خطّ المَوْضِع

2849
(reference) line
F ligne de référence *f.*
E línea de referencia *f.*
D Basislinie *f.*
R нулевая линия *ж.*
A خطّ المَرْجِع

2850 [L]
(report) line
F ligne de compte-rendu
  f.
E línea de informe *f.*
D Meldelinie *f.*
R рубеж отправки
  докладов *м.*
A خط الإلتِحاق

2851
(rhumb) line
F loxodromie *f.*
E línea de rumbo *f.;* línea
  loxodrómica *f.*
D Kompasslinie *f.*
R локсодромия *ж.*
A خَط مُتَسَاوِي المَيْل

177

2852 [L]

(safety) line

F marquage de sécurité
   m.
E sendero seguro *m.*
D Sicherheitslinie *f.*
R рубеж безопасности
   м.
A خط الامان

2853 [A,L]

(static) line

F sangle d'ouverture
   automatique *f.*
E cable de apertura
   automática *m.*
D selbsttätige Aufziehleine
   f.
R вытяжная верёвка *ж.*
A الحبل القراري

2854 [L]

(thrust) line

F axe de progression *m.*
E dirección de ataque *f.*
D Angriffsachse *f.*
R ось наступления *ж.*
A مُحْوَر الدَّفْع

2855

line abreast

F ligne (de front) *f.*
E línea frontal *f.*
D Dwarslinie *f.*
R строй фронта *м.*
A خَطّ الإصْطِفاف *m*

2856

linear

F linéaire
E lineal; (de) longitud
D linear *adj.;* Linear-
   komp.; Längs- *komp.*
R линейный
A خَطّي

178

2857

linear defense

F défense linéaire *f.*
E defensa lineal (frontal)
   f.
D lineare Verteidigung *f.*
R линейная оборона *ж.*
A الدِّفاع الخَطّي *m*

2858

linear scale

F échelle linéaire *f.*
E escala lineal *f.*
D lineare Skala *f.*
R линейная шкала *ж.*
A مِقْياس خَطّي *m*

2859

line astern

F colonne *f.*
E columna *f.*
D Kiellinie *f.*
R строй кильватера *м.*
A خَطّ الى الخَلْف

2860 [L]

line of departure

F ligne de débouche *f.*
E línea de partida *f.*
D Ablauflinie *f.*
R исходный рубеж *м.;*
   исходная линия *ж.*
A خط الخروج

2861 [S]

line of departure

F ligne de projection *f.*
E línea de proyección *f.*
D Abgangsrichtung *f.*
R исходный рубеж *м.;*
   исходная линия *ж.*
A خَطّ الخُرُوج

2862 [S]

line of elevation

F ligne de tir *f.*
E línea de tiro *f.*
D Höhenlinie *f.*
R линия возвышения *ж.*
A خط الارتفاع

2863

line of position

F ligne de position *f.*
E línea de situación *f.*
D Ziellinie *f.*
R линия цели *ж.*
A خَطّ المَوْضِع

2864 [S]

line of sight

F ligne de mire *f.;* ligne
   de visée *f.*
E línea de mira *f.*
D Visierlinie *f.*
R линия прицеливания
   ж.
A خط البصر

2865 [R]

line of sight

F vue optique *f.*
E vista óptica *f.*
D Richtlinie *f.*
R линия зрения *ж.*
A خط البصر

2866

line of supply

F échelon de soutien *m.;*
   système de soutien *m.*
E línea de abastecimiento
   f.; línea de suministro
   f.
D Versorgungssystem
   neut.
R путь снабжения *м.;*
   путь подвоза *м.*
A خَطّ التَّمْوين

2867 [L]

line of withdrawal

F cheminement de repli
   m.
E camino de retiro *m.*
D Rückzugslinie *f.*
R путь отхода *м.*
A خط الإنسحاب

2868 [M]
liner
F paquebot *m.*
E buque de línea *m.*
D Linienschiff *neut.*
R лайнер *м.*;
   пассажирский
   пароход *м.*
A بَاخِرة *f*

2869 [L]
lines
F ligne avant *f.*
E primera línea *f.*
D vorderen Linien
   eigener/feindlicher
   Truppen
R рубеж *м.*; линия *ж.*
A الخطوط الامامية

2870 [L]
**lines of
   communication**
F lignes de
   communication *f. pl.*
E líneas de comunicación
   f. pl.
D Verbindungslinien *f.*
   (pl.)
R пути сообщения *м.*
   мн. ч.
A خطوط المواصلات

2871 [L]
line up *v.*
F mettre en ligne
E formar en línea
D aufstellen *tr.*
R выстраивать(ся)
A صَفّ

2872 [E]
line up *v.*
F aligner
E alinear
D ausrichten *tr.*
R присоединять(ся)
A صَفّ

2873 [E]
lining
F garniture *f.;* revêtement
   m.
E guarnición *f.;*
   revestimiento *m.*
D Belag *m.*
R подкладка *ж.*; обивка
   ж.
A بِطَانَة *f*

2874 [R]
link *v.*
F relier
E enlazar
D verbinden
R связывать(ся)
A إتَّصَل بِ

2875 [R]
link *n.*
F liaison *f.*
E conexión *f.*
D Verbindung *f.*
R радиорелейная линия
   ж.; связь *ж.*
A اتَّصَال *m*

2876 [R]
(data) link
F transmission des
   données *f.*
E transmisión de datos *f.*
D automatische
   Datenübertragungs-
   verbindung *f.*
R автоматическая
   передача данных *ж.*
A وصلة معطيات

2877
linkage
F liaison *f.*
E sistema articulado *m.*
D Verkettung *f.*
R сцепление *ср.*;
   соединение *ср.*
A مَجْمُوعَة وُصَل *f*

2878
liquid *adj. n.*
F liquide *adj. n. m.*
E líquido *adj. s. m.*
D flüssig *adj.;* Flüssigkeit
   f.
R жидкий *прилаг.*;
   жидкость *ж.*
A سَائِل *m*

2879
liquid propellant
F propergol liquide *m.*
E líquido propulsor *m.*
D Flüssigkeitstreibstoff *m.*
R жидкое топливо *ср.*
A وقود سائل *m*

2880 [M]
list *n.*
F bande *f.;* gîte *f.*
E escora *f.*
D Schlagseite *f.*
R крен *м.*
A مَيْل *m*

2881 [U]
list *n.*
F liste *f.*
E lista *f.*
D Liste *f.*
R список *м.*
A جَدْوَل *m*

2882 [M]
list *v.*
F prendre de la gîte
E escorar
D krängen
R (на)крениться
A مَال

2883 [H]
(casualty) list
F état des pertes *m.*
E número de bajas *m.*
D Verwundetenrolle *f.*
R список раненных *м.*
A بيان الخسائر

179

2884

(target) list

F liste d'objectifs *f.*
E lista de objetivos *f.*
D Zielliste *f.*
R список целей *м.;*
распределение целей
м.
A جدول اهداف

2885

listen

F écouter
E escuchar
D horchen
R слушать;
подслушивать
A اسْتَمَع

2886 [H]

litter

F brancard *m.*
E camilla *f.;* litera *f.*
D (Feld-) Trage *f.*
R носилки *ж. мн. ч.*
A نَقَّالة *f*

2887

litter bearer

F brancardier *m.*
E camillero *m.*
D Krankenträger *m.*
R носильщик *м.*
A نَقَّال *m*

2888

litter patient

F malade couché *m.*
E herido transportado en
camilla *m.*
D bettlägeriger Patient *m.*
R носилочный раненый
м.
A مَريض النَّقَّالة *m*

180

2889

littoral

F littoral *m.*
E litoral *m.*
D Küstenstreifen *m.*
R побережье *ср.;*
приморье *ср.*
A سَاحِلي

2890 [E]

live

F en charge; sous tension
E conectado
D stromführend
R (под) напряжением
A مُكَهْرَب

2891 [B,S]

live

F chargé
E activado; cargado
D scharf
R боевой; действующий
A حَيّ

2892

live firing

F tir à balles réelles *m.*
E tiro con balas *m.*
D Scharfschiessen *neut.*
R боевая стрельба *ж.*
A الرمي بذخيرة حيّة

2893 [G,S]

load *n.*

F charge *f.*
E carga *f.*
D Ladung *f.*
R заряд *м.*
A حُمُولَة *f*

2894 [S]

load *v.*

F charger
E cargar
D laden *untr.*
R заряжать
A مـلأ

2895 [A,E,M,V]

load *n.*

F charge *f.*
E carga *f.*
D Belastung [E] *f.;* Last
[AMV] *f.;*
R груз [AMV] *м.;*
нагрузка [E] *ж.*
A شحن

2896 [A,E,M,V]

load *v.*

F charger
E cargar
D laden; aufladen [AEMV];
beladen [AMV]
R грузить; нагружать
A حَمَّل

2897 [A,M]

(allowable cargo) load

F charge maximum
autorisée *f.*
E carga máxima *f.*
D zulässige
Höchstbelastung *f.*
R максимально-
допустимая нагрузка
ж.
A الحمولة القصوى

2898

(basic) load

F dotation initiale *f.*
E dotación mínima *f.*
D Ausgangswerkstoff *m.*
R наличные боеприпасы
м. мн. ч.;
положенные
боеприпасы *м. мн. ч.*
A مُرتَّب الذخيرة *m* ـ الوحدة
النارية *f*

2899

(palletized unit) load

F charge standard sur
palette *f.*
E carga standard en
plataforma *f.*
D palettierte
Sammelpackung *f.*
R груз на поддонах *м.*
A حُمُولات مَجْمُوعَة على مِنَصَّة

2900

(prescribed) load
F dotation prescrit *f.*
E dotación reglamentaria *f.*
D vorgeschriebener (Material-) Ausrüstungsbestand *m.*
R установленное снабжение части *ср.*
A الحمولة المخصصة

2901

(safe working) load
F charge pratique limite *f.*
E carga práctica de seguridad *f.*
D zulässige Belastung *f.*
R допускаемая нагрузка *ж.*
A حَمُولة تَشْغِيل مَأْمُونة

2902 [A]

(sling) load
F capacité d'emport sous élingue *f.*
E capacidad de llevar carga bajo eslinga *f.*
D Geschirrlast *f.*
R грузоподъёмность вертолёта *ж.*
A الحمولة المعلقة القصوى

2903 [A]

(slung) load
F charge sous élingue *f.*
E carga llevada bajo eslinga *f.*
D Geschirrlast *f.*
R подвесной груз *м.*
A الحمولة المعلقة

2904

(standard) load
F chargement standard *m.*
E carga reglementaria *f.*
D Standardlast *f.*
R стандартный груз *м.*
A حَمُولة قِيَاسِيَّة *f*

2905

(underslung) load
F charge sous élingue *f.*
E carga transportada debajo de helicóptero *f.*
D Hängeladung *f.*
R подвесной груз *ср.*
A حمولة معلقة

2906

(unit) load
F charge normalisée *f.*
E carga por unidad *f.*
D Gemischtbeladung *f.;* Sammelpackung *f.*
R партия груза для части
A وَحْدَة الحِمْل

2907

loaded
F chargé
E cargado
D geladen
R заряженный; погруженный
A مُحَمَّل

2908 [A]

loading
F chargement *m.*
E carga *f.*
D Belastung *f.*
R погрузочный коэффициент *м.*
A عامل تحميل القدرة

2909 [S]

(breech) loading *adj.*
F chargé par la culasse
E (de) retrocarga
D Waffe mit Hinterladung *f.*
R заряженный с казённой части
A يُملأ من المغلق

2910

(convoy) loading
F chargement par convoi *m.*
E carga para un convoy *f.*
D Geleitzugbeladung *f.*
R погрузка по конвойным эшелонам *ж.*
A تَحْمِيل القَافِلة

2911

(tactical) loading
F chargement tactique *m.*
E carga táctica *f.*
D Gemischtbeladung *f.;* Sammelpackung *f.*
R тактическая погрузка *ж.*
A تَحْمِيل تَعْبَوِي

2912

(unit) loading
F chargement par unité constituée *m.*
E embarque por unidades completas *m.*
D Verladung einer Truppeneinheit *f.*
R прогрузка части на одно средство транспорта *ж.*
A تَحْمِيل الوَحْدَة

2913 [M]

(vertical) loading
F chargement vertical *m.*
E cargamento vertical *m.*
D Vertikalladung *f.*
R вертикальная погрузка *ж.;* погрузка с воздуха *ж.*
A تحميل أفقي

2914

load spreader
F répartiteur de charges *m.*
E distribuidora de cargas *f.*
D Belastungsverteiler *m.*
R распределитель нагрузки *м.*
A ناشِرة الوَزْن *f*

2915

loan *n.*
F prêt *m.*
E préstamo *m.*
D Anleihe *f.*
R заём *м.*
A قَرْض *m*

2916

loan *v.*
F prêter
E prestar
D verleihen
R давать взаимы; ссужать
A أقْرَض

2917

loan personnel
F personnel détaché (auprès de...) *m.*
E personal prestado *m.*
D Leihpersonal *neut.*
R временно приданный личный состав *м.*
A أفْرَاد مُسْتَعِيرون *m*

2918

local
F local
E local
D örtlich *adj.;* Orts-*komp.*
R местный; локальный
A مَحَلِّي *m*

2919 [L]

(tactical) locality
F zone d'intérêt tactique *f.*
E zona de importancia táctica *f.*
D (taktisches) Gebiet *neut.*
R местность *ж.*
A نقطة تعبوية *f*

2920

localize *v.*
F localiser
E localizar
D begrenzen
R ограничивать
A حصر في موضع

2921

local purchase
F achat local *m.*
E compra local *f.*
D örtlicher Ankauf *m.*
R местные закупки *ж. мн. ч.*
A شِرَاء مَحَلِّي *m*

2922

local time
F heure locale *f.*
E hora local *f.*
D Ortszeit *f.*
R местное время *ср.*
A تَوْقيت مَحَلِّي *m*

2923

locate
F localiser
E localizar
D örtlich festlegen; peilen [AMRS]; stellen
R обнаруживать; располагать
A حَدَّد مَكَاناً

2924

locate *v.*
F repérer; situer
E localizar; colocar
D orten; verlegen
R обнаруживать; размещать
A حدِّد مكانا

2925 [S]

locating battery
F batterie d'acquisition d'objectifs *f.*
E batería de adquisición de blancos *f.*
D aufklärende Batterie *f.*
R батарея, отыскивающая целей *ж.*
A سرية تحديد الهدف *f*

2926

location
F emplacement *m.;* position *f.*
E situación *f.;* posición *f.*
D Lage *f.;* Platz *m.;* Stellung *f.*
R местонахождение *ср.;* обнаруживание *ср.*
A مَوْقِع *m*

2927 [U]

lock *n.*
F serrure *f.*
E cerradura *f.*
D Schloss *neut.*
R замок *м.*
A قُفْل *m*

2928 [E]

lock *n.*
F blocage *m.*
E traba *f.*
D Sperrvorrichtung *f.*
R стопор *м.;* чека *ж.*
A قُفْل *m*

2929 [U]
lock *v.*
F fermer à clef
E cerrar con llave
D verschliessen
R запирать на замок
A أغْلَقَ

2930 [E]
lock *v.*
F enclencher
E trabar
D sperren
R запирать
A ثَبَّت

2931 [S]
lock *n.*
F platine *f.*
E cierre *m.*
D Verschluss *m.*
R запирающий
механизм *м.*; затвор
м.
A سَدّ *m*

2932 [M]
lock *n.*
F écluse *f.*
E esclusa *f.*
D Schleuse *f.*
R шлюз *м.*
A رِتاج *m*

2933 [G,R,S]
lock on *adj.*
F (à) verrouillage (radar)
E (a) localización
D anhängefähig
R захватывающий;
захватный
A متابع الهَدَف بالرَّادار

2934 [G,R,S]
lock on (to) *v.*
F accrocher; verrouiller
(à)
E captarse (a)
D anhängen *tr.*
R ловить (цель);
захватывать
A تابَع الهَدَف بالرَّادار

2935 [M]
lock through *v.*
F écluser
E (hacer) pasar por una
esclusa
D durchschleusen *tr.*
R шлюзовать
A مَرَّ بالسَّدّ

2936
locomotive
F locomotive *f.*
E locomotora *f.*
D Lokomotive *f.*
R локомотив *м.*
A قاطِرَة

2937
(to make a) lodgement
F (s')installer dans une
position avantageuse
E ganar una posición
ventajosa
D Halt *m.*
R закрепляться на
захваченной позиции
A إستولى على

2938 [R]
log *v.*
F repérer
E localizar
D erfassen
R записывать
A سَجَّل

2939 [A,E,M,L,R]
log *n.*
F journal (de bord) [AM]
m.; loch [M] *m.*
E diario (de navegación)
m.; registro de
navegación *m.*
D Bordbuch [A] *neut.;*
Logbuch [M] *neut.;*
R вахтенный журнал [M]
м.; бортовой журнал
[A] *м.*
A سِجِل *m*

2940 [E,L,R]
log *n.*
F carnet *m.*
E cuaderno de trabajo [E]
m.
D Logbuch [ER] *neut.;*
Tagebuch [ELR] *neut.*
R журнал *м.*; журнал
радиостанции [R] *м.*
A خَشَب *m*

2941
log *v.*
F porter (au journal
etc...)
E registrar
D eintragen *tr.;* loggen
R вносить в журнал
A سَجَّل

2942
logarithm
F logarithme *m.*
E logaritmo *m.*
D Logarithmus *m.*
R логарифм *м.*
A لُوغارِثْمَة *f*

2943
logistic(al)
F logistique
E logístico
D logistisch
R тыловой;
интендантский
A إداري

2944
logistical command
F groupement logistique
m.
E mando logístico *m.*
D logistische
Kommandobehörde *f.*
R командование тыла
ср.
A قِيَادَة إداريَّة *f*

2945
logistical constraint
F contrainte logistique *f.*
E dificultad logística *f.*
D logistischer Fehlbetrag *m.*
R недостатки в материально-техническом обеспечении *м. мн. ч.*
A قيد اداري

2946
logistical estimate
F bilan logistique *m.*
E cálculo logístico *m.;* apreciación (estimación) logística *f.*
D Beurteilung der Versorgungslage *f.*
R оценка возможностей тыла *ж.*
A تَقْدِير إدَاري *m*

2947
logistics
F logistique *f.*
E logística *f.*
D Logistik *f.*
R тыл и снабжение *м.;* техника штабной службы *ж.*
A الشُّؤُون الإدَارِيَّة *f*

2948
long
F long
E largo
D lang
R длинный; долгий
A طويل

2949
longitude
F longitude *f.*
E longitud *f.*
D Länge *f.*
R долгота *ж.;* длина *ж.*
A خَطّ الطُّول *m*

2950
longitudinal
F longitudinal
E longitudinal
D Längs- *komp.*
R продольный
A طُولي

2951
longitudinal axis
F axe longitudinal *m.*
E eje longitudinal *m.*
D Längsachse *f.*
R продольная ось *ж.*
A مِحْوَر طُولي *m*

2952
long-wave *adj.*
F (à) ondes longues
E (de) onda larga
D Langwellen- *komp.*
R длинноволновой
A طَويل الموجة

2953 [L]
lookout *n.*
F guetteur *m.*
E vígia *m.*
D Wachtposten *m.*
R наблюдатель *м.*
A مُرَاقِب *m*

2954 [M]
lookout *n.*
F homme de veille *m.*
E vígia *m.*
D Ausguck *m.;*
R наблюдатель *м.*
A مُرَاقِب *m*

2955
loophole
F meurtrière *f.;* créneau *m.*
E aspillera *f.*
D Lücke *f.*
R бойница *ж.*
A فُتحة رمي

2956 [U]
loose
F détaché
E suelto
D frei
R свободный; шатающийся
A غَيْر مُحْكَم

2957 [E]
loose
F desserré
E flojo; suelto
D lose
R холостой *м.*
A رَخْو *m*

2958 [A,M,V]
loose (cargo)
F (en) vrac
E (a) granel
D Schütt- *komp.*
R незакреплённый
A سَائِب

2959 [U]
loosen
F relâcher
E desatar
D (auf)lockern
R развязывать(ся); отпускать
A أرْخَى

2960 [E]
loosen
F desserrer
E aflojar
D (auf)lockern
R развязывать(ся); отпускать
A أرْخَى

2961
loot *v.*
F piller
E saquear *m.*
D plündern
R грабить
A سلب

2962
lorry
F camion *m.*
E camión
D Lastwagen *m.*
R грузовик *м.*
A حافلة شحن

2963 [A,M,L,U]
lose
F perdre
E perder
D verlieren
R терять
A فَقَد

2964
lose contact *v.*
F perdre le contact
E perder contacto
D Kontakt verlieren
R терять контакт
A فقد الإتّصال

2965
lose ground *v.*
F perdre du terrain
E perder terreno
D (an) Boden verlieren
R отставать
A خسِر أرضا

2966
losses *pl.*
F pertes *f. pl.*
E bajas *f. pl.;* perdidas *f. pl.*
D Verluste *m. pl.*
R потери *ж. мн. ч.*
A خَسَائِر *f*

2967 [S]
lost
F non vu
E perdido
D verloren
R не виден
A مَفْقُود

2968 [S]
lot
F lot *m.*
E lote *m.*
D Los *neut.*
R партия *ж.*
A فِرْزَة *f*

2969
loudhailer
F porte-voix *m. inv.*
E portavoz *m.*
D Megaphon *neut.*
R звукоусилитель *м.*
A مِجْهار ذو بُوق *m*

2970
loudspeaker
F haut-parleur *m.*
E altavoz *m.*
D Lautsprecher *m.*
R громкоговоритель *м.*
A مِجْهَار ـ مُكَبِر الصوت *m*

2971
louse *n.*
F pou *m.*
E piojo *m.*
D Laus *f.*
R вошь *ж.*
A قَمْلة *f*

2972
low
F bas
E bajo
D nieder; tief (beides auch *komp.*)
R низкий; малый
A مُنْخَفِض

2973 [N]
low airburst
F explosion à basse altitude *f.*
E explosión a baja cota *f.*
D niedriger *m.;* Luftsprengpunkt
R воздушный взрыв на малой высоте *м.*
A إنْفِجَار جَوِّي مُنْخَفِض *m*

2974
low-altitude bombing
F bombardement à basse altitude *m.*
E bombardeo a baja cota *m.*
D Skip-Bombenabwurf-verfahren *neut.*
R бомбометание с малых высот *ср.*
A قَصْف على إرْتِفَاع مُنْخَفِض

2975
low altitude flying
F vol à basse altitude *m.*
E vuelo a baja cota *m.*
D Tiefflug *m.*
R полёт на малой высоте *м.*
A طَيَران على إرْتِفَاع مُنْخَفِض

2976
lower *adj.*
F inférieur
E inferior
D Unter- *komp.;* untere *adj.*
R нижний; низший
A أَسْفَل

2977
lower *v.*
F amener [M]; (a)baisser; descendre; diminuer
E bajar; lanzar [M]; disminuir
D ausfahren [A] *tr.;* fieren [M] *v.;* herabsetzen *tr.*
R спускать; опускать; снижать
A خَفَض

2978 [R]
low frequencies
F fréquences basses *f. pl.*
E bajas frecuencias *f. pl.*
D untere Frequenzbereiche *f.*
R низкие частоты *ж. мн. ч.*
A ترددات منخفضة *f*

185

2979

low loader
F porte-engins *m. inv.*
E portamáquinas *m. inv.*
D Tieflader *m.*
R низкорамочный
   (полу)прицеп *м.*
A شَحّان مُنْخَفِض *m*

2980

low oblique (photo)
F photo oblique basse *f.*
E fotografía oblicua baja
   f.
D Steilluftbild *neut.*
R низкий перспективный
   аэроснимок *м.*
A تَصْوِير مَائِل قَلِيل المَيْل

2981

low speed
F marche lente *f.*
E velocidad baja *f.*
D langsame Fahrt *f.*
R малая скорость *ж.*
A سُرْعَة مُنْخَفِضَة *f*

2982

low water
F marée basse *f.*
E bajamar *f.*
D Ebbe *f.*
R малая вода *ж.*
A ماء الجَزْر

2983

lox
F oxygène liquide *m.*
E oxígeno líquido *m.*
D Flüssigsauerstoff *m.*
R жидкий кислород *м.*
A أُكْسِجِين سَائِل *m*

2984 [M]

lubber line
F ligne de foi *f.*
E línea de fe *f.*
D Peilstrich *m.*
R курсовая черта *ж.*
A خط اليقين

2985

lubricant
F lubrifiant
E lubricante *m.*
D Schmierstoff *m.*
R смазка *ж.*; смазочное
   вещество *ср.*
A زيت تشحيم *m*

2986

lubricate
F graisser; lubrifier
E engraser; lubricar
D (ab-)schmieren *tr.*
R смазывать
A زَيَّت

2987

lubrication
F graissage *m.;*
   lubrification *f.*
E engrase *m.;* lubricación
   f.
D Schmierung *f.*
R смазывание *ср.*
A تَزْيِيت *m*

2988

lug *n.*
F oreille *f.*
E orejeta *f.*
D Ansatz *m.*
R ушко *ср.*; развилка
   ж.; глазок *м.*
A لِسَان ـ طَرَف تَوصِيل *m*

2989

lung
F poumon *m.*
E pulmón *m.*
D Lunge *f.*
R лёгкое *ср.*
A رِئَة *f*

M

2990 [P]

MACE
F gaz lacrymogène *m.*
E gas lacrimógeno *m.*
D Tränengas *neut.*
R слезоточивый газ *м.*
A غاز مضاد للمشاغبين

2991

machine *n.*
F machine *f.*
E máquina *f.*
D Maschine *f.*
R машина *ж.*; станок *м.*
A مَكَنَة *f*

2992

machine gun
F mitrailleuse *f.*
E ametralladora *f.*
D Maschinengewehr *neut.*
R пулемёт *м.*
A رَشَاش *m*

2993

mach number
F nombre de mach *m.*
E número de mach *m.*
D Machzahl *f.*
R число Маха *ср.*;
   число-M *ср.*
A العَدَد المَاخِي *m*

2994 [M]

magazine
F soute (aux poudres) *f.*
E santabárbara *f.*
D Munitionsbunker *m.*
R (артиллерийский)
   погреб *м.*
A مَخْزَن *m*

2995 [S]

magazine

F magasin *m.*

E cargador *m.*

D Magazin *neut.*

R магазин *м.;*
магазинная коробка
ж.

A مَخْزَن *m*

2996 [L]

magazine

F dépôt de munitions *m.*

E depósito de munición
m.

D Munitionslager *neut.*

R склад боеприпасов *м.*

A مَخْزَن *m*

2997 [S]

magazine (small arms)

F chargeur *m.*

E cargador *m.*

D Magazin *neut.*

R магазин *м.*

A مخزن *m*

2998

magnet

F aimant *m.*

E imán *m.*

D Magnet *m.*

R магнит *м.*

A مَغْنَطِيس *m*

2999

magnetic

F magnétique

E magnético

D Magnet- *komp.;*
magnetisch *adj.*

R магнитный

A مَغْنَطِيسي *m*

3000

magnetic bearing

F relèvement magnétique
m.

E marcación magnética *f.*

D magnetische Peilung *f.*

R магнитный азимут *м.;*
пеленг *м.*

A الإِتِّجَاه الزَّاوِي المَغْنَطِيسي *m*

3001

magnetic compass

F compas magnétique *m.*

E brújula magnética *f.*

D (Magnet-)Kompass *m.*

R компас *м.*

A بُوصَلة مَغْنَطِيسِيَّة *f*

3002

magnetic pole

F pôle magnétique *m.*

E polo magnético *m.*

D Magnetpol *m.*

R магнитный полюс *м.*

A قُطْب مَغْنَطِيسي *m*

3003

magnetic storm

F orage magnétique *m.*

E tormenta magnética *f.*

D magnetischer Sturm *m.*

R магнитная буря *ж.*

A عاصِفة مَغْنَطِيسِيَّة *f*

3004

magnetic variation

F déclinaison magnétique
f.

E declinación magnética
f.

D magnetische Deviation
[L] *f.;* Missweisung
[AM] *f.*

R магнитное склонение
ср.

A التَّغَيُّر المَغْنَطِيسي *m*

3005

magnification

F grossissement *m.*

E aumento *m.*

D Vergrösserung *f.*

R увеличение *ср.;*
усиление *ср.*

A تَكْبِير *m*

3006

magnifier

F loupe *f.*

E lupa *f.*

D Lupe *f.*

R усилитель *м.;* лупа *ж.*

A مُكَبِّرة *f*

3007

magnify

F grossir

E aumentar

D vergrössern

R преувеличивать

A كَبَّر

3008

mail *n.*

F courrier *m.;* poste *f.*

E correo *m.*

D Post *f.*

R почта *ж.*

A بَرِيد *m*

3009

mail *v.*

F expédier; poster

E enviar por correo

D (in der Post)
(ab-)senden *(tr.) untr.*

R посылать по почте

A أَرْسَل بالبريد

3010

main

F principal

E principal

D Haupt- *komp.;*
hauptsächlich *adj.*

R главный; основной

A رَئِيسي

187

3011
main airfield
F terrain principal *m.*
E campo (de aviación)
   principal *m.*
D Hauptflugplatz *m.*
R главный аэродром *м.*
A المَطَار الرَّئيسي *m*

3012
main attack
F attaque principale *m.*
E ataque principal *m.*
D Hauptangriff *m.*
R главный удар *м.*
A الهُجُوم الرَّئيسي *m*

3013
main base
F base principale *f.*
E base principal *f.*
D Hauptstützpunkt *m.*
R главная база *ж.*
A القَاعِدة الرَّئيسيّة *f*

3014
main battle area
F zone de bataille
   principale *f.*
E zona principal de
   batalla *f.*
D Hauptgefechtsfeld *neut.*
R главное поле боя *ср.*
A ساحة المعركة الرئيسية *f*

3015
main body
F gros *m.*
E grueso *m.*
D Haupttrupp *m.*
R главные силы *ж. мн.
   ч.*
A القِسْم الاكْبَر *m*

188

3016 [L]
(tactical) main guard
F gros de l'avant-garde
E guardia principal *f.*
D Hauptsicherung *f.*
R главные силы
   авангарда *ж. мн. ч.*
A الطَّليعة *f*

3017
mainland
F continent *m.*
E continente *m.*
D Festland *neut.*
R материк *м.*
A قَارَّة *f*

3018 [U]
maintain
F maintenir
E mantener
D aufrechterhalten *tr.*
R поддерживать
A حَافظ على

3019 [E]
maintain
F entretenir
E entretener
D instandhalten *tr.;*
   warten *v.*
R обслуживать
A صَان

3020 [A,L,M]
maintenance
F maintenance (de
   matériel) *f.*
E conservación *f.;*
   mantenimiento *m.*
D Materialerhaltung *f.*
R ремонт *м.;*
   обслуживание *ср.;*
   техническое
   обеспечение *ср.*
A صِيَانة *f*

3021 [E]
maintenance
F entretien *m.*
E entretenimiento *m.*
D Instandhaltung *f.;*
   Wartung *f.*
R текущий ремонт *м.;*
   исправление *ср.*
A صِيَانة *f*

3022
(corrective)
   maintenance
F réparation legère *f.*
E reparación *f.*
D Instandsetzung *f.*
R исправительный
   ремонт *м.*
A الصِّيَانة التَّصْحيحية *f*

3023 [L,P]
maintenance (of law
   and order)
F maintien de l'ordre *m.*
E conservación (del orden
   público) *f.*
D Aufrechterhaltung *f.*
R поддержание
   (правопорядка) *ср.*
A حِفْظ النظام

3024
(preventive)
   maintenance
F entretien préventif *m.*
E mantenimiento
   preventivo *f.*
D vorbeugende Wartung
   f.
R профилактический
   ремонт *м.*
A الصِّيَانة الوِقَائيَّة *f*

3025 [A,L,M,V]
(routine) maintenance
F entretien courant *m.*
E conservación *f.*
D (routinemässige)
   Wartung *f.*
R текущее техническое
   обслуживание *ср.*
A الصيانة العادية

3026 [A,L,M,V]

(scheduled)
maintenance
F entretien périodique *m.*
E entretenimiento *m.*
D (planmässige) Wartung
  f.
R плановое техническое
  обслуживание *cp.*
A صِيَانَة مُخَطَّطَة *f*

3027

maintenance area
F zone des soutiens *f.*
E zona de conservación *f.*
D Versorgungsraum *m.*
R район технического
  обслуживания *м.*
A مِنْطَقَة الصِيانة *f*

3028

major *adj.*
F majeur
E mayor; principal
D gross *adj.;* Haupt-
  komp.
R главный; основной
A رَئِيسي . كبير

3029

major assembly
F ensemble majeur *m.*
E montaje principal *m.*
D Hauptbaugruppe *f.;*
  Hauptgruppe *f.*
R главный сбор *м.;*
  агрегат *м.*
A مَجْمُوعة رَئِيسيَّة / كَبِيرَة *f*

3030

make for
F (se) diriger vers
E dirigirse a
D losgehen auf *tr.*
R направляться (на)
A إتَّجَـه إلى

3031 [P]

make inquiries
F enquêter; prendre des
  renseignements
E estar investigando
D Erhebungen anstellen
  tr.
R наводить справки;
  расследовать
A حَقَّقَ

3032

male
F mâle
E macho
D männlich
R мужской
A ذَكَر

3033

malfunction *v.*
F être défectueux; mal
  fonctionner
E averiarse
D versagen
R неправильно
  срабатывать
A ساء أَدَاؤه

3034

malfunction *n.*
F défectuosité *f.;* mauvais
  fonctionnement *m.*
E avería *f.;*
  entorpecimiento *m.*
D Funktionsstörung *f.*
R неисправность *ж.;*
  задержка *ж.*
A سُوء الأداء *m*

3035

malinger
F faire le malade; tirer au
  flanc
E fingirse enfermo
D (sich) krank stellen
R притворяться
  больным
A تَمَارَض

3036 [A,M,V]

man *v.*
F armer
E tripular
D bemannen
R укомплектовывать
  (личным составом)
A زَوَّد بِالرِّجَال

3037 [L]

man *v.*
F garnir
E guarnecer
D besetzen
R занимать
A زَوَّد بِالرِّجَال

3038 [S]

man *v.*
F servir une pièce
E servir; guarnecer
D bemannen
R укомплектовывать
A زَوَّد بِالرِّجَال

3039 [A,M,V,U]

manage
F diriger
E conducir [AV];
  gobernar[M];
  dirigir[U]
D leiten [U]; führen
  [AMV]
R управлять;
  руководить
A أَدَار

3040

(man) management
F gestion des personnels
  f.
E administración (de
  personal) *f.*
D Personalführung *f.*
R управление (личным
  составом) *cp.*
A إدارة

3041

mandatory
F obligatoire
E obligatorio
D verbindlich
R обязательный;
  принудительный
A إِجْبَارِي

3042

maneuver v.
F manœuvre
E (hacer) maniobrar
D lenken; manövrieren
R маневрировать;
  проводить манёвры
A نَاوَر

3043

maneuver n.
F manœuvre f.
E maniobra f.
D Manöver
R манёвр м.
A مُنَاوَرَة f

3044 [A]

(escape) maneuver
F manoeuvre d'évasion f.
E maniobra de evasión f.
D taktische Bewegung f.
R уклонение ср.
A مناورة النجاة

3045

maneuverable
F maniable
E maniobrable
D manövrierbar
R манёвренный
A تَسْهُل المُنَاوَرَة به

3046

maneuvering force
F force de manœuvre f.
E fuerza de maniobra f.
D bewegliche
  Einsatzkräfte f. pl.
R манёвренная группа
  ж.
A قُوَّة مُنَاوَرَة f

190

3047

maneuvers pl.
F manœuvre f.
E maniobras f. pl.
D Manöver neut.;
  (Feld-)Übungen f. pl.
R манёвры м. мн. ч.
A مُنَاوَرَات f

3048

manhandle v.
F manutentionner
E mover a brazo
D (durch) Menschenkraft
  bewegen
R тащить
A نقل باليد

3049 [A,M]

manifest n.
F manifeste m.
E manifiesto m.
D Ladeliste f.
R манифест м.
A بَيَان الشَّحْنَة

3050 [E]

manifold
F collecteur d'admission
  et d'échappement m.
E colector de escape m.
D Verteilerrohr neut.
R коллектор м.
A مجمِّع

3051 [L,M]

**manipulative
electronic
deception**
F manipulations de trafic
  (déception) f. pl.
E manipulación de
  propias emisiones
  para confundir al
  enemigo f.
D (durch) Manipulation
  herbeigeführte
  elektronische
  Täuschung
R радиодезинформация
  ж.
A الخُدعة بتداول الاشارات
  الالكترونية

3052

**man movable
(manportable)**
F portatif
E portátil
D tragbar
R переносный;
  портативный
A قَابِل للحَمْل

3053

manpower
F main d'œuvre f.
E mano de obra f.
D Arbeitskräfte f. pl.
R живая сила ж.
A القُدْرَة البَشَرِيَّة

3054

manpower
F main d'oeuvre f.
E personal m.
D verfügbare Kampfkraft
  f.
R живая сила ж.
A عدد الافراد المتيسرين
  للعملية

3055
manpower-intensive
F (à) forte proportion de main d'oeuvre
E intensivo en efectivos (en la mano de obra)
D arbeitsaufwendig
R интенсивный по живой силе
A تَرْكيز القُدْرَة البَشَريَّة

3056 [V]
mantlet
F masque; bouclier (char)
E mantelete m.
D Blende f.
R мантелет м.; щит м.
A بَاب المَرْكَبَة m

3057
manual adj. n.
F manuel adj. n. m.
E manual adj. s. m.
D Handbuch neut.; Handkomp.
R ручной прилаг.; устав м.; наставление ср.
A يَدَوِي

3058 [G]
manual command guidance
F guidage par télécommande manuelle m.
E teledirección manual f.
D Handfernlenkung f.
R ручное командное управление ср.; ручное командное наведение ср.
A تَوْجِيه بالتَّحَكُّم اليَدَوِي

3059 [A,M,V]
manual control
F commande manuelle f.
E mando manual m.
D Handbedienung f.
R ручное управление ср.
A تَحَكُّم يَدَوِي

3060
map n.
F carte f.
E mapa (geográfico) m.; plano (topográfico) m.
D Karte f.
R карта ж.
A خَريطة f

3061
map v.
F dresser une carte de
E levantar un plano de
D kartographisch darstellen tr.
R наносить на карту
A رَسَم خَريطة

3062
(battle) map
F plan directeur m.
E mapa de batalla m.
D Feldkarte f.
R схема ж.
A خَريطة المَعركة f

3063
map case
F porte-carte m.
E cartera portamapas f.
D Kartenmappe f.
R портфель м.
A مَحْفَظة الخَرائط f

3064
map reference
F référence cartographique f.
E referencia cartográfica f.
D Koordinaten f. pl.
R координаты по карте ж. мн. ч.
A مَرْجِع الخَريطة m

3065
map room
F chambre des cartes f.
E cámara de mapas f.
D Kartenraum m.
R штабный отдел м.; прицеп [V] м.
A حُجْرَة الخَرائط f

3066 [P]
marauder
F maraudeur m.
E saqueador m.
D Plünderer m.
R мародёр м.
A سارق m

3067
march n.
F marche f.
E marcha f.
D Marsch m.
R марш м.; походное движение ср.
A سَيْر m

3068
march v.
F marcher
E marchar
D marschieren
R маршировать; двигаться походным порядком
A سار

3069
marchpast
F défilé m.
E desfile militar m.
D Parademarsch m.
R торжественный марш м.
A عرض m

191

3070

margin

F marge *f.*

E margen *m.*

D Rand, Spielraum *m. m.*

R поле *ср.*; край *м.*;
  запас *м.*

A هَامِش *m*

3071

marginal

F marginal

E marginal

D begrenzt; geringfügig

R написанный на полях;
  зарамочный

A هَامِشي

3072

marginal data *pl.*

F données marginales *f.
  pl.*

E datos marginales *f. pl.*

D Zusatzinformationen *f.
  pl.*

R зарамочное
  оформление карты
  ср.

A مُعْطَيَات تَفْسِيرِيَّة *f*

3073

marine *adj.*

F marin; maritime

E marino, marítimo *m.*

D Meeres- *komp.;* See-
  komp.

R морской; судовой

A بَحْري

3074

marines *n. pl.*

F marines *m. pl.*

E infantería de marina *f.*

D Marineinfanterie *f.*

R морская пехота *ж.*

A مُشَاة البَحْرِيَّة *m*

3075

maritime

F maritime

E marítimo

D See- *komp.*

R морской

A بَحْرِي

3076

mark *v.*

F marquer

E marcar

D markieren

R отмечать; обозначать

A دَلّ ـ حَدَّد

3077

marker

F marqueur *m.*

E señal *f.*

D Markierungszeichen
  neut.

R указатель *м.*;
  ориентир *м.*

A دَلِيل *m*

3078 [L,M]

(gap) marker

F marqueur d'extrémité
  de trouée de mine *m.*

E señal del hueco en un
  campo minado *f.*

D Breschenmarkierung *f.*

R указатель (минного
  поля) *м.*

A علامة حدود الثغرة

3079 [L]

(lane) marker

F panneau de marquage
  de cheminement *m.*

E señal que jalona el
  camino por un campo
  minado *f.*

D Gassenmarkierung *f.*

R (маршрутный)
  указатель *м.*

A علامة الممر ـ دليل الممر

3080 [A,L,M]

(laser) marker

F marqueur laser *m.*

E adquisición del blanco
  por laser *f.*

D Laserzielerfassung *f.*

R (лазерный) индикатор
  (цели) *м.*

A دليل ليزر

3081

(parade) marker

F homme de base *m.*

E guía *m.*

D Flügelmann *m.*

R линейный

A دليل

3082 [R]

(range) marker

F marqueur de distance
  m.

E marcador de distancia
  m.

D Entfernungsmess-
  markierung *f.*

R калибрационная
  отметка дальности
  ж.

A عَلَامَة المَدَى *f*

3083

mark on the map *v.*

F porter sur la carte

E indicar en el mapa

D markieren

R наносить на карту

A وضع علامة على الخريطة

3084 [A,L]

mark out

F baliser sur la carte [A];
  indiquer sur la carte
  [L]

E marcar

D markieren bezeichnen

R обозначать

A وضع العلامات

3085

marksman
F tireur d'élite *m.*
E tirador de segunda *m.*
D Scharfschütze *m.*
R отличный стрелок *м.*;
   меткий стрелок *м.*
A رَام مَاهِر *m*

3086

maroon *n.*
F fusée à pétard *f.*
E petardo *m.*
D Kanonenschlag *m.*
R петарда *ж.*; хлопушка
   ж.
A مُفَرْقَعَات التَّحْذِير *f*

3087

maroon *v.*
F abandonner
E abandonar; aislar
D aussetzen *tr.*
R высаживать (на
   остров); оставлять
A هَجَر

3088

marooned
F abandonné
E aislado
D isoliert
R высаженный (на
   острове);
   оставшийся
A مَهْجُور

3089

marquee
F grande tente *f.*
E entoldado *m.*
D grosses Zelt *neut.*
R шатёр *м.*
A سرادق *m*

3090

marsh
F marais *m.*
E pantano *m.*
D Sumpf *m.*
R болото *ср.*
A مستنقع *m*

3091

marshal *v.*
F ranger; rassembler
E formar; ordenar; reunir
D aufstellen *tr.;* ordnen *v.*
R сосредоточивать;
   приводить в
   порядок
A رَتَّب ـ حَشَد

3092

(aircraft) marshaller
F personnel de
   manoeuvre des avions
   au sol *m.*
E indicador de dirrección
   m.
D (Flugzeug-)
   Bodenkontrolleur *m.*
R сортировщик *м.*
A دليل (الطائرات)

3093

marshalling
F agencement *m.*
E reunión *f.*
D Bereitstellen *neut.;*
   Versammeln *neut.*
R сосредоточение *ср.*
A تَرْتِيب ـ حَشْد *m*

3094

marshalling area
F zone des préparatifs *f.*
E zona de reunión *f.;*
   patio de maniobra *m.*
D Bereitstellungsraum *m.*
R район сосредоточения
   м.
A مِنْطَقة التَّصْنِيف *f*

3095

mask *v.*
F masquer
E enmascarar
D maskieren
R маскировать;
   скрывать
A سَتَر

3096 [M,S]

mask clearance
F angle de surplomb *m.*
E flecha de la trayectoria
   f.
D (der) Zwischenraum
   eines Geschosses im
   Flug über ein Objekt
   m.; Deckungswinkel
R просвет *м.*
A اجتناب أي حاجز في المحرك

3097

mass *n.*
F masse *f.*
E masa *f.*
D Masse *f.*
R массирование *ср.*;
   масса *ж.*
A كُتْلَة *f*

3098

mass *v.*
F concentrer
E concentrar
D massieren *untr.;*
   zusammenfassen *tr.*
R массировать;
   сосредоточивать(ся)
A كَتَّل

3099

massacre
F massacre *m.*
E carnicería *f.*
D Massaker *neut.*
R резня *ж.*
A مَذْبحة *f*

3100

mass casualties *pl.*
F pertes massives *f. pl.*
E perdidas masivas *f. pl.*
D Massenverluste *m. pl.*
R массовые потери *ж.*
   мн. ч.
A خَسَائِر ضَخْمَة *f*

3101 [M]
mast
F mât *m.*
E palo *m.*
D Mast *m.*
R мачта *ж.*
A صار *m*

3102 [R]
mast
F pylône *m.*
E torre *f.*
D Mast *m.*
R мачта *ж.*
A صار

3103
master *v.*
F maîtriser
E dominar
D beherrschen
R осваивать;
   господствовать
A سَيْطَر

3104 [M]
master *n.*
F capitaine *m.*
E capitán *m.*
D Kapitän *m.*
R капитан *м.*; командир
   м.
A رُبَّان *m*

3105
master *adj.*
F -maître (suffixe)
E maestro; principal
   (sufijo)
D Haupt- [EU] *komp.;*
   Kontroll- [AEM]
   komp.; Muster- [R]
   komp.
R главный; основной
A رَئِيسِي

194

3106
master film
F film original *m.*
E película original *f.*
D Musterfilm *m.*
R основная плёнка *ж.*
A فِيلْم رَئِيسِي *m*

3107 [A,T]
master plot
F plot de photos
   aériennes *m.*
E esquema de superficie
   fotografiada *m.*
D Luftaufklärungskarte
   (eng.Begriff zu
   werden) *f.*
R главный планшет *м.*
A التحديد الرئيسي *m*

3108
master switch
F interrupteur général *m.*
E interruptor principal *m.*
D Hauptschalter *m.;*
   Trennschalter *m.*
R главный выключатель
   м.
A مِفْتَاح رَئِيسِي *m*

3109 [A,L,M]
mastery (command or
   control)
F maîtrise *f.;* domination
   f.
E dominación *f.*
D Herrschaft *f.*
R управление *ср.*
A سيادة

3110
mastery (of a skill)
F connaissance
   approfondie *f.*
E maestría *f.*
D Beherrschung (einer
   Geschicklichkeit) *f.*
R освоение *ср.*
A حُسْن الاستخدام

3111 [A,M]
mat
F grille [A] *f.;* paillet [M]
   m.
E emparrillado [A] *m.;*
   palete [M] *m.*
D Landematte [A] *f.;*
   Leckmatte [M] *f.*
R (плетёный) мат *м.;*
   циновка *ж.;*
   рогожка *ж.*
A حَصِير *m*

3112
mat(t)(e) *adj.*
F mat
E mate
D matt *adj.;* glanzlos *adj.*
R матовый
A غير لامع

3113 [M]
mate
F second *m.*
E maestre *m.;* segundo *m.*
D Maat *m.*
R старшина *м.;*
   помощник капитана
   м.
A مُسَاعِد *m*

3114
material
F matériau *m.;* matière *f.*
E material *m. f.*
D Material *m.*
R материал *м.;*
   вещество *ср.*
A مَادَّة *f*

3115
material handling
   equipment
F équipement de
   manutention *m.*
E equipo de manipulación
   (de matérial) *m.*
D Verladevorrichtungen *f.*
   pl.
R погрузочно-
   разгрузочное
   оборудование *ср.*
A أَجْهِزَة لِتَحْرِيك المُعَدَّات *f*

3116

matériel

F matériel *m.*
E material *m.*
D Kriegsmaterial *m.*
R материальная часть
   ж.; имущества *ср.*
   мн. ч.
A مُعَدَّات *f*

3117

maximum *adj. n.*

F maximum *adj. inv. n.*
   m.; limite *adj.*
E máximo *adj. s. m.*
D grösste *adj.;* Höchst-
   komp.; Maximum *s.*
   neut.
R максимальный
   прилаг.; максимум
   м.
A الأَقْصَى

3118 [N]

maximum permissible dose

F dose maximum
   admissible *f.*
E dosis máxima aceptable
   f.
D höchstzulässige Dosis *f.*
R предельно
   допускаемая доза
   облучения *ж.*
A الجُرْعَة القُصْوَى المَسْمُوح
   بها *f*

3119

M-Day (mobilization)

F jour M (de
   mobilisation) *m.*
E día M (de movilización)
   f.
D M-Tag
   (Mobilmachungs-) *m.*
R день начала
   мобилизации *м.*
A يَوْم التعبِئة *m*

3120 [A,M]

meaconing

F transplexion *f.*
E emisiones enemigas
   engañosas para
   confundir
   la navegación de
   aviones o de buques
   f. pl.
D Täuschsignale (vom
   Feind übermittelt) *f.*
   pl.
R сбивание (пеленга) *ср.*
A ارسال اشارات الملاحة
   الكاذبة

3121

meal

F repas *m.*
E comida *f.*
D Essen *neut.*
R еда *ж.*
A وجبة *f*

3122

mean *adj.*

F moyen
E medio
D durchschnittlich;
   mittlerer
R средний
A مُعَدَّل *m*

3123

mean distance between failures (MDBF)

F distance moyenne entre
   pannes *f.*
E distancia media entre
   fallas *f.*
D durchschnittlicher
   Abstand
   zwischenVersagen
   (meist eng.
   Abkürzung)
R среднее расстояние
   между поломками
   ср.
A مُتَوَسِّط البُعْد بين
   التَّوَقُّفَات *m*

3124

means

F moyens *m. pl.*
E medios *m. pl.*
D Mittel *neut.*
R средство *ср.*
A وسائل

3125

mean sea level (MSL)

F niveau moyen de la mer
   m.
E nivel medio del mar
D Normalnull (NN) *neut.*
R нормальный уровень
   воды *м.*; межень *ж.*
A سَطْح البَحْر *m*

3126

mean time between failures (MTBF)

F temps moyen entre
   défectuosités *m.*
E tiempo medio entre
   fallos *m.*
D durchschnittliche Zeit
   zwischenVersagen
   (meist eng.
   Abkürzung)
R среднее время между
   поломками *ср.*;
   средний срок между
   поломками *м.*
A مُتَوَسِّط الزَّمَن بَيْن
   التَّوَقُّفَات *m*

3127

measure *v.*

F mesurer
E medir
D messen
R измерять; мерить
A قَاس

3128

measurement

F mesure *f.*
E medida *f.*
D Abmessung *f.*
R измерение *ср.*;
   размеры *ж. мн. ч.*
A قِيَاس *m*

3129
mechanic
F mécanicien *m.*
E mecánico *m.*
D Mechaniker *m.*
R механик *м.*
A عَامِل ميكانيكي ـ علم
الميكانيكا *m*

3130
mechanism
F mécanisme *m.*
E mecanismo *m.*
D Mechanismus *m.*
R механизм *м.*
A آلِيَّة *f*

3131
(breech) mechanism
F mécanisme de culasse *m.*
E mecanismo de cierre *m.*
D Verschlussmechanismus *m.*
R затворный механизм *м.*
A آلية المِغْلاق *f*

3132 [S]
(counter-recoil) mechanism
F récupérateur *m.*
E dispositivo contraretroceso *m.*
D Vorholmechanismus *neut.*
R механизм *м.*
A جهاز الارتداد *m*

3133 [S]
(firing) mechanism
F mécanisme de mise à feu *m.*
E dispositivo de tiro *m.*
D Abschussmechanismus *neut.*
R (стреляющий) механизм *м.*
A آلية الرمي *f*

196

3134
(self-destruct) mechanism
F mécanisme d'autodestruction *m.*
E dispositivo auto-destructor *m.*
D Selbstzerstörungs-mechanismus *m.*
R самоликвидация *ж.*; самоликвидатор *м.*
A جهاز مدمِّر ذاتيا

3135 [E]
(servo-)mechanism
F servomécanisme *m.*
E servomecanismo *m.*
D Servomechanismus *m.*
R сервомеханизм *м.*
A مضاعف

3136 [E]
(timing) mechanism
F minuterie *f.*
E relojería *f.*
D Einstellungsmechanismus *neut.*
R (регулирующий) механизм *м.*
A آلية التوقيت *f*

3137
(trip) mechanism
F allumeur à traction *m.*
E mecanismo de disparo *m.*
D Auslösemechanismus *m.*
R спотыкач *м.*
A جهاز تسييب

3138
mechanize
F mécaniser
E mecanizar
D mechanisieren [E]; motorisieren [L]
R механизировать
A حوَّل إلى آلة

3139
mechanized infantry
F infanterie mécanisée *f.*
E infantería mecanizada *f.*
D Panzergrenadiere *m. pl.*
R мотопехотный *прилаг.*; мотопехота *ж.*
A مُشاة آلِيَّة *f*

3140 [A,H]
medevac *adj. n.*
F (d')évacuation sanitaire par air (évasan) *(adj.) n. f.*
E evacuación de heridos *f.*
D Verwundetenabtransport per Hubschrauber *m.*
R медицинская эвакуация *ж.*
A اخلاء المرضى بطائرة

3141
median *adj. n.*
F médian *adj. n. m.*
E mediano *adj. s. m.*
D Mittel *komp.*; mittlerer *adj.*; Mittelwert *m.*
R средний *прилаг.*; медиана *ж.*
A أوْسَط *m*

3142
median lethal dose (LD 50)
F dose létale moyenne *f.*
E dosis letal media *f.*
D mittlere Lethaldosis *f.*
R средняя смертельная доза *ж.*
A نصف جُرعة مُميتة *f*

3143
mediate *v.*
F servir d'intermédiaire entre
E mediar
D vermitteln
R посредничать
A توسط

3144
medical
F médical
E médico
D ärztlich *adj.;* Sanitäts-
komp.
R (медицинско-)
санитарный
A طِبِّي

3145
medical treatment
F traitement médical *m.*
E tratamiento médico *m.*
D ärztliche Behandlung *f.*
R лечение *ср.;*
медицинская
помощь *ж.*
A مُعَالَجَة طِبِّيَّة *f*

3146
medical unit
F unité médicale *f.*
E unidad de sanidad *f.*
D Sanitätseinheit *f.*
R санитарная часть *ж.*
A وَحْدَة طِبِّيَّة *f*

3147
medicament
F médicament *m.*
E medicamento *m.*
D Arzneimittel *neut.*
R лекарство *ср.*
A دَوَاء *m*

3148
medium
F moyen
E mediano
D mittlerer; mittelschwer
[MV]
R средний
A مُتَوَسِّط *m*

3149 [R]
medium frequency
F fréquence moyenne *f.*
E frecuencia media *f.*
D Mittelfrequenz *f.*
R средняя волна *ж.*
A تردد متوسط

3150
meet *v.*
F rencontrer
E encontrar
D treffen
R встречать(ся)
A قَابَل

3151
meeting
F réunion *f.*
E reunión *f.*
D Tagung *f.;* Treffen *neut.*
R встреча *ж.;* сбор *м.*
A اِجْتِمَاع *m*

3152
meeting engagement
F combat de rencontre *m.*
E combate de encuentro
m.
D Begegnungsgefecht
neut.
R встречный бой *м.*
A تنفيذ الوعد *m*

3153 [R]
megahertz
F mégahertz *m.*
E megahertz *m.*
D Megahertz *neut.*
R мегагерц *м.*
A ميكاهرتز

3154
megaton
F mégatonne *f.*
E megatón *m.*
D Megatonne *f.*
R мегатонна *ж.;*
мегатон *м.*
A ميكاطن *m*

3155
megawatt
F mégawatt *m.*
E megavatio *m.*
D Megawatt *neut.*
R мегаватт *м.*
A ميكاوَط *m*

3156
memorandum
F mémorandum *m.*
E apunte *m.*
D Vermerk *m.*
R справка *ж.*
A مُذكِّرة

3157 [D]
memory
F mémoire *f.*
E memoria *f.*
D Widerrufvermögen
neut.
R память *ж.*
A ذاكرة

3158 [M]
mercantile
F marchand
E mercante
D Handels- *komp.*
R (морско-)торговый;
коммерческий
A تِجَارى

3159 [T]
Mercator's projection
F projection de mercator
f.
E proyección Mercator *f.*
D Merkatorprojektion *f.*
R меркаторская
проекция *ж.*
A أَرْتِسَام مُرْكَاتور *m*

3160
mercury
F mercure *m.*
E mercurio *m.*
D Quecksilber *neut.*
R ртуть *ж.*
A زِئْبَق *m*

3161 [T]
meridian
F méridien *m.*
E meridiano *m.*
D Meridian *m.*
R меридиан *м.*
A خَطّ زَوَال *m*

3162 [A,L,M]
(officers') mess
F mess des officiers *m.;*
 carré des officiers *m.*
E club de oficiales *m.*
D Offiziersmesse *f.;*
 Offizierskasino *neut.*
R (Сов.) офицерский
 клуб-столовая *м.;*
 офицерское собрание
 ср.
A نادي الضباط *m*

3163 [U]
message
F message *m.*
E mensaje *m.*
D Nachricht *f.*
R сообщение *ср.;*
 донесение *ср.*
A رِسَالَة *f*

3164 [R]
message
F communication *f.*
E transmisión *f.*
D (Funk-) Spruch *m.*
R радиограмма *ж.;*
 «радио» *ср.*
A بَرْقِيَّة *f*

3165
message center
F bureau des messages *m.*
E centro de mensajes *m.*
D Meldekopf *m.*
R пункт сбора
 донесений *м.*
A مَكْتَب الإِشَارَة *m*

3166
mess deck
F poste d'équipage *m.*
E sollado de marinería *m.*
D Messe *f.*
R жилая палуба *ж.*
A سَطْح سَكَن النُّوتِيِّين (في
 سَفِينَة) *m*

3167 [A,L,M]
messenger
F estafette *f.;* planton *m.*
E mensajero *m.*
D Bote [A] *m.;* Melder
 [LM] *m.*
R связной *м.*
A رسول

3168 [A,L,M]
mess hall (enlisted
 men)
F salle de mess *f.;*
 réfectoire *m.;* carré
 [M] *m.*
E cantina *f.*
D Speisesaal [AL] *m.;*
 Messe [M] *f.*
R столовая *ж.*
A قاعة الطعام *f*

3169
metal
F métal *m.*
E metal *m.*
D Metall *neut.*
R металл *м.*
A فِلِزّ *m*

3170
metal fatigue
F fatigue du métal *f.*
E fatiga del metal *f.*
D Metallermüdung *f.*
R металлоусталость *ж.*
A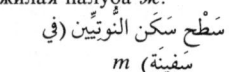
 m

3171
metallic
F métallique
E metálico
D Metall- *komp.*
R металлический
A مَعْدَنِي

3172
meteorological
 message
F message météorologique
 m.
E parte meteorológico *m.*
D Wettermeldung *f.*
R метеосводка *ж.;*
 «метео» *ср.*
A نَشْرَة جَوَّيَّة *f*

3173
meteorological
F météorologique
E meteorológico
D meteorologisch
R метеорологический
A جَوِّي

3174
meteorological data
F données
 météorologiques *f. pl.*
E datos meteorológicos
 m. f.
D meteorologische
 Unterlagen *f. pl.*
R метеорологические
 данные *ж. мн. ч.*
A مُعْطَيَات جَوَّيَّة *f*

3175
meteorology
F météorologie *f.*
E meteorología *f.*
D Meteorologie *f.;*
 Wetterkunde *f.*
R метеорология *ж.*
A عِلْم الظَّوَاهِر الجَوَّيَّة *m*

3176
(linear) meter *n.*
F mètre (linéaire) *m.*
E metro (lineal) *m.*
D Meter *neut.*
R метр *м.*
A مِتْر *m*

3177
metric
F métrique
E métrico
D metrisch
R метрический
A مِتْرِي

3178 [I,P]
microfilm
F microfilm *m.*
E microfilm *m.*
D Mikrofilm *m.*
R микрофильм *м.*
A فلم مُصغّر

3179
microphone
F microphone *m.*
E micrófono *m.*
D Mikrophon *neut.*
R микрофон *м.*
A ميكروفون *m*

3180
microprocessor
F microprocesseur *m.*
E microprocesador *m.*
D Mikroverarbeitungsgerät *neut.*
R микропроцессор *м.*
A آلة حاسبة صغيرة

3181
mid-course guidance
F guidage en vol *m.*
E dirección en vuelo *f.*
D Fernlenkung *f.*
R наведение на маршевом участке полёта *ср.*
A التّوْجِيه في نِصْف المَسافة

3182
middle *n.*
F centre *m.;* milieu *m.*
E centro *m.;* medio *m.*
D Mitte *f.*
R середина *ж.*
A وَسَط *m*

3183
middle *adj.*
F central; intermédiaire
E central; intermedio; medio
D Mittel- *komp.;* mittler *adj.;* Zwischen- *komp.*
R средний; срединный
A أوْسَط

3184
middle channel
F milieu de chenal *m.*
E centro de canal *m.*
D Fahrwassermitte *f.*
R средний фарватер *м.*
A الطّرِيق الأوْسَط *m*

3185
middle watch
F quart de minuit *m.*
E media guardia *f.*
D Mittelwache *f.*
R ночная вахта *ж.*
A نَوْبَة الحِرَاسة الوُسْطى *f*

3186
midget submarine
F sous-marin de poche *m.*
E submarino de bolsillo *m.*
D Klein-U-Boot *neut.*
R сверхмалая подлодка *ж.*
A غَوّاصَة صَغيرة *f*

3187
mil
F millième *m.*
E milésimo *f.*
D Strich *m.*
R тысячная *ж.*
A مِلِّيم *m*

3188
(nautical) mile
F mille marin *m.*
E milla marina *f.*
D Seemeile *f.*
R морская миля *ж.*
A مِيل بَحْرِي *m*

3189
(statute) mile
F mille terrestre *m.*
E milla (terrestre) *f.*
D Meile *f.*
R миля *ж.*
A ميل بري

3190
militant *n.*
F militant *m.*
E militante *m. f.*
D Streiter *m.*
R воинствующий *м.*
A مُقَاتِل *m*

3191
military
F militaire
E militar
D Militär- *komp.;* militärisch *adj.*
R военный
A عَسْكَرِي

3192
military assistance
F assistance militaire *f.*
E apoyo militar *m.*
D Militärunterstützung *f.*
R военная помощь *ж.*
A مُسَاعَدَة عَسْكَرِيَّة *f*

3193
military attaché
F attaché militaire *m.*
E agregado militar *m.*
D Militärattaché *m.*
R военный атташе *м.*
A ملحق عسكري *m*

3194 [T]
military crest
F crête militaire *f.*
E cresta militar *f.*
D Scheitelpunkt *m.*
R (боевой) гребень *м.*
A الذروة العسكرية

199

3195

military government
F gouvernement militaire *m.*
E gobierno militar *m.*
D Militärregierung *f.*
R военная
   администрация *ж.*
A حُكُومَة عَسْكَرِيَّة *f*

3196

military grid
F carroyage militaire *m.*
E cuadrícula militar *f.*
D militärisches
   Kartengitter *neut.*
R прямоугольная
   (координатная)
   сетка *ж.*
A نِظَام تَرْبِيع الخَرَائِط
   العَسْكَرِيَّة

3197

military law
F législation militaire *f.*
E derecho militar *m.*
D Kriegsstandrecht *neut.*
R военное право *ср.*
A القانون العسكري *m*

3198

military load
   classification
   (MLC)
F classe de chargement
   militaire *m.*
E clasificación militar *f.*
D Militärlastenklasse
   (meist eng.
   Abkürzung) *f.*
R классификация моста
   ж.; классификация
   боевой части *ж.*
A تَصْنِيف الحِمْل العَسْكَرِي *m*

3199

military police
F police militaire *f.*
E policía militar *f.*
D Feldjäger *m. pl.*
R военная полиция *ж.*
A شُرْطَة عَسْكَرِيَّة *f*

3200

militia
F milice *f.*
E milicia *f.*
D Miliz *f.*
R ополчение *ср.*
A حَرَس وَطَنِي *m*

3201

mine *n.*
F mine *f.*
E mina *f.*
D Mine *f.*
R мина *ж.*
A لُغْم *m*

3202

mine *v.*
F miner
E minar
D verminen
R минировать
A لَغَم

3203

(acoustic) mine
F mine (acoustique) *f.*
E mina (acústica) *f.*
D Schallmine *f.*
R (акустическая) мина
   ж.
A لغم صوتي

3204

(activated) mine
F mine à détonateur
   auxiliaire *f.*
E mina (activada) *f.*
D Mine mit Zweitzünder
   f.
R (активизированная)
   мина *ж.*
A لغم مجهز بصاعق اضافي

3205

(actuated) mine
F mine armée *f.*
E mina (accionada) *f.*
D scharf gemachte Mine
R (снаряжённая) мина
   ж.
A لغم مسلح

3206

(antipersonnel) mine
F mine antipersonnel *f.*
E mina contra personal *f.*
D Schützenmine *f.*
R (противопехотная)
   мина *м.*
A لغم ضد الاشخاص

3207

(antipersonnel) mine
F mine antipersonnel *f.*
E mina contra personal *f.*
D Schützenmine *f.*
R противопехотная
   мина *ж.*
A لُغْم مُضَادّ لِلاشْخاص *m*

3208

(bottom) mine
F mine de fond *f.*
E mina (de fondo) *f.*
D Tiefwassermine (auf
   dem Meeresboden
   liegend) *f.*
R (противодесантная)
   мина *ж.*
A لغم (قاع)

3209

(claymore) mine
F mine à effet dirigé *f.*
E mina (contra personal
   de tipo claymore) *f.*
D Streumine *f.*
R (осколочная) мина *ж.*
A لغم (كليمور)

3210

(clock) mine
F mine à retard *f.*
E mina reloj *f.*
D Uhrmine *f.*
R (механическая) мина
   ж.
A لغم (ساعة)

3211 [M]

(creeping) mine

F mine rampante *f.*
E mina móbil *f.*
D Kriechmine *f.*
R (ползучая) мина *ж.*
A لغم (زاحف)

3212

(delayed) mine

F mine à retard *f.*
E mina (de acción retardada) *f.*
D Mine mit Verzögerungszünder *f.*
R мина (замедленного действия) *ж.*
A لغم (متأخر الفِعل)

3213

(disarmed) mine

F mine neutralisée *f.*
E mina (desarmada) *f.*
D entschärfte Mine *f.*
R (обезвреженная) мина *ж.*
A لغم مُبْطَل

3214

(drifting) mine

F mine dérivante *f.*
E mina (a la deriva) *f.*
D Treibmine *f.*
R (плавучая) мина *ж.*
A لغم عائم

3215

(drill) mine

F fausse mine *f.*
E mina (de instrucción) *f.*
D Drillmine *f.*
R мина-болванка *ж.*
A لغم (تدريب)

3216

(homing) mine

F mine à tête chercheuse *f.*
E mina (buscadora) *f.*
D zielansteurende Mine *f.*
R (самонаводящаяся) мина *ж.*
A لغم موجّه

3217

(horizontal action) mine

F mine à effet horizontal *f.*
E mina (de acción horizontal) *f.*
D Streumine (horizontaler Wirkung) *f.*
R мина (горизонтального действия) *ж.*
A لغم افقي الفعل

3218

(inert) mine

F mine inerte *f.*
E mina (inactiva) *f.*
D (nicht scharf gemachte) Mine *f.*
R (обезвреженная) мина *ж.*
A لغم باطل

3219

(land) mine

F mine terrestre *f.*
E mina terrestre *f.*
D Landmine *f.*
R (наземная) мина *ж.*
A لُغْم أَرْضِي *m*

3220

(limpet) mine

F mine ventouse *f.*
E mina magnética *f.*
D Haftmine *f.*
R присасывающийся (подрывной) заряд *м.*
A قُنْبُلة مُلْتَصِقة *f*

3221

(magnetic) mine

F mine magnétique *f.*
E mina magnética *f.*
D Magnetmine *f.*
R магнитная мина *ж.*
A لُغْم مَغْنِطيسي *m*

3222 [M]

(mobile) mine

F mine autopropulsée *f.*
E mina autopropulsada *f.*
D Selbsuchmine *f.*
R самоходная мина *ж.*
A لُغْم مُتَحَرِّك *m*

3223

(pressure) mine

F mine à dépression *f.*
E mina (de presión) *f.*
D Tretmine *f.*
R мина (нажимного действия) *ж.*
A لغم ضغطي

3224

mine clearance *n.*

F déminage *m.*
E levantamiento de minas *m.*
D Minenräumung *f.*
R разминирование *ср.*
A إزالة ألغام *f*

3225 [L]

mine clearance *adj.*

F (de) déminage
E destructor de minas
D Minenräum- *komp.*
R разминированный
A إزَالة أَلْغَام *f*

3226

mine countermeasures
pl.

F mesures de défense
contre les mines *f. pl.*

E medidas contra minas *f.
pl.*

D Minenabwehrmass-
nahmen *f. pl.*

R противоминные
действия *ср. мн. ч.*

A تَدَابِيرُ مُضَادَّة لِلْألْغَام *f*

3227

mined area

F zone minée *f.*

E zona minada *f.*

D Minensperre *f.*

R заминированный
участок *м.*; минное
поле *ср.*

A مِنْطَقَة مَلْغُومة *f*

3228

mine disposal

F déminage *m.*

E levantamiento de minas
m.

D Minenvernichtung *f.*

R обезвреживание мин
ср.

A التَّخَلُّص من الأَلْغَام *m*

3229

minefield

F champ de mines *m.*

E campo de minas *m.*

D Minenfeld *neut.*

R минное поле *ср.*

A حَقْل ألْغَام *m*

3230

(barrier) minefield

F champ de mines de
cloisonnement *m.*

E (campo) minado *m.*

D Minenfeld *neut.*;
Minensperre *f.*

R минное поле *ср.*

A حقل الغام حاجز

3231

(nuisance) minefield

F champ de mines de
harcèlement *m.*

E campo de minas de
hostigamiento *m.*

D Störminenfeld *neut.*

R беспокоящее минное
поле *ср.*

A حقل الغام ازعاج

3232

minefield record

F plan de pose de mines
m.

E registro de minas *m.*

D Minenplan *m.*

R формуляр минного
поля *м.*

A سِجِل حَقْل الأَلْغَام *m*

3233 [L]

minelayer

F poseur de mines *m.*;
mouilleur de mines

E tendedor (sembrador)
de minas *m.*

D Minenleger *m.*

R минный раскладчик
м.; минный
заградитель *м.*

A زَارِعَة ألْغَام *f*

3234

mine row

F rangée de mines *f.*

E hilera de minas *f.*

D Minenreihe *f.*

R минный ряд *м.*

A صَفّ ألْغَام *m*

3235 [M]

minesweeping

F dragage de mines *m.*

E dragar minas

D Minenräumung *f.*

R траление *ср.*

A كَسْح الالغام

3236

minimize

F minimiser; réduire au
minimum

E minimizar

D (auf ein) Minimum
herabsetzen *tr.*

R доводить до
минимума;
преуменьшать

A قَلَّل

3237 [R]

"Minimize!"

F "Prudence!"
(international);
"Trafic réduit!"

E "¡Reducir al minimo!"

D „Betriebsbeschränkung!"
(eng. Begriff zu
verwenden) *f.*

R "Молчать!";
"Прекратите
болтать!"

A قَلَّل

3238

minimum *adj. n.*

F minimum *adj. inv. n. m.*

E mínimo *adj. s. m.*

D Mindest- *komp.*;
minimal *adj.*;
Minimum *neut.*

R минимальный *прилаг.*;
минимум *м.*

A الأدْنَى

3239 [N]

**minimum burst
altitude**

F hauteur minimum
d'explosion *f.*

E altura mínima de
explosión *f.*

D Mindestdetonationshöhe
f.

R минимальная высота
взрыва *ж.*

A أدْنَى إرْتِفَاع لِلإنْفِجَار

3240 [P]
minimum force
F force minimum *f.*
E fuerza mínima *f.*
D minimale Stärke *f.*
R минимальная сила *ж.*
A أَصْغَرُ قُوَّةٍ *f*

3241 [A]
minimum safe altitude
F altitude minimum de
sécurité *f.*
E altura mínima segura *f.*
D Mindestflughöhe *f.*
R минимальная
безопасная высота
ж.
A اَدْنَى إِرْتِفَاعٍ لِلْأَمَان

3242 [N]
minimum safe distance
F distance minimum de
sécurité *f.*
E distancia mínima
segura *f.*
D Sicherheitsabstand *m.*
R минимальное
безопасное удаление
ср.
A أَدْنَى مَسَافَةٍ لِلْأَمان

3243
Ministry of Defence
F Ministère de la Défense
Nationale *m.*
E Ministerio de Defensa
m.
D Verteidigungsministerium
neut.
R Министерство
Обороны *ср.*
A وزارة الدفاع

3244
minor *adj.*
F mineur
E menor
D geringfügig; kleiner
R меньший; мелкий
A صَغِير

3245
minute *n.*
F minute *f.*
E minuto *m.*
D Minute *f.*
R минута *ж.*
A دَقِيقَة *f*

3246
minutes
F procès-verbal *m.*
E actas *f. pl.*
D (Sitzungs-)Protokoll
neut.
R протокол *м.*
A مَحْضَرُ الجَلسة *m*

3247
mirror *n.*
F miroir *m.*
E espejo *m.*
D Spiegel *m.*
R зеркало *ср.*
A مِرْآة *f*

3248 [V]
misfire *n.*
F raté *m.*
E fallo del encendido *m.*
D Fehlzündung *f.*
R незапуск *м.*; перебой
м.
A خَلَل الإِشْعَال *m*

3249 [S]
misfire *v.*
F rater
E fallar el tiro
D versagen
R давать осечку; не
взрываться
A كَذَب الرَّمْي

3250 [S]
misfire *n.*
F raté *m.*
E fallo del disparo *m.*
D Versager *m.*
R осечка *ж.*; незапуск *м.*
A كِذْبَة رَمْي *f*

3251 [P]
missile
F projectile *m.*
E objeto arrojadizo *m.*
D Wurfgeschoss *neut.*
R снаряд *м.*
A مَقْذُوف *m*

3252 [G]
missile
F engin *m.;* missile *m.*
E misil *m.*
D (ferngesteuerte) Rakete
f.
R ракета *ж.*; снаряд *м.*
A صَارُوخ *m*

3253
**(air-launched ballistic)
missile**
F engin balistique lancé
d'aéronef *m.*
E misil (balístico lanzado
de avión) *m.*
D ballistische Bordrakete
f.
R ракета (авиационная
баллистическая) *ж.*
A صاروخ قِذافي مطلق من الجو

3254
(air-to-air) missile
F engin air-air *m.*
E misil (aire-aire) *m.*
D Bord-Bordrakete *f.*
R ракета (воздух-воздух)
ж.
A صاروخ جو ـ جو

3255
(air-to-ground) missile
F engin air-sol *m.*
E misil (aire-suelo) *m.*
D Bord-Bodenrakete *f.*
R ракета (воздух-земля)
ж.
A صاروخ جو ـ ارض

203

3256

(air-to-underwater) missile

F missile anti-sous-marin lancé d'un aéronef *m.*

E misil (contra-submarino lanzado de avión) *m.*

D Bord-Unterwasserrakete *f.*

R ракета (воздух-море) *ж.*

A صاروخ (جو إلى تحت الماء)

3257

(antiaircraft) missile

F missile antiaérien *m.*

E misil (contra-avión) *m.*

D Flakrakete *f.*

R зенитная ракета *ж.*

A صاروخ ضد الطائرات

3258

(antiballistic) missile

F missile anti-missile *m.*

E misil (contra misil balístico) *m.*

D Raketenabwehr-flugkörper *m.*

R антиракета *ж.*

A صاروخ ضد المقذوفات القذافية

3259

(antitank guided) missile

F engin (téléguidé) antichar *m.*

E proyectil (teledirigido) contracarro *m.*

D Panzerabwehrrakete *f.*

R противотанковый управляемый реактивный снаряд (ПТУРС) *м.*

A صاروخ مُوَجَّه مُضَادَ للدَّبَابَات *m*

3260

(ballistic) missile

F missile balistique *m.*

E proyectil balístico *m.*

D Rakete *f.*

R баллистическая ракета *ж.*

A صاروخ مَقْذُوفِي *m*

3261

(cruise) missile

F missile de croisière *m.*

E misil de crucero *m.*

D Marschflugkörper *m.*

R крылатая ракета *ж.*

A صَارُوخ عابِر للقارات *m*

3262

(free flight) missile

F missile libre *m.*

E misil libre *m.*

D ungelenkte Rakete *f.*

R неуправляемая (баллистическая) ракета *ж.*

A صاروخ ذُو طَيَرَان حُرّ *m*

3263

(guided) missile

F engin téléguidé *m.;* missile *m.*

E proyectil guiado *m.*

D Fernlenkrakete *f.*

R управляемая ракета *ж.*

A صَارُوخ مُوَجَّه *m*

3264

(homing) missile

F engin autoguidé *m.*

E misil (buscador) *m.*

D zielansteuernde Rakete *f.*

R (самоуправляемый) снаряд *м.;* (самоуправляемая) ракета *ж.*

A صاروخ موجه

3265

(intercontinental ballistic) missile (ICBM)

F missile balistique intercontinental *m.*

E proyectil balístico intercontinental *m.*

D Interkontinentalrakete *f.*

R межконтинентальная баллистическая ракета *ж.*

A صَارُوخ عابِر لِلقَارَات *m*

3266

(intermediate range ballistic) missile (IRBM)

F missile balistique de portée intermédiaire *m.*

E misil (balístico de alcance medio) *m.*

D Mittelstreckenrakete *f.*

R (управляемая) ракета (средней дальности) *ж.*

A صاروخ قذافي متوسط المدى

3267

(launch and leave) missile

F missile "lance et n'y pense plus" *m.*

E misil (autónomo) *m.*

D Rakete (ungelenkt nach dem Abschuss) *f.*

R (самонаводящийся) снаряд *м.*

A صاروخ للاطلاق والترك

3268

(medium range ballistic) missile

F missile balistique à moyenne portée *f.*

E misil balístico de alcance medio *m.*

D Mittelstreckenrakete *f.*

R баллистическая ракета средней дальности *ж.*

A *m* مَقْذُوف مُتَوَسِّط المَدَى

3269

(sea-skimming) missile

F missile volant au ras de l'eau *m.*

E misil (de vuelo rasante) *m.*

D Rakete (über das Meer streichend) *f.*

R (моребреющий) снаряд *ж.*

A صاروخ يطير فوق سطح الماء

3270 [G]

(seeker) missile

F engin chercheur auto-directeur *m.*

E mísil buscador *m.*

D Zielsuchflugkörper *m.*

R самонаводящаяся ракета *ж.*

A صاروخ باحث عن الهدف

3271

(stand-off) missile

F missile auto-guidé *m.*

E misil (de alcance medio) *m.*

D Abstandflugkörper *m.*

R (наземный/воздушный самоуправляемый) снаряд *м.*

A صاروخ مطلق من مدى بعيد

3272

(submarine-launched ballistic) missile

F engin lancé par sous-marin

E mísil balístico de lanzamiento-submarino *m.*

D U-Boot-Rakete *f.*

R баллистическая ракета подводного базирования *ж.*; подлочная ракета *ж.*

A صاروخ مَقْذُوف مِن غَوَّاصَة

3273

(surface-to-surface) missile

F missile surface-surface *m.*

E misil (buque-buque) *m.*

D Boden-Bodenrakete *f.*

R ракета (земля-земля) *ж.*; снаряд (земля-земля) *м.*

A صاروخ ارض ـ ارض

3274

(tube-launched optically-tracked wire-guided) missile (TOW)

F missile filoguidé *m.*

E misil TOW *m.*

D Rakete (TOW) *f.;* (Rohrabschuss; optisch verfolgt; mit Drahtlenkung)

R ПТУРС (противотанковый управляемый реактивный снаряд) типа TOW *м.*

A صاروخ مطلق من انبوب ـ متعقب بصرياً ـ موجه بأسلاك

3275

(underwater-to-air) missile

F missile sous-marin-air *m.*

E mísil submarino-aire

D Unterwasser-Luftflugkörper *m*

R подлодочная зенитная ракета *ж.*

A مقذوف من تحت سطح الماء إلى الجو

3276

missing in action (MIA)

F disparu au combat

E desparecido en combate

D vermisst (im Krieg) *adj.;* Vermisste(r) *m.*

R пропавший без вести

A مفقود في القتال

3277

mission

F mission *f.*

E misión *f.*

D Auftrag *m.*

R (боевая) задача *ж.*

A مُهِمّة *f*

3278

(call) mission

F mission sur demande urgente *f.*

E ataque aéreo pedido a corto plazo *m.*

D Fliegereinsatz (auf Abruf) *m.*

R (вызов на срочный) воздушный налёт *м.*

A مهمّة *f*

3279

(military) mission

F mission *f.*

E misión *f.*

D Militärmission *f.*

R (военная) миссия *ж.*

A بَعْثة *f*

205

3280 [L]

(retain) mission

F mission de conservation
   de terrain *f.*
E misión de conservación
   de terreno *f.*
D Sicherungsauftrag *m.*
R удерживающий бой *м.*
A مهمة الإعاقة

3281

(search) mission

F mission de recherche *f.*
E misión de exploración
   f.
D Suchauftrag *m.*
R вылет на поиск *м.*
A مُهِمَّة تَفْتِيش

3282 [A]

**(immediate) mission
   request**

F demande de mission
   d'appui aérien urgente
   f.
E misión no prevista *f.*
D unmittelbarer
   Luftauftrag *m.*
R требование срочного
   воздушного налёта
   ср.
A طلب لضرب جوي فوراً

3283

**(preplanned) mission
   request**

F demande d'appui aérien
   (mission planifiée) *f.*
E petición prevista *f.*
D vorgeplante
   Anforderung von
   Luftunterstützung *f.*
R плановый вызов (огня
   м.; воздушного
   налёта)
A طلب مهمة مخططه

3284

miss the target *v.*

F rater/manquer la cible
E errar el blanco
D (das) Ziel verfehlen
R промахиваться
A أخطأ الهدف

3285

mist

F brume *f.*
E niebla *f.*
D Dunst *m.*
R туман *м.*
A ضباب *m*

3286

mixed

F mixte
E mixto
D gemischt *adj.;* Misch-
   komp.
R смешанный
A مُخْتَلِف ـ مُخْتَلَط

3287

mixture

F mélange *m.*
E mezcla *f.*
D Gemisch *neut.*
R смешивание *ср.;* смесь
   ж.
A مَزِيج *m*

3288 [P]

mob *n.*

F cohue *f.*
E turba *f.*
D Pöbel *m.*
R сборище *ср.*
A غوغاء *m*

3289

mobile

F mobile
E móvil
D beweglich
R подвижной;
   мобильный
A مُتَحَرِّك

3290

mobile defense

F défense mobile *f.*
E defensa móvil *f.*
D bewegliche Verteidigung
   f.
R подвижная оборона
   ж.; активная
   оборона *ж.*
A دِفَاع مُتَحَرِّك *m*

3291 [M]

mobile support group

F groupe de support
   logistique *m.*
E grupo de
   abastacimiento *m.*
D bewegliche
   Versorgungsgruppe *f.*
R подвижная группа
   поддержки *ж.*
A جَمَاعَة سَنْد مُتَنَقِّلَة *f*

3292

mobile warfare

F guerre de mouvement *f.*
E guerra de movimiento
   f.
D Bewegungskrieg *m.*
R манёвренная война *ж.*
A حَرْب الحَرَكَة *f*

3293

mobility

F mobilité *f.*
E movilidad *f.*
D Beweglichkeit *f.*
R подвижность *ж.;*
   мобильность *ж.;*
   манёвренность *ж.*
A سُهُولة الحَرَكَة *f*

3294

mobilization

F mobilisation *f.*
E movilización *f.*
D Mobilmachung *f.*
R мобилизация *ж.*
A نَفِير ـ تَعْبِئَة ـ حَشْد *m*

3295
mobilize
F mobiliser
E movilizar
D mobil machen
R мобилизовать
A أَنْفَرَ ـ عبأَ ـ حَشَدَ

3296
mock-up
F maquette f.
E maqueta f.
D Modell neut.
R макет м.
A نَمُوذَج بالْحَجَم الْحَقِيقِي m

3297
model n.
F modèle m.
E modelo m.
D Modell neut.
R модель ж.; макет м.
A نَمُوذَج m

3298
(advanced development) model
F prototype m.
E prototipo m.
D Modell (Prototyp) neut. m.
R прототип м.
A تَطْوِير مُتَقَدَّم

3299
model (on) v.
F modeler (sur)
E modelar (sobre)
D einrichten (nach) tr.; modellieren v.
R создавать по образцу
A إِمْتَثَلَ ـ مَثَّلَ

3300
(production) model
F modèle de série
E producto m.
D Modell (Serienfertigung) neut. f.
R модель ж.
A نَمُوذَج إِنْتَاج

3301
mode of transport
F mode de transport m.
E medio de transporte m.
D Transportweise f.
R вид транспорта м.
A طَرِيقَة نَقْل f

3302
moderate v.
F (se) modérer
E calmarse; moderarse
D nachlassen tr.
R умерять; смягчать
A عَدَّل

3303
moderate adj.
F modéré
E moderado
D mässig
R умеренный
A مُعْتَدِل

3304
modern
F moderne
E moderno
D modern
R современный
A حَدِيث

3305
modernization
F modernisation f.
E modernización f.
D Modernisierung f.
R модернизация ж.
A تَعْصِير m

3306
modernize
F moderniser
E modernizar
D modernisieren
R модернизировать
A عَصَّرَ

3307
modification
F modification f.
E modificación f.
D Abänderung f.
R модификация ж.
A تَعْدِيل m

3308
modify
F modifier
E modificar
D abändern tr.; umsteuern [DR] tr.
R модифицировать; изменять
A عَدَّل

3309
modular
F modulaire
E modular
D Einheits- komp.
R модульный
A تشكل من وحدات

3310 [R]
modulation
F modulation f.
E modulación f.
D Modulation f.
R модуляция ж.
A تضمين

207

3311

(amplitude)
modulation (AM)
F modulation d'amplitude *f.*
E modulación de amplitud *f.*
D Amplitudenmodulation *f.*
R (амплитудная) модуляция *ж.*
A تضمين الذَّروة

3312

(frequency)
modulation (FM)
F modulation de fréquence *f.*
E modulación (de frecuencia) *f.*
D Frequenzmodulation *f.*
R (частотная) модуляция *ж.*
A تضمين التردد

3313

(pulse code)
modulation
F modulation par impulsions et codage *f.*
E modulación (de pulsación) *f.*
D Impulsmodulation *f.*
R (импульсная) модуляция *ж.*
A تضمين النَّبْض

3314

Molotov cocktail
(petrol bomb)
F cocktail Molotov *m.*
E cóctel Molotov *m.*
D „Molotov cocktail" (eng. Begriff) *neut*; Benzinbombe *f.*
R зажигательная бомба *ж.*
A قنبلة بنزين

208

3315 [E,R]

monitor *n.*
F moniteur *m.*
E monitor *m.*
D Monitor *m.*
R контрольный аппарат *м.*; радиоперехватчик *м.*
A جِهَاز مُرَاقَبَة *m*

3316

monitor *v.*
F contrôler; surveiller
E controlar; escuchar; vigilar
D abhören *tr.;* überwachen *untr.*
R контролировать; подслушивать
A رَاقَب

3317

monsoon
F mousson *f.*
E monzón *m. f.*
D Monsun *m.*
R муссон *м.*; дождливый сезон *м.*
A رِيح مَوْسِمِيَّة *f*

3318

moon
F lune *f.*
E luna *f.*
D Mond *m.*
R луна *ж.*
A قَمَر *m*

3319

moor *v.*
F accoster; amarrer
E amarrar
D anlegen *tr.;* festmachen *tr.*
R ставить на якорь; становиться на якорь; причаливать
A أرْسَى

3320

moor alongside
F accoster
E atracar de costado
D anlegen (an) *tr.*
R пришвартовать(ся)
A رَبط على طول . . .

3321

mooring
F amarrage *m.*
E amarradero *m.*
D Liegeplatz *m.*
R постановка на якорь *ж.*; причал *м.*
A مَرْبَط *m*

3322

mooring buoy
F bouée de corps-mort *f.*
E boya de amarre *f.*
D Ankerboje *f.*
R рейдовая бочка *ж.*
A عَوَّامَة إرْسَاء *f*

3323 [L]

mopping up
F nettoyage *m.*
E limpieza *f.*
D Säuberung *f.*
R прочёсывание *ср.*; очистка *ж.*
A تَطْهِير *m*

3324

morale
F moral *m.*
E moral *f.*
D innere Haltung *f.*
R моральное состояние *ср.*
A مَعْنَوِيَّات *f*

3325

moratorium
F moratorium *m.*
E moratoria *f.*
D Moratorium *m.*
R мораторий *м.*
A تأجيل دفع الديون *m*

3326
morning watch
F quart de 4h à 8h
E cuarto de diana *m.*
D Morgenwache *f.*
R утренняя вахта *ж.*
A نَوْبَة الحِرَاسَة صَبَاحاً *f*

3327
Morse code
F morse *m.*
E alfabeto Morse *m.*
D Morsealphabet *neut.*
R азбука Морзе *ж.*
A رُموز إِشارَات مورس *f*

3328
mortar
F mortier *m.*
E mortero *m.*
D Mörser *m.*
R миномёт *м.*
A هَاوُن *m*

3329
mortar *v.*
F bombarder aux mortiers
E bombardear con
   morteros
D (mit) Mörserfeuer
   belegen
R обстреливать из
   миномёта
A قَصَف بِالهَاوُن

3330
mortar bomb
F obus de mortier *m.*
E granada de mortero
D Mörsergranate *f.*
R (миномётная) мина
   ж.
A قُنْبَلَة هَاوُن *f*

3331
mosaic
F mosaique *f.*
E mosaico *m.*
D Reihenluftbild *neut.*
R фотоплан *м.*;
   фотосхема *ж.*
A فُسَيْفِسَاء *m*

3332
mosquito
F moustique *m.*
E mosquito *m.*
D Stechmücke *f.*
R москит *м.*
A بَعُوض *m*

3333
mosquito net
F moustiquaire *f.*
E mosquitero *m.*
D Moskitonetz *neut.*
R сетка от москитов *ж.*
A نَامُوسِيَّة *f*

3334 [E,M]
mothball *v.*
F mettre en cocon
E sacar de servicio
D „mothball" (eng. Begriff);
   Ausrüstungsgegen-
   stände stillegen
R хранить
A وضع جهازاً محفوظاً في مكان
   واق

3335 [M]
mothership
F navire ravitailleur *m.*
E buque de
   abastecimiento *m.*
D Mutterschiff *neut.*
R матка *ж.*; плавучая
   база *ж.*
A السفينة الأم

3336
motor *n.*
F moteur *m.*
E motor *m.*
D Motor *m.*
R мотор *м.*; двигатель
   м.
A مُحَرِّك

3337 [G]
(rocket) motor
F propulseur *m.*
E reactor *m.*
D Raketenmotor *m.*
R двигатель *м.*
A محرك صاروخي

3338
motorcycle
F motocyclette *f.*
E motocicleta *f.*
D Krad (=Kraftrad)
   neut.
R мотоцикл *м.*
A دَرَّاجَة ذات مُحَرِّك *f*

3339
motorized unit
F unité motorisée *f.*
E unidad motorizada *f.*
D motorisierte Einheit *f.*
R моторизованная часть
   ж.
A وَحْدَة آلِيَّة *f*

3340
mound
F tertre *m.*
E montículo *m.*
D Erdhügel *m.*
R холм *м.*; курган *м.*
A تَـلْ *m* - ربوة *f*

3341 [V]
mount *v.*
F monter
E montar
D einsteigen *tr.*
R садиться
A رَكِبَ

3342 [E]
mount *v.*
F monter
E montar
D montieren
R устанавливать;
   монтировать
A رَكَّب

3343 [A,M,L]
mount *v.*
F préparer
E montar
D bereitstellen *tr.*
R предпринимать;
   организовывать
A رَكَّب

209

3344
mountain *adj.*
F (de) montagne
E (de) montaña
D Gebirgs- *komp.*
R горный
A جَبَلِي

3345
mountain *n.*
F montagne *f.*
E montaña *f.*
D Berg *m.*
R гора *ж.*
A جَبَل *m*

3346
mountainous
F montagneux
E montañoso
D bergig
R гористый
A جَبَلِي

3347 [V]
mounted
F monté; porté
E montado
D aufgesessen
R посаженный
A مُرَكَّب

3348
(turret) mounted
F sous-tourelle
E montado en torreta
D (am) Turm versehen;
befestigt (z.B.
Maschinengewehr)
R установленный
A مركب في البرج

3349
(vehicle) mounted *adj.*
F (de) bord
E montado (instalado) en
vehículo
D Bord- *komp.*
R моторизованный;
посаженный на
машине
A مُرَكَّب عَلى عَرَبَة

3350 [A,M,L]
mounting
F préparatifs *m. pl.*
E preparativos *m. pl.*
D Bereitstellung *f.*
R подготовка *ж.*
A جِهَاز تَثْبِيت *m*

3351 [S]
mounting
F affût *m.*
E montaje *m.*
D Lafette *f.*
R установка *ж.*
A حَاضِن *m*

3352 [A,M,L]
mounting area
F zone d'embarquement
f.
E zona de embarco *f.*
D Einschiffungsraum [M]
m.; Verladeraum
[AL] *m.*
R район сосредоточения
и подготовки войск
м.
A مِنْطَقَة التَّرْكِيب *f*

3353
movable
F mobile
E movible
D beweglich
R подвижной;
переносный
A قَابِل لِلتَّحَرُّك

3354
move *v.*
F (se) déplacer
E mover(se)
D (sich) bewegen
R двигать(ся)
A نَقَل

3355
move (of a unit)
F déplacement *m.*
E desplazarse
D Verlegung (einer
Einheit) *f.*
R перевоз части *м.*
A انتقال الوحدة

3356
movement
F mouvement *m.*
E movimiento *m.*
D Marsch [L] *m.;*
Transport [AMV] *m.;*
Verlegung *f.*
R движение *ср.;* ход *м.*
A إنْتِقَال *m*

3357
movement control
F organisation des
mouvements *f.*
E control de la
organización del
movimiento *m.*
D Marschüberwachung *f.*
R управление
движением *ср.;*
перевозкой *м.*
A تَنْظِيم السَّيْر *m*

3358
movement priority
F priorité de mouvement
f.
E prioridad de
movimiento *f.*
D Verlegungspriorität *f.*
R очередность перевозки
ж.; очередность
движения *ж.*
A أفْضَلِيَّة السَّيْر *f*

3359

movement restriction
F restriction de la
circulation *f.*
E restricción de
movimientos *f.*
D Verkehrsbeschränkung
f.
R ограничение перевозки
ср.; ограничение
движения *ср.*
A تَحْدِيد السَّيْر *m*

3360

movement to contact
F marche à l'ennemi *f.*
E avance hacia el
enemigo *m.*
D Vormarsch *m.*
R подход *м.*
A التقدم حتى الاشتباك

3361 [E,V]

muffler
F silencieux *m.*
E silenciador *m.*
D Auspufftopf *m.*
R глушитель *м.*
A خَافِت لِلصَّوْت ـ كَاتِم
للصوت *m*

3362

(in) mufti
F (en) civil
E vestido de paisano
D Zivilkleidung *f.;* (in)
Zivil
R штатская одежда (в
штатском) *ж.*
A باللباس المدني

3363

mule
F mulet *m.*
E mulo *m.*
D Maultier *neut.*
R мул *м.*
A بغل *m*

3364

muleteer
F muletier *m.*
E mulatero *m.*
D Maultiertreiber *m.*
R погонщик мулов *м.*
A بغّال *m*

3365 [S]

multibarrel *adj.*
F (à) canons multiples
E multitubo *inv.*
D mehrläufig
R многоствольный
A مُتَعَدِّد المَوَاسِير

3366

multicapable
F polyvalent
E adaptable
D Mehrzweck- *komp.*
R многоспособный
A متعدد القدرات

3367

multifuel
F polycarburant
E polivalente en
combustible
D Vielstoff- (z.B. Motor)
komp.
R многотопливный
A متعدد الوقود

3368

multilateral
F multilatéral
E multilátero
D multilateral
R многосторонний
A مُتَعَدِّد الأَضْلاع

3369

multinational
F multinational
E multinacional
D multinational
R многонациональный
A مُتَعَدِّد الجِنْسِيَّة

3370

multiple *adj.*
F multiple
E múltiplo
D Mehr- *komp.;* mehrfach
komp.
R многократный;
составной; складной
A مُتَعَدِّد

3371

multiple employment
n.
F souplesse d'emploi *f.*
E empleo múltiple *m.*
D mehrfacher Einsatz *m.*
R многоспособность *ж.*
A استخدام متعدد

3372 [R]

multiplex *adj.*
F multiplex *inv.*
E múltiplex *inv.*
D Multiplex- *komp.*
R многоканальный
A مُتَعَدِّد العَنَاصر

3373

multipurpose *adj.*
F polyvalent; multi-usages
E adaptable
D Mehrzweck- *komp.*
R многоцелевый
A متعدد الاغراض

3374

munitions *pl.*
F munitions *f. pl.*
E municiones *f. pl.*
D Munition *f.*
R военные запасы *м. мн.*
ч.; имущества *ср.*
мн. ч.
A مُعَدَّات حَرْبِيَّة *f*

211

3375 [P]
murder *n.*
F meurtre *m.*
E asesinato *m.*
D Mord *s. m.*
R убийство *ср.*
A قتل *m*

3376
murderer
F meurtrier *m.*
E asesino *m.*
D Mörder *m.*
R убийца *м. ж.*
A قاتل *m*

3377 [N]
mushroom cloud
F champignon atomique *m.*
E nube atómica *f.*
D Pilzwolke *f.*
R грибовидное облако *ср.*
A سحابة عش الغراب *f*

3378 [S]
musketry
F instruction de tir *f.*
E tiro de armas *m.*
D Schiessunterricht *m.*
R стрелковое дело *ср.*
A فن الرمي *m*

3379 [A,L,M]
muster *v.*
F rassembler
E juntar para pasar revista
D mustern
R набирать
A جَمَع *m*

3380 [A,L,M]
muster *n.*
F rassemblement *m.*
E revista *f.*
D Musterung *f.*
R сбор *м.*; скопление *ср.*
A اجتماع *m*

3381
mutilated
F mutilé
E mutilado
D verstümmelt
R увечный
A مبتور

3382
mutiny *n.*
F mutinerie *f.*
E motín *m.*
D Meuterei *f.*
R бунт *м.*; мятеж *м.*
A عصيان *m*

3383
mutiny *v.*
F (se) mutiner
E amotinarse
D meutern
R бунтовать(ся)
A عَصَى *m*

3384
mutual
F mutuel
E mutuo
D gegenseitig
R взаимный; обоюдный
A مُتَبَادَل *m*

3385 [S]
muzzle
F bouche *f.*
E boca *f.*
D Mündung *f.*
R дуло *ср.*
A فُوهَة *f*

3386
muzzle boresight
F simbleau de bouche *m.*
E centrador de boca *m.*
D Justierkreuz *neut.*
R дульный визир *м.*
A نَاظِمَة الفُوهَة *f*

3387
muzzle flash
F lueur de bouche *m.*
E fogonazo *m.*
D Mündungsfeuer *neut.*
R дульное пламя *ср.*
A وَمِيض الفُوهَة *m*

N

3388
NAAFI
F économat *m.*
E cantina *f.*
D Kantine *f.*
R войсковая лавка *ж.*
A نادي الجنود *m*

3389
nadir
F nadir *m.*
E nadir *m.*
D Nadir *m.;* Fusspunkt *m.*
R надир *м.*
A نَظِير *m*

3390
napalm *adj. n.*
F (au) napalm (*adj.*) *n. m.*
E (de) napalm (*adj.*) s. m.
D Napalm(-) *neut.* (*komp.*)
R напалмовый *прилаг.*; напалм *м.*
A نَابَالِم *m*

3391
nap-of-the-earth (NOE) flying
F vol en rase-mottes *m.*
E vuelo a ras del suelo *m.*
D Konturenflug *m.*
R бреющий полёт *м.*
A الطَّيْرَان بالقُرْب من سَطْح الأَرْض

212

3392

narrow
F étroit
E estrecho
D eng
R узкий
A ضَيِّق

3393

narrow gauge railway
F chemin de fer à voie
    étroite m.
E ferrocarril de vía
    estrecha m.
D Kleinspurbahn f.
R узкоколейная
    железная дорога ж.
A سكة حديد ضيقة

3394

nation
F nation f.
E nación f.
D Nation f.
R народ м.; нация ж.
A أُمَّة f

3395

national adj. n.
F national adj. n. m.;
    ressortissant n. m.
E nacional adj. s. m. f.
D national adj.;
    Staatsangehörige(r)
    m.
R народный прилаг.;
    национальный
    прилаг.; подданный
    м.; гражданин м.
A وَطَنِي

3396

national component
F contingent national m.
E contingente nacional m.
D nationaler Anteil neut.
R национальный
    контингент м.; часть
    ж.
A العُنْصُر الوَطَني m

3397

national infrastructure
F infrastructure nationale
    f.
E infraestructura nacional
    f.
D nationale Infrastruktur
    f.
R национальная
    инфраструктура ж.
A الأُسُس الوَطَنِيَّة f

3398

nationality
F nationalité f.
E nacionalidad f.
D Staatsangehörigkeit f.;
R национальность ж.;
    подданство ср.
A جِنْسِيَّة f

3399

**National Liberation
Front**
F Front National de
    Libération m.
E Frente National de
    Liberacíon m.
D Nationale
    Freiheitsfront f.
R Фронт (для)
    Народного
    Освобождения м.
A جبهة التحرير القومية f

3400

NATO
F OTAN
E OTAN f.
D NATO f.
R Североатлантический
    союз м.; НАТО м.;
    натовский прилаг.
A منظمة حلف شمال الاطلسي

3401

natural adj.
F naturel
E natural
D Natur- komp.; natürlich
    adj.
R естественный;
    природный
A طَبِيعِي

3402

natural barrier
F barrière naturelle f.
E barrera natural f.
D natürliche Sperre f.
R естественное
    заграждение ср.
A مَانِع طَبِيعِي m

3403 [H,P]

natural death
F mort naturelle f.
E muerte por causas
    naturales f.
D natürlicher Tod m.
R естественная смерть
    ж.
A الموت الطبيعي

3404 [T]

natural feature
F détail topographique m.
E característica naturel f.
D natürliches Merkmal
    neut.
R естественный предмет
    м.
A عَارِضَة طَبِيعِيَّة f

3405

nautical
F nautique
E náutico
D See- komp.;
    seemännisch adj.
R морской; мореходный
A بَحْري

213

3406

nautical twilight

F crépuscule nautique *m.*

E crepúsculo náutico *m.*

D nautisches Zwielicht *neut.*

R морские сумерки *ж. мн. ч.*

A الشَّفَق البَحْري *m*

3407

naval

F naval

E naval

D Marine- *komp.*

R военно-морской

A بَحْرِي

3408

Naval academy

F école navale *f.*

E escuela naval militar *f.*

D Marine-Akademie *f.*

R военно-морское училище *ср.*; Академия Военно-Морского Флота *ж.*

A كلية بحرية

3409

naval action

F combat sur mer *m.*

E combate naval *m.*

D Seekampf *m.*

R морской бой *м.*

A قتال بحري

3410

naval assault group

F groupement naval d'assaut *m.*

E grupo naval de asalto *m.*

D Landungsverband *m.*

R морская десантная группа *ж.*

A جَمَاعَة إقْتِحَام بَحْرِيَّة *f*

3411

naval beach group

F groupement naval de plage *m.*

E grupo naval de playa *m.*

D Strandmeistergruppe *f.*

R морская высадочная группа *ж.*

A مَجْمُوعَة الشَّاطِىء البَحْرِيَّة *f*

3412

naval support area

F zone de soutien naval *f.*

E zona de apoyo naval *f.*

D Seeversorgungsraum *m. m.*

R район поддержки военно-морскими силами *м.*

A مِنْطَقَة السَّنْد البَحْري *f*

3413 [A,M]

navigable

F navigable

E gobernable

D lenkbar

R лётный; судоходный

A صَالِح لسَّيْر السُّفُن أو الطَّائِرات

3414 [T]

navigable

F (en) état de naviguer

E navegable

D schiffbar

R судоходный

A قَابِل لِلمِلاَحَة

3415

navigation

F navigation *f.*

E navegación *f.*

D Navigation *f.*

R навигация *ж.*

A مِلاَحَة *f*

3416

navigational grid

F grille de navigation *m.*

E cuadrícula de navegación *f.*

D Navigationsgitter *f.*

R навигационная (координатная) сетка *ж.*

A تَرْبِيع المِلاَحَة *m*

3417

navigation head

F point de transbordement *m.*

E cabeza de etapa fluvial *f.*

D Navigationsmarke *f.*; Umladungspunkt *m.*

R пристань снабжения *ж.*; конечный пункт разгрузки *м.*

A رَأس جِسْرِ المِلاَحَة *m*

3418 [A,M]

navigation lights *pl.*

F feux de route *m. pl.*

E luces de navegación *f. pl.*

D Befeuerung [A] *f.*; Positionslampen [M] *f. pl.*

R аэронавигационные огни [A] *м. мн. ч.*; ходовые огни [M] *м. мн. ч.*

A أنْوَار المِلاَحَة *f*

3419

navigation mode

F contrôle automatique de navigation *m.*

E control automático de navegación *m.*

D Selbststeuerung durch automatische Peilung (meist eng. Begriff) *f.*

R система самонаведения *ж.*

A طَرِيقَة المِلاَحَة *f*

3420 [A,M]

navigator *m.*

F (officier) navigateur
  [(M)A] *m.*
E official de navegación
  [M]; navigante [A]
D Navigator [A] *m.;*
  Navigationsoffizier
  [M] *m.*
R навигатор [A] *м.;*
  штурман [M] *м.*
A مَلَّاح *m*

3421

navy

F marine de guerre *f.*
E marina de guerra *f.*
D Kriegsmarine *f.*
R военно-морской флот
  (ВМФ) *м.*
A البَحْرِيَّة *f*

3422

neap tide

F mortes eaux *f. pl.*
E marea muerta *f.*
D Nippflut *f.*
R квадратурный прилив
  м.
A المَدّ الأَصْغَر *m*

3423 [H]

neck

F cou *m.*
E cuello *m.*
D Hals *m.*
R шея *ж.*
A عُنْق *m*

3424 [H]

neck

F cou *m.*
E cuello *m.*
D Hals *m.*
R шея *ж.*
A عُنْق *m*

3425

negative *adj. n.*

F négatif *adj. n. m.*
E negativo *adj. s. m.*
D negativ *adj.;* Negativ
  neut.
R отрицательный;
  негативный
A سَلْبِي

3426

negative answer

F réponse négative *f.*
E respuesta negativa *f.*
D Verneinung *f.*
R отрицательный ответ
  м.
A جَواب سَلْبِي

3427 [S]

negligent discharge

F décharge involontaire
  d'une arme
  (punissable) *f.*
E disparo fortuito
  (punible) *m.*
D nachlässiger Abschuss
  m.
R небрежный выстрел *м.*
A الرمي بالاهمال

3428 [N]

**negligible risk
(nuclear)**

F risque négligeable
  (nucléaire) *m.*
E riesgo insignificante *m.*
D Mindestgefahr *f.;*
  Mindestrisiko *neut.*
R незначительный риск
  м.
A خطر يمكن اهماله

3429

negotiate

F négocier
E negociar
D verhandeln
R вести переговоры
A فَاوَض

3430

negotiate an obstacle

F franchir un obstacle
E pasar un obstáculo
D überwinden ein
  Hindermiss *untr.*
R преодолевать
  заграждение
A اقتحم مانعا

3431

negotiation(s) *pl.*

F négociations *f.* (*pl.*)
E negociación(es) *f.* (*pl.*)
D Verhandlungen *f.* (*pl.*)
R переговоры *м. мн. ч.*
A مُفَاوَضات *f*

3432

negotiator

F négociateur *m.*
E negociador *m.*
D Verhändler *m.*
R ведущий переговоры
  м.
A مُفَاوِض *m*

3433

neighboring unit

F unité voisine *f.*
E unidad inmediata *f.*
D flankierende Einheiten
  f.
R сосед *м.;* соседное
  подразделение *ср.*
A وحدة مجاورة *f*

3434

nerve

F nerf *m.*
E nervio *m.*
D Nerv *m.*
R нерв *м.*
A عَصَب *m*

3435
nerve agent
F agent (neurotoxique) *m.*
E agente nervioso *m.*
D Nervenkampfstoff *m.*
R отравляющее
   вещество (ОВ)
   нервно-
   паралитического
   действия *ср.*
A عَامِل مُثِيرِ لِلأعْصَاب *m*

3436
(machine-gun) nest
F nid de mitrailleuses *m.*
E nido de ametralladoras
   m.
D MG-Stellung *f.*
R (пулемётное) гнездо
   ср.
A عشّ الرشاشة

3437 [R]
net *v.*
F mettre en réseau;
   syntoniser (à)
E sintonizar (a)
D einstellen (auf) *tr.*
R устанавливать сеть
A نَصَب شَبَكَة

3438 [R,U]
net *n.*
F filet *m.;* réseau [R] *m.*
E red *f.*
D Netz *neut.*
R (радио-)сеть *ж.*; сетка
   ж.
A شَبَكَة *f*

3439 [R]
(command) net
F réseau de transmissions
   m.
E red (de mando) *f.*
D Führungsnetz *neut.*
R (командная)
   радиосеть *ж.*
A شَبَكَة القِيَادة *f*

3440 [R]
(rear-link) net
F réseau de liaison *m.*
E enlace con escalón
   superior *m.*
D Funknetz (zum
   rückwärtigen Stab)
   neut.
R радиосеть *ж.*
A شبكة الاتصالات إلى الخلف

3441
(warning) net
F réseau d'alerte *m.*
E red de alarma *m.*
D Warnnetz *neut.*
R сеть оповещения *ж.*
A شَبَكَة إنْذَار

3442
net authentication
F authentification de
   réseau *f.*
E autentificación de red *f.*
D Netzauthentisierung *f.*
R позывные сети *м. мн.*
   ч.
A إثْبَات شَرْعِيَّة الشَّبَكَة

3443 [R]
netting call
F signal d'accord en
   réseau *m.*
E llamada general *f.*
D Funknetzabruf *m.*
R вступление в связь *ср.*
A نِداء الشبكة

3444 [T]
(communications)
   network
F réseau *m.*
E red *f.*
D Verbindungsnetz *neut.*
R сеть (сообщений) *ж.*
A شبكة المواصلات

3445 [A,M]
(fixer) network
F réseau de détermination
   de position *m.;* réseau
   de trigonométrie *m.*
E red de fijación *f.*
D Peilnetz *neut.*
R сеть
   радиопеленгаторных
   постов *ж.*
A شَبَكَة رَادْيو لتَعْيِين النُّقْطَة *f*

3446
(transport) network
F réseau de transport *m.*
E red de transporte *m.*
D Verkehrsnetz *neut.*
R маршрутная сеть *ж.*
A شبكة نقل

3447
neutral
F neutre
E neutral
D neutral
R нейтральный
A محايد

3448
neutralization
F neutralisation *f.*
E neutralización *f.*
D Neutralisierung [S] *f.;*
   Lähmung [ALM] *f.*
R подавление (огнём)
   ср.; нейтрализация
   ж.
A إبْطَال *m*

3449
neutralize *v.*
F neutraliser
E neutralizar
D ausschalten *tr.*
R подавлять огнём
A أبْطَل

3450

neutron
F neutron *m.*
E neutrón *m.*
D Neutron *neut.*
R нейтрон *м.*
A بِيُوتْرُون *m*

3451

neutron-induced activity
F radioactivité induite par les neutrons *f.*
E actividad inducida por los neutrones *f.*
D neutroninduzierte Aktivität *f.*
R нейтронно-индуктированная радиоактивность *ж.*
A نَشاط مِن التَّفَاعُل النِّيوتْرُوني

3452

next-of-kin
F plus proche parent *m.*
E parientes más próximas *m. f.*
D Angehörige *m. f.*
R родные *мн. ч.*
A اقرب الاقرباء

3453 [N]

N-hour
F heure N *f.*
E hora N *f.*
D Stunde N *f.*
R час-Н *м.*
A سَاعَة أن

3454

nickname
F appellation conventionnelle *f.*
E apodo *m.*
D Spitzname *m.;* Deckname *m.*
R условное наименование *ср.;* кличка *ж.*
A لقب

3455

night
F nuit *f.*
E noche *f.*
D Nacht *f.*
R ночь *ж.*
A لَيْل *m*

3456

night adaptation
F adaptation à la nuit *f.*
E adaptación a la noche *f.*
D Nachtadaption *f.*
R ночная адаптация глаз *ж.*
A تَكَيُّف مَع اللَّيْل *m*

3457 [R]

night effect
F effet de nuit *m.*
E efecto de noche *m.*
D Nachteffekt *m.*
R ночной эффект *м.;* ночное воздействие *ср.*
A الظَّاهِرة اللَّيْلِيَّة *f* / تَأْثِير اللَّيل *m*

3458 [A]

night fighter
F chasseur(s)-intercepteur(s) de nuit *m. (pl.)*
E caza nocturno *m.*
D Nachtjäger *m.*
R ночной истребитель *м.*
A مُقَاتِلة لَيْلِيَّة *f*

3459

night fighting
F combat de nuit *m.*
E combate nocturno *m.*
D Nachtkampf *m.*
R ночной бой *м.*
A القِتَال اللَّيْلِي *m*

3460

night fighting aid
F moyen d'aide au combat de nuit *m.*
E dispositivo de combate nocturno *m.*
D Nachtkampfhilfsmittel *neut.*
R средство обеспечения боевых действий ночью *ср.*
A أَدَاة مُسَاعِدَة لِلقِتَال اللَّيْلِي *f*

3461

night glasses
F jumelles d'observation de nuit *f. pl.*
E gafas nocturnas *f. pl.*
D Nacht(fern)glas *neut.*
R ночной бинокль *м.*
A مِنْظار الرَّصْد اللَّيْلِي *m*

3462

night vision
F vision nocturne *f.*
E visión nocturna *f.*
D Nachtsicht *f.*
R ночное видение *ср.*
A الرُّؤْية اللَّيْلِيَّة *f*

3463

night vision system
F système d'observation nocturne *m.*
E sistema de visión nocturna *m.*
D Nachtsichtgerät *neut.*
R прибор ночного видения *м.*
A جَهَاز الرُّؤْية اللَّيْلِيَّة *m*

3464

night watch
F quart de nuit *m.*
E cuarto de noche *m.*
D Nachtwache *f.*
R ночная вахта *ж.*
A حِرَاسَة لَيْلِيَّة *f*

217

3465

no fire area (NFA)

F zone d'interdiction de tir *f.*

E área de fuego prohibido *f.*

D Feuerverbotsgebiet *neut.*

R запретная зона (для стрельбы) *ж.*

A منطقة عدم الرمي

3466 [P]

no-go area

F zone à éviter *f.*

E zona peligrosa *f.*

D Sperrgebiet *neut.*

R запретный район *м.*

A مَنْطِقَة حَرَام *f*

3467 [R,U]

noise

F bruit (de fond) *m.*

E ruído *m.*

D Geräusch [U] *neut.;* Störung [R] *f.*

R шум *м.;* звук *м.*

A ضَوْضَاء *m*

3468

no man's land

F no man's land

E tierra de nadie *f.*

D Niemandsland *neut.*

R ничья земля *ж.;* ничейная местность *ж.*

A أَرْض حرام *f*

3469

nominal

F nominal

E nominal

D Nenn- *komp.;* nominal *adj.;* nominal- *komp.*

R номинальный; условный

A إسمي - رَمزي

3470

nominate

F désigner; nommer

E nombrar

D ernennen

R назначать

A عَيَّنَ - رَشَّحَ

3471

nonbattle casualties

F pertes hors combat *f. pl.*

E bajas por accidente *f. pl.*

D Ausfälle ohne Feindeinwirkung *m. pl.*

R небоевые потери *ж. мн. ч.*

A خَسَائِر خَارِج المَعْرَكَة *f*

3472

nonbelligerent *adj.*

F non-belligérant

E no combatiente

D nicht-kriegführend

R несражающийся

A غير محارب

3473

noneffective *adj. n.*

F inapte *adj. n. m. f.*

E incapacitado *adj. s. m.*

D dienstunfähig(-e,-er) *adj.(s. m.)*

R негодный к службе

A غَيْر مُؤَثِّر

3474 [C]

nonpersistent agent

F agent fugace *m.*

E agresivo fugaz *m.*

D Luftkampfstoff *m.*

R нестойкое отравляющее вещество (OB) *ср.*

A عَامِل غَيْر مُدَاوِم *m*

3475 [L,M]

nonscheduled unit

F unité de réserve *f.*

E unidad en reserva *f.*

D Reserve-Einheit *f.*

R часть резерва *ж.*

A وَحْدَة غَيْر مُدْرَجَة *f*

3476

nonskid

F antidérapant

E antideslizante

D schleuderfrei

R нескользящий

A ضد الانزلاق

3477

nonvital cargo

F cargaison non essentielle *f.*

E carga no esencial *f.*

D nicht lebenswichtige Ladung *f.*

R второстепенный груз *м.*

A حُمُولَة غَيْر حَيَّوِيَّة *f*

3478

noon

F midi *m.*

E mediodía *m.*

D Mittag *m.*

R полдень *м.*

A ظُهْر *m*

3479

normal

F normal

E normal

D normal *adj.;* normal- *komp.*

R нормальный; обыкновенный

A عَادِي

3480 [S]
normal impact
F impact normal *m.*
E impacto normal *m.*
D Normalaufschlag *m.*
R попадание по
   нормали
A إِصَابَة عَادِيَّة

3481
normalize
F normaliser
E normalizar
D normalisieren
R нормализовать
A جَعَلَه سَوِيًّا

3482
north *adj. n.*
F (du) nord *(adj.) n. m.*
E (del) norte *(adj.) s. m.*
D Nord(en) *komp. (s. m.)*
   nördlich *adj.*
R северный *прилаг.*;
   север *м.*
A شِمَال *m*

3483
northbound
F (en) direction du nord
E hacia el norte
D (nach) Norden
   unterwegs
R направляющийся на
   север
A موجه إلى الشمال

3484
northing *n.*
F ordonnée nord *f.*
E hacia el norte *f.*;
   ordenada del norte
D Nordwert *m.*
R северное отшествие
   ср.
A خط عرض شمالي *m*

3485
northwards
F vers le nord
E hacia el norte
D (nach) Norden;
   nördlich
R (на) север; (к) северу
A في إتِّجَاه الشمال

3486 [G,S]
nose *adj. n.*
F (d')ogive *(adj.) n. f.;*
E (de) ojiva *(adj.) s. f.*
D Spitze(n-) *f. (komp.)*
R носовой *прилаг.*; нос
   м.
A مُقَدَّمَة *f*

3487 [G,S]
nosecone
F ogive *f.*
E ojiva *f.*
D Spitze *f.*
R носовой конус *м.*;
   головная часть *ж.*
A المَخْرُوط الأَمَامِي *m*

3488
nosedive
F piqué *m.*
E picado *m.*
D Sturzflug *m.*
R (крутое) пикирование
   ср.; пике *ср.*
A طَيْرَان إنْقِضَاضِي *m*

3489
notation
F notation *f.;* numération
   f.
E notación *f.*
D Bezeichnung *f.*
R нотация *ж.*;
   примечание *ср.*
A تَدْوِين *m*

3490
notch (sight)
F cran de mire *m.*
E muesca *f.*
D Visierkimme *f.*
R целик *м.*
A حَزّ *m*

3491
nothing to report
F (rien à) signaler (RAS)
E (sin) novedad
D (nichts zu) melden
R (нечего) докладывать
A تقرير بلا شيء

3492
notice *n.*
F avis *m.;* notification *f.*
E anuncio *m.;* aviso *m.*
D Ankündigung *f.*
R извещение *ср.*;
   уведомление *ср.*
A إعْلَان *m*

3493
**notice (board, plate
etc.)** *n.*
F affiche *f.*
E aviso *m.;* letrero *m.*
D Aushang *m.*
R объявление *ср.*;
   бюллетень *м.*
A شَارَة *f*

3494
noticeboard
F tableau d'affichage *m.*
E tablón de anuncios *m.*
D schwarzes Brett *neut.*
R доска для объявлений
   ж.
A لَوْحَة إعْلَانَات *f*

3495

notices to airmen (NOTAM) *pl.*

F avis aux navigateurs aériens *m. pl.*

E avisos a los navegantes aéreos *m. pl.*

D Nachrichten für Flieger *f. pl.*

R извещения лётчикам *ср. мн. ч.*

A تَنْبِيهَات لِلطَّيَّارِين *f*

3496

notices to mariners

F avis aux navigateurs *m.*

E avisos a los navegadores *m. pl.*

D Nachrichten für Seefahrer *f. pl.*

R извещения мореплавателям *ср. мн. ч.*

A تَنْبِيهَات لِلمَلَّاحِين *f*

3497 [A,L,M]

notice to move (at ... hours)

F (à ... heures) préavis de mise en route *m.*

E estar listo para el movimiento (dentro de ....horas)

D Verlegungsmeldung (auf....Abruf) *f.*

R наготове к перевозке

A الانذار للحركة

3498

notification

F notification *f.*

E aviso *m.*

D Mitteilung *f.*

R оповещение *ср.*

A إعلام

220

3499

nozzle

F bec *m.; jet m.; lance f.*

E tobera *f.; boquilla (de manga) f.*

D Düse *f.;* Schlauchendstück *neut.*

R насадок *м.;* форсунка *ж.*

A نَضَّاحَة *f*

3500

nuclear

F nucléaire

E nuclear

D Atom- *komp.;* Kern- *komp.*

R ядерный

A نَوَوِي

3501

nuclear, biological and chemical (NBC)

F nucléaire, biologique et chimique (NBC)

E nuclear, bacteriológica e química (NBQ)

D atomar, biologisch und chemisch (ABC)

R ядерний, биологический и химический

A نَوَوِي جُرْثُومِي كِيمَاوِي

3502

nuclear airburst

F explosion nucléaire aérienne *f.*

E explosión nuclear en el aire *f.*

D Luftdetonation *f.*

R воздушный ядерный взрыв *м.*

A إنْفِجَار نَوَوِي جَوِّي *f*

3503

nuclear cloud

F nuage radioactif *m.*

E nube nuclear *m.*

D (nukleare) Detonationswolke *f.*

R (ядерная) грибовидная туча *ж.*

A سَحَابة نَوَوِيَّة *f*

3504

nuclear damage

F dommage nucléaire *m.*

E daño nuclear *m.*

D Atomschäden *m. pl.*

R ядерное разрушение *ср.;* поражение *ср.*

A التَّدْمِير النَّوَوِي *m*

3505

nuclear damage assessment

F évaluation de dommages nucléaires *f.*

E evaluación de daños nucleares *f.*

D Atomschädenbewertung *f.*

R оценка ядерного поражения *ж.*

A تَقْدِير التَّدْمِير النَّوَوِي *m*

3506

nuclear defense

F défense nucléaire *f.*

E defensa nuclear *f.*

D atomare Verteidigung *f.*

R противоядерная защита *ж.;* противоядерная оборона *ж.*

A الدِّفَاع النَّوَوِي *m*

3507

nuclear device

F charge nucléaire *f.*

E carga nuclear *f.*

D Atomsprengkörper *m.*

R ядерное оружие *ср.*

A جِهَاز نَوَوِي *m*

3508

nuclear incident
F incident nucléaire *m.*
E incidente nuclear *m.*
D atomarer Zwischenfall *m.*
R происшествие с ядерным оружием *ср.*
A حَادِث نَوَوِي طَارِىء *m*

3509 [P]

nuclear power
F puissance nucléaire *f.*
E potencia nuclear *f.*
D Atommacht *f.*
R ядерная энергия *ж.*
A قُدْرَة نَوَوِيَّة *f*

3510

nuclear power plant
F groupe moteur nucléaire *m.*
E grupo motor nuclear *m.*
D Atomantrieb *m.*
R ядерная/атомная электростанция *ж.*
A مَحَطَّة تَوْلِيد طَاقَة نَوَوِيَّة *f*

3511

nuclear radiation
F rayonnement nucléaire *m.*
E radiación nuclear *f.*
D Kernstrahlung *f.*
R ядерное излучение *ср.*
A إِشْعَاع نَوَوِي *m*

3512

nuclear reactor
F réacteur nucléaire *m.*
E reactor nuclear *m.*
D Kernreaktor *m.*
R ядерный реактор *м.*
A مُفَاعِل نَوَوِي *m*

3513

nuclear surface burst
F explosion nucléaire en surface *f.*
E explosión nuclear de superficie *f.*
D Bodendetonation *f.*
R наземный ядерный взрыв *м.*
A تَفْجِير نَوَوِي على سَطْح الأَرْض

3514

nuclear underground burst
F explosion nucléaire souterraine *f.*
E explosión nuclear subterránea *f.*
D Untergrunddetonation *f.*
R подземный ядерный взрыв *м.*
A تَفْجِير نَوَوِي تَحْت سَطْح الأَرْض

3515

nuclear underwater burst
F explosion nucléaire sous-marine *f.*
E explosión nuclear submarina *f.*
D Unterwasserdetonation *f.*
R подводный ядерный взрыв *м.*
A تَفْجِير نَوَوِي تَحْت سَطْح المَاء

3516

nuclear weapon(s) accident
F accident d'arme(s) nucléaire(s) *m.*
E accidente de arma(s) nuclear(es) *m.*
D Atomwaffenunfall *m.*
R происшествие с ядерным оружием *ср.*
A حَادِث بِسِلَاح نَوَوِي

3517

nuclear yield
F puissance *f.*
E potencia *f.*
D Detonationswert *m.*
R мощность ядерного оружия *ж.*
A الحَاصِل النَّوَوِي *m*

3518

number *v.*
F numéroter
E numerar
D nummerieren
R нумеровать; считать
A رَقَّم . عَدَّ

3519

number (figure) *n.*
F chiffre *m.*
E cifra *f.*
D Ziffer *f.*
R цифра *ж.*
A رقم

3520

number (quantity) *n.*
F nombre *m.*
E número *m.*
D Anzahl *m.*
R количество *ср.*; число *ср.*
A عَدَد *m*

3521

number (serial) *n.*
F numéro *m.*
E número *m.*
D Nummer *f.*
R номер *м.*
A رقم

3522

nurse *v.*
F soigner
E cuidar (a)
D pflegen
R ухаживать (за)
A مَرَّض

3523

(male) nurse

F infirmier *m.*
E enfermero *m.*
D Krankenpfleger *m.;*
   Krankenschwester *f.*
R санитар(ка) *м.*(*ж.*);
   медсестра *ж.*
A (مُمَرِّض) مُمَرِّضَة

3524

nut

F écrou *m.*
E tuerca *f.*
D Mutter(n) *f.*
R гайка *ж.*
A صَمُولة *f*

3525

nylon *adj. n.*

F (en) nylon (*adj.*) *n. m.*
E (de) nilón (*adj.*) *s. m.*
D Nylon(-) *neut.* (*komp.*)
R нейлоновый *прилаг.*;
   нейлон *м.*
A نِيلُون

O

3526

oar

F rame *f.*
E remo *f.*
D Ruder *neut.*
R весло *ср.*
A مِجْداف *m*

3527

oarlocks *pl.*

F tolets *m. pl.*
E escalameras *f. pl.*
D Runzeln *f. pl.*
R уключины *ж. мн. ч.*
A حَامِلات مِجْداف *f*

3528

obey

F obéir (à)
E obedecer; responder
D (jemanden) gehorchen
R повиноваться;
   подчиняться
A أَطَاع

3529

object *v.*

F faire objection (à)
E oponerse (a)
D Einwand erheben
R возражать;
   протестовать
A إحْتَجّ

3530

object *n.*

F objet *m.*
E objeto *m.*
D Gegenstand *m.*
R объект *м.*; цель *ж.*;
   предмет *м.*
A غَرَض *m*

3531

objection

F objection *f.*
E objeción *f.*
D Einwand *m.*
R возражение *ср.*;
   протест *м.*
A إحْتِجاج *m*

3532

objective

F objectif *m.*
E objetivo *m.*
D Objektiv *neut.*
R объект *м.*; цель *ж.*;
   линза объектива *ж.*
A هَدَف *m*

3533 [S]

objective area

F zone à l'objectif battue
   par le feu *f.*
E zona del objetivo *f.*
D Beschussgebiet *neut.*
R район объекта *м.*
A مِنْطَقَة الهَدَف *f*

3534 [L]

objective area

F zone de l'objectif *f.*
E zona del objetivo *f.;*
   zona a batir *f.*
D Zielraum *m.*
R район цели *м.*; район
   высадки десанта *м.*
A مِنْطَقَة الهَدَف *f*

3535

(conscientious)
   objector

F objecteur de conscience
   m.
E pacifista (que se niega a
   tomar) las armas *m. f.*
D Kriegsdienst-
   verweigerer *m.*
R отказывающийся (от
   военной службы)
A مُسْتَنْكِف ضَمِيرِيا

3536

oblique

F oblique
E oblicuo
D schräg
R косой; наклонный;
   перспективный
A مائِل ـ مُنْحَرِف

3537

obscuration

F obscurcissement *m.*
E oscurecimiento *m.*
D Verdunkelung *f.*
R невыясненность *ж.*
A الاظلام

3538
observation
F observation *f.*
E observación *f.*
D Beobachtung *f.*
R наблюдение *ср.*
A رَصْد *m*

3539
observe
F surveiller
E observar
D beobachten
R наблюдать
A رَصَد

3540
observed fire chart
F plan des tirs observés *m.*
E plano artillero de tiro observado *m.*
D Zieltafel für beobachtetes Schiessen *m.*
R таблица поправок (огня) *ж.*
A لَوْحة النَّار المَرْصُودة *f*

3541
observer
F observateur *m.*
E observador *m.*
D Beobachter [SU] *m.;* Orter [A] *m.*
R наблюдатель *м.*
A رَاصِد ـ مُرَاقِب

3542
obsolescent
F (en) désuétude; vieillissant
E (en) desuso
D veraltend
R устаревающий
A في طريق الزَّوَال

3543
obsolete
F déclassé; suranné
E (en) desuso; obsoleto
D veraltet
R устарелый
A قَدِيم *m*

3544
obstacle
F obstacle *m.*
E obstáculo *m.*
D Hindernis *neut.*
R препятствие *ср.*; заграждение *ср.*; преграда *ж.*
A حَاجِز *m*

3545
(cultural) obstacle
F obstacle artificiel *m.*
E obstáculo artificial *m.*
D Hindernis (künstlich) *neut.*
R заграждение *ср.*
A مانع إصْطِنَاعي

3546
(existing) obstacle
F obstacle existant *m.*
E obstáculo natural *m.*
D Hindernis (natürlich) *neut.*
R преграда *ж.*
A مانع موجود

3547
(flanking) obstacle
F obstacle passif de flanc *m.*
E obstáculo en un flanco *m.*
D Flankhindernis *neut.*
R фланкирующее заграждение *ср.*
A مانع جناحي *m*

3548
(impassable) obstacle
F obstacle infranchissable *m.*
E obstáculo infranqueable *m.*
D Hindernis (unpassierbar) *neut.*
R (непроходимое) препятствие *ср.*
A حاجز لا يمكن عبوره

3549
(natural) obstacle
F obstacle naturel *m.*
E obstáculo natural *m.*
D Hindernis (natürlich) *neut.*
R преграда *ж.*
A حَاجِز طَبِيعِي *m*

3550
(protective) obstacle
F obstacle passif *m.*
E obstáculo protector *m.*
D Schutzhindernis *neut.*
R (заградительное) препятствие *ср.*
A مانع وِقَائي

3551
(rear) obstacle
F obstacle dans la zone arrière *m.*
E obstáculo de profundidad *m.*
D Hindernis (im rückwärtigen Kampfgebiet) *neut.*
R (тыловое защитное) заграждение *ср.*
A مانع خلفي

3552
(standard) obstacle
F obstacle de série *m.*
E obstáculo estandar *m.*
D Hindernis (ständig) *neut.*
R (инженерное) заграждение *ср.*
A مانع عادي

3553

(tactical) obstacle

F obstacle actif *m.*
E obstáculo táctico *m.*
D Hindernis (taktisch) *neut.*
R (тактическое инженерное) заграждение *ср.*
A مانع تعبوي

3554

obstruct

F boucher; encombrer
E cerrar; obstruir
D hemmen; sperren
R заграждать; преграждать
A عَاق

3555

obstruction

F encombrement *m.*
E obstrucción *f.*
D Hindernis *neut.*
R препятствие *ср.*; заграждение *ср.*
A سَدّ ـ عَائِق *m*

3556 [A,M]

occulting

F (à) occultations
E (de) ocultaciones
D unterbrochen
R затмевающий (огонь, маяк)
A إسْتِتَار *m*

3557

occupation

F occupation *f.*
E ocupación *f.*
D Besetzung *f.*
R оккупация *ж.*; занятие *ср.*
A إحْتِلَال *m*

224

3558 [P]

occupied territory

F territoire occupé *m.*
E territorio ocupado *m.*
D besetztes Gebiet *neut.*
R оккупированная территория *ж.*
A أرْض مُحْتَلَة *f*

3559

occupy

F garnir; occuper
E ocupar
D besetzen
R оккупировать; занимать
A إحْتَلّ

3560

occupy a position

F occuper une position
E ocupar una posición
D (eine) Stellung beziehen
R занимать позицию
A استولى على

3561

ocean

F océan *m.*
E océano *m.*
D Ozean *m.*
R океан *м.*
A مُحِيط *m*

3562

ocean convoy

F convoi océanique *m.*
E convoy oceánico *m.*
D Hochseegeleitzug *m.*
R океанский конвой *м.*
A قَافِلَة بَحْرِيَّة *f*

3563

oceangoing *adj.*

F (au) long cours
E (de) alta mar
D Hochsee- *komp.*
R океанский
A بَحْرِي ـ عَابِر المُحِيطَات *f*

3564

ocean manifest

F manifeste maritime *m.*
E manifiesto marítimo *m.*
D Seeladeverzeichnis *neut.*
R океанский манифест *м.*
A بيان بحري

3565

oceanography

F océanographie *f.*
E oceanografía *f.*
D Ozeanographie *f.*
R океанография *ж.*
A جغرافية المحيطات

3566

octane number

F indice d'octane *m.*
E octanaje *m.*
D Oktanzahl *f.*
R октановое число *ср.*
A العدد الاوكتاني

3567

octant

F octant *m.*
E octante *m.*
D Oktant *m.*
R октант *м.*
A مِنْقَل ـ ثُمْن مُحِيط (مساحة) الدائرة *m*

3568

ocular *adj. n.*

F oculaire *adj. n. m.*
E ocular *adj. s. m.*
D Augen- *komp.*; Okular *s. neut.*
R глазной; окулярный; окуляр
A عَيْنِي

3569

odd number

F nombre impair *m.*
E número impar *m.*
D ungerade Zahl *f.*
R нечётное число *ср.*
A عدد فردي

3570

offensive *adj.*

F offensif

E ofensivo

D offensiv *adj.;* Offensiv-
komp.

R наступательный

A تَعَرُّضي ـ هجومي

3571

offensive *n.*

F offensive *f.*

E ofensiva *f.*

D Offensive *f.*

R наступление *ср.*

A تَعَرُّض ـ هُجْوم *m*

3572

office

F bureau *m.*

E oficina *f.*

D Büro *neut.;* Dienststelle
f.

R контора *ж.;* бюро *ср.*

A مَكْتَب *m*

3573

officer

F officier *m.*

E oficial *m.*

D Offizier *m.*

R офицер *м.*

A ضَابِط *m*

3574

(air liaison) officer

F officier de liaison (air)
m.

E oficial de enlace aire *m.*

D Luftverbindungsoffizier
m.

R офицер воздушной
связи *м.*

A ضابط ارتباط جوي

3575

(air movement) officer

F officier régulateur *m.*

E oficial de movimiento
aéreo *m.*

D Lufttransportoffizier *m.*

R офицер воздушных
перевозок *м.*

A ضابط نقل جوي

3576

(billeting) officer

F Major de cantonnement
m.

E oficial de alojamiento
m.

D Quartieroffizier *m.*

R офицер
расквартирования *м.*

A ضابط الايواء

3577

(bomb disposal) officer

F artificier *m.*

E desactivador (de
explosivos) *m.*

D Bombenentschärfungs-
offizier *m.;*
Kampfmittelbeseiti-
gungsoffizier *m.*

R офицер по
обезвреживанию
бомб *м.*

A ضابط التخلص من القنابل

3578

(catering) officer

F officier d'ordinaire *m.*

E oficial de
abastecimiento *m.*

D Proviantoffizier *m.*

R офицер-хозяйственник
м.

A ضابط الإعاشة

3579

(commanding) officer

F chef de corps *m.*

E jefe *m.*

D Einheitsführer *m.;*
Kommandeur *m.*

R командир *м.*

A قائِد وَحْدة *m*

3580

(dental) officer

F dentiste militaire *m.*

E oficial dentista *m.*

D Truppenzahnarzt *m.*

R военный зубной врач
м.

A ضابط أسنان

3581

(duty) officer

F officier de permanence
m.

E oficial del día, oficial de
servicio *m.*

D Offizier vom Dienst *m.*

R дежурный офицер *м.*

A الضابط المناوب

3582

(escort) officer

F officier accompagnateur
m.

E oficial de escolta *m.*

D Begleitoffizier *m.*

R офицер конвоя *м.*

A ضابط للحراسة

3583

(executive) officer

F officier adjoint *m.*

E segundo oficial *m.*

D Verwaltungsoffizier *m.*

R заместитель
командира *м.*

A ضابط منفذ

3584 [M]

(flag) officer

F officier supérieure
navale

E almirante *m.;* ayudante

D Flaggoffizier *m.*

R заместитель
командира по ... *м.*

A ضابط من مرتبة القادة

3585

(forward observation) officer (FOO)

F officier observateur d'artillerie *m.*

E observador avanzado *m.*

D vorgeschobener Beobachtungsoffizier *m.*

R передовой артиллерийский наблюдатель *м.*

A ضابط الرصد الامامي

3586

(gun position) officer (GPO)

F Lieutenant de tir *m.*

E oficial de artillería *m.*

D Offizier der Geschützstellung *m.*

R старший (офицер батареи) *м.*

A ضابط موقع الرمي

3587

(ground liasion) officer

F officier de liaison de l'armée de terre *m.*

E oficial de enlace en tierra *m.*

D Heeresverbindungs- offizier *m.*

R офицер связи сухопутных войск *м.*

A ضَابِط إِرْتَباط أَرْضِي *m*

3588

(instructor) officer

F officier instructeur *m.*

E instructor *m.*

D Ausbildungsoffizier *m.*

R офицер-инструктор *м.*

A ضابط مدرب

3589

(intelligence) officer

F officier de renseignement *m.*

E oficial de información *m.*

D Nachrichtenoffizier *m.*

R офицер разведки *м.*

A ضابط الاستخبارات

3590

(junior) officer

F officier subalterne *m.*

E oficial subalterno *m.*

D dienstjüngerer Offizier *m.*

R младший офицер *м.*

A ضابط (صغير)

3591

(liaison) officer

F officier de liaison *m.*

E oficial de enlace *m.*

D Verbindungsoffizier *m.*

R офицер связи *м.*

A ضَابِط إِرْتَباط *m*

3592

(medical) officer

F médecin militaire *m.*

E médico *m.*

D Sanitätsoffizier *m.*

R военный врач *м.*

A ضابط طبي

3593

(messing) officer

F officier d'ordinaire *m.*

E oficial de rancho *m.*

D Proviantoffizier *m.*

R офицер-хозяйственник *м.*

A ضابط المطعم

3594

(motor transport) officer (MTO)

F officier auto *m.*

E oficial de transportes *m.*

D Transportoffizier *m.*

R заместитель командира по автотранспорту *м.*

A ضابط النقل

3595

(naval) officer

F officier de marine *m.*

E oficial de marina *m.*

D Marineoffizier *m.*

R офицер военно-морского флота (ВМФ) *м.*

A ضَابِط بَحْري *m*

3596

(noncommissioned) officer

F sous-officier *m.*

E suboficial *m.*

D Unteroffizier *m.*

R сержант *м.*

A ضَابِط صَفّ *m*

3597

(operations) officer

F officier du 3e Bureau *m.*

E oficial de operaciones *m.*

D Operationsoffizier *m.*

R дежурный офицер по военным действиям *м.*

A ضابط العمليات

3598

(orderly) officer

F officier de permanence *m.*

E oficial del día *m.*

D Offizier vom Dienst *m.*

R дежурный офицер *м.*

A ضابط الخفر

3599

(ordnance) officer

F officier du matériel *m.*
E oficial de material (de guerra) *m.*
D Offizier der Nachschubtruppe *m.*
R офицер службы снабжения *м.*
A ضَابِط العَتاد والذخيرة *m*

3600

(staff) officer

F officier d'état-major *m.*
E official de estado mayor *m.*
D Stabsoffizier *m.*
R штабной офицер *м.*
A ضَابِط رُكْن

3601

(subaltern) officer

F officier subalterne *m.*
E alférez *m.*
D Subalternoffizier *m.*
R младший лейтенант *м.*
A ضَابِط عون

3602

(supply) officer

F officier d'approvisionnements *m.*
E oficial de intendencia *m.*
D Nachschubsoffizier *m.*
R офицер по снабжению *м.*
A ضَابِط التموين

3603

officer of the day (OD)

F officier de jour *m.*
E oficial del día *m.*
D Offizier vom Dienst (OvD) *m.*
R дежурный офицер *м.*
A ضَابِط اليَوْم *m*

3604

officer of the guard

F officier de la garde *m.*
E oficial de la guardia *m.*
D Wachoffizier *m.*
R командир караула *м.*
A ضابط الحراسة

3605

officer of the watch

F officier de quart *m.*
E oficial de cuarto *m.*
D Wachoffizier *m.*
R вахтенный офицер *м.*; командир вахты *м.*
A ضَابِط الحِرَاسَة *m*

3606

official *adj.*

F officiel *m.*
E oficial
D amtlich *adj.;* Amts-*komp.*
R служебный; официальный
A رَسْمِي

3607

official *n.*

F fonctionnaire *m.*
E funcionario *m.*
D Beamte *m.*
R должностное лицо *ср.*; служащий *м.*
A مُوَظَّف *m*

3608 [M]

(in the) offing

F (au) large
E cerca
D (in) Aussicht
R (в) виду берега
A عرض البحر غير بعيد

3609

Off limits!

F Hors limites!
E ¡Zona vedada!; ¡Fuera de límites!
D Betreten verboten!
R вход военнослужащим воспрещён
A مَحْظُور على الجُنْد

3610 [A,L]

off-load *v.*

F décharger
E descargar
D abladen *tr.*
R выгружать
A أنزل

3611 [E]

offset *n.*

F décalage *m.;* décentrage *m.*
E desvío *m.*
D Versetzung *f.*
R смещение *ср.*; офсет *м.*
A مُعادِل . موازن

3612 [U]

offset *n.*

F compensation *f.*
E compensación *f.*
D Ausgleich *m.*
R возмещение *ср.*; отвод *м.*; офсет *м.*
A حَيْد

3613 [U]

offset *v.*

F compenser
E compensar
D ausgleichen *tr.*
R возмещать; печатать офсетным способом
A حاد

227

3614
offset bombing
F bombardement en
   déport *m.*
E bombardeo por visado
   indirecto *m.*
D versetztes
   Bombenabwurf-
   verfahren *neut.*
R бомбометание с
   косовым
   прицеливанием *ср.*
A القَصْف الغير مُباشِر *m*

3615 [N]
offset distance
F distance de
   décentrement *f.*
E distancia de desvío *f.*
D Mindestabstand *m.*
R отклонение эпицентра
   (от цели) *ср.*
A مَسَافَة الإزَاحَة *f*

3616
offshore *adj.*
F au large
E (de) altura
D (in) Küstennähe
R близко от берега
A في عُرْض البَحْر

3617
offshore wind
F brise de terre *f.*
E terral *m.;* viento de
   tierra *m.*
D ablandiger Wind *m.*
R ветер с берега *м.*
A نَسِيم البَرّ *m*

3618
oil *v.*
F graisser
E lubricar
D schmieren
R смазывать
A زَيَّت

228

3619
oil *n.*
F huile *f.*
E aceite *m.*
D Öl *m.*
R масло *ср.*
A زَيْت *m*

3620
oil *adj.*
F pétrolier
E petrolero
D Erdöl- *komp.*
R нефтяной
A زَيْتي

3621
(diesel) oil
F gas-oil *m.*
E gas-oil *m.*
D Dieselöl *neut.*
R дизельное топливо *ср.*
A وقود زيت الديزل *m*

3622
(fuel) oil
F mazout *m.*
E aceite pesado *m.*
D Heizöl *neut.*
R горючее *ср.*
A مَازُوت *m*

3623
(natural) oil *n.*
F pétrole *m.*
E petróleo *m.*
D Erdöl *m.*
R нефть *ж.*
A نفط

3624
oil rig
F plateforme petrolière *f.*
E torre de perforación *f.*
D Bohrturm *m.*
R (плавучая) нефтяная
   буровая платформа
   ж.
A جِهَاز الحَفْر *m*

3625
oilskins *pl.*
F ciré *m.*
E traje de encerado *m.*
D Ölzeug *neut.*
R дождевое платье *ср.*
A مُشَمَّعَات *f*

3626
olive drab(s) *adj. (n.pl.)*
F (uniforme de couleur)
   gris-vert (olive) (*n.
   m.*) *adj.*
E verde-oliva *adj.;*
   uniforme(s) verde
   oliva *m. (pl.)*
D olivengrün *adj.;*
   Olivengrün
   (*amerikanische
   Uniform) neut.*
R оливковый *прилаг.;*
   повседневная форма
   ж.
A بزة من لون الزيتون

3627
on board
F à bord
E (a) bordo
D (an) Bord
R (на) борту
A على ظَهْر . . .

3628
on-board spares *pl.*
F pièces de rechange de
   bord *f. pl.*
E piezas de recambio de a
   bordo *f. pl.*
D Bordersatzteile *neut. pl.*
R бортовые запасные
   части (запчасти) *ж.
   мн. ч.*
A قِطَع تَبْدِيل على ظَهْر . . .

3629
on call
F à la demande
E (a) petición *m.*
D (auf) Abruf
R (по) вызову
A تَحْت الطَّلَب

3630

on deck

F (sur) le pont
E (en) cubierta
D (auf) Deck
R (на) палубе
A على السَّطْح

3631 [R]

one time system

F système à clé
  nonréutilisable *m.*
E sistema no repetible *m.*
D Einmal(schlüssel)system
  neut.
R одноразовая
  шифровальная
  система *ж.*
A نِظام رُموز يُسْتَعْمَل مَرَّة
  وَاحِدة

3632 [L]

one up

F un(e) en tête; en
  premier échelon
E (en) cuña
D einer(-e,-es) voraus
R (по) одному; (на) одну
  степень выше
A وَاحِد إلى الأمام

3633

on limits

F accès permis *m.*
E dentro de los límites
D Zutritt gestattet
R незапрещённый
A غير محظور على الجند

3634

ooze *n.*

F vase *f.*
E cieno *m.*
D Schlick *m.*
R ил *м.*; тина *ж.*
A نَزَّ ـ رَشَحَ *m*

3635 [U]

open *adj.*

F ouvert
E abierto
D offen
R открытый
A مَفْتوح

3636

open *v.*

F ouvrir
E abrir
D aufmachen *tr.*; öffnen
  untr.
R открывать
A فَتَح

3637 [T]

open *adj.*

F ras
E raso
D frei
R открытый
A مَكْشوف

3638

(in the) open *n.*

F (à) découvert
E (al) raso
D (im) Freien
R (на) свежем воздухе
A في أرض مكشوفة

3639

open city

F ville ouverte *f.*
E ciudad abierta *f.*
D entmilitarisierte Stadt
  (einem Angriff nicht
  ausgesetzt) *f.*
R открытый город *м.*
A مدينة مفتوحة

3640

open fire

F ouvrir le feu
E abrir el fuego
D Feuer eröffnen
R открывать огонь
A أَطْلَق النّار

3641

opening *n.*

F brèche *f.*; ouverture *f.*
E abertura *f.*; brecha *f.*
D Lücke *f.*; Öffnung *f.*
R отверстие *ср.*; щель
  ж.
A ثُغْرَة *f*

3642

(delayed) opening *n.*

F (saut) à ouverture
  commandée *m.*
E apertura retardada *f.*
D verzögerte Eröffnung
  (eines Fallschirms) *f.*
R замедленное открытие
  зонта *ср.*
A فتح متأخر

3643

(static line) opening *n.*

F ouverture automatique
  f.
E apertura automática *f.*
D statische Eröffnung
  (eines Fallschirms) *f.*
R открытие вытяжным
  фалом *ср.*
A الفتح بالحبل القراري

3644

open order *adj.*

F (en) ordre dispersé
E orden abierto *m.*
D (in) geöffneter Ordnung
R расчленённый строй
  м.
A في ترتيب مفتوح

3645 [A,L,M]

operate *v.*

F opérer
E operar
D operieren
R действовать
A قام بعمليات

3646 [E,R]

operate *v.*

F fonctionner
E manejar
D funktionieren
R работать
A عَمِل ـ شَغَّلَ

3647

operate *v.*

F manipuler
E funcionar
D bedienen
R действовать; работать
A قَام بِعَمَلِيَّة

3648 [H]

operate (on someone for)

F opérer (quelqu'un de)
E operar (a uno de algo)
D (jemanden (-em etwas)) operieren
R оперировать
A قام بعملية

3649

(gas) operated

F fonctionnant par emprunt des gaz
E accionado (por gas)
D (von) Gas getrieben
R действуемый газом
A يُشغَّل بالغاز

3650

operating lever

F levier de commande *m.*
E palanca de mando *f.*
D Bedienungshebel *m.*
R ручка управления *ж.*; рычаг управления *м.*
A عَتَلَة التَّشْغِيل *f*

230

3651

operation

F opération *f.*
E operación *f.*
D Operation [H] *f.*; Unternehmung [AML] *f.*
R операция *ж.*; действие *ср.*
A عَمَلِيَّة *f*

3652

(airmobile) operation

F opération héliportée *f.*
E operación con tropas aerotransportodas *f.*
D luftbewegliche Operation *f.*
R (воздушнодесантное) действие *ср.*
A عملية بالحركة الجوية

3653

(amphibious) operation

F opération amphibie *f.*
E operación anfibia *f.*
D amphibische Operation *f.*
R морская десантная операция *ж.*
A عَمَلِيَات بَرْمَائِية

3654

(area interdiction) operation

F opération d'interdiction *f.*
E operación de interdicción de zona *f.*
D Gebiets-Sperroperation *f.*
R (воздушные) действия по воспрещению местности *ср. мн. ч.*
A عملية تحريم المنطقة

3655 [M]

(boarding) operation

F manoeuvre d'abordage *f.*
E operación de abordaje *f.*; operación de vista y reconocimiento *f.*
D Enteroperation *f.*
R взятие на абордаж *ср.*
A عملية اقتحام السفينة

3656 [I]

(clandestine) operation

F mission clandestine *f.*
E operación clandestina *f.*
D Geheimoperation *f.*
R (тайные) действия *ср. мн. ч.*; (подрывные) действия *ср. мн. ч.*
A عملية سرية

3657

(combined) operation

F opération combinée (interalliée) *f.*
E operación combinada (de fuerzas aliadas) *f.*
D verbündete Operation *f.*
R совместные (десантные) действия *ср. мн. ч.*
A عَمَلِية مُشْتَرِكة *f*

3658 [L,P]

(cordon and search) operation

F opération de bouclage et de ratissage *f.*
E operación de acordonamiento y búsqueda *f.*
D Sperr- und Suchoperation *f.*
R плановые поиски *ср. мн. ч.*
A عَمَلِية المُحَاصَرة والبَحْث

3659 [A]
(counterair) operation
F opération de supériorité
  aérienne *f.*
E operación de
  superioridad aérea *f.*
D Fliegergegenangriff *m.*
R противовоздушные
  действия *ср. мн. ч.*
A عملية مضادة جوية

3660
(counterbarrier)
  operation
F opérations de
  destruction d'un
  barrage *f. pl.*
E maniobra por
  contrarrestar un
  obstáculo *f.*
D Hindernisräumung *f.*
R операция по
  подавлению преград
  ж.
A عملية مكافحة الموانع

3661
(countermobility)
  operation
F (des) opérations contre-
  mobilité *f. pl.*
E operación de
  contramobilidad *f.*
D Beweglichkeit-
  begrenzungoperation
  f.
R сооружение
  препятствий и
  преград *ср.*
A عملية مكافحة التحركات *f*

3662
(intervention)
  operation
F opération d'intervention
  rapide (hors limites) *f.*
E operación de
  intervención *f.*
D Vermittelungsoperation
  f.
R интервенция *ж.*
A عملية التدخل

3663
(joint) operation
F opération combinée
  interalliée *f.*
E operación combinada
  m.
D verbündete Operation *f.*
R совместные действия
  ср. мн. ч.
A عَمَلِيَّة مُشْتَرَكَة *f*

3664
(peacekeeping)
  operation *n.*
F opération de maintien
  de la paix *f.*
E operación pacificadora
  f.
D Operation zur
  Aufrechterhaltung des
  Friedens
R действия по
  поддержанию мира
  ср. мн. ч.
A عَمَلِيَّة لِلحِفْظ على الأَمْن

3665
(policing) operation
F opération de police *f.*
E (acción de) limpieza *f.*
D Säuberungsoperation *f.*
R действия по
  поддержке порядка
  ср. мн. ч.
A عَمَلِيَّة لِلمُحَافظة على
  النِّظام ـ عملية بوليسية

3666 [L]
(screening) operation
F opération de ratissage
  et de bouclage *f.*
E operación de cobertura
  f.
D Abschirmungsoperation
  f.
R изнурительно-
  защищающие
  действия *ср. мн. ч.*
A عملية ساترة

3667
(search and destroy)
  operation
F (opération de) ratissage
  et (de) nettoyage *f.*
E operación de búsqueda
  y destrucción *f.*
D Such- und
  Zerstöroperation *f.*
R действия обнаружения
  и уничтожения
  противника *ср. мн.*
  ч.
A عملية البحث والتدمير

3668 [L]
(single envelopment)
  operation
F opération de
  débordement *f.*
E operación de
  envolvimiento *f.*
D (einseitige) Umfassung
  f.
R обход по одному
  флангу *м.*
A عملية إلتفاف حول جناح
  واحد

3669 [E]
operational
F en état de fonctionner
E operacional
D einsatzfähig
R действующий;
  рабочий
A مُتَعَلِّق بِالعَمَلِيَّات

3670
operational airfield
F terrain (d'aviation)
  opérationnel *m.*
E campo (de aviación)
  operativo *m.*
D Einsatzflugplatz *m.*
R оперативный
  аэродром *м.*;
  активный аэродром
  м.
A مَطَار لِلعَمَلِيَّات *m*

231

3671

**operational
characteristic**

F caractéristique de
fonctionnement *f.*

E característica operativa
f.

D Leistungsmerkmal *neut.*

R эксплуатационная
характеристика *ж.*

A مُمَيِّزَة لِلعَمَلِيَّات

3672

operational command

F commandement
opérationnel *m.*

E mando operativo *m.*

D an bestimmten
Auftrag
gebundene Befehls-
gewalt (meist
eng.Begriff) *f.*

R оперативное
командование *ср.*

A القِيَادَة لِلعَمَلِيَّات

3673

operational control

F contrôle opérationnel
m.

E control operativo *m.*

D allgemeine
Befehlsgewalt (meist
eng. Begriff) *f.*

R оперативное
управление *ср.*

A المُرَاقَبَة لِلعَمَلِيَّات

3674

operational level

F niveau opérationnel *m.*

E nivel operativo *m.*

D operativer Bereich *m.*

R оперативный уровень
м.

A الإِرْتِفَاع الفِعْلِي

3675

operational mobility

F mobilité opérationnelle
f.

E movilidad operativa *f.*

D operative Beweglichkeit
f.

R боевая мобильность.
ж.; оперативная
мобильность *ж.*

A قَابِلِيَّة الحَرَكَة لِلعَمَلِيَّات

3676

**operational
requirement**

F besoin opérationnel *m.*

E requisito operativo *m.*

D Militärforderung *f.*

R боевое требование *ср.*

A مُقْتَضَى العَمَلِيَّات *m*

3677

operational reserve

F réserve opérationnelle *f.*

E reserva operativa *f.*

D örtliche Reserve *f.*

R боевой резерв *м.*;
оперативный резерв
м.

A إِحْتِيَاطِي العَمَلِيَّات *m*

3678

operational stocks *pl.*

F stocks opérationnels *m.*
pl.

E existencias operativas

D Kriegsvorrat *m.*

R боевые запасы
имуществ *м. мн. ч.*

A مَخْزُونَات العَمَلِيَّات *f*

3679 [A,L,M]

(combined) operations

F opérations interarmes *f.*
pl.

E operaciones conjuntas
f. pl.

D interalliierte Operation
f.

R совместные действия
ср. мн. ч.

A عَمَلِيات مُشْتَرِكة

3680

**(limited visibility)
operations**

F opérations de nuit *f.*
pl.; opérations
exécutées dans des
conditions de
mauvaise visibilité *f.*
pl.

E operaciones de noche *f.*
pl.; operaciones
realizadas en malas
condiciones de
visibilidad *f. pl.*

D Operationen bei
begrenzter
Sichtbarkeit

R ограниченная
видимость *ж.*

A العمليات في أحوال الرؤية
المحدودة

3681 [L]

(security) operations

F opérations de sécurité
(couverture et
jalonnement) *f. pl.*

E operaciones de
seguridad y cobertura
f. pl.

D Sicherungsoperationen
f. pl.

R действия сохранения и
безопасности *ср. мн.*
ч.

A عملية الحراسة والتغطية

3682

(special) operations

F opérations commando
f. pl.

E operaciones especiales

D Sonderoperation *f.*

R действия
специального
назначения *ср. мн. ч.*

A عمليات خاصة

3683

(survivability) operations

F opérations de capacité de survie *f.*

E operaciones de capacidad de supervivencia *f. pl.*

D Pioniersicherungs-operationen *f. pl.*

R инженерное защитное сооружение *ср.*

A عمليات البقاء

3684

(tactical) operations

F opérations tactiques *f. pl.*

E operaciones tácticas *f. pl.*

D taktische Führung *f.*

R тактические боевые действия *ср. мн. ч.*

A عمليات تعبوية

3685

(tactical air) operations

F opérations aériennes tactiques *f. pl.*

E operaciones aéreas tácticas *f. pl.*

D taktische Luftoperationen *f. pl.*

R тактические воздушные действия *ср. мн. ч.*

A عمليات جوية تعبوية

3686

operations map

F plan directeur *m.*

E mapa de operaciones *m.*

D Operationskarte *f.*

R рабочая карта *ж.*

A خريطة العمليات

3687 [R]

operator

F radio *m.*

E operador *m.;* radiotelefonista *m.*

D Funker *m.*

R связист *м.*; радист *м.*

A عَامِل *m*

3688

opponent *n.*

F adversaire *m.*

E adversario *m.*

D Gegner *m.*

R противник *м.*

A خصم *m*

3689

opportunity

F occasion *f.*

E oportunidad *f.*

D Gelegenheit *f.*

R удобный случай *м.*; возможность *ж.*

A فُرْصَة *f*

3690

opposite *adj.*

F contraire; opposé

E (de) enfrente; opuesto

D entgegengesetzt; gegenüberliegend

R противоположный

A مُقَابِل

3691

opposite number

F homologue *m.*

E correspondiente *m.;* equivalente (en grado)

D Gegenüber *m. neut.*

R занимающий соответствующий пост

A شَخْص يُشَغِّل وَظِيفَة مُمَاثِلة

3692

opposition

F opposition *f.*

E oposición *f.*

D Widerstand *m.*

R сопротивление *ср.*; оппозиция *ж*

A مُعَارَضَة *ر*

3693

optical axis

F axe optique *m.*

E eje óptico *m.*

D optische Achse *f.;* Sehachse *f.*

R оптическая ось *ж.*

A المِحْوَر البَصَرِي *m*

3694

optics

F optique *f.*

E óptica *f.*

D Optik *f.*

R оптика *ж.*

A بَصَرِيَّات *f*

3695

optimum *adj. n.*

F optimum *adj. n. m.*

E óptimo *adj. s. m.*

D optimal *adj.;* Optimum *s. neut.*

R оптимальный *прилаг.*; благоприятное условие *ср.*

A الأَمْثَل *m*

3696

option

F option *f.*

E opción *f.*

D Möglichkeit *f.*

R выбор *м.*; вариант *м.*

A خِيَار *m*

3697

optional

F facultatif

E facultativo

D beliebig

R необязательный

A خِيَاري

3698

optronics
F optronique *f.*
E optrónica *f.*
D Optronik *f.*
R оптроника *ж.*
A عِلْم البَصَرِيَّات *m*

3699

orbit *n.*
F orbite *f.*
E órbita *f.*
D (Umlauf-)Bahn *f.*
R орбита *ж.*
A مَدَار *m*

3700

orbit *v.*
F orbiter
E orbitar
D umkreisen (Satellit)
R совершать полёт по
   орбите
A دار

3701

orbital
F orbital
E orbital
D Bahn- *komp.*
R орбитальный
A مَدَارِي

3702

orbiting satellite
F satellite en orbite *m.*
E satélite en órbita *m.*
D Satellit in
   Erdumlaufbahn *m.*
R спутник на орбите *м.*
A قَمْر صِناعِي مَدَارِي *m*

3703

order *n.*
F ordre *m.*
E orden *m. f.*
D Befehl *m.*
R приказ *м.*
A أمْر *m*

234

3704 [A,M,L]

order *v.*
F ordonner (à, de)
E mandar (a)
D befehlen;
   kommandieren
R приказывать
A أمَر

3705

(administrative) order
F ordre logistique *m.*
E orden logística *f.*
D Verwaltungsbefehl *m.*
R указание *ср.*
A امر إداري

3706

**(complete equipment
   fighting) order
   (CEFO)**
F tenue de combat
   complète *f.*
E equipo completo de
   combate *m.*
D Einsatzbefehl (mit
   voller
   Kampfausrüstung) *m.*
R (полное боевое)
   снаряжение *ср.*
A الاجهزة الكاملة للقتال

3707

(executive) order
F ordre d'exécution *f.*
E orden de ejecución *m.*
D Ausführungsbefehl
   (meist eng. Begriff) *m.*
R приказ выполнения *м.*
A أمْر تَنْفِيذِي *m*

3708

(fire) order
F ordre de tir *m.*
E orden de fuego *m.*
D Feuerbefehl *m.*
R приказ об открытии
   огня *м.*
A أمْر الرَّمْي *m*

3709

(grounding) order
F ordre d'interdiction de
   vol *m.*
E orden que prohibe los
   vuelos *f.*
D Flugverbotsbefehl *m.*
R запрет (полёта) *м.*
A أمر منع الطيران

3710

(movement) order
F ordre de mouvement *m.*
E orden de movimiento
   m.
D Marschbefehl *m.*
R приказ на перевозку
   м.; движение *ср.*
A أمْر التَّنَقّل *m*

3711

(on) order
F (sur) ordre
E (al) recibir el aviso
D (auf) Befehl
R (по) приказу
A عند تلقي الأمر

3712

(operation) order
F ordre d'opération *m.;*
   emploi tactique des
   moyens *m.*
E orden de operaciones *f.*
D Operationsbefehl *m.*
R боевой приказ *м.*
A أمْر العَمَلِيَّات *m*

3713

(out of) order
F (en) dérangement; (en)
   panne
E estropeado
D nicht in Ordnung
R неисправный
A مختَلّ

3714

(standing) order
F ordre permanent *m.;*
consigne permanente
f.
E orden de régimen
interior *f.*
D ständiger Befehl *m.*
R (постоянный) приказ
м.
A تعليمات دائمة

3715

(warning) order
F ordre d'avertissement
m.
E orden de alerta *f.*
D Vorbefehl *m.*
R предварительное
указание *ср.*
A أمر إنذاري *m*

3716

"Order arms!"
F "Reposez armes!"
E "¡Descansen armas!"
D „Gewehr ab!"
R „К ноге!"
A جنباً سلاح

3717

order into action *v.*
F engager dans le combat
E enviar a las tropas a la
batalla
D Einsatzbefehl geben
R направлять в бой
A أمر الهجوم

3718

orderly *n.*
F infirmier *m.;* planton
m.
E enfermero *m.;*
ordenanza *m.*
D Krankenpfleger [H] *m.;*
Ordonnanz *m.*
R вестовой *м.;*
ординарец *м.*
A حَاجِب ـ مُنَسِّق *m*

3719

order of the day
F ordre du jour *m.*
E orden del día *f.*
D Tagesbefehl *m.*
R приказ по части *м.*
A أمر يومي

3720

orders group (O Gp)
F briefing *m.*
E (el) acto de recibir la
orden *m.*
D Lagebesprechung *f.*
R сбор *м.*
A جماعة القيادة

3721 [S]

ordnance
F artillerie *f.*
E artillería *f.*
D Geschütze *neut. pl.*
R артиллерия *ж.*
A مُعَدَّات مِدْفَعِيَّة *f*

3722 [A,M,L]

ordnance
F matériel *m.*
E material (de guerra) *m.*
D Material *neut.*
R вооружение *ср.;*
боевая техника *ж.*
A العتاد والذخيرة

3723

(explosive) ordnance
F explosifs et munitions
m. pl. f. pl.
E material explosivo *m.*
D Feldzeugmaterial *neut.*
R артиллерийско-
техническое
снабжение *ср.*
A مُعَدَّات مُتَفَجِّرَة *f*

3724

ordnance depot
F dépôt du matériel *m.*
E depósito de material
(de guerra) *m.*
D Gerätedepot *neut.*
R склад артиллерийско-
технического
имущества *м.*
A مُسْتَوْدَع العتاد والذخيرة *m*

3725

organic
F organique
E orgánico
D einheitseigen [AML];
organisch
R штатный
A عُضْوِي

3726 [A,M,L,U]

organization
F organisation *f.*
E organización *f.*
D Gliederung *f.*
R организация *ж.;*
(боевое)
распределение *ср.*
A تَنْظِيم *m*

3727

organization chart
F organigramme *m.*
E cuadro orgánico *m.*
D Gliederungsübersicht *f.*
R схема организации *ж.*
A خَرِيطَة تَنْظِيم *f*

3728 [L]

**organization of
defense (objective)**
F organisation de la
défense *f.*
E organización de la
defensa *f.*
D Organisation der
Verteidigung *f.*
R оборудование
обороны *ср.*
A تَنْظِيم الدِّفَاع *m*

235

3729 [L]
organization of the ground
F organisation du terrain *f.*
E organización del terreno *f.*
D Geländeausnutzung *f.*
R оборудование местности *ср.*
A تَنْظِيم الأرْض *m*

3730
organize
F organiser
E organizar
D aufstellen [AML] *tr.;* ausbauen[L] *tr.;* gliedern[AML] *;* organisieren[U] .
R организовывать; оборудовать
A نَظَّم

3731 [S]
orient *v.*
F orienter
E orientar
D einrichten *tr.*
R наводить
A وَجَّه

3732 [T,U]
orient *v.*
F orienter
E orientar
D orientieren *tr.*
R ориентировать(ся)
A وَجَّه

3733
orientation *f.*
F orientation *f.*
E orientación *f.*
D Orientierung *f.*
R ориентировка *ж.;* наводка *ж.*
A تَوْجِيه *m*

3734
origin
F origine *f.*
E origen *m.*
D Nullpunkt [E] *m.;* Ursprung [U] *m.*
R источник *м.;* начало *ср.*
A أصْل *m*

3735 [R]
originator
F autorité d'origine *f.;* expéditeur *m.*
E expedidor *m.*
D Aufgeber *m.*
R отправитель *м.*
A مُسَبِّب ـ مُبْتَكِر *m*

3736
oscillate
F (faire) osciller
E (hacer) oscilar
D schwingen (lassen)
R качать(ся); колебаться
A تَذَبْذَب

3737
oscillating
F oscillateur(-trice)
E oscilante
D schwingend
R качающийся
A مُتَذَبْذِب

3738
oscillation
F oscillation *f.*
E oscilación *f.*
D Schwingung *f.*
R качание *ср.;* колебание *ср.*
A ذَبْذَبَة *f*

3739
oscillator
F oscillateur *m.*
E oscilador *m.*
D Oszillator *m.*
R вибратор *м.;* осциллятор *м.*
A مُذَبْذِب *m*

3740
oscillograph
F oscillographe *m.*
E oscilógrafo *m.*
D Oszillograph *m.*
R осциллограф *м.*
A مُسَجِّل الذَّبْذَبَة *m*

3741
oscilloscope
F oscilloscope *m.*
E osciloscopio *m.*
D Oszilloskop *m.*
R осциллоскоп *м.*
A مِكْشَاف الذَّبْذَبَة *m*

3742 [R]
"Out!"
F "Terminé!"
E "¡Fin!"
D „Ende!"
R "конец связи"; "С К" (эсс-ках)
A إنْتَهِي

3743
outboard *adj.*
F extérieur; hors bord
E fuera-bordo
D aussenbord(s)
R (за) бортом; подвесной
A جَانِبِي

3744
outboard motor
F moteur hors-bord *m.*
E motor fuera-bordo *m.*
D Aussenbordmotor *m.*
R крыльевой двигатель *м.;* подвесной мотор *м.*
A مُحَرِّك زَوْرَق *m*

3745
outer
F extérieur
E exterior
D Aussen- *komp.;* äusserer *adj.*
R внешний; наружный
A خَارِجِي

236

3746

outflank
F déborder
E desbordar
D umfassen
R обходить фланг
A إِلْتَفَ حَوْل

3747

outgoing unit
F unité relevée
E unidad de salida *f.*
D (ausziehende) Einheit *f.*
R (отходящее)
   подразделение *ср.*
A وَحْدة مُغَادِرة

3748

outgun *v.*
F posséder une puissance
   de feu supérieure
E exceder al enemigo en
   artillería
D schneller schiessen an
   Schuss ; übertreffen
   untr.
R превосходить по
   огневой силе
A تفوق على غيره في نيران
   المدافع

3749

outline *v.*
F ébaucher; esquisser
E bosquejar; prefigurar
D umreissen
R наметить в общих
   чертах
A رَسَم

3750

outline(s) *n.(pl.)*
F profil *m.;* grandes
   lignes *f. pl.*
E contorno *m.;* idea
   general *f.;* trazado
   general *m.*
D (grobe) Umriss(e) *m.*
   (pl.)
R очертание *ср.;* черты
   ж. мн. ч.
A جَانِبِيَّة *f*

3751

outline of operation
F idée de manoeuvre *f.*
E trazado general de la
   operación *m.*
D Operationsumriss *m.*
R вводная *ж.*
A الخطة الاولية للعملية

3752

outpost
F avant-poste *m.*
E puesto avanzado *m.*
D (Gefechts-)Vorposten
   m.
R аванпост *м.;*
   сторожевая застава
   ж.
A نُقْطَة خَارِجِيَّة *f*

3753 [E]

output
F rendement *m.*
E rendimiento *m.*
D Leistung *f.*
R отдача *ж.*
A قُدْرَة *f*

3754 [U]

output
F production *f.*
E producción *f.*
D Ertrag *m.*
R производство *ср.;*
   выпуск *м.*
A نِتَاج *m*

3755 [D]

(computer) output
F sortie *f.*
E salida (del ordenador)
   f.
D (Arbeits-)Leistung *f.*
R вывод *м.*
A الخرج *m*

3756

outrank *v.*
F avoir un rang supérieur
E ser de grado superior
D (im) Rang höherstehen
   tr.
R быть старше
A كان أعلى رتبةً

3757 [R]

outstation
F station secondaire *f.*
E estación satélite *f.*
D Aussenstation *f.*
R абонент *м.*
A مركز مقابل

3758 [S]

over *n.*
F coup long *m.*
E tiro largo *m.*
D Weitschuss *m.*
R перелёт *м.*
A طَلْقَة مُجَاوِزة *f*

3759 [R]

"Over!"
F "Parlez!"
E "¡Cambio!"
D „Kommen!"
R "Приём!"
A حَوِّل

3760

overboard
F à la mer (homme); par-
   dessus le bord (jeter
   qch.)
E al agua (hombre); por
   la borda (echar algo)
D über Bord
R (за) бортом
A خَارِج السَّفِينَة

3761

overcast *adj.*
F bouché
E encapotado
D bewölkt
R пасмурный; хмурый
A غَائِم

237

3762

overflight
F survol *m.*
E vuelo de
reconocimiento *m.*
D Überflug *m.*
R перелёт *м.*
A طيران فوق منطقة معينة

3763 [A,M,V]

overhaul *v.*
F rattraper; dépasser
E alcanzar
D überholen *untr.;*
einholen *tr.*
R обгонять
A تجاوز

3764 [E]

overhaul *v.*
F réviser
E revisar
D überholen (ein Gerät)
untr.
R ремонтировать
A أصْلَح

3765 [A,M,V]

overhaul *v.*
F rattraper
E alcanzar
D einholen *tr.*
R догонять
A أصْلَح

3766

overhaul *n.*
F révision *f.*
E revisión general *f.*
D Überholung *f.*
R подробный осмотр *м.;*
текущий ремонт *м.*
A إصْلاَح *m*

3767

overhead *adj.*
F aérien; au dessus
E aéreo; (en) cabeza
D Luft- *komp.;*
überirdisch *adj.*
R верхний; воздушный
A فَوْق

238

3768

overhead clearance
F hauteur libre *f.;* tirant
d'air *m.*
E luz libre *f.*
D lichte Höhe *f.*
R верхний клиренс *м.*
A الخُلُوص فَوْقه

3769

overheat
F échauffer
E recalentarse
D heisslaufen *tr.*
R перегревать(ся)
A أفْرَط الإحْمَاء

3770

overkill
F surabondance *f.*
E exceso de poder
destructor *m.*
D (mehr) Waffenwirkung
als nötig
R многократное
уничтожение *ср.;*
перегиб *м.*
A قَتْل بَالِغ *m*

3771

overland *adv.*
F (par voie de) terre
E por tierra
D über Land
R сухопутный
A فوق اليابسة

3772

overlap *n.*
F recouvrement *m.*
E traslapo *m.*
D Überschneidung *f.*
R перекрытие *ср.;*
затягивание *ср.*
A تَرَاكُب *m*

3773

overlay
F calque *m.*
E superpuesto *m.*
D Planpause *f.*
R калька *ж.*
A غِطَاء

3774

(operations) overlay
F calque renseigné *m.*
E superponible *m.*
D Auflegekarte *f.*
R шаблон *м.*
A شفاف

3775

(target) overlay
F calque d'objectifs *m.*
E superponible de
objetivos *m.*
D Planpause *f.*
R схема целей *ж.*
A شَفَاف الأهْدَاف

3776

overlook *v.*
F donner sur; dominer
E dominar
D Aussicht gewähren auf
R господствовать
A أشْرَف ـ أطَلّ

3777

overnight pass
F titre de permission
d'une nuit *m.*
E licencia de una noche *f.*
D Kurzurlaub (für die
Nacht)
R отпускной билет (на
(...) часов).
A اجازة ليلية

3778 [A]

overpass *n.*
F survol *m.*
E sobrevuelo *m.*
D Überflug *m.*
R перелёт *м.*
A طيران فوق الهدف

3779
overpower
F vaincre; subjuguer
E dominar
D überwältigen
R перегрузка ж.
A غَلَب

3780 [N]
overpressure
F surpression f.
E sobrepresión f.
D Überdruck m.
R повышение давления
    ср.; сверхдавление
    ср.
A ضَغْط زَائِد m

3781
overprint v.
F surimprimer
E sobrecargar
D überdrucken tr.
R перепечатывать на
    карте
A تَرَاكُب الصُّوَر

3782
overprint n.
F surimpression f.
E sobrecarga f.
D Überdruck m.
R перепечатка на карте
    ж.
A تَرَاكُب الصُّوَر m

3783
overrule
F annuler une décision
E anular; rechazar
D überstimmen
R отвергать
A ساد

3784 [L]
overrun
F envahir; submerger
E invadir
D durchstossen;
    herfallen über tr.
R опустошать
A إكْتَسَح

3785
overseas adj.
F (d')outre-mer
E (de) ultramar
D Übersee- komp.
R заморский
A وَرَاء البِحَار

3786 [A]
overshoot n.
F remise des gaz f.
E planeo largo m.
D Hinausschiessen neut.
R перелёт м.
A مُجَاوَزَة المَدْرَج

3787 [S]
overshoot n.
F tir long m.
E tiro largo m.
D über das Ziel
R промах м.
A مُجَاوَزَة الهَدَف f

3788 [A]
overshoot v.
F remettre les gaz
E pasar por encima
D durchstarten
R перелетать
A جاوز المَدْرَج

3789 [S]
overshoot v.
F tirer long
E tirar largo
D über das Ziel
    hinausschiessen
R промахиваться
A جاوز الهَدَف

3790 [A,H,M]
overtake
F rattraper
E alcanzar
D einholen tr.
R догонять
A تَجَاوَز

3791 [V]
overtake
F doubler
E adelantar
D überholen
R обгонять
A تَجَاوَز

3792 [M]
overturn
F (faire) chavirer
E (hacer) zozobrar
D kentern (lassen)
R опрокидываться;
    перевёртываться
A إنْقَلَب

3793 [P]
overturn
F renverser
E derrumbar
D niederwerfen tr.
R совершить переворот
A قَلَب

3794 [A,V,U]
overturn
F capoter [AV]; (se)
    renverser [U]
E volcar
D (sich) überschlagen
    [A]; umkippen
    (lassen) [VU]
R опрокидывать(ся)
A إنْقَلَب

3795 [L]
overwatch
F mission de couverture
    éloignée f.
E vigilancia f.
D Überwachung f.;
    Unterstützungs-
    bereitschaft f.
R тактическое защитное
    наблюдение ср.;
    готовность к
    тактическому
    защитному
    обеспечению ж.
A في حالة الانتباه لسند وحدة
    أخرى بالنيران

3796

(bounding) overwatch

F manoeuvre en
  perroquet *f.*
E vigilancia por saltos *f.*
D vorstossende
  Überwachung *f.*
R тактическое
  обеспечение *ср.*
A الوحدة المساندة تتقدم
       بقفزات

3797

(travelling) overwatch

F surveillance eloignée
  par les éléments
  arrière *f.;*
  déplacement en
  sécurité *m.*
E vigilancia móvil *f.*
D bewegliche
  Überwachung *f.*
R готовность к
  оказанию
  тактического
  защитного
  обеспечения с ходу
A سير الوحدة والاشتباك مع
       العدو المتوقع

3798 [E]

oxy-acetylene cutter

F chalumeau
  oxyacétylénique *m.*
E soplete oxiacetilénico
  m.
D Schneidbrenner *m.*
R газовый резчик *м.*
A مشعل قطع بالاكسجين
       والسيتيلين

3799

oxygen

F oxygène *m.*
E oxígeno *m.*
D Sauerstoff *m.*
R кислород *м.*
A أُكْسِجِين *m*

240

3800

oxygen cylinder

F bouteille d'oxygène *f.*
E cilindro de oxígeno *m.*
D Sauerstofflasche *f.*
R баллон с кислородом
  м.
A أُسْطُوَانَة أُكْسِيجِين *f*

3801

oxygen mask

F masque à oxygène *m.*
E máscara de oxígeno *f.*
D Sauerstoffmaske *f.*
R кислородная маска *ж.*
A قناع اكسيجين

P

3802 [H]

pace

F pas *m.*
E paso *m.*
D Schritt *m.*
R шаг *м.*
A خُطْوَة *f*

3803 [L]

pace

F vitesse de marche *f.*
E cadencia del paso *f.*
D Marschtempo *neut.*
R темп марша *м.*
A سُرْعَة السَّيْر *f*

3804 [A,M,L,V]

pace *n.*

F allure *f.*
E velocidad *f.*
D Geschwindigkeit *f.*
R скорость *ж.*
A سُرْعَة *f*

3805 [A,L,M]

pace *n.*

F rythme *m.*
E ritmo *m.*
D Tempo *neut.*
R темп *м.*
A خُطْوَة *f*

3806

pace setter

F guide *m.*
E marcador de pasos *m.*
D Schrittmacher *m.*
R задающий темп *м.*
A مُنَظِّم السُّرْعَة *m*

3807

pack *n.*

F paquetage *m.*
E carga *f.;* mochila *f.*
D Gepäck *neut.;* (Trag-)
  Last *f.*
R сумка *ж.;* ранец *м.*
A حَقِيبَة *f*

3808

pack *v.*

F emballer
E embalar
D verpacken
R упаковывать;
  тюковать
A حَمَل

3809

(back) pack

F sac à dos *m.*
E mochila *f.*
D Tornister *m.*
R вещевой мешок *м.*
A حقيبة

3810

(power) pack

F groupe motopropulseur
  m.
E motor *m.*
D (Triebwerk-)Anlage
  (*neut.*) *f.*
R двигатель (танка) *м.*
A مجموعة توليد الطاقة

3811

pack animal

F bête de somme *f.*
E animal de carga *m.*
D Lasttier *neut.*
R вьючное животное *ср.*
A حَيَوَان تَحْمِيل *m*

3812

pack a parachute *v.*

F plier un parachute
E preparar una paracaída
D packen falten (ein Fallshirm)
R запаковать парашют
A عبأ المظلة

3813

(landing) pad

F aire d'atterrissage verticale *f.*
E plataforma de aterrizaje *f.*
D Landerampe *f.*
R (посадочная) площадка *ж.*
A سُطيح الهبوط

3814

(launching) pad

F plate-forme de lancement *f.*
E plataforma de lanzamiento *f.*
D Abschussrampe *f.*
R (стартовая) площадка *ж.*
A سُطيح القَذْف

3815 [R]

(one-time) pad

F bloc à clé nonreutilisable *m.*
E clave no repetible *f.*
D (Einmal-)Schlüsselblock *m.*
R одноразовый шифровальный блокнот (шифроблокнот) *м.*
A دَفْتَر رُمُوز يُسْتَعْمَل مَرَّة وَاحِدة فَقَط

3816

padding

F bourre *f.*
E relleno *m.*
D Polsterung *f.*
R набивка *ж.*
A حشْو

3817

paddle *v.*

F pagayer
E remar con canalete
D paddeln
R грести гребками
A جَذَف ـ جَدَّف

3818

paddle *n.*

F pagaie *f.*
E canalete *m.;* remo *m.*
D Paddel *neut.*
R гребок *м.*
A مِجْذاف *m*

3819

padre

F aumônier *m.*
E capellán militar *m.;* Padre
D Pfarrer *m.*
R полковой священник *м.*
A كاهن

3820

pain

F douleur *f.*
E dolor *m.*
D Schmerz *m.*
R боль *ж.*
A أَلَم *m*

3821

painkiller

F analgésique *m.*
E calmante *m.*
D schmerzstillendes Mittel *neut.*
R болеутоляющее средство *ср.*
A دواء ضد الالم

3822

paint *n.*

F peinture *f.*
E pintura *f.*
D Farbe *f.*
R краска *ж.*
A دِهان *m*

3823

paint *v.*

F peindre
E pintar
D anstreichen *tr.*
R расписывать
A دَهَن

3824 [M]

painter

F amarre *f.;* haussière *f.*
E amarra *f.*
D Fangleine *f.*
R (носовой) фалинь *м.*
A حَبْل صَغِير *m*

3825

pallet

F palette *f.*
E plataforma *f.*
D Palette *f.*
R поддон *м.*
A مِنَصّة *f*

3826

palletize

F palettiser
E cargar sobre plataformas
D palettieren
R ставить (груз) на поддоны
A وَضَع على مِنَصّة

3827

(instruction) pamphlet

F document d'instruction *m.*
E folleto de instrucción *m.*
D Merkblatt *neut.*
R проспект *м.;* брошюра *ж.*
A كُرّاسة

241

3828 [E]
panel
F panneau *m.*
E tablero *m.*
D Schalttafel *f.*
R пульт *м.*; консоль *м.*
A لَوْحَة *f*

3829 [A,L]
panel
F panneau *m.*
E panel *m.*
D Schalttafel *f.*
R панель *м.*; щиток *м.*
A لَوْحَة *f*

3830
(control) panel
F tableau de commande *m.*
E tablero de instrumentos *m.*
D Bedienungspult *neut.;* Schaltbrett *neut.*
R пульт управления *м.*
A لَوْح الإِدَارة *m*

3831 [A,E,V]
(instrument) panel
F tableau de bord *m.*
E tablero *m.*
D Armaturenbrett *neut.*
R пульт *м.*
A لَوْحَة القِيَادة *f*

3832 [A,L]
(marking) panel
F panneau de signalisation *m.*
E balizamiento *m.*
D Signaltafel *f.*
R опозновательное полотнище *ср.*
A لَوْحَة تَشْخِيصْ *f*

242

3833 [L]
(splash) panel
F bouclier de protection contre les vagues (véhicule amphibie) *m.*
E panel de navegación *m.*
D Wellen-Schutzblech *neut.*
R щит *м.*; щитка *ж.*
A واقية الامواج

3834
panel code
F code de panneaux *m.*
E código de paneles *m.*
D Signaltuchkode *m.*
R код-полотнище *м.*
A لَوْحة الأَسْمَاء الرَّمْزِيَّة *f*

3835
pannier
F pannier de bât *m.*
E alforja *f.*
D Tragkorb *m.*
R корзина *ж.*
A سَفَط

3836
panoramic
F panoramique
E panorámico
D Rundsicht *komp.*
R панорамный
A بَانُورَامِي

3837
parachute *n.*
F parachute *m.*
E paracaídas *m. inv.*
D Fallschirm *m.*
R парашют *м.*
A مِظَلَّة *f*

3838
parachute *v.*
F descendre en parachute; parachuter
E lanzar(se) en paracaídas
D abspringen; abwerfen
R парашютировать; сбрасывать(ся) на парашюте
A هَبَط بِمَظَلَّة

3839
(brake) parachute
F parachute de freinage *m.*
E paracaídas de freno *m.*
D Bremsschirm *m.*
R (парашютный) тормоз *м.*
A مظلة كابحة

3840
(cargo) parachute
F parachute à matériel *m.*
E paracaídas de carga *m.*
D Frachtgutfallschirm *m.*
R (грузовой) парашют *м.*
A مظلة شحن

3841
(chest pack) parachute
F parachute ventral *m.*
E paracaídas de pecho *m.;* paracaídas ventral *m.*
D Brustfallschirm *m.*
R (дополнительный) парашют *м.*
A مظلة صدرية

3842

(extraction) parachute

F parachute extracteur *m.*

E paracaídas extractor;
paracaídas pilotillo

D Ausziehschirm; (ein
kleiner) Schirm der
den Hauptschirm
herauszieht

R (вытяжной) парашют
м.

A مِظَلَّة إسْتِخْرَاجِية *f*

3843

(free fall) parachute

F parachute à ouverture
commandée *m.*

E paracaídas de caída
libre *m.*

D Sportfallschirm *m.*

R (управляемый)
парашют *м.*

A مظلة للاسقاط الحّر

3844

**(self-opening)
parachute**

F parachute à ouverture
automatique *m.*

E paracaídas de apertura
automática *m.*

D selbstöffnender
Fallschirm *m.*

R (саморазвёртывающий)
парашют *м.*

A مظلة ذاتية الانفتاح

3845

parachuting

F parachutage *m.*

E saltar en paracaídas

D (mit dem) Fallschirm
abspringen

R прыгание с
парашютом *ср.*

A قفز بمظلات

3846

**(delayed drop)
parachuting**

F parachutage à
ouverture retardée *m.*

E salto de apertura
retardada *m.*

D Fallschirmspringen
(verzögerter
Absprung) *neut.*

R прыжок с
замедленным
развёртыванием
парашюта *м.*

A هبوط متأخر بمظلات

3847

**(high altitude low
opening)
parachuting
(HALO)**

F largage à haute altitude
ouverture à basse
altitude *m.*

E salto a gran altura y
apertura baja *m.*

D fallschirmspringen (aus
hoher Flughöhe
mitFallschirmeröffnung
aus neidriger Höhe)
neut.

R прыжок с больших
высот с
замедленным
развёртыванием
парашюта *м.*

A القفز من ارتفاع عال وانفتاح
المظلة على ارتفاع
منخفض

3848

parachutist

F parachutiste *m.*

E paracaidista *m.*

D Fallschirmspringer *m.*

R парашютист *м.*;
десантник *м.*

A مظلي

3849

parade *v.*

F défiler; parader

E formar en parada (para
revista)

D vorbei marschieren

R строить(ся)

A عَرَض

3850

parade *n.*

F défilé *m.;* prise d'armes
f.

E defile *m.;* parada *f.*

D Parade *f.*

R парад *м.*

A عَرْض *m*

3851

(muster) parade

F rassemblement *m.;*
appel *m.*

E revista *f.*

D Appell *m.*

R построение *ср.*

A عرض الجمع

3852

(sick) parade

F visite médicale *m.*

E reconocimiento médico
m.

D Krankenappell *m.*

R амбулаторный приём
м.

A تفتيش المرضى

3853

parados

F parados *m.*

E parados *m.*

D Parados *m.*

R парадос *м.*

A سُترة خلفية في التحصينات

243

3854

paradrop

F parachutage *m.*

E lanzamiento con paracaídas *m.*

D Absprung *m.;* Abwurf *m.*

R выброска *ж.;* сбрасывание *ср.*

A هُبُوط بمظلّات *m*

3855

paraffin

F paraffine *f.*

E petróleo *m.*

D Paraffinöl *neut.*

R парафин *м.*

A بارافين

3856

parallax

F parallaxe *f.*

E paralaje *f.*

D Parallaxe *f.*

R параллакс *м.*

A إخْتِلاَف المَنْظَر *m*

3857

parallel *adj.*

F parallèle

E paralelo

D parallel

R параллельный

A مُتَوازٍ

3858

parallel rulers

F règle cras *f.*

E transportador *m.*

D Parallellineal *neut.*

R параллельные линейки *ж. мн. ч.*

A مسطرة رسم خطوط متوازية

244

3859

parallel sheaf

F faisceau parallèle *m.*

E haz paralelo *m.*

D Feuer mit parallelen Rohren *neut.*

R параллельный веер *м.*

A حزمة متوازية

3860

paramedic

F médecin parachutiste *m.*

E médico *m.*

D Arzt bei der Fallschirmtruppe *m.*

R врач, приданный воздушнодесантным войскам *м.*

A طبيب مظلي

3861

paramilitary *adj.*

F paramilitaire *adj.*

E paramilitar

D halbmilitärisch

R спасательные команды *ж. мн. ч.;* милиционерный *прилаг.*

A شبه عسكري

3862

parapet

F parapet *m.;* épaulement *m.*

E parapeto *m.*

D Wall *m.*

R бруствер *м.*

A سُترة خندق

3863

pararaft

F canot pneumatique parachutable *m.*

E paracaídas-bolsa neumática *m.*

D (ein mit dem) Fallschirm abwerfbares Schlauchboot *neut.*

R (парашютный) надувной плотик *м.*

A طوف قابل للانزال بمظلة

3864

paratrooper

F soldat parachutiste *m.*

E paracaidista *m.*

D Fallschirmjäger *m.*

R десантник *м.*

A جندي مظلي

3865

paratroops *pl.*

F troupes parachutistes *f. pl.*

E tropas paracaidistas *f. pl.*

D Fallschirmjäger *m. pl.*

R парашютнодесантные войска *ср. мн. ч.*

A مِظَلِّيُون *m*

3866

paravane

F paravane *f.*

E paraván *m.*

D Minenabweiser *m.*

R параван *м.*

A جرَّافه

3867

parent unit

F unité d'appartenance *f.*

E unidad de pertenencia *f.*

D Stammtruppenteil *m.*

R основная часть *ж.*

A وَحْدة أَساسِيَّة *f*

3868

parka
F parka *f.*
E parka *m.*
D Parka(s) *m.*
R парка *ж.*
A مِعْطَف ثَقِيل (باركا) *m*

3869

parking brake
F frein de parking *m.*
E freno de mano *m.*
D Parkbremse *f.*
R ручной тормоз *м.*
A كَابِحَة الوُقُوف *f*

3870

parley *v.*
F (être en) pourparlers avec
E parlamentar
D verhandeln
R вести переговоры
A فاوض

3871

parole
F parole *f.*
E palabra de honor *f.*
D Parole *f.*
R честное слово *ср.*
A وعـد *m*

3872 [E]

part
F pièce *f.*
E pieza *f.*
D Teil *m.*
R часть *ж.*; деталь *м.*
A قِطْعَة *f*

3873 [U]

part
F part *f.*
E parte *f.*
D Teil *m.*
R часть *ж.*
A جُزْء *m*

3874

(spare) part(s)
F pièce(s) de rechange *f.*
E recambio *m.*
D Ersatzteil(e) *m. (pl.)*
R запчасти *ж. мн. ч.*
A قِطَع الغِيار

3875

partial
F partiel
E parcial
D Teil- *komp.*; teilweise *adj.*
R частичный
A جُزْئِي

3876

particle
F particule *f.*
E partícula *f.*
D Teilchen *neut.*
R частица *ж.*
A دَقِيقَة ـ ذَرَّة *f*

3877

partisan *n.*
F partisan *m.*
E partisano *m.*
D Partisan *m.*
R партизан *м.*
A نصير

3878 [A,M,L]

party
F détachement *m.*
E destacamento *m.*
D Kommando *neut.*
R команда *ж.*; группа *ж.*
A جَمَاعَة *f*

3879 [P]

party
F parti *m.*
E partido *m.*
D Partei *f.*
R партия *ж.*
A حِزْب *m*

3880

(advance) party
F détachement précurseur *m.*
E grupo de vanguardia *m.*
D Vortrupp *m.*
R передовая группа *ж.*
A جَمَاعَة مُتَقَدِّمَة *f*

3881

(boarding) party
F groupe d'abordage *m.*
E destacamento de abordaje *m.*
D Enterkommando *neut.*
R призовая команда *ж.*
A مَفْرَزَة تَفْتِيش سَفِينة أو إقْتِحَامها

3882

(burial) party
F piquet d'honneurs funèbres *m.*
E destacamento de entierro *m.*
D Beerdigungstrupp *m.*
R похоронная команда *ж.*
A جماعة الدفن

3883

(demolition) party
F groupe de destruction *m.*
E destacamento de demolición *m.*
D Sprengtrupp *m.*
R команда разрушения *ж.*
A جماعة التدمير

3884

(demolition firing) party
F équipe chargée des destructions *f.*
E destacamento de demolición *m.*
D Sprengkommando *neut.*
R команда подрывников *ж.*
A جَمَاعَة تَدْمِير *f*

3885

(fatigue) party
F détachement de corvée *m.*
E destacamento de trabajo *m.*
D Arbeitskommando *neut.*
R рабочая команда *ж.*
A جَمَاعَة شُغْل *f*

3886

(firing) party
F peloton d'exécution *m.*
E pelotón de ejecución *m.*
D Ehrensalutkommando *neut.;* Exekutionskommando *neut.*
R салютная команда *ж.;* расстрельная команда *ж.*
A جماعة تنفيذ الاعدام

3887

(landing) party
F détachement de débarquement *m.*
E destacamento de desembarco *m.*
D Landungskommando *neut.;* Strandkommando *neut.*
R десантная партия *ж.;* команда *ж.*
A مُفْرَزَة إنْزَال *f*

3888

(left-behind) party
F élément dépassé *m.;* élément laissé sur les arrières *m.*
E destacamento retrasado de información *m.;* de vigilancia
D zurückgelassener Spähtrupp *m.*
R группа, действующая в тылу противника *ж.*
A جماعة باقِيَة وَرَاء خُطُوط العَدو

3889

(quartering) party
F détachement précurseur *m.*
E partida de alojamiento *f.*
D Unterbringungstrupp *m.*
R передовая группа расквартирования штаба *ж.*
A جماعة الايواء *f*

3890

(raiding) party
F commando *m.*
E partida de incursión *f.*
D Überfallkommando *neut.*
R диверсионная группа *ж.*
A جَمَاعَة مُغِيرة *f*

3891

(rear) party
F détachement postcurseur *m.*
E destacamento de retaguardia *m.*
D Nachkommando *neut.*
R тыльная походная застава (тпз) *ж.*
A جَمَاعَة المُؤَخَّرَة *f*

3892

(search) party
F expédition de secours *f.*
E destacamento de busquada (de exploración) *m.*
D Rettungsmannschaft *f.*
R поисковая группа *ж.*
A جَمَاعَة تَفْتِيش

3893

(shore) party
F élément de plage *m.*
E destacamento de playa *m.*
D Strandkommando *neut.;* Uferkommando *neut.*
R береговая команда *ж.;* береговой отряд *м.*
A جَمَاعَة تَنْظِيم الشَّاطِىء

3894 [T]

pass *n.*
F col *m.*
E puerto *m.*
D Pass *m.*
R перевал *м.*
A مَمَرّ *m*

3895 [A]

pass *n.*
F passage *m.*
E pasaje *m.*
D Vorbeiflug *m.*
R заход на цель *м.*
A مَمَرّ *m*

3896 [A,M,L]

pass *n.*
F permission *f.*
E permiso *m.*
D Urlaubschein *m.*
R путёвка *ж.*
A جِهَاز *m*

3897 [P]

pass *n.*
F laisser-passer *m. inv.*
E pase *m.*
D Ausweis *m.*
R пропуск *м.*
A بِطَاقَة مُرُور *f*

3898

passable
F franchissable
E franqueable; vadeable
D passierbar
R проходимый
A صالح للمرور

246

3899 [A,M]

passage
F passage *m.*
E pasaje *m.*
D Passage *f.*
R проход *м.*; переход *м.*
A عُبُور-مُرُور *m*

3900

passage lanes
F couloirs de dépassement *m. pl.*
E vías de paso *f. pl.*
D Passierschneise *f.*
R пути прохождения *м. мн. ч.*; пути продвижения *м. мн. ч.*
A طرق المرور

3901

passage of command
F transmission de responsabilité *f.*
E transferencia de mando *f.*
D Verantwortungswechsel *m.*
R передача командования *ж.*
A تبادل القيادة

3902

passage of lines (forward and rear)
F manoeuvre de dépassement *f.;* manoeuvre de recueil *f.*
E paso de línea *m.;* paso de escalón *m.*
D Liniendurchlass (durch die vorgeschobenen und rückwärtigen Linien) *m.*
R (передовое) прохождение *ср.*; (тыловое) прохождение *ср.*
A اجتياز الخطوط

3903 [A,M,V]

passenger
F passager [AM] *m.;* voyageur [V] *m.*
E pasajero [AM] *m.;* viajero [V] *m.*
D Fahrgast [MV] *m.;* Fluggast [A] *m.*
R пассажир *м.*
A رَاكِب *m*

3904

"Pass friend!"
F "Passez!"
E "¡Adelante!"; "¡Pase!"
D „Sie dürfen passieren!"
R "Проходите!"
A مُرَّ يا صَدِيق

3905

passive air defense
F défense passive aérienne *f.*
E defensa antiaérea pasiva *f.*
D passive Luftverteidigung *f.*
R пассивная противовоздушная оборона (ПВО) *ж.*
A دِفَاع جَوِّي سَلْبِي *m*

3906

passive homing guidance
F guidage passif *m.*
E dirección buscadora pasiva *f.*
D passive Zielsuchlenkung *f.*
R пассивное самонаведение *ср.*
A جِهَاز تَوْجِيه سَلْبِي *m*

3907 [V]

pass time
F durée d'écoulement *f.*
E duración de pasaje *f.*
D Durchlaufzeit *f.*
R время прохождения *ср.*
A زَمَن المُرُور *m*

3908

password
F mot de passe *m.*
E consigna *f.*
D Kennwort *neut.*
R пароль *м.*
A كَلِمَة المُرُور *f*

3909 [T]

path
F sentier *m.*
E sendero *m.*
D Pfad *m.*
R тропа *ж.*
A درب *m*

3910 [H]

patient *n.*
F malade *m. f.*; patient *m.*
E paciente *m. f.*
D Patient *m.*
R пациент *м.*; больной *м.*
A مَرِيض *m*

3911

patrol *n.*
F patrouille *f.*
E patrulla *f.*
D Streife *f.*
R дозор *м.*
A دَوْرِيَّة *f*

3912

patrol *v.*
F patrouiller
E patrullar
D patrouillieren
R охранять дозорами; патрулировать
A قَام بِدَوْرِيَّة

3913

(border) patrol
F patrouille de frontière *f.*
E patrulla fronteriza *f.*
D Grenzstreife *f.*
R пограничный дозор *м.*
A دَوْرِية الحُدود *f*

3914

(combat) patrol
F patrouille de combat *f.*
E patrulla de combate *f.*
D Kampfpatrouille *f.*
R (боевой) дозор *м.*
A دَوْرِية قِتال *f*

3915 [M]

(inshore) patrol
F défense littorale *f.*
E defensa litoral *f.*;
   patrulla de aguas
   limítrofes
D Küstenschutz *m.*
R прибрежный дозор *м.*
A دَوْرِية سَاحِليَّة

3916

(reconnaissance)
   patrol
F patrouille de
   reconnaissance *f.*
E patrulla de
   reconocimiento *f.*
D Spähtrupp *m.*
R (разведывательный)
   дозор *м.*
A دورية استطلاع

3917 [M]

(shore) patrol
F patrouille de police
   militaire (navale) *f.*
E patrulla en tierra *f.*
D Hafenpolizei *f.*
R береговой дозор *м.*
A دورية الشرطة البحرية

3918

(snatch) patrol
F patrouille d'enlèvement
   de prisonnier *f.*
E patrulla de captura *f.*
D Schnappatrouille *f.*
R поисковая группа *ж.*
A دورية خطف الاسرى

3919

(standing) patrol
F sonnette *f.*
E patrulla de vigilancia *f.*
D stehender Spähtrupp *m.*
R неподвижный дозор
   м.; патрулирующий
   полёт *м.*
A دَوْرِيَّة ثَابِتَة

3920

(tank hunting) patrol
F combat rapproché anti-
   chars (CRAC) *m.*
E patrulla de cazacarros
   f.
D Panzersuchpatrouille *f.*
R противотанковый
   разведдозор *м.*
A دورية صيد الدبابات

3921

patrol boat
F patrouilleur *m.*
E barco patrullero *m.*
D Wachboot *neut.*
R дозорный катер *м.*;
   сторожевой катер *м.*
A سَفِينَة دَوْرِيَّة *f*

3922

pattern bombing
F bombardement par
   vagues *m.*
E bombardeo sistemático
   m.
D Flächenbombardierung
   f.
R групповое
   бомбометание *ср.*
A قَصْف مُنَظَّم *m*

3923

pay *v.*
F solder
E pagar
D besolden
R платить
A دَفَع

3924

pay *n.*
F solde *f.*
E sueldo *m.*
D Sold *m.*
R зарплата *ж.*
A رَاتِب *m*

3925

payload
F charge utile *f.*
E carga útil *f.*
D Nutzlast *f.*
R бомбовый груз *м.*;
   боевой заряд *м.*
A وَزْن المَلَّاحِين والآلات

3926

paymaster
F trésorier *m.*;
   commissaire *m.*
E habilitado *m.*; oficial
   pagador *m.*
D Zahlmeister *m.*
R кассир *м.*; казначей *м.*
A أمِين صُنْدُوق *m*

3927 [M]

pay out
F (laisser) filer
E ir dando
D fieren
R травить (трос)
A أرْخَى

3928 [A,L]

P day
F jour P *m.*
E día P *m.*
D Tag P *m.*
R день достижения
   полной мощности *м.*
A يَوْم دَفْع الرَّاتِب *m*

248

3929

peace
F paix *f.*
E paz *f.*
D Friede *m.*
R мир *м.*
A سِلم *m*

3930

peacekeeping force
F force de maintien de la paix *f.*
E fuerza(s) de pacificación *f.* (*pl.*)
D Truppen zur Aufrechterhaltung des Friedens
R войска по поддержанию мира *ср. мн. ч.;* силы по поддержанию мира *ж. мн. ч.*
A قُوَّة حَافِظَة عَلى الأَمْن

3931

peacetime *n.*
F temps de paix *m.*
E tiempo de paz *m.*
D Frieden *m.*
R мирное время *ср.*
A زَمَن السِّلْم *m*

3932

peacetime establishment
F tableau d'effectifs du temps de paix *m.*
E estadillo del tiempo de paz *m.*
D Friedensstärke *f.*
R численность сил мирного времени *ж.*
A إقرار السِّلْم *m* ـ ملاك السلم

3933

peak *n.*
F cime *f.;* pic *m.*
E cumbre *m.;* pico *m.*
D Gipfel *m.;* Spitze *f.*
R пик *м.;* максимум *м.;* разгар *м.*
A ذُرْوَة *f*

3934

peak *adj.*
F (de) crête; (de) pointe
E (em) cresta; (en) punta
D Höchst- *komp.*; Spitzen- *komp.*
R максимальный; высший
A ذُرْوِي

3935

peak overpressure
F surpression de crête *f.*
E sobrepresión máxima *f.*
D Spitzenüberdruck *m.*
R пиковое давление *ср.*
A الضَّغْط الذُّرْوِي

3936 [E,V]

pedal *n.*
F pédale *f.*
E pedal *m.*
D Pedal *neut.*
R педаль *ж.*
A دوّاسة *f*

3937

pellet
F grain de plomb *m.*
E perdigón *m.*
D Kügelchen *neut.*
R дробинка *ж.*
A بُنْدَقة *f*

3938 [M]

(submarine) pen
F abri *m.*
E abrigo *m.*
D (U-Boot-)Bunker *m.*
R укрытие для подлодок *ср.*
A مَلْجَأ *m* (غَوَّاصَة)

3939

pencil beam
F pinceau lumineux *m.*
E rayo delgado *m.*
D Strahlenbündel *neut.*
R узкий луч *м.;* луч точного наведения *м.*
A حُزْمَة نُور ضَيِّقَة *f*

3940

penetrate [A,L,S]
F pénétrer
E penetrar
D einbrechen [L] *tr.;* einfliegen [A] *tr.;* durchschlagen [S] *untr.*
R проникать; прорывать
A إخْتَرَق

3941

penetration [A,L,S]
F pénétration *f.;* rupture [L] *f.*
E penetración *f.*
D Einbruch [L] *m.;* Einflug [A] *m.;* Durchschlag [S] *m.*
R проникание *ср.;* прорыв *м.*
A إخْتِرَاق *m*

3942

peninsula
F péninsule *f.*
E península *f.*
D Halbinsel *f.*
R полуостров *м.*
A شِبْه جَزيرَة *m*

3943 [M]

pennant
F flamme *f.*
E bandería *f.;* gallardete *m.*
D Wimpel *m.*
R вымпел *м.*
A رَاية *f*

3944

(on) pension
F pensionné
E jubilado
D (im) Ruhestand
R (быть на) пенсии;
пенсионный
A متقاعد

3945

peppill
F stimulant m.
E píldora anti-fatiga
D Aufputschungsmittel
neut.
R возбуждающее
средство cp.
A حبة نشاط

3946

percussion cap
F amorce f.
E cápsula fulminante f.
D Zündkapsel f.
R ударный капсюль м.
A كَبْسُولة f

3947

perforate
F perforer
E perforar
D durchlöchern untr.
R пробивать;
просверливать
A ثَقَب

3948

perforation
F perforation f.
E perforación f.
D Perforierung [U] f.;
Durchlöcherung [S] f.
R просверливание cp.;
отверстие cp.
A ثَقَب m

250

3949 [E]

performance
F rendement m.
E rendimiento m.
D Leistung f.
R характеристика ж.;
эксплуатационные
качества cp. мн. ч.
A نَتَائِج قِيَاسِيَّة f / أداء m

3950

perigee
F périgée m.
E perigeo m.
D Perigäum neut.;
Erdnähe f.
R перигей м.
A نُقْطَة الحَضِيض f

3951

perimeter
F périmètre m.
E perímetro m.
D Umkreis [U] m.;
Verteidigungsgürtel
[L] m.
R периметр м.; внешняя
граница обороны ж.
A مُحِيط ـ نِطاق m

3952

perimeter defense
F défense périphérique f.
E defensa perimétrica f.
D Rundumverteidigung f.
R круговая оборона ж.
A دِفَاع مُحِيطي m

3953

period
F délai m.; période f.
E período m.; plazo m.
D Zeitraum m.
R период м.; срок м.
A مُدَّة f

3954

periodic
F périodique
E periódico
D periodisch adv.
R периодический
A دَوْرِي

3955

periscope
F périscope m.
E periscopio m.
D Periskop neut.
R перископ м.
A مَرْقَب الغَوَّاصَة m

3956 [M]

(at) periscope depth
F (en) plongée
(immersion)
périscopique f.
E (en) profundidad de
periscopio f.
D (an)Periskoptiefe f.
R (на) перископной
глубине
A المَرْقَب صَاعِد

3957

periscopic
F périscopique
E periscópico
D Periskop- komp.
R перископический
A مَرْقَبي

3958

permanent
F permanent
E permanente
D ständig adj.
R постоянный
A دَائِم

3959 [R]

permanent echo
F écho permanent m.
E eco permanente m.
D Festzeichen neut.
R постоянное отражение
cp.
A صَدَى دائِم m

3960

permission

F autorisation *f.*
E autorización *f.*
D Genehmigung *f.*
R разрешение *cp.*
A رُخْصَة *f* ـ إذن *m*

3961

permit *v.*

F permettre
E permitir
D zulassen *tr.*
R разрешать; позволять
A سَمَح

3962

permit *n.*

F permis *m.*
E autorización *f.; pase m.*
D Durchlasschein *m.*
R пропуск *м.;*
разрешение *cp.*
A إجَازَة *f*

3963

perpendicular *adj. n.*

F perpendiculaire *adj. n. f.*
E perpendicular *adj. s. f.*
D perpendikular *adj. (komp);* Senkrechte *s. f.*
R перпендикуляр *м.;*
перпендикулярный *прилаг.*
A عَمُودي *m* عَمُود

3964 [C]

persistent agent

F gaz persistant *m.*
E agresivo persistente *m.*
D Geländekampfstoff *m.*
R стойкое отравляющее вещество *cp.*
A عَامِل ثَابِت *m*

3965

personal effects *pl.*

F effets personnels *m. pl.*
E efectos personales *m. pl.*
D persönliche Habe *f.*
R личные вещи *ж. мн. ч.*
A مملوكات شخصية

3966

personnel

F personnel *m.*
E personal *m.*
D Personal *neut.*
R личный состав *м.*
A أَفْرَاد

3967

personnel services *pl.*

F service du personnel *m.*
E servicio de personal *m.*
D Personalabteilung *f.*
R администрация *ж.;*
интендантская служба *ж.;*
интендантство *cp.*
A مَصَالِح الأَفْرَاد

3968

(in) perspective *(adj.)* *n.*

F (en) perspective *(adj.) n. f.*
E (en) perspectiva *(adj.) s. f.*
D Perspektive *s. f.;* perspektivisch *adj.*
R перспектива *ж.;* вид *м.;* (в) виду
A مَنْظُوري مَنْظُور *m*

3969

petroleum

F pétrole *m.*
E petróleo *m.*
D Erdöl *m.*
R нефть *ж.*
A بَنْزِين *m*

3970

petroleum oil and lubricants (POL)

F carburants et lubrifiants *m. pl.*
E carburantes y lubricantes *m. pl.*
D Treib-und Schmierstoffe *m. pl.*
R горючее и смазочные материалы (ГСМ) *cp. м. мн. ч.*
A البَنْزِين والزُّيُوت

3971

phase *v.*

F découper en phases
E desarrollar en fases
D (in) Phasen einteilen *tr.*
R вводить по этапам; принимать по этапам
A بَلَّغ

3972

phase *n.*

F phase *f.*
E fase *f.*
D Phase *f.*
R этап *м.;* фаза *ж.*
A مَرْحَلَة *f*

3973

phased out

F retiré du service
E retirado
D abgewickelt; aufgelöst
R выведенный
A مُخْرَج مِن برنَامِج تَدْرِيجِياً

3974

phase out *v.*

F déclasser; retirer progressivement
E retirar
D abwickeln *tr.;* auflösen *tr.*
R выводить
A اخرج من برنامج تدريجيا

251

3975

phenomenon
F phénomène *m.*
E fenómeno *m.*
D Phänomen *neut.*
R явление *ср.*
A ظَاهِرَة *f*

3976

phonetic alphabet
F alphabet phonétique *m.*
E alfabeto fonético *m.*
D phonetisches Alphabet
   neut.
R условный
   фонетический
   алфавит *м.*
A أَبْجَدِيَّة صَوْتِيَّة *f*

3977 [R]

phonetic spelling
F épellation phonétique *f.*
E deletreo fonético *m.*
D phonetische Schrift *f.*
R фонетическое
   правописание *ср.*;
   буквальное
   правописание *ср.*
A التَّهْجِئَة الصَّوْتِيَّة *f*

3978

phony
F faux
E simulado
D Schein- *komp.*
R ложный
A كَاذِب

3979 [C]

phosgene
F phosgène *m.*
E fosgeno *m.*
D Phosgen *neut.*
R фосген *м.*
A فُسْجِين *m*

3980 [C,S]

phosphorus
F phosphore *m.*
E fósforo *m.*
D Phosphor *m.*
R фосфор *м.*
A فُوسْفور *m*

3981

photocell
F cellule photo-électrique
   f.
E fotocélula *f.*
D Photozelle *f.*
R фотоэлемент *м.*
A خَلِيَّة كَهْرُبَائِيَّة ضَوْئِيَّة *f*

3982

photoflash bomb
F bombe photo-éclair *f.*
E bomba de iluminación
   f.
D Blitzlichtbombe *f.*
R фотобомба *ж.*
A قُنْبُلَة نَارِيَّة لِلتَّصْوِير فِي اللَّيل

3983

photogrammetry
F photogrammétrie *f.*
E fotogrametría *f.*
D Photogrammetrie *f.*
R фотограмметрия *ж.*
A مِسَاحَة تَصْوِيرِيَّة *f*

3984

(air) photograph
F photo aérienne *f.*
E fotografía aérea *f.*
D Luftaufnahme *f.*
R аэрофотоснимок *м.*
A صُورَة جَوِّيَّة *f*

3985

**(oblique air)
photograph**
F photo aérienne oblique
   f.
E fotografía aérea oblicua
   f.
D Schrägluftaufnahme *f.*
R планово-перспективный
   аэрофотоснимок *м.*
A صُورَة جَوِّيَّة مَائِلة *f*

3986

**(vertical air)
photograph**
F photo aérienne verticale
   f.
E fotografía aérea vertical
   f.
D Vertikalluftaufnahme *f.*
R плановый аэроснимок
   м.
A صورة جوية عمودية

3987

(air) photography
F photographie aérienne
   f.
E fotografía aérea *f.*
D Luftbildwesen *neut.*
R аэрофотосъёмка *ж.*
A تصوير جوي

3988

**(continuous strip)
photography**
F photographie en
   défilement continu *f.*
E itinerario fotográfico *m.*
D Durchlaufphotographie
   f.
R маршрутная
   аэрофотосъёмка *ж.*
A التصوير بشريط مستمر

3989

photomap

F photo-carte *f.*

E fotomapa *m.*

D (Luft-)Bildkarte *f.*

R (воздушная)
фотокарта *ж.*

A خَرِيطَة تَصْوِيرِيَّة *f*

3990

photoreconnaissance

F reconnaissance
photographique *f.*

E reconocimiento
fotográfico *m.*

D (Luft-)Bildaufklärung *f.*

R (воздушная)
(аэро-)фоторазведка
х.

A اسْتِطْلَاع تَصْوِيري *m*

3991

P hour

F heure-P *f.*

E hora P *f.*

D Stunde P *f.*

R время выброски *ср.*;
час-П *м.*

A ساعة (پي)

3992

physical

F physique

E físico

D physikalisch [E] *adj.;*
physisch [U]

R физический

A بَدَني

3993

physical education

F éducation physique *f.*

E educación física *f.*

D Sport *m.*

R физкультура *ж.*

A تعليم الرِّياضَة البَدَنِيَّة *m*

3994

physical examination

F examen médical *m.*

E reconocimiento médico
m.

D ärztliche Untersuchung
f.

R медицинский осмотр
м.

A فحص بدني

3995 [L]

picket

F piquet *m.*

E piquete *m.*

D Feldwache *f.*

R (сторожевая) застава
ж.

A جماعة *f* / وتد *m*

3996

(fire) picket

F piquet d'incendie *m.*

E destacamento de
contraincendios *m.*

D Brandwache *f.*

R пожарная команда *ж.*

A مُفْرَزة الاطفاء

3997

(radar) picket

F piquet radar *m.*

E defensa de radar *f.*

D Radarposten *m.*

R радиолокационный
дозор *м.*

A مفرزة رادار

3998 [A]

pick up *v.*

F relever

E recoger

D aufnehmen *tr.*

R обнаруживать;
перехватывать

A إلْتَقَط من الأرْض

3999 [R]

pick up *v.*

F repérer

E captar

D aufgreifen *tr.*

R принимать

A إلْتَقَط

4000 [A]

pickup point/field

F point/champ de
relèvement *m. m.*

E punto/campo de
recogida *m. m.*

D Aufnahmepunkt/-platz
m. m.

R место погрузки *ср.*

A نُقْطة أو مَيْدَان الإلْتِقاط

4001 [M]

pier

F jetée *f.;* quai *m.*

E embarcadero *m.*

D Landesteg *m.*

R пирс *м.*; мол *м.*

A رَصِيف *m*

4002 [B]

pier

F pile (de pont) *f.*

E estribo *m.*

D Brückenpfeiler *m.*

R бык *м.*; волнолом *м.*

A عَمُود *m*

4003

pierce

F percer

E penetrar

D durchschlagen *untr;*
durchstossen *untr.*

R пробивать; проникать

A خَرَق

4004

pill

F pilule *f.*

E píldora *f.*

D Tablette *f.*

R пилюля *ж.*

A حَبَّة *f*

253

4005

(smoke) pillar *n.*

F colonne (de fumée) *f.*
E columna (de humo) *f.*
D Rauchwolke *f.*
R столб *м.*
A عَمُود دُخَانِي *m*

4006

pillbox

F blockhaus *m.;* casemate *f.*
E fortín *m.*
D Bunker *m.*
R ДОС (долговременное оборонительное сооружение) *м.* *(ср.)*; ДОТ *м.*; ЗОС *м.*
A مِتْرَسَة لِلْمَدَافِع *f*

4007 [V]

pillion

F siège arrière *m.*
E asiento de atrás *m.*
D Soziussitz *m.*
R заднее сиденье *ср.*
A مَقْعَد خَلْفِي *m*

4008 [E]

pilot *adj.*

F pilote *inv.*
E piloto *inv.*
D Versuchs- *komp.*
R опытный; предварительный
A تَجْرِيبِي

4009 [A,M]

pilot *v.*

F piloter
E pilotar [A]; practicar [M]
D führen [A]; lotsen [M]
R вести; управлять
A أَرْشَد

4010 [A,M]

pilot *n.*

F pilote *m.*
E piloto [A] *m.;* practico [M] *m.*
D Flugzeugführer [A] *m.;* Lotse [M] *m.*
R лётчик [A] *м.*; лоцман [M] *м.*
A مُرْشِد *m*

4011 [M]

pilotage

F pilotage *m.*
E practicaje *m.*
D Küstennavigation *f.;* Lotsen *neut.*
R проводка судов *ж.*; лоцманское дело *ср.*
A إِرْشَاد *m*

4012

pilot boat

F bateau pilote *m.*
E bote de práctico *m.*
D Lotsenboot *neut.*
R лоцманское судно *ср.*
A زَوْرَق مُرْشِد *m*

4013

pilot flag

F pavillon-pilote *m.*
E bandera de pedir práctico *f.*
D Lotsenflagge *f.*
R лоцманский флаг *м.*
A راية إِسْتِدْعَاء المُرْشِد *f*

4014

pilot flame

F veilleuse *f.*
E llama piloto *f.*
D Zündflamme *f.*
R воспламенитель *м.*
A مَشْعَل تِلْقَائِي

4015

pilot light

F lampe témoin *f.*
E luz piloto *f.*
D Kontrollampe *f.*
R индикаторный огонь *м.*
A مِصْبَاح إِرْشَاد

4016

(firing) pin

F percuteur *m.*
E percutor *m.*
D Schlagbolzen *m.*
R боёк *м.*
A إِبْرَة الرَّمْي *f*

4017

(safety) pin

F goupille de sécurité *f.*
E aguja del seguro *f.*
D Sicherungssplint *m.*
R предохранительный шплинт *м.*
A سقاطة الامان

4018 [L]

pincer movement

F manœuvre en tenailles *f.*
E movimiento de pinza *m.*
D Zangenbewegung *f.*
R двойной охват *м.*; клещи *ж. мн. ч.*
A تَطْوِيق من الجَانِبَيْنْ

4019 [L]

pin down

F clouer au sol
E copar
D niederhalten *tr.*
R сковывать огнём; подавлять огнём
A ثَبَّت العَدُوّ

4020

pinnace
F chaloupe *f.*
E pinaza *f.*
D Pinasse *f.*
R пинас *м.*
A ‏زورق‏ *m*

4021

pinpoint *v.*
F indiquer exactement
E indicar con precisión
D genau festlegen *tr.*
R точно определять;
   указывать
A ‏عَيَّنَ مَوْضِعاً بِدِقَّة‏

4022

pinpoint *adj.*
F ponctuel
E (con) precisión
D Punkt- *komp.*
R точечный
A ‏نُقْطِي‏

4023

pinpoint *n.*
F repère *m.*
E punto preciso *m.*
D genauer Standort *m.*
R точечный ориентир *м.*
A ‏نُقْطَة مُعْلَمة‏ *f*

4024 [L]

pioneer
F pionnier *m.;* sapeur *m.*
E zapador *m.*
D Pionier *m.*
R сапёр *м.*
A ‏مهندس عسكري‏ *m*

4025 [R]

pip
F plot radar *m.*
E marcador *m.*
D Signal (Echo-, Morse-
   usw.) *neut.;* Zeichen
   neut.
R импульс *м.;* точка *ж.*
A ‏نُقطه‏ *f*

4026

pipe *n.*
F tuyau *m.*
E tubo *m.*
D Rohr *neut.*
R труба *ж.;*
   трубопровод *м.*
A ‏أُنْبُوب‏ *m*

4027

pipe *v.*
F canaliser
E conducir en cañerías
D (durch) Rohre leiten
R снабжать трубой
A ‏أَجْرَى في أَنابِيب‏

4028

**pipeline (gas, oil,
   water)**
F canalisation *f.;* gazoduc
   m.; oléoduc *m.*
E gaseoducto *m.;*
   oleoducto *m.;* tubería
   f.
D Fernleitung
   (Gas,Öl,Wasser) *f.;*
   Rohrleitung *f.*
R (газо-, нефте-,
   водо-)провод *м.*
A ‏خَطّ أَنابِيب‏ *m*

4029

pistol
F pistolet *m.*
E pistola *f.*
D Pistole *f.*
R пистолет *м.*
A ‏مُسَدَّس‏ *m*

4030

(flare) pistol
F pistolet lance-fusées *m.*
E pistola de señales *f.*
D Leuchtpistole *f.*
R ракетница *ж.*
A ‏مسدس للاشارة‏

4031

(machine) pistol
F pistolet mitrailleur *m.*
E pistola automática *f.*
D Maschinenpistole *f.*
R ручной пулемёт *м.*
A ‏رشاش قصير‏

4032

(signal) pistol
F pistolet de signalisation
   m.
E pistola de señales *f.*
D Leuchtpistole *f.*
R сигнальный
   (ракетный) пистолет
   м.
A ‏مُسَدَّس إِشَارَة‏

4033

(Verey light) pistol
F pistolet lance-fusées *m.*
E pistola Verey de señales
   f.
D Leuchtpistole *f.*
R сигнальная ракета *ж.*
A ‏مسدس اشارات ضوئية‏

4034

(inspection) pit
F fosse de réparations *f.*
E foso de inspección *m.*
D Inspektionsgrube *f.*
R смотровая яма *ж.*
A ‏حفرة فحص‏ *f*

4035

(machine-gun) pit
F nid de mitrailleuses *m.*
E pozo de ametralladoras
   m.
D Maschinengewehrgraben
   m.
R стрелковая ячейка *ж.*
A ‏حفرة رشاش‏ *f*

4036 [A]
pitch *v.*
F tanguer
E cabecear
D kippen
R наклоняться
A تَمَوَّج

4037 [M]
pitch *v.*
F tanguer
E cabecear
D stampfen [M]
R подвергаться килевой
  качке
A تَمَوَّج

4038 [E]
pitch
F pas (d'une vis etc...) *m.*
E paso *m.*
D Steigung *f.*
R шаг *м.*
A تَمَوَّج *m*

4039 [A,M]
pitch *n.*
F tangage *m.*
E cabezada *f.*
D Längsneigung [A] *f.;*
  Stampfen [M] *neut.*
R наклон *м.;* тангаж *м.*
A مَيْل *m*

4040 [E]
pivot *n.*
F pivot
E pivote *m.*
D Drehzapfen *m.*
R ось *ж.*
A قُطْب ـ مِحور *m*

4041 [L,U]
pivot *n.*
F pivot *m.*
E eje *m.*
D Drehpunkt *m.*
R точка вращения *ж.;*
  опорная точка *ж.*
A قُطْب ـ مِحور *m*

4042
pivot (about) *v.*
F (faire) pivoter (sur)
E girar (sobre)
D (sich) drehen (um)
R вертеться; вращаться
A دار حَوْل (قُطْب) مِحور

4043
pivotal *adj.*
F pivotant
E central *adj.*
D Schlüssel- *komp.*
R центральный; осевой
A مِحْوَرِي

4044 [T]
plain *n.*
F plaine *f.*
E llano *m.*
D Ebene *f.*
R равнина *ж.*
A سَهْل *m*

4045
(in) plain clothes (*adj.*)
n.
F (en) civil (*adj.*) *n. m.*
E (en) traje de calle (de
  paisano) (*adj.*) *s. m.*
D (in) Zivil (*adj.*) *s. neut.*
R штатская одежда *ж.;*
  в штатских
A (ب) اللِّبَاس المَدَنِي *m*

4046 [R]
plain text
F texte en clair *m.*
E texto claro *m.*
D Klartext *m.*
R открытый текст *м.;*
  незашифрованный
  текст *м.*
A مَتْن وَاضِح *m*

4047
plan *n.*
F plan *m.*
E plano *m.*
D Plan *m.*
R план *м.;* схема *ж.*
A خُطَّة *f*

4048
plan *v.*
F planifier; projeter
E planificar; proyectar
D planen
R составлять план;
  планировать
A خَطَّط

4049
(contingency) plan
F plan de contingence *m.*
E plan de emergencia *m.*
D Ausweichplan *m.*
R вторичный план *м.*
A خطة احتمالية

4050
(cover) plan
F plan de dissimulation
  m.
E plan de encubrimiento
  m.
D Deckungsplan *m.*
R план обеспечения *м.*
A خطة الاخفاء

4051
(deception) plan
F plan de déception *m.*
E plan de decepción *m.*
D Täuschungsplan *m.*
R дезинформация *ж.*
A خطة الخِداع

4052
(emplaning) plan
F plan d'enlèvement *m.*
E plan de embarque *m.*
D Flugzeugverladungsplan
  m.
R план погрузки *м.*
A خطة تحميل الطائرة

4053
(fire) plan
F plan d'emploi des feux
   m.
E plan de fuegos *m.*
D Feuerplan *m.*
R план огня *м.*
A خِطَّة النِيرَان *f*

4054
(flight) plan
F plan de vol *m.*
E plan de vuelo *m.*
D Flugplan *m.*
R план полёта *м.*
A خِطَّة الطَّيْرَان *f*

4055 [L]
(ground tactical) plan
F plan tactique au sol *m.*
E plan táctico de
   operaciones *m.*
D taktischer Plan für den
   Bodenkampf *m.*
R план боевых действий
   м.
A خطة تعبوية للعمليات البرية

4056 [A,L]
(landing) plan
F plan d'atterrissage
   (troupes aéromobiles)
   m.
E plan de aterrizaje (de
   tropas
   aerotransportadas)
D Landungsplan *m.*
R план высадки десанта
   м.
A خطة الانزال *f*

4057
(loading) plan
F plan de chargement *m.*
E plan de carga *m.*
D Ladeplan *m.*
R план погрузки *м.*
A خِطَّة التَّحْميل *f*

4058
(master) plan
F plan-maître *m.*
E plano maestro *m.*
D Generalplan *m.*
R общий план *м.*
A خِطَّة رَئيسِيَّة *f*

4059
(operation) plan
F plan d'opération *m.*
E plano de operación *m.*
D Operationsplan *m.*
R оперативный план *м.*
A خِطَّة العَمَلِيَّات *f*

4060
(outline) plan
F avant-projet *m.;*
   ébauche de
   conception *f.*
E plano a grandes rasgos
   m.
D Grobplan *m.*
R краткий план *м.*
A خِطَّة أَوَّلِيَّة *f*

4061
(restrictive fire) plan
F plan de tir restrictif
E plano de tiro restrictivo
   m.
D begrenztes Feuerverbot
   für Steilfeuerwaffen
   neut.
R ограниченный план
   ведения огня *м.*
A خُطَّة نِيرَان تَضْيِيقِيَّة

4062 [A,M]
plane *v.*
F planer
E planear
D (sich) abheben [M] *tr.;*
   gleiten [A]
R скользить;
   планировать
A إنْطَلَق بِمُسْتَوَى سَطْح الماء

4063
plane *adj.*
F plan
E plano *adj.*
D flach
R плоский
A مُسْتَوٍ

4064
plane *n.*
F plan *m.*
E nivel [U] *m.;* plano [ES]
   m.
D Fläche [ES] *f.;* Niveau
   [U] *neut.;* Tragfläche
   [A] *f.*
R плоскость *ж.;* уровень
   м.; крыло [A] *ср.*
A سَطْح *m*

4065
planet
F planète *f.*
E planeta *f.*
D Planet *m.*
R планета *ж.*
A سَيَّار ـ كوكب *m*

4066
planimetric map
F carte planimétrique *f.*
E plano planimétrico *m.*
D planimetrische Karte *f.*
R контурная карта *ж.*
A خَريطة لِقِياس المَسَاحَات غَيْر
   المُنْتَظَمَة

4067
plank
F planche *f.*
E tablón *m.*
D Planke *f.*
R доска *ж.*
A لَوْح خَشَب

4068
plankton
F plancton *m.*
E plankton *m.*
D Plankton *neut.*
R планктон *м.*
A عَوالِق أَحْيَائِيَّة ـ كائِنات مائية
   دقيقة *f*

4069

planned

F planifié; prévu
E planeado; previsto
D geplant; vorgesehen
R планированный;
   плановый
A مُخَطَّط

4070

(logistic) planning factor

F facteur de planification
   logistique *m.*
E factor de planificación
   logística *m.*
D Planungsfaktor *m.*
R плановый
   коэффициент *м.*
A عَامِل لِلتَّخْطِيط الإِدَارِي

4071

plant

F équipement *m.*
E equipo *m.*
D Aggregat *neut.*
R установка *ж.*
A معدات الهندسة *f*

4072

plant a bomb *v.*

F poser une bombe
E colocar una bomba
D Minen legen
R поставлять бомбу
A وضع قنبلة

4073

plasma

F plasma *m.*
E plasma *m.*
D Plasma *m.*
R плазма *ж.*
A بَلَازْمَا *f*

258

4074

plastic *adj. n.*

F (matière) plastique (*n.
   f.*) adj.
E plástico *adj. s. m.*
D Kunststoff(-) *m.*
   (*komp.*)
R пластический *прилаг.*;
   пластичный *прилаг.*;
   пластмасса *ж.*
A لَدْن ـ بِلاستيك *m*

4075

plastic explosive

F (explosif) plastic *m.*
E explosivo plástico *m.*
D plastischer Sprengstoff
   m.
R пластичное
   взрывчатое вещество
   ср.
A مُتَفَجِّر لَدْن *m*

4076

plasticizer

F plastifiant *m.*
E plastificador *m.*
D Weichmacher *m.*
R пластификатор *м.*
A مُلَدِّن *m*

4077

(armor) plate

F plaque de blindage *f.*
E blindaje *m.*
D Plattenpanzerung *f.*
R бронеплита *ж.*
A صفيح الدَّرْع *m*

4078 [S]

(base) plate

F plaque de base *f.*
E placa base *f.*
D Bodenplatte *f.*
R опорная плита *ж.*
A صفيحة القاعدة *f*

4079

(butt) plate

F plaque de couche de
   crosse *f.*
E culata *f.*
D Kolbenkappe *f.*
R затылок *м.*
A صفيحة الاخمص *f*

4080

(deck) plate

F tôle de pont *f.*
E plancha de cubierta *f.*
D Decksplatte *f.*
R лист настилки палубы
   м.
A لَوْح سَطْحِي *m*

4081

(glacis) plate

F blindage glacis *m.*
E plato blindado
   inclinado *m.*
D Glacispanzerung *f.*
R гласис *м.*
A منحدر مُقْبِل

4082

(pressure) plate

F plateau de pression *m.*
E placa de presión *f.*
D Druckkappe *f.*
R нажимная плита *ж.*
A صفيحة الضغط

4083 [T]

plateau

F plateau *m.*
E meseta *f.*
D Hochebene *f.*
R плоскогорье *ср.*;
   плато *ср.*
A نَجْد *m* هَضَبَة *f*

4084 [A]

platform

F plate-forme *f.*
E plataforma *f.*
D Plattform *f.*
R платформа *ж.*
A مِنَصَّة *f*

4085 [A]
platform drop
F largage lourd *m.*
E lanzamiento pesado *m.*
D Plattform-Abwurf *m.*
R тяжёловесное
сбрасывание *ср.*
A إِسْقَاط مِنَصَّة

4086 [A,M,V]
plating *n.*
F blindage *m.*
E blindaje *m.*
D Panzerung *f.*
R броневая плита *ж.*
A أَلْوَاح تَغْطِيَة

4087
**Plimsoll line; Plimsoll
mark**
F ligne de Plimsoll *f.;*
marque de Plimsoll *f.*
E línea Plimsoll *f.;* marca
Plimsoll *f.*
D Höchstladelinie *f.;*
Höchstlademarke *f.*
R грузовая марка *ж.*
A خَطّ بلمسول / خط الشحن
على السفينة

4088 [P]
plot *v.*
F conspirer
E conspirar
D (sich) verschwören
R составлять заговор;
замышлять
A تَآمَر

4089 [A,L,M,R,S]
plot *n.*
F tracé *m.*
E trazado *m.*
D Standort *m.;* Zielort [S]
m.
R засечка цели *ж.;*
трасса *ж.;* график
м.
A مُخَطَّط *m*

4090 [P]
plot *n.*
F complot *m.*
E complot *m.*
D Verschwörung *f.*
R заговор *м.*
A مُؤَامَرَة *f*

4091 [A,L,M,R,S]
plot *v.*
F rapporter; tracer
E trazar
D eintragen *tr.*
R наносить;
прокладывать (курс)
A عَيِّن مَوَاقِع النُّقْطَة

4092 [S]
plotting board
F tableau de tracé
E tablero *m.*
D Feuerleitplatte *f.*
R планшет *м.*
A لوحة بيانية لاهداف المدفعية

4093 [B,M]
plug *n.*
F bouchon *m.*
E tapón *m.*
D Stöpsel *m.*
R затвор *м.;* пробка *ж.*
м.; стопор
A سِدَادة

4094 [B,M]
plug *v.*
F boucher
E tapar
D zustopfen *tr.*
R затыкать
A سَدّ

4095 [E,R]
(spark) plug *n.*
F fiche *f.;*
E clavija *f.*
D Stecker *m.*
R штепсель *м.*
A شَمْعَة الإِشْعَال

4096 [V]
(spark) plug *n.*
F bougie *f.*
E bujía *f.*
D Zündkerze *f.*
R запальная свеча *ж.*
A شَمْعَة الإِشْعَال

4097
plug adaptor
F adapteur *m.*
E adaptador *m.*
D Übergangsstecker *m.*
R адаптор *м.;*
переходник *м.*
A مُهَايِىء لِلْقَابِس

4098
plug in *v.*
F brancher
E conectar
D anschliessen *tr.*
R вставлять (штепсель);
соединять
A وَصَل

4099
plunder *v.*
F piller
E saquear
D plündern
R грабить
A سَلَب

259

4100

plutonium

F plutonium *m.*
E plutónio *m.*
D Plutonium *m.*
R плутоний *м.*
A البلوتونيوم *m*

4101

plywood

F contre-plaqué *m.*
E madera contraplacada *f.*
D Sperrholz *neut.*
R клеёная фанера *ж.*
A خَشَب رَقائقِي *m*

4102

pneumatic

F pneumatique
E neumático
D Druckluft- *komp.;* Luft-*komp.;* pneumatisch *adj.*
R надувной; пневматический
A مُختَصّ بِالهَواء والغازات

4103 [A,H,L,U]

pocket

F poche [HLU] *f.;* trou [A] *m.*
E bolsa [ALU] *f.;* bolsillo [H] *m.*
D Luftloch [A] *neut.;* Tasche[HU] *f.;* Widerstandsnest[L] *neut.*
R карман *м.;* мешок *м.;* очаг *м.;* яма *ж.*
A جَيْب *m*

4104

pod

F nacelle *f.;* point d'attache des armes de bord *m.*
E barquilla *f.*
D Hülse *f.*
R (отделяемый) отсек *м.*
A حجيرة طولية تحت جناح الطائرة

4105 [L]

point *n.*

F pointe *f.*
E punta *f.*
D Spitze *f.*
R головной дозор *м.*
A نُقْطة *f*

4106 [T,U]

point *n.*

F point *m.*
E punto *m.*
D Punkt *m.*
R пункт *м.;* точка *ж.*
A نُقْطة

4107

(air control) point (ACP)

F point de contrôle aérien *m.*
E punto de control de operaciones aéreas *m.*
D Luftkommandostelle *f.*
R аэродромный пункт управления *м.*
A نُقطة سيطرة جوية *f*

4108

(aiming) point

F point de visée *m.*
E punto de puntería *m.*
D Zielpunkt *m.*
R точка прицеливания *ж.;* точка наводки *ж.*
A نقطة التسديد

4109 [S]

point (at) *v.*

F braquer (sur)
E apuntar (a, con)
D richten
R направлять; наводить
A سَدَّد

4110 [A]

(bomb release) point

F point de largage des bombes *m.*
E punto de suelta de la bomba *m.*
D Bombenabwurfpunkt *m.*
R точка сбрасывания *ж.*
A خَطّ / نُقْطة إلْقَاء القَنَابِل

4111

(border crossing) point

F poste frontière *m.*
E puesto fronterizo *m.*
D Grenzübergang *m.*
R пограничный контрольный пункт *м.;* контрольно-пропускной пункт *м.*
A نُقْطة عُبور الحُدود *f*

4112 [E]

(breaking) point

F limite de résistance *f.*
E límite de ruptura *m.*
D Bruchgrenze *f.*
R предельное напряжение *ср.*
A نُقْطة الكَسْر

4113

(casualty collection) point

F section de ramassage *f.*
E nido de heridos *m.*
D Verwundetensammel-punkt *m.*
R пункт сбора раненых *м.*
A مركز جمع الخسائر

4114

(casualty evacuation) point

F poste d' évacuation des morts et des blessés *f.*

E punto de evacuación de heridos *m.*

D Verwundetentransport-punkt *m.*

R пункт эвакуации раненых *м.*

A نقطة اخلاء الخسائر

4115 [E]

(charging) point

F prise de ravitaillement *f.*

E puesto de carga *m.*

D Batterieladestelle *f.*

R пункт зарядки *м.*

A نُقْطَة الشَّحْن *f*

4116 [V]

(check) point

F poste de controle *f.*

E puesto de control *m.*

D Kontrollpunkt *m.;* kritischer Punkt *m.*

R контрольно-пропускной пункт (КПП) *м.*

A نقطة الفحص

4117

(contact) point

F point de jonction *m.*

E punto de contacto (de unión) *m.*

D Treffpunkt *m.*

R пункт соприкосновения *м.;* район соприкосновения *м.*

A نُقْطَة التَّمَاسّ *f*

4118

(control) point

F contrôle *m.*

E punto de control *m.*

D Kontrollpunkt *m.;* Meldestelle *f.*

R пункт регулирования *м.;* контрольный пункт *м.*

A نُقْطَة المُرَاقَبَة *f*

4119 [L]

(critical) point

F pcint sensible *m.*

E punto crítico *m.*

D Belastungsgrenze *f.*

R критический пункт *м.*

A نقطة حَرِجَة

4120

(crossing) point

F point de passage *m.*

E punto de paso *m.*

D Übergangsstelle *f.*

R КПП *м.;* переезд *м.*

A نقطة العبور

4121

(decimal) point

F virgule *f.*

E coma *f.*

D Komma *neut.*

R запятая *ж.*

A الصِفر العُشْرِي

4122 [A]

(departure) point

F point d'origine *m.*

E punto de partida *m.*

D Ausgangspunkt *m.*

R исходный пункт *м.*

A نُقْطَة (البَدْء/ الرَّحِيل) *f*

4123

(distribution) point

F point de distribution *m.*

E punto de distribución *m.*

D Ausgabestelle *f.*

R пункт распределения *м.*

A نُقْطَة تَوْزِيع *f*

4124 [A]

(filler) point

F point de ravitaillement *m.*

E enchufe de recarga *m.*

D (Ein-)Füllöffnung *f.*

R заправочный пункт *м.*

A نُقْطَة المَلْء *f*

4125 [E,P]

(flash) point

F point d'ignition [E]; (sur le) point d'exploser [P]

E punto de inflamación *m.*

D Entzündungspunkt *m.*

R точка возгорания *ж.*

A نُقْطَة (الإلْتِهَاب/ الوَمِيض) *f*

4126

(forward arming and refuelling) point

F point avancé de ravitaillement de munitions et de carburants (hélicoptère) *m.*

E punto avanzado de municionamiento y repostaje *m.*

D vorgeschobener Stützpunkt für Treibstoff und Munition *m.*

R передовой пункт снаряжения и заправки *м.*

A نقطة أمامية للتسليح والتموين بالوقود

4127 [A]

(holding) point

F point d'attente *m.*

E punto de espera *m.*

D Wartepunkt *m.*

R пункт ожидания *м.*

A نُقْطَة تَجَمُّع *f*

4128 [S]

(impact) point

F point d'impact *m.*

E punto de incidencia *m.*

D Treffpunkt *m.*

R точка попадания *ж.*; точка приземления *ж.*

A نُقْطَة الإِصَابَة *f*

4129 [A]

(intercept) point

F point d'interception *m.*

E punto de intercepción *m.*

D Abfangpunkt *m.*

R пункт перехвата *м.*

A نُقْطَة الإِعْتِرَاض *f*

4130 [L]

(junction) point

F point de jonction *m.*

E punto de juntura *m.;* conexión *f.*

D Schnittpunkt *m.*

R стык *м.*

A نُقْطَة إِتِّصَال *f*

4131 [L]

(key) point

F point sensible *m.*

E punto clave *m.*

D Schlüsselpunkt *m.*

R ключевой пункт *м.*

A نُقْطَة حَيَوِيَّة *f*

4132

(landing) point

F point d'atterrissage *m.;* point de débarquement *m.*

E punto de aterrizaje *m.*

D Landepunkt *m.*

R пункт высадки *м.*

A نُقْطَة إِنْزَال *f*

262

4133

(lashing down) point

F point d'arrimage *m.*

E punto de sujeción (de ajuste, de atadura) *m.*

D Befestigungspunkt *m.*

R закрепительный пункт *м.*; люверс *м.*

A حَلْقَة الرَّبْط *f*

4134

(loading) point

F point de chargement *m.*

E punto de embarque *m.*

D Ladepunkt *m.*

R пункт погрузки *м.*

A نُقْطَة التَّحْمِيل *f*

4135 [A]

(offset) point

F point futur *m.*

E punto de intersección *m.*

D versetzter Zielpunkt *m.*

R вспомогательная точка *ж.*

A نُقْطَة حَائِدة *f*

4136

(offset aiming) point

F point de viseé décalée *m.*

E punto de puntería compensada *m.*

D abweichender Zielpunkt *m.*

R косая точка наводки *ж.*

A نقطة تسديد غير مباشر

4137

(orbit) point

F point d'attente *m.*

E punto de escalonar (de espera) *m.*

D Warteraum *m.*

R выжидательный пункт *м.*

A نُقْطَة مَدَار *f*

4138

(orienting) point

F point de repère *m.*

E punto de orientación *m.*

D Richtpunkt *m.*

R репер *м.*; отметка *ж.*; ориентирная точка *ж.*

A نُقْطَة التَّوْجِيه *f*

4139

(passage) point

F point de passage *m.*

E punto de paso *m.*

D Durchlasspunkt *m.*

R точка прохождения *ж.*; пункт прохождения *м.*

A نقطة اجتياز الخطوط

4140 [A]

(pop-up) point

F point de début d'attaque *m.*

E pop-up point *m.*

D Steigpunkt (eines Flugzeuges) *m.*

R пункт перехода к атаке *м.*

A نقطة الصعود من ارتفاع منخفض لبدء الهجوم

4141

(prisoner of war collecting) point

F centre de regroupement de prisonniers de guerre *m.*

E punto de concentración de prisioneros de guerra *m.*

D (Kriegsgefangenen-) Sammelstelle *f.*

R пункт сбора военнопленных *м.*

A نُقْطَة جَمْع أَسْرَى الحَرْب

4142 [A]
(pull-up) point
F point de cabré *m.*
E pullup point *m.*
D Anhaltepunkt *m.*
R пункт перехода к
     набору высоты *м.*
A نقطة الشبوب

4143 [T]
(reference) point
F point de référence *m.*
E punto de referencia *m.*
D Orientierungspunkt *m.*
R отметка *ж.*
A نقطة مَرْجِع

4144 [S]
(registration) point
F point de réglage *m.*
E blanco auxiliar *m.*
D Einschiesspunkt *m.*
R пристрелочный репер
     м.
A نُقْطَة التَّسْجِيل

4145 [A]
(release) point
F point de largage *m.;*
     point de dislocation
E punto de suelta *m.*
D Bombenpunkt *m.;*
     Abwurfpunkt *m.*
R пункт сбрасывание *м.*
A نقطة الالقاء

4146 [L,V]
(start, starting) point
F point initial *m.*
E punto de partida *m.*
D Ablaufpunkt *m.*
R исходный пункт *м.;*
     рубеж *м.*
A خَطّ الإنْطِلَاق

4147 [L]
(strong) point
F centre de résistance *m.*
E punto de resistencia *m.*
D Stützpunkt *m.*
R опорный пункт *м.*
A نُقْطَة حَصِينة *f*

4148
(supply) point
F point de ravitaillement
     m.
E punto de
     abastecimiento *m.*
D Versorgungspunkt *m.*
R пункт снабжения *м.*
A نُقْطَة تَمْوِين

4149 [S]
(surveyed) point
F point de pièce *m.*
E (exacta) localización de
     un emplazamiento *f.*
D vermessene
     Kanonenstellung *f.*
R ориентированная
     точка *ж.*
A نقطة ممسوحة ــ نقطة مخططة

4150 [A]
(target approach)
     point
F point d'orientation *m.*
E punto de orientación *m.*
D Ablaufpunkt für
     Heeres-
     Nahunterstützung *m.*
R точка захода на цель
     ж.
A نُقْطَة الإقْتِرَاب من الهَدَف

4151
(target reference)
     point
F point de réglage *m.*
E punto de referencia de
     objetivo *m.*
D Zielorientierungspunkt
     m.
R отметка *ж.*
A نقطة الدلالة على الهدف

4152
(tie-down) point
F point d'arrimage *m.*
E punto de amarre *m.*
D Befestigungspunkt *m.*
R глазок *м.;* ушко *ср.*
A نُقْطَة رَبْط

4153 [T]
(trig) point
F point géodésique *m.*
E vértice geodésico *m.*
D Kartenvermessungspunkt
     m.
R геодезический пункт
     м.
A نقطة التثليث

4154 [A]
(turn-in) point
F point de début
     d'attaque *m.*
E punto de picado *m.*
D Luftangriff-
     Ausgangspunkt *m.*
R пункт перехода к
     атаке *м.*
A نُقْطَة انْعِطَاف

4155 [U]
(turning) point
F tournant *m.*
E punto decisivo *m.*
D Wendepunkt *m.*
R поворотный пункт *м.*
A نُقْطَة تَحَوُّل

4156
(vantage) point
F point tactique *m.;*
     position avantageuse
     f.
E posición ventajosa *f.*
D Aussichtspunkt *m.*
R (выгодная) позиция
     ж.
A موقع ممتاز

4157 [L]
(vital) point
F point d'importance
     tactique *m.*
E punto vital *m.*
D Schlüsselpunkt *m.*
R важнейший пункт *м.;*
     первостепенный
     пункт *м.*
A نقطة حيوية

4158

(vulnerable) point

F point sensible *m.*
E punto vulnerable *m.*
D empfindlicher Punkt *m.*
R уязвимая точка *ж.*;
    позиция *ж.*
A نُقْطَة مُعَرَّضَة لِلْخَطَر

4159

(watering) point

F point de distribution
    d'eau *m.*
E punto de aguada *m.*
D Wasserstelle *f.*
R водораспредели-
    тельный пункт *м.*;
    пункт
    водоснабжения *м.*
A نُقْطَة تَوْزِيع الماء

4160

point blank *adj.*

F (à) bout portant
E (a) quemarropa
D direkt; nahe
R прямой наводкой
A عن كَثَب

4161 [E]

pointer

F aiguille *f.*
E indicador *m.*
D Zeiger *m.*
R стрелка *ж.*; указатель
    м.
A إِبْرَة *f*

4162

point of burst

F point d'explosion *m.*
E punto de explosión *m.*
D Detonationspunkt *m.*
R точка разрыва *ж.*
A نُقْطَة الإِنْفِجَار

264

4163

point of departure

F point de départ *m.*
E punto de partida *m.*
D Ausgangspunkt *m.*
R исходный пункт *м.*
A نُقْطَة الإِنْطَلَاق

4164 [S]

point of impact

F point d'impact *m.*
E punto de impacto *m.*
D Auftreffpunkt *m.*
R точка попадания *ж.*
A نُقْطَة الصَدم

4165 [A]

point of impact

F point d'atterrissage *m.*
E punto de llegada *m.*
D Landepunkt *m.*
R точка попадания *ж.*
A نُقْطَة الإِصَابة

4166

**(mean) point of
impact (MPI)**

F point moyen des
    impacts *m.*
E punto medio de
    impacto *m.*
D mittlerer Treffpunkt *m.*
R средняя точка
    попадания *ж.*
A نُقْطَة الإِصَابَة الوَسَطِيَّة *f*

4167

point of no return

F point de non retour *m.*
E punto de no regreso *m.*
D Umkehrgrenzpunkt *m.*
R "пункт невозврата" *м.*
A نُقْطَة اللَّا عودَة

4168

point out

F indiquer
E indicar
D (auf etwas) hinweisen
    tr.
R указывать;
    показывать
A أَشَار

4169

(cardinal) points *pl.*

F points cardinaux *m. pl.*
E puntos cardinales *m. pl.*
D Himmelsrichtungen *f.*
    pl.
R главные румбы *м. мн.*
    ч.; страны света *ж.*
    мн. ч.
A الجِهَات الأَرْبَع

4170

(lubrication) points

F graisseurs *m. pl.*
E puestos de engrase *m.*
    pl.
D Schmierpunkt *m.*
R пункты смазывания *м.*
    мн. ч.
A نقط التزييت

4171

polar *adj.*

F polaire
E polar
D polar *adj.*; Polar- *komp.*
R полярный
A قُطْبِي

4172

polar coordinates *pl.*

F coordonnées polaires *f.*
    pl.
E coordenadas polares *f.*
    pl.
D Polarkoordinaten *f. pl.*
R полярные координаты
    ж. мн. ч.
A الإِحْدَاثِيَّات القُطْبِيَّة *f*

4173
polaroid camera
F appareil-photo
  "polaroid" *m.*
E cámara polaroid *f.*
D Polaroidkamera *f.*
R фотоаппарат
  Поларойд *м.*
A آلة تصوير بولارويد

4174 [E,T]
pole
F pôle *m.*
E polo *m.*
D Pol *m.*
R полюс *м.*
A قُطْب *m*

4175 [B,U]
pole
F échalas *m.*
E poste *m.*
D Stange *f.*
R столб *м.*
A عُمُود / عِماد

4176
police *v.*
F maintenir l'ordre;
  patrouiller
E mantener el orden
  público; vigilar
D (die) Ordnung
  aufrechthalten *tr.;*
  schützen
R охранять порядок;
  поддерживать
  порядок
A رَاقَب

4177
police *n.*
F police *f.*
E policía *f.*
D Polizei *f.*
R полиция *ж.;*
  полицейские *м. мн.*
  ч.
A شُرْطَة *f* / بوليس *m*

4178
policy
F ligne de conduite *f.;*
  politique *f.*
E normas de conducta *f.*
  pl.
D Grundsatz *m.;* Politik *f.*
R политика *ж.*
A سِياسَة *f*

4179
politician
F homme politique *m.*
E político *m.*
D Politiker *m.*
R политический деятель
  м.
A سياسي *m*

4180
polluted water
F eau contaminée *f.;* eau
  non-potable *f.*
E agua contaminada *f.*
D verseuchtes Wasser
  neut.
R загрязнённая вода *ж.*
A ماء ملوَّث *m*

4181
polystyrene *n.*
F polystyrène *m.*
E poliestireno
D Styropor *neut.*
R полистирол *м.*
A بوليستيرين *m*

4182 [T]
pond
F étang *m.;* mare *f.*
E charca *f.;* estanque *m.*
D Teich *m.*
R пруд *м.*
A بركة *f*

4183 [B]
pontoon
F ponton *m.*
E pontón *m.*
D Ponton *m.*
R понтон *м.*
A مَرْكَب ـ صَنْدَل *m*

4184
pool *v.*
F mettre en pool (en
  commun)
E mancomunar
D zusammenlegen *tr.*
R объединять;
  складывать
A شَارَك

4185 [V]
pool *n.*
F pool (de véhicules) *m.*
E centro común (de
  vehículos) *m.*
D Fahrbereitschaft *f.*
R автобаза *ж.*
A مَعِين المَرْكَبات

4186
population
F population *f.*
E población *f.*
D Bevölkerung *f.*
R население *ср.*
A سُكَّان

4187
port *n.*
F port *m.*
E puerto *m.*
D Hafen *m.*
R гавань *ж.;* порт *м.*
A مِيناء *m*

4188 [M]
port *adj.*
F (de) bâbord
E (a) babor
D backbord; Backbord-
  komp.
R левый; левого борта;
  налево
A يَسَار السفينة

4189
portable
F portatif
E portátil
D tragbar
R переносный;
     портативный
A نَقَّالِي

4190
portage
F portage *m.*
E porteo *m.*
D Portage *f.*
R переноска *ж.*
A حمل الزوارق

4191
port area
F zone portuaire *f.*
E zona portuaria *f.*
D Hafengebiet *neut.*
R портовый район *м.*
A مِنْطقة المِيناء

4192
**port throughput
   capacity**
F capacité portuaire *f.*
E capacidad de puerto *f.*
D Hafen-Entladekapazität
     f.
R портовый
     грузооборот *м.*
A طاقة المِيناء

4193 [A,L,M,U,V]
position *n.*
F position *f.*
E posición [ALMUV] *f;*
     situación [U] *f.*
D Position [AMUV] *f.;*
     Stellung [LU] *f.*
R позиция *ж.;*
     местоположение *ср.*
A مَوْضِع *m*

4194
(alternate) position
F position de rechange *f.*
E puesto eventual *m.*
D Ausweichstellung *f.*
R запасная позиция *ж.*
A موقع تَبادُلي *m*

4195
(assembly) position *n.*
F position d'attente *f.*
E zona de reunión *f.*
D Truppensammelstelle *f.*
R пункт сбора *м.;*
     выжидательная
     позиция *ж.*
A موضع الاجتماع

4196
(attack) position
F zone de déploiement *f.*
E posición de ataque *f.*
D Angriffsstellung *f.*
R исходное положение
     ср.
A موضع الاقتحام

4197 [S]
(base plate) position
F emplacement de
     batterie d'un mortier
     m.
E posición de la placa
     base *f.*
D Bodenplattenstellung *f.*
R миномётная огневая
     позиция *ж.*
A موضع صفيحة القاعدة

4198
(blocking) position
F position d'arrêt *f.*
E posición de bloqueo *f.*
D Sperrstellung *f.*
R заградительное
     положение *ср.*
A موضع حاجز

4199
(defilade) position
F position défilée *f.*
E posición desenfilada *f.*
D gedeckte Stellung *f.*
R (естественное)
     укрытие *ср.*
A موضع مستور من الجانب

4200
(dominating) position
F position dominante *f.*
E posición dominante *f.*
D beherrschende Stellung
     f.
R (господствующая)
     позиция *ж.*
A موضع اشراف

4201
(dummy) position
F (fausse) position *f.*
E posición simulada *f.*
D Scheinstellung *f.*
R (ложная огневая)
     позиция *ж.*
A موضع هيكلي

4202
(firing) position
F position de tir *f.*
E posición de fuego *f.*
D Feuerstellung
R положение для
     стрельбы *ср.*
A موضع الرمي

4203
**(hulldown firing)
   position**
F position de tir à
     défilement de caisse *f.*
E posición en desenfilada
     de casco *f.*
D (teilgedeckte) Stellung *f.*
R скрытым корпусом
A موضع الرمي بالبدن مستور

4204

(key) position
F position-clé *f.*
E posición clave
(estratégica, vital) *f.*
D Schlüsselstellung *f.*
R ключевая позиция *ж.*
A مَوْضِع حَيَوي *m*

4205

(kneeling) position
F position à genou *f.*
E posición de rodillas *f.*
D Kniestellung *f.*
R положение с колена
ср.
A وضع الجثو

4206

(overwatch) position
F position dominante *f.*
E posición de vigilancia *f.*
D Überwachungsstellung
f.
R наготове к
(тактическому
защитному
обеспечению)
A موضع الانتباه

4207 [S]

(predicted) position
F position prédite *f.*
E posición de predicción
f.
D Vorhaltepunkt *m.*
R упреждённая точка *ж.*
A مَوْضِع مُتوقع

4208

(pre-reconnoitred)
position
F position reconnue à
l'avance *f.*
E posición previamente
reconocida *f.*
D vorerkundete Stellung *f.*
R (заранее
рекогносцированная)
позиция *ж.*
A موضع مستطلع متقدماً

4209

(present) position
F position du moment *f.*
E posición actual *f.*
D gegenwärtige Stellung *f.*
R (настоящая) позиция
ж.
A الموضع الحالي

4210

(prone) position
F positicn de tireur
couché *f.*
E posición de tendido *f.*
D Bauchlage *f.;* Anschlag
liegend *m.*
R положение ничком *ср.*
A وضع الانبطاح

4211 [A]

(ready) position
F position d'attente *f.*
E posición de apresto *f.*
D Bereithalteplatz *m.*
R выжидательное
положение *ср.;*
изготовка *ж.*
A الموضع المُهَيَّأ للإقْلاع

4212

(turret-down) position
F défilement de tourelle
m.
E desenfilada de torreta *f.*
D Deckungsstellung *f.*
R башней укрытой
A وِضعة البرج المستور

4213

position defense
F défense ferme *f.*
E defensa en posición
(organizada) *f.*
D Verteidigung aus
Stellungen *f.*
R позиционная оборона
ж.
A دِفَاع ثَابِت *m*

4214

position error
F écart de position *m.*
E desvío de posición *m.*
D Ortungsfehler *m.*
R отклонение [A] *ж.;*
погрешность
местоположения [M]
ср.
A خَطَأ المَوْضِع

4215

position forces *v.*
F déployer des forces
E disponer a las tropas
D Truppen stationieren
untr.
R развёртывать силы;
развёртывать войска
A وَضع القوات

4216

positioning
F mise en batterie [S] *f.;*
mise en place [L] *f.*
E disposición *f.*
D stationieren
R развёртывание *ср.*
A وضع

4217 [A,M,V]

position light
F feu de position *m.*
E luz de situación *f.*
D Positionslampe *f.*
R кильватерный огонь
м.; позиционный
огонь *м.*
A ضَوْء المَوْضِع

4218

positive *adj. n.*
F positif *adj. n. m.*
E positivo *adj. s. m.*
D positiv *adj. s. neut.*
R положительный;
позитивный
A مُوجَب *m*

267

4219 [A,L,M]

post *n.*

F garnison *f.*
E guarnición *f.*
D Garnison *f.*
R гарнизон *м.*; пост *м.*;
   пункт *м.*
A مَرْكَز *m*

4220 [L]

post *v.*

F poster
E apostar
D (als Posten) aufstellen
   tr.
R назначать на
   должность
A وَضَع

4221 [L]

post *n.*

F poste *m.*
E puesto *m.*
D Posten(bereich) *m.*
R должность *ж.*; пост
   м.
A بريد

4222 [A,L,M]

post *v.*

F affecter
E destinar
D versetzen
R расставлять; ставить
A عَيَّن

4223

(aiming) post

F repère *m.*
E soporte del arma *m.*
D Zielpfosten *m.*
R прицельная веха *ж.*
A وتد التسديد

4224

(air command) post

F poste de
   commandement en
   vol *f.*
E puesto de mando aéreo
D Luftbefehlsstelle *f.*
R воздушный
   командный пункт
   (ВКП) *м.*
A مركز قيادة جوية

4225

(air observation) post

F poste d' observation
   aérienne *f.*
E puesto de observación
   aérea
D Luftbeobachtungsstelle
   f.
R воздушный
   наблюдательный
   пункт (ВНП) *м.*
A مرصد جوي

4226

(air staging) post

F escale aérienne *f.*
E campo de aterrizaje de
   tránsito *m.*
D Zwischenlandungs-
   flughafen *m.*
R передовой
   (промежуточный)
   аэродром *м.*
A قَاعِدَة تَرْحِيل جَوِّي *f*

4227

(command) post

F poste de
   commandement *f.*
E puesto de mando *m.*
D Befehlsstelle *f.*;
   Gefechtsstand *m.*
R командный пункт
   (КП) *м.*; командно-
   наблюдательный
   пункт (КНП) *м.*
A مَرْكَز القِيَادة *m*

4228

(listening) post

F poste d'écoute *f.*
E puesto escucha *m.*
D Horchposten *m.*
R пост подслушивания
   м.
A مَرْكَز تَصَنُّت *m*

4229

(lookout) post

F poste de surveillance *f.*
E puesto de observación
   m.
D Beobachtungsposten
   m.; Spähposten *m.*
R наблюдательный пост
   м.
A مركز الكشّاف

4230

**(observation) post
(OP)**

F observatoire *m.;* poste
   d'observation *m.*
E observatorio *m.*
D Beobachtungstelle (B-
   Stelle) *f.*
R наблюдательный
   пункт (НП) *м.*
A مَرْكَز رَصْد *m*

4231 [R]

(reporting) post

F poste de détection *m.*
E estación de detección *f.*
D Ortungsstelle *f.*
R пункт оповещения *м.*
A مَرْكَز الإتِّصَال

4232

(sign) post

F poteau indicateur *m.*
E indicador (de dirección)
   m.
D Anweisungsschild *neut.*
R указатель маршрута
   м.
A عُمود الدَلَّلة

4233

(traffic) post
F poste de régulation *m.*
E punto de regulación *m.*
D Verkehrsleitpunkt *m.*
R пост регулирования
  движения *м.*; пост
  регулировщиков *м.*
A مَرْكَز المرور

4234

posting
F affectation *f.*
E destino *m.*
D Versetzung *f.*
R перевод *м.*
A التعيين في . . .

4235

(service) post office
F vaguemestre *m.*
E correos *m.*
D Postdienst
R полевая почта *ж.*
A مكتب البريد

4236

**(regimental) post
  orderly**
F vaguemestre *m.*
E suboficial de correos *m.*
D Postbote (des
  Bataillons) *m.*
R дневальный почтовой
  конторы *м.*
A موزع البريد

4237

(kill) potential
F potentiel de destruction
  m.
E capacidad (de
  destrucción) *f.*
D Abschuss- *komp.*;
  Zerstörungs- *komp.*; -
  Möglichkeit *komp.*
R способность (к
  поражению) *ж.*
A احتمالية القتل

4238

(ammunition) pouch
F cartouchière
E cartuchera *f.*
D Patronentasche *f.*
R патронная сумка *ж.*
A جَعْبَة (الخَرْطُوش) جِراب
  m (الذخيرة)

4239 [E,P,U]

power *n.*
F pouvoir [U] *m.*;
  puissance [EP] *f.*
E poder [U] *m.*; potencia
  [EP] *f.*
D Leistung [E] *f.*; Macht
  [PU] *f.*
R сила [U] *ж.*;
  мощность [E] *ж.*;
  власть [P] *ж.*
A قُدْرَة ـ طاقة *f*

4240

**(armor piercing or
  penetrating)
  power**
F pouvoir perforant *m.*
E potencia de penetración
  f.
D Durchschlagskraft *f.*
R (пробивная) сила *ж.*
A قدرة خرق الدروع

4241

(brake horse) power
F puissance au frein *f.*
E potencia al freno *f.*
D Pferdestärke *f.*
R индикаторная
  мощность *ж.*
A القدرة المكبحية

4242

powerboat
F vedette *f.*
E lancha motora *f.*;
  motorbote *m.*
D Motorboot *neut.*
R моторный катер *м.*
A زَوْرَق ذُو مُحَرَّك

4243

powered
F (à) moteur; automoteur
E automóvil; con motor;
  mecánico
D Kraft- *komp.*; Motor-
  komp.
R моторный;
  самоходный
A ذُو مُحَرَّك

4244

power-weight ratio
F puissance *f.*; spécifique
  rapport puissance
  poids *m.*
E potencia específica *f.*
D Leistungsgewicht *neut.*
R удельная мощность
  ж.
A نِسْبَة القُدْرَة إلى الوَزْن

4245

practice *adj.*
F (d')entraînement;
  (d')exercice;
  (d')instruction
E (de) ejercicio
D Übungs- *komp.*
R учебный
A لِلتَّمْرين

4246

practice *v.*
F (s')exercer(à)
E ensayarse (a)
D ausüben *tr.*; üben
R тренировать(ся)
A تَمَرَّن

4247

prearranged
F préparé à l'avance
E convenido; previsto
D vorbereitet; vereinbart
R плановый; заранее
  подготовленный
A مُدَبَّر

4248 [R]

precedence
F préséance *f.*; priorité *f.*
E precendencia *f.*
D Dringlichkeitsstufe *f.*
R степень срочности *ж.*
A حَقّ التَقَدُّم

4249

precession
F précession *f.*
E precesión *f.*
D Präzession *f.*
R прецессия *ж.*
A *m* سَبْق

4250

precision
F précision *f.*
E precisión *f.*
D Präzision *f.*
R точность *ж.*
A دقة

4251 [S]

predictor
F appareil de préparation de tir *m.*
E predictor *m.*
D Feuerleitrechengerät *neut.*
R упредительный механизм *м.*
A جهاز التنبؤ

4252

preemptive
F préventif
E preventivo
D vorbeugend *adj.*; Vorbeugungs- *komp.*
R упреждающий
A بالشُفْعَة

4253

preflight *adj.*
F avant le vol
E preparatorio de vuelo
D vor dem Flug
R предполётный
A قَبْل الطَيْرَان

4254

preliminary *adj. n.*
F préliminaire *adj. n. m.*
E preliminar *adj. s. m.*
D Vor- *komp.*; vorläufig *adj.*; Vorbereitung *s. f.*
R предварительный
A تَمْهِيدِي

4255 [S]

premature (explosion) *n.*
F explosion prématurée *f.*
E explosión prematura *f.*
D Frühzünder *m.*
R преждевременный (взрыв) *м.*
A إِنْفِجَار سَابِق لِلأَوَان

4256 [U]

preparation
F préparation *f.*
E preparación *f.*
D Vorbereitung *f.*
R подготовка *ж.*
A *m* تَحْضِير

4257 [S]

preparation
F tir de préparation *m.*
E tiro de preparación *m.*
D Vorbereitungsfeuer *neut.*
R (артиллерийская) подготовка *ж.*
A *m* إِعْدَاد

4258

prepare
F préparer
E preparar
D vorbereiten *tr.*
R приготовлять(ся)
A أَعَدّ

4259

preplanned
F (à) temps; prévu
E previsto
D (im) voraus geplant
R (заранее) спланированный
A مُدَبَّر سَابِقاً

4260

pre-position *v.*
F mettre en place (préalablement)
E situar de antemano
D (im) voraus in den Einsatzraum bringen
R заранее устанавливать
A وَضَع تَمْهِيدِيّا

4261 [H]

prescribe
F prescrire (à)
E recetar
D verschreiben
R предписывать
A وَصَف

4262 [H]

prescription
F ordonnance *f.*
E receta *f.*
D Verordnung *f.*
R предписание *ср.*; рецепт *м.*
A *f* وَصْفَة

4263

present *adj.*
F actuel; présent
E actual; presente
D answesend; gegenwärtig
R настоящий; присутствующий
A حَالِي

4264

presentation
F exposition *f.*
E exposición *f.*
D Darstellung *f.*
R представление *ср.*
A *m* عَرْض

4265
preset *v.*
F pré-régler
E arreglar por adelantado
D vorhereinstellen *tr.*
R заранее наводить;
    программировать
A سَبَق ضَبْطُه

4266
preset *adj.*
F pré-réglé
E preestablecido
D vorprogrammiert
R заранее наведённый;
    запрограммиров-
    анный
A ضَبَط مُتَقَدِّماً

4267 [A,L,M]
press *v.*
F serrer de près
E acosar
D drängen
R давить; сильно
    теснить
A ضَغَط . شَدَّد

4268
press *adj. n.*
F (de) presse (*adj.*) *n. f.*
E (de) prensa (*adj*) *s. f.*
D Presse(-) *f.* (*komp.*)
R пресса *ж.*; газетный
    прилаг.
A الصَّحافة *f* صُحُفِي

4269
press ahead
F presser le pas
E apretar el paso
D weiterdrängen *tr.*
R продвигаться;
    стремиться
A دَفَع إلى الأَمام

4270
press for
F insister sur
E presionar (sobre, por)
D drängen auf
R требовать; настаивать
    (на)
A سَرَّع

4271
pressure
F pression *f.*
E presión *f.*
D Druck *m.*
R давление *ср.*; сжатие
    ср.; нажим *м.*
A ضَغْط *m*

4272
pressure altitude
F altitude-pression *f.*
E altura barométrica
    standard *f.*
D Druckhöhe *f.*
R барометрическая
    высота *ж.*; высота
    по давлению *ж.*
A الإِرْتِفاع الضَّغْطِي *m*

4273
pressure suit
F vêtement de
    pressurisation *m.*
E traje de presión *m.*
D Druckanzug *m.*
R герметичный костюм
    м.; высотный
    скафандр *м.*
A لِباس الضَّغْط *m*

4274 [A]
pressurized cabin
F cabine étanche *f.;*
    cabine pressurisée *f.*
E cabina a presión *f.*
D Druckkabine *f.*
R гермокабина *ж.*
A مَقْصُورَة مُكَيَّفَة الضَّغْط

4275 [B]
prestressed concrete
F béton précontraint *m.*
E hormigón pretensado
    m.
D Spannbeton *m.*
R напряжённый бетон *м.*
A خرسانة مسلحة تحت الضغط

4276
prevent
F empêcher
E impedir
D verhindern
R предотвращать
A مَنَع

4277
prevention
F prévention *f.*
E prevención *f.*
D Vorbeugung *f.*
R предотвращение *ср.*
A مَانِع *m*

4278
preventive medicine
F médecine préventive *f.*
E medicina preventiva *f.*
D Gesundheitsvorsorge *f.*
R профилактическая
    медицина *ж.*
A الطِّب الوِقائِي *m*

4279
primary cell
F pile *f.*
E pila *f.*
D Primärzelle *f.*
R первичный элемент
    м.; клетка *ж.*
A خَلِيَّة ابتِدائِيَّة *f*

4280
prime *v.*
F amorcer
E cebar
D füllen; laden
R воспламенять;
    подкачивать
A وضع الكبسولة

4281 [E,S,V]
prime *v.*
F amorcer
E cebar
D scharfmachen [S] *tr.*;
   (eine Pumpe)
   anlassen [E] *tr.*;
   (Benzin) einspritzen
   [V] *tr.*
R подкачивать [EV];
   заряжать [S]
A وَضَع الكُبْسُولَة

4282
prime mover
F tracteur *m.*
E tractor *m.*
D Zugmaschine *f.*
R тягач *м.*
A مُحَرِّك اسَاسِي *m*

4283 [S]
primer
F amorce *f.*
E cebo *m.*
D Sprengkapsel *f.*
R капсюль *м.*;
   (воспламенительная)
   трубка *ж.*
A كُبْسُولَة *f* . شَعِيلَة *f*

4284
print *n.*
F épreuve *f.*; tirage *m.*
E positiva *f.*
D Abzug *m.*
R опечаток *м.*; печать *ж.*
A نُسْخَة مَطْبُوعَة *f*

4285 [D,R]
printout
F données de sortie *f. pl.*
E datos de salida *m. pl.*
D Ausdruck *m.*
R распечатка *ж.*
A طَبْع *m*

4286
(order of) priority
F (ordre de) priorité *f.*
E (orden de) prioridad(es)
   f.
D Dringlichkeitsreihenfolge
   f.
R очередность *ж.*;
   очередь срочности
   ж.
A (نِظَام) الأُولُويات

4287 [S]
priority of fires
F priorité des feux *f.*
E prioridad de fuegos *f.*
D Beschusspriorität *f.*
R распределение целей
   ср.
A اولوية السند بالبنزين

4288
priority of support
F priorité d'appui et de
   soutien *f.*
E prioridad de apoyo *f.*
D Versorgungsvorrang *m.*
R очерёдность
   обеспечения огнём *ж.*
A اولوية السند

4289 [S]
priority of target
F objectif de priorité
   absolue *m.*
E objetivo prioritario *m.*
D Vorrangziel *neut.*
R главная цель *ж.*
A الهدف الرئيسي

4290
prior permission
F autorisation préalable *f.*
E autorización previa *f.*
D Vorausgenehmigung *f.*
R прежнее разрешение *ср.*
A إجَازَة سَابِقَة *f* / إذن
   سابق *m*

4291
prison
F prison *f.*
E prisión *f.*
D Gefängnis *neut.*
R тюрьма *ж.*
A سِجْن *m*

4292 [P]
prisoner
F prisonnier *m.*
E detenido *m.*
D Arrestant *m.*
R заключённый *м.*
A سَجِين *m*

4293
(take) prisoner *v.*
F (faire) prisonnier
E tomar prisionero;
   apresar
D gefangennehmen *tr.*
R брать в плен
A أسر

4294
**prisoner of war
   (POW)**
F prisonnier de guerre *m.*
E prisionero de guerra *m.*
D Kriegsgefangene(r) *m.*
R военнопленный *м.*
A أسِير *m*

4295
prisoner of war camp
F camp de prisonniers de
   guerre *m.*
E campamento de
   prisioneros de guerra
   m.
D Kriegsgefangenenlager
   neut.
R лагерь военнопленных
   м.
A مُعَسْكَر أُسْرَى الحَرْب

4296

probability
F probabilité *f.*
E probabilidad *f.*
D Wahrscheinlichkeit *f.*
R вероятность *ж.*
A إِحْتِمَال *m*

4297

probable error
F écart probable *m.*
E desvío probable *m.*
D wahrscheinliche Abweichung
R вероятное отклонение *ср.*
A الْخَطَأ الْمُحْتَمَل *m*

4298 [A]

(in-flight refuelling) probe
F perche de ravitaillement en vol *f.*
E percha de reabastecimiento *m.*
D Flugauftankung *f.*
R (заправочный) щуп *м.*
A مسبار للتموين بالوقود في الطيران

4299 [U]

procedure
F procédé *m.*
E procedimiento *m.*
D Verfahren *neut.*
R метод *м.*; процедура *ж.*
A سِيَاق *m*

4300 [R]

procedure
F procédure (radio) *f.*
E código radiotelefónico *m.*
D Funkbetriebsverfahren *neut.*
R радиосвязная работа *ж.*
A أُسْلُوب *m*

4301

(standard operating) procedures (SOPs)
F instructions permanentes *f. pl.*
E procedimientos operativos normales *m. pl.*
D ständige Dienstanweisung *f.*
R уставной порядок *м.*
A الاوامر المستديمة

4302 [A]

procedure turn
F virage conventionnel *m.*
E viraje convencional *m.*
D Umkehrflug *m.*
R очередной вираж *м.*; поворот *м.*
A إِنْعِطَاف مُدَبَّر *m*

4303

proceed to a location *v.*
F (se) déplacer vers
E ir a una posición
D vorgehen *tr.*
R продвигаться на местонахождение
A إتجه إلى . . . *A*

4304

process *v.*
F traiter
E tratar
D bearbeiten; verarbeiten
R обрабатывать
A حوَّل . عَالِج

4305 [P]

process *v.*
F trier
E registrar
D registrieren
R оформлять
A حوَّل

4306

process *n.*
F processus *m.*
E proceso *m.*
D Vorgang *m.*
R процесс *м.*
A طَرِيقَة *f*

4307 [D,E,P,U]

processing
F traitement [DEU] *m.;* triage [P] *m.*
E filiación [P] *f.;* tratamiento [DEU] *m.*
D Verarbeitung [DE] *f.;* Registrierung [P] *f.;* Bearbeitung [U] *f.*
R обработка [DEU] *ж.;* оформление [P] *ср.;* подготовка *ж.*
A تَحْوِيل . مُعَالَجَة *f m*

4308

(automatic data) processing (ADP)
F traitement automatique des données *m.;* informatique *f.*
E procesamiento automático de datos *m.*
D automatische Datenverarbeitung *f.*
R обработка данных *ж.*
A تَحْلِيل الْمُعْطِيَات تِلْقَائِياً

4309 [I]

(information) processing
F traitement du renseignement *m.*
E procesamiento *m.*
D Informationsbearbeitung *f.*
R обработка сведений *ж.*
A تحويل المعلومات

273

4310
(word) processing
F traitement des mots *m.*
E procesar información *v.*
D Sprachenverarbeitung *f.*
R обработка слов *ж.*
A تسجيل الكلمات في حاسبة آلية

4311
procure
F approvisionner
E adquirir
D beschaffen
R доставлять;
обеспечивать
A نال / حصل

4312
procurement
F approvisionnement *m.*
E adquisición *f.*
D Beschaffung *f.*
R заготовка и получение
ж. ср.
A نيل / حُصول *m*

4313 [E]
produce
F fabriquer
E fabricar
D herstellen *tr.*
R производить
A أنْتَج

4314
production
F production *f.*
E producción *f.*
D Produktion *f.*
R производство *ср.;*
изготовление *ср.*
A إنْتاج *m*

4315
proficiency
F compétence *f.*
E habilidad *f.*
D Fertigkeit *f.*
R умение *ср.*
A كفاية

4316 [P]
profile
F profil *m.*
E descripción detallada *f.*
D Kurzbiographie *f.*
R профиль *м.*
A صورة جانبية

4317
proforma
F formulaire *m.*
E hoja *f.*
D Formular *neut.*
R плановые донесения
ср. мн. ч.
A صِيغَة *f*

4318 [D,S,U]
program *n.*
F programme *m.*
E programa *m.*
D Programm *neut.*
R программа *ж.*
A بَرْنامَج *m*

4319 [D]
programer
F programmeur *m.*
E programador *m.*
D Programmierer *m.*
R програмист *м.*
A مدير البرنامج

4320 [D]
programing language
F langage de
programmation *m.*
E lenguaje de
programación *m.*
D programmieren; ein
Programm aufstellen
R програмный язык *м.;*
оформление
программы *ср.*
A ادارة البرنامج

4321 [L]
progress *v.*
F avancer
E avanzar
D vorrücken
R продвигаться
A تَقَدَّم

4322 [U]
progress *v.*
F (faire le) suivi (de)
E perseguir
D bearbeiten
R развиваться
A تَقَدَّم

4323
progress *n.*
F progrès *m. pl.*
E progresos *m. pl.*
D Fortschritt *m.*
R прогресс *м.;* развитие
ср.
A تَقَدُّم *m*

4324 [P]
progressive *n.*
F progressiste *m. f.*
E progresista *m. f.*
D Progressive(r) *m. f.*
R прогрессивный
A مُؤْمِن بالاصْلاح الإجْتِماعي

4325 [U]
progressive *adj.*
F progressif
E progresivo
D fortschreitend
R прогрессивный
A تَقَدُّمِي *m*

4326
prohibit
F interdire
E prohibir
D verbieten
R запрещать
A حَظَر

4327 [A,M]
prohibited area
F zone interdite *f.*
E zona prohibida *f.*
D Sperrgebiet *neut.*
R запретная зона *ж.*
A مِنْطَقَة مُحَرَّمَة *f*

4328
project *n.*
F projet *m.*
E proyecto *m.*
D Projekt *neut.;*
Vorhaben *neut.*
R проект *м.;* план *м.*
A مَشْرُوع *m*

4329
projectile
F projectile *m.*
E proyectil *m.*
D Geschoss *neut.*
R снаряд *м.*
A مَقْذُوف *m*

4330
projecting
F (en) saillie
E saliente
D vorspringend
R выступающий;
выдающийся
A بَارِز

4331
projection
F saillie *f.*
E saliente *m.*
D Vorsprung *f.*
R выступ *м.;*
проектирование *ср.*
A بُرُوز ـ قَذْف *m*

4332
prolong
F prolonger
E extender (espacio);
prolongar (tiempo)
D verlängern
R продолжать;
продлевать
A أطال

4333
prolongation
F prolongement (espace)
m.; prolongation
(temps) *f.*
E extensión (espacio) *f.;*
prolongación (tiempo)
f.
D Verlängerung *f.*
R продление *ср.;*
продолжение *ср.*
A إطَالَة *f*

4334 [T]
prominent feature
F point haut *m.*
E rasgo dominante *m.*
D markanter
Geländepunkt *m.*
R местный предмет *м.*
A عارضة بارزة *f*

4335 [T]
promontary
F promontoire *m.*
E promontorio *m.*
D Vorgebirge *neut.*
R мыс *м.*
A رعن *m*

4336
promotion
F promotion *f.*
E ascenso *m.*
D Beförderung *f.*
R присвоение звания *ср.*
A ترقية *f*

4337
propaganda
F propagande *f.*
E propaganda *f.*
D Propaganda *f.*
R пропаганда *ж.*
A دِعَايَة *f*

4338 [A,E,M,V]
propel *v.*
F commander [E];
propulser [AMV]
E impulsar [E]; propulsar
[AMV]
D treiben
R продвигать (вперёд);
толкать
A دَفَع

4339
propellant *n.*
F propergol *m.*
E propulsor *m.*
D Treibstoff *m.*
R метательное BB *ср.*
A دَافِعَة *f*

4340
propeller
F hélice *f.*
E hélice *f.*
D Propeller *m.*
R движитель *м.;*
(гребной) винт *м.*
A مِرْوَحَة *f*

4341
propelling charge
F charge propulsive *f.*
E carga de proyección *f.*
D Treibladung *f.*
R метательный заряд *м.*
A شُحْنَة دَافِعَة *f*

4342
propulsion
F propulsion *f.*
E propulsión *f.*
D Antrieb *m.*
R продвижение *ср.;*
движущая сила *ж.*
A دَفْع *m*

275

4343 [P]

prosecute *v.*
F poursuivre
E procesar
D gerichtlich vorgehen *tr.*
R вести судебное
  преследование
A قاضى

4344 [P]

prosecutor
F plaignant *m.;* procureur
  m.
E acusador *m.*
D Anklagevertreter *m.*
R прокурор *м.*
A النائب العام

4345

protect (against)
F protéger (contre)
E proteger (contra)
D abschirmen *tr.;*
  schützen (vor)
R защищать (от)
A حَمَى (من)

4346 [A,M,U,V]

protection
F blindage [AMV] *m.;*
  défense[M] *f.;*
  protection *f.*
E blindaje [AMV] *m.;*
  protección [U] *f.*
D Abschirmung *f.;*
  Panzerung [AMV] *f.;*
  Schutz *m.*
R защита *ж.;* охранение
  ср.
A حِمَايَة *f*

4347

protective
F protecteur
E protector(a)
D Schutz- *komp.*
R защитный
A واقٍ

276

4348

protective clothing
F vêtement protecteur *m.*
E ropa protectora *f.*
D Schutzkleidung *f.*
R защитная одежда *ж.*
A مَلَابِس وَاقِيَة *f*

4349

protocol
F protocole *m.*
E protocolo *m.*
D Protokoll
  (Verhandlungs-
  protokoll) *neut.*
R протокол *м.*
A مراسم

4350

proton
F proton *m.*
E protón *m.*
D Proton *neut.*
R протон *м.*
A بروتون *m*

4351

prototype *n.*
F prototype *m.*
E prototipo *m.*
D Ausgangsbautyp *m.*
R прототип *м.*
A الطراز البدئي

4352

protractor
F rapporteur *m.*
E transportador *m.*
D Gradbogen *m.*
R транспортир *м.;*
  протрактор *м.*
A مِنْقَلَة *f* مقياس الزوايا *m*

4353

provide
F fournir
E proporcionar; proveer
D versorgen
R снабжать;
  обеспечивать;
  доставлять
A زَوَّد

4354

provisional
F temporaire
E provisional
D vorläufig
R временный; условный
A مُوَقَّت

4355

provisions *pl.*
F vivres *m. pl.*
E víveres *m. pl.*
D Proviant *m.*
R продукты питания *м.*
  мн. ч.
A أرْزَاق *f* ـ تَموين *m*

4356

provost *adj.*
F prévôtal
E prebostal
D Feldjäger- *komp.*
R военный полицейский
  м.
A شُرْطَة عَسْكَرِيَّة *f*

4357 [P]

provost marshal
F prévôt *m.*
E capitán preboste *m.*
D Kommandeur (der
  Militärpolizei) *m.*
R начальник военной
  полиции *м.*
A قائد الشرطة العسكرية *m*

4358

prowler guard
F garde mobile *f.*
E guardia de ronda *m. pl.*
D Streifwache *f.*
R патрулирующий
  часовой *м.*
A حارس جائل *m*

4359

proword
F terme de procédure *m.*
E palabra de servicio *f.*
D Betriebswort *neut.*
R условное слово *ср.*
A كَلِمَة مُقَدَّمَة *f*

4360

psychiatric
F psychiatrique
E psiquiátrico
D psychiatrisch
R психиатрический
A نَفْسِي

4361

psychological
F psychologique
E psicológico
D psychologisch
R психологический
A مُخْتَصّ بِعِلْم النَّفْس

4362

public *adj. n.*
F public *adj. n. m.*
E público *adj. s. m.*
D öffentlich *adj.;*
   Öffentlichkeit *s. f.*
R общественный;
   публичный
A عَامّ . الجُمْهُور *m*

4363

public information
F information publique *f.*
E información pública *f.*
D öffentliche Information
   f.
R общественная
   информация *ж.*
A مَعْلُومَات عَامَّة *f*

4364

public relations (PR)
F public-relations *f. pl.*
E relaciones públicas *f. pl.*
D Beziehungen zur
   Öffentlichkeit *f. pl.*
R общественная
   информация *ж.*
A علاقات عامة

4365

public utilities
F services publics *m. pl.*
E servicios públicos *m. pl.*
D öffentliche
   Einrichtungen *f. pl.*
R коммунальные
   сооружения *ср. мн.*
   ч.; услуги *ж. мн. ч.*
A خِدَمَات عَامَّة *f*

4366 [M,S,U]

pull *v.*
F presser [S]; ramer[M];
   tirer[U]
E apretar [S]; remar[M];
   tirar[U]
D abdrücken [S];
   rudern[M]; ziehen[U]
R тянуть; тащить
A سَحَب

4367 [B,E,M]

pulley
F moufle [BM] *f.;* poulie
   [E] *f.*
E apareo [BM] *m.;* polea
   [E] *f.*
D Flaschenzug [BM] *m.;*
   Rolle [E] *f.*
R шкив *м.;* ролик *м.;*
   блок *м.*
A بَكْرَة *f*

4368 [L]

pull out *v.*
F (se) retirer
E retirar(se)
D zurückziehen *tr.*
R отступать
A سحب

4369

pull-ring
F anneau de la goupille
   d'une grenade *m.*
E aro de seguridad *m.*
D Ziehring *m.*
R вытяжное кольцо *ср.*
A حلقة النزع

4370 [S]

pull-through
F ficelle de nettoyage *f.*
E baquetón *m.*
D Reinigungskette *f.*
R протирка *ж.*
A حبل التنظيف

4371 [A]

pull up
F cabrer
E encabritar
D abheben *tr.*
R резко подниматься
A صَعَد

4372 [D,R]

pulse *adj.*
F (à) impulsions
E (a, de) impulsos *m.*
D Impuls- *komp.*
R импульсный
A نَبْضِي

4373 [H,D,R]

pulse *n.*
F pouls [H] *m.;* impulsion
   [DR] *f.*
E pulso *m.*
D Impuls [DR] *m.;* Puls
   [H] *m.*
R импульс *м.*
A نَبْض *m*

4374

pulse-jet engine
F pulsoréacteur *m.*
E pulsoreactor *m.*
D IL-Triebwerk
   (intermittierende
   Luftstrahl-) *neut.*
R пульсирующий
   воздушно-
   реактивный
   двигатель (ПВРД)
   м.
A مُحَرِّك نَافُوري نَبْضِي *m*

4375
pump *n.*
F pompe *f.*
E bomba *f.;* pompa *f.*
D Pumpe *f.*
R насос *м.*
A مِضَخَّة *f*

4376
pump *v.*
F pomper
E bombear; sacar (etc.)
   con una bomba
D pumpen
R накачивать; качать
   насосом
A ضَخَّ

4377 [S]
pump-action *adj.*
F (fusil à) chargement par
   levier
E (de) repetición
D Stosswirkung *f.*
R полуавтоматический
A بندقية تعمل كمنفاخ الدراجة

4378
pump up
F gonfler
E inflar
D aufpumpen *tr.*
R накачивать; надувать
A (ضَخَّ) وَرَفَعَ بِمِضَخَّة

4379 [V]
puncture *n.*
F crevaison *f.*
E pinchazo *m.*
D Reifenpanne *f.*
R прокол *м.;* пробой *м.*
A ثَقْب *m*

4380
punish *v.*
F punir
E castigar
D bestrafen
R наказывать
A عاقب

4381
punishable
F punissable
E castigable
D strafbar
R наказуемый
A يُعَاقَب

4382
punishment
F punition *f.*
E castigo *m.*
D Strafe *f.*
R наказание *ср.*
A عقوبة

4383
punitive
F punitif
E punitivo
D strafend
R карательный
A عقابي

4384 [E]
purchase
F prise *f.*
E agarre *m.;*
   apalancamiento *m.*
D Hebelkraft *f.;*
   Hebevorrichtung *f.*
R точка опоры *ж.*
A (قوة) قَبْضَة *f*

4385
pursue
F poursuivre
E (dar) caz [AM];
   perseguir
D verfolgen
R преследовать
A لَاحَق

4386
pursuit
F poursuite *f.*
E caza *f.*
D Verfolgung *f.*
R преследование *ср.*
A مُلَاحَقَة *f*

4387
push *n.*
F avance *f.;* poussée *f.*
E avance *m.;* empuje *m.*
D Stoss *m.;* Vorstoss *m.*
R толчок *м.;* удар *м.*
A دَفْعَة *f*

4388
pushbutton *n.*
F poussoir *m.*
E pulsador *m.*
D Drucktaste *f.*
R кнопка *ж.*
A زِرّ كَبَّاس *m*

4389 [L]
push on
F pousser
E avanzar
D vorstossen *tr.*
R спешить; ускорять
A دَفَع إلى الأَمَام

4390
puttee
F bande molletière *f.*
E polaina *f.*
D Wickelgamasche *f.*
R обмотка *ж.*
A جُرْمُوق . لفافة الساق

4391
pyrotechnics *pl.*
F artifices *m. pl.*
E artificios pirotécnicos
   m. pl.
D pyrotechnische
   Gegenstände *m. sing.*
R пиротехнические
   вещества *ср. мн. ч.*
A سِهَام نارِيَّة *f*

Q

4392 [S]
quadrant *n.*
F quadrant *m.*
E nivel de puntería *m.*
D Quadrant *m.*
R квадрант *м.*
A مِسْوَاة التَّسْدِيد

4393 [S]
quadrant elevation
F angle au niveau *m.*
E ángulo de tiro *m.*
D Aufsatzwinkel *m.*
R угол возвышения *м.*
A زَاوِيَة المِحْوَر

4394
qualification
F qualification *f.;* titre *m.*
E título *m.*
D Befähigung *f.;* Qualifikation *f.*
R квалификация *ж.;* специальность *ж.*
A أَهْلِيَّة *f*

4395
qualified
F breveté
E titulado
D geeignet; qualifiziert
R квалифицированный
A مُؤَهَّل

4396
quality assurance
F contrôle de qualité *m.*
E control de calidad *m.*
D Qualitätskontrolle *f.*
R учёт (высокого) качества *м.;* обеспечение (высокого) качества *ср.*
A مُرَاقَبة الجَوْدَة

4397 [E]
quantum
F quantum *m.*
E quantum *m.*
D Quant(en) *neut.*
R квант *м.;* квантовый прилаг.
A كَمّ *m*

4398
(in) quarantine
F (en) quarantaine *f.*
E (en) cuarentena *f.*
D (unter) Quarantäne *f.*
R карантин *м.;* (в) карантине
A (في) العَزْل *m*

4399 [L,U]
quarter *v.*
F cantonner
E acuartelar
D einquartieren *tr.*
R размещать; помещать
A أَقَام . أَسْكَن

4400
quarterdeck
F plage arrière *f.*
E alcázar *m.*
D Achterdeck *neut.*
R ют *м.;* шканцы *м. мн. ч.*
A طَرَف المُؤَخَّر . مركز الضباط *m*

4401 [L]
quartermaster
F intendant *m.*
E intendente *m.*
D Quartiermeister *m.*
R квартирмейстер *м.;* интендант *м.*
A ضَابِط تَمْوِين *m*

4402 [M]
quartermaster
F maître timonier *m.*
E timonero *m.*
D Steuermann
R старшина-рулевой *м.*
A مُدِير السُّكَّان في السَّفِينَة

4403
quartermaster corps
F service de l'intendance *m.*
E servicio de intendencia *m.*
D Nachschubtruppe *f.*
R интендантская служба *ж.*
A سِلَاح التَّمْوِين *m*

4404 [L]
quarters *pl.*
F logement *m.*
E alojamiento *m.*
D Quartier *neut.*
R квартиры *ж. мн. ч.;* размещение *ср.*
A تُكْنَة . مَسَاكِن *f*

4405 [M]
(crew's) quarters
F poste d'équipage *f.*
E camareta (de la tripulación) *f.*
D Quartier *neut.;* (Mannschafts-) Unterkunft *f.*
R помещение *ср.*
A مراكز النوتية

4406 [L]
(general) quarters
F branle-bas de combat *m.*
E alerta general *m.*
D Einsatzalarm *m.;* Bereitschaftsalarm *m.*
R боевые посты *м. мн. ч.*
A استعداد للقتال

4407
quay
F quai *m.*
E muelle *m.*
D Kai *m.*
R набережная *ж.;* причал *м.;* мол *м.*
A رَصِيف *m*

279

4408

quell a riot

F réprimer une émeute
E dominar un disturbio
D unterdrücken (einen
 Aufstand) *untr.*
R подавлять мятеж
A قمع

4409 [E,R,U]

quench

F amortir [RU];
 éteindre[U];
 tremper[E]
E apagar [RU]; templar
 [E]
D abschrecken [E];
 löschen [RU]
R гасить; тушить
A أَطْفَأَ

4410 [A,M]

quick-flashing light

F feu scintillant *m.*
E luz centelleante *f.*
D Funkelfeuer *neut.*
R частопроблесковый
 огонь *м.*
A مِصْبَاح سَرِيع الوَمِيض

4411

quicksand

F sable(s) mouvant(s) *m.*
 (pl.)
E arena movediza *f.*
D Treibsand *m.*
R зыбучие пески *м. мн.*
 ч.
A رَمْل مُتَحَرِّك *m*

4412

quick time

F pas accéléré *m.*
E ritmo de paso *m.*
D schnelles Marschtempo
 neut.
R походный шаг *м.*
A خطى متسارعة *f*

4413

quota

F quote-part *f.*
E cuota *f.*
D Anteil *m.*
R доля *ж.;* квота *ж.;*
 норма *ж.*
A نِصَاب *m*

4414

quotient

F quotient *m.*
E cociente *m.*
D Quotient *m.*
R коэффициент *м.*
A خَارِج القِسْمَة

R

4415

(ammunition) rack

F magasin à munitions *m.*
E estante de munición *m.*
D Munitionsständer *m.*
R стеллаж *м.*
A مسند الذخيرة *m*

4416

(arms) rack

F ratelier d'armes *m.*
E armero *m.*
D Waffenständer *m.*
R стойка *ж.*
A مسند الأسلحة *m*

4417 [A]

(rocket) rack

F pylône pour roquettes
 m.
E jaula (de cohetes) *f.*
D Raketenrüstsatz *m.*
R ракетодержатель *м.*
A حامل صاروخ *m*

4418 [N]

rad

F rad *m.*
E rad *m.*
D Rad *neut.*
R рад *м.*
A وَحْدَة قِيَاس الجُرْعَة
 الشُّعَاعِيَّة *f*

4419

radar *adj. n.*

F radar *adj. inv. n. m.*
E radar
D Radar(-) *s. m. (komp.)*
R радиолокационный
 прилаг.;
 радиолокационная
 станция (РЛС) *ж.*
A رادار *m*

4420

(acquisition) radar

F radar d'acquisition *m.*
E radar de adquisición *m.*
D Radarauffassung *f.*
R радиолокатор
 обнаружения цели
 м.
A رادار التفتيش

4421

**(advance early
 warning) radar**

F radar d'alerte avancée
 m.
E radar avanzado de
 alerta previa *m.*
D Frühwarnradar *m.*
R радиолокационная
 станция (РЛС)
 дальнего
 обнаружения *ж.*
A رادار الانذار

4422

**(continuous wave
 acquisition) radar**

F radar d'acquisition à
 ondes entretenues *m.*
E radar de adquisición de
 onda continua *m.*
D standige
 Radarwellenauffassung
 f.
R радиолокатор с
 непрерывным
 излучением *м.*
A رادار تفتيش ذوموجات
 مضمنة

4423

(Doppler) radar

F radar Doppler *m.*

E radar Doppler *m.*

D Dopplerradar *neut.*

R доплеровская радиолокационная станция (РЛС) *ж.*

A رادار دُوبْلِر *m*

4424

(frequency jumping) radar

F radar à fréquence variable *m.*

E radar con saltos de frecuencia *m.*

D frequenzhüpfender Radar *m.*

R частото-поворотливая РЛС *ж.*

A رادار قافز الترددات

4425

(look-down) radar

F radar d'observation verticale *f.*

E radar vertical *m.*

D Abwärtssichtradar *neut.*

R РЛС вертикального обзора

A رَادَار طَائِرَة يراقب نَحْوَ الأَرْض

4426

(mortar locating) radar

F radar anti-mortier *m.*

E radar de localización de morteros *m.*

D Mörserortungsradar *m.*

R контрминомётный радиолокатор *м.*

A رادار تحديد أماكن الهاونات

4427

(phased array) radar

F radar à balayage électronique *m.*

E radar de sistema de fases *m.*

D elektronischer Suchradar *m.*

R РЛС с фазово-антенной системой *ж.*

A رادار متطاور الصف

4428

(search) radar

F radar de recherche *m.*

E radar de exploración *m.*

D Suchradar *m.*

R поисковый радиолокатор *м.*

A رادار باحث

4429

(sidelooking airborne) radar (SLAR)

F radar aéroporté à antenne latérale *m.*

E radar aerotransportado de lóbulo oblicuo *m.*

D seitlich gerichteter Bordradar *m.*

R воздушная РЛС бокового действия *ж.*

A رادار جوي يبحث إلى الجانب

4430

(surveillance) radar

F radar de veille *m.*

E radar de vigilancia *m.*

D Überwachungsradar *m.*

R обзорный радиолокатор *м.*

A رادار الرصد

4431

radar camouflage

F camouflage-radar *m.*

E camuflaje radar *m.*

D (Radar-)Tarnung *f.*

R противорадиолокационная маскировка *ж.*

A تَمْوِيه الرَّادَار

4432

radar clutter

F interférence radar *f.*

E interferencia radar *f.*

D Radarstörung *f.*

R местные помехи *ж. мн. ч.*

A تَشْوِيش الرَّادَار

4433

radar countermeasures *pl.*

F contre-mesures radar *f. pl.*

E contramedidas radar *f. pl.*

D Radar-Gegenmassnahmen *f. pl.*

R противорадиолокационные меры *ж. мн. ч.*

A تَدَابِير الرَّادَار المُضَادَّة *f*

4434

radar coverage

F couverture radar *f.*

E cobertura radar *f.*

D Radarerfassungsbereich *m.*

R радиолокационное наблюдение *ср.*

A تَغْطِيَة الرَّادَار *f*

4435

radar echo

F écho radar *m.*

E eco radar *m.*

D Radarecho *neut.*

R отражённый (радиолокационный) сигнал *м.*

A صدى الرَّادَار *m*

4436

radar locating
F détection radar *f.*
E localización radar *f.*
D Radarpeilung *f.*
R радиолокация *ж.*
A تحديد المكان بالرادار

4437

radar rangefinding
F télémétrie radar *f.*
E telemetria radar *f.*
D Funkmessverfahren *neut.*
R радиолокационное определение дальности *ср.*
A تحديد المدى بالرادار

4438

radar screen
F écran radar *m.*
E pantalla de radar *f.*
D Radarschirm *m.*
R индикаторный экран РЛС *м.*
A لوحة رادار *f*

4439

radar silence
F silence radar *m.*
E silencio de radar *m.*
D Radarstille *f.*
R прекращение работы РЛС *ср.*
A صمت الرادار *m*

4440

radiac detector
F détecteur radiologique *m.*
E detector radiológico *m.*
D Strahlungsmesser *neut.*
R радиометр *м.*
A جهاز رادْيَك *m*

4441

radiac dosimeter
F dosimètre de radiation *m.*
E dosímetro radiológico *m.*
D Strahlungsmesser
R дозиметр *м.*
A مقياس جرعات رادياك

4442

radial
F radiale
E radial
D radial; strahlenförmig
R радиальный
A شُعاعي - إشعاعي *m*

4443

radiation dose
F dose d'irradiation *f.*
E dosis de radiación *f.*
D Strahlendosis *f.*
R доза (радиоактивного) облучения *ж.*
A مقدار الجرْعة الاشعاعية *m*

4444

radiation dose rate
F intensité de la dose absorbée *f.*
E intensidad de la dosis radiactiva *f.*
D Strahlendosisleistung *f.*
R мощность дозы облучения *ж.*
A معدّل الجرْعة الاشعاعية *m*

4445

radiation sickness
F maladie des rayons *f.*
E enfermedad de radiación *f.*
D Strahlungskrankheit *f.*
R лучевая болезнь *ж.*
A المرْض الاشعاعي *m*

4446

radiator
F radiateur *m.*
E radiador *m.*
D Heizkörper [BU] *m.;* Kühler [EV] *m.*
R радиатор *м.;* батарея *ж.*
A مُشَعّ *m*

4447

radio *adj. n.*
F radio *adj. inv. n. f.*
E radio *adj. inv. s. f.*
D Radio(-) *s. neut.* (*komp.*)
R радио- *прилаг.;* радио *ср.*
A لاسلْكي . راديو *m*

4448

radioactive
F radioactif
E radi(o)activo
D radioaktiv
R радиоактивный
A ذو فاعلية إشعاعية

4449

radio approach aids *pl.*
F aides radio d'approche *m. pl.*
E ayudas radio de acercamiento *f. pl.*
D Funkanflugeinrichtungen *f. pl.*
R радионавигационные средства *ср. мн. ч.*
A معدّات الإقتراب اللاسلكيّة *f*

4450

radio channel
F canal radio *m.*
E canal radio *m.*
D Radiokanal *m.*
R радиодорожка *ж.*
A قناة لاسلْكيّة *f*

4451
radio control
F téléguidage *m.*
E teledirección *f.*
D Funkfernsteuerung *f.*
R радиоуправление *ср.*
A تَحَكُّم بالرَّادُيُو *m*

4452
radio deception
F mesures de déception radio *f. pl.*
E radiotransmisión engañosa *f.*
D Funktäuschung *f.*
R радиообман *м.*
A الخِدَاع اللاسِلْكِي *m*

4453 [A,M,R]
radio direction finding (RDF)
F radiogoniométrie *f.*
E radiogoniometría *f.*
D Funkortung [R] *f.;* Funkpeilung [AM] *f.*
R радиопеленгация *ж.*
A تَحْدِيد الإتِّجَاه بالرَّادُيُو

4454
radioisotope
F radio-isotope *m.*
E radioisotopo *m.*
D Radioisotop *neut.*
R радиоизотоп *м.*
A نَظِير مُشِعّ *m*

4455
radio location
F radio repérage *m.*
E radio goniometría *f.*
D Funkauffassung *f.*
R радиолокация *ж.*
A استدلال بالراديو

4456
radiological
F radiologique
E radiológico
D radiologisch
R радиологический; радиоактивный
A اشْعَاعِي

4457
radio navigation
F radionavigation *f.*
E navegación radiogoniométrica *f.*
D Funknavigation *f.*
R радионавигация *ж.*
A مِلاَحَة لاسِلْكِيَّة *f*

4458
radio operator
F radio *m.*
E radiotelefonista *m.*
D Funker *m.*
R радист *м.*; связист *м.*
A عَامِل لاَسِلْكِي *m*

4459
radio rangefinding
F radiotélémétrie *f.*
E radiotelemetría *f.*
D Funkmessverfahren *neut.*
R радиолокационное определение дальности *ср.*
A تَحْدِيد المَدَى بالأَسِلْكِي *m*

4460
radio recognition
F identification radioélectrique *f.*
E identificaciónradio-eléctrica *f.;* reconocimientoradio-eléctrico *m.*
D Funkkennung *f.*
R радиоопознавание *ср.*
A التَّمْيِيز اللاَّسِلْكِي *m*

4461
radio relay system
F câble hertzien *m.*
E red de retransmisión *m.;* radio relé *m.*
D Richtfunknetz *neut.*
R система радиорелейной связи *ж.*
A نِظَام الترحيل اللاّسِلْكِي

4462
radio silence
F silence radio *m.*
E silencio de radio *m.*
D Funk(sende)verbot *neut.*
R радиомолчание *ср.*
A الصَّمْت اللاسِلْكِي *m*

4463 [R]
radio telegraphy
F radiotélégraphie *f.*
E radiotelegrafía *f.*
D Funktelegraphie *f.*
R радиотелеграфия *ж.*
A بَرَق راديوي *m*

4464
radio telephony
F radiotéléphonie *f.*
E radiotelefonía *f.*
D Sprechfunk *m.*
R радиосвязь *ж.*
A مهاتفة لاسلكية *f*

4465
radio transceiver
F émetteur-récepteur *m.*
E transceptor *m.*
D Senderempfänger *m.*
R (радио)приёмо-передатчик *м.*
A جِهَاز مُرْسِل مُسْتَقْبِل *m*

4466
radio transmitter
F émetteur *m.*
E radiotransmisor *m.*
D Sender *m.*
R радиопередатчик *м.*
A جهاز إرْسَال *m*

4467
radius
F rayon *m.*
E radio *m.*
D Radius *m.*
R радиус *м.*
A نِصْف القُطْر

283

4468

(effective lethal) radius

F rayon létal efficace *m.*
E radio *m.*
D wirksamer und tödlicher Radius *m.*
R радиус сплошного поражения *м.*
A نصف القطر المميت والفعال

4469 [A,V]

(minimum turning) radius

F rayon de virage minimum *m.*
E radio mínimo de giro *m.*
D Mindestdrehkreis *m.*
R минимальный радиус поворота
A نصف قطر الدوران الاصغر

4470 [V]

(turn) radius

F rayon de braquage *m.*
E radio de giro *m.*
D Wenderadius *m.*
R радиус разворота *м.*
A نِصْف قُطْر الدَوَران

4471

radius of action

F rayon d'action *m.*
E radio de acción *m.*
D Aktionsradius [AM] *neut.;* Fahrbereich [V] *m.*
R радиус действий *м.*
A نِصْف المَدَى

4472 [N]

radius of damage

F rayon d'effets *m.*
E radio de daños (de efectos) *m.*
D Wirkungsradius *m.*
R радиус поражения *м.*
A مَدَى التَّدْمِير

4473 [N]

radius of safety

F rayon de sécurité *m.*
E radio de seguridad *m.*
D Sicherheitsradius *m.*
R дальность безопасного удаления *ж.*
A مَدَى الأَمَان

4474 [N]

radius of vulnerability

F rayon de vulnérabilité aux effets nucléaires *m.*
E radio de vulnerabilidad *m.*
D Verwundbarkeitskreis *m.;* Risikoradius *m.*
R радиус уязвимости *м.*
A نصف قطر التعرض

4475

radome

F radôme *m.*
E radomo *m.*
D Antennendom *m.*
R колпак *м.;* кожух *м.;* (носовой) обтекатель антенны *м.*
A قُبَّة هَوَائِي الرَّادَار *f*

4476

raft *n.*

F radeau *m.*
E balsa *f.*
D Floss *neut.*
R плот *м.;* паром *м.*
A طَوْف *m*

4477 [M]

(life) raft

F radeau de sauvetage *m.*
E balsa salvavida *f.*
D Rettungsboot *neut.*
R (спасательный) плотик *м.*
A طوف النجاة

4478 [L]

raid *v.*

F faire un raid
E realizar una incursión
D überfallen *tr.*
R производить налёт
A أَغَار

4479 [L]

raid *n.*

F coup de main *m.;* raid *m.*
E incursión *f.*
D Überfall *m.*
R налёт *м.;* нападение *ср.;* рейд [M] *м.*
A غَارَة *f*

4480

(air) raid

F bombardement aérien *m.*
E bombardeo aéreo *m.*
D Luftangriff *m.;* Fliegerangriff *m.*
R воздушный налёт *м.*
A غارة جوية

4481

raider

F commando *m.*
E bombardero *m.*
D Stürmer *m.*
R мародёр *м.*
A مغير

4482

railhead

F tête de ligne *f.*
E cabeza de etapa ferroviaria *f.*
D Endstation *f.*
R железнодорожный конечный пункт *м.;* станция снабжения *ж.*
A رَأْس سِكَّة حديد

4483
railroad *n.*
F chemin de fer *m.*
E ferrocarril *m.*
D Eisenbahn *f.*
R железная дорога *ж.*
A سِكَّة حديد

4484
railroad *adj.*
F ferroviaire *adj.*
E ferroviario
D Eisenbahn- *komp.*
R железнодорожный
A سِكَّة الحَدِيد

4485 [N]
rainout
F pluie radioactive *f.*
E lluvia radiactiva *f.*
D radioaktiver
   Niederschlag *m.*
R радиоактивный дождь
   м.
A إنعدام الرُّؤية بسبب المطر

4486
raise a blocade *v.*
F lever le blocus
E bloquear
D (eine) Blockade
   aufheben
R деблокировать
A رفع الحصار

4487 [A]
raise the
   undercarriage *v.*
F rentrer le train
E levantar el tren de
   aterrizaje
D (das) Fahrgestell
   einfahren
R поднимать шасси
A رفع عجلات الهبوط

4488 [S]
rake *v.*
F enfiler
E barrer
D bestreichen
R обстреливать
   (продольным огнём)
A جَرَف

4489
rally troops *v.*
F rallier les troupes
E rehacerse
D Truppen wieder
   sammeln; ordnen
R собирать войска
A جمع القطع

4490 [S]
ramjet engine
F statoréacteur *m.*
E motor autoreactor *m.*
D Staustrahltriebwerk
   neut.
R прямоточный
   воздушнореактивный
   двигатель (пврд) *м.*
A مُحَرِّك نَفَاث تَضَاغُطي *m*

4491
ramp
F rampe *f.*
E rampa *f.*
D Rampe *f.*
R скат *м.*; наклонная
   установка *ж.*
A مَزْلَقَة *f*

4492
rampart
F rempart *m.*
E terraplén *m.*
D Schutzwall *m.*
R вал *м.*
A مِتراس *m*

4493
ramrod
F refouloir *m.;* baguette
   (de fusil) *f.*
E baqueta *f.*
D Ladestock *m.;*
   Reinigungsstock *m.*
R шомпол *м.;*
   прибойник *м.*
A قضيب التنظيف *m*

4494
random
F (au) hasard
E (al) azar, aleatorio
D zufällig
R случайный;
   беспорядочный
A عَرَضاً ـ عَشوائي

4495
R and R
F (cantonnement de)
   repos *m.*
E recreo educativo del
   soldado (RES) *m.*
D Ruhe und Erholung *f.*
   f.
R отпуск *м.;* отдых и
   выздоровление *м.*
   ср.
A الاستراحة والاسترجاع

4496 [R]
range *n.*
F portée *f.*
E alcance *m.*
D Funksendeweite *f.*
R дальность *ж.*
A مدى العمل

4497 [A,M,V]
range *n.*
F autonomie *f.*
E autonomía *f.*
D Fahrbereich [MV] *m.;*
   Flugbereich [A] *m.*
R дальность действия
   ж.
A مَدَى *m*

285

4498 [S]

range *v.*

F régler le tir
E medir la distancia;
 regular el tiro
D einschiessen
R пристреливаться по
 дальности
A أَحكم الرمي

4499 [R]

(day) range

F portée diurne *f.*
E alcance de día *m.*
D Tagessendeweite *f.*
R дальность
 действительного
 огня днём *ж.*
A المدى في النهار

4500 [R]

(frequency) range

F gamme de fréquences *f.*
E gama de frecuencias *f.*
D Frequenzweite *f.*
R диапазон *м.*
A سُلّم الترددات

4501 [S]

(long) range *adj.*

F à longue portée
E (de) largo alcance
D Fernkampf- *komp.*
R дальнобойный
A بَعيد المَدَى

4502 [A,M,V]

(long) range *adj.*

F à long rayon d'action
E (de) gran radio de
 acción
D Fern- *komp.;*
 Langstrecken- *komp.*
R дальнобойный;
 дальнего действия
A بَعيد المَدَى

4503 [R]

(long) range *adj.*

F à longue portée
E (de) largo alcance
D Fern- *komp.*
R дальний;
 дальнобойный
A بعيد المدى

4504 [S]

(maximum) range

F portée maximum *f.*
E máximo alcance *m.*
D grösste
 Schussentfernung *f.*
R максимальная
 дальность *ж.*
A المَدَى الأَقْصَى *m*

4505 [A,M,V]

(maximum) range

F rayon d'action
 maximum *m.*
E radio de acción
 máximo *m.*
D grösste Reichweite *f.*
R максимальная
 дальность действия
 ж.; максимальная
 дальность полёта *ж.*
A المَدَى الأَقْصَى *m*

4506 [S]

(maximum) range

F portée *f.*
E alcance *m.*
D (Höchst-)
 Schussweite *f.*
R (максимальная)
 дальность *ж.*
A المَدَى الأَقْصَى *m*

4507

**(maximum effective)
 range**

F portée efficace
 maximale *f.*
E alcance máximo eficaz
 m.
D wirksame
 Höchstschussent-
 fernung *f.*
R максимальная
 дальность
 действительного
 огня *ж.*
A المَدَى الأَقْصَى المُؤَثِّر *m*

4508

(minimum) range

F portée minimum *f.*
E alcance mínimo *m.*
D Mindestentfernung *f.*
R минимальная
 дальность *ж.*
A المدى الادنى

4509

(out of) range

F hors de portée
E fuera de alcance
D ausser Schussweite
R вне досягаемости
A خارج المدى

4510

(point blank) range

F à bout portant
E (a) quemarropa; (a)
 boca de jarro
D Kernschussweite *f.*
R прямая наводка *ж.*
A مدّى عن كثب

4511 [S]

(short) range *adj.*

F (à) courte portée
E (de) corto alcance
D Nahkampf- *komp.*
R ближний;
 (...)ближнего
 действия
A قَصير المَدَى

4512 [A,M,V]
(short) range *adj.*
F (d')autonomie limitée
E (de) autonomía
  reducida
D Nahverkehrs- *komp.*
R (...)ближнего действия
A قَصِيرِ المَدَى

4513
(slant) range
F distance oblique *f.*
E distancia-objectivo *f.*
D Schrägentfernung *f.*
R наклонная дальность
  ж.
A المَدَى المَائِل *m*

4514 [S]
(target) range
F distance *f.*
E distancia *f.*
D (Ziel-)Entfernung *f.*
R дальность *ж.*;
  полигон *м.*
A مَدَى (الهدف)

4515 [S]
range by bracketing *v.*
F régler le tir à la
  fourchette
E medir la distancia por
  horquillamiento
D richten; durch
  eingabeln
R отыскивать боковую
  вилку
A صحّح الرمي بالحصر

4516
range finding
F télémétrie *f.*
E telemetria *f.*
D Entfernungsmessung *f.*
R определение
  дальности *ср.*
A تَحْدِيد المَدَى

4517 [S]
range for elevation *v.*
F régler en hauteur
E medir la distancia por
  elevación
D (auf) Höhe richten
R пристреливать на угол
  возвышения
A صحّح الرمي بزاوية الارتفاع

4518 [S]
range for fuze *v.*
F régler l'hauteur
  d'éclatement
E medir la distancia por
  graduación de la
  espoleta
D (auf) Detonation
  richten
R пристреливать по
  трубочной
  дальности
A ضبط الصمامة بحسب المدى

4519 [S]
range for line *v.*
F régler en direction
E medir la distancia por
  líneas
D (auf) Länge richten
R отыскивать основное
  направление
A صحّح الرمي بحسب الخط

4520
ranger
F soldat commando
  (Etats Unis) *m.*
E soldado de fuerzas
  especiales
  estadounidenses *m.*
D Soldat der
  Kommandotruppe *m.*
R "рэнджер" *м.*;
  военнослужащий
  американских войск
  специального
  назначения *м.*
A جندي من المغاوير

4521 [R]
range resolution
F pouvoir séparateur en
  portée *m.*
E poder separador en
  alcance *m.*
D Auslösungsvermögen
  neut.
R разрешающая
  способность по
  дальности (РСД) *ж.*
A تَحْلِيل المَدَى

4522 [S]
range spread
F tir échelonné *m.*
E distribución en alcance
  f.; tiro escalanado *m.*
D Feuerverteilung in
  Schussrichtung *f.*
R рассеивание по
  дальности *ср.*
A انْتِشَار المَدَى العَمِيق

4523
range to *v.*
F avoir une portée de
E tener un alcance de
D richten (auf)
R иметь дальнобойность
  (...) метров
A مدى السلاح حتى ...

4524
rank
F grade *m.*
E grado *m.*
D Dienstgrad *m.*
R воинское звание *ср.*
A رتبة *f*

4525
rank and file
F hommes de troupe *m.*
  pl.
E soldados rasos *m. pl.*
D Mannschaftsstand *m.*
R личный состав *м.*
A الجنود

287

4526
(all) ranks
F militaires de tous
  grades *m. pl.*
E todos *m. pl.*
D Mannschaftsstand *m.*
R личный состав *м.*
A جميع الرتب

4527
(other) ranks (GB)
F sous-officiers et
  hommes de troupe *m.
  pl.*
E soldados *m. pl.*
D Unteroffiziere und
  Mannschaften *m. pl.
  f. pl.*
R личный состав *м.*
A الرتب الأخرى

4528
ransom
F rançon *f.*
E rescate *m.*
D Lösegeld *neut.*
R выкуп *м.*
A فِدْيَة *f*

4529
**rapid deployment
  force**
F force d'intervention
  rapide *f.*
E fuerza de intervención
  inmediata *f.*
D schnelle Einsatztruppe
  f.
R войска быстрого
  развёртывания *ср.
  мн. ч.*
A القوة السريعة الانتشار *f*

4530 [T]
rapids *pl.*
F rapides *m. pl.*
E rapidos *m. pl.*
D Stromschnellen *f. pl.*
R речной порог *м.;*
  стремнина *ж.*
A جَرْيَة سَرِيعَة لِلنَّهر *f*

4531 [L]
**rate of advance
  (march)**
F vitesse de progression *f.*
E velocidad de progresión
  f.
D Vormarsch-
  geschwindigkeit *f.*
R темп наступления *м.*
A سُرْعَة التَّقَدُّم

4532
rate of climb
F vitesse ascensionnelle *f.*
E régimen ascensional *m.*
D Aufstieggeschwindigkeit
  f.
R скороподъёмность *ж.*
A مُعَدَّل الصُّعُود

4533
rate of fire
F cadence (de tir) *f.*
E cadencia (de tiro) *f.*
D Feuergeschwindigkeit
  f.; Feuerkadenz *f.*
R темп огня *м.*
A مُعَدَّل إطْلَاق النِّيرَان

4534 [I]
**(intelligence
  evaluation) rating**
F cotation d'un
  renseignement *f.*
E evaluación *f.*
D Nachrichtenbewertung
  f.
R (разведывательная)
  оценка *ж.*
A تقييم *m*

4535 [M]
ratings (GB)
F matelots et petits
  grades *m. pl.*
E marinería *f.*
D Matrosen *m. pl.*
R матросы *м. мн. ч.*
A نوتية البحرية العسكرية

4536
ratio
F rapport *m.*
E razón *f.*
D Verhältnis *neut.*
R отношение *ср.;*
  пропорция *ж.*
A نِسْبَة *f*

4537
ration *v.*
F rationner (c.à.d.
  limiter); mettre à la
  ration
E racionar
D rationieren (d.h.
  beschränken);
  verpflegen
R выдавать паёк;
  ограничивать
A خَصَّص الأرْزَاق

4538
ration *n.*
F ration *f.*
E ración *f.*
D Ration *f.*
R паёк *м.*
A ارْزَاق ـ حصة *f*

4539
ravine
F ravin *m.*
E barranco *m.*
D Schlucht *f.*
R овраг *м.;* лощина *ж.*
A وَهْدَة *f*

4540
raw troops
F troupes non-aguerris *m.
  pl.*
E tropas inexpertas *f. pl.*
D unerfahrene Truppen
  pl.
R новобранцы *м. мн. ч.*
A جنود مستجدون

4541 [M,T]
reach *n.*
F bief (canal) *m.;* section
   f.
E extensión *f.*
D Flusstrecke *f.;*
   Kanalabschnitt *m.;*
   Meeresarm *m.*
R плёс *м.;* колено реки
   ср.
A إمْتِداد

4542
reach *v.*
F arriver à
E llegar a
D erreichen
R достигать; доходить
   (до)
A وَصَل إلى

4543
react
F réagir
E reaccionar
D reagieren (auf)
R реагировать;
   воздействовать
A تَفَاعَل

4544
reaction
F réaction *f.*
E reacción *f.*
D Reaktion *f.*
R реакция *ж.;*
   реагирование *ср.*
A رَدَّ فِعْل

4545
reaction time
F temps de réaction *m.*
E tiempo de reacción *m.*
D Reaktionzeit *f.*
R время готовности *ср.*
A الفَاصِل الزَّمَنِي لِرَدّ الفِعْل

4546
reactivate
F réactiver
E reactivar
D reaktivieren
R снова
   сформировывать
A أعَاد تَنْشِيط

4547 [N]
reactor
F réacteur *m.*
E reactor *m.*
D Reaktor *m.*
R реактор *м.;* атомный
   котёл *м.*
A مُفَاعِل *m*

4548
reactor core
F noyau du réacteur *m.*
E núcleo de reactor *m.*
D Reaktorkern *m.*
R ядро реактора *ср.*
A قَلْب المُفَاعِل

4549
reactor vessel
F coque du réacteur *f.*
E recipiente de reactor *m.*
D Reaktorgehäuse *neut.*
R реакторный котёл *м.*
A وِعَاء المُفَاعِل

4550
readiness
F disponibilité *f.*
E disponibilidad *f.*
D Bereitschaft *f.*
R готовность *ж.*
A إسْتِعْدَاد *m*

4551
(combat) readiness
F disponibilité
   opérationnelle *f.*
E disponibilidad para
   combate *f.*
D Einsatzbereitschaft *f.*
R боеготовность *ж.*
A الإسْتِعْدَاد لِلقِتَال

4552
(immediate) readiness
F alerte immédiate *f.;*
   disponibilité
   instantanée *f.*
E disposición inmediata *f.*
D volle Einsatzbereitschaft
   f.
R (срочная/полная)
   готовность *ж.*
A حَالَة الإنْذَار الفَوْرِي

4553
**(immediate
   operational)
   readiness**
F situation parée à
   combattre *f.*
E (estado de) alerta
   immediato *m.*
D unmittelbare
   Operationsbereitschaft
   f.
R (срочная)
   боеготовность *ж.*
A في حَالَة الانذار

4554
(operational) readiness
F disponibilité
   opérationnelle *f.*
E (estado de) alerta *m.*
D Operationsbereitschaft
   f.
R боеготовность *ж.*
A الإسْتِعْدَاد لِلعَمَلِيَّات

4555
(barometric) reading
F hauteur barométrique *f.*
E lectura barométrica *f.*
D Barometerstand *m.*
R барометрическое
   показание *ср.*
A قِرَاءة بارومترية *f*

289

4556 [E]

(instrument) reading
F indication *f.*
E indicación *f.*
D Instrumentenangabe *f.*
R отметка *ж.*;
показание *ср.*
A الرقم الذي تشير اليه الأداة

4557

ready
F prêt
E dispuesto; listo;
preparado
D bereit
R готовый
A مُسْتَعِدّ

4558

reallocate
F réattribuer
E reasignar
D neu zuteilen *tr.*
R переподчинять
A أعاد التَّخْصِيص

4559

reallocation
F réattribution *f.*
E reasignación *f.*
D Neuzuteilung *f.*
R переподчинение *ср.*
A إعَادة التَّخْصِيص

4560

real time *adj.*
F (en) temps réel
E (en) tiempo real
D Echtzeit- *komp.*
R реальновременный; (в)
реальном масштабе
времени
A الوَقْت الحَقِيقي *f m*

4561

rear *adj.*
F arrière *inv.*
E atrás; detrás; trasera
D hinter *adj.;* hinter-
komp.
R тыловой
A خَلْفِي

290

4562

rear area
F zone arrière *f.*
E zona de retaguardia *f.*
D rückwärtiges Gebiet
neut.
R тыл *м.*; тыловой
район *м.*
A مِنْطَقَة خَلْفِيَّة *f*

4563

rearguard
F arrière-garde *f.*
E retaguardia *f.*
D Nachtruppen *f. pl.*
R арьергард *м.*
A مُؤْخَرَة *f*

4564

rearm *v.*
F réarmer
E rearmar
D wiederaufrüsten *tr.;*
wiederbewaffnen *untr.*
R перевооружать
A اعاد التسليح

4565

rearsight
F cran de mire de la
hausse
E alza posterior *f.*
D Kimme *f.*
R прицел *м.*
A مسددة خلفية *f*

4566

rebel *n.*
F rebelle *m. f.*
E rebelde *m. f.*
D Rebell *m.*
R повстанец *м.*
A ثائر *m*

4567 [A]

recall *v.*
F rappeler
E llamar
D widerrufen
R объявлять отбой
A إسْتَدْعَى

4568 [A]

recall *n.*
F rappel *m.*
E llamada *f.*
D Widerruf *m.*
R отбой *м.*; отозвание
ср.
A إسْتِدْعَاء *m*

4569

recapture *v.*
F reprendre
E recobrar; reconquistar
D zurückerobern *tr.;*
wieder ergreifen
R отбывать; перебрать в
плен
A إسترد

4570 [R]

receipt
F accusé de réception *m.*
E acuse de recibo *m.*
D Empfangsbestätigung *f.*
R подтверждение
принятия *ср.*
A تَسَلُّم *m*

4571

receive
F recevoir
E recibir
D empfangen
R принимать; получать
A تَسَلُّم

4572 [R]

receiver
F récepteur *m.*
E receptor *m.*
D Empfänger *m.*
R (радио)приёмник *м.*
A مُسْتَقْبِل *m*

4573

reception
F réception *f.*
E recepción *f.*
D Empfang *m.*
R (радио)приём *м.*
A إسْتِقْبَال *m*

4574

reciprocal *adj. n.*
F réciproque *adj. n. f.*
E recíproco(-a) *adj. (s. f.)*
D gegenseitig [U] *adj.;*
  reziprok[E] *adj.;*
  reziproker Wert[E]
  s. m.
R взаимный; обоюдный
A مُتَبَادَل *m*

4575

(dead) reckoning
F navigation à l'estime *f.*
E navegación de estima *f.*
D gegisstes Besteck *neut.*
R (навигационное)
  счисление *ср.*
A تَقْدِير المَوْضِع *m*

4576

reclassify a document
  v.
F changer la classification
  d'un document
E reclasificar
D neu klassifizieren
R засекречивать по
  высшей степени
A أعاد التصنيف

4577

recognition
F reconnaissance *f.*
E reconocimiento *m.*
D Erkennung *f.*
R опознавание *ср.*
A مَعْرِفَة *f* ـ تَعَرُّف *m*

4578

recognize
F reconnaître
E reconocer
D erkennen
R признавать; узнавать
A تَعَرُّف

4579

recoil *n.*
F recul *m.*
E retroceso *m.*
D Rücklauf *m.*
R откат *м.;* отдача *ж.*
A إرْتِدَاد *m*

4580

recoil *v.*
F reculer
E retroceder
D zurücklaufen *tr.*
R откатываться
A إرْتَدَّ

4581

recommend
F recommander
E recomendar
D empfehlen
R представлять;
  рекомендовать
A وَصَّى

4582

recommendation
F recommandation *f.*
E recomendación *f.*
D Empfehlung *f.*
R представление *ср.;*
  рекомендация *ж.*
A تَوْصِيَة *f*

4583

reconnaissance
F reconnaissance *f.*
E reconocimiento *m.*
D Aufklärung *f.*
R разведка *ж.;*
  рекогносцировка *ж.*
A إسْتِطْلَاع *m*

4584

(air) reconnaissance
F reconnaissance aérienne
  f.
E reconocimiento aéreo
  m.
D Luftaufklärung *f.*
R воздушная разведка
  ж.
A إسْتِطْلَاع جَوِّي *m*

4585 [A]

(armed)
  reconnaissance
F reconnaissance armée *f.*
E reconocimiento armado
  m.
D bewaffnete
  Luftaufklärung *f.*
R боевая разведка *ж.*
A إسْتِطْلَاع مُسَلَّح

4586

(long range)
  reconnaissance
F reconnaissance
  profonde *f.;*
  découverte *f.*
E reconocimiento en
  profundidad *m.*
D Fernaufklärung *f.*
R дальняя разведка *ж.*
A استطلاع لمدى بعيد

4587

(maritime)
  reconnaissance
F reconnaissance
  maritime *f.*
E reconocimiento
  marítimo *m.*
D Seeaufklärung *f.*
R береговая разведка *ж.*
A إسْتِطْلَاع بَحْرِي

291

4588

(route) reconnaissance

F reconnaissance de route
  f.
E reconocimiento del
  itinerario *m.*
D Marschstrassen-
  erkundung *f.*
R разведка маршрута
  ж.; рекогносцировка
  маршрута *ж.*
A إسْتِطْلاَع الطَّرِيق

4589

(zone) reconnaissance

F reconnaissance de zone
  f.
E reconocimiento de una
  zona *m.*
D Gebietsaufklärung *f.*
R разведка района *ж.*
A إستكشاف المنطقة

4590

reconnaissance by fire

F reconnaissance par la
  voie des feux *f.*
E reconocimiento por el
  fuego *m.*
D Aufklärung (durch
  Beschuss) *f.*
R разведка огнём *ж.*
A الاستطلاع بالرمي

4591

**reconnaissance in
  force**

F reconnaissance en force
  f.
E reconocimiento en gran
  número *m.*
D Aufklärung (durch
  Gewalt) *f.*
R разведка боем *ж.*
A استطلاع تعرضي

292

4592

reconnoiter

F reconnaître
E reconocer
D erkunden
R производить разведку;
  рекогносцировать
A إسْتَطْلَع

4593 [S]

record as a target *v.*

F enregistrer comme cible
E registrar un blanco
D (als) Ziel erfassen
R записывать целью
A سجّل كهدف

4594 [A]

recover

F recueillir
E recuperar; restablecer
D wiedergewinnen *tr.*
R выводить (из фигуры);
  принимать (на
  палубу)
A أصْلَح

4595 [V,U]

recover

F dépanner [V];
  récupérer [U]
E recuperar
D bergen
R собирать;
  восстанавливать
A أصْلَح

4596 [V]

recoverable

F récupérable
E recuperable
D wiedererlangbar
R эвакуируемый
A قابل للاصلاح

4597 [A]

recovery

F recueil *m.*
E restablecimiento *m.*
D Wiedergewinnung *f.*
R выведение *ср.*;
  принятие *ср.*
A إصلاح *m*

4598 [V,U]

recovery

F dépannage [V] *m.;*
  récupération [U] *f.*
E recuperación *f.*
D Bergung *f.*
R собрание *ср.*;
  восстановление *ср.*
A إصلاح *m*

4599

recruit *v.*

F recruter
E reclutar
D rekrutieren
R комплектовать
A جَنَّدَ

4600

recruit *n.*

F recrue *f.*
E recluta *m.*
D Rekrut *m.*
R новобранец *м.*; рекрут
  м.
A مُجَنَّد *m*

4601

recruitment

F recrutement *m.*
E reclutamiento *m.*
D Einberufung *f.*
R вербовка *ж.*
A تَجْنِيد *m*

4602 [E]

rectification

F rectification *f.*
E rectificación *f.*
D Gleichrichtung *f.*
R детектирование *ср.*
A تَقْوِيم *m*

4603 [E]

rectifier

F redresseur *m.*
E rectificador *m.*
D Gleichrichter *m.*
R выпрямитель *м.*;
   детектор *м.*
A مُعَدِّلة *f* ـ مُقوِّمْ *m*

4604

recuperate

F se rétablir
E restablecerse
D (sich) erholen
R поправляться;
   оправляться
A تعافى ـ إسترد

4605

recuperation

F rétablissement *m.*
E restablecimiento *m.*
D Erholung *f.*
R выздоровление *ср.*
A تعافى ـ إسترداد *m*

4606

red alert

F alerte rouge *f.*
E alerta rojo *m.*
D Alarmstufe rot *f.*
R тревога 'красная' *ж.*
A إنْذار أحمر *m*

4607

Red Crescent

F (le) Croissant Rouge *m.*
E Media Luna Roja *f.*
D Roter Halbmond *m.*
R Красный Полумесяц
   м.
A الهلال الاحمر *m*

4608

Red Cross

F (la) Croix Rouge *f.*
E Cruz Roja *f.*
D Rotes Kreuz *neut.*
R Красный Крест *м.*
A الصليب الاحمر *m*

4609

redeploy

F redéployer
E cambiar de frente;
   redesplegar; transferir
D verlegen
R перегруппировывать;
   производить
   перегруппировку
A اعاد (التوزيع ـ الإنتشار)

4610

redeployment airfield

F terrain de
   redéploiement *m.*
E campo de redespliegue
   m.
D Verlegungsflugplatz *m.*
R запасный аэродром
   для перебазирования
   м.
A مطار إعَادة (التوزيع ـ
   الإنتشار)

4611

redirect

F faire suivre
E reexpedir
D nachsenden *tr.*
R перенацеливать
   (огонь)
A أعاد التَّوْجيه

4612

reduce

F réduire
E reducir
D vermindern
R сокращать;
   уменьшать
A خَفَّض . صَغَّر

4613 [S]

reduced charge

F charge réduite *f.*
E carga reducida *f.*
D verminderte Ladung *f.*
R уменьшенный заряд
   м.
A حُشْوَة (ناقِصَة / مُخَفَّضة) *f*

4614 [S]

(flash) reducer

F réducteur de lueur de
   bouche *m.*
E cubrefuegos *m. inv.*
D Mündungsfeuerdämpfer
   m.
R пламегаситель *м.*
A مانعة الوميض

4615

reduction

F réduction *f.*
E reducción *f.*
D Verminderung *f.*
R сокращение *ср.*;
   уменьшение *ср.*;
   снижение *ср.*
A تَخْفِيض *m* تَصْغِير *m*

4616

reduction in rank

F rétrogradation *f.*
E reducción de grado *f.*
D Degradierung *f.*
R понижение в звании
   ср.
A تخفيض الرتبة

4617 [M]

reef

F écueil *m.;* récif *m.*
E rizo escollo *m.*
D Riff *neut.*
R риф *м.*
A رصيف صخري

4618 [A]

reenter

F rentrer
E reentrar
D wieder eintreten *tr.*
R возвращаться (в
   атмосферу)
A رَجَع

293

4619 [T]
reentrant
F indentation *f.;* rentrant
   m.
E reentrante *m.*
D Öffnung *f.*
R входящая долина *ж.*
A وادٍ ـ أخدود *m*

4620 [A]
reentry
F rentrée *f.*
E reentrada *f.*
D Wiedereintritt *m.*
R возврат *м.;*
   возвращение *ср.*
A رَجْعَة *f*

4621
(grid) reference
F coordonnées
   hectométriques *f. pl.*
E coordenadas *f. pl.*
D (Gitter) Kartenpunkt
   m.
R координаты *ж. мн. ч.;*
   отметка по решётке
   ж.
A مرجع الخريطة

4622
reference datum
F plan de référence *m:*
E plano de referencia *m.*
D Bezugsebene *f.*
R исходная плоскость
   ж.
A مُعْطَى المَرْجِع ـ مرجع
   المعلومات

4623
reference number
F numéro de référence *m.*
E número de referencia
   m.
D Bezugsnummer *f.*
R отметка (высоты) *ж.;*
   относительный
   номер *м.*
A رَقْم المَرْجِع

4624 [M]
refit *v.*
F radouber
E embonar; restaurar
D depot instandsetzen *tr.*
R ремонтировать;
   чинить
A جَدَّد

4625 [M]
refit *n.*
F radoub *m.*
E embonada *f.*
D Depot-Instandsetzung *f.*
R ремонт *м.;* починка
   ж.
A تَجْدِيد *m*

4626
reflect
F réfléchir
E reflejar
D nachdenken *tr.;*
   widerspiegeln *tr.*
R отражать
A إِنْعَكَس

4627
reflection
F réflexion *f.*
E reflejo (imagen) *m.;*
   reflexión *f.*
D Reflexion *f.;* Spiegelung
   f.
R отражение *ср.;*
   отблеск *м.*
A إِنْعِكَاس *m*

4628 [V]
reflector
F catadioptre *m.*
E captafaros *m. inv.*
D Rückstrahler *m.*
R рефлектор *м.*
A عَاكِس *m*

4629 [E,U]
reflector
F réflecteur *m.*
E reflector *m.*
D Reflektor [E] *m.;*
   Spiegel [U] *m.*
R рефлектор *м.;*
   отражатель *м.*
A عَاكِس *m*

4630
refract
F réfracter
E refractar
D ablenken *tr.;* brechen
R преломлять
A إِنْكَسَر

4631
refraction
F réfraction *f.*
E refracción *f.*
D Brechung *f.*
R преломление *ср.;*
   рефракция *ж.*
A إِنْكِسَار *m*

4632
refuel
F ravitailler (en
   carburant)
E reaprovisionar (de
   combustible)
D (auf)tanken (*tr.*)
R дозаправлять
   (топливом)
A مَوَّن بالوُقُود

4633 [A]
refueling
F ravitaillement en
   carburant *m.*
E reaprovisionamiento de
   combustible
D Tanken *neut.*
R (воздушная)
   дозаправка *ж.*
A تَمْوِين بالوُقُود

4634 [A]

(hot) refueling

F ravitaillement avec
   moteur et rotor
   tournant *f.*

E repostaje *m.*

D Auftanken (bei laufenden
   Motoren) *neut.*

R срочная заправка *ж.*

A التمـوِين بالوقود والمحرك
   يُشَغَّل

4635

refuge

F (lieu de) refuge *m.*

E (lugar de) refugio *m.*

D Zuflucht *f.;*
   Zufluchtsort *m.*

R укрытие *cp.;* убежище
   cp.

A مَلْجأ *m*

4636

(seek) refuge

F (chercher) refuge

E (buscar) refugio

D Zuflucht (suchen) *untr.*

R (искать) убежища

A (إلى) إلْتَجَأ

4637

refugee

F refugié *m.*

E refugiado *m.*

D Flüchtling *m. f.*

R беженец *м.*

A لاَجِىء *m*

4638

refuge port

F port de salut *m.*

E puerto de refugio *m.*

D Schutzhafen *m.*

R порт-убежище *cp.;*
   гавань *ж.*

A مَرْفَأ الإلْتِجَاء

4639

regiment

F régiment *m.*

E regimiento *m.*

D Regiment *neut.*

R полк *м.*

A كتيبة

4640 [P,U]

register *v.*

F enregistrer

E registrar

D eintragen *tr.*

R регистрировать;
   заносить в список

A سَجَّل

4641 [T]

register *n.*

F contrôle de repérage *m.*

E registro *m.*

D Verzeichnis *m.*

R список *м.;* опись *ж.*

A سِجِل *m*

4642 [S]

register *v.*

F régler le tir (sur point
   de repère)

E reglar el tiro (sobre
   blanco auxiliar)

D (sich) einschiessen *tr.*

R пристреливать

A سَجَّل

4643

registered document

F publication enregistrée
   f.

E documento registrado
   m.

D Verschlussache *f.*

R заказной документ *м.*

A وَثِيقَة مُسَجَّلة *f*

4644 [S]

(ground) registration

F repérage du terrain *m.*

E registro de objetivos *m.*

D Bodenerfassung *f.*

R пристрелка по реперу
   ж.

A تسجيل الأرْض

4645

regroup

F (se) regrouper

E reagrupar(se)

D umgruppieren *tr.*

R перегруппировывать

A أعاد التَّجْمِيع

4646

regular *adj.*

F régulier

E regular

D regelmässig

R регулярный; кадровый
   (солдат)

A نِظامِي

4647

Regular Army

F armée de métier *f.*

E ejército permanente *m.*

D stehendes Heer *neut.*

R регулярная армия *ж.*

A جيش نظامي

4648 [E]

regulate

F régler

E regular

D einstellen *tr.*

R регулировать

A نَظَّم

4649 [E]

regulation

F ajustage *m.;* réglage *m.*

E regulación *f.*

D Einstellung *f.*

R регулирование *cp.*

A تَنْظِيم *m*

4650 [P]
regulation(s) *(pl.)*
F réglementation *f.*
E reglamento(s) *m.* *(pl.)*
D Vorschrift(en) *f.* *(pl.)*
R устав *м.*; инструкция
   ж.
A أَنْظِمَة *f*

4651
(security) regulations
F règles de sécurité *f. pl.*
E reglamentos de
   seguridad *m. pl.*
D Sicherheitsvorschriften
   f. pl.
R устав о безопасности
   м.
A تعليمات الأمن

4652
(traffic) regulations
F réglementation de la
   circulation *f.*
E reglamentos de la
   circulación *m. pl.*
D Verkehrsvorschriften *f.*
   pl.
R устав о движении *м.*
A أنظمة السَّير

4653 [E]
regulator
F régulateur *m.*
E regulador *m.*
D Regler *m.*
R регулятор *м.*
A مُنَظِّم *m*

4654
rehabilitate
F réhabiliter
E rehabilitar
D auffrischen [AML] *tr.*;
   rehabilitieren [U]
R реабилитировать;
   исправлять
A رَدّ الإعْتِبَار

296

4655
rehabilitation
F réhabilitation *f.*
E rehabilitación *f.*
D Rehabilitierung
R реабилитация *ж.*;
   восстановление *ср.*
A رَدُّ الإعْتِبَار

4656 [E,U]
rehabilitation
F remise en état *f.*
E repaso general *m.*
D Wiederherrstellung *f.*
R ремонт *м.*
A تصحيح

4657
rehearsal
F répétition *f.*
E ensayo *m.*
D Probe *f.*
R репетиция *ж.*;
   повторение *ср.*
A تَجْرِبة *f*

4658 [S]
reheater
F dispositif de post-
   combustion *m.*
E dispositivo de
   poscombustión *m.*
D Nachbrenner *m.*
R форсажёр *м.*
A مُعِيد التَّسْخِين

4659 [A,E,M,L,U]
reinforce
F renforcer
E reforzar
D verstärken
R усиливать; укреплять
A عَزَّز

4660 [B]
reinforce
F armer (béton)
E armar (hormigón)
D armieren (Beton)
R укреплять
A قَوَّى

4661
reinforced concrete
F béton armé *m.*
E hormigón armado *m.*;
   cemento armado *m.*
D Stahlbeton *m.*
R железобетон *м.*
A خَرَسَانة مُسَلَّحَة *f*

4662
reinforcement(s) *pl.*
F renforcement *m.*;
   renforts *m. pl.*
E reforzamiento *m.*;
   refuerzos *m. pl.*
D Verstärkung(en) *f. (pl.)*
R дополнительные
   войска *ср. мн. ч.*;
   подкрепление *ср.*
A تَعْزِيزَات *f*

4663 [E]
reject *n.*
F (pièce de) rebut *m.*
E cosa rechazada *f.*
D Ausschuss *m.*
R отброска *ж.*; отбросы
   м. мн. ч.
A رَفْض *m*

4664
reject *v.*
F mettre au rebut;
   rejeter
E rechazar
D ausmustern *tr.*;
   verwerfen
R отбрасывать;
   отказывать
A رَفَض

4665 [E]
relation(s) *(pl.)*
F rapport *m.*
E relación(es) *f. (pl.)*
D Verhältnis *neut.*
R отношения *ср. мн. ч.*;
   соотношения *ср. мн.*
   ч.
A عَلاَقَات *f*

4666 [P]

relation(s) *pl.*

F relations *f. pl.*
E relación(es) *f. (pl.)*
D Beziehungen *f. pl.*
R отношения *ср. мн. ч.;*
соотношения *ср. мн. ч.*
A عَلاَقَات

4667

relative bearing

F relèvement relatif *m.*
E marcación relativa *f.*
D Peilwinkel *m.*
R курсовой угол *м.*
A الإتِّجَاه الزَّاوي النِّسْبِي *m*

4668

relative biological effectiveness (RBE)

F efficacité biologique relative *f.*
E eficacia biológica relativa *f.*
D relative biologische Wirkung *f.*
R относительная биологическая эффективность *ж.*
A الفَعَّالِيَّة الأَحْيَائِيَّة النِّسْبِيَّة *f*

4669

relative humidity

F humidité relative *f.*
E humedad relativa *f.*
D (relative) Luftfeuchtigkeit *f.*
R относительная влажность *ж.*
A رُطُوبَة نِسْبِيَّة *f*

4670 [P]

relax

F relâcher
E mitigar
D lockern
R смягчать; уменьшать напряжение
A أَهْمَل - إسْتَرْخى

4671 [R]

relay *v.*

F transmettre
E retransmitir
D übertragen (mit Relais) *untr.*
R ретранслировать
A رحّل

4672 [R]

relay *n.*

F relais radio *m.*
E retransmisión *f.*
D Funkrelais *neut.*
R трансляция *ж.*
A ترحيل

4673 [E]

release *v.*

F déclencher
E disparar
D auslösen *tr.*
R разблокировать; расцеплять
A أرْخَى

4674 [A,G]

release *v.*

F lancer; larguer [A]
E lanzar [A]; soltar [G]
D abschiessen *tr.;* abwerfen *tr.*
R запускать; отпускать
A إطْلَق

4675 [C]

release *v.*

F dégager
E despedir
D abblasen *tr.*
R выделять
A سَيَّب

4676 [P,U]

release *v.*

F libérer [P]; publier
E publicar; soltar [P]
D freigeben *tr.*
R освобождать
A أطْلَق سَرَاح

4677 [P]

release *n.*

F libération *f.*
E liberación *f.*
D Befreiung *f.*
R освобождение *ср.*
A إطْلَاق السَّرَاح

4678 [A]

release (load)

F largage de la charge *m.*
E descargo *m.*
D Ausklinkung *f.*
R сбрасывание *ср.*
A إسقاط

4679

(parachute) release

F dispositif d'ouverture de parachute *m.*
E dispositivo de apertura *m.*
D Fallschirmauslöser *m.*
R защёлка *ж.*
A جهاز فتح المظلة

4680

(quick) release

F dégrafage rapide *m.*
E dispositivo de apertura rápida *m.*
D Auslösevorrichtung *f.*
R (быстродействующая) скоба *ж.*
A جهاز تسييب سريع

4681 [N]
releasing commander
F commandant
approbateur *m.*
E comandante autorizado
m.
D freigebender
Truppenführer *m.*
R разрешающий
начальник *м.*
A قَائِد الإِطْلاق

4682 [E]
reliability
F fiabilité *f.;* régularité de
marche *f.*
E fiabilidad *f.;* seguridad
de funcionamiento *f.*
D Zuverlässigkeit *f.*
R надёжность *ж.;*
прочность *ж.*
A أَهْلِيَّة الثِّقة

4683
reliability (of
intelligence)
F valeur *f.*
E valor *m.*
D Zuverlässigkeitsgrad der
Nachrichten *m.*
R достоверность
сведений *ж.*
A أهلية الاستخبارات

4684 [E]
reliable
F (d'un fonctionnement)
sûr
E confiable; seguro
D zuverlässig
R надёжный; прочный
A يُرْكَن إِلَيْه

4685 [L,M]
relief
F relève *m.*
E relevo *m.*
D Ablösung *f.*
R смена *ж.*
A إِسْعَاف *m*

4686 [T]
relief
F relief *m.*
E relieve *m.*
D Relief *neut.*
R рельеф *м.*
A بُرُوز *m*

4687 [A,L,M,U]
relief
F secours *m.*
E socorro *m.*
D Hilfe *f.*
R помощь *ж.*
A تَبْديل *m*

4688
relief map
F carte en relief *f.*
E plano (mapa) en relieve
m.
D Reliefkarte *f.*
R рельефная карта *ж.*
A خَرِيطة بَارِزَة *f*

4689 [L,M]
relieve
F dégager (ville etc...);
relever
E descercar (ciudad etc.);
relevar
D ablösen; entsetzen
(Stadt usw.)
R освобождать (город);
производить смену;
сменять
A بَدَّل

4690 [H,U]
relieve
F soulager
E aliviar
D lindern
R облегчать; приходить
на помощь
A أَسْعَف

4691
reload
F recharger
E recargar
D umladen [AMV];
wieder laden [S]
R перезаряжать;
перегружать
A اعاد المَلْء ـ أعاد التحميل

4692
relocate a unit
F déplacer; replacer
E transladar una unidad
(a otro sitio)
D (eine) Einheit verlegen
R перемещать часть
A اعاد تحديد موقع الوحدة

4693
remote *adj.*
F lointain
E (a) distancia; remoto
D abgelegen
R дальний;
дистанционный
A مِنْ بُعْد

4694
remote control *n.*
F télécommande *f.*
E control remoto *m.;*
telecontrol *m.*
D Fernlenkung *f.*
R телеуправление *ср.;*
дистанционное
управление(ДУ) *ср.;*
телерегулирование
ср.
A السَّيْطَرة من بُعْد

4695
remote control *adj.*
F télécommandé
E (a) control remoto;
teledirigido
D Fernlenk- *komp.*
R телеуправляемый
A سَيْطَرة من بُعْد

4696

remote ground sensors
F détecteurs au sol
   télécommandés *m.*
E detectores terrestres
   teledirigidos *m. pl.*
D fernbedienter
   Bodensensor *m.*
R дистанционные
   наземные детекторы
   м. мн. ч.
A اجهزة ارضية للاحساس من
   بُعد

4697 [D,R]

remote terminal
F terminal éloigné *m.*
E equipo terminal a
   distancia *m.*
D Fernbediengerät *neut.*
R удалённая позиция *ж.*
A مِرْبَط بَعِيد *m*

4698

render safe *v.*
F mettre hors d'état de
   fonctionner
E desarmar; descargar
D entschärfen
R поставлять (оружие)
   на предохранитель;
   обезвреживать
   (бомбу)
A أَمَّن

4699

rendezvous
F point de rassemblement
   m.
E punto de reunión *m.*
D Versammlungsplatz *m.*
R сбор *м.*; встреча *ж.*
A مَوْعِد *m*

4700

renegade *n.*
F renégat *m.*
E renegado *m.*
D Renegat *m.*
R изменник *м.*
A مُرْتَدّ

4701

renew
F renouveler
E renovar
D erneuern
R обновлять;
   восстанавливать
A جَدَّد

4702

renovate
F remettre à neuf
E renovar
D instandsetzen *tr.;*
   renovieren
R починять;
   восстанавливать
A أَصْلَح

4703

reorganization
F réorganisation *f.*
E reorganización *f.*
D Neugliederung *f.;*
   Umgruppierung *f.*
R реорганизация *ж.*
A إِعَادة التَّنْظِيم

4704

reorganize
F (se) réorganiser
E reorganizar(se)
D neugliedern *tr.;*
   umgruppieren *tr.*
R реорганизовывать
A اعاد التَّنْظِيم

4705

repair *n.*
F réparation *f.*
E reparación *f.*
D Instandsetzung *f.;*
   Reparatur *f.*
R ремонт *м.*; починка
   ж.
A إِصْلَاح

4706

repair *v.*
F réparer
E reparar
D instandsetzen *tr.;*
   reparieren
R ремонтировать;
   починять
A أَصْلَح

4707

repair and recovery *n.*
F réparation et
   récupération *f. f.*
E reparación y
   recuperación *f.*
D Reparatur und Bergung
   f. f.
R восстановление *ср.*
A الاصلاح

4708

reparations
F réparations *f. pl.*
E reparaciones *f. pl.*
D Entschädigung *f.*
R возмещения *ср. мн. ч.*
A تعويضات *f*

4709

repeat *v.*
F répéter
E repetir
D wiederholen
R повторять
A تِكْرار *m*

4710 [S]

"Repeat!"
F "Mêmes éléments!"
E "¡Repita!"
D „Die gleiche
   Entfernung!"
R "Повторите!"
A كَرِّر

299

4711 [D]
repeater
F répéteur *m.*
E repetidor *m.*
D Zwischenverstärker *m.*
R повторитель *м.*;
усилитель *м.*
A مُعيد *m*

4712 [R]
repeater
F station relais *f.*
E estación de
retransmisión *f.*
D Relaisstelle *f.*
R повторитель *м.*;
усилитель *м.*
A مُعيد *m*

4713
repeating firearm
F arme à répétition *f.*
E arma automática *f.*
D automatische Waffe *f.*;
Mehrladewaffe *f.*
R магазинное оружие *ср.*
A بندقية تلقائية

4714 [S]
"Repeat range!"
F "Même portée!" *f.*
E "¡Mismo alcance!" *m.*
D „Gleiche
Schussentfernung!" *f.*
R "Дальность по
прежнему!" *ж.*
A كرر المدى

4715
repel an attack
F repousser une attaque
E rechazar un ataque
D Angriff zurückschlagen
R отражать атаку
A صدّ هجوما

300

4716
(insect) repellent
F crème anti-insecte *f.*;
aérosol insecticide *m.*
E loción para ahuyentar
los insectos *f.*
D (Insekten-)
Gegenmittel *neut.*
R средство от
насекомых *ср.*
A مُنَفِّر

4717
replace
F remplacer
E reemplazar; reponer
D ersetzen
R заменять; замещать
A بَدَّل

4718 [E]
replacement
F pièce de rechange *f.*
E pieza de recambio *f.*
D Ersatzteil *m.*
R замена *ж.*;
возмещение *ср.*
A تَعْويض *m*

4719
replacement factor
F taux de remplacement
m.
E factor de recambio *m.*
D geschätzter Ausfall *m.*
R коэффициент
возмещения потерь
м.
A عَامِل تَقْدِير التَّعْويض

4720 [H]
replacements *pl.*
F recomplètement *m.*
E recambios *m. pl.*;
reemplazos *m. pl.*
D Personalersatz *m.*
R пополнения в личном
составе *ср. мн. ч.*
A تَعْويضات *f*

4721
replenishment
F réapprovisionnement *m.*
E reaprovisionamiento *m.*
D Nachschub *m.*
R пополнение запаса *ср.*
A إعَادَة التَّمْوين

4722
reply *v.*
F répondre
E contestar; responder
D (be-)antworten;
erwidern
R отвечать
A أجَاب

4723
reply *n.*
F réponse *f.*
E contraseña *f.*
D Antwort *f.*
R ответ *м.*
A جَوَاب *m*

4724 [R]
reply *n.*
F réponse *f.*
E contraseña *f.*
D Rückmeldung *f.*
R ответ *м.*
A جَوَاب *m*

4725
report *v.*
F rapporter
E hacer un informe
D berichten über
R доносить; делать
донесение
A عَمَل تقريراً

4726

report *v.*

F signaler; rendre compte de
E informar (acerca de)
D melden
R докладывать
A كتب تقريراً

4727

report *v.*

F (se) présenter
E presentarse
D (sich) melden
R заявляться; появляться
A حضر

4728 [U]

report

F compte rendu *m.;* procès-verbal *m.*
E informe *m.;* parte *m.*
D Bericht *m.;* Meldung *f.*
R донесение *ср.*; сообщение *ср.*; доклад *м.*
A تَقْرِير *m*

4729 [S]

report

F détonation *f.*
E estampido *m.*
D Knall *m.*
R звук взрыва *м.*
A تَقْرِير *m*

4730

(absentee) report

F bulletin d'appel *m.*
E parte de ausentes *m.*
D Abwesendenmeldung *f.*
R донесение об отсутствующих *ср.*
A بيان الغائبين

4731 [I]

(amplifying) report

F compte rendu supplémentaire de contact *m.*
E informe aclaratorio *m.*
D Ergänzungsmeldung *f.*
R разработающее донесение *ср.*
A تقرير اضافي

4732

(bombing) report

F compte rendu de bombardement *m.*
E informe de bombardeo *m.*
D Bombenschädenbericht *m.*
R доклад о результатах бомбометания *м.*
A تَقْرِير القَصْف *m*

4733

(casualty) report

F état des pertes *m.*
E parte de bajas *m.*
D Verwundetenmeldung *f.*
R донесение о поражённых *ср.*
A بيان الخسائر

4734

(confidential) report

F notes annuelles *f. pl.*
E informe reservado *m.*
D (vertraulicher) Bericht *m.*
R аттестация *ж.*
A تقرير خصوصي

4735

(inflight) report

F compte rendu en vol *m.*
E parte al vuelo *m.*
D Lagemeldung aus der Luft (meist eng. Begriff) *f.*
R донесение при выходе на цель *ср.*; доклад при выходе на цель *м.*
A تَقْرِير أَثْنَاء الطَّيْرَان

4736

(intelligence) report

F compte rendu de renseignement *m.*
E informe de inteligencia *m.*
D Nachrichtenmeldung *f.*
R разведывательное донесение *ср.*
A تَقْرِير إِسْتِخْبَارَات *m*

4737

(nil) report

F état néant *m.*
E ningún informe *m.*
D Fehlanzeige *f.*
R доклад неизменной обстановки *м.*
A تقرير بلا شيء

4738 [S]

(shelling) report

F compte rendu de bombardement *m.*
E informe de bombardeo *m.*
D Beschussmeldung *f.*
R донесение об обстреле *ср.*
A تقرير قصف

4739

(sighting) report

F compte rendu d'observation *m.*
E informe de observación *m.*
D Beobachtungsmeldung *f.*
R доклад об обнаружении *м.*
A تَقْرِير كَشْف

4740

(situation) report

F compte rendu de situation *m.*
E informe de la situación *m.*
D Lagebericht *m.*
R доклад об обстановке *м.*
A تَقْرِير المَوْقِف

4741

(strength) report
F situation d'effectifs *f.*
E parte del estado de
 fuerzas *m.*
D Kampfstärkenmeldung
 f.
R списочная
 численность *ж.*
A بيان القوة

4742

(toxic) report
F compte rendu d'attaque
 par agent toxique *m.*
E informe de agresivo *m.*
D Kampfstoffaufklärungs-
 bericht *m.*
R доклад (донесение) о
 применении ОВ *м.*
 (*ср.*)
A تقرير السُّمِّيَّة

4743

(weather) report
F bulletin météorologique
 m.
E boletín meteorológico
 m.
D Wetterbericht *m.*
R метеосводка *ж.*
A نَشْرَة جَوَّيَّة *f*

4744

report for duty *v.*
F (se) présenter à son
 poste
E presentarse en el puesto
D (sich zum Dienst)
 melden
R заявляться на службу
A حضر الخدمة

4745 [R]

report into a net *v.*
F entrer dans le réseau
E entrar en la malla
D (sich einem Funknetz)
 anschliessen *tr.*
R присоединяться к сети
A التحق بشبكة لاسلكية

302

4746 [S]

"Report ready!"
F "Annoncez prêt!"
E "¡Listo!"
D „(sich schussbereit)
 melden!"
R "Доложите о
 готовности!"
A اخبر باستعداد المدافع

4747

report sick *v.*
F (se faire porter) malade
E (pasar) reconocimiento
D (sich) krank melden
R сводка больных *ж.*
A بلغ عن مرضه

4748

represent *v.*
F représenter
E representar
D vertreten
R представлять
A ناب عن . . .

4749

representative *n.*
F représentant *m.*
E representante *m.*
D Vertreter *m.*
R представитель *м.*
A وكيل / نائب

4750 [T]

representative fraction
F échelle numérique *f.*
E escala numérica *f.*
D (Karten-)Massstab *m.*
R численный масштаб
 м.
A كَسْر مُمَثِّل *m*

4751

reprimand *v.*
F réprimander
E reprender
D Verweis erteilen
R (давать) выговор *м.*
A وبّخ

4752

reprisals
F représailles *f. pl.*
E represalias *f. pl.*
D Vergeltungsmassnahmen
 f. pl.
R репрессалия *ж.*
A أعمال انتقامية

4753

repulse an attack *v.*
F repousser une attaque
E rechazar un ataque
D (einen Angriff)
 zurückschlagen *tr.*
R отражать атаку
A صدّ هجوماً

4754 [A,S]

request *n.*
F demande *f.*
E solicitud *f.*
D (Luft-)Anforderung *f.*
R требование *ср.*; заявка
 ж.
A طَلَب *m*

4755 [A,S]

request *v.*
F demander
E pedir
D anfordern *tr.*
R требовать
A طَلَب

4756

require
F nécessiter
E necesitar
D verlangen
R нуждаться; требовать
A طَلَب

4757 [U]

requirement
F besoin *m.*
E requisito *m.*
D Bedürfnis *neut.*
R потребность *ж.*;
 нужда *ж.*
A حَاجَة *f*

4758 [A,M,L,U]

requisition v.

F demander
E pedir
D anfordern tr.
R реквизировать
A صَادِر

4759

requisition n.

F demande f.; réquisition f.
E pedido m.
D Anforderung f.
R заявка ж.; реквизиция ж.
A مُصَادَرَة f

4760 [P]

requisition v.

F réquisitionner
E requisar
D beschlagnahmen
R истребовать
A صَادِر

4761

requisitioning

F réquisition f.
E requisa f.
D Beschlagnahme f.
R истребование ср.; реквизиция ж.
A مُصَادَرَة f

4762

rescue v.

F sauver
E salvar
D retten
R спасать
A أَنْقَذ

4763

rescue adj. n.

F (de) sauvetage (adj.) n. m.
E (de) salvamento (adj.) s. m.
D Rettung(s-) s. f. (komp.)
R спасательные операции ж. мн. ч.
A إِنْقَاذ m

4764

(operations) research

F recherche opérationelle f.
E investigación operacional f.
D Operationsforschung f.
R исследование ср.
A البحوث للعمليات

4765

research and development (R and D)

F conception et réalisation f. f.
E investigación y desarollo f.
D Forschung und Entwicklung f. f.
R научно-техническая разработка ж.
A البحوث والتطورات

4766 [T]

resection

F intersection inverse f.
E intersección inversa f.
D Rückwärtseinschnitt m.
R обратная засечка ж.
A تَقَاطُع خَلْفِي

4767

reserve adj. n.

F (de) réserve (adj.) n. f.
E (de) reserva (adj.) s. f.
D Reserve(-) s. f. (komp.)
R резерв(ный) м. (прилаг.); запас(ный) м. (прилаг.)
A إِحْتِيَاطِي m

4768

reserved demolition

F destruction réservée f.
E demolición reservada f.
D offenzuhaltene Sperre f.
R подготовленное разрушение ср.
A تَدْمِير مُخَصَّص m

4769

reserve forces pl.

F forces de réserve f. pl.
E fuerzas de reserva f. pl.
D Reserveverbände m. pl.
R резерв м.; войска резерва ср. мн. ч.
A إِحْتِيَاطَات f

4770

reserve stocks pl.

F stocks de réserve m.
E abastecimientos de reserva m. pl.
D Reservebestände m. pl.
R резервный запас м.
A إِمْدَادَات إِحْتِيَاطِيَّة f

4771 [T]

reservoir

F réservoir m.
E depósito m.; embalse m.
D Reservoir neut.
R водохранилище ср.
A خَزَّان

303

4772

resident unit

F unité permanente *f.*

E unidad residente *f.*

D Stammeinheit *f.*

R постоянная часть *ж.*

A وَحْدَة مُقِيمَة *f*

4773 [C,N]

residual contamination

F contamination
résiduelle *f.*

E contaminación residual
f.

D Restverseuchung *f.*

R остаточное заражение
ср.

A تَلَوُّث مُتَخَلِّف *m*

4774 [N]

residual radiation

F rayonnement résiduel
m.

E radiación residual *f.*

D Reststrahlung *f.*

R остаточная радиация
ж.

A اِشْعَاع نَوَوِي مُتَخَلِّف *m*

4775

**(main line of)
resistance**

F ligne principale de
résistance *f.*

E línea principal de
resistencia *f.*

D Hauptwiderstandslinie
f.

R главная полоса
обороны *ж.*

A خَطّ المُقَاوَمَة الرَّئِيسِي *m*

304

4776

**(main point of)
resistance**

F (noyau de) résistance
m.

E núcleo de resistencia *m.*

D Hauptwiderstandspunkt
m.

R главная точка
сопротивления *ж.*

A النقطة الرئيسية للمقاومة

4777

(organized) resistance

F résistance organisée *f.*

E resistencia organizada *f.*

D organisierter
Widerstand *m.*

R организованное
сопротивление *ср.*

A مُقَاوَمَة مُنَظَّمَة *f*

4778

(token) resistance

F résistance symbolique

E resistencia simbólica *f.*

D Scheinwiderstand *m.*

R видимость
сопротивления *ж.*

A مقاومة تظاهرية

4779 [E,R]

resolution

F pouvoir séparateur *m.*

E resolución *f.*

D Auflösungsvermögen
neut.

R разложение *ср.*

A تَحْلِيل *m*

4780

resources

F ressources *f. pl.*

E recursos *m. pl.*

D Mittel *neut. pl.*

R ресурсы *м. мн. ч.;*
запасы *м. мн. ч.*

A موارد *f*

4781 [C]

respirator

F masque à gaz *m.*

E respirador *m.*

D ABC-Schutzmaske *f.*

R противогаз *м.*

A قِنَاع التَّنَفُّس

4782 [R]

responder

F répondeur *m.*

E respondedor *m.*

D Responder *m.*

R ответчик *м.*

A وحدة مستجيبة

4783

(flexible) response

F riposte graduée *f.*

E reacción graduada *f.*

D (flexibele) Reaktion *f.*

R гибкий ответ *м.*

A إِسْتِجَابَة مَرِنة *f*

4784

(graduated) response

F riposte graduée *f.*

E respuesta gradual *f.*

D angepasste Reaktion
(meist eng.Begriff) *f.*

R дифференцированные
ответные действия
ср. мн. ч.

A إِسْتِجَابَة تدريجية *f*

4785 [N]

(target) response

F effet sur l'objectif *m.*

E efecto sobre el objetivo

D Wirkung im Ziel *f.*

R воздействие на цель
ср.; эффект на цель
ср.

A رَدّ الفِعْل لِلهَدَف

4786

responsibility

F responsabilité *f.*

E responsabilidad *f.*

D Verantwortung *f.*

R ответственность *ж.*

A مسؤولية *f*

4787 [R]

responsor

F répondeur *m.*
E respondedor
D Bewerter *m.*
R ответчик *м.*
A المجيب (آلة الكترونية)

4788 [S]

"Rest!"

F "Repos!"
E "¡Descanso!"
D „Feuerpause!" (ohne Batterie)
R "Отдых!"
A إسْتَرِح

4789

(aiming) rest

F chevalet de pointage *m.*
E soporte del arma *m.*
D Zielstütze *f.*
R (прицельный) станок *м.*
A متكأ التصويب *m*

4790

(butt) rest

F bout de crosse *m.;* support de crosse *m.*
E soporte de culatín *m.*
D Kolbenplatte *f.*
R (прикладный) упор *м.*
A حامل الاخمص *m*

4791

(cheek) rest

F appui-joue *m.*
E soporte de mejilla *m.*
D Backenauflage *f.*
R плечо приклада *ср.*
A خدّ الاخمص *m*

4792

(firing) rest

F appui de tir *m.*
E soporte de tiro
D Schießstütze *f.*
R (прицельный) станок *м.*
A مسند الرمي *m*

4793

(knife) rest

F cheval de frise *m.*
E caballo de frisa *m.*
D Sperrbalken *m.*
R рогатка *ж.*
A صَهَوات *f*

4794

restitution

F restitution *f.*
E restitución *f.*
D Entschädigung *f.*
R возвращение *ср.;* восстановление *ср.*
A إرْتِداد *m*

4795 [A,M]

restraint factor

F coefficient d'arrimage *m.*
E coeficiente de restricción *m.*
D Befestigungsfaktor *m.*
R ограничительный фактор *м.*
A عَامِل لِلتَّقْييد

4796

restrict

F restreindre
E limitar; restringir
D einschränken *tr.*
R ограничивать; сдерживать
A حَدَّد

4797

restricted air cargo

F fret aérien dangereux *m.*
E carga aérea peligrosa *f.*
D beschränkte Luftfracht *f.*
R ограниченный груз самолёта *м.*
A شحنة جوية محددة *f*

4798 [A]

restricted area

F zone réglementée *f.*
E zona restringida *f.*
D Sperrgebiet *neut.*
R ограниченная зона *ж.;* запретная зона *ж.*
A مِنْطَقَة مَحْظُورَة *f*

4799

restricted information

F information à diffusion restreinte *f.*
E información restringida *f.*
D vertrauliche Information *f.*
R секретные сведения *ср. мн. ч.;* сведения для служебного пользования *ср. мн. ч.*
A مَعْلُومَات مَحْظُورَة *f*

4800

(exercise) restrictions

F conventions de manoeuvres *f. pl.*
E limitaciones de maniobras *f. pl.*
D Übungsbeschränkungen *f. pl.*
R ограничение *ср. мн. ч.*
A تقييدات التمرين

4801

restrictive fire area (RFA)

F zone de tir prescrit *f.*
E zona de fuego limitado *f.*
D (einbeschränkter) Feuerwirkungsbereich *m.*
R район ограничения артиллерийского огня *м.*
A منطقة رمي مقيَّدة

4802

resupply *n.*

F ravitaillement *m.*
E reabastecimiento *m.;*
reaprovisionamiento
m.
D Anschlussversorgung *f.*
R переснаряжение *ср.*
A تموين

4803

resupply *v.*

F recompléter
E reabastacer
D (die Depots)
wiederauffüllen *tr.*
R пополнение запаса *ср.;*
подвоз *м.*
A اعاد التَّموين

4804

(automatic) resupply

F ravitaillement
automatique *m.*
E reaprovisionamiento
automático *m.*
D (automatische)
Ergänzung *f.*
R (очередное)
переснабжение *ср.*
A التموين التلقائي

4805

(initial) resupply

F ravitaillement initial *m.*
E abastecimiento inicial
m.
D Versorgungsbeginn *m.*
R первоначальное
пополнение *ср.;*
первоначальный
подвоз *м.*
A تموين أوّلي *m*

4806

(vertical) resupply

F ravitaillement par
hélicoptère *m.*
E reaprovisionamiento
por helicóptero *m.*
D Luftanschlussversorgung
f.
R переснаряжение
вертолётом *ср.*
A التموين بطائرة عمودية

4807

resupply on demand

F ravitaillement sur
demande *m.*
E reaprovisionamiento a
petición *m.*
D (auf Antrag) ergänzen
R срочное
переснабжение *ср.*
A التموين عند الطلب

4808

retaliation

F revanche *f.;* représailles
f. pl.
E desquite *m.;* revancha *f.*
D Vergeltung *f.*
R возмездие *ср.;*
ответный удар *м.*
A إنتقام

4809

retard *v.*

F retarder
E retardar
D verzögern
R задерживать
(нападение)
A أخّر

4810

reticle

F réticule
E retículo *m.*
D Strichplatte *f.*
R сетка (оптического
прибора) *ж.*
A شُبَيْكَة *f*

4811

retinue

F suite *f.*
E séquito *m.*
D Gefolge *neut.*
R свита *ж.*
A الحاشية *f*

4812

retire

F (se) replier
E retirarse
D (sich) zurückziehen *tr.*
R отходить; отступать
A إنْسَحَب

4813

**retire from active
service**

F (prendre sa) retraite; (se
mettre à la) retraite
E retirarse de servicio
activo
D (in den Ruhestand)
treten; pensionieren
R выходить в отставку
A تقَاعَد

4814

retractable

F escamotable
E replegable
D einziehbar
R убирающийся
A يَنْضَم

4815

retreat *v.*

F battre en retraite
E batirse en retirada
D (sich) zurückziehen *tr.*
R отступать; отходить
A تَرَاجَع

4816

retreat *n.*

F retraite *f.*
E retirada *f.*
D Rückzug *m.*
R отступление *ср.;* отход
м.
A تَقَهقُر *m*

4817 [D]

(information) retrieval

F recherche documentaire
  f.
E sacar (información)
D Informationswiedergabe
  f.
R находка данных *ж.*;
  возвращение данных
  ср.
A استعادة (المعلومات)

4818

retrofire *v.*

F ralentir en actionnant
  ses rétrofusées
E frenarse por retrocohete
D (Bremsrakete) zünden
R включать тормозную
  ракету
A اشعل صاروخاً كابحاً

4819

retrofit

F modification
  retrospective *f.*
E modificación
  retrospectiva *f.*
D Formänderung *f.*
R модификация *ж.*
A تَعْدِيل *m*

4820

retrograde movement

F mouvement rétrograde
  m.
E movimiento retrogrado
  m.
D Rückzugbewegung *f.*
R отступательные
  действия *ср. мн. ч.*
A حَرَكَة التَّرَاجُع

4821

retrorocket

F rétro-fusée *f.*
E retrocohete *m.*
D Bremsrakete *f.*
R тормозная ракета *ж.*
A صاروخ كابح *m*

4822

return

F rapport (état/situation)
  m.
E informe *m.*
D Erstattung *f.*; -
  Erstattung *komp.*
R сводка *ж.*
A بيان

4823 [R]

(ground) return

F écho de sol *m.*
E eco de tierra *m.*
D Bodenwiederempfang
  m.
R отражение от
  местности *ср.*
A صدّى أرضي

4824

(radar) return

F écho radar *m.*
E eco radar *m.*
D Radarwiederempfang
  m.
R местные помехи *ж.*
  мн. ч.
A صدّى راداري

4825

return fire *v.*

F riposter
E replicar el fuego
D (Feuer) erwidern
R отвечать на огонь
A رد على النيران

4826

reveille

F réveil *m.*
E torque de diana *m.*
D Wecken *neut.*
R подъём *м.*
A استيقاظ

4827 [V]

reverse *v.*

F faire marche arrière
E (dar) marcha atrás (a)
D rückwärts fahren
R давать задний ход;
  давать обратный
  ход
A سَار إلَى الوَرَاء

4828

revet

F revêtir
E revestir
D verkleiden
R облицовывать
A كَسَا

4829

revetment

F revêtement *m.*
E revestimiento *m.*
D Verkleidung *f.*
R облицовка *ж.*;
  обшивка *ж.*
A تَكْسِيَة *f*

4830

review *n.*

F revue *f.*
E revista *f.*
D Parade *f.*
R смотр *м.*; парад *м.*
A استعراض

4831 [H]

revive *v.*

F réanimer
E resucitar; reanimar
D wiederbeleben
R восстанавливать
A أَحْيَا

4832 [P]

revolt *v.*

F (se) révolter
E rebelarse
D revoltieren
R восставать
A ثَار

4833
revolt *n.*
F révolte *f.*
E rebelión *f.;* revuelta *f.*
D Aufstand *m.*
R восстание *ср.;* мятеж
м.
A ثَوْرَة *f*

4834 [P]
revolution
F révolution *f.*
E revolución *f.*
D Revolution *f.*
R революция *ж.*
A ثَوْرَة *f*

4835
revolutions per minute
(rpm) *pl.*
F tours par minute
(t/min) *m. pl.*
E revoluciones por
minuto (rpm) *f. pl.*
D Umdrehungen pro
Minute (U/Min.) *f.*
pl.
R (№) оборотов в
минуту
A دَوْرَات في الدَّقِيقَة

4836
revolver
F revolver *m.*
E revólver *m.*
D Revolver *m.*
R револьвер *м.*
A مُسَدَّس *m*

4837
rev up
F emballer le moteur
E girar el motor
D (den Motor auf)
Touren bringen
R увеличивать число
оборотов; заводить
A زوّد دورات المحرك

4838 [P]
reward *n.*
F récompense *f.*
E recompensa *f.*
D Belohnung *f.*
R награда *ж.;*
вознаграждение *ср.*
A مُكَافَأَة *f*

4839 [E]
rheostat
F rhéostat *m.*
E reóstato *m.*
D Rheostat *m.*
R реостат *м.*
A مُعدِّلة *f*

4840
(medal) ribbon
F décoration militaire *f.*
E galón *m.*
D Ordensband *neut.*
R (орденская) лента *ж.*
A شَريط (نوط)

4841
ricochet *v.*
F ricocher
E rebotar
D abprallen *tr.*
R (бить) рикошетом
A تَنَطَّط ـ إِرتد

4842
ricochet *n.*
F ricochet *m.*
E rebote *m.*
D Abprall(er) *m.*
R рикошет *м.*
A إِرتداد *m*

4843 [T]
ride *n.*
F piste *f.*
E vereda *f.*
D Schneise *f.*
R аллея *ж.*
A مَمَرّ بغابة

4844
ride at anchor *v.*
F être à l'ancre
E estar al ancla
D (vor) Anker liegen
R стоять на якоре
A أَلْقَى المِرْسَاة وتَوَقَّف

4845 [T]
ridge *n.*
F arête *f.;* crête *f.*
E cadena *f.;* cresta *f.*
D (Berg-) Kamm *m.;*
Rücken *m.*
R гребень *м.;* хребет *м.*
A ذِرْوَة *f*

4846
rifle
F fusil *m.*
E fusil *m.*
D Gewehr *neut.*
R винтовка *ж.*
A بُنْدُقِيَّة *f*

4847
(automatic) rifle
F fusil automatique *m.*
E fusil automático *m.*
D Sturmgewehr *neut.*
R автомат *м.*
A بندقية تلقائية

4848
(clip-fed) rifle
F fusil à répétition *m.*
E fusil de repetición *m.*
D Gewehr mit
Rahmenzuführung
neut.
R (магазинная) винтовка
ж.
A بندقية ذات مخزن

4849

(grenade) rifle

F fusil lance-grenade *m.*
E fusil lanzagranadas *m.*
D Gewehrgranate *f.*
R (гранатомётная)
   винтовка *ж.*
A قنبلة بندقية

4850

(recoilless) rifle

F canon sans recul *m.*
E fusil sin retroceso *m.*
D (rückstossfreies)
   Gewehr *neut.*
R (безоткатное) орудие
   ср.
A مدفع عديم الارتداد

4851

(scope-mounted) rifle

F fusil à lunette *m.*
E fusil provisto de visor
   m.
D Gewehr mit
   Zielfernrohr *neut.*
R винтовка с
   телескопическим
   прицелом *ж.*
A بندقية ذات مِقراب

4852

(sniper's) rifle

F fusil de tireur d'élite *m.*
E fusil de tirador
   apostado *m.*
D Scharfschützengewehr
   neut.
R (снайперская)
   винтовка *ж.*
A بندقية القنَّاص

4853 [S]

rifled bore

F âme rayée *f.*
E ánima rayada *f.*
D gezogener Lauf *m.*
R нарезной канал ствола
   м.
A جَوْف مُحَلْزَن *m*

4854

rifleman

F grenadier-voltigeur
   (GV) *m.;* fantassin *m.*
E fusilero *m.*
D Schütze *m.*
R стрелок *м.*
A رامي

4855

rifling

F rayure *m.*
E rayado *m.*
D Züge *m. pl.*
R нарезка *ж.*
A حَلْزَنَة

4856 [E]

(test) rig

F banc d'essai *m.*
E equipo de pruebas *m.*
D Probebank *f.*
R испытательное
   оборудование *ср.*
A جهاز اختبار

4857

(parachute) rigger

F plieur de parachute *m.*
E aparejador de
   paracaídas *m.*
D Fallschirmpacker *m.*
R сборщик парашютов
   м.
A مُركِّب المظلّات

4858 [M]

rigging

F gréement *m.*
E aparejo *m.*
D Takelage *f.*
R такелаж *м.;* снасти *ж.*
   мн. ч.
A عُدّة السفينة *f*

4859

rigging lines

F suspentes *f. pl.*
E cuerdas de sustentación
   f. pl.
D Fallschirmleinen *f. pl.*
R парашютное
   снаряжение *ср.*
A حِبال المظلّة *f*

4860

right *adj.*

F correct
E correcto
D richtig
R правый
A صَحيح

4861

right *n.*

F droit
E derecho *m.*
D Recht *neut.*
R право *ср.*
A حَقّ *m*

4862

right (on, to the right)
   adj.

F droit (à (la) droite)
E derecho (a la derecha)
D recht(s)
R направо; справа
A يَمِين

4863

right-handed

F droitier
E (persona) que usa la
   mano derecha; de
   derecha
D rechtshändig;
   rechtsgängig
R пользующийся правой
   рукой; правая
   нарезка [E]
A يَميني

309

4864 [M]

right of search

F droit de visite *m.*

E derecho de visita *m.*

D Durchsuchungsrecht *neut.*

R право на обыск *ср.*

A حَقّ التَّفْتِيش

4865 [M]

right of stopping

F droit d'arraisonnement *m.*

E derecho de detención *m.*

D Anhalterecht *neut.*

R право остановки *ср.*; право на остановку *ср.*

A حَقّ التَّوْقِيف

4866 [M,T,V]

right of way

F priorité [MV] *f.;* servitude de passage [T] *f.*

E derecho de paso [MV] *m.*; servidumbre de paso [T] *f.*

D Wegerecht [MT] *neut.;* Vorfahrt [V] *f.*

R право прохода *ср.*; право проезда *ср.*

A أَفْضَلِيَّة المُرُور

4867

rigid

F rigide

E rígido

D stabil

R жёсткий; негибкий

A صُلْب

4868 [I]

(espionage) ring

F réseau d'espionnage *m.*

E grupo de espionaje *m.*

D Spionagering *m.*

R банда *ж.*; клика *ж.*

A عصابة جاسوسية

4869

(safety pin) ring

F anneau de goupille *m.*

E aro de seguro *m.*

D Sicherungssplintring *m.*

R предохранительный шплинт *м.*

A حَلْقة الأمان

4870 [P]

ringleader

F meneur *m.*

E cabecilla *f.*

D Rädelsführer *m.*

R главарь *м.*

A زعيم حركة

4871

riot *n.*

F émeute *f.*

E motín *m.*

D Aufruhr *m.*

R бунт *м.*; мятеж *м.*

A هَيَاج *m* شَغَب *m*

4872

riot *v.*

F (faire une) émeute

E amotinarse

D meutern; randalieren

R бунтовать

A هَاج

4873

riot control

F maintien de l'ordre *m.*

E acción antisubversiva *f.*

D Aufruhr-bekämpfung *f.*

R действия по подавлению беспорядков *ср. мн. ч.*

A ضَبْط الشَّغَب

4874 [C]

riot control agent

F agent (chimique) utilisé en maintien de l'ordre (p. ex. lacrymogène) *m.*

E agente antisubversivo *m.*

D Aufruhrbekämpfungs-mittel *neut.*

R отравляющее вещество для контрбеспорядочных действий *ср.*; слезоточивый газ *ср. м.*

A عَامل كِيمَاوِي لضَبْط الشَّغَب

4875

rioter

F émeutier *m.*

E amotinado *m.*

D Aufrührer *m.* Unruhestifter *m.*

R мятежник *м.*; бунтовщик *м.*

A مُشَاغِب *m*

4876

riot gun

F fusil à canon tronçonné *m.*

E escopeta recortada *f.*

D kurzläufige Flinte *f.*

R полицейская винтовка *ж.*

A بُنْدُقِيَّة لِضَبْط الشَّغَب

4877

rioting

F émeutes *f. pl.*

E disturbios *m. pl.*

D Aufruhr *f.;* Unruhen *f. pl.*

R беспорядки *м. мн. ч.*

A مُشَاغَبَة *f*

4878

riot squad

F équipe d'intervention f.
E brigada anti-disturbios f.
D Überfallkommando neut.
R противоповстанческий полицейский отряд м.
A قوات مكافحة الشغب

4879

ripcord

F poignée d'ouverture f.
E cable de apertura m.
D Aufziehkabel neut.
R вытяжной трос м.
A حبل الفتح m

4880 [T]

rise

F côte f.; pente f.
E cuesta f.; pendiente f.
D Steigung f.
R возвышенность ж.
A مرتفع m

4881 [M]

rise

F flux m.
E altura de la marea f.
D Fluthöhe f.
R подъём воды м.
A ارتفاع المد m

4882 [P]

rising n.

F émeute f.
E alzamiento m.; levantamiento m.; sublevación f.
D Aufstand m.
R восстание cp.
A ثَوْرَة f

4883

risk n.

F risque m.
E riesgo m.
D Risiko neut.
R риск м.
A مُغَامَرَة f

4884

risk v.

F risquer
E arriesgar(se)
D wagen
R рисковать
A غَامَر

4885

river adj.

F fluvial
E fluvial
D Fluss- komp.
R речной
A نَهْري

4886

river n.

F fleuve m.; rivière f.
E río m.
D Fluss m.
R река ж.
A نَهْر m

4887

river crossing

F franchissement de rivière m.
E paso de río m.
D Übergang über ein Gewässer m.
R переправа ж.; форсирование cp.
A عبور الانهار

4888

rivet v.

F river (riveter)
E remacher
D (ver-)nieten
R заклёпывать
A دَسَر ـ بَرشَم

4889

rivet n.

F rivet m.
E remache m.; roblón m.
D Niete f.
R заклёпка ж.
A دِسَار ـ مُسمَار بِرشَام m

4890 [M]

road

F rade f.
E rada f.
D Reede f.
R рейд м.
A مَرْسَى m

4891 [T]

road

F route f.
E camino m.
D Strasse f.
R дорога ж.
A طَرِيق f, m

4892

(access) road

F route d'accès f.
E camino de acceso m.
D Zufahrtsstrasse
R подъездной путь м.
A طريق الدخول

4893

(hilly) road

F chemin à fortes pentes m.
E camino de fuertes pendientes m.
D Bergstrasse f.
R холмистая дорога ж.; горная дорога ж.
A طَرِيق جَبَلِية f

311

4894

(trunk) road
F grande route *f.*
E carretera *f.;* ruta
  principal *f.*
D Hauptstrasse *f.*
R шоссе *ср.;* шоссейная
  дорога *ж.;*
  магистраль *м.*
A طَرِيق رَئِيسِي *m*

4895

road block
F barrage *m.*
E barricada *f.*
D Strassensperre *f.*
R дорожная застава *ж.;*
  дорожный
  контрольно-
  пропускной пункт
  (КПП) *м.*
A مِتْراس الطَّرِيق

4896

road capacity
F capacité de route *f.*
E capacidad (de tránsito)
  de camino *f.*
D Strassenkapazität *f.*
R пропускная
  способность дорог
  ж.
A سَعَة الطَّرِيق

4897

road clear
F route ouverte *f.*
E camino libre *m.*
D Strasse frei
R дорога открыта *ж.*
A طَرِيق مَفْتُوح

4898

road clearance time
F durée d'encombrement
  (circulation routière)
  f.
E duración del despejo de
  caminos *f.*
D Strassenfreigabezeit *f.*
R срок расчистки дорог
  м.
A ساعة العبور

312

4899

road clearing
F ouverture de route *f.*
E despejo de caminos *m.*
D Strassenräumung *f.*
R расчистка дорог *ж.*
A فَتْح الطَّرِيق

4900

road closed
F route fermée *f.*
E camino cerrado *m.;*
  carretera cortada *f.*
D Strasse gesperrt *f.*
R дорога закрыта *ж.*
A قَطْع في الطَّرِيق ـ طريق
  مقفول

4901

road junction
F bifurcation *f.*
E empalme de caminos *m.*
D (Strassen-)Kreuzung *f.*
R стык дорог *м.*
A مُلْتَقَى طُرُق

4902

road marking
F balisage *m.*
E marca de carreteras *f.*
D Strassenmarkierung *f.*
R маркировка дороги
  ж.
A اشارات الطرق

4903

road space
F longueur de colonne *f.*
E fondo de marcha *m.*
D Marschlänge *f.*
R глубина колонны *ж.*
A فُرْجَة الطَّرِيق

4904 [M]

roadstead
F rade *f.*
E rada *f.*
D Reede *f.*
R рейд *м.*
A مرسى

4905

road time
F durée de trajet *f.*
E duración del desfile *f.*
D Marschzeit *f.*
R время прохождения
  по дороге *ср.*
A مُدَّة المُرُور

4906 [V]

roadworthy
F apte à rouler
E listo para marchar
D betriebsicher
R дорожноспособный
A صَالِح لِلطَّرِيق

4907

robbery
F vol *m.*
E robo *m.*
D Raub *m.*
R кража *ж.;* грабёж *м.*
A سرقة *f*

4908

rock *v.*
F balancer
E balancear(se)
D schaukeln
R качать(ся);
  колебать(ся)
A أَرَجَّحَ ـ هَزَّ

4909

rock *n.*
F roche *f.*
E roca *f.*
D Fels [L] *m.;* Klippe [M]
  f.
R скала *ж.;* утёс *м.*
A صَخْر *m*

4910

rocket
F fusée *f.*
E cohete *m.*
D Rakete *f.*
R ракета *ж.*
A صَارُوخ *m*

4911

(anti-submarine) rocket

F roquette ASM *f.*

E cohete contra-submarinos *m.*

D (U-Boot-) Abwehrrakete *f.*

R (противолодочный) снаряд *м.*

A صاروخ مضاد للغوّاصات

4912

(liquid-fuelled) rocket

F fusée à liquide(s) *f.;* fusée à combustible(s) *f.*

E cohete a líquido(s) *m.*

D Flüssigkeitsrakete *f.*

R жидкостный ракетный снаряд (ЖРС) *м.*

A صَارُوخ ذُو وُقُود سَائِل

4913

rocket-assisted projectile

F projectile à fusée *m.*

E proyectil ayudado por cohete *m.*

D (mit) Raketen nachbeschleunigtes Geschoss *neut.*

R ракетный снаряд *м.*

A قَذِيفَة صَارُوخِيَّة *f*

4914

rocket launcher

F lance-roquette *m.*

E lanzacohetes *m. inv.*

D Raketenwerfer *m.*

R реактивный гранатомёт (ручной) *м.*; пусковая установка *ж.*

A قَاذِفَة صَوَارِيخ

4915

rocket propulsion

F propulsion à fusée *f.*

E propulsión por reacción *f.*

D Raketenantrieb *m.*

R ракетная движущая сила *ж.*

A دَفْع صَارُوخِي *m*

4916 [T]

rocky *adj.*

F rocheux (montagne); rocailleux (sentier)

E rocoso

D felsig

R скалистый

A صخري

4917

"Roger!"

F "Reçu!"

E "¡Roger!"; "¡Conforme!"; "¡Recibido!"

D „Verstanden!"

R "Понял!"

A مَفْهُوم

4918 [P]

rogues' gallery

F (collection de) photographies de repris de justice *f.*

E álbum de fotos de personas buscadas *m.*

D Verbrecheralbum *neut.*

R альбом фотоснимков преступников *м.*

A مجموعة من صور المجرمين

4919

role

F rôle *m.*

E papel *m.*

D Rolle *f.*

R роль *ж.*

A دَوْر *m*

4920 [A,M]

roll *v.*

F rouler

E balancearse

D rollen; schlingern

R катить(ся)

A تَقَلَّب . تَمَايَل

4921 [B,E]

roll *v.*

F cylindrer; laminer; rouler

E apisonar [B]; laminar [E]

D walzen

R вращать(ся)

A مَلَّس

4922 [U]

roll *n.*

F liste *f.*

E lista *f.*

D Liste *f.*

R вращение *ср.*; крен *м.*

A سِجْل *m*

4923 [A]

roll *n.*

F tonneau *m.*

E tonel *m.*

D Rolle *f.*

R крен *м.*

A تَقَلُّب *m*

4924 [M]

roll *n.*

F (coup de) roulis *m.*

E balanceo *m.*

D Schlingern *neut.*

R (килевая) качка *ж.*

A تَمَايُل *m*

4925

(nominal) roll

F état nominatif *m.*

E lista de personal *f.*

D Namensliste *f.*

R (именной) список *м.*

A سجل أسمى

313

4926
(pay) roll
F feuille de solde *f.*
E nómina *f.*
D Lohnliste *f.*
R платёжная ведомость
   ж.
A جدول الرواتب

4927
roll call
F appel *m.*
E lista *f.*
D Appell *m.*
R перекличка *ж.*
A تَفَقُّد *m*

4928 [B,E]
roller
F rouleau-compresseur [B]
   m.; rouleau [E] *m.*
E apisonadora [B] *f.;*
   rodillo [E] *m.*
D Rolle [E] *f.;* Walze [BE]
   f.
R ролик *м.;* вал *м.*
A بَكَرَة *f*

4929
(barbed wire) roller
F tambour de fil barbelé
   m.
E rodillo (para alambre
   de púas) *m.*
D Stacheldrahttrommel *f.*
R ролик *м.;* вал *м.*
A بكرة اسلاك شائكة

4930
roller conveyor
F transporteur à rouleaux
   m.
E transportador de
   rodillos *m.*
D Rollenbahn *f.*
R роликовый конвейер
   м.
A نَاقِلة باسطوانات دوارة

4931
rolling stock
F matériel roulant *m.*
E material rodante *m.*
D rollendes Material *neut.*
R подвижной состав *м.*
A دَوَارِج ـ عربات سكة حديد *f*

4932
**roll-on roll-off
(RORO)** *adj.*
F roll on roll off; (navire)
   transroulier *m.*
E transbordador RORO
   adj.
D Durchfahr *komp.*
R автовоз РОРО *м.*
A سفينة يمكن السيارات تسير
   داخلها من الرصيف

4933
roll up *v.*
F (faire) replier; déborder
E arrollar
D einkesseln *tr.;* aufrollen
   tr.
R свёртывать
A التف على موضع العدو
   واستغل

4934
romer
F équerre de report de
   point *f.*
E escalímetro *m.*
D Gittermessgerät *neut.*
R координатная мерка
   ж.
A آلة للاحداثيات التربيعية

4935
(chart) room
F chambre des cartes *f.*
E cuarto de derrota *m.*
D Kartenzimmer *neut.*
R штурманская рубка
   ж.
A حُجْرَة الخَرَائِط *f*

4936 [A,M,U]
(control) room
F salle de contrôle *f.*
E sala de control *f.*
D Kommandoraum *m.*
R пункт управления *м.*
A حُجْرة الإدارة *f*

4937
(engine) room
F salle des machines *f.*
E sala de máquinas *f.*
D Motorraum *m.*
R машинное отделение
   ср.
A حُجْرة المُحَرِّكَات *f*

4938 [M]
(engine) room
F chambre des machines
   f.
E cámara de máquinas *f.*
D Maschinenraum *m.*
R машинное отделение
   ср.
A حجرة المُحَرِّكَات *f*

4939
(operations) room
F salle d'opération *f.*
E sala de operaciones *f.*
D Operationszentrale *f.*
R командный пункт
   (КП) *м.*
A حُجْرَة العَمَلِيَات *f*

4940 [L]
(orderly) room
F salle des rapports *f.;*
   bureau du secrétariat
   régimentaire *m.*
E oficina *f.*
D Wachstube *f.*
R канцелярия *ж.*
A حجر الاوامر

4941 [A,L]

(situation) room

F salle des cartes *f.*

E sala de operaciones *f.*

D Lagezimmer *neut.*

R главный командный
пункт *м.*

A حجرة خريطة الموقف

4942 [A,M,L]

room for maneuver

F espace de manoeuvre
m.

E espacio para hacer
maniobra *m.*

D Bewegungsfreiheit *f.*

R манёвроспособная
местность *ж.*

A مجال للمناورة

4943 [U]

rope

F corde *f.*

E cuerda *f.*

D Seil *neut.*

R трос *м.*; верёвка *ж.*

A حَبْل *m*

4944 [R]

rope (chaff)

F plaquettes de brouillage
f. pl.

E (ristra de) virutas *f. pl.*

D Düppel *m.;*
Düppelstreifen *m. pl.*

R металлизированные
отражатели *м. мн. ч.*

A قَش *m*

4945

rope ladder

F échelle de corde *f.*

E escala de cuerda *f.*

D Strickleiter *f.*

R шторм-трап *м.*

A سُلَّم حِبَال *m*

4946

roster

F tableau de service *m.*

E lista *f.*

D Dienstplan *m.*

R расписание *ср.*; список
нарядов *м.*

A جَدْوَل الخدمة *m*

4947

rotary

F rotatif

E rotativo

D Dreh- *komp.;* rotierend
adj.

R ротационный;
вращательный

A دَوَّار

4948

rotate

F (faire) tourner

E (hacer) girar

D (sich) drehen

R вращать(ся)

A دَار

4949

rotation

F rotation *f.*

E rotación *f.*

D (Um-)Drehung *f.;*
Rotation *f.*

R вращение *ср.*; ротация
ж.

A دَوَرَان *m*

4950

rotor

F rotor *m.*

E rotor *m.*

D Drehflügel [A] *m.;*
Rotor [AE] *m.*

R ротор *м.*

A دَوَّار *m*

4951

rotor blade

F pale de rotor *f.*

E pala de rotor *f.*

D Rotorblatt *neut.*

R лопасть винта *ж.;*
ротор *м.*

A رِيشَة دَوَّاره *f*

4952 [M]

rough

F houleux

E picado

D (stark) bewegt

R беспокойный; бурный

A هَائِج

4953 [T]

rough

F inégal; cahoteux
(chemin)

E desigual (camino);
escabroso

D uneben

R пересечённый

A وَعْر

4954 [S]

round *n.*

F coup complet *m.*

E tiro *m.*

D Schuss *m.*

R пуля *ж.*; снаряд *м.*

A طَلْقَة *f*

4955

round *adj.*

F rond

E redondo

D rund

R круглый

A مُسْتَدِير

4956

(baton) round

F balle en caoutchouc *f.*

E cilíndrico de caucho *m.*

D Stumpfgeschoss *neut.*

R резиновая пуля *ж.*

A طَلْقَة (هراوية . مطاط) *f*

4957

(cased) round

F cartouche à douille *f.*
E munición con vaina *f.*
D Kartuschmunition *f.*
R гильзовый снаряд *м.*
A طَلَقَة مُغَلَّفة *f*

4958 [S]

(ranging) round

F coup de réglage *m.*
E tiro telemétrico *m.*
D Einschiessgeschoss *neut.*
R (дальномерный) выстрел *м.*
A طلقة تحديد المدى

4959

(spent) round

F cartouche utilisée *f.;* cartouche explosée *f.*
E (bala) fría *f.*
D (matte) Kugel
R стреляный патрон *м.*
A طلقة ميتة

4960

rounds-per-gun

F coup par pièce *m.*
E (número de) disparos o proyectiles por cañón
D Schuss pro Waffe *m.*
R очередь *ж.*
A الطلقات لكل مدفع

4961

rout *v.*

F (mettre en) déroute
E derrotar
D vertreiben
R громить
A هــزم

4962

route

F itinéraire *m.*
E itinerario *m.;* ruta *f.*
D Route *f.;* Strecke *f.*
R маршрут *м.;* курс *м.;* путь *м.*
A طَريق *f , m*

4963

(air) route

F route aérienne *f.*
E ruta aérea *f.*
D Flugstrecke *f.*
R воздушная трасса *ж.*
A خَطّ جَوّي *m*

4964

(alternative) route

F itinéraire de rechange *m.*
E ruta de emergencia *f.*
D Ausweichstrecke *f.*
R путь обхода *м.*
A طريق بديل

4965

(bug-out) route

F voie d'évacuation d'urgence *f.*
E ruta de retiro de emergencia *f.*
D Ausweichstrecke *f.*
R путь отхода *м.*
A طريق الانسحاب

4966

(convoy) route

F route de convoi *m.*
E ruta del convoy *f.*
D Geleitzugweg *m.*
R маршрут конвоя *м.*
A طَريق القافِلة

4967

(great circle) route

F route orthodromique *f.*
E ruta ortodrómica *f.*
D Grosskreisroute *f.*
R ортодромический курс *м.*
A مَسِير دَائِرة عُظْمَى

4968

(main supply) route

F itinéraire principal de ravitaillement *m.*
E vía principal de abastacimiento *f.*
D Hauptnachschubstrasse *f.*
R главный путь подвоза *м.;* главный путь снабжения *м.*
A مِحْوَر التَّمْوِين *m*

4969

(reserved) route

F itinéraire réservé *m.*
E itinerario reservado *m.*
D reservierte Militärstrasse *f.*
R забронированный маршрут *м.*
A طَرِيق مُخَصَّص

4970 [A]

(trunk) route

F voie aérienne principale *f.*
E ruta aérea principal *f.*
D Langstrecken-Luftweg *m.*
R магистральная трасса *ж.*
A طَرِيق رَئِيسي *m*

4971

route card (map)

F carte routière *f.*
E mapa itinerario *m.*
D Marschkarte *f.*
R маршрутная карта *ж.;* путевая карта *ж.*
A بِطَاقَة الطَّرِيق

4972

route classification

F classification d'un itinéraire *f.*

E clasificación del itinerario *f.*

D Strassenbelastbarkeits-bewertung *f.*

R классификация дорог *ж.*

A تَصْنِيف الطُّرُق

4973

routine

F routine *f.*

E rutina *f.*

D Gewohnheit *f.;* Routine *f.*

R (установленный) распорядок *м.*

A مُعْتَاد

4974

roving gun

F pièce nomade *f.*

E cañón autónomo

D Einzelkanone *f.*

R самостоятельное орудие *ср.*

A مدفع متحرك

4975 [M]

row *v.*

F ramer

E remar

D rudern

R грести

A جَذَف

4976

row (line) *n.*

F rangée *f.*

E hilera *f.*

D Reihe *f.*

R ряд *м.*

A صَفّ

4977

rub *v.*

F frotter

E frotar

D reiben

R тереть(ся)

A دَلَك

4978

rubber

F caoutchouc *m.*

E caucho *m.*

D Gummi *neut.*

R резина *ж.*; каучук *м.*

A مَطَاط *m*

4979

rucksack

F sac à dos *m.*

E mochila alpina *f.*

D Rucksack *m.*

R рюкзак *м.*; походный мешок *м.*

A كِيس جَبَلي *m* ـ جربنديّة *f*

4980 [A,M]

rudder

F gouvernail (de direction [A]) *m.*

E timón [M] (de dirección [A]) *m.*

D Seitenruder [A] *neut.;* Steuerruder [M] *neut.*

R руль *м.*

A سُكَّان *m* ـ دَفّة *f*

4981 [M]

(diving) rudder

F barre de plongée *f.*

E timón de picado *m.*

D Tauchruder *neut.*

R руль *м.*; гидрокрыло *ср.*

A دفّة الانقضاض

4982 [T]

rugged

F accidenté

E escabroso

D uneben

R пересечённый

A خَشِن

4983

ruins *pl.*

F ruines *f. pl.*

E ruinas *f. pl.*

D Ruinen *f. pl.*

R развалина *ж.*

A أَخْرِبَة *f*

4984 [P]

rule

F autorité *f.*

E autoridad *f.*

D Herrschaft *f.*

R власть *ж.*; господство *ср.*

A حَكَم

4985 [E,U]

ruler

F règle *f.*

E regla *f.*

D Lineal *neut.*

R линейка *ж.*

A مِسْطَرَة *f*

4986

rules

F règles *m. pl.;* règlements

E reglamentos *m. pl.*

D Regeln *f. pl.*

R правила *ср. мн. ч.*

A قوانين/قواعد *f*

4987 [S]

rules of engagement

F règles d'engagement (antiaérien) *f. pl.*

E reglas de tiro antiaéreo *f. pl.*

D Feuerregelung für Flug-u. Fliegerabwehr *f.*

R порядок ведения огня *м.*; ведения в бой

A قَوَاعِد الإشْتِبَاك

4988 [P]

rummage *v.*

F fouiller

E registrar a fondo

D durchsuchen *untr.*

R обыскивать

A فتش بدقة

4989
rung
F échelon *m.*
E escalón *m.*
D Sprosse *f.*
R ступенька *ж.*
A دَرَجَة *f*

4990 [L]
runner
F agent de liaison *m.*
E ordenanza *m.*
D Meldegänger *m.*
R посыльный
A سَاعٍ *m*

4991
running engine
F moteur en marche *m.*
E motor en marcha *m.*
D laufender Motor *m.*
R работающий
  двигатель *м.*
A مُحَرِّك مُشْتَغِل *m*

4992
running light
F feux de bord [M] *m.*
  pl.; lanterne [V] *f.*
E luz *f.;* faro *m.*
D Fahrbeleuchtung *f.*
R навигационный огонь
  [M] *м.;* подфарник
  [V] *м.*
A نور تحديد الوضع

4993
runway
F piste (d'envoi) *f.*
E pista (de despegue) *f.*
D Startbahn *f.*
R взлётно-посадочная
  полоса *ж.*
A مَدْرَج *m*

4994 [E]
rupture *n.*
F rupture *f.*
E ruptura *f.*
D Bruch *m.*
R перелом *м.;* пролом
  м.
A تَشَقُّق *m*

4995
rupture force
F force de rupture *f.*
E fuerza de ruptura *f.*
D Stosstrupp *m.*
R прорывная сила *ж.*
A قوة الاختراق

4996
ruse
F ruse *f.*
E ardid *m.*
D List *f.*
R уловка *ж.;* хитрость
  ж.
A خُدْعَة *f*

4997
rush a position *v.*
F foncer sur une position
  ennemie
E arremeter contra una
  posición
D (eine Stellung) stürmen
R бросаться на позицию
A هجم باندفاع

4998
rust *n.*
F rouille *f.*
E herrumbro *f.;* oxidación
  f.
D Rost *m.*
R ржавчина *ж.*
A صَدَأ *m*

4999
rust *v.*
F (se) rouiller
E aherrumbrarse;
  oxidarse
D (ver)rosten
R ржаветь
A صَدِىء

5000
rust inhibitor
F antirouille *m. inv.*
E antioxidante *m.*
D Rostschutzmittel *neut.*
R антикоррозийное
  средство *ср.*
A مَانِع لِلصَّدَأ

5001
rustproof
F inoxydable
E inoxidable
D rostfrei
R нержавеющий
A مُضَاد لِلصَّدَأ

5002
rusty
F rouillé
E herrumbroso; oxidado
D rostig
R ржавый
A مُصْدَأ ـ صَدِىء

S

5003 [S]
sabot
F sabot *m.*
E sabot (término francés)
D Geschossführungspring
  m.
R поддон *м.*
A نعل القذيفة

5004

(armor-piercing discarding) sabot (APDS)

F obus perforant à sabot détachable *m.*
E munición APDS *f.*
D (panzerbrechendes) Geschoss *neut.*
R подкалиберный бронебойный снаряд с поддоном APDS *м.*
A ذخيرة عِيار مُصغر خارق للدروع *f*

5005

sabotage *n.*

F sabotage *m.*
E sabotaje *m.*
D Sabotage *f.*
R диверсия *ж.*; саботаж *м.*
A تَخْريب سِرّي *m*

5006

sabotage *v.*

F saboter
E sabotear
D sabotieren
R саботировать
A خَرَّب

5007

saboteur

F saboteur *m.*
E saboteador *m.*
D Saboteur *m.*
R диверсант *м.*
A مُخَرِّب

5008

sack *v.*

F piller
E saquear
D ausplündern *tr.*
R грабить
A نهب

5009

saddle *n.*

F selle *f.*
E silla *f.* ; sillín *m.*
D Sattel *m.*
R седло *ср.*
A سَرْج

5010 [T]

saddle *n.*

F selle *f.*
E collado *m.*
D Bergsattel *m.*
R седловина *ж.*
A مرتفع يصل بين قمتين

5011

(pack) saddle

F bât *m.*
E alforja *f.*
D Tornistersattel *m.*
R вьючное седло *ср.*
A سرج التحميل

5012

safe conduct

F sauf-conduit *m.*
E salvoconducto *m.*
D freies Geleit *neut.*
R охранное свидетельство *ср.*
A جَوَاز المُرُور

5013 [S]

safety catch

F cran de sûreté *m.*
E seguro *m.*
D Sicherung *f.*
R предохранитель *м.*
A مِسْمَار الأَمَان

5014

safety distance

F distance de sécurité *f.*
E distancia de seguridad *f.*
D Sicherheitsabstand *m.*
R безопасная дистанция *ж.*
A مَسَافة الأَمَان

5015

safety factor

F facteur de securité *m.*
E coeficiente de seguridad *m.*
D Sicherheitskoeffizient *m.*
R коэффициент безопасности *м.*
A عَامِل الأَمَان

5016 [M]

safety lane

F chenal de sécurité *m.*
E ruta de seguridad *f.*
D sichere Schiffahrtsroute *f.*
R безопасный проход *м.*
A مَمَرّ الأمان

5017

safety limit

F limite de sécurité *f.*
E límite de seguridad *m.*
D Sicherheitsgrenze *f.*
R граница безопасной зоны *ж.*; рубеж безопасной зоны *м.*
A حَدّ الأمان

5018

safety valve

F soupape de sûreté
E válvula de seguridad *f.*
D Sicherheitsventil *neut.*
R предохранительный клапан *м.*
A صِمَام الأمان

5019

sail (for)

F faire route (sur)
E zarpar (para)
D auslaufen (nach) *tr.*
R отправлять(ся)
A أَبْحَر (إلى)

319

5020

sail (leave port)
F sortir
E salir
D auslaufen *tr.*
R отплывать
A أَقْلَعَ

5021 [M]

(submarine) sail
F kiosque *m.*
E torreta (de un submarino) *f.*
D Kommandoturm *m.*
R боевая рубка *ж.*
A برج القيادة (في غواصة)

5022

sailing orders
F ordre d'appareillage *m.*
E instrucciones de navegación *f. pl.*
D Auslaufbefehl *m.*
R приказ о выходе в море *м.*
A أَوَامِر الإِقْلَاع

5023

sailing time
F heure de départ *f.*
E hora de salida *f.*
D Abfahrtszeit *f.*
R время отплытия *ср.*
A مُدَّة الإِبْحَار

5024

sailor
F matelot *m.*
E marinero *m.*
D Matrose *m.*
R матрос *м.*; моряк *м.*
A بَحَّار

5025

salient *n.*
F saillant *m.*
E saliente *m.*
D Keil *m.*
R выступ *м.*; клин *м.*
A بَارِز

5026

saline *adj.*
F salin
E salino
D Salz- *komp.;* salzig *adj.*
R соляной; солевой
A مِلْجِي

5027

salt water *adj. n.*
F (d')eau salée *(adj.) n. f.*
E (de) agua salada *(adj.) s. f.*
D Salzwasser(-) *neut. (komp.)*
R солёная вода *ж.;* солёный *прилаг.*
A مَاء مَلِح

5028

salute *v.*
F faire un salut (à qn.)
E saludar
D salutieren; grüssen
R отдавать честь
A حيَّا

5029

(take the) salute
F (passer les troupes en) revue
E tomar el saludo
D (die) Parade abnehmen *tr.*
R принимать честь
A إستعرض

5030 [S]

salute of ... guns *n.*
F salve de .... coups de canon *f.*
E salva de ....cañonazos *f.*
D Salut von .... Schuss *m.*
R салют *м.*
A تحية بالمدافع

5031 [E]

salvage *v.*
F récupérer
E recuperar
D sammeln
R собирать; вывозить
A إسْتَعَاد

5032 [M]

salvage *v.*
F sauver
E salvar
D bergen
R спасать
A أنْقَذَ

5033 [L,M]

salvage (clearance) *n.*
F déblaiement et renflouage *m.*
E limpieza *f.*
D Räumung *f.*
R расчистка *ж.;* очистка *ж.*
A تَخْلِيص *m*

5034 [E]

salvage (reclamation) *n.*
F récupération *f.*
E recuperación *f.*
D Abschub *m.;* Sammeln *neut.*
R сбор *м.;* эвакуация *ж.;* ремонт *м.*
A إسْتِعَادة *f*

5035 [M]

salvage (rescue) *n.*
F sauvetage *m.*
E salvamento *m.*
D Bergung *f.*
R спасение *ср.*
A إنْقَاذ *m*

5036

salvage tug
F remorqueur de
 sauvetage *m.*
E remolcador de
 salvamento *m.*
D Bergungsschlepper *m.*
R спасательный буксир
 м.
A قَاطِرَة إِنْقَاذ

5037

sanction *v.*
F sanctionner
E sancionar
D billigen
R одобрять
A صدَّق على

5038

sanctions *n. pl.*
F sanctions *f. pl.*
E sanciones *f. pl.*
D Sanktionen *f. pl.*
R санкции *ж. мн. ч.*
A عقوبات

5039

sand *n.*
F sable *m.*
E arena *f.*
D Sand *m.*
R песок *м.*
A رَمْل *m*

5040

sandbag *n.*
F sac de sable *m.*
E saco terrero *m.*
D Sandsack *m.*
R мешок с песком *м.;*
 земленосный мешок
 м.
A كِيس رَمْل

5041

sandbag *v.*
F renforcer une position
 avec des sacs de sable
E proteger con sacos de
 arena
D (mit) Sandsäcken
 befestigen
R защищать мешками с
 песком
A كسا باكياس رمل

5042 [T,M]

sand bank
F banc de sable [T] *m.;*
 barre [M] *f.*
E banco de arena *m.*
D Sandbank *f.*
R песчаная отмель *ж.*
A رصيف رملي

5043

sandstorm
F tempête de sable *f.*
E tempestad de arena *f.;*
 tormento de arena *m.*
D Sandsturm *m.*
R песчаная буря *ж.;*
 самум *м.*
A عَاصِفَة رَمْلِيَّة *f*

5044

sandy
F sablonneux
E arenoso
D sandig
R песчаный; песочный
A رَمْلي

5045 [B]

sap *v.*
F saper
E zapar
D untergraben
R подкапывать;
 подрывать
A نَقَب

5046 [A,L,M]

sash
F écharpe *f.*
E fajín *m.*
D Schärpe *f.*
R кушак *м.*
A حزام الخدمة

5047

satellite *adj. n.*
F satellite *adj. n. m.*
E satélite *adj. sing. m.*
D Hilfs- *komp.;* Satellit *m.*
R спутник *м.;* -спутник
 прилаг.
A تَابِع . قَمْر *m*

5048 [E,S,U]

saturate
F saturer
E saturar
D sättigen [EU]; schwer
 beschiessen [S]
R насыщать(ся);
 подавлять огнём
A أشْبَع

5049 [E,S,U]

saturation
F saturation *f.*
E saturación *f.*
D Sättigung *f.*
R насыщение *ср.;*
 подавление (огнём)
 ср.
A إشْبَاع *m*

5050 [D]

save *v.*
F mémoriser
E conservar
D speichern
R складывать
A ادّخر

5051

saw *v.*
F scier
E serrar
D sägen
R пилить(ся)
A نَشَر

5052 [P]
sawn off shotgun
F carabine à canon scié f.
E escopeta (de cañón acortado) f.
D (kurzgesägte) Schrotflinte f.
R ружьё с пропиленными стволами ср.
A بندقية صيد ذات سبطانة مقصّرة

5053
(bayonet) scabbard
F fourreau de baïonnette m.
E vaina (de bayoneta) f.
D Bajonettscheide f.
R ножны ж. мн. ч.
A غِمْد الحربة

5054 [E]
scale v.
F désincruster
E desincrustar
D (den) Kesselstein entfernen
R очищать от накипи
A قَاس

5055 [L]
scale v.
F ecalader
E escalar
D erklettern
R взбираться (на)
A تَسلَّق

5056 [B,T]
scale v.
F porter (etc...) à l'échelle
E escalar
D massstabgerecht machen
R изображать в определённом масштабе
A دَرَج

5057 [B,E,T]
scale n.
F échelle f.
E escala f.
D Massstab m.
R масштаб м.; шкала ж.
A مِقْياس m

5058 [R]
scan v.
F balayer
E explorar; escrutar
D abtasten tr.
R производить поиск; наблюдать
A مَسَح

5059
scan(ning) n.
F balayage m.
E exploración f.
D Abtasten neut.
R поиск м.; развёртка ж.; сканирование ср.
A مَسْح m

5060 [P]
scar
F cicatrice f.
E cicatriz f.
D Narbe f.
R рубец м.
A ندبة

5061 [R,S]
scatter n.
F dispersion f.
E dispersión f.
D Streuung f.
R рассеяние ср.
A تَناشُر m

5062
scatter v.
F (se) disperser; (s') écarter
E dispersar(se)
D (sich) streuen
R рассеивать(ся)
A تَناشَر

5063
schedule v.
F dresser le programme (de)
E fijar la hora (de); proyectar
D festsetzen tr.; (zeitlich) planen
R планировать
A نَظِّم جَدْوَلًا

5064
schedule n.
F horaire m.; planning m.
E horario m.; programa m.
D Plan m.; Zeittafel f.
R расписание ср.; план м.; график м.
A جَدْوَل m

5065
(convoy) schedule
F plan de convoi m.
E programa de convoy m.
D Geleitzugfahrplan m.
R план конвоя м.; расписание конвоя ср.
A تَوْقيت القافلة

5066 [V]
(servicing) schedule
F périodicité d'entretien f.
E programa de conservación m.
D (planmässiges) Fahrbereitmachen neut.
R график технического обслуживания м.
A جدول الصيانة

5067 [A,U]
scheduled
F régulier
E según horario
D fahrplanmässig
R плановый
A مُحَدَّد المِيعاد

5068 [S]
scheduled
F (à) l'horaire
E previsto
D Plan- *komp.*
R плановый
A مُعَيَّن

5069 [L]
(demolition) scheme
F plan de destruction *m.*
E plan de demolición *m.*
D Sprengplan *m.*
R план подрыва *м.*
A خطة التدمير

5070
(exchange) scheme
F convention d'échange *f.*
E acuerdo de canje *m.*
D Austauschplan *m.*
R план обмена *м.*
A خِطَّة تَبَادُل *f*

5071 [L]
scheme of maneuver
F conception de
    manoeuvre *f.*
E plan de maniobra *m.*
D Angriffsplan *m.*
R план движения *м.*
A خطة المناورة

5072
school
F école *f.*
E escuela *f.*
D Schule *f.*
R училище *ср.*; школа
    ж.
A مَدْرَسَة *f*

5073
science
F science *f.*
E ciencia *f.*
D (Natur-) Wissenschaft *f.*
R наука *ж.*
A عِلْم *m*

5074
scientific
F scientifique
E científico
D wissenschaftlich
R научный
A عِلْمِي

5075
scientist
F savant *m.*
E científico *m. f.*
D Wissenschaftler *m.*
R учёный
A عَالِم *m*

5076
scout *n.*
F éclaireur *m.*
E explorador *m.*
D Späher *m.*
R разведчик *м.*
A كَشَّاف *m*

5077
scout *v.*
F aller en reconnaissance
E explorar
D aufklären *tr.*
R разведывать;
    рекогносцировать
A كَشَف

5078
scout car
F véhicule blindé de
    reconnaissance *m.*
E carro *m.*
D Spähwagen *m.*
R разведывательная
    машина *ж.*
A سيارة استكشاف

5079
(long range) scouting
F découverte *f.*
E reconocimiento a larga
    distancia *m.*
D Fernspähung *f.*
R дальняя разведка *ж.*
A استطلاع

5080 [A,G]
scramble *n.*
F décollage immédiat *m.*
E despegue inmediato *m.*
D Alarmstart *m.*
R взлёт по тревоге *м.*
A البدء في الطَّيْران

5081 [R]
scramble *v.*
F brouiller
E cifrar; desmodular
D verwürfeln
R зашифровывать
A مَنَع مِن الإسْتِرَاق التليفوني

5082 [A,G]
scramble *v.*
F décoller immédiatement
E despegar
    inmediatamente
D (einen) Alarmstart
    durchführen
R взлетать по тревоге
A شَرَع في الطَّيْران

5083 [R]
scrambler
F brouilleur
E criptógrafo *m.*;
    desmodulador *m.*
D Verwürf(e)lungs-
    vorrichtung *f.*
R шифровальное
    устройство *ср.*
A مَانِع من الإسْتِرَاق التليفوني

5084
scrap *v.*
F envoyer à la ferraille
E reducir a chatarra
D verschrotten
R отдавать на слом
A نَفَى ـ الغى

323

5085 [E]
scrap *n.*
F déchets *m. pl.;* ferraille *f.*
E chatarra *f.*
D Abfälle *m. pl.;* Schrott *m.*
R металлолом *м.*
A نُفَايَة *f*

5086
scree
F éboulis *m.*
E ladera de cantos rodados *f.*
D Geröll *neut.*
R (каменистая) осыпь *ж.*
A رُكَام *m*

5087 [L,M]
screen *n.*
F écran *m.;* protection *f.*
E cobertura *f.*
D Sicherungsschleier *m.*
R прикрытие *ср.;* охранение *ср.*
A سِتَار *m*

5088 [S]
screen *n.*
F rideau (de fumée) *m.*
E cortina (de humo *f.;* de niebla)
D Nebelwand *f.;* Sperrfeuer *neut.*
R завеса *ж.;* маска *ж.*
A سِتَار

5089 [E,R]
screen *v.*
F abriter; blinder
E abrigar; blindar; cubrir
D abschirmen *tr.;* bewehren
R экранировать
A حَجَب

5090 [L,M]
screen *v.*
F assurer la sûreté; protéger
E proteger
D sichern
R прикрывать; охранять
A سَتَر

5091 [E,R]
screen *n.*
F écran *m.*
E pantalla *f.*
D Abschirmung *f.;* (Bild-)Schirm *m.*
R экран *м.*
A لَوْحَة *f*

5092 [S]
screen *v.*
F jalonner; masquer
E ocultar
D abschirmen *tr.;* einnebeln *tr.*
R прикрывать завесой; прикрывать огнём
A سَتَر

5093 [I]
screen *v.*
F trier; (faire le) tri de
E investigar
D durchsieben
R прочёсывать
A فحص المشبوهين

5094 [M]
(antisubmarine) screen
F écran de protection ASM *m.*
E cortina contra-submarinos *f.*
D Sicherung gegen-U-Boote *f.;* Geleitschutz *m.*
R противолодочное прикрытие *ср.*
A حاجز ضد الغواصات

5095 [V]
(flotation) screen
F jupe de flottaison *f.*
E pantalla de flotación *f.*
D Schwimmbalg *m.*
R юбка плавучести *ж.*
A جِهَاز تَعْويم (دبابة) *m*

5096 [L,M]
(security) screen
F protection *f.*
E cobertura de protección *f.*
D Sicherungsschleier *m.*
R охранение *ср.;* прикрытие *ср.*
A سِتَار أَمْن

5097
(smoke) screen
F rideau de fumée *m.*
E cortina de humo *f.*
D Nebelwand *f.*
R дымовая завеса *ж.*
A حِجَاب دُخَان

5098
screw *v.*
F visser
E atornillar
D schrauben
R привинчивать
A شَدَّ اللَّوْلَب

5099 [A,E,M]
screw *n.*
F hélice [AM] *f.;* vis [E] *f.*
E hélice [AM] *f.;* tornillo [E] *m.*
D Schraube *f.*
R винт *м.*
A لَوْلَب *m*

5100 [T]
scrub
F broussailles *f. pl.*
E maleza *f.*
D Gestrüpp *neut.*
R кустарник *м.*
A شُجَيْرَات قَصِيرَة *f*

5101
scupper *n.*
F dalot *m.*
E imbornal *m.*
D Speigatt *neut.*
R шпигат *м.*;
  водопроток *м.*
A بالوعة *f*

5102 [M]
scuttle *v.*
F saborder
E echar a pique
D anbohren *tr.*
R затоплять
A خَرَق سَفِينَة

5103 [M]
scuttle *n.*
F hublot *m.*
E barrenar
D Springluke *f.*
R люк *м.*
A كوة

5104
sea
F mer *f.*
E mar
D Meer *neut.;* See *f.*
R море *ср.*
A بَحْر *m*

5105 [M]
sea anchor
F ancre flottante *f.*
E ancla flotante *f.*
D Treibanker *m.*
R якорь *м.*
A مرساة طافية

5106
sea bed
F fond de la mer *m.*
E fondo marino *m.*
D Meeresboden *m.*
R дно моря *ср.*
A قاع البحر

5107
seaborne
F transporté par mer
E transportado por mar
D (auf dem) Seeweg
  befördert
R морской;
  перевозимый морем
A مَنْقُول بَحْرًا

5108
sea cock
F robinet de prise d'eau
  m.
E grifo de mar *m.*
D Bodenventil *neut.*
R кингстон *м.*;
  забортный клапан
  м.
A مِحْبَس مَاء البَحْر

5109
seagoing
F (de) haute mer
E (de) alta mar
D Hochsee- *komp.*
R мореходный; морской
A بَحْرِيّ

5110 [U]
seal *v.*
F cacheter; plomber;
  sceller
E cerrar; emplomar; sellar
D versiegeln
R ставить печать
A خَتَم

5111 [U]
seal *n.*
F cachet *m.;* plomb *m.;*
  sceau *m.*
E sello (de plomo) *m.*
D Siegel *m.;* Versiegelung
  f.
R печать *ж.*
A خِتْم *m*

5112 [E]
seal *n.*
F (dispositif d') étanchéité
  f.
E junta *f.;* sello *m.*
D Dichtung *f.*
R обтюратор *м.*
A وُصْلَة *f*

5113 [E]
seal *v.*
F boucher; rendre étanche
E cerrar; obturar
D abdichten *tr.*
R обтюрировать
A سَدّ

5114
sealed orders
F ordres cachetés *m. pl.*
E instrucciones selladas *f.*
  pl.
D versiegelte Befehle *m.*
  pl.
R запечатанные приказы
  м. мн. ч.
A أَوَامِر مَخْتُومة *f*

5115 [T]
sea level
F niveau de la mer *m.*
E nivel del mar *m.*
D Meereshöhe *f.*
R уровень моря *м.*
A مستوى البحر

5116 [E,M,U]
seam *n.*
F couture [U] *f.;*
  soudure[E] *f.;*
  joint[M] *m.*
E costura [MU] *f.;*
  soldadura [E] *f.*
D Naht *f.*
R шов *м.;* паз [ME] *м.;*
  спай [ME] *м.*
A خياطة ـ شَقّ *m*

325

5117
seamanship
F matelotage *m.*
E náutica *f.*
D Seemanstum *neut.*
R морская практика *ж.*
A مَهَارَات البَحْر

5118
sea mark
F balise *f.;* marque de
　　mer *f.*
E marcaciones marítimas
　　f. pl.
D Farbmarkierung *f.;*
　　Seezeichen *neut.*
R навигационный знак
　　м.
A مَنَار *m*

5119
seaplane
F hydravion *m.*
E hidroavión *m.*
D Wasserflugzeug *neut.*
R гидросамолёт *м.*
A طائرة مائية

5120
sear *n.*
F gâchette *f.*
E fiador *m.*
D Spannstück *neut.*
R спусковой рычаг *м.*
A لفحة الحر

5121
search *v.*
F rechercher; visiter
E explorar [AM];
　　cachear[P];
　　registrar[P]
D durchsuchen [P] *untr.;*
　　suchen
R искать; обыскивать
A بَحَث

5122 [A,M]
search *n.*
F recherche *f.*
E exploración *f.*
D Suche *f.*
R поиск *м.*
A إِرْتِيَاد *m* كَشْف *m*

5123 [P]
search *n.*
F visite *f.*
E registro *m.*
D Durchsuchung *f.*
R обыск *м.*
A تَفْتِيش *m*

5124
search and rescue
F recherche et sauvetage
　　f. m.
E exploración y rescate *f.*
　　m.
D Such- und
　　Rettungsoperation
　　komp. f.
R поиск и спасение *м.*
　　ср.
A البَحْث والإِنْقَاذ

5125
searched channel
F chenal exploré *m.*
E canal explorado *m.*
D geräumter Kanal *m.*
R обследованный
　　фарватер *м.*
A مَمَرّ مُفَتَّش *m*

5126 [P]
searcher
F fouilleur *m.*
E explorador *m.*
　　investigador *m.*
D Durchsuchende(r) *m.*
R обыскивающий *м.*
A مُفَتِّش *m*

5127
searchlight
F projecteur *m.*
E proyector *m.*
D Scheinwerfer *m.*
R прожектор *м.*
A نُور كَاشِف

5128
seasick(ness) *adj. n.*
F (ayant le) mal de mer
　　(adj.) n. m.
E mareado *adj.;* mareo *s.*
　　m.
D seekrank(heit) *adj. (s. f.)*
R морская болезнь *ж.*
A دُوَار البَحْر

5129
seat
F siège *m.*
E asiento *m.*
D Sitz *m.*
R сиденье *ср.;* место *ср.*
A مَقْعَد *m*

5130
sea wall
F digue *f.*
E rompeolas *m.;* dique
　　(marítimo) *m.*
D Deich *m.*
R дамба *ж.;* волнолом
　　м.
A سَدّ بَحري

5131
seaward(s)
F vers le large
E hacia el mar; mar a
　　dentro
D seewärts
R (направленный) к
　　морю
A نَحْو البَحْر

5132

seaworthy
(seaworthiness)
adj. (n.)

F apte (aptitude) à
  naviguer *adj. (n. f.)*
E (en) condiciones de
  navegar *(adj.) s. f.*
  pl.; marinero *adj.*
D seetüchtig *adj.;*
  Seetüchtigkeit *s. f.*
R мореходный *прилаг.;*
  мореходность *сущ.*
  ж.
A صَلَاحِيَّة الإِبْحَار

5133

second *n.*

F seconde *f.*
E segundo *m.*
D Sekunde *f.*
R секунда *ж.*
A ثَانِيَة *f*

5134

secondary

F auxiliaire; secondaire
E auxiliar; secundario
D Hilfs- *komp.;* Neben-
  komp.; sekundar *adj.*
R вторичный;
  второстепенный
A ثَانَوِي

5135

secondary radiation

F radiation secondaire *f.*
E radiación secundaria *f.*
D Sekundarstrahlung *f.*
R вторичное излучение
  ср.
A اِشْعَاع ثَانَوِي *m*

5136

secondment

F (être) détaché
E agregación *f.*
D Abstellung *f.*
R прикомандирование
  ср.
A إعارة

5137

second strike
capability

F capacité de deuxième
  frappe *f.*
E capacidad de segundo
  ataque *f.*
D Zweitschlagkapazität
  (meist eng. Begriff) *f.*
R способность к
  ответному удару *ж.*
A مَقْدِرَة الضَّرْبَة الثَّانِيَّة

5138

secrecy

F discrétion *f.*
E reserva *f.*
D Geheimhaltung *f.*
R скрытность *ж.;*
  секретность *ж.*
A سِرِّيَّة

5139

secret *adj. n.*

F secret *adj. n. m.*
E secreto *s. m.*
D geheim(nis) *adj. (s.*
  neut.)
R тайный *прилаг.;* тайна
  сущ. ж.; секретный
  прилаг.; секрет *сущ.*
  м.
A سِرِّي . سِرّ *m*

5140 [P,U]

secretary

F secrétaire *m. f.*
E ministro [P] *m.;*
  secretario [U] *m.*
D Sekretär *m.*
R секретарь *м.*
A أَمِين *m*

5141

secretary-general

F directeur-général *m.;*
  secrétaire-général *m.*
E secretario general *m.*
D Generalsekretär *m.;*
  Direktor *m.*
R генеральный-секретарь
  м.
A أَمِين عَامّ *m*

5142 [E]

section

F coupe *f.;* section *f.*
E sección *f.*
D (Quer-)Schnitt *m.;* Teil
  m.
R часть *ж.;* отсек *м.*
A مَقْطَع *m*

5143 [B]

(bridge) section

F section de pontage *f.*
E sección de pontoneros
  f.
D Brückenabschnitt *m.*
R понтонный блок *м.*
A مَقْطَع جسر

5144 [B]

(cross) section

F section transversale
  d'un pont *f.*
E sección transversal *f.*
D Querschnitt *m.*
R сечение *ср.*
A مقطع عرضي

5145

(fitter) section

F équipe de dépannage *f.*
E destacamento de
  mecánicos *m.*
D Instandsetzungstrupp
R слесарно-механическое
  отделение *ср.*
A حضيرة صيانة

5146

sector
F secteur *m.*
E sector *m.*
D Abschnitt *m.;* Sektor *m.*
R участок *м.;* сектор *м.*
A قِطاع *m*

5147

sector controller
F contrôleur de secteur *m.*
E director de sector *m.*
D Abschnittskontrolleur *m.;* Sektorkontrolleur *m.*
R офицер по наведению средства ПВО *м.*
A مُدير القِطاع

5148

sector of fire
F secteur de tir *m.*
E sector de tiro *m.*
D Zielabschnitt *m.*
R сектор обстрела *м.*
A قِطاع رَمْي

5149 [R]

sector scan
F recherche sectorielle *f.*
E exploración de sector *f.*
D Sektorenabtastung *f.*
R секторный поиск *м.*
A مَسْح القِطاع

5150 [B,M]

secure *adj.*
F saisi
E asegurado
D festgemacht
R закреплённый
A آمِن

328

5151 [L,P]

secure *adj.*
F sûr
E seguro
D sicher
R безопасный; защищённый
A مَحْمِي

5152 [L,M]

secure *v.*
F attacher; saisir
E asegurar
D befestigen [L]; fest machen [M]
R закреплять
A حَمِي

5153 [P]

secure *v.*
F mettre en sûreté
E cerrar
D sicherstellen *tr.*
R держать в надёжном месте
A أَمَّن

5154 [I,L,P]

security
F sécurité *f.*
E seguridad *f.*
D Geheimhaltung [IP] *f.;* Sicherheit [L] *f.*
R охранение *ср.;* безопасность *ж.*
A أَمْن *m*

5155 [I,P]

(close) security
F sécurité rapprochée *f.*
E guardia personal *f.*
D (enge) Sicherheit *f.*
R сохранность *ж.*
A أمن قريب

5156

(communications) security
F sécurité de transmissions *f.*
E seguridad de transmisiones *f.*
D Verbindungssicherheit *f.*
R безопасность связи *ж.*
A أمن المواصلات *m*

5157 [L]

(local) security
F forces et moyens de sûreté locale *f. pl. m. pl.*
E fuerzas de segurdad local *f. pl.*
D Nahsicherung *f.*
R ближнее охранение *ср.*
A الأمن المحلي

5158

(physical) security
F sécurité physique *f.*
E seguridad física *f.*
D (personelle) Sicherheit *f.*
R непосредственное охранение *ср.*
A الأمن المادّي *m*

5159

(signal) security
F sécurité des transmissions *f.*
E seguridad de transmisiones *f.*
D Meldesicherheit *f.*
R скрытность связи *ж.*
A أمن الاشارة

5160

security certification
F certificat de sécurité *m.*
E certificado de seguridad *m.*
D Sicherheitsbescheinigung *f.*
R удостоверение безопасности *ср.*
A تَصْدِيق الأمْن

5161

security check
F contrôle de sécurité *m.*
E control de seguridad *m.*
D Sicherheitsüberprüfung
  f.
R проверка
  безопасности *ж.*;
  проверка
  секретности *ж.*
A فَحْض الأمْن

5162

security classification
F classification de sécurité
  f.
E clasificación de
  seguridad *f.*
D Geheimhaltungsstufe *f.*
R гриф секретности *м.*
A تصْنِيف الأمْن

5163

security clearance
F habilitation de sécurité
  f.
E nivel de seguridad *m.*
D Sicherheitsbescheid *m.*
R допуск к секретной
  работе *м.*
A تَفْوِيض الأمْن

5164 [H]

(under) sedation
F (sous) sédation
E tranquilizado
D (unter dem Einfluss
  von)
  Beruhigungsmittel
R покой (под покоем) *м.*
A مُسَكِّن

5165

seismic
F sismique
E sísmico
D seismisch
R сейсмический
A زَلْزَالِي

5166

seismograph
F sismographe *m.*
E sismógrafo *m.*
D Seismograph *m.*
R сейсмограф *м.*
A مَرْسَمَة الزَّلازِل *f*

5167 [A,L,M,P]

seize
F capturer
E ocupar
D (er)greifen
R хватать; захватывать
A إسْتَوْلَى على

5168 [E]

seize
F coincer
E agarrotarse
D (sich) festfressen *tr.*
R заедать
A تعطل

5169 [A,L,M,P]

seizure
F prise *f.;* capture *f.*
E captura *f.*
D Ergreifung *f.*
R захват *м.*
A إستيلاء

5170

select (from)
F choisir (parmi)
E elegir; escoger;
  seleccionar (entre)
D (aus)wählen (aus) *tr.*
R выбирать (из);
  отбирать
A إختار (من)

5171

selection
F choix *m.*
E selección
D Auswahl *f.*
R выбор *м.*; отбор *м.*
A إخْتِيَار *m*

5172

selective
F sélecteur; sélectif
E selectivo
D Auswahl- [U] *komp.;*
  trennscharf [R] *adj.*
R отборный;
  селективный
A إخْتِيَارِي

5173

self-cocking
F automatique
E automático
D selbstspannend;
  automatisch
R самовзводный
A نصب تلقائي

5174

self-defense
F légitime défense *f.*
E defensa propia *f.*
D Selbstverteidigung *f.*
R самооборона *ж.*
A دفاع عن النفس

5175

self-propelled
F automoteur
E autopropulsado
D Selbstfahr- *komp.*
R самоходный
A ذَاتِي الحَرَكَة

5176

self-righting
F auto-élévateur;
  autoinclinable
E autoenderezante;
  autodirigido
D selbstaufrichtend
R остойчивый;
  самовыпрямляю-
  щийся
A الاسْتَعَادة الذَّاتِيَّة

329

5177

(combat) self-sufficiency

F autonomie de combat *f.*
E autosuficiencia en combate *f.*
D Kampfunabhängigkeit *f.*
R самостоятельность *ж.*
A ذاتي التموين (في القتال)

5178

semiactive homing

F autoguidage semi-actif *m.*
E guía semiactiva *f.*
D halbautomatische Zielsuchung (auch eng. Begriff) *f.*
R полуактивное самонаведение *ср.*
A تَوْجِيه نِصْف إِيْجابي

5179

semiautomatic

F semiautomatique
E semiautomática
D halbautomatisch
R полуавтоматический
A شِبْه تِلْقَائي

5180

semifixed ammunition

F munition à douille séparée *f.*
E munición semifija *f.*
D Aufsteckmunition *f.*
R унитарные боеприпасы с переменными зарядами *м. мн. ч.*
A ذَخِيرة نِصْف ثابِتة

5181

semitrailer

F semi-remorque *f.*
E semiremolque *m.*
D Auflieger *m.;* Sattelanhänger *m.*
R полуприцеп *м.*
A نِصْف مَقْطُورة

5182 [R]

send *v.*

F émettre
E transmitir
D senden
R передавать
A أرسل

5183

senior

F (de grade) supérieur
E superior; más antiguo
D ranghöher
R старший
A الأَقْدام

5184

seniority

F ancienneté *f.*
E antigüedad *f.*
D höheres Dienstalter *neut.;* höherer Dienstgrad *m.*
R старшинство *ср.*
A أَقْدَمِيَّة *f*

5185 [H]

sense *n.*

F sens *m.*
E sentido *m.*
D Sinn *m.*
R чувство *ср.*
A حَاسَّة *f*

5186 [E,H]

sense *v.*

F capter [E]; sentir [H]
E captar [E]; sentir [H]
D auffassen [E]; empfinden [H]
R ощущать; чувствовать
A أَحَسّ

5187

sensitive

F sensible
E delicado
D empfindlich
R чувствительный
A حَسَّاس

5188

sensitivity

F sensibilité *f.*
E sensibilidad *f.*
D Empfindlichkeit *f.*
R чувствительность *ж.*
A حَسَاسِيَّة *f*

5189

sensor

F détecteur
E detector *m.*
D Auffassungsgerät (meist eng. Begriff) *neut.*
R датчик *м.;* чувствительный элемент *м.*
A جِهَاز إِحْسَاس *m*

5190

(unattended ground) sensor

F détecteur au sol télécommandé *m.*
E sensor terrestre autónomo *m.*
D (unabhängiger) Bodensensor *m.*
R самостоятельный наземный элемент *м.*
A جهاز تلقائي ارضي للإحساس

5191

sentry

F factionnaire *m. f.;* sentinelle (en campagne)
E centinela *m.*
D Posten *m.*
R часовой
A حَارِس *m*

5192

(to stand) sentry

F (être de) garde; (être en) sentinelle
E (estar de) guardia
D Wache halten
R нести полевой караул
A قام بالحراسة

5193

sentry box
F guérite *f.*
E garita de centinela *f.*
D Schilderhaus *neut.*
R будка часового *ж.*
A مظلة الحارس

5194

separate *adj.*
F détaché; indépendant; séparé
E distinto; independiente; separado
D getrennt; selbstständig
R отдельный; индивидуальный
A مُنْفَصِل

5195

separate *v.*
F séparer
E separar
D trennen
R отделять(ся); разделять(ся)
A فَرَّق

5196

sequence
F succession *f.*
E sucesión *f.*
D Reihenfolge *f.*
R последовательность *ж.*; порядок *м.*
A تَسَلْسُل *m*

5197

sequential
F séquentiel
E consecutivo
D Folge- *komp.*; Reihenfolge- *komp.*
R последовательный
A مُتَتَابِع

5198

serial
F numéro d'ordre *m.*
E serie *f.;* número de orden (en una marcha) *m.*
D Marschgruppe *f.;* Transportgruppe *f.*
R маршевый эшелон *м.*
A مُتَسَلْسِل *m*

5199

serial number
F numéro de fabrication *m.;* numéro de série *m.*
E número de serie *m.*
D laufende Nummer *f.*
R личный номер *м.*; порядковый номер *м.*
A رَقْم مُسَلْسَل

5200

series
F série *f.*
E serie *f.*
D Serie *f.*
R серия *ж.*; ряд *м.*
A سِلْسِلَة *f*

5201

seriously wounded *adj.*
F grièvement blessé
E gravemente herido
D schwerverwundet
R тяжело раненый
A مجروح بخطورة

5202

service *adj.*
F militaire
E militar
D Dienst- *komp.*
R военный
A نظامي

5203

service *n.*
F service *m.*
E servicio militar *m.*
D (im) Dienst (sein)
R служба *ж.*
A خدمة *f*

5204

service *n.*
F armée *f.*
E ejército *m.*
D Teilstreitkraft *f.*
R род войск *м.*; служба *ж.*
A خِدْمَة *f*

5205 [S]

service *adj.*
F réel
E (de) guerre
D scharf
R боевой
A نِظَامي . حَرْبي

5206 [A,E,V]

service *v.*
F entretenir
E mantener
D pflegen; warten [V]
R заправлять; смазывать [V]
A صان

5207

(intelligence) service
F service de renseignements *m.*
E servicio de información *m.*
D Nachrichtendienst *m.*
R разведывательная служба *ж.*
A مصلحة الاستخبارات

5208

(in the) service

F (dans) les forces
   armeés; militaire
E (en el) servicio militar
   m.
D Wehrmachtsangehörige(r)
   m.
R служба ж.; (в/на
   службе)
A في الخدمة

5209 [A,L,M]

(logistic) service *adj.*

F logistique
E logístico
D Nachschub- *komp.;*
   Versorgungs- *komp.*
R обслуживающий;
   интендантский;
   тыловой
A خِدْمة (إدَارية)

5210

(ordnance) service

F service du matériel *m.*
E servicio de material (de
   guerra) *m.*
D Nachschubtruppe *f.*
R служба снабжения *ж.*
A إدَارة العتاد والذخيرة *f*

5211

(out of) service

F hors d'usage
E no funcionar
D ausser Dienst
R из строя
A خارج الخدمة

5212

(welfare) service

F service d'assistance
   sociale *m.*
E servicio de asistencia *m.*
D Betreuungsdienst *m.*
R служба
   культурно-бытового
   обслуживания *ж.*
A خِدْمَة التّرْفِيه

5213

serviceability

F disponibilité *f.*
E utilidad *f.*
D Verwendbarkeit *f.*
R исправное состояние
   ср.
A صَلاحِيَّة الخِدْمَة

5214

service club

F club militaire *m.*
E club militar *m.*
D Soldatenklub *m.*
R клуб военнослужащих
   м.
A نادي عسكري

5215

serviceman

F militaire *m.*
E militar *m.*
D Soldat *m.*
R военнослужащий *м.*
A عسكري

5216

service stripe

F galon de service actif
   m.
E galón de servicio
   militar *m.*
D Dienststreifen *m.*
R нашивка *ж.*
A شريط الخدمة

5217

servicing

F entretien *m.*
E entretenimiento *m.*
D Pflege *f.;* Wartung *f.*
R (техническое)
   обслуживание *ср.*
A صِيَانة *f*

5218

set (a fuze) *v.*

F régler (une fusée)
E graduar (un espoleta)
D (einen Zünder)
   einstellen
R устанавливать
   (взрыватель)
A جَهَّز (صِمَامة)

5219

(demolition) set

F dispositif de destruction
   m.
E juego de equipo de
   demolición *m.*
D Sprengmaterialien *neut.*
   pl.
R подрывной комплект
   м.; подрывное
   устройство *ср.*
A جِهَاز تَدْمِير *m*

5220

(igniter) set

F bouchon-allumeur *m.*
E equipo (de encendido)
   m.
D Zündsatz *m.*
R запал *м.*
A جهاز إشعال

5221

set (of tools etc.) *n.*

F jeu *m.*
E juego
D Satz *m.*
R комплект *м.;* агрегат
   м.
A مَجْمُوعَة *f*

5222

(radio) set

F poste de radio *f.*
E aparato (de radio) *m.*
D Funkgerät *neut.*
R радиоприёмник *м.*
A جهاز راديو

5223
set a map
F orienter la carte
E orientar un mapa
D (eine Karte) richten
R ориентировать карту
A وجّه خريطة

5224 [S]
setback
F recul *m.*
E reacción de inercia *f.*
D Rückschlag *m.*
R отход *м.;* неудача *ж.*
A تَراجُع *m*

5225
set course for
F mettre le cap vers
E hacer rumbo
D (etwas) ansteuern *tr.*
R прокладывать курс
(на)
A إتّجه إلى

5226
set off *v.*
F faire exploser
E (hacer) estallar
D detonieren
R взрывать
A أطلق

5227
set-piece *adj.*
F classique
E clásico
D Muster *komp.*
R планированный
A مُدبّر

5228 [T]
settlement
F centre *m.;* village *m.*
E pueblo *m.*
D Siedlung *f.*
R поселение *ср.*
A مستعمرة

5229
settling shot
F coup pour asseoir une
pièce *m.*
E disparo de
asentamiento *m.*
D Festschiessen *neut.*
R осадительный выстрел
м.
A طَلْقة التّثْبِيت

5230
set up an ambush
F tendre une embuscade
E preparar una
emboscada
D (hinterhalt) legen
R устраивать засаду
A نصب كميناً

5231
sewer
F égout *m.*
E alcantarilla *f.*
D Abwasserkanal *m.*
R сточная труба *ж.*
A أنبوب المجاري

5232
sextant
F sextant *m.*
E sextante *m.*
D Sextant *m.*
R секстан *м.;* секстант
м.
A سُدْسِيَّة *f*

5233
sextant angle
F angle de sextant *m.*
E ángulo de sextante *m.*
D Sextantwinkel *m.*
R угол по секстанту *м.*
A زَاوِية السُّدْسِيَّة

5234
sextant error
F écart de sextant *m.*
E desvío de sextante *m.*
D Sextantabweichung *f.*
R ошибка по секстанту
ж.; отклонение по
секстанту *ср.*
A خَطَأ السُّدْسِيَّة

5235
shackle *v.*
F mailler
E trabar
D anschäkeln *tr.*
R (при)соединять скобой
A قَيّد

5236
shackle *n.*
F maillon d'attache *m.*
E grillete *m.;* abrazadera
f.
D Schäkel *m.*
R скоба *ж.;* шекл *м.*
A حَلْقة *f* ـ قَيْد *m* ـ صَفْد *m*

5237
shade *v.*
F abriter de (soleil); voiler
(lumière)
E poner pantalla a (luz);
proteger contra (sol)
D abschirmen (Licht) *tr.;*
beschatten (Sonne)
R заслонять (от света);
затенять
A اظَلّ

5238
(in the) shade *n.*
F (à l')ombre *f.*
E (a la) sombra *f.*
D (im) Schatten *neut.*
R (в) тени
A (في) الظِّلّ *m*

333

5239 [A,M,P]
shadow v.
F suivre; veiller
E perseguir; vigilar [P]
D beschatten;
   überwachen *untr.*
R выслеживать; следовать
A تَعَقَّب

5240 [A,M]
shadow(er) n.
F (avion *m.*; navire *m.*)
   suiveur *m.*
E perseguidor *m.*
D Überwacher *m.*
R выслеживающий
A مُتَعَقِّب *m*

5241 [B]
shaft
F puits *m.*
E pozo *m.*
D Schacht *f.*; Schaft *m.*
R шахта *ж.*
A بِثْر *m*

5242 [E]
shaft
F arbre *m.*
E árbol *m.*; eje *m.*; lanza
   f.
D Schaft *m.*; Welle *f.*
R ось *ж.*; шпиндель *м.*
A جِذْع *m*

5243
(main rotor) shaft
F arbre rotor principal *m.*
E eje (del rotor principal)
   m.
D Hauptrotorwelle *f.*
R несущий винт *м.*
A جِذْع الدوّار الرئيسي

5244
(propeller) shaft
F arbre de transmission
   m.
E eje de transmisión *m.*
D Kardanwelle *f.*
R вал винта *м.*
A جذع الادارة

5245 [E,U]
shallow
F plat
E llano
D flach
R мелкий; плоский
A سَطْحي

5246 [M]
shallow adj.
F peu profond
E poco profundo
D seicht
R мелкий
A قَليل العُمْق

5247
shallow(s) n. pl.
F bas-fond *m.*; haut-fond
   m.
E bajos *m. pl.*
D Untiefe *f.*
R мелководье *ср.*
A مِياه غَيْر عَميقة *f*

5248
shaped charge adj. n.
F (à) charge creuse *(adj.)*
   n. *f.*
E (a) carga perforante
   (adj.) s. *f.*
D Hohlladung(s-) s. *f.*
   (komp.)
R кумулятивный заряд
   м.
A حُشْوَة مُشَكَّله *f*

5249 [E]
sharp adj.
F aigu; net; tranchant
E cortante; nítido (foto
   etc); puntiagudo
D scharf; spitz
R резкий; острый
A حَادّ

5250
sharpen v.
F aiguiser
E afilar
D schärfen
R точить; обострять
A سَنَّ

5251
sharpshooter
F tireur d'élite *m.*
E tirador selecto *m.*
D Scharfschütze *m.*
R меткий стрелок *м.*;
   снайпер *м.*
A رام ماهِر *m*

5252
shatterproof adj.
F (de) sécurité (verre)
E inastillable
D bruchfest
R небьющийся
A غَيْر قابل للكسر

5253 [S]
sheaf
F faisceau *m.*
E haz *m.*
D Feuergarbe *f.*
R веер *м.*; сноп *м.*
A حُزْمَة *f*

5254
sheath
F gaine *f.*
E vaina *f.*
D Armierung *f.*; Scheide
   f.
R ножны *ж. мн. ч.*
A غِمْد *m*

5255 [T]
sheer adj.
F (à) pic
E escarpado; abrupto
D steil
R отвесный
A عَمُودي

334

5256
(ground) sheet
F tapis de sol *m.*
E tela impermeable *f.*
D Zeltbahn *f.*
R плащ-палатка *ж.*
A مِشْمَع أَرْضِي *m*

5257
(map) sheet
F feuille de carte *f.*
E hoja *f.*
D Kartenauszug *m.*
R карта *ж.*
A صحيفة خريطة

5258
(military conduct) sheet
F carnet de punition *m.*
E hoja (de premios y castigos) *f.*
D Führungsunterlage *f.*
R карточка взысканий поощрений *ж.*
A سجل العقوبات

5259 [M]
sheet anchor
F ancre de veille *f.*
E ancla de la esperanza *f.*
D Notanker *m.*
R запасный становой якорь *м.*
A مرساة الامان

5260
sheet metal
F tôle *f.*
E chapa; hoja (estanada)
D Blech *neut.*
R листовой металл *м.*
A صَفِيحَة مَعْدِن

5261 [M]
shelf
F banc *m.;* bas-fond *m.;* haut-fond *m.*
E plataforma *f.*
D Schelf *m.*
R шельф *м.*
A رَصِيف صَخْرِي *m*

5262
shelflife
F durée de conservation *f.*
E duración *f.;* tiempo de vida *m.*
D Lagerbeständigkeit *f.*
R срок хранения *м.*
A عُمْر الخَزْن

5263 [S]
shell *n.*
F obus *m.*
E granada *f.*
D Geschoss *neut.*
R снаряд *м.*
A قَذِيفَة *f*

5264 [S]
shell *v.*
F bombarder
E bombardear
D beschiessen
R обстреливать
A قَصَف

5265
(anti-tank) shell
F obus anti-char *m.*
E proyectil (contracarros) *m.*
D Panzerabwehrgeschoss *neut.*
R противотанковый снаряд *м.*
A قذيفة مضادة للدبابات

5266
(armor-piercing) shell
F obus perforant *m.*
E proyectil (perforante) *m.*
D (panzerbrechendes) Geschoss *neut.*
R бронебойный снаряд *м.*
A قذيفة خارقة

5267
(base ejection) shell
F obus à éjection par le culot *m.*
E proyectil (de expulsión de culote) *m.*
D Zapfenauswurfgeschoss *neut.*
R снаряд с вышибным дном *м.*
A قذيفة (قَذْفها بعقِبها)

5268
(canister) shell
F obus *m.*
E proyectil (de bote de metralla) *m.*
D Geschosskanister *m.*
R картечный снаряд *м.*
A قذيفة شظايا

5269
(chemical) shell
F obus chimique *m.*
E granada química *f.;* proyectil químico *m.*
D Gasgranate *f.*
R химический снаряд *м.*
A قَذِيفَة كِيمِيَائية *f*

5270
(high explosive squash head) shell (HESH)
F obus explosif à tête d'écrasement *m.*
E proyectil HESH *m.*
D Weichkopfgeschoss *neut.*
R снаряд с расплющивающейся головной части *м.*
A قذيفة شديدة الانفجار ذات رأس مهروس

5271

(illuminating) shell
F obus éclairant *m.*
E proyectil de iluminación *m.*
D Leuchtgeschoss *neut.*
R осветительный снаряд *м.*
A قذيفة مضيئة

5272

(smoke) shell
F obus fumigène *m.*
E proyectil fumígena *m.*
D Nebelgeschoss *neut.*
R дымовой снаряд *м.*
A قذيفة دخان

5273

(target indicating) shell
F obus marqueur *m.*
E proyectil indicadora de objetivos *m.*
D Zielanzeigegeschoss *neut.*
R целеуказательны снаряд *м.*
A قذيفة تشير إلى الهدف

5274

shellfire
F bombardement par obus *m.;* bombardement d'artillerie *m.*
E fuego de artillería *m.*
D Granatfeuer *neut.*
R артиллерийский огонь *м.*
A رمي بقذائف

5275

shell-shocked
F commotionné
E (que padece) neurosis de guerra
D (unter einer) Kriegsneurose *f.*
R страдающийся от контузии
A مصاب بصدمة القذائف

5276

shelter *v.*
F (s')abriter
E abrigar(se)
D beschützen (vor); Schutz suchen
R укрывать(ся)
A أَلْجَأَ ـ لَجَأَ

5277

shelter *n.*
F abri *m.;* refuge *m.*
E abrigo *m.;* refugio *m.*
D Bunker [L] *m.;* Zufluchtsort *m.*
R убежище *ср.;* блиндаж *м.*
A مَلْجَأ *m*

5278 [M,T]

sheltered
F abrité
E abrigado
D geschützt
R укрытый; прикрытый
A محمي

5279

shelter-tent
F bivouac *m.*
E vivaque *m.*
D Zeltschutz *m.*
R бивак *м.*
A خيمة ملجأ

5280 [M,T]

shelve *v.*
F (s')incliner
E bajar [M]; estar en declive [T]
D (sich) neigen
R отлого спускаться
A إنْحَدَر

5281 [P]

shield *n.*
F bouclier *m.*
E escudo *m.*
D Schild *m.*
R щит *м.*
A تُرْس *m*

5282

shield *n.*
F pare-... *c.;* tôle protectrice *f.*
E capa protectora [E] *f.;* blindaje *m.*
D Schutz *m.;* Schutzblech *neut.*
R щит *м.;* защита *ж.*
A سِتَار *m*

5283

shield (from) *v.*
F blinder; protéger (contre)
E blinder; proteger (contra)
D abschirmen *tr.;* beschützen (vor)
R защищать (от)
A حَمَى

5284

(heat) shield
F écran thermique *m.*
E escudo térmico *m.*
D Hitzeschutz *m.*
R тепловой экран *м.*
A دِرْع حراري

5285 [N]

shielding
F écran de protection *m.*
E blindaje *m.*
D Abschirmung *f.*
R защита *ж.;* защитная толща *ж.*
A سِتَار *m*

5286

(radar) shift
F variation de centrage (radar) *f.*
E transporte del radar *m.*
D (Radar) neu richten
R сдвиг *м.;* фаз *м.*
A نقل الرادار

5287
(work) shift *n.*
F (travail par) équipe(s) *f.*
E (trabajo por) turno *m.*
D Schicht (-arbeit) *f.*
R смена *ж.*; сменная
работа *ж.*
A مُنَاوَبَة *f*

5288
shift fire *v.*
F déplacer le tir
E transportar el tiro
D Feuer verlegen
R водить доворот
A نقل النيران

5289
shift gear *v.*
F changer de vitesse
E cambiar de marcha
D (um)schalten *(tr.)*
R переключать передачу
A بَدَّل السُّرْعَة

5290
ship
F navire *m.*
E buque *m.*
D Schiff *neut.*
R корабль *м.*; судно *ср.*
A سَفِينة *f*

5291
ship's papers
F papiers de bord *m. pl.*
E documentación del
buque *f.*
D Schiffspapiere *f. pl.*
R капитанские
документы *м. мн. ч.*
A وثائق السفينة

5292
ship's time
F heure du bord *f.*
E hora de a bordo *f.*
D Schiffzeit *f.*
R судовое время *ср.*;
местное время *ср.*
A وَقْت السَّفِينة

5293
(container) ship
F bateau porte-conteneurs
m.
E buque de contenedores
m.
D Containerschiff *neut.*
R контейнерное
(грузовое) судно *ср.*
A سَفِينة شَحْن *f*

5294
(decoy) ship
F navire-piège *m.*
E buque señuelo *m.*
D U-Boot-Falle *f.*
R (противолодочное)
судно-ловушка *ср.*
A سَفِينة (خَادِعة) هيكَلِيَّة *f*

5295
(depot) ship
F ravitailleur *m.*
E buque nodriza *m.*
D Versorgungsschiff *neut.*
R плавучая база *ж.*
A سَفِينة مُسْتَوْدَع *f*

5296
(flag) ship
F vaisseau amiral *m.*
E buque almirante *m.*
D Flaggschiff *neut.*
R флагманский корабль
м.
A سفينة القيادة

5297
(heavy lift) ship
F navire gros porteur *m.*
E buque transporte de
cargas pesadas *m.*
D Schwerguttransporter
m.
R тяжёлое грузовое
судно *ср.*
A سَفِينة نَاقِلة *f*

5298
(landing) ship
F bâtiment de
débarquement *m.*
E buque de desembarco
m.
D Landungsschiff *neut.*
R высадочное судно *ср.*
A سفينة انزال

5299
(parent) ship
F ravitailleur *m.*
E buque matriz (de
abastecimiento) *m.*
D Mutterschiff *s.*
R плавучая база *ж.*
A سَفِينة أَسَاسِيَّة *f*

5300
(merchant) ship(ping)
F navire(s) marchand(s)
m. (pl.)
E buque(s) (mercante(s))
m. (pl.)
D Handelsschiff(e) *neut.*
(pl.)
R торговое судоходство
ср.; торговые суда
ср. мн. ч.
A بَحْرِيَّة تِجَارِيَّة *f*

5301
(replenishment) ship
F ravitailleur *m.*
E buque de
aprovisionamiento *m.*
D Vessorgungsschiff *neut.*
R судно переснаряжения
ср.
A سفينة تموين

5302
shipboard *adj.*
F (de) bord
E (a) bordo
D (an) Bord
R бортовой
A على متن السفينة

5303

shipload
F chargement *m.*
E cargamento (entero) *m.*
D (volle) Schiffladung *f.*
R судовой груз *м.*
A حُمُولة *f* شُحْنَة *f*

5304

shipment
F embarquement *m.;*
expédition *f.*
E embarque *m.;* envío *m.*
D Verladung *f.;* Versand
m.
R погрузка *ж.;*
перевозка *ж.*
A شَحْن *m*

5305

shipped
F installé à bord
E montado
D (an) Bord
R бортовой
A مركّب

5306

shipping
F bâtiments *m. pl.*
E buques *m. pl.*
D (Handels-)Flotte *f.;*
Schiffahrt *f.*
R суда *ср. мн. ч.;*
погрузка *ж.;*
судоходство *ср.*
A بَحْرِيَّة تِجَارِيَّة *f*

5307

shipping lane
F route maritime *f.*
E ruta de navegación *f.*
D Schiffahrtsstrasse *f.*
R морской путь *м.*
A مَمَرّ بَحْري

338

5308

ship-to-ship transfer
F transbordement *m.*
E transbordo *m.*
D Schiff-zu-Schiff
Verladung *f.*
R перегрузка
корабль-корабль *ж.*
A نَقْل مِن سَفِينة إلى أُخْرَى

5309

**ship-to-shore
movement**
F mouvement navire-
rivage *m.*
E movimiento buque-
orilla *m.*
D Anlanden *neut.*
R высадка (десанта) *ж.*
A نَقْل مِن سَفِينة إلى الشَّاطِىء

5310

shipwreck
F naufrage *m.*
E naufragio *m.*
D Schiffbruch *m.*
R кораблекрушение *ср.*
A حُطام السفينة

5311

shipyard
F chantier (naval) *m.*
E astillero *m.*
D Werft *f.*
R верфь *ж.;*
судостроительный
завод *м.*
A تِرسانة *f*

5312 [T]

shoal
F banc de sable *m.;* écueil
m.
E banco de arena *m.*
D Schwarm *m.*
R мелководье *ср.*
A قليل العمق

5313

shock
F choc *m.*
E choque *m.*
D Schock *m.;* Stoss *m.*
R удар *м.;* толчок *м.*
A صَدْمَة *f*

5314

(electric) shock
F décharge électrique *f.*
E shock eléctrico *m.*
D Elektroschock *m.*
R удар *м.;* толчок *м.*
A صدمة كهربائية

5315

(ground) shock
F onde de choc terrestre
f.
E onda explosiva terrestre
f.
D Bodenstoss *m.*
R наземный скачок
уплотнения *м.*
A صدمة ارضية

5316

shock absorber
F amortisseur *m.*
E amortiguador *m.;*
parachoques *m.*
D Stossdämpfer *m.*
R амортизатор *м.*
A جهاز إمتصاص الصدمات

5317

shock action
F action de choc *f.*
E acción de choque *f.*
D Schockwirkung durch
Feuer und/oder
Bewegung *f.*
R ударное действие *ср.*
A عَمَلِيَّة صَدم

5318
shock front
F front de choc *m.*
E frente de choque *m.*
D Schockwelle *f.*
R фронт ударной волны
  м.
A جِبْهَة الصَّدم

5319
shock-proof
F anti-choc
E (a) prueba de choques
D schocksicher
R ударностойкий
A مقاوم الصدمات

5320
shock troops *pl.*
F troupes de choc *f. pl.*
E tropas de choque *f. pl.*
D Angriffstruppen *f. pl.*
R ударные войска *ср.*
  мн. ч.
A قوات الصّدم *f*

5321
shoot *v.*
F fusiller; tirer
E disparar; fusilar; tirar
D erschiessen; schiessen
R стрелять; вести огонь
A رمَى

5322
shoot and scoot
  technique
F décrochage sitôt tiré *m.*
E disparar una arma y
  moverse rápidamente
  a otra posición de tiro
D Schiess- und
  Laufverfahren *neut.*
R ведение огня при
  быстром
  развёртывании и
  перемещении
A فن الرمي والهرب

5323
(snap) shooting
F tir instinctif *m.*
E fuego (instintivo) *m.*
D Schnellschiessen *neut.*
R стрельба навскидку
  ж.
A الرمي الخاطف

5324 [M]
shore
F côte *f.;* rivage *m.*
E costa *f.;* orilla *f.*
D Küste [M] *f.;* Ufer *neut.*
R берег *м.;* побережье
  ср.
A شَاطِىء *m*

5325
(hostile) shore
F rive ennemie *f.*
E ribera enemiga *f.*
D (feindliches) Ufer *neut.*
R чужой берег *м.*
A شاطىء العدو

5326
shore bombardment
F tir contre la terre *m.*
E bombardeo de la costa
  m.
D Küstenbeschuss *m.*
R артподдержка
  (морского) десанта
  ж.; обстреливание
  берега *ср.*
A قَصْف سَاحِلي

5327
shore leave
F permission à terre *f.*
E permiso de paseo en
  tierra *m.*
D Landurlaub *m.*
R отпуск на берег
A أجازة نزول الى البرّ

5328 [T]
shore line
F bord de la mer *m.*
E línea de la playa *f.*
D Küstenlinie *f.*
R побережье *ср.*
A خط الساحل

5329 [B]
shoring
F étai *m.*
E puntales *m. pl.;* entibos
  m. pl.
D Spreizholz *neut.*
R крепление *ср.*
A دعم (الابنية)

5330 [S]
short *adj. n.*
F court *adj. n. m.*
E corto
D kurz *adj.;* Kurzschuss *s.*
  m.
R короткий *прилаг.;*
  краткий *прилаг.;*
  недолёт *сущ. м.*
A قَاصِر (طَلقَة قَاصِرَه)

5331
shortage
F manque *m.*
E falta *f.*
D Mangel *m.*
R недостаток *м.*
A نَقْص

5332
short circuit *n.*
F court-circuit *m.*
E corto circuito *m.*
D Kurzschluss *m.*
R (короткое) замыкание
  ср.
A دائرة قصيرة

339

5333
short-handed
F (à) court de personnel
E falto de personal
D knapp an
  Arbeitskräften
R нуждающийся
A عجز في قوة الأفراد

5334
short-term
F (à) court terme
E (a) plazo corto
D kurzfristig
R краткосрочный
A قصير الأجل

5335 [R]
short-wave
F (sur) ondes courtes f. pl.
E (de) onda corta
D Kurzwelle(-) f.(komp.)
R коротковолновый
A موجة قصيرة

5336
shot n.
F coup de feu m.
E tiro m.
D Schuss m.
R выстрел м.;
  пристрелка ж.;
  запуск м.
A طَلْقة f

5337 [S]
"Shot!" ("On the way!")
F "Coup parti!"
E "¡Disparado!"
D „Abgefeuert!"
R "Выстрел!" "Пуск!" [G]
A مَرْمى

5338
shotgun
F fusil de chasse m.
E escopeta f.
D Schrotflinte f.
R ружьё ср.
A بُنْدُقِيَّة الصَّيْد

5339
shoulder strap
F bretelle f.; patte
  d'épaule f.
E hombrera f.
D Schultergurt m.
R погон м.; плечевой
A سَيْر الكَتْف

5340
(mechanical) shovel
F pelle mécanique f.
E excavadora con pala f.
D Bagger m.
R (механическая) лопата
  ж.
A مجرفة آلية

5341
shrapnel n.
F éclats d'obus m. pl.
E metralla f.
D Schrapnell neut.
R шрапнель ж.
A قنبلة شظايا

5342
shunting
F (opération de) triage f.;
  (manoeuvres
  d')aiguillage f. pl.
E maniobras f. pl.
D Rangieren neut.
R маневрирование ср.;
  шунтирование ср.
A المناورة

5343
shut v.
F fermer
E cerrar
D (ver-)schliessen;
  zumachen tr.
R закрывать
A أقفل

5344 [E]
shut down v.
F couper
E parar
D abschalten tr.
R прекращать работу;
  выключать
A وَقَف

5345 [E]
shutter
F obturateur m.
E obturador m.
D Verschluss m.
R затвор м.; заслонка
  ж.
A سِدَادة f غَلَق m

5346
shuttle v.
F faire la navette
E transportar en viajes de
  retorno
D pendeln
R перевозить
A أقْبَل وأدْبَر

5347
shuttle movement
F mouvement en navette
  m.
E transporte de ida y
  vuelta m.
D Pendelverkehr m.
R авто- составной;
  аэроперевозка ж.
A حَرَكة مُتَنَاوِبَة النَّقْل

5348
sick
F malade
E enfermo
D krank
R больной
A مَرِيض

5349

sick bay

F infirmerie *f.*

E enfermería *f.*

D (Schiffs-)Lazarett *neut.*

R лазарет *м.*

A جَنَاح المَرْضَى

5350

sick call

F appel de consultants *m.*

E reconocimiento médico *m.*

D Krankenappell *m.*

R вызов больных (к врачу) *м.*

A تَفَقُّد المَرْضَى

5351

side *adj.*

F latéral

E lateral

D Seiten- *komp.;* seitlich *adj.*

R боковой; побочный (вопрос)

A جَانِبي

5352

side *n.*

F côte *m.*

E lado *m.*

D Seite *f.*

R сторона *ж.*; бок *м.*; край *м.*

A جَانِب *m*

5353

side arms *pl.*

F armes blanches *f. pl.*

E armas de cinturón *f. pl.*

D Seitenwaffen *f. pl.*

R личное оружие *ср.*

A أَسْلِحَة تُحْمَل على الجنب

5354 [A,M,V]

sidelights *pl.*

F feux de position *m.*

E luces de situación [AM] *f. pl.;* luces laterales [V] *f. pl.*

D Begrenzungslichter [V] *neut. pl.;* Seitenlampen [AM] *f. pl.*

R боковые фонари *м.* *мн. ч.*

A أَنْوار جانِبِيَّة *f*

5355

(railway) siding

F voie de garage *f.*

E apartadero *m.*

D Rangiergleis *neut.*

R запасный путь *м.*

A سكة اضافية

5356

siege *n.*

F (état de) siège *m.*

E sitio *m.;* cerco *m.*

D Belagerung *f.*

R осада *ж.*

A حصار *m*

5357 [S]

sight *v.*

F pointer

E apuntar

D richten; zielen

R прицеливаться; наводить

A سَدَّد

5358 [U]

sight *v.*

F apercevoir

E ver

D erblicken

R обнаруживать

A نَظَر

5359 [S]

sight *n.*

F visée *f.*

E mira *f.;* visor *m.*

D Visier (-einrichtung) *neut.* (*f.*)

R прицел *м.;* визир *м.*

A مُسَدِّدَة *f*

5360

(aerial) sight

F mire aérienne *f.*

E alza aérea *f.*

D Visierantenne *f.*

R воздушный прицел *м.*

A مسددة جوية

5361

(back) sight

F cran de mire de la hausse *m.*

E alza posterior *f.*

D Rückvisier *neut.*

R целик *м.;* прицел *м.*

A مسددة خلفية

5362

(battle) sight

F hausse de combat *f.*

E alza abatida *f.*

D Nahkampfvisier *neut.*

R постоянный прицел *м.*

A مسددة قتالية

5363

(leaf) sight

F hausse à charnière *f.*

E alza abatible *f.*

D Klappvisier *neut.*

R (складной) прицел *м.*

A مُسَدِّدَة ذات صَفِيحَة *f*

5364 [S]

(night) sight

F appareil de visée
nocturne *m.*

E visor nocturno *m.*

D Nachtzieleinrichtung *f.*

R ночной
(инфракрасный)
прицел *м.*

A مُسَدَّدَة لَيْلِيَّة *f*

5365

(peep) sight

F hausse à oeilleton *f.*

E mira dióptica *f.*

D Guckvisier *neut.*

R смотровой прицел *м.*

A مسددة التدقيق

5366

(rear) sight

F cran de mire de la
hausse *m.*

E alza *f.*

D Kimme *f.*

R целик *м.*; прицел *м.*

A مسددة خلفية

5367

(reflex) sight

F viseur reflex *m.*

E mira de reflexión *f.*

D Spiegelvisier *neut.*

R зеркальный прицел *м.*

A مُسَدَّدَة إنْعِكَاسِيَّة *f*

5368

(ring) sight

F oeilleton de visée *m.*

E alza de anillo *f.*

D Ringvisier *neut.*

R кольцевой прицел *м.*

A مسددة ذات حَدَقة

342

5369

(tachometric) sight

F viseur tachymétrique *m.*

E mira tacométrica *f.*

D Vorhaltevisiereinrichtung
f.

R тахометрический
визир *м.*

A مسددة تاكومترية

5370

(telescopic) sight

F viseur télescopique *m.*

E goniómetro telescópico
m.

D Zielfernrohr *neut.*

R телескопический
прицел *м.*;
оптический прицел
м.

A مُسَدَّدَة تليسكوبية

5371

sighter

F balle d'essai *f.*

E disparo de armas
portátiles de
verificación del
goniómetro *m.*

D Visierschuss *m.*

R контрольный выстрел
м.

A طلقة تجريب

5372

sighting shot

F coup de réglage *m.*

E disparo de verificación
m.

D Einschiessschuss *m.*

R пристрелочный
выстрел *м.*

A طَلْقة تَجْرِيب

5373 [R]

(call) sign

F indicatif d'appel *m.*

E indicativo de llamada
m.

D Rufzeichen *neut.*

R позывной *м.*

A رَمْز النِّداء *m*

5374

(conventional) sign

F signe conventionnel *m.*

E señal convencional *f.*

D Kartensymbol *neut.*

R условный знак *м.*

A اشارة اصطلاحية

5375 [M]

**(international call)
sign**

F indicatif (d'appel)
international *m.*

E indicativo internacional
m.

D internationales
Rufzeichen *neut.*

R международный
позывной (сигнал)
м.; международный
свод *м.*

A إشارة نداء دولية

5376

(net call) sign

F indicatif (d'appel) de
réseau *m.*

E indicativo de red *m.*

D Netzrufzeichen *neut.*

R позывной (сигнал)
сети *м.*

A عَلَامَة نِداء الشَّبَكَة *f*

5377 [R]

**(procedure) sign
(prosign)**

F signe de procédure *f.*

E contraseña de servicio
f.

D Betriebszeichen *neut.*

R служебный сигнал *м.*

A إشارة لاسلكية مُنتظمة *f*

5378

(visual call) sign

F indicatif (d'appel) visuel *m.*

E indicativo visual *m.*

D optisches Rufzeichen *neut.*

R зрительный позывной (сигнал) *м.*

A رَمْزِنِدَاء بَصْرِي

5379

signal *v.*

F donner un signal; signaler

E hacer señales

D signalisieren

R сигнализировать; передавать; давать сигнал; давать знак

A أشار

5380 [R,U]

signal *n.*

F signal *m.;* transmission [R] *f.*

E señal [U] *f.;* transmisión [R] *f.*

D Signal *neut.;* (Funk-) Spruch [R] *m.*

R сигнал *м.;* передача [R] *ж.;* радиограмма [R] *ж.*

A إشَارَة *f*

5381 [GR]

(command destruct) signal

F signal de destruction télécommandé *m.*

E señal

D Vernichtungssignal *neut.*

R сигнал-команда по самоликвидации *ж.*

A اشارة المراقبة للتدمير

5382

(distress) signal

F signal de détresse *m.*

E señal de socorro *f.*

D Notsignal *neut.*

R сигнал бедствия *м.*

A إسْتِغَاثَة *f*

5383

signal (end of message)

F signal fin de communication *m.*

E señal de fin *f.*

D Spruchendesignal *neut.*

R сигнал о конце связи *м.;* СК

A اشارة نهاية البرقية

5384

(fog) signal

F signal de brume *m.*

E señal de niebla *f.*

D Nebelsignal *neut.*

R туманный сигнал *м.*

A إشَارَة الضَّبَاب *f*

5385

(ground) signal

F signal de trafic (air) *m.*

E señal terrestre *f.*

D Boden-Signal *neut.*

R наземный сигнал *м.*

A إشَارَة أَرْضِيَّة *f*

5386

(modulating) signal

F signal modulateur *m.*

E frecuencia modulada *f.*

D Modulationsignal *neut.*

R модулирующий сигнал *м.*

A اشارة التضمين

5387

(recognition) signal

F signal de reconnaissance *m.*

E señal de reconocimiento *f.;* consigna *f.*

D Erkennungssignal *neut.*

R сигнал опознания *м.*

A اشارة التمييز

5388

signal center

F centre de transmission *m.*

E centro de transmisiones *m.*

D Fernmeldezentrale *f.*

R узел связи *м.*

A مَرْكَز الإِشَارَة

5389

signal corps

F service de transmissions *m.*

E servicio de transmisiones *m.*

D Fernmeldetruppe *f.*

R служба связи *ж.;* войска связи *ср. мн. ч.*

A سِلَاح الإِشَارَة

5390

signal flag

F fanion de signalisation *m.*

E bandera de señales *f.*

D Signalflagge *f.*

R сигнальный флажок *м.*

A عَلَم إشَارَة

5391

signaller

F transmetteur *m.*

E señalero *m.*

D Melder *m.*

R связист *м.*

A عامل إشارة

5392 [A,R,S,V]
signature
F signature de détection *f.*
E señal de identificación *f.*
D typisches Signal *neut.*
R характеристика *ж.*
A سِمَة مُمَيِّزة *f*

5393 [U]
signature
F signature *f.*
E firma *f.*
D Unterschrift *f.*
R подпись *ж.*
A إمْضَاء *m*

5394
(sound) signature
F signature sonore *f.*
E reflexión acústica *f.*
D Kennton *m.*
R звуковое отражение *ср.*
A امضاء صوتي

5395
(weapon) signature
F signature d'une arme *f.*
E efectos identificadores del arma *m. pl.*
D Schusseigenschaft *f.*
R характеристика оружия *ж.*
A مميز السلاح

5396
signboard
F panneau *m.*
E letrero *m.*
D Aushängeschild *neut.*
R указатель *м.*; вывеска *ж.*
A لوحة الدلالة

5397
(road) signing
F fléchage et jalonnement *m. m.*
E señalamiento del camino *m.*
D Strassenbeschilderung *f.*
R указывание маршрутов *ср.*
A وضع اشارات الطرق

5398
sign up *v.*
F s'engager
E alistar
D (sich) verpflichten
R поступать (в армию)
A دخل في الخدمة

5399
silence
F silence *m.*
E silencio *m.*
D Schweigen *neut.*
R молчание *ср.*
A سُكُوت *m*

5400 [S]
silencer
F silencieux *m.*
E silenciador *m.*
D Schalldämpfer *m.*
R глушитель *м.*
A خَافِت *m*

5401 [V]
silencer
F silencieux *m.*
E silenciador *m.*
D Auspufftopf *m.*
R глушитель *м.*
A خافت

5402
silent
F silencieux
E silencioso
D schweigend
R бесшумный; безмолвный
A سَاكِت

5403
silent hours
F heures hors service *f. pl.*
E silencio *m.*
D Ruhezeit *f.*
R тёмное время *ср.*
A ساعات الليل

5404
silhouette
F silhouette *f.*
E silueta *f.*
D Silhouette *f.*
R силуэт *м.*
A شَبَح *m*

5405
(to be) silhouetted
F découper; (se) silhouetter
E destacarse
D (sich) abheben *tr.*
R быть вырисованный (на фоне горизонта); вырисовывать
A بَرَز بِشَكْل شَبَح

5406 [G]
silo
F silo *m.*
E silo *m.*
D Silo *m.*
R стартовая шахта *ж.*
A مستودع أسطواني

5407
silt
F vase *f.*
E sedimento *m.*
D Schlamm *m.*
R ил *м.*; тина *ж.*; наносы *м. мн. ч.*
A غِرْين *m* طَمْي *m*

5408
silt up
F (s')envaser
E obstruir(se) con
  sedimentos
D verschlammen
R засорять(ся) илом
A إِمْتَلأَ بِالطَّمْي

5409
simulate
F simuler
E simular
D nachbilden [E] *tr.;*
  vortäuschen *tr.*
R имитировать;
  симулировать
A تَظاهَرَ بِ

5410
simulator
F simulateur *m.*
E simulador *m.*
D Nachbildner *m.;*
  Simulator *m.*
R имитирующее
  устройство *ср.;*
  тренажер *м.*
A مُحاكٍ *m*

5411 [E]
single
F mono- *c.;* unique *adj.*
E mono- *c;* único *adj.*
D Ein- *komp.;* einfach
  adj.; Einfach- *komp.;*
  einzeln *adj.*
R простой; одно-
A مُفْرَد

5412
single file
F (en) file indienne *f.*
E fila de a uno *f.*
D Einzelreihe *f.*
R гуськом
A رتل أُحادي

5413
sink *v.*
F couler; mouiller (corps-
  mort; mine)
E hundir(se)
D sinken; versenken
R (за)тонуть; (за)топить
A غَرِق

5414
siphon *v.*
F siphonner
E sacar con sifón
D absaugen *tr.*
R переливать через
  сифон
A ثَعَب ـ أفرغ بالمص

5415
siren
F sirène *f.*
E sirena *f.*
D Sirene *f.*
R сирена *ж.;* гудок *м.*
A صُفَّارة الإنذار *f*

5416 [B,S,U]
site *n.*
F chantier [B] *m.;*
  emplacement *m.;*
  endroit *m.*
E lugar [U] *m.;* sitio [U]
  m.; situación [SU] *f.;*
  solar [B] *m.*
D Baustelle [B] *f.;*
  Standort [U] *m.;*
  Stellung [S] *f.*
R место *ср.;* участок *м.*
A مَوْضِع *m*

5417
(crossing) site
F lieu de franchissement
  m.
E punto de paso *m.*
D Übergangstelle *f.*
R пункт переправы *м.*
A موقع عبور *m*

5418 [G,N]
(hardened) site
F site protégé *m.*
E lugar protegido *m.*
D befestigte Fläche *f.*
R укреплённая
  площадка *ж.*
A مَوْضِع صَلْد *m*

5419 [A,L]
(landing) site
F site d'atterrissage *m.*
E lugar de desembarco
D Landestelle *f.*
R площадка десанта *ж.*
A مكان الهبوط

5420 [G]
(launching) site
F site de lancement *m.*
E lugar de lanzamiento *f.*
D Abschussraum *m.*
R стартовая площадка
  ж.
A قاعِدَة إطْلاق *f*

5421
situation
F situation *f.*
E situación *f.*
D Lage [ALM] *f.;*
  Standort *m.*
R обстановка *ж.;*
  положение *ср.*
A مَوْقِف *m*

5422
situation map
F carte de situation *f.*
E plano de la situación
  m.
D Lagekarte *f.*
R карта обстановки *ж.*
A خَريطة المَوْقِف

5423 [A,L,M,P]

situation map

F carte renseignée *f.;* plan directeur *m.*
E mapa de operaciones *m.*
D Lagekarte *f.*
R карта боевой обстановки *ж.*
A خريطة الموقف

5424

sketch a plan *v.*

F esquisser un plan
E hacer un dibujo
D skizzieren
R рисовать кроӄи
A رَسَم مخططاً

5425

sketch-map

F croquis du terrain *m.*
E dibujo *m.*
D Skizzenkarte *f.*
R схема *ж.;* отчётная карточка *ж.*
A خريطة تخطيطية

5426

ski *v.*

F skier
E esquiar
D schilaufen *tr.*
R ходить на лыжах; идти на лыжах
A تَزَحْلَق

5427

ski *n.*

F ski *m.*
E esquí *m.*
D Schi *m.*
R лыжа *ж.*
A زَحَّافة *f*

5428 [V]

skid *v.*

F déraper
E derrapar
D rutschen; schleudern
R заносить; буксовать
A إنْزَلَق

5429 [V]

skid *n.*

F dérapage *m.*
E derrape *m.*
D Rutschen *neut.;* Schleudern *neut.*
R занос *м.;* буксование *ср.*
A إنْزِلَاق *m*

5430 [A,E,M]

skid *n.*

F patin [AE] *m.;* traineau [M] *m.*
E patín [AE] *m.;* varadera [M] *f.*
D Kufe [A] *f.;* Schlitten [EM] *m.*
R хвостовой костыль *м.;* полоз *м.;* рельс *м.*
A زَحَّافة *f*

5431

skidproof

F antidérapant
E (a) prueba de patinazos
D schleudersicher
R нескользящий
A لا ينزلق

5432

(cross country) skiing

F ski de fond *m.*
E esquí nórdico *m.*
D Geländeschilauf *m.*
R ходьба на лыжах *ж.*
A التزحلق عبر الاراضي

5433

(downhill) skiing

F ski alpin *m.*
E esquí alpino *m.*
D Abfahrtschilauf *m.*
R горнолыжный спорт *м.*
A التزحلق (منحدراً)

5434

skill-at-arms

F (habilité au) maniement des armes *m.*
E manejo de armas *m.*
D Waffengeschicklichkeit *f.*
R стрелковое мастерство *ср.*
A مهارة في استعمال الاسلحة

5435

skim *v.*

F raser
E pasar rasando
D fliegen; gleiten
R скользить (по)
A طار فَوْق سَطْح المَاء

5436

skin

F peau *f.*
E piel *f.*
D Haut *f.*
R кожа *ж.*
A جِلْد *m*

5437

skin diver

F plongeur autonome *m.*
E bucendor autónomo *m.*
D Schwimmtaucher *m.*
R лёгководолаз *м.*
A غَاطِس *m*

5438

skip bombing

F bombardement par ricochet *m.*
E bombardeo de rebote *m.*
D Abpraller-Bombenwurf *m.*
R бомбометание с рикошетованием бомб *ср.*
A قَصْف إنْقِضَاضي فَصْعُود عَمُودي

5439 [R]

(zone) skip distance
F zone des ricochets f.
E distancia (zona) de salto f.
D tote Zone f.
R расстояние скачка ср.; величина мёртвой зоны ж.
A مَسَافَة التَّفْوِيت

5440

ski pole
F bâton de ski m.
E bastón de esquiar m.
D Schistock m.
R лыжная палка ж.
A عَمُود تَزَحْلُق

5441 [M]

skipper
F capitaine m.
E patrón m.; capitán m.
D Kapitän m.
R командир корабля м.
A رُبّان

5442

skirmish v.
F escarmoucher
E escaramuzar
D plänkeln
R перестреливаться
A نَاوَش

5443

skirmish n.
F escarmouche f.
E escaramuza f.
D Scharmützel neut.
R стычка ж.; схватка ж.
A مُنَاوَشَة f

5444

skirmisher
F tirailleur m.
E escaramuzador m.
D Plänkler m.
R стрелок м.
A مناوش

5445

skirt n.
F jupe de flottaison f.
E faldones m. pl.
D Weste f.
R юбка ж.; фальшборт м.
A حافة / طرف

5446

skirt v.
F contourner; longer
E bordear; rodear
D umgehen untr.
R обходить; идти вдоль края
A دَار حَوْل ـ سار بِمُحَاذاة

5447

ski troops pl.
F troupes à skis f.; troupes de montagne f.
E tropas esquiadoras f. pl.
D Schitruppen f. pl.
R лыжные войска ср. мн. ч.
A قِطَع التَّزَحْلُق ـ قوات التزحلق

5448

sky
F ciel m.
E cielo m.
D Himmel m.
R небо ср.
A سَمَاء f and m

5449

skydiving
F saut à ouverture retardée m.
E paracaída acrobática f.
D Fallschirmspringen neut.
R парашютизм м.
A إسقاط حرّ

5450

skyline
F horizon m.
E horizonte m.
D Horizont m.
R горизонт м.; линия горизонта ж.
A أُفُق m

5451 [R]

skywave
F onde ionosphérique f.
E onda de espacio f.
D Raumwelle f.
R отражённая радиоволна ж.
A موجة سماوية

5452 [B,E]

slab charge
F charge platte f.
E carga explosiva
D Plattenladung f.
R шашка ж.
A عجينة متفجرة

5453

slack adj. n.
F desserré adj.; mou adj. n. m.
E flojo adj. s. m.
D lose adj.; Lose s. neut.
R слабый прилаг.; ненатянутый прилаг.; слабина сущ. ж.
A مُرْتَخ ـ راكد m

5454

slacken
F desserrer; (se) détendre
E aflojar(se)
D erlahmen [U]; fieren[M]; lockern[E]
R ослаблять; замедлять
A حَلّ ـ تَمَهَّل

347

5455
slack water
F étale *m.*
E agua firma *f.*
D Stauwasser *neut.*
R стояние прилива *ср.;*
    стояние отлива *ср.*
A بَحْر ثَابِت المُسْتَوَى

5456
slant *v.*
F (s')incliner
E inclinarse
D schräg stellen (sein)
R направлять вкось;
    идти вкось;
    наклонять(ся)
A مَال

5457
slant *n.*
F inclinaison *f.*
E inclinación *f.*
D Neigung *f.*
R склон *м.;* уклон *м.*
A مَيْل *m*

5458
slanting
F incliné
E inclinado
D schräg
R косой; наклонный
A مَائِل

5459
slash *v.*
F entailler
E hostigar
D schlitzen
R рубить
A قطع

5460
sledge
F traineau *m.*
E trineo *m.*
D Schlitten *m.*
R сани *ж. мн. ч.;* нарты
    ж. мн. ч.
A زَحَّافَة *f*

5461
sleeping bag
F sac de couchage *m.*
E saco-manta *m.;* saco de
    dormir *m.*
D Schlafsack *m.*
R спальный мешок *ср.*
A كيس للنوم بداخله

5462 [E]
sleeve
F manchon *m.*
E manguito *m.*
D Muffe *f.*
R муфта *ж.;* втулка *ж.;*
    гильза *ж.*
A كُمّ *m*

5463
slice (logistic etc.)
F tranche *f.*
E parte (logística) *f.*
D Anteil *m.*
R доля (части) *ж.*
A قِطْعَة *f*

5464 [E]
slide *n.*
F coulisse (mobile) *f.;*
    glissière (fixe) *f.*
E corredera *f.*
D Rutschbahn (fest) *m.;*
    Schieber (beweglich)
    m.
R золотник *м.;* ползун
    м.
A مِزْلَقَة *f* زَلَّاقَة *f*

5465
slide *v.*
F (faire) glisser
E deslizar(se)
D gleiten (lassen)
R скользить
A زَلِقَ

5466
(photo) slide *n.*
F diapositive *f.*
E diapositiva *f.*
D Dia(positiv) *neut.*
R диапозитив *м.;*
    кассета *ж.*
A شَرِيحَة *f*

5467
slide rule
F règle à calculer *f.*
E regla de cálculo *f.*
D Rechenschieber *m.*
R логарифмическая
    линейка *ж.;* счётная
    линейка *ж.*
A مسطرة حاسبة

5468 [B,E,M]
sling *v.*
F élinguer
E eslingar
D umschlingen
R подвешивать;
    подвязывать
A عَلَّق

5469 [S]
sling *n.*
F bandoulière *f.;* bretelle
    f.
E bandolera *f.;* portafusil
    m.
D Riemen *m.*
R ремень *м.*
A جَمَالَة *f*

5470 [B,E,M]
sling *n.*
F élingue *f.*
E eslinga *f.*
D Gurt *m.*
R подвязка *ж.;* строп
    м.; петля *ж.*
A حَبْل رَفْع

5471
slip v.
F glisser
E deslizarse
D gleiten; schlüpfen
R скользить
A إنْزَلَق

5472 [E]
slip n.
F glissement m.
E deslizamiento m.
D Schlupf m.
R скольжение ср.
A إنْزِلَاق m

5473
slipknot
F nœud coulant m.
E nudo corredizo m.
D Laufknoten m.;
Ziehknoten m.
R скользящий узел м.
A عُقْدَة زَالِقَة

5474
slippery
F glissant
E resbaladizo
D schlüpfrig
R скользкий
A زَلِق

5475 [A,V]
slipstream
F sillage (d'air) m.
E torbellino m.
D Propellerbö f.
R воздушный поток м.;
спутная струя ж.
A رِيح المِرْوَحَة

5476
slipway
F cale m.; slip
E grada f.
D Slipp neut.
R стапель м.; слип м.
A رَصِيف الجَرّ والسَّحْب

5477
(gun) slit
F embrasure f.;
meurtrière f.
E tronera f.
D Schießspalte f.
R прорез м.
A كُوَّة المِدْفع

5478
(forward) slope
F glacis m.
E glacis m.
D Vorderhang m.
R передний скат м.
A مُنْحَدِر أمامي m

5479 [T]
(gradual) slope
F pente douce f.
E pendiente gradual f.
D leichte Neigung f.
R постепенный скат м.
A مِيل تدريجي m

5480
(reverse) slope
F contre-pente f.
E contrapendiente f.
D Hinterhang m.
R обратный скат м.
A منحدر خلفي

5481 [S]
(extractor) slot
F fenêtre d'extraction f.
E ventana de expulsión f.
D Auszieherluke f.
R выбрасывательный
паз м.
A مَجرى اللقاف

5482
slow
F lent
E lento
D langsam
R медленный; тихий
A بَطِيء

5483
slow down (up)
F ralentir
E moderar la marcha de
D (sich) verlangsamen
R замедлять(ся)
A أَبْطَأ

5484
sluice
F écluse f.
E esclusa f.
D Schleuse f.
R шлюз м.; затвор м.
A سَدّ ذُو بَوَّابَات

5485
small arms pl.
F armes portatives f. pl.
E armas portátiles f. pl.
D Handfeuerwaffen f. pl.
R ручное оружие ср.;
стрелковое оружие
ср.
A أَسْلِحَة صَغِيرة f

5486
small-bore adj.
F (à) petit calibre
E (de) calibre reducido
D kleinkaliber
R малокалиберный
A صغير العيار

5487
"smart" bomb
F projectile laser-guidé ou
vidéoguidé
E proyectil inteligente m.
D Smart-geschoss (video-
oder lasergelenkters)
Geschoss m.
R бомба с лазерным
наведением ж.
A قنبلة موجهة بالليزر

5488 [A,S,P]
smoke *adj.*
F fumigène
E fumígeno
D Nebel- *komp.*
R дымовой
A دُخَاني

5489
smoke *n.*
F fumée *f.*
E humo *m.*
D Nebel [S] *m.;* Rauch [U] *m.*
R дым *м.*
A دُخَان *m*

5490
(identification) smoke
F fumeé d'identification *f.*
E humo indicador *m.*
D Erkennungsrauch *m.*
R опозновательный дым *м.*
A دخان التمييز

5491
(obscuration) smoke
F écran de fumée *m.*
E humo de oscurecimiento *m.*
D Verdunkelungsrauch *m.*
R маскирующий дым *м.*
A دخان إظلام

5492
(screening) smoke
F écran de fumeé *m.*
E humo de ocultación *m.*
D Deckungsnebel *m.*
R маскирующий дым *м.*
A حجاب دخان

5493
smoke pot
F fumigène *m.*
E aparato productor de humos *f.*
D Rauchtopf *m.*
R дымовая шашка *ж.*
A علبة دخان

5494
smoothbore *adj.*
F (à) âme lisse
E (de) ánima lisa
D (mit) glattem Rohr
R гладкоствольный
A جَوْف أَمْلَس

5495
smuggle
F faire de la contrabande
E pasar de contrabando
D schmuggeln
R провозить контрабандой; заниматься контрабандой
A هَرَّب

5496
smuggler
F contrebandier *m.*
E contrabandista *m. f.*
D Schmuggler *m.*
R контрабандист *м.*
A مُهَرِّب *m*

5497 [A]
snake mode
F contrôle de manoeuvre d'identification *m.*
E control de maniobra de identificación
D Suchflugbahn *f.*
R манёвр "змейка" *м.*
A طَرِيقَة الحَوِيَّة ـ التحرك بالزجزاج

5498
snapshot
F tir instinctif *m.*
E tiro rápido *m.*
D Schnellschuss *m.*
R выстрел без прицела *м.*
A خطفة

5499
sneak attack
F attaque-surprise *f.*
E ataque de sorpresa *m.*
D Überraschungsangriff *m.*
R внезапная атака *ж.*
A غارة خفية

5500 [L,P]
sniffer
F renifleur *m.*
E detector de explosivos *m.*
D Schnüffeler *m.*
R нюхатель *м.*
A آلة تحديد مكان المتفجرات بتحليل الهواء

5501 [P]
sniffer dog
F chien de recherche *m.*
E perro detector de explosivos *m.*
D Schnüffelhund *m.*
R собака-искатель *м.*
A كلب يشم المتفجرات

5502
sniper
F tireur isolé *m.*
E francotirador *m.*
D Scharfschütze *m.*
R снайпер *м.;* меткий стрелок *м.*
A قَنَّاص

5503
sniperscope
F viseur à lunette *m.*
E alza óptica de tirador *f.*
D Infrarotvisier *neut.*
R снайперский телескоп *м.*
A مرقب القنّاص

5504

snooperscope

F lunette à infrarouge *f.*
E alza óptica infrarroja *f.*
D Infrarotvisier mit Bildwandler *neut.*
R инфракрасный усилитель изображений *м.*
A غاص بالمنشاق

5505 [M,V]

snorkel *v.*

F naviguer au schnorkel [M]; traverser en submersion [V]
E navegar a snorkel [M]; atravesar a snorkel [V]
D schnorcheln
R переправляться с шноркелем
A غَاص المِنْشَاق

5506

snorkel *n.*

F schnorkel *m.*
E tubo snorkel (de respiración) *m.*
D Schnorchel *m.*
R шноркель *м.*
A مِنْشَاق ـ أنبوب التنفس *m*

5507

snow *n.*

F neige *f.*
E nieve *f.*
D Schnee *m.*
R снег *м.*
A ثَلْج *m*

5508

snow *v.*

F neiger
E nevar
D schneien
R сыпаться снегом; идёт снег
A هَبَط الثَّلْج

5509

snow blindness

F cécité des neiges *f.*
E ceguera causada por los reflejos de la nieve *f.*
D Schneeblindheit *f.*
R слепота от снега
A عمى الثلج

5510

snowdrift

F congère *f.*
E montón de nieve *m.*
D Schneewehe *f.*
R сугроб *м.*
A كومة ثلج

5511

snow hole

F trou dans la neige *m.*
E cueva *f.*
D Schneeloch *neut.*
R укрытие-нора *ж.*
A حفرة ثلج

5512

snowline

F limite des neiges éternelles *f.*
E límite de las nieves perpetuas *m.*
D Schneegrenze *f.*
R снеговая линия *ж.*; предел вечного снега *м.*
A خَطّ الثَّلْج

5513

snowmobile

F auto-neige *f.*
E vehículo para la nieve *m.*
D Schneemobil *neut.*
R снегоход *м.*
A زحّافة الثلج

5514

snow plow

F chasse-neige *m. inv.*
E quitanieves *m.*
D Schneepflug *m.*
R снеговой плуг *м.*; снегоочиститель *м.*
A مِحْفَار ثَلْج

5515

snow shoes

F raquettes *f. pl.*
E raquetas de nieve *f. pl.*
D Schneeschuhe *m. pl.*
R снегоступы *мн. ч.*
A نعل التزلج

5516

snow tire

F pneu à clou *m.*
E neumático de nieve *m.*
D Schneereifen *m. pl.*
R снегоходная шина *ж.*
A إطار ثلج

5517

soften up

F affaiblir
E debilitar
D zermürben
R ослаблять оборону; обрабатывать оборону
A أَضْعَف

5518 [D]

software

F logiciel *m.*
E software *m.*
D Software *neut.*
R программное обеспечение *ср.*
A اللوازم لوضع برنامج الآلة الحاسبة

5519

soil

F terre *f.*
E suelo *m.*
D Boden *m.*
R почва *ж.*; земля *ж.*
A تُرَاب *m* ـ تُرَبَه *f*

351

5520

soil shear strength (soil bearing pressure)
F résistance des sols *f.*
E resistencia del suelo *f.*
D Scherfestigkeit des Bodens *f.*
R прочность почвы на сдвиг *ж.*; прочность почвы на срезывание *ж.*
A قوة تَحَمُّل للتُرْبَة

5521

solar
F solaire
E solar
D Solar- *komp.;* Sonnen- *komp.*
R солнечный
A شَمْسِي

5522

solder *v.*
F souder
E soldar
D löten
R паять; спаивать
A لحم

5523

soldier
F soldat *m.*
E soldado *m.*
D Soldat *m.*
R солдат *м.*; военнослужащий *м.*
A جُنْدِي *m*

5524

(foot) soldier
F fantassin *m.*
E infante *m.*
D Fußsoldat *m.*
R пехотинец *м.*
A جندي مُشاة *m*

352

5525

(regular) soldier
F soldat de métier *m.;* officier de carrière *m.*
E soldado (profesional) *m.*
D Berufssoldat *m.*
R солдат *м.*; воин *м.*
A جندي نظامي

5526

soldierly
F (de) militaire
E militar
D militärisch
R воинский
A عسكري

5527

soldiery
F soldats *m. pl.*
E soldadesca *f.*
D Militär *neut.*
R солдаты *м. мн. ч.;* солдатня *ж.*
A الجند

5528

solid-state *adj.*
F transistorisé
E (de) estado sólido
D fest *adj.;* Fest- *komp.*
R твёрдый; (в) твёрдом состоянии
A حالة الصلابة

5529

solitary confinement
F (au) régime cellulaire *m.*
E (estar) incomunicado
D Einzelhaft *f.*
R одиночное заключение *ср.*
A سجن منفرد

5530

solvent *n.*
F solvant *m.*
E solvente *m.*
D Lösungsmittel *neut.*
R растворитель *м.*
A مُذِيب *m*

5531

sonar
F sonar *m.*
E sonar *m.*
D Sonar *neut.*
R сонар *м.;* гидролокатор *м.*
A سُونَار *m*

5532

(dipping) sonar
F sonar immersif *m.*
E sonar submergido *m.*
D Tauchsonar *neut.*
R гидролокатор *м.;* сонар *м.*
A سونار معلق بطائرة عمودية

5533

(dunking) sonar
F sonar d'hélicoptère *m.;* sonar immergeable
E sonar suspendido de helicóptero *m.*
D Tunksonar *neut.*
R гидролокатор (на) буксире *м.*
A سونار مقطور خلف سفينة

5534

(passive) sonar
F sonar passif *m.*
E sonar pasivo *m.*
D passives Sonar *neut.*
R пассивная гидролокация *ж.*
A سُونَار سَلْبِي

5535

sonobuoy

F radiobouée f.

E boya acústica f.;
  sonoboya f.

D Sonoboje f.

R гидроакустический
  буй м.; звуковой буй
  м.

A طَافِيَة صَوْتِيَّة رَادِيَّة f

5536 [A]

sortie

F sortie f.

E salida f.

D Einsatz m.

R самолётовылет м.;
  боевой вылет м.;
  вылазка ж.

A إِنْطِلاَقَه ـ طَلْعة f

5537 [U]

sound n.

F son m.

E sonido m.

D Schall neut.

R звук м.

A صَوْت m

5538 [M]

sound n.

F détroit m.

E estrecho m.

D Meerenge f.

R пролив м.; бухта ж.

A مَمَر مَائِي m

5539 [M]

sound v.

F sonder

E sondar

D loten

R измерять глубину;
  бросать лот

A سَبَر عُمق

5540

sound barrier

F mur du son m.

E (barrera del) sonido f.

D Schallmauer f.

R звуковой барьер м.

A الحَاجِز الصوتي

5541

sounding

F sondage m.

E sondeo m.

D Loten neut.

R измерение глубины ср.

A سَبْر عُمق m

5542

sound proofing

F insonorisation f.

E insonorización f.

D Schalldichtung f.

R звукоизоляция ж.

A تخفيف الصوت

5543

sound ranging

F repérage par le son m.

E fonolocalización f.

D Schallrichten neut.

R звукометрия ж.

A تحديد المدى بالصوت

5544

soundwave

F onde sonore f.

E onda acústica f.

D Schallwelle f.

R звуковая волна ж.

A مَوْجَه صَوْتِيَّة

5545

south(ern) adj. n.

F (au, du) sud (adj.) n.
  m.;

E (al, del) sur (adj.) s. m.;

D Süd- komp.; Süden s.
  m.; südlich adj.

R южный прилаг.; юг
  сущ. м.

A جَنُوبِي . جَنُوب m

5546 [U]

span n.

F portée f.

E vano (techo) m.

D Spannweite f.

R размах м.

A عَرْض m

5547 [B]

span n.

F portée f.; travée (pont)
  f.

E arcada (puente) f.; luz
  f.

D Brückenbogen m.;
  Spannweite f.

R пролёт м.

A حَامِلَة جِسر

5548 [B]

span v.

F franchir

E cruzar

D überbrücken untr.

R перекидывать (мост);
  соединять

A جَسَّر

5549 [A]

span n.

F envergure f.

E envergadura f.

D Spannweite f.

R размах м.

A جِسر

5550

**(ditch) spanning
  ability**

F franchissement de fosse
  m.

E capacidad de cruzar
  (zanjas) f.

D Spannfähigkeit f.

R способность
  преодолевать
  (канаву) ж.

A قدرة الاجتياز

5551 [M]

spar *n.*
F espar *m.*
E palo *m.; mastil m.*
D Rundholz *m.*
R рангоут *м.;* бревно
  ср.; шест *м.*
A عَارِضَة ـ سارية *f*

5552

spare wheel
F roue de secours *f.*
E rueda de repuesta *f.*
D Ersatzrad *neut.*
R запасное колесо *ср.*
A عجلة احتياطية

5553

spark *n.*
F étincelle *f.*
E chispa *adj.*
D Funke *m.*
R искра *ж.*
A شَرَارَة *f*

5554

spat
F guêtre *f.*
E polaina corta *f.*
D Gamasche *f.*
R гетры *мн. ч.*
A غطاء لحماية رسخ القدم

5555

spearhead an attack
  v.
F marcher en pointe; (être
  le) fer de lance de
  l'attaque
E conducir un ataque *v.*
D (die) Spitze bilden
R действовать в первом
  эшелоне;
  действовать клином
  атаки
A قاد رأس سهم الهجوم

354

5556

spearhead group
F bataillon d'intervention
  m.
E grupo de vanguardia *m.*
D Angriffsspitze *f.*
R острие клина *ср.;*
  передовой отряд *м.*
A مجموعة رأس المقدمة

5557

special activities *pl.*
F services spéciaux *m. pl.*
E servicios especiales *m.*
  pl.
D Sonderdienste *m. pl.*
R специальные действия
  ср. мн. ч.;
  спецдействия *ср. мн.*
  ч.
A أَعْمَال خَاصَّة *f*

5558

special ammunition
F munition spéciale *f.*
E munición especial *f.*
D Sondermunition *f.*
R специальные
  боеприпасы *м. мн. ч.*
A ذَخِيرَة خَاصَّة *f*

5559

special forces *pl.*
F forces spéciales *f. pl.*
E fuerzas especiales *f. pl.*
D Sonderverbände *m. pl.*
R войска специального
  назначения *ср. мн.*
  ч.; диверсионные
  войска *ср. мн. ч.*
A قُوَّات خَاصَّة *f*

5560

specialist *n.*
F spécialiste *m. f.*
E especialista *m. f.*
D Fachmann *m. ;*
  Sachverständige(r) *m.*
R специалист *м.*
A إِخْتِصَاصِي *m*

5561

specialize
F (se) spécialiser
E especializarse
D (sich) spezialisieren
R специализировать(ся)
A تَخَصَّص

5562

specification
F spécification *f.*
E especificación *f.*
D Spezifikation *f.*
R спецификация *ж.*
A تَعْيِين *m* ـ مواصَفات *f*

5563

specific gravity
F poids spécifique *m.*
E peso específico *m.*
D spezifisches Gewich
  neut.
R удельный вес *м.*
A الثَّقْل النَّوْعِي *m*

5564

specify
F préciser
E especificar
D festlegen *tr.*
R точно определять;
  устанавливать
A وَاصَف ـ عَيَّن ـ حَدَّد

5565

spectrum
F spectre *m.*
E espectro *m.*
D Spektrum *neut.*
R спектр *м.;* диапазон
  м.
A طَيْف *m*

5566

speed
F vitesse *f.*
E velocidad *f.*
D Geschwindigkeit *f.*
R скорость *ж.;* темп *м.;*
  ход *м.*
A سُرْعَة *f*

5567
(climbing) speed
F vitesse ascensionnelle *f.*
E velocidad de ascenso *f.*
D Steiggeschwindigkeit *f.*
R скорость подъёма *ж.*
A سرعة التسلق

5568
(convoy) speed
F vitesse de convoi *f.*
E velocidad de convoy *f.*
D Geleitzuggeschwindigkeit *f.*
R скорость конвоя *ж.*
A سُرْعَة القافلة *f*

5569
(economic cruising) speed
F vitesse de croisière économique *f.*
E velocidad económica de crucero *f.*
D Marschgeschwindigkeit *f.*
R эксплуатационная скорость *ж.*
A سرعة المحرك الاقتصادية

5570
(ground) speed
F vitesse sol *f.*
E velocidad absoluta *f.*
D Geschwindigkeit über Grund *f.*
R путевая скорость *ж.*
A السُرْعَة بِالنِّسْبَة إلى الأَرْض

5571
(jump) speed
F vitesse de largage *f.*
E velocidad de salto *f.*
D Absprungsgeschwindigkeit *f.*
R безопасная скорость для прыжка *ж.*
A سُرْعَة القَفْز *f*

5572 [V]
(maximum sustained) speed
F vitesse maximum de croisière *f.*
E velocidad máxima de crucero *f.*
D höchste Marschgeschwindigkeit *f.*
R наибольшая скорость хода *ж.*
A السُّرْعَة القُصْوَى *f*

5573
(wind) speed
F vitesse du vent *f.*
E velocidad de viento *f.*
D Windgeschwindigkeit *f.*
R скорость ветра *ж.*
A سُرْعَة الرِّيح

5574
speed limit
F vitesse maximale permise *f.*
E velocidad máxima permitida *f.*
D Geschwindigkeitsgrenze *f.*
R дозволенная скорость *ж.*
A حد السرعة المباحة

5575
speed of sound
F vitesse du son *f.*
E velocidad del sonido *f.*
D Schallgeschwindigkeit *f.*
R скорость звука *ж.*
A سُرْعَة الصَّوْت

5576
spell ("I spell")
F "J'épelle"
E "Deletreo"
D „(Ich) buchstabiere"
R "Даю по буквам"
A اهجي

5577
sphere
F sphère *f.*
E esfera *f.*
D Bereich *m.;* Kugel *f.*
R сфера *ж.;* шар *м.*
A (كُرَة) *f* جِسْم كُرَويّ *m*

5578
spherical
F sphérique
E esférico
D kugelförmig
R сферический; шарообразный
A كُرَوي

5579
spillage
F débordement *m.*
E líquido derramado *m.*
D Überlauf *m.*
R утечка *ж.;* утруска *ж.*
A أنْسِكَاب *m*

5580
spill over
F déborder
E derramar; desbordarse
D überlaufen (lassen) *tr.*
R разливаться; поливать
A إنْسَكَب

5581 [A]
spin *n.*
F vrille *f.*
E barrena *f.*
D Trudeln *neut.*
R штопор *м.*
A دَوَّامَة *f*

5582 [E,U]
spin *v.*
F tourner
E girar
D wirbeln
R вращать(ся); крутить(ся)
A دَار

355

5583 [T]
spinney
F boqueteau *m.*
E bosquecillo *m.*
D Dickicht *neut.*
R рощица *ж.*
A غيضة صغيرة الاشجار

5584
spin-stabilized
F stabilisé par rotation
E establilizado por
   rotación
D drallstabilisiert
R стабилизируемый
   вращением
A ذو إسْتِقَرار الدَّوَران

5585
spiral *adj.*
F spiral
E espiral
D Spiral- *komp.*
R спиральный;
   винтообразный
A لَوْلَبِي

5586 [T]
spit
F digue *m.;* langue de
   sable *f.*
E lengua *f.*
D Landzunge *f.*
R коса *ж.;* отмель *м.*
A لِسَان أَرْضِي

5587 [A,S]
"Splash!"
F "Arrivée!" [S]; "Coup
   au but!" [A]
E "¡Objetivo destruido!"
D „Aufschlag!" [A];
   „Treffer!" [S]
R "Всплеск!"
A تَنَاثُر

356

5588 [V]
splashboard
F protège-embruns *m. inv.*
E guardabarros *m. ins.*
D Schwallbrett *neut.*
R крыло *ср.;* фальшборт
   м.
A وَاقِيَة الوَحْل

5589
splashdown
F amerrissage *m.*
E amerizaje *m.*
D Anwassern *neut.*
R приводнение *ср.*
A هُبُوط في البَحْر

5590
splint
F éclisse *f.*
E tablilla *f.*
D Schiene *f.*
R лубок *м.;* шина *ж.*
A جبيرة

5591
splinter *n.*
F éclat *m.*
E casco *m.*
D Splitter *m.*
R осколок *м.*
A شَظِيَّة *f*

5592
split up
F (se) diviser
E separarse
D (sich) trennen
R разделять(ся)
A قَسَّم

5593 [B]
spoil *n.*
F déblais *m. pl.*
E tierra excavada *f.*
D Bodenaushub *m.*
R вынутая земля *ж.;*
   пустая порода *ж.*
A حَفِير *m*

5594
spoiling attack
F harcèlement *m.*
E acción de
   hostigamiento *f.*
D Störangriff *m.*
R упреждающий удар *м.*
A هُجُوم الإحْبَاط

5595
spokesman
F porte-parole *m.*
E portavoz *m.*
D Sprecher *m.*
R представитель *м.*
A ناطِق

5596
spot *v.*
F observer
E observar
D beobachten
R обнаруживать
A إسْتَكْشَف

5597 [N]
(hot) spot
F zone chaude *f.*
E (alta) radiación *f.*
D Bestrahlungspunkt *m.*
R место сильного
   излечения *ср.;*
   "горячее пятно" *ср.*
A نقطة إشعاع شديد

5598 [P]
spot check
F contrôle intermittent *m.*
E inspeccion al azar *f.*
D Stichprobe *f.*
R выборочная проверка
   ж.
A تحقيق جزئي

5599 [T]
spot elevation
F point coté *m.*
E cota (del terreno) *f.*
D Höhenpunkt *m.*
R отметка высоты *ж.*
A إرْتِفَاع نُقْطَة

5600

spotlight

F rayon de projecteur m.; projecteur m.
E faro (orientable) m.
D Scheinwerferlicht neut.
R подвижная фара ж.
A ضوء موضعي

5601 [S]

spotter plane

F avion de réglage du tir m.
E avión observador m.
D Beobachtungsflugzeug neut.
R самолёт-корректировщик м.
A طائرة رصد

5602

spotting

F observation du tir f.
E observación f.
D Beobachtung f.
R корректировка ж.
A رصد الرمي

5603 [A]

spot turn

F tour sur place f.
E girar sobre sí mismo v.
D Punktdrehung f.
R поворот на месте м.
A إنعطاف نقطي

5604

spray n.

F pulvérisation f.
E pulverización f.
D Spritzen neut.
R брызги м. мн. ч.; разлёт (осколков) м.; поливка ж.
A نشر

5605

spray attack

F attaque par répandage f.
E ataque diseminado m.
D Giftregenangriff m.
R поливка отравляющих веществ (ОВ) ж.
A هُجُوم بِنَشْر المَوَادّ الكِيمَاوِيَّة

5606 [N]

spray dome

F dôme d'écume m.
E chorro nuclear m.
D Wasserfontaine f.
R водяной столб м.
A قُبَّة رَشّ

5607 [S]

spread of shot

F dispersion f.
E dispersión del tiro f.
D Schussbereich m.
R рассредоточение огня ср.
A إنتشار الرمي

5608 [E]

spring n.

F ressort m.
E ballesta [V] f.; resorte m.
D Feder f.
R пружина ж.
A نَابِض m

5609 [T]

(water) spring n.

F source f.
E fuente f.
D Brunnen m.
R источник м.
A عَيْن f

5610

spring-loaded

F (à) ressort
E (a) resorte
D (unter) Federspannung
R пружинный
A مُحَمَّل بِنَابِض

5611

spring tide

F marée de vive eau f.
E marea viva f.
D Springflut f.
R сизигийный прилив м.
A المَدّ الأكبر

5612 [B,E,M]

sprinkler system

F installation d'extinction automatique d'incendie f.
E sistema de regaderas m.
D Feuerlöschanlage f.
R противопожарная система ж.
A مِرَشَّة f

5613 [T]

spur

F éperon m.
E espolón m.
D Gebirgsvorsprung m.
R отрог горы м.; уступ горы м.
A نتوء

5614

spy n.

F espion m.
E espía m.(f.)
D Spion(in) m.
R шпион м.
A جَاسُوس m

5615

spy (on) v.

F espionner (quelqu'un)
E espiar
D spionieren (auf)
R шпионить; следить (за)
A تَجَسَّس (على)

5616
squad
F escouade *f.;* groupe *m.*
E escuadra *m.;*
   destacamento
   (pequeño)
D Gruppe *f.;* (kleines)
   Kommando *neut.*
R отделение *ср.;*
   команда *ж.*
A جَمَاعة *f*

5617 [M]
squad leader
F chef de groupe *m.;* chef
   de pièce *m.*
E jefe de pelotón *m.;* jefe
   de escuadra *m.*
D Truppführer *m.*
R командир отряда *м.*
A قائد الجماعة

5618
squall
F grain *m.*
E chubasco *m.;* ráfaga (de
   viento) *f.*
D Bö *f.*
R шквал *м.*
A هَبَّة الرِّيح

5619
square *adj. n.*
F carré *adj. n. m.*
E cuadrado *adj. s. m.*
D Quadrat [E] *neut.;*
   Viereck *m.;* viereckig
   adj.
R квадратный *прилаг.;*
   квадрат *м.*
A مُرَبَّع *m*

5620 [S]
squash head
F (obus explosif à) tête
   d'écrasement *f.*
E cabeza de guerra
   HESH *f.*
D Weichkopf *komp.*
R расплющивающаяся
   головная часть *ж.*
A رأس مهروس

5621
stab *v.*
F poignarder
E apuñalar
D (er)stechen
R вонзать; наносить
   удар
A طَعَن

5622
stability
F stabilité *f.*
E estabilidad *f.*
D Stabilität *f.*
R устойчивость *ж.;*
   стабильность *ж.*
A إِسْتِقْرَار *m*

5623
stabilize
F stabiliser
E estabilizar
D stabilisieren
R стабилизировать
A أَقَرَّ

5624
stabilizer *n.*
F stabilisateur *m.*
E estabilizador *m.*
D Höhenflosse [A] *f.;*
   Stabilisator *m.*
R стабилизатор *м.*
A مُقِرّ *m*

5625
stable *adj.*
F stable
E estable
D stabil
R устойчивый; стойкий
A مُسْتَقِرّ

5626
staff *adj. n.*
F (de l')état-major (*adj.*)
   n. m.
E (de) estado mayor
   (*adj.*) *s. m.*
D Stab(s-) *m. (komp.)*
R штабной *прилаг.;*
   штаб *сущ. м.*
A أَرْكان *f*

5627
(combined) staff
F état-major interallié *m.*
E estado mayor conjunto
   m.
D (vereinter) Stab *m.*
R главный штаб *м.*
A أركان مشتركة

5628
(general) staff
F état-major *m.*
E estado mayor *m.*
D Generalstab *m.*
R генеральный штаб *м.*
A الأَرْكَان العَامَة *f*

5629
(integrated) staff
F état-major intégré *m.*
E estado mayor integrado
   m.
D integrierter Stab *m.*
R сводный штаб *м.*
A هَيْئَة أَرْكَان مُوَحَّدَة *f*

5630
(joint) staff
F état-major interarmées
   m.
E estado mayor de
   conjunto (EMACON)
   m.
D gemeinsamer Stab *m.*
R объединённый штаб
   м.
A أَرْكَان مُشْتَرَكَة *f*

5631

(joint operations) staff
F centre combiné
d'opérations *m.*
E estado mayor de
operaciones conjunto
m.
D (verbündeter)
Operationsstab *m.*
R объединённый штаб
м.; Ставка *ж.*
A أركان العمليات المشتركة

5632

(parallel) staff
F état-major parallèle *m.*
E estado mayor paralelo
m.
D Parallelstab *m.*
R параллельный штаб
м.
A أَرْكان مُتَوازِيَة *f*

5633

(permanent) staff
F personnel permanent *m.*
E personal permanente *m.*
D Stammpersonal *neut.*
R личный состав в
кадрах *м.*
A طاقم دائِم *m*

5634

(personal) staff
F état major particulier
m.
E ayudantes de campo *m.*
pl.
D persönlicher Stab *m.*
R личный штаб *м.*
A ضُبّاط مُرافِقون *m*

5635

(planning) staff
F état-major de
planification *m.*
E estado mayor de
planificación *m.*
D Planungsstab *m.*
R штаб оперативного
планирования *м.*
A أرْكان التَّخْطيط / أفراد
التخطيط

5636

staff car
F voiture de liaison *f.*
E automóvil del comando
m.; automóvil del
Estado Mayor *m.*
D Stabsfahrzeug *neut.*
R офицерская машина
ж.
A عربة أركان

5637

staff duty
F service d'état-major *m.*
E servicio de estado
mayor *m.*
D Stabsdienst *m.*
R штабное дежурство
ср.
A واجبات الأركان

5638

staff estimate
F étude de facteurs *f.*
E estimación de oficial del
Estado Mayor *f.*
D Stabsschätzung *f.*
R штабная оценка *ж.*
A تقدير الأركان

5639

staffing
F encadrement d' unité
m.
E (número de) personal
m.
D Stabseinteilung *f.*
R состав *м.*
A أعمال الأركان

5640

stage
F étage *f.*
E escalón *m.*
D Stufe *f.*
R ступень *ж.*
A طبقة

5641 [G]

(booster) stage
F phase d'accélération *f.*
E etapa de impulsión *f.*
D Beschleunigungsstufe *f.*
R разгонная ступень *ж.*
A طبقة التعزيز

5642

(landing) stage
F débarcadère *m.*
E desembarcadero *m.*
D Landestag *m.*
R пристань *ж.*
A رصيف الإنزال

5643

stage (movement) *n.*
F étape *f.*
E etapa *f.;* fase *f.*
D Etappe *f.*
R фаза *ж.;* этап *м.*
A مَرْحَلَة *f*

5644

staging area
F zone de mise en
condition *f.;* zone
étape *f.*
E zona de
estacionamiento *f.*
D Rastraum *m.*
R район сосредоточения
м.
A مِنْطَقَة المَراحِل

5645 [P]

stake out *v.*
F garder sous surveillance
continue
E observar
clandestinamente
D überwachen
R держать под
надзором
A راقب

359

5646

stalemate
F impasse *f.*
E punto muerto *m.*
D Patt *neut.;* matt *adj.*
R тупик *м.*
A موقف غامض

5647 [A]

stall *v.*
F (causer une) perte de vitesse
E parar(se)
D (Motor) abwürgen *tr.*
R терять скорость
A نقصت السرعة

5648

stamina
F endurance *f.*
E resistencia *f.;* vigor *m.*
D Lebenskraft *f.*
R выдержка *ж.*
A تجلد

5649 [B,M]

stanchion
F étai [B] *m.;* chandelier [M] *m.*
E montante [B] *m.;* candalero [M] *m.*
D Stütze *f.*
R подпорка [B] *ж.;* пиллерс [M] *м.*
A دِعَامَة *f*

5650 [E]

standard *n.*
F norme *f.*
E norma *f.*
D Norm *f.*
R норма *ж.;* эталон *м.*
A قِيَاس *m*

5651

standard *adj.*
F standard *inv.*
E standard
D Standard- *komp.*
R стандартный; типовой
A قِيَاسي

360

5652

standard bearer
F porte-étendard *m.*
E abanderado *m.*
D Fahnenträger *m.*
R знаменосец *м.*
A حامل العلم

5653

standard deviation
F écart type *m.*
E desvío normal *m.*
D Normalabweichung *f.*
R нормальное отклонение *ср.*
A الإنْحِرَاف القِيَاسِي *m*

5654

standardization
F standardisation *f.*
E estandardización *f.*
D Standardisierung *f.*
R стандартизация *ж.*
A توحيد *m*

5655

standardize
F standardiser
E estandardizar
D standardisieren
R стандартизировать
A وَحَّد

5656

stand by *v.*
F (se) tenir paré (à) [M]; (se) tenir prêt (à) [ALU]
E estar listo (para); estar a la espera
D (in) Bereitschaft sein (zu)
R быть наготове; быть готовым
A إحْتَاط

5657

stand-by *adj.*
F (de) réserve; (de) secours
E (de) emergencia
D Bereitschafts- *komp.*
R запасный; резервный
A احتياطي

5658

stand fast *v.*
F tenir bon
E mantenerse firme
D (sich) festlegen *tr.*
R стоять на месте
A تَوَقَّف - صَمَّدَ

5659

standing operating procedure (SOP)
F instructions permanentes *f. pl.;* procédure opérationnelle permanente (POP) *f.*
E reglas permanentes de servicio *f. pl.*
D ständige Führungsweisung (bzw. Dienstanweisung) *f.*
R постоянно действующая инструкция *ж.*
A أوَامِر ثَابِتة *f*

5660

stand off *v.*
F courir au large
E apartarse
D (die) Küste meiden
R удаляться от берега; держаться от берега
A تَوَقَّف بعيداً عن ...

5661 [M]

stand out
F gagner le large
E hacerse a la mar
D (in) See gehen
R выходить в море
A أقْلَع

5662

stand-to *n.*

F état d'alerte *m.*

E alerta *f.*

D Alarm *m.*

R боеготовность *ж.*

A إنْذَار *m*

5663

stand to *v.*

F (se) tenir sous les armes

E estar sobre las armas

D (in) Alarmstand sein

R быть в боеготовности

A تَأَهَّب

5664

star

F étoile *f.*

E estrella *f.*

D Stern *m.*

R звезда *ж.;* звездочка (форма) *ж.*

A نَجْم *m*

5665

(pole) star

F étoile polaire *f.*

E estrella polar *f.*

D Polarstern *m.*

R Полярная звезда *ж.*

A النجم القُطْبي

5666

starboard *adj. n.*

F (à) tribord *(adj.) n. m.*

E (a) estribor

D Steuerbord *adj. s. neut.*

R правого борта *прилаг.;* правый борт *сущ. м.*

A مَيْمَنَة *f*

5667 [A,M,V]

start *v.*

F démarrer; mettre en marche

E arrancar [MV]; poner en marcha

D anlassen (Motor) *tr.;* (in) Gang setzen

R пускать(ся) в ход; начинать(ся); заводить; запускать

A إنْطَلَق ـ بدأ

5668

starter

F démarreur *m.*

E arranque *m.*

D Anlasser *m.*

R стартер *м.*

A مُشَغِّل

5669 [P,U]

state *n.*

F état *m.*

E estado *m.*

D Staat [P] *m.;* Zustand [U] *m.*

R государство [P] *ср.;* положение *ср.;* состояние *ср.*

A دَوْلَة *f*

5670

(parade) state

F état des effectifs *m.*

E estadillo de formación *m.*

D Paradestand *m.*

R парадное положение *ср.*

A بيان حالة الوحدة

5671

stateless person

F apatride *m. f.*

E apatrida *m. f.;* desnacionalizado *m.*

D Staatenlose *f.*

R перемещенное лицо *ср.;* беспаспортное лицо *ср.*

A لَاجِىء *m* (شَخْص بِلا وَطَن)

5672

state of readiness

F stade d'alerte *m.*

E estado de preparación *m.*

D Bereitschaftsstufe *f.*

R степень (бое)готовности *ж.;* боеготовность *ж.*

A حَالَة الإسْتِعْدَاد

5673

state of readiness: armed

F stade d'alerte armé *m.*

E estado de preparación armada *m.*

D Zündbereitschaft '2' (entsichert) *f.*

R готовность объекта к подрыву (боевая) *ж.;* готовность 2 (два) [G] *ж.*

A حَالَة الإسْتِعْدَاد ـ كَامِل التَّسْلِيح

5674

state of readiness: safe

F stade d'alerte non-amorcé (sécurité)

E estado de preparación de seguridad *m.*

D Zündbereitschaft '1' (gesichert) *f.*

R готовность объекта к подрыву (безопасная) *ж.;* готовность 1 (один) [G] *ж.*

A حَالَة الإسْتِعْدَاد المَأْمُونَة

5675
state of war
F état de guerre *m.*
E estado de guerra *m.*
D Kriegszustand *m.*
R военное время *ср.*
A حالة الحرب

5676 [L,U]
static *adj.*
F permanent; statique
E permanente; estático
D Standort- *komp.;*
  statisch *adj.*
R стационарный;
  неподвижный
A قَرَارِي ـ سَاكِن ـ ثابِت

5677 [R]
static *n.*
F parasites *m. pl.*
E parásitos *m. pl.*
D Störungen *f. pl.*
R атмосферные помехи
  ж. мн. ч.
A تَشْوِيش *m*

5678 [A,L,M]
station *v.*
F poster
E estacionar
D stationieren
R размещать;
  распределять
A رَكَز

5679
station *n.*
F base *f.;* poste *m.;*
  station *f.*
E apostadero *m.;* base *f.*
D Base *f.;* Standort *m.*
R гарнизон *м.;* база *ж.;*
  пост *м.*
A مَحَطَّة *f* مَرْكَز *m*

362

5680 [M]
station *n.*
F poste *m.*
E puesto *m.*
D Position *f.*
R стоянка *ж.;* место в
  строю *ср.*
A مَرْكَز *m*

5681 [V]
station *n.*
F gare *f.*
E estación *f.*
D Bahnhof *m.*
R вокзал *м.;* станция *ж.*
A مَحَطَّة *f*

5682 [R]
station *n.*
F émetteur *m.;* poste *m.*
E emisora *f.;* puesto *m.*
D Funkstelle *f.;* Sender *m.*
R радиостанция *ж.*
A مَحَطَّة *f*

5683
(advanced dressing)
  station
F poste de triage et de
  premiers soins *f.*
E hospital de campaña
  avanzado *m.*
D (vorgeschobener)
  Verbandsposten *m.*
R перевязочный пункт
  м.
A محطة إسعاف متقدمة

5684
(aid) station
F poste de secours *f.*
E puesto de socorro *m.*
D Truppenverbandsplatz
  m.
R медицинский пункт *м.*
A محطة إسعاف

5685 [M]
(conning) station
F poste de manoeuvre *f.*
E puesto de mando *m.*
D Kommandoturm *m.*
R станция управления
  ж.
A محطة قيادة

5686
(direction finding)
  station
F poste de
  radiogoniométrie *f.*
E estacion
  radiogoniométrica *f.*
D Ortungsstandort *m.*
R пеленгаторная
  станция *ж.*
A محطة تحديد الإتجاه

5687
(dressing) station
F poste de triage et de
  premiers soins *f.*
E puesto de socorro *m.*
D Verbandstelle *f.*
R перевязочный пункт
  м.
A مركز اسعاف مُتَقَدِّم

5688
(filling) station
F poste d'essence *m.*
E estación de servicio *f.*
D Tankstelle *f.*
R заправочная станция
  ж.
A مَحَطَّة بَنْزِين *f*

5689
(fire) station
F caserne de pompiers *f.*
E parque de bomberos *m.*
D Feuerwehrwache *f.*
R пожарная станция *ж.*
A مَخْفَر الاطْفَاء

5690

(gasoline) station

F poste d'essence *f.*
E gasolinera *f.*
D Tankstelle *f.*
R заправочная станция
 ж.
A محطة بنزين

5691

(net control) station

F station directrice *f.*
E estación directora *f.*
D Kreisleitstelle *f.*
R главная станция сети
 ж.
A مَحَطَّة السَّيْطَرَة عَلَى الشَّبَكَة *f*

5692

(police) station

F poste de police *f.*
E comisaría *f.*
D Polizeiwache *f.*
R полицейский участок
 м.
A مخفر شرطة

5693 [R]

(relay) station

F station relais *f.*
E estación de
 retransmisión *f.*
D Relaisstation *f.*
R радиорелейная
 станция *ж.*;
 ретрансляционная
 станция *ж.*
A مَحَطَّة التَّرْحِيل

5694

(take) station *v.*

F (prendre sa) poste
E colocarse (en su puesto)
D (Stellung) nehmen
R занимать место;
 занимать позицию
A استعد

5695

stationary

F stationnaire
E estacionario; inmóvil
D feststehend;
 (an)gehalten
R неподвижный;
 стационарный
A ثَابِت

5696

stationed at

F (en) poste à
E destinado en
D stationniert
R дислоцирующий
A مقيم في

5697

**stations ("Action
 stations!")**

F "A vos postes!"
E puestos de combate *m.*
 pl.; "¡Acción!" *f.*
D „Kampfstationen!"
R "К бою!"
A مراكز القتال

5698

statistical

F statistique
E estadístico
D statistisch
R статистический
A إِحْصَائِي

5699

statistics

F statistique(s) *f. (pl.)*
E estadística(s) *f. (pl.)*
D Statistik *f.*
R статистика *ж.*
A إِحْصَائِيَّات *f*

5700

status

F statut *m.*
E categoría *f.*
D Zustand *m.*
R состояние *ср.*;
 положение *ср.*
A وَضْع *m* ـ حالة *f*

5701

**(air defense weapon
 control) status**

F consigne de tir *f.*
E tipo de posición de las
 armas *m.*
D Feuerregelung für
 Flugabwehr
R режим управления
 оружием *м.*
A وضع السيطرة على الأسلحة
 المضادة للطائرات

5702

(equipment) status

F condition de
 fonctionnement *f.*
E situación de equipos *f.*
D Ausrüstungszustand *m.*
R состояние *ср.*;
 положение *ср.*
A وضع الجهاز

5703

(fuel) status

F carburant restant *m.*
E situación de
 combustible *f.*
D Treibstoffzustand *m.*
R наличие топлива *ср.*
A وضع الوقود

5704

(radiation) status

F taux de radiation *m.*
E cantidad de radiación *f.*
D Strahlungsgrad *m.*
R оценка радиационной
 обстановке *ж.*
A مقياس الاشعاع

5705

(readiness) status

F état de préparation *m.*

E situación (de la unidad) *f.*

D Bereitschaftszustand *m.*

R готовность *ж.*

A وضع الإستعداد

5706 [G]

(weapon control) status

F consigne de tir *f.*

E tipo de posición de las armas *m.*

D Waffenbereitschafts-zustand

R готовность *ж.*

A وضع الإستعداد لإطلاق المقذوف

5707

stay behind forces

F éléments clandestins dépassés par l'ennemi *m. pl.*

E patrulla retardadora *f.*

D (zurückgelassene) Kräfte *f.*

R (НАТО) диверсионные группы *ж. мн. ч.;* (Сов.) войска специального назначения (спецназ) *ср. мн. ч. (м.)*

A قوات خاصة تبقى في منطقة بعد استيلاء العدو عليها

5708

steady state

F état constant *m.*

E estado constante *m.*

D Dauerzustand *m.*

R постоянное положение *ср.*

A حَالَة إسْتِقْرَار

5709

steam *adj. n.*

F (à) vapeur *(adj.) n. f.*

E (de) vapor *(adj.) s. m.*

D Dampf(-) *s. m. (komp.)*

R паровой *прилаг.;* пар *сущ. м.*

A بُخَارِي . بُخَار *m*

5710

steamship

F (navire à) vapeur *m.*

E (buque de) vapor *m.*

D Dampfer *m.*

R пароход *м.*

A بَاخِرَة *f*

5711

steel *adj. n.*

F (d') (en) acier *(adj.) n. m.*

E (de) acero *(adj.) s. m.*

D Stahl(-) *s. m. (komp.)*

R стальной *прилаг.;* сталь *сущ. ж.*

A فُولَاذِي . فُولَاذ *m*

5712

steep *adj.*

F (à) pic; fort

E (a) pico; fuerte

D steil

R крутой

A حَادّ

5713

steer *v.*

F conduire; gouverner [M]

E conducir; gobernar [M]

D lenken; steuern

R управлять; водить

A وَجَّه

5714

steerage way

F vitesse minimale de manoeuvre *f.*

E gobierno *m.*

D Steuervermögen *neut.*

R ход, достаточный для управления *м.*

A ما يَكْفِي مِن السُّرْعَة لِجَعْل السَّفِينة تَحْت سَيْطَرَة الدَّفَة

5715

(power) steering

F direction assistée *f.*

E servodirección *f.*

D Kraftsteuerung *f.*

R рулевое управление с усилителем *ср.*

A مِقْود آلي

5716

(in) step

F (au) pas *m.*

E (al) paso *m.*

D (im Gleich-)Schritt *m.*

R шаг *м.;* (в) ногу

A على الخُطْوَة

5717 [E,U]

step up *v.*

F augmenter [U]; survolter [E]

E aumentar [U]; elevar [E]

D hochtransformieren [E]; vermehren [U]

R увеличивать(ся); повышать(ся)

A مَدّ الى الأمَام

5718

stereo(scopic) vision

F vision stéréoscopique *f.*

E vista estereoscópica *f.*

D stereoskopische Sicht *f.*

R стереоскопическое видение *ср.*

A رُؤْية مُجَسَّمَة *f*

5719
stern *n.*
F arrière *m.;* poupe *f.*
E popa *f.*
D Heck *neut.*
R корма *ж.*
A مؤخِّرة السَّفِينة

5720
stern anchor
F ancre de poupe *m.*
E ancla de popa *f.*
D Heckanker *m.*
R кормовой якорь *м.*
A مِرْسَاة مؤخِّرة السفينة

5721
stern light
F feu de poupe *m.*
E luz de popa *f.*
D Hecklampe *f.*
R гакабортный огонь *м.;*
   кормовой огонь *м.*
A نُور مؤخِّرة السفينة

5722
stevedore
F aconier *m.;* arrimeur *m.*
E estibador *m.*
D Stauer *m.*
R портовой грузчик *м.*
A مُرَتِّب الحُمُولة

5723 [A]
(joy) stick
F manche à balai *f.*
E palanca de gobierno *f.*
D Steuerknüppel *m.*
R ручка управления *ж.;*
   рычаг управления *м.*
A عصا القِيادة

5724 [A]
stick (paratroops)
F stick *m.*
E grupo de salto *m.*
D Absprunggruppe *f.*
R группа
   (парашютистов) *ж.*
A نَسَق *m*

5725 [A]
stick of bombs
F chapelet de bombes *m.*
E grupo de bombas *m.*
D Bombengruppe *f.*
R серия бомб *ж.*
A قَصْف نَسَق قَنَابِل

5726 [S]
stock
F crosse *f.*
E caja *f.*
D Schaft *m.*
R ложа *ж.*
A مِقْبَض ـ كَعْب *m*

5727 [U]
stock(s) *n. (pl.)*
F stock *m.*
E existencias *f. pl.;*
   depósito *m.*
D Bevorratung *f.*
R запас(ы) *м. (мн. ч.)*
A مَخْزُون *m* مَخْزُونَات *f*

5728
stock control
F contrôle de stocks *m.*
E control de existencias
   m.
D Bevorratungskontrolle
   f.
R контроль запасов *м.*
A مُرَاقَبَة المَخْزُون

5729
stock level
F niveau de stocks *m.*
E nivel de existencias *m.*
D Bevorratungshöhe *f.*
R норма содержания *ж.;*
   размер запаса *м.*
A مُسْتَوَى المَخْزُونَات

5730
stockpile *n.*
F stock de réserve *m.*
E acopio *m.*
D Vorratreserve *f.*
R запас *м.;* резерв *м.*
A كُدْس *m*

5731
stockpile *v.*
F constituer des stocks de
   réserve
E acopiar
D ansammeln *tr.*
R накапливать (запасы)
A كَدَّس

5732
stone *n.*
F pierre *f.*
E piedra *f.*
D Stein *m.*
R камень *м.*
A حَجَر *m*

5733
stone *v.*
F lapider
E apedrear
D steinigen
R побивать камнями
A رَجَم

5734
stop *v.*
F (s')arrêter
E parar(se)
D (an)halten *(tr.)*
R останавливать(ся)
A وَقَف

5735
"Stop!"
F "Stop!"
E "¡Alto!"
D „Halt!"
R "Стой!"; "Стоп!"
A قِفْ

5736

stopcock
F robinet de fermeture *m.*
E llave de cierre *f.*
D Absperrhahn *m.*
R запорный кран *м.*
A مِحْبَس *m*

5737 [S]

stoppage
F arrêt *m.;* interruption *f.*
E interrupción *f.*
D Ladehemmung *f.*
R задержка *ж.*
A وُقُوف *m*

5738

stopwatch
F chronomètre *m.*
E cronómetro *m.*
D Stoppuhr *f.*
R секундомер *м.*
A عَدَّاد الثَّوَانِي ـ ساعة توقيت

5739

storage
F stockage *m.;*
emmagasinage *m.*
E almacenaje *m.;* depósito *m.*
D Lagerung *f.*
R хранение *ср.*
A خزن

5740

storage capacity
F capacité de stockage *f.;*
capacité de mémoire [D]
E capacidad de almacenaje *f.*
D Fassungsvermögen *m.*
R вместимость *ж.*
A سعة الخزن

5741 [B]

store *n.*
F magasin *m.*
E almacén *m.*
D Lagerraum *m.*
R склад *м.;* магазин *м.*
A مَخْزَن *m*

366

5742

(ammunition) store
F soute à munitions *f.*
E depósito de municiones *m.*
D Munitionslager *neut.*
R склад боеприпасов *м.*
A مستودع الذخيرة

5743

(clothing) store
F magasin d'habillement *m.*
E depósito de vestuario *m.*
D Kleidungslager *neut.*
R вещевой склад *м.*
A مستودع الألبسة

5744

store lorry
F camion-magasin *m.*
E camión de suministros *m.*
D (motorisiertes) Lager *neut.*
R грузовик со складом готовых имуществ *м.*
A حافلة نقل المؤن

5745

stores
F approvisionnements *m. pl.*
E suministros *m. f.*
D Versorgungsgüter *f. pl.*
R материальные средства *ср. мн. ч.*
A المعدّات

5746

store up
F amasser
E acumular
D ansammeln *tr.*
R запасать; откладывать
A خَزَن

5747

storm *v.*
F emporter d'assaut
E asaltar
D stürmen
R штурмовать; стремительно атаковать
A إقتحم

5748

storm (Force 10)
F tempête *f.*
E temporal *m.*
D (schwerer) Sturm *m.*
R (жестокий) шторм *м.*
A عَاصِفَة *f*

5749 [B]

story
F étage *m.*
E piso *m.*
D Stock (-werk) *m. (neut.)*
R этаж *м.*
A طَابِق *m*

5750

stow
F arrimer
E estibar
D verstauen
R укладывать; штивать
A رَتَّب

5751

stowage diagram
F plan d'arrimage *m.*
E plano de estiba *m.*
D Stauplan *m.*
R план штивки *м.;* схема штивки *ж.*
A رَسَم التَّرْتِيب

5752 [S]

straddle *v.*
F encadrer
E encuadrar
D (ein)decken *(tr.)*
R накрывать
A حَصَر

5753
strafe *v.*
F mitrailler [L];
  bombarder [A]
E cañonear
D (im Tiefflug) angreifen
  tr.
R обстреливать с
  воздуха
A قصف

5754
straggle
F traîner
E rezagarse
D (von der Truppe)
  abkommen *tr.*
R отставать(ся); идти
  вразброд
A تَخَلَّف

5755
straggler
F isolé *m.* traînard *m.*
E rezagado *m.*
D Nachzügler *m.*
  Versprengter *m.*
R отставший
A مُتَخَلِّف *m*

5756
strait
F détroit *m.*
E estrecho *m.*
D (Meer-)Enge *f.*
R пролив *м.*; узкость *ж.*
A مَضِيق *m*

5757 [M]
stranded
F échoué
E encallado
D gestrandet
R сидящий на мели;
  выброшенный на
  берег
A مَجْنُوح

5758
strap *n.*
F courroie *f.*
E correa *f.*
D Riemen *m.*
R ремень *м.*
A حبل رفع *m*

5759
strap *v.*
F sangler
E atar con correa
D festschnallen *tr.*
R стягивать ремнем
A رَبَط بِسَيْر

5760
strap
F élingue *f.*; courroie
  d'attache *f.*
E eslinga *f.*
D Gurt *m.*
R штроп *м.*
A جِزَام *m*

5761
strategic
F stratégique
E estratégico
D strategisch
R стратегический
A إستراتيجي

5762
strategic concentration
F concentration
  stratégique *f.*
E concentración
  estratégica *f.*
D strategische
  Zusammenfassung *f.*
R стратегическое
  сосредоточение *ср.*
A حَشْد إستراتيجي

5763
strategic concept
F conception stratégique
  f.
E concepto estratégico *m.*
D strategische Konzeption
  f.
R стратегический
  замысел *м.*
A مَفْهُوم إستراتيجي

5764
strategic deterrent
F force de dissuasion
  stratégique *f.*
E fuerza de disuasión *f.;*
  estratégica *f.*
D strategische
  Abschreckungswaffe
  f.
R стратегическое
  средство устрашения
  ср.
A رادع إستراتيجي *m*

5765
strategic reserve
F réserve stratégique *f.*
E reserva estratégica *f.*
D strategische Reserve *f.*
R стратегический резерв
  м.; оперативный
  резерв *м.*
A احتياط استراتيجي *m*

5766
strategy
F stratégie *f.*
E estrategia *f.*
D Strategie *f.*
R стратегия *ж.*
A فَنّ الإستراتيجية

5767
stratosphere
F stratosphère *f.*
E estratosfera *f.*
D Stratosphäre *f.*
R стратосфера *ж.*
A طَبَقَة الجَوّ العَالِيَة

5768 [T]

stream *n.*

F ruisseau *m.*
E arroyo *m.*
D Bach *m.*
R речка *ж.*; ручей *м.*
A مَجْرَى *m*

5769 [E,U]

stream *n.*

F courant *m.*
E corriente *m.*
D Strom *m.;* (Strömung
  f.)
R струя *ж.*
A تَيَّار مائي *m*

5770

streamlined

F aérodynamique
E aerodinámico
D stromlinienförmig
R обтекаемый
A على شكل انسيابي

5771

street fighting

F combats de rues *m. pl.*
E combate en las calles
  m.
D Strassenkampf *m.*
R уличные бои *м. мн. ч.*
A قتال الشوارع

5772

strength

F effectif(s) [AML] *m.*
  (pl.); force [EU] *f.*
E efectivos [AML] *m. pl.;*
  fuerza [EU] *f.*
D Leistung [E] *f.;* Stärke
  [ALMU] *f.*
R сила *ж.*; численность
  ж.
A قُوَّة ـ طاقة *f*

5773

(actual) strength

F effectif réalisé *m.*
E estado de efectivos *m.*
D Ist-Stärke *f.*
R настоящая
  численность *ж.*
A الملاك الفعلي

5774

(authorised) strength

F effectifs théoriques *m.*
  pl.
E personal en plantilla
D (geplante)
  Truppenstärke *f.*
R штатная численность
  ж.
A الملاك ـ ملاك الوحدة

5775

(full) strength

F effectifs complets *m. pl.*
E plantilla al completo *f.*
D (volle) Truppenstärke *f.*
R укомплектованность
  ж.
A الملاك الكامل

5776

(signal) strength

F qualité de transmission
  f.
E potencia de señal *f.*
D Signalstärke *f.*
R сила сигнала *ж.*
A قوة الإشارة

5777

(strike off) strength *v.*

F rayer de l'effectif
E borrar de la plantilla
D ausstreichen *tr.*
R вычёркивать
  численности
A شَطَب من السجلات

5778

strengthen *v.*

F consolider
E fortalecer
D verstärken
R укреплять
A عزّز

5779 [E]

stress *v.*

F charger
E cargar
D beanspruchen
R подвергать
  напряжению
A أجْهَد

5780 [U]

stress *n.*

F insistance *f.*
E énfasis *m.*
D Betonung *f.*
R ударение *ср.*; значение
  ср.
A ضَغْط *m*

5781 [U]

stress *v.*

F souligner
E subrayar
D betonen
R подчёркивать
A ضَغْط

5782 [E]

stress *n.*

F charge *f.*
E carga *f.*
D Belastung *f.*
R напряжение *ср.*;
  нажим *м.*
A جُهْد *m*

5783 [H]

stress *n.*

F stress *m.*
E tensión *f.*
D Stress *m.*
R стресс *м.*
A ضَغْط *m*

5784

stretcher
F brancard *m.*
E camilla *f.*
D Tragbahre *f.*
R носилки *ж. мн. ч.*
A نَقَّالة

5785

stretcher bearer
F brancardier *m.*
E camillero *m.*
D Krankenträger *m.*
R санитар-носильщик *м.*
A حامل نقالة

5786 [A,L,M]

strike
F attaque *f.;* frappe *f.*
E ataque *m.*
D Angriff *m.*
R удар *м.;* налёт *м.;*
    атака *ж.*
A ضَرْبَة *f* أَثَر *m*

5787 [F,P]

strike
F grève *f.*
E huelga *m.*
D Streik *m.*
R забастовка *ж.*
A إضْرَاب *m*

5788 [N]

(single weapon) strike
F tir unitaire *m.*
E ataque con un disparo
    nuclear *m.*
D Einzelschlagwaffe *f.*
R (ядерный) удар с
    одной ракетой *ср.*
A ضربة بسلاح واحد

5789

strike force
F force de frappe *f.*
E escalón de ataque *m.*
D Schlagkräfte *f. pl.*
R ударные войска *ср.*
    мн. ч.
A قوة ضاربة

5790

(air) strip
F piste d'atterrissage *f.*
E franja de aterrizaje *f.*
D Landeplatz *m.*
R взлётно-посадочная
    полоса *ж.*
A أرض هبوط

5791 [E]

strip (down)
F démonter
E despiezar
D abmontieren *tr.*
R демонтировать
A فَكَّ

5792

(flight) strip
F piste temporaire d'envol
    et d'atterrissage *f.*
E franja de aterrizaje *f.*
D Landestreifen *m.*
R взлётно-посадочная
    полоса *ж.*
A شُقَّة هبوط *f*

5793 [A]

(landing) strip
F piste d'atterrissage *f.*
E franja de aterrizaje *f.*
D Behelfslandeplatz *m.*
R взлётно-посадочная
    полоса *ж.*
A شُقَّة هُبُوط

5794

(mine) strip
F rangée double de mines *f.*
E hilera doble de mines *f.*
D Minen-Doppelreihe *f.*
R минная полоса *ж.*
A شفة ألغام *f* ـ شَريط
    ألغام *m*

5795

stripe
F chevron *m.;* galon *m.*
E divisa *f.;* galón *m.*
D (Dienstgrad-)
    Abzeichen *neut. pl.;*
    Paspelierung *f.*
R нашивка (форма) *ж.;*
    полоса *ж.*
A شَارَة *f* ـ شريط *m*

5796

**strip mosaic (strip
    plot)**
F boucle mosaique photo
    f.
E tira de mosaico
    fotográfico *f.*
D Bildreihe *f.*
R монтаж аэроснимков
    м.
A شَريط الصُّوَر

5797

(butt) stroke
F coup de crosse *m.*
E golpe con la culata de
    un fusil *m.*
D Kolbenschlag *m.*
R удар прикладом *м.*
A ضربة اخمص

5798 [A,L,M,U]

strong
F fort (de)
E fuerte [U]; (de
    ...)efectivos [ALM]
D stark
R сильный; крепкий;
    прочный
A قَوِّي

5799

stronghold
F place forte *f.*
E plaza fuerte *f.*
D Festung *f.*
R крепость *ж.;* оплот *м.*
A مَوْقِع حَصِين *m*

369

5800 [A,B,M]
strut *n.*
F entretoise *f.*
E riostra *f.*
D Strebe *f.*
R подкос *м.*; подпорка
 ж.; раскос *м.*;
 распорка *ж.*
A دِعَامَة *f*

5801
(military) student
F stagiaire *m.*
E estudiante militar
D Militärstudent *m.*
R курсант *м.*
A طالب عسكري

5802
subatomic
F subatomique
E subatómico
D subatomar
R податомный
A دُون الذَّرِّي

5803
subcaliber
F sous-calibré
E (de) calibre reducido
D Kleinkaliber- (z.B.
 Ausbildungs-
 vorrichtung *f.*);
 Unterkaliber- (z.B.
 Geschoss *n.*)[S]
 komp.
R подкалиберный
A عِيَار مُصَغَّر

5804
subcritical
F sous-critique
E subcrítico
D unterkritisch
R докритический
A دُون الحَرِج

370

5805
subject
F objet d'un document *m.*
E tema *m.*
D Thema *neut.*
R повод *м.*; тема *ж.*
A موضوع

5806
subkiloton
F inférieur au
 kilotonne; sous-
 kilotonnique
E sub-kilotónico
D Subkilotonnen- *komp.*
R подкилотонный
A أَدْنَى من كِيلُوطَنّ

5807
submachine carbine
F pistolet mitrailleur *m.*
E pistola ametralladora
 f.; metralleta *f.;*
 subfusil *m.*
D Maschinenpistole *f.*
R автомат *м.*
A رَشَاش قَصِير *m*

5808
submarine
F sous-marin *m.*
E submarino *m.*
D U-Boot *neut.*
R подводная лодка *ж.*;
 подлодка *ж.*
A غَوَّاصَة *f*

5809
submariner
F sous-marinier *m.*
E marinero de submarino
 m.
D U-Bootfahrer *m.*
R подводник *м.*
A نُوتِي غَوَّاصَة

5810
submerge
F plonger
E sumergir
D tauchen
R затоплять(ся);
 погружать(ся)
A غَمَرَ

5811 [M,V]
submerged crossing
F passage en submersion
 [V] *m;* passage en
 plongée [M] *m.*
E crucero submarino [M]
 m.; travesía a
 snorkel[V] *f.*
D Unterwasserfahren
 neut.
R переправа под водой
 ж.; переправа по
 дну *ж.*
A عُبُور مَغْمُور

5812
submersible *adj.*
F submersible
E sumergible
D Tauch- *komp.*
R подводный;
 погружающийся
A قَابِل لِلغَمْر

5813 [P,U]
submission
F soumission *f.*
E sumisión *f.*
D Unterwerfung [P] *f.;*
 Vorlage [U] *f.*
R подчинение *ср.*;
 подавление *ср.*;
 представление
 (докум.) *ср.*
A تَقْدِيم

5814 [P,U]
submit
F (se) soumettre
E someter(se)
D (sich) unterwerfen [P];
vorlegen [U]
R подчиняться;
представлять
A قَدَّم

5815
subordinate adj. n.
F subordonné adj. n. m.
E subordinado adj. s. m.
D unterstellt(e-r) adj. (s. m.)
R подчинённый прилаг.;
подчинённый сущ. м.
A مَرْؤُوس m

5816
subsequent operation phase
F phase d'exploitation f.
E fase secundaria de operación f.
D Nachoperationsphase f.
R последующий этап действий м.
A مرحلة تالية للعملية

5817 [B]
subside
F (s')affaisser
E hundirse
D sinken
R оседать
A هَبَط

5818 [B]
subsidence
F affaissement m.
E hundimiento m.;
socavón m.
D (Boden-)Senkung f.
R оседание ср.
A هُبُوط m

5819
subsidiary adj.
F auxiliaire
E auxiliar; subsidiario
D Hilfs- komp.
R вспомогательный
A إضافِي ـ ثانوي

5820
subsistence
F subsistance f.
E subsistencia f.
D Unterhalt m.
R пропитание ср.;
довольствие ср.
A إعَاشَة f

5821
subsonic
F subsonique
E subsónico
D Unterschall- komp.
R дозвуковой
A دُون سُرْعَة الصَّوْت

5822 [C]
substance
F substance f.
E sustancia f.
D Substanz f.
R вещество ср.
A مَادَّة f

5823
substantive rank
F grade à titre définitif f.
E empleo activo m.
D Dienstgrad mit Patent m.
R действительный чин м.; действительное воинское звание ср.
A رتبة فعلية f

5824 [H]
substitute n.
F remplaçant m.
E suplente m. f.
D Stellvertreter m.
R заместитель м.
A نَائِب m

5825 [E,U]
substitute n.
F succédané m.
E sucedáneo m.; sustituto m.
D Ersatzstoff m.
R замена ж.
A بَدِيل

5826
substitute adj.
F succédané
E sucedáneo; sustituto
D Ausweich- komp.;
Ersatz- komp.
R заменяющий;
замещающий
A بَدِيل

5827
sub-unit
F sous-groupement m.
E unidad subordinada f.
D Teileinheit f.
R подразделение ср.
A وحدة صغيرة

5828
subversion
F subversion f.
E subversión f.
D Umsturz m.
R диверсия ж.;
свержение ср.
A أعْمَال الهَدْم

5829
subversive adj. n.
F subversif adj. n. m.
E subversivo adj. s. m.
D umstürzlerisch(-e-r) adj. (s.m.); subversiv
R подрывной;
диверсионный
A هَدَّام

5830

subvert

F renverser
E subvertir
D (um)stürzen;
   untergraben
R ниспровергать;
   подрывать
A هَدَم

5831

sub-zero *adj.*

F (au) dessous de zéro
E bajo cero
D unter Null
R ниженулевой
A تحت الصفر

5832

successive positions

F positions successives *f.*
   pl.
E posiciones sucesivas *f.*
   pl.
D (folgende) Stellungen *f.*
   pl.
R позиции, находящиеся
   одна за другой *ж.*
   мн. ч.
A مواضع متتالية

5833 [E]

suction pump

F pompe aspirante *f.*
E bomba de succión *f.*
D Saugpumpe *f.*
R всасывающий насос
   м.
A مضخة ماصّة

5834

suffer losses *v.*

F subir des pertes
E sufrir bajas
D Verluste erleiden
R поносить потери *ж.*
   мн. ч.
A تكبد خسائر

5835

sum *n.*

F montant *m.*
E total *m.*
D Betrag *m.*
R сумма *ж.*; итог *м.*
A مَجْمُوع *m*

5836

sum *v.*

F additionner
E sumar
D summieren
R подводить итог
A أجْمَل

5837

summarize

F résumer
E resumir
D zusammenfassen *tr.*
R суммировать;
   резюмировать
A لَخَّص

5838

summary

F résumé *m.*
E resumen *m.*
D Zusammenfassung *f.*
R резюме *ср.*; сводка *ж.*
A خلاصة *f*

5839

sump

F carter *m.*; puisard *m.*
E colector de aceite *m.*
D (Öl) Wanne *f.*
R отстойник *м.*
A خزان الزيت

5840

sun

F soleil *m.*
E sol *m.*
D Sonne *f.*
R солнце *ср.*
A شَمْس *f*

5841

sun compass

F compas solaire *m.*
E brújula solar *m.*
D Sonnenkompass *m.*
R солнечный компас *м.*
A بُوصَلَة شَمْسِيَّة *f*

5842

sunglasses *pl.*

F lunettes fumées *f. pl.*
E gafas de sol *f. pl.*
D Sonnenbrille *f.*
R солнцезащитные очки
   м. мн. ч.; тёмные
   очки *м. мн. ч.*
A نَظَّارَات الشَّمْس

5843 [T]

sunken *adj.*

F creux
E hundido
D versunken
R осевший
A غائر

5844

sunrise

F lever du soleil *m.*
E aurora *f.;* salida del sol
   f.
D Sonnenaufgang *m.*
R восход солнца *м.*;
   утренняя заря *ж.*
A شُرُوق الشَّمْس

5845

sunset

F coucher du soleil *m.*
E ocaso *m.;* puesta del sol
   f.
D Sonnenuntergang *m.*
R заход солнца *м.*;
   вечерняя заря *ж.*
A غُرُوب الشَّمْس

5846 [S]
supercharge *adj. n.*
F (à) charge additionnelle
   (adj.) n. f.
E (de) sobrecarga *(adj.) s.*
   f.
D (mit) verstärkte(r)
   Ladung (bzw. eng.
   Begriff) *(adj.) s. f.*
R усиленный заряд *м.*
A فَرْط التَّغْذِيَة . زِيَادَة الحُشْوَة

5847 [S,U]
superimpose
F superposer
E sobreponer
D überlagern
R переносить; наносить
A رَكِبَ على

5848
superior *adj. n.*
F supérieur *adj. n. m.*
E superior *adj. s. m.*
D überlegen *adj.;*
   vorgesetzt *adj.;*
   Vorgesetzte(r) *s. m.*
R высший;
   превосходный;
   старший
A أعْلى

5849
superiority
F supériorité *f.*
E superioridad *f.*
D Überlegenheit *f.*
R превосходство *ср.;*
   старшинство *ср.*
A تَفَوُّق *m*

5850
supernumerary *n.*
F supernuméraire *m. pl.*
E supernumerario *m.*
D Supernumerar *m.*
R сверхкомплектный *м.*
A فائض عن العدد

5851
supersonic
F supersonique
E supersónico
D Überschall- *komp.*
R сверхзвуковой
A أعْلى مِن سُرْعَة الصّوت

5852
superstructure
F superstructure *f.*
E superestructura *f.*
D Aufbau(ten) *m.*
R надстройка *ж.*
A فوق السطح

5853
supervise
F surveiller
E supervisar
D beaufsichtigen
R заведовать; надзирать
A اشْرَف على

5854
supervisor
F surveillant *m.*
E supervisor *m.*
D Aufseher *m.*
R надзиратель *м.;*
   контролёр *м.*
A مُشْرِف *m*

5855
supplement *v.*
F ajouter
E complementar
D ergänzen
R пополнять; добавлять
A ألحَق

5856
supplement *n.*
F supplément *m.*
E suplemento *m.*
D Zusatz *m.*
R приложение *ср.;*
   дополнение *ср.*
A مُلْحَق *m*

5857
supplementary
F supplémentaire
E complementario;
   suplementario
D Ausweich- *komp.;*
   Zusatz- *komp.;*
   zusätzlich *adj.*
R дополнительный
A إلْحَاقِي

5858
supplier
F approvisionneur *n. m.*
   adj.
   ravitailleur *n. m. adj.*
E proveedor *m.*
D Lieferant *m.*
R доставляющий
   (средств, провианта)
   м.
A مُورِّد *m*

5859
supplies *pl.*
F ravitaillement *m.*
E abastecimientos *f. pl.*
D Nachschub *m.;*
   Versorgungsgüter
   neut. pl.
R снабжение *ср.;*
   материальные
   средства *ср. мн. ч.;*
   провиант *м.*
A تَمْوِينَات *f*

5860
(accompanying)
   supplies
F ravitaillement et
   dotation de combat
   m.
E suministros de
   acompañamiento *m.*
   pl.
D Begleitgüter *neut. pl.*
R готовые
   материальные
   средства *ср. мн. ч.*
A المعدات المرافقة

5861

(nonexpendable) supplies

F matériel non-consommable *m.*

E material no fungible *m.*

D Nichtverbrauchsgüter *neut. pl.*

R нерасходуемые предметы снабжения *м. мн. ч.*

A مُعَدّات غَيْر قَابِلة لِلصَّرَف *f*

5862

(expendable) supplies and materials *pl.*

F approvisionnements et articles consommables *m. pl.*

E materiales y suministros consumibles *m. pl.*

D (Mengen- u. Einzel-) Verbrauchsgüter *neut. pl.*

R расходуемый материал снабжения *м.*

A مُعَدّات تُسْتَهْلَك *f*

5863

supply *v.*

F fournir; ravitailler

E abastecer

D versorgen

R снабжать; доставлять; обеспечивать

A مَوَّن

5864

(air) supply

F ravitaillement par air *m.*

E abastecimiento aéreo *m.*

D Luftversorgung *f.*

R снабжение по воздуху *cp.*

A تَمْوين جَوّي *m*

374

5865

(follow up) supply

F ravitaillement initial *m.*

E reabastecimiento aerotransportado *m.*

D unmittelbare Versorgung *f.*

R немедленное снабжение по воздуху *cp.*

A التموين اللاحق

5866

(forward) supply

F soutien de l'avant *m.;* ravitaillement de l'avant *m.*

E suministro avanzado *m.*

D Direktversorgung *f.*

R передовое снабжение *cp.*; прямое снабжение *cp.*

A التَّمْوين الأمامي *m*

5867

(one day's) supply

F jour de ravitaillement *m.*

E día de abastacimiento *m.*

D Tagesnachschub *m.*

R суточная норма снабжения *ж.*

A تَعْيينات يوم وَاحِد

5868

(power) supply point

F prise de courant *f.*

E punto de suministro de energía eléctrica *m.*

D Starkstromanschluss *m.*

R подача электропитания *ж.*

A وَقْبة مَأْخَذ التَّيَار

5869

supply point distribution

F ravitaillement au point de ravitaillement *m.*

E distribución por puntos de abastecimiento *f.*

D Lieferung durch Versorgungspunkte

R выдача из пунктов снабжения *ж.*

A نقطة توزيع التموينات

5870 [S]

(required) supply rate

F taux de ravitaillement requis *m.*

E régimen necesario de abastecimiento *m.*

D Nachschubbedarfs-quote *f.*

R требуемая норма снабжения *ж.*

A مُعَدّل التَّمْوين المَطْلوب

5871

(air) support

F appui aérien *m.*

E apoyo aéreo *m.*

D Luftunterstützung *f.*

R воздушная поддержка *ж.*

A سند جَوّي *m*

5872

(close) support

F appui rapproché *m.*

E apoyo inmediato *m.*

D Nahunterstützung *f.*

R непосредственная поддержка *ж.*

A سَنْد مُبَاشِر *m*
تعاون وثيق *m*

5873

(close air) support

F appui aérien rapproché *m.*

E apoyo inmediato aéreo *m.*

D (Luftnah)unterstützung *f.*

R авиационная поддержка войск *ж.*

A السند الجوي القريب

5874

(combat) support

F appui *m.*

E apoyo de combate *m.*

D (Kampf)unterstützung *f.*

R боевая поддержка *ж.*; взаимодействие *ср.*

A المساندة القتالية *f*

5875 [A,L,M,]

(combat air) support

F appui aérien rapproché *m.*

E apoyo aéreo de combate *m.*

D Luftunterstützung (im Gefecht) *f.*

R боевая авиационная поддержка *ж.*

A مساندة القتال الجوي *f*

5876

(combat service) support

F soutien *m.*

E apoyo logístico de combate *m.*

D (Nachschubs) unterstützung *f.*

R материально-техническое обеспечение *ср.*

A القوات المساندة للقتال

5877 [S]

(direct) support

F appui direct *m.*

E apoyo directo *m.*

D (unmittelbare) Unterstützung *f.*

R непосредственная поддержка *ж.*

A سند مباشر

5878

(floating base) support

F bâtiments de soutien logistique *m. pl.*

E buques de abastecimiento *m. pl.*

D Nachschubschiffsgruppe *f.*

R пополнение кораблей на рейде *ср.*

A حَامِل سَائِب *m*

5879 [S]

(general) support

F action d'ensemble *f.*

E apoyo general (de artillería) *m.*

D (allgemeine) Unterstützung *f.*

R планированная артиллерийская поддержка *ж.*

A سَنْد عَام *m*

5880

(immediate air) support

F appui aérien urgent *m.*

E apoyo inmediato aéreo *m.*

D (unmittelbare) Luftunterstützung *f.*

R срочная авиационная поддержка *ж.*

A سند جوي فوري

5881

(impromptu) support

F appui inopiné *m.*

E apoyo imprevisto *m.*

D (ungeplante) Unterstützung *f.*

R импровизированная поддержка *ж.*

A سند ارتجالي

5882

(indirect) support

F appui indirect *m.*

E apoyo indirecto *m.*

D indirekte Unterstützung *f.*

R косвенная поддержка *ж.*

A سَنْد غَيْر مُبَاشِر *m*

5883 [A]

(indirect air) support

F appui aérien indirect *m.*

E apoyo aéreo indirecto *m.*

D indirekte Luftunterstützung *f.*

R косвенная авиаподдержка *ж.*

A سَنْد جَوِي غَيْر مُبَاشِر *m*

5884

support(ing) *adj.*

F d'appui

E (de) apoyo

D Unterstützungs- *komp.*

R поддерживающий; ...поддержки

A سَانِد

5885

(integrated) support

F soutien intégré *m.*

E apoyo integrado *m.*

D integrierte Unterstützung *f.*

R объединённая поддержка *ж.*

A سَنْد مُتَكَامِل *m*

5886

(logistical) support

F soutien logistique *m.*
E apoyo logístico *m.*
D Versorgung *f.*
R снабжение *ср.*;
материально-
техническое
обеспечение *ср.*
A سَنْد اداري *m*

5887

(mutual) support

F appuis réciproques *m. pl.*
E apoyo mutuo *m.*
D (gegenseitige) Unterstützung *f.*
R взаимодействие
A سند متبادل

5888

(naval gunfire) support

F appui de feu naval *m.*
E apoyo de fuego naval *m.*
D Schiffsartillerie-
unterstützung *f.*
R корабельная
артиллерийская
поддержка
(артподдержка) *ж.*
A سَنْد مِدْفَعِيَّة البَحْرِيَّة *m*

5889

(offensive air) support

F appui aérien offensif *m.*
E apoyo aéreo ofensivo *m.*
D offensive Luftunterstützung *f.*
R наступательное
авиационное
обеспечение *ср.*
A السند الجوي التعرضي *m*

5890

(service) support

F soutien *m.*
E apoyo logístico de combate *m.*
D (logistische) Kampfunterstützung *f.*
R материально-
техническое
обеспечение *ср.*
A سند الخدمة

5891

(tactical) support

F appui *m.*
E apoyo de combate *m.*
D (taktische) Kampfunterstützung *f.*
R боевая поддержка *ж.*;
взаимодействие *ср.*
A السند التعبوي

5892

(tactical air) support

F appui aérien tactique *m.*
E apoyo aéreo táctico *m.*
D taktische Luftunterstützung *f.*
R тактическая
воздушная
поддержка *ж.*
A سَنْد جَوِّي تَعْبَوِي

5893

supporting distance

F distance de sécurité et d'appui réciproque *f.*
E distancia de apoyo recíproco *f.*
D Unterstützungs-
entfernung *f.*
R указанное расстояние
взаимного действия *ср.*
A مسافة السند

5894

support unit

F unité d'appui *f.;* unité de soutien *f.*
E unidad de apoyo *f.*
D Kampfunterstützungs-
einheit *f.;*
Reservetruppen *f. pl.*
R поддерживающее
подразделение *ср.*
A وَحْدة مُسَانَدة

5895

(flash) suppressor

F dispositif anti-lueur *m.*
E amortiguador de fogonazo *m.*
D Mündungsfeuerdämpfer *m.*
R пламегаситель *м.*
A مَانِع الوَمِيض *m*

5896

(heat) suppressor

F suppresseur thermique *m.*
E supresor de calor (para evitar la detección) *m.*
D Wärmezerstreuer *m.*
R жароглушитель *м.*
A مانع الحرارة

5897

suppress riots *v.*

F réprimer des émeutes
E suprimir disturbios
D unterdrücken *untr.*
R подавлять бунты;
подавлять мятежи
A احمد شغبا

5898

surf

F déferlement *m.;* ressac *m.*
E rompientes *f. pl.*
D Brandung *f.*
R прибой *м.;* буруны *м. мн. ч.*
A أَمْوَاج الشَّاطِىء

5899

surface *n.*

F surface *f.*
E superficie *f.*
D Oberfläche *f.*
R поверхность *ж.*
A سَطْح *m*

5900 [M]

surface *v.*

F faire surface
E salir a la superficie
D auftauchen *tr.*
R всплывать
A صَعَدَ إلى السَّطْح

5901 [N]

surface burst

F explosion au sol *f.*
E explosión superficial *f.*
D Bodendetonation *f.*
R наземный взрыв *м.*
A تَفْجِير على السَّطْح

5902 [G]

surface-to-air *adj.*

F sol-air *inv.*
E tierra-aire *inv.*
D Boden-Luft- *komp.;*
  Flugabwehr- *komp.;*
  Fla- *komp.*
R земля-воздух
A أرْض ـ جَوّ

5903 [G]

surface-to-subsurface
  adj.

F antisousmarins *inv.;*
  mer-mer
E superficie-submarino
  inv.
D (schiffsgestützte) U-
  Boot-Bekämpfungs-
  komp.
R противолодочный
A أرْض ـ تَحْت الأرْض

5904

surface-to-surface *adj.*

F sol-sol *inv.;* surface-
  surface *inv.*
E superficie-superficie *inv.*
D Boden-Boden- *komp.*
R земля-земля
A أرْض ـ أرْض

5905 [M]

surge *n.*

F houle *f.*
E oleaje *m.*
D Woge *f.*
R зыбь *ж.;* качка *ж.*
A مَوْجَه *f*

5906 [M]

surge *v.*

F (se) soulever
E hervir
D wogen
R волноваться;
  вздыматься
A مَاج

5907

surgeon

F chirurgien *m.*
E cirujano *m.*
D Chirurg *m.*
R хирург *м.*
A جَرَّاح *m*

5908

surgery

F chirurgie *f.*
E cirurgía *f.*
D Chirurgie *f.*
R хирургия *ж.*
A جِرَاحَة *f*

5909

surgical

F chirurgical
E quirúrgico
D chirurgisch
R хирургический
A جِرَاحِي

5910

surplus *adj. n.*

F surplus *adj. inv. n. m.*
E excedente *adj. s. m.*
D Überschuss(-) *s. m.*
  (komp.)
R излишний *прилаг.;*
  излишек *м.;* остаток
  м.
A فَائِض *m*

5911

surprise *v.*

F surprendre
E sorprender
D überraschen *untr.*
R заставать врасплох
A بَاغَت

5912

surprise *adj. n.*

F (par) surprise *(adj.) n.*
  f.
E (de) sorpresa *(adj.) s. f.*
D Überraschung(s-) *s. f.*
  (komp.)
R внезапный *прилаг.;*
  внезапность *ж.*
A مُفَاجَأة *f*

5913

surprise attack

F attaque par surprise *f.*
E ataque de sorpresa *m.*
D Überraschungsangriff
  m.
R внезапное нападение
  ср.
A هُجُوم مُبَاغِت *m*

377

5914 [C]

surprise dosage attack

F attaque chimique à dose surprise *f.*

E ataque químico a dosis desprevenida *f.*

D Angriffsvorbereitung durch überraschenden C-Waffeneinsatz *f.*

R внезапная поливка *ж.*; внезапный химический налёт *м.*

A هُجُوم كِيمَاوِي مُبَاغِت

5915

surrender *n.*

F reddition *f.*

E capitulación *f.*

D Übergabe *f.*

R скача *ж.*; капитуляция *ж.*

A إسْتِسْلَام *m*

5916

surrender *v.*

F (se) rendre

E rendir(se)

D kapitalieren; übergeben

R сдаваться; уступать

A إسْتَسْلَم

5917

surround *v.*

F cerner

E cercar

D einkesseln *tr.*

R окружать

A أَحَاط بِ

5918

surveillance *adj. n.*

F (de) surveillance *(adj.)* *n. f.*

E (de) vigilancia *(adj.) s. f.*

D Überwachung(s-) *s. f.* *(komp.)*

R наблюдение *ср.*; надзор *м.*

A مُرَاقَبَة *f*

378

5919 [T]

survey *v.*

F relever

E apear

D vermessen

R производить топосъёмку

A مَسَح

5920 [T]

survey *adj.*

F topographique

E topográfico

D Vermessungs- *komp.*

R топографический

A مِسَاحَة *f*

5921 [T]

survey *n.*

F levée *f.*

E levantamiento topográfico *m.*

D Vermessung *f.*

R топографическая съёмка *ж.*; топосъёмка *ж.*

A مِسَاحَة *f*

5922 [C]

(chemical) survey

F reconnaissance chimique *f.*

E reconocimiento químico *m.*

D (chemische) Überwachung *f.*

R химическое обследование *ср.*; химическая разведка *ж.*

A مساحة كيميائية

5923

(damage) survey

F expertise des dégâts *f.*

E evaluación de averías *f.*

D Schadenbegutachtung *f.*

R оценка поражения

A مساحة العطب

5924

(plane-table) survey

F levé à la planchette *m.*

E reconocimiento de plancheta *m.*

D Kartenvermessung *f.*

R мензульная съёмка *ж.*

A لوحة مساحة

5925

(road) survey

F levé d'itinéraire *m.*

E reconocimiento (para planificar una carreterra) *m.*

D Strassenvermessung *f.*

R маршрутная съёмка *ж.*

A مسح الطرق

5926

surveyed *adj.*

F déterminé par travail topographique

E medido

D vermessen

R промеренный

A ممسوح

5927 [A,V]

survivability

F capacité de survie *f.*

E capacidad de supervivencia *f.*

D Überlebensfähigkeit *f.*

R живучесть *ж.*

A القدرة على البقاء

5928

survival *adj. n.*

F (de) survie *(adj.) n. f.*

E (de) supervivencia *(adj.) s. f.*

D Überleben(s-) *s. neut.* *(komp.)*

R спасательный; аварийный

A البَقَاء *m*

5929

survive

F survivre

E sobrevivir

D überleben *untr.*

R выживать; уцелеть

A بَقِيَ حَيًّا

5930

survivor

F survivant *m.*

E sobreviviente *m.*

D Überlebende(r) *m.*

R уцелевший *м.*;
оставшийся в живых

A باقٍ على الحياة

5931 [I,P]

suspect *v.*

F soupçonner

E sospechar

D verdächtigen

R подозревать

A اشتبه

5932 [I,P]

suspect *n.*

F suspect *m.*

E sospechoso *m.*

D Verdächtige(r) *m.*

R подозреваемый *м.*

A مشبوه *m*

5933

suspend (cease)

F suspendre

E suspender

D einstellen *tr.;*
verschieben

R прекращать

A وَقَف

5934 [U,V]

suspension

F suspension *f.*

E suspensión *f.*

D Aufhängung [V] *f.;*
Einstellung [U] *f.*

R прекращение *ср.*;
подвес(ка) [V] *м.(ж.)*

A تَعْلِيق *m*

5935

suspension bridge

F pont suspendu *m.*

E puente colgante *m.*

D Hängebrücke *f.*

R висячий мост *м.*;
подвесной мост *м.*

A جِسْر مُعَلَّق *m*

5936

**(parachute) suspension
lines**

F suspentes *m. pl.*

E cuerdas de suspensión
f. pl.

D Fallschirmtragseile
neut. pl.

R подвесные верёвки *ж.
мн. ч.*

A حبال التعليق

5937

**suspension of
hostilities**

F cessation des hostilités
f.; suspension des
hostilités *f.*

E suspensión de armas *f.*

D Einstellung der
Feindseligkeiten *f.*

R перемирие *ср.*

A توقف الاعمال العدوانية

5938 [A]

suspension strop

F raccord d'élingue *m.*

E conexión de eslinga *f.*

D Ausgleichsstück *neut.*

R компенсатор *м.*

A سَيْر تَعْلِيق *m*

5939

sustained rate of fire

F cadence normal
pratique *f.*

E cadencia normal *f.*

D Dauerfeuerkadenz *f.*

R практическая
скорострельность *ж.*

A مُعَدِّل الرَّمْي الغَزِير

5940

swamp *n.*

F marais *m.*

E pantano *m.*

D Moor *neut.*

R болото *ср. ж.*; топь

A مُسْتَنْقَع *m*

5941 [R]

sweep *n.*

F balayage *m.*

E abarcadura *m.*

D Abtastung *f.*

R развёртка *ж.*; поиск
м.; облучение *ср.*

A إمْتِداد *m*

5942 [R]

sweep *v.*

F balayer

E abarcar

D abtasten *tr.*

R развёртывать поиск
ж.; проводить поиск
м.; облучать *ср.*

A إمْتَدَ

5943 [S]

sweep *v.*

F balayer

E barrer

D bestreichen

R обстреливать;
простреливать

A حَصَد

379

5944 [M]
sweep (mines) *v.*
F draguer
E barrer
D räumen
R тралить
A كَسَح

5945 [M]
swell *n.*
F houle *f.*
E marejada *f.*
D Dünung *f.*
R волнение *ср.*; зыбь *ж.*
A تَمَوُّج *m*

5946
swept channel
F chenal dragué *m.*
E canal barrido *m.*
D geräumtes Fahrwasser
  neut.
R протраленный
  фарватер *м.*
A مَمَرّ مائي مَكْسُوح

5947
swim capability
F aptitude au
  franchissement des
  cours d'eau *f.*
E capacidad anfibia *f.*
D Schwimmfähigkeit *f.*
R способность плавать
  ж.; плавучесть *ж.*
A مقدرة العوم

5948 [L]
swing *v.*
F changer la direction
E girar
D (ein-)schwenken *tr.*
R перебрасывать(ся)
A دَار

5949
swing bridge
F pont tournant *m.*
E puente giratorio *m.*
D Drehbrücke *f.*
R поворотный мост *м.*
A جِسْر دَوَّار *m*

5950 [L,U]
switch *v.*
F changer
E cambiar
D schalten; umstellen *tr.*
R переносить;
  перецеливать;
  переключать
A بَدَّل

5951 [E]
switch *n.*
F commutateur *m.;*
  interrupteur *m.*
E conmutador *m.;*
  interruptor *m.*
D Schalter *m.*
R выключатель *м.;*
  переключатель *м.*
A قَاطِع *m* مِفْتَاح *m*

5952 [E]
switch off *v.*
F couper
E quitar
D abschalten *tr.*
R выключать
A قَطَع

5953 [E]
switch on *v.*
F allumer
E encender
D einschalten *tr.*
R включать
A فَتَح

5954
swivel *v.*
F pivoter
E girar
D schwenken
R вертеться; вращаться
A تَحَرُّك حَوْل مِحْوَر

5955
swivel *n.*
F émerillon *m.*
E eslabón *m.*
D Drehgelenk *neut.*
R вертлюг *м.*
A قُطْب *m*

5956 [S]
swivel gun
F canon à pivot *m.*
E cañón de colisa *m.*
D Drehzapfen-MG *neut.*
R орудие на
  вертлюжной
  установке *ср.*
A مِدْفع مُرَكَّب عَلَى قُطْب

5957
symbol
F symbole *m.*
E símbolo *m.*
D Symbol *neut.;* Zeichen
  neut.
R символ *м.;* знак *м.*
A رَمْز *m*

5958
**sympathetic
  detonation**
F détonation par
  sympathie *f.*
E detonacion por
  simpatía *f.*
D Leitfeuerzündung *f.*
R симпатичная
  детонация *ж.;*
  ответная детонация
  ж.
A انفجار بالتَّأْثِير

5959

synchronize

F synchroniser
E sincronizar
D synchronisieren
R синхронизировать;
  сверять (часы)
A زَامَن

5960

synchronous

F synchrone
E sincrónico
D gleichzeitig; synchron
R синхронный;
  одновременный
A تَزَامُنِي

5961

system

F système *m.*
E sistema *m.*
D System *m.*
R система *ж.*
A نِظَام *m*

5962

systems analysis

F analyse de systèmes *f.*
E análisis de sistemas *m.*
D Systemanalyse *f.*
R системный анализ *м.*;
  оценка систем *ж.*;
  оценка комплектов
  ж.
A تَحْلِيل نِظَامِي *m*

T

5963

(chart) table

F table *f.;* tableau *m.*
E tabla *f.*
D Tabelle *f.;* Tafel *f.*
R таблица *ж.;* планшет
  м.
A لَوْحَة خَرَائِط *m*

5964

(movement) table

F tableau des
  mouvements *m.*
E tablero de movimiento
  m.
D Marschtabelle *f.*
R план перевозки *м.;*
  план движения *м.*
A جَدْوَل التَّنَقُّل *m*

5965

(tide) table

F annuaire des marées *m.*
E tabla de mareas *f.*
D Gezeitentafel *f.*
R табель времён
  полных вод *м.;*
  таблица приливов
  ж.
A جَدْوَل المَدّ والجَزْر

5966

**table of organization
and equipment
(TOE)**

F tableau d'effectifs et de
  dotation (TED) *m.*
E cuadro orgánico y de
  equipo *m.*
D Stärke- und
  Ausrüstungsnachweis
  m.
R штатно-
  организационное
  расписание и табель
  имущества *ср. м.*
A جدْول التَّنْظِيم والتَّجْهِيزَات

5967 [H]

tablet

F comprimé *m.*
E pastilla *f.*
D Tablette *f.*
R таблетка *ж.*
A قُرُص دواء

5968

tabular

F tabulaire
E tabular
D Tabellen- *komp.*
R табличный
A جَدْوَلِي

5969

tachometer

F tachymètre *m.*
E tacómetro *m.*
D Tachometer *neut.*
R тахометр *м.*
A مقياس السرعة

5970 [B,M]

tackle

F palans *m. pl.*
E aparejo *m.*
D Takel *neut.*
R снасть *ж.;* такелаж *м.*
A آلة رافعة

5971

tactical

F tactique
E táctico
D taktisch
R тактический;
  оперативный
A تَعْبَوِي

5972

**tactical air control
center**

F centre de contrôle
  aérien tactique *m.*
E centro de control aéreo
  táctico *m.*
D taktische
  Fliegerleitzentrale *f.*
R центр управления
  тактической
  авиацией *м.*
A مَرْكَز السَّيْطَرَة الجَوِّيَّة التَّعْبَوِي

5973

tactical air controller

F contrôleur aérien
    tactique *m*.

E controlador aéreo
    táctico *m*.

D taktischer
    Fliegerleitoffizier *m*.

R начальник (офицер)
    центра управления
    тактической
    авиацией *м*.

A مُسَيْطِر جَوِّي تَعْبَوِي

5974

tactical air force

F force aérienne tactique
    f.

E aviación táctica *f*.

D taktische Luftflotte *f*.

R тактические
    военно-воздушные
    силы (ВВС) *ж. мн.
    ч*.

A قُوَّة جَوِّيَّة تَعْبَوِيَّة

5975

tactical airlift

F aérotransport tactique
    m.

E puente aéreo táctico
    m.

D taktischer Lufttransport
    m.

R тактическая перевозка
    ж.

A عملية ـ نقل جوية تعبوية

5976

tactical control

F contrôle tactique *m*.

E control táctico *m*.

D Taktische Kontrolle *f*.

R боевое управление *ср*.

A السَّيْطَرَة التَّعْبَوِيَّة

382

5977

**tactical nuclear
    weapons** *pl*.

F armes nucléaires
    tactiques *f. pl*.

E armas nucleares tácticas
    f. pl.

D taktische A-Waffen *f.
    pl*.

R тактическое ядерное
    оружие *ср*.

A أَسْلِحَة نَوَوِيَّة تَعْبَوِيَّة

5978

tactical reserve

F réserve tactique *f*.

E reserva táctica *f*.

D taktische Reserve *f*.

R тактический резерв *м*.

A إِحْتِيَاط تَعْبَوِي

5979

tactician

F tacticien *m*.

E táctico *m*.

D Taktiker *m*.

R тактик *м*.

A خبير بالتعبئة

5980

tactics

F tactique *f*.

E táctica *f*.

D Taktik *f*.

R тактика *ж*.

A تَعْبِئَة *f*

5981

(dog) tag

F plaque d'identité *f*.

E ficha de identidad *f*.

D Erkennungsmarke *f*.

R личный знак *м*.

A قُرص الهوية *m*

5982

(identity) tag

F plaque d'identité *f*.

E chapa de identidad *f*.

D Erkennungsmarke *f*.

R личный
    (опознавательный)
    знак *м*.

A قرص هوية

5983

tailboard

F hayon *m*.

E tablero posterior *m*.

D Ladeklappe *f*.

R откидной борт *м*.

A باب خلفي

5984

tailgate

F hayon *m*.

E tablero posterior *m*.

D Ladeklappe *f*.

R откидной борт *м*.

A باب خلفي

5985

tail light

F feu arrière *m*.

E luz de cola *f*.

D Schlusslicht *neut*.

R хвостовой фонарь *м*.

A مِصْبَاح خَلْفِي

5986

tail pipe

F tuyau d'échappement
    m.

E tubo de escape *m*.

D Schlussrohr *neut*.

R выхлопное сопло *ср*.

A انبوب العادم

5987

tail plane

F plan fixe horizontal *m*.

E plano de cola *m*.

D Höhenflosse *f*.

R стабилизатор *м*.

A سَطْح الذَّيْل

5988 [A,M]
tail wind
F vent arrière *m.*
E viento de cola *m.*
D Rückenwind *m.*
R попутный ветер *м.*
A ريح خلفية

5989
take off *n.*
F décollage *m.*
E despegue *m.*
D Start *m.*
R взлёт *м.*
A إقْلَاع

5990
take off
F décoller
E despegar
D starten
R взлетать
A أقْلَع

5991 [A]
(jump) take off
F décollage sauté *m.*
E despegue de salto *m.*
D Sprungstart *m.*
R взлёт прыжком
A الاقلاع من رصيف مائل

5992 [A]
(running) take off
F décollage roulé *m.*
E despegue normal *m.*
D Start *m.*
R (обычный) взлёт *м.*
A الاقلاع من مدرج عادي

5993 [A]
(stream) take off
F décollage en série *m.*
E despegue en serie *m.*
D Reihenstart *m.*
R (серийный) взлёт *м.*
A الاقلاع بالتسلسل

5994
(short) take off and landing (STOL) *adj.*
F (à) décollage et atterrissage courts
E (de) despegue y aterrizaje cortos
D kurz-Start-und-Lande-*komp.*
R взлёт и посадка с коротким разбегом и прибегом *м. ж.*
A قصير الاقلاع والهبوط

5995 [A,S,I,P]
"take out" *v.*
F détruire [AS]; enlever [IP]
E destruir [AS]; tomar [IP]
D ausschalten *tr.*
R уничтожать; брать в плен
A قبض على ـ دمر هدفا

5996
talcum powder
F talc *m.*
E (polvo de) talco *m.*
D Körperpuder *m.*
R тальк *м.*
A مسحوق الطلق

5997
talk down *v.*
F aider à atterrir par radio-contrôle
E controlar el aterrizaje de un avión desde tierra
D (ein Flugzeug) heruntersprechen *tr.*
R передавать указания лётчику по радио при заходе на посадку
A ارشد الطيار في عملية الهبوط

5998
talks *pl.*
F pourparlers *m. pl.*
E conversación *f.;* parlamento *m.*
D Verhandlungen *f. pl.*
R переговоры *м. мн. ч.*
A مُحَادَثَات *f*

5999 [S]
tampion
F tape de canon *f.*
E tapabocas *m. pl.*
D Mündungspropfen *m.*
R дульная пробка *ж.*
A سِدادة

6000 [A,E,V,U]
tank
F réservoir *m.*
E depósito *m.*
D Behälter *m.;* Tank(s) *m. (pl.)*
R бак *м.;* цистерна *ж.*
A خَزَّان *m*

6001 [V]
tank
F char *m.*
E carro de combate *m.*
D Kampfpanzer *m.*
R танк *м.*
A دَبّابَة *f*

6002
(amphibious) tank
F char amphibie *m.*
E carro de combate anfibio *m.*
D Schwimmpanzer *m.*
R плавающий танк *м.*
A دَبَابَة بَرْمَائِية *f*

6003
(bridge laying) tank
F char poseur de pont *m.*
E carro tiendepuentes *m.*
D Brückenlegepanzer *m.*
R танковый мостоукладчик (ТМУ) *м.*
A دبابة تجسير

383

6004

(fuel) tank

F réservoir de carburant *m.*
E depósito de combustible *m.*
D Kraftstoffbehälter *m.*
R топливный бак *м.*
A خَزَّان وُقُود *m*

6005

(jettison) tank

F réservoir largable *m.*
E depósito lanzable *m.*
D abwerfbarer Zusatztank *m.*
R сбрасываемый топливный бак *м.*
A خَزَّان الإِسْقاط *m*

6006

(main battle) tank

F char de bataille *m.*
E carro de combate *m.*
D Kampfpanzer *m.*
R (основной) танк *м.*
A دبابة قتالية رئيسية

6007

(positioned or dug-in) tank

F char embossé *m.*
E carro (enterrado y emplazado) *m.*
D Panzer (in Deckungsstellung)
R (окопанный) танк *м.*
A دبابة مخندقة

6008

(self-sealing) tank

F réservoir à obturation automatique *m.*
E depósito de obturación automática *m.*
D selbstabdichtender Kraftstoffbehälter *m.*
R самозатягивающаяся бочка *ж.*
A خزان وقود قادر للسد تلقائيا

384

6009

(self-sealing fuel) tank

F réservoir auto obturateur *m.*
E depósito de combustible deobturación automática *m.*
D (selbstverdichtender) Treibstoffbehälter *m.*
R (самозатягивающийся топливный) бак *м.*
A خَزَّان تِلْقَائِي القَفل

6010 [A]

(slip) tank

F réservoir largable *m.*
E depósito de combustible lanzable *m.*
D (abwerfbarer) Treibstoffbehälter *m.*
R (сбрасываемый топливный) бак *м.*
A خزان الاسقاط

6011

tank crew

F équipage de char *m.*
E tripulación de carro de combate *f.*
D Panzerbesatzung *f.*
R экипаж танка *м.*
A طاقِم دَبَّابَة

6012

tank destroyer

F chasseur de chars *m.*
E destructor de carros *m.*
D Jagdpanzer *m.* (-Kanone *f;* -Rakete *f.*)
R самоходное противотанковое орудие *ср.;* противотанковая самоходка *ж.*
A قَانِصَة لِلدَّبَّابَات

6013 [A]

tanker

F avion ravitailleur *m.*
E avión de reaprovisionamiento *m.*
D Tankflugzeug *neut.*
R самолёт-заправщик *м.*
A نَاقِلة نَفْط

6014

(oil) tanker

F pétrolier *m.*
E petrolero *m.*
D Tanker *m.*
R танкер *м.;* нефтеналивное судно *ср.*
A نَاقِلة زَيْت

6015

tank gun

F canon de bord *m.;* canon de char *m.*
E cañón de carro *m.*
D (Panzer-) Bordkanone *f.*
R танковая пушка *ж.*
A مِدْفَع دَبَّابَة

6016

tank truck

F camion-citerne *m.*
E camión cisterna *m.*
D Tankwagen *m.*
R цистерна *ж.*
A شَاحِنَة لِلدَّبَّابَات

6017 [R]

tape *v.*

F enregistrer
E grabar
D aufnehmen *tr.*
R записывать на плёнке
A سَجِّل

6018 [D,R]
tape *n.*
F bande (magnétique) *f.;*
 ruban (magnétique)
 m.
E cinta magnetofónica *f.*
D Tonband *neut.*
R плёнка *ж.;* лента *ж.*
A شَرِيط *m*

6019
tape recorder
F magnétophone *m.*
E magnetofón *m.*
D Tonbandgerät *neut.*
R магнитофон *м.*
A مُسَجِّل *m*

6020
tape recording
F enregistrement *m.*
E grabación en cinta *f.*
D Tonbandaufnahme *f.*
R плёнка *ж.;*
 магнитофонная
 запись *ж.*
A تسجيل شريطي

6021
tar
F goudron *m.*
E alquitrán *m.*
D Teer *m.*
R смола *ж.*
A قَطِرَان *m*

6022
target *n.*
F cible *f.;* objectif *m.*
E blanco *m.;* objetivo *m.*
D Ziel *neut.*
R цель *ж.*
A هَدَف *m*

6023 [S]
"Target!"
F "Coup au but!"
E "¡Blanco!"; "¡Objetivo
 batido!"
D „Treffer!"
R "Цель!"
A هدف ! *m*

6024
(area) target
F objectif non ponctuel
 m.
E objetivo extenso *m.*
D Flächenziel *neut.*
R (площадная) цель *ж.*
A هَدَف مِنْطَقَة *m*

6025
(denial) target
F zone d'interdiction *f.*
E defensa a toda costa *f.*
D (gesichtertes) Ziel *neut.*
R цель (воспрещения)
A هدف مانع

6026
(hard) target
F objectif résistant *m.*
E objetivo acorazado *m.*
D (hartes) Ziel *neut.*
R (твёрдая) цель *ж.*
A هدف صُلب

6027
(impromptu) target
F objectif inopiné *m.*
E objetivo imprevisto *m.*
D (Gelegenheits-)Ziel
 neut.
R (неплановая) цель *ж.*
A هدف مباغت

6028
(linear) target
F objectif linéaire *m.*
E objetivo lineal *m.*
D Breitenziel
 neut.
R удлинённая цель *ж.*
A هَدَف خَطِّي *m*

6029
(live) target
F cible réelle *f.*
E objetivo real *m.*
D (lebendes) Ziel *neut.*
R (боевая) цель *ж.;*
 (живая) цель *ж.*
A هدف حي

6030
(moving) target
F objectif mouvant *m.*
E objetivo móvil *m.*
D (bewegliches) Ziel *neut.*
R движущаяся цель *ж.*
A غَرَض مُتَحَرِّك *m*

6031
(on call) target
F objectif à la demande
 m.
E objetivo inmediato *m.*
D Ziel (auf Abruf) *neut.*
R (срочная) цель *ж.*
A هدف مطلوب

6032
(opportunity) target
F objectif inopiné *m.*
E objetivo imprevisto *m.*
D (Gelegenheits-)Ziel
 neut.
R (неплановая) цель *ж.*
A هدف مباغت

6033
(planned) target
F objectif prévu *m.*
E objetivo previsto *m.*
D (geplantes) Ziel *neut.*
R (плановая) цель *ж.*
A هدف مخطط

6034
(point) target
F objectif ponctuel *m.*
E objetivo aislado *m.*
D (Punkt-)Ziel *neut.*
R (одиночная) цель *ж.*
A هدف نقطي

6035
(pop-up) target
F cible à éclipse *f.*
E objetivo sorpresa *m.*
D (plötzlich
 auftauchendes) Ziel
 neut.
R появляющаяся
 мишень *ж.*
A هدف تدريب لرمي الخطف

6036

(reserve demolition) target

F ouvrage à destruction réservée

E objetivo constituído por una demolición reservada *m.*

D Vorbehaltssprengobjekt *neut.*

R (запасная) цель подрывания *ж.*

A هدف مهدد للتدمير

6037 [S]

(scheduled) target

F objectif à battre à l'horaire

E objetivo programado *m.*

D Fahrplanziel *neut.*

R плановая цель *ж.*

A هدف معين

6038

(secondary) target

F objectif secondaire *m.*

E objetivo secundario *m.*

D Ausweichziel *neut.*

R запасная цель *ж.*

A هَدَف ثَانوِي *m*

6039

(soft) target

F objectif peu résistant *m.;* cible facile *f.*

E objetivo vulnerable *m.*

D (weiches) Ziel *neut.*

R (незащитная) цель *ж.*

A هَدَف ضَعِيف

6040

(subsurface) target

F objectif immergé *m.*

E objetivo submarino *m.*

D (Unterwasser-)Ziel *neut.*

R (подводная) цель *ж.*

A هدف تحت سطح الماء

6041

(tow) target

F cible aérienne remorquée *f.*

E blanco remolcado *m.*

D (Schlepp-)Ziel *neut.*

R (буксируемая) мишень *ж.*

A هدف مقطور

6042

target acquisition

F acquisition d'objectif *f.*

E adquisición de objetivo *f.*

D Zielerfassung *f.*

R целеуказание *ср.*

A تَحْدِيد الهَدَف

6043 [N]

target analysis

F analyse d'objectifs *f.*

E análisis de objetivos *m.*

D Zielanalyse *f.*

R оценка цели *ж.;* анализ цели *м.*

A تَحْلِيل الهَدَف

6044

target area

F zone d'objectif *f.*

E zona-objetivo *f.*

D Zielraum *m.*

R район цели *м.*

A مِنْطَقَة الهَدَف

6045

target data *pl.*

F éléments de tir *m. pl.*

E elementos de tiro *m. pl.*

D Zielangaben *f. pl.*

R данные о цели *ж. мн. ч.*

A معلومات الهَدَف

6046

target date

F date d'exécution *f.*

E fecha tope *f.*

D Termin *m.*

R дата удара *ж.;* дата начала действий *ж.*

A مِيعَاد *m*

6047

target discrimination

F discrimination d'objectifs *f.*

E resolución de objetivos *f.*

D Zielauflösungsvermögen *neut.*

R выделение целей *ср.*

A تَمِييز الأهْدَاف

6048

target dossier

F dossier d'objectifs *m.*

E expediente de objetivos *m.*

D Zielunterlagen *f. pl.*

R дело целей *ср.*

A دَفْتَر إسْتِخْبَارَات عن الهَدَف

6049 [S]

target grid procedure

F procédé grille d'objectifs *m.*

E procedimiento cuadriculado de objetivo *m.*

D Zielgitterverfahren *neut.*

R корректирование по координатам *ср.*

A أُسْلُوب تَرْبِيع الهَدَف

6050

target of opportunity

F objectif inopiné *m.*

E objetivo de oportunidad *m.*

D Gelegenheitsziel (beobachtetes Schiessen) *neut.*

R неплановая цель *ж.*

A هَدَف عَارِض *m*

6051

tarmac

F goudron *m.*
E alquitranado *m.*
D Asphalt *m.*
R гудрон *ср.*; асфальт
   ср.
A حَصْبَاء مُقَيَّرة لِرَصْف الطُّرُق

6052

tarpaulin

F bâche *f.;* prélart [M] *m.*
E encerado *m.*
D Plane *f.*
R брезент *м.*
A نَسِيج مُشَمَّع *m*

6053

task *n.*

F tâche *f.*
E tarea *f.*
D Aufgabe *f.*
R задача *ж.*; задание *ср.*
A وَاجِب *m*

6054

task *v.*

F charger quelqu'un (de);
   donner pour mission
   (de)
E encargar;
   responsibilizar (a uno
   de)
D beaufgaben
R ставить задачу
A كَلَّف بِعَمَل

6055

task force

F groupement *m.*
E agrupación táctica *f.*
D Einsatzverband *m.*
R (временная)
   оперативная группа
   ж.
A قُوَّة وَاجِب مُعَيَّن

6056

(company) task force

F sous-groupement *m.*
E agrupación mixta de
   combate *f.*
D Kampfgruppe *f.*
R боевая группа *ж.*
A سرية واجب معين

6057

(joint) task force

F groupement interarmées
   m.
E agrupación conjunta *f.*
D (gemeinsamer)
   Gefechtsverband *m.*
R (общевойсковая)
   боевая группа *ж.*
A قوة مشتركة لواجب معين

6058

tattoo

F tatouage *m.*
E tatuaje *m.*
D Tätowierung *f.*
R татуировка *ж.*
A دقة العودة

6059

tattoo

F fête militaire *f.*
E gran espectáculo militar
   m.
D (festliche) Abendparade
   f.
R торжественная заря
   ж.
A معرض ـ مظاهرة عسكرية

6060

tattoo

F extinction des feux *f.*
E retreta *f.*
D Zapfenstreich *m.*
R вечерняя заря *ж.*
A وشم

6061

taut

F tendu
E tenso
D gespannt; straff
R тугой
A مُوَتَّر ـ مشدود

6062 [A]

taxi *v.*

F rouler
E rodar
D rollen
R рулить
A دَرَج عَلى الأرْض

6063

taxiway

F piste de roulement (de
   circulation) *f.*
E pista de rodaje *f.*
D Rollbahn *f.*
R рулежная полоса *ж.*;
   дорожка *ж.*
A مِدْرَجَة جَانِبِيَّة *f*

6064

team

F équipe *f.*
E equipo *m.*
D Team *neut.*
R команда *ж.*; расчёт
   м.; экипаж *м.*
A زُمْرَة *f*

6065

(assault) team

F équipe choc *f.*
E equipo de asalto *m.*
D Angriffstrupp *m.*
R (штурмовая) группа
   ж.
A فريق اقتحام

387

6066

(central planning) team

F groupe central de planification *m.*

E equipo central de planificación

D zentraler Planungsstab *m.*

R центральная группа планирования *ж.*

A فَرِيق التَّخْطِيط المَرْكَزِي

6067

(combat engineer) team

F groupement d'emploi du génie *m.*

E grupo de ingenieros de combate *m.*

D Kampfpioniertrupp *m.*

R (инженерно-сапёрная) команда *ж.*

A زمرة هندسة القتال

6068 [L]

(combined arms) team

F sous-groupement *m.*

E agrupamiento táctico *m.*

D verbündete Kampfgruppe

R боевая группа соединённых войск *ж.*

A مجموعة الأسلحة المُشتركة

6069

(company) team

F sous-groupement

E agrupación mixta de combate *f.*

D (verstärkte) Kompanie *f.*

R боевая группа *ж.*

A زمرة قتال

6070

(field surgical) team

F équipe chirurgicale mobile *f.*

E equipo quirúrgico de campaña *m.*

D Chirurgengruppe *f.*

R (полевая санитарная) команда *ж.*; полевой госпиталь *м.*

A زمرة جراحية ميدانية

6071

(Forward Air Control) Team

F poste de guidage avancée *f.*

E grupo avanzado de aviación *m.*

D vorgeschobener Lufteinsatztrupp *m.*

R команда передовых авианаводчиков *ж.*

A زمرة أمامية للسيطرة الجوية

6072

(pathfinder) team

F équipe d'orienteurs-marqueurs *f.*

E equipo de señaladores-guías *m.*

D Zielmarkierungs-kommando *neut.*

R следопытная группа *ж.*; команда *ж.*

A زُمْرة كَشافة *f*

6073

(tank killer) team

F équipe CRAC (combat rapproché anti-char) *f.*

E grupo de antitanquistas *m.*

D Panzerabwehrtrupp *m.*

R (противотанковое) отделение *ср.*

A زمرة مضادة للدبابات

6074

technical

F technique

E técnico

D technisch

R технический

A فَنِّي

6075

technical specification

F spécification technique *f.*

E especificación técnica *f.*

D (technische) Forderungen *f. pl.*

R технические требования; *ср. мн. ч.* технические условия *ср. мн. ч.*; текнические спецификации *ж. мн. ч.*

A وَصْف فَنِّي *m*

6076

technician

F technicien *m.*

E técnico *m.*

D Techniker *m.*

R техник *м.*

A تِقْني

6077

telecommunication(s) n. (pl.)

F télécommunication(s) *f. (pl.)*

E telecomunicaciónes *f. pl.;* transmisiones *f. pl.*

D Fernmeldetechnik *f.;* Fernmeldeübertra-gung(en) *f. (pl.)*

R дальняя связь *ж.*; средства связи *ср. мн. ч.*

A إتِّصالات مِن بُعْد

6078

telegraphy
F télégraphie *f.*
E telegrafia *f.*
D Telegraphie *f.*
R телеграфия *ж.*
A بَرْق

6079

telemetry
F télémétrie *f.*
E telemetría *f.*
D Entfernungsmessung *f.;*
  Telemetrie *f.*
R телеметрия *ж.*
A قِيَاس البُعْد

6080

telephone *v.*
F téléphoner (à)
E telefonear (a)
D anrufen *tr.;* telefonieren
R звонить
A تكلم في التليفون

6081

telephone *n.*
F téléphone *m.*
E teléfono *m.*
D Fernsprecher *m.*
R телефон *м.*
A هَاتِف ـ تليفون *m*

6082

(tank) telephone
F téléphone extérieure *f.*
E teléfono del carro *m.*
D Panzersprechgerät *neut.*
R телефон танка *м.*
A هاتف دبابة

6083

telephone call
F appel *m.;* coup de
  téléphone *m.*
E llamada telefónica *f.*
D Anruf *m.*
R вызов (по телефону)
  м.
A نِداء هَاتِفي *m*

6084

telephone exchange
F central téléphonique *m.*
E central (telefónica) *f.*
D (Fernsprech-)
  Vermittlung *f.;*
  Zentrale *f.*
R телефонная станция
  ж.; центральная *ж.*
A بَدَّالة هَاتِفِيَّة *f*

6085

telephone number
F numéro de téléphone
  m.
E número de teléfono *m.*
D Fernsprechnummer *f.;*
  Telefonnummer *f.*
R телефонный номер *м.*
A رَقْم هَاتِفي ـ رقم التليفون

6086

teleprinter
F téléimprimeur *m.*
E teleimpresor *m.*
D Fernschreiber *m.*
R буквопечатающий
  телеграфный
  аппарат *м.*
A طابعة برقية

6087

telescope *n.*
F télescope *m.*
E telescopio *m.*
D Fernrohr *neut.*
R телескоп *м.*
A مِقْرَاب ـ تِليسكوب *m*

6088

telescope *v.*
F (s')emboîter; (se)
  télescoper
E comprimir; meterse
  dentro de;
  telescopar(se)
D (sich)
  ineinanderschieben
  tr.; verkürzen
R складывать(ся)
A تَدَاخَل

6089

telescopic
F coulissant; télescopique
E extensible; telescópico
D Teleskop- *komp.;*
  teleskopisch *adj.*
R телескопический
A مُتَدَاخِل ـ تِليسكوبي

6090

teletype
F télétype *m.*
E teletipo *m.*
D Fernschreiber *m.;*
  Fernschreib- *komp.*
R радиотелепринтер *м.*
A كتابة برقية

6091

televise
F téléviser
E televisar
D (durch Fernsehsender)
  übertragen *untr.*
R передавать по
  телевидению
A بَثَّ بِالتِلْفِزْيُون

6092

television camera
F caméra de télévision *f.*
E cámara de televisión *f.*
D Fernsehkamera *f.*
R телевизионная камера
  ж.
A كاميرا تِلْيفزيونية *f*

6093

television monitor
F poste de télévision *m.;*
  téléviseur *m.*
E monitor *m.;* receptor
  (de televisión) *m.;*
  televisor *m.*
D Fernsehgerät *neut.;*
  Monitor *m.*
R (телевизионный)
  монитор *м.*
A جِهَاز تِلْفِزْيُون *m*

6094

temperate
F tempéré
E templado
D gemässigt
R умеренный
A مُعْتَدَل

6095

temperature
F température *f.*
E temperatura *f.*
D Temperatur *f.*
R температура *ж.*
A دَرَجَة الْحَرَارَة *f*

6096 [E]

templet
F gabarit *m.*
E plantilla *f.;* patrón *m.*
D Schablone *f.*
R лекало *ср.*
A مِعْيَار

6097

temporary
F (de) fortune; provisoire;
    temporaire
E interino; provisional;
    temporal
D vorläufig; zeitweilig
R временный
A مُؤَقَّت

6098

(submarine) tender
F bâtiment de servitude
    m.
E buque nodriza para
    submarinos *m.*
D Tender *m.*
R плавучая база
    подлодок *ж.*
A ممونة الغواصات

390

6099

tense *adj.*
F tendu
E tenso
D gespannt
R напряжённый;
    натянутый
A مُتَوَتِّر

6100

tensile
F (de) traction
E tensor
D Zug- *komp.*
R растяжимый
A قابِل لِلْمَدّ

6101

tension
F tension *f.*
E tensión *f.*
D Spannung *f.*
R напряжение *ср.;*
    растяжение *ср.*
A تَوَتُّر *m*

6102

tent
F tente *f.*
E tienda *f.*
D Zelt *neut.*
R палатка *ж.*
A خَيْمة *f*

6103

tentacle
F poste de liaison *m.*
E puesto de enlace *m.*
D Verbindungstrupp *m.*
R подвижный пост связи
    м.
A مِجَسّ *m*

6104

tentage
F (des) tentes *f. pl.*
E tiendas de campaña *f.*
    pl.
D Zeltausrüstung *f.*
R палаточное
    имущество *ср.*
A معدات التخييم

6105 [A,V]

terminal *n.*
F aérogare [A] *f.;*
    terminus [V] *m.*
E terminal [A] *f.;* término
    [V] *m.*
D Einfangsgebäude [A]
    neut.; Endbahnhof [V]
    m.
R аэровокзал [A] *м.;*
    конечный пункт [V]
    м.
A آخَر مَحَطَّة

6106 [R]

terminal *n.*
F terminal *m.*
E terminal *f.*
D Empfänger bzw. Sender
    m. m.
R клемма *ж.*
A مِرْبَط

6107

terminal *adj.*
F final
E terminal
D End- *komp.*
R конечный;
    терминальный;
    предсмертный
A نِهائِي

6108 [E]

terminal *n.*
F borne *f.*
E borne *m.*
D Klemme *f.*
R клемма *ж.*
A مِرْبَط *m*

6109 [D]

terminal *n.*
F sortie *f.;* terminal *m.*
E salida *f.;* terminal *f.*
D Ausgang *m.;* Terminal
    m.
R ввод *м.;* вывод *м.*
A مِرْبَط

6110

terminal effect

F effets de feu sur
l'objectif *m. pl.*

E efectos reales *m. f.*

D Endphasenwirkung *f.*

R окончательное
действие *ср.*

A التأثير النهائي

6111 [G]

terminal guidance

F guidage de fin de
trajectoire *m.*

E guía terminal *f.*

D Endphasenlenkung *f.*

R конечное наведение на
цель *ср.*

A تَوْجِيه نِهائِي *m*

6112

terminal leave

F congés de fin de service
m. pl.

E permiso final *m.*

D Entlassungsurlaub *m.*

R отпуск по случаю
выхода в резервы *м.*

A اجازة نهائية

6113

terminate

F (se) terminer

E terminar(se)

D abschliessen *tr.*

R кончать; завершать

A أنْهَى

6114

terrace

F terre-plein(s) *m. (pl.)*

E terraplén *m.*

D Terrasse *f.*

R терраса *ж.*

A سَطْح

6115

terrain

F terrain *m.*

E terreno *m.*

D Gelände *neut.*

R местность *ж.*

A أرْض *f* تَضَارِيس *f*

6116

(dominant) terrain

F terrain en surplomb *m.*

E terreno dominante *m.*

D beherrschendes Gelände
neut.

R господствующая
местность *ж.*

A منطقة حاكمة *f*

6117

(key) terrain

F position-clef *f.*

E clave estratégica *f.*

D Schlüsselgelände *neut.*

R (ключевая) местность
ж.

A أرْض حَيَوِيَّة *f*

6118

(trafficable) terrain

F terrain practicable *m.*

E terreno transitable por
vehículo *m.*

D (fahrbares) Gelände
neut.

R (проходимая)
местность *ж.*

A ارض صالحة للسيارات

6119 [A]

**terrain avoidance
system**

F fonction "évitement du
sol" *f.*

E sistema de esquivar el
terreno *m.*

D Hindernisanzeige für
Tiefflug *f.*

R высотомерная РЛС
ж.

A جِهَاز لاجْتِنَاب تَضَارِيس
الأرْض

6120 [A]

**terrain clearance
system**

F fonction "découpe" *f.*

E sistema de vuelo a
altura constante *m.*

D (automatische)
Höhensteuerung für
Tiefflug *f.*

R РЛС определения
местности (при
маловысотном
полёте) *ж.*

A جِهَاز يُخَطِّي تَضَارِيس
الأرْض

6121 [A]

**terrain following
system**

F fonction "suivi du
terrain" *f.*

E sistema de adaptación
al terreno *m.*

D Konturenflughilfe *f.*

R контурноопределяющая
РЛС *ж.*

A جِهَاز يَتْبَع تَضَارِيس الأرْض

6122

terrain reinforcement

F travaux de mobilité, de
contre-mobilité et
d'enfouissement *m. pl.*

E refuerzo del terreno *m.*

D (befestigtes) Gelände
neut.

R оборонительная
подготовка
местности *ж.*

A تعزيز الأرض (بالحواجز
والتحصينات)

6123

territory

F territoire *m.*

E territorio *m.*

D Gebiet *neut.*

R территория *ж.*; земля
ж.

A إقْلِيم *m*

6124

terrorism

F terrorisme *m.*
E terrorismo *m.*
D Terrorismus *m.*
R терроризм *м.*
A إرْهَاب *m*

6125

terrorist *adj. n.*

F terroriste *adj. n. m. f.*
E terrorista *adj. s. m. f.*
D terroristisch *adj.;*
Terrorist- *komp.;*
Terrorist *s. m.*
R террористический
прилаг.;
террорист(ка) *м.(ж.)*
A إرْهَابِي *m*

6126

terrorize

F terroriser
E aterrorizar
D terrorisieren
R терроризировать
A أرْهَب

6127

test *adj.*

F d'essai(s)
E (de) prueba(s)
D Prüf- *komp.;* Test- [H]
komp.
R испытательный;
пробный
A تَجْرِيبِي

6128

test *v.*

F essayer; tester; vérifier
E ensayar; probar;
verificar
D erproben; kontrollieren;
prüfen
R испытывать;
проверять;
пробовать
A جَرَّب

6129

test *n.*

F épreuve *f.;* essai *m.;*
test *m.*
E ensayo *m.;* prueba *f.*
D Probe *f.;* Prüfung *f.;*
Test(s) *m.*
R испытание *ср.;*
проверка *ж.*
A تَجْرِبَة *f*

6130

(basic fitness) test

F contrôle de la valeur
physique individuel
m.; test de cooper
E prueba de aptitud física
f.
D körperliche
Tauglichkeitsprüfung
f.
R испытание основной
пригодности *ср.*
A اختبار اللياقة البدنية

6131

**(enlistment screening)
test**

F test d'aptitude à
l'engagement *m.*
E test de selección de
reclutas *m.*
D Einstellungsprüfung *f.*
R приёмная
контрольная
проверка
новобранцев *ж.*
A اختبار المجندين

6132

(flight) test

F vol d'essai *m.*
E vuelo de prueba *m.*
D Probeflug *m.*
R испытательный полёт
м.
A اختبار طيران *m*

6133

**(operational
acceptance) test**

F test d'aptitude
opérationelle *m.*
E prueba operacional de
recepción *f.*
D Annahmeprobe *f.*
R (приёмное) испытание
ср.
A اختبار الصلاحية للعمليات

6134 [G]

(pre-launch) test

F essai des circuits de la
phase de
prélancement *m.*
E pruebas preparatorias
de vuelo *f. pl.*
D Abschussprobe *f.*
R (предпусковой)
осмотр *м.*
A اختبار قبل الاطلاق

6135 [A,E]

(run-up) test

F essai au point fixe *m.*
E prueba en tierra *f.*
D statische Motorenprobe
f.
R (стендовое) испытание
ср.
A تجربة المحرك

6136

test bench

F banc d'essai *m.*
E aparato de pruebas *m.*
D Probebank *f.*
R испытательный стенд
м.
A دكّة التجربة

6137

test pilot

F pilote d'essais *m.*
E piloto de pruebas *m.*
D Testpilot *m.*
R лётчик-испытатель *м.*
A طيار التجربة

6138

theater (of operations)

F théâtre (d'opérations) *m.*

E teatro (de operaciones) *m.*

D (Kriegs-)Theater *neut.*

R район боевых действий; театр боевых действий

A مَسْرَح العَمَلِيَّات *m*

6139

theodolite

F théodolite *m.*

E teodolito *m.*

D Theodolit *m.*

R теодолит *м.*

A آلة قِيَاس زَوَايَا الإِرْتِفَاع

6140

thermal *adj. n.*

F thermique *adj. n. m.*

E térmico *adj.*

D thermisch *adj.;* Wärme-komp.; Warmluftströmung *s. f.*

R термический *прилаг.;* тепловой *прилаг.;* тёплая струя *ж.*

A حَرَارِي

6141 [N]

thermal exposure

F effet thermique *m.*

E efecto térmico *m.*

D Hitzewirkung

R тепловое облучение *ср.*

A تَعَرُّض حَرَارِي *m*

6142

thermal imager

F dispositif d'image thermique *m.*

E dispositivo de imagen térmica *m.*

D Wärmebildgerät *neut.*

R электроннооптический тепловой пеленгатор *м.;* индикатор *м.*

A مُصَوِّر حَرَارِي *m*

6143

thermal insulation

F calorifugeage [E] *m.;* isolation thermique *f.*

E aislamiento térmico *m.;* calorifugaje [E] *m.*

D Wärmeschutz *m.*

R теплоизоляция *ж.;* термическая изоляция *ж.*

A عَزْل حَرَارِي *m*

6144

thermal radiation

F rayonnement thermique *m.*

E radiación térmica *f.*

D Hitzestrahlung *f.*

R тепловое излучение *ср.*

A إِشْعَاع حَرَارِي *m*

6145

thermometer

F thermomètre *m.*

E termómetro *m.*

D Temperaturmesser *m.*

R термометр *м.*

A مِقْيَاس الحرارة

6146

thermonuclear

F thermonucléaire

E termonuclear

D thermonuklear

R термоядерный

A نَوَوِي حَرَارِي

6147

thickening

F renforcer une unité

E refuerzo *m.*

D Verstärken *neut.*

R подкрепление *ср.*

A تعزيز القطع في الدفاع

6148 [T]

thicket

F fourré *m.*

E matorral *m.*

D Dickicht *neut.*

R чаща *ж.*

A حِرْش

6149

thinning out

F opération dégraissement *f.*

E reducción *f.*

D Verringern *neut.*

R прореживание *ср.*

A تخفيف القطع في القتال

6150

"This is …"

F "Ici…"

E "Aquí…"

D „Hier ist…"

R "Я"

A هذا . . . ' واحد '

6151

threat

F menace *f.*

E amenaza *f.*

D Drohung *f.*

R угроза *ж.*

A تَهْدِيد *m*

6152

(enemy) threat

F ennemi éventuel *m.*

E amenaza *f.*

D Feind *m.*

R угроза *ж.;* вероятный противник *м.*

A التهديد

6153

threaten

F menacer (quelqu'un de)
E amenazar
D (be)drohen
R угрожать
A هدَّد

6154 [M]

three-mile limit

F limite des eaux
    territoriales *f.*
E límite de tres millas *m.*
D Dreimeilengrenze *f.*
R трёхмильный рубеж
    м.
A حد ثلاثة أميال

6155

throttle lever

F manette des gaz *f.*
E estrangulador *m.;*
    palanca de gases *f.*
D Gashebel *m.*
R рычаг управления
    газом *м.*
A ذِراع الصِّمام الخانِق

6156

throw *v.*

F jeter
E echar; lanzar
D werfen
R метать; бросать
A رمَى

6157

throw back an attack
    v.

F repousser une attaque
E rechazar un ataque
D (einen Angriff)
    zurückstossen *tr.*
R отражать атаку
A رد هجوما

6158

(flame) thrower

F lance-flammes *m.*
E lanzallamas *m.*
D Flammenwerfer *m.*
R огнемёт *м.*
A قاذفة لهب

6159 [L]

thrust *n.*

F poussée *f.*
E arremetida *f.*
D Vorstoss *m.*
R удар *м.*; продвижение
    ср.
A دَفْع *m*

6160 [L]

thrust *v.*

F pousser en avant
E arremeter; empujar
D vorstossen *tr.*
R (стремительно)
    продвигаться
A دَفَع

6161

tidal

F (à) marée
E (de) marea
D Gezeiten- *komp.;* Tide-
    komp.
R приливо-отливный
A مُتَعَلِّق بالمَدّ والجَزْر

6162

tidal basin

F bassin à flot *m.*
E dique de marea *m.*
D Tidebecken *neut.*
R приливный бассейн *м.*
A حَوْض المَدّ والجَزْر

6163

tide

F marée *f.*
E marea *f.*
D Gezeit *f.;* Tide *f.*
R прилив *м.*; отлив *м.*
A المَدّ والجَزْر

6164

tideway

F lit de marée *m.*
E canal de marea *m.*
D Stromstrich *m.*
R направление течения
    прилива (отлива) *ср.*
A مَجْرى المَدّ والجَزْر

6165

tie *v.*

F faire un nœud; lier
E atar; hacer un nudo
D binden; Knoten machen
R завязывать;
    привязывать
A شَدَّ

6166

tie down *n.*

F (matières d') arrimage
    m.
E (materiales de) amarre
    m.;
D Befestigung
    (- smaterial) *f. (neut.)*
R прихватка *ж.*
A رَبْط *m*

6167

tie down *v.*

F arrimer; assujettir; fixer
    [L]
E amarrar; atracar [M];
    sujetar [LU]
D befestigen
R привязывать
A رَبَط

6168

tie down troops *v.*

F fixer
E mantener tropas
    distraídas del cuerpo
    principal
D (Truppen) niederhalten
    tr.
R привязывать войска
A شغل قطع العدو

6169

(weapons) tight!

F tir restreint!
E (armas) en posción
D Waffen (mit bedingter Feuererlaubnis) *f. pl.*
R разрешается стрелять только по точному опознанию самолёта
A الاشتباك مع طائرات العدو التي تمييزها أكيد

6170

tiller

F barre *f.*
E caña del timón *f.*
D Ruderpinne
R румпель *м.*
A عَجَلَة السُّكَّان ـ ذراع الدَّفة

6171

tilt *v.*

F (s')incliner
E inclinar(se)
D schräg sein; stellen
R наклонять(ся)
A مَال

6172

tilt angle

F angle d'inclinaison *m.*
E ángulo de inclinación *m.*
D Verkantungswinkel *m.*
R угол наклона *м.*
A زَاوِيَة المَيْل

6173

timber

F bois *m. pl.;* madrier *m.*
E madera *f.*
D Nutzholz *neut.*
R лесоматериалы *м. мн. ч.*
A خَشَب *m*

6174

timberline

F limite forestière *f.*
E límite forestal *m.*
D Baumgrenze *f.*
R верхняя граница леса *ж.*
A خَطّ الأَشْجَار

6175

time *n.*

F temps *m.*
E tiempo *m.*
D Zeit *f.*
R время *ср.*
A وَقْت *m*

6176

time bomb

F bombe à retardement *f.*
E bomba de retardo *f.*
D Zeitbombe *f.*
R бомба замедленного действия *ж.*
A قُنْبُلَة مَوْقُوتة *f*

6177

time check

F synchronisation de l'heure *f.*
E comprobación de la hora *f.*
D Uhrzeiteinstellung *f.*
R проверка времени *ж.*
A تحقيق الوقت

6178

time for execution

F délai d'exécution *m.*
E tiempo concedido (para una acción) *m.*
D Operationsfrist *f.*
R указанное время *ср.*
A الزمن المسموح للتنفيذ

6179 [S]

time of flight

F durée de trajet *f.*
E duración (de la trayectoria) *f.*
D Flugzeit *f.*
R время полёта *ср.;* продолжительность полёта *ж.*
A مُدَّة الطيران

6180

time of origin

F heure de sortie d'un document *f.*
E hora de origen *f.*
D Ursprungszeit *f.*
R время передачи *ср.*
A وقت التوقيع

6181 [R]

time of receipt

F heure de réception *f.*
E hora de llegada (de recibo) *f.*
D Empfangzeit *f.*
R время принятия *ср.*
A وَقْت التَّسَلُّم

6182 [S]

time-on-target method

F heure sur l'objectif *f.*
E concentración simultánea *f.*
D Feuerzusammenfassung im Ziel *f.*
R сосредоточение огневых налётов *ср.*
A طَرِيقة الوَقْت على الهَدَف

6183 [A]

time over target

F heure sur l'objectif *f.*
E tiempo sobre el objetivo *m.;* TOT *m.*
D Zeit über Ziel
R указанное время налёта *ср.*
A الوقت المتوقع فوق الهدف

6184

timetable

F horaire *m.*
E horario *m.*
D Zeittabelle *f.*
R расписание *ср.*
A جدول التوقيت

6185 [I,P]

tip-off *n.*

F avertissement par une
  dénonciation *m.*
E aviso *m.*
D Tip *m.*
R намёк *м.*
A قدّم معلومات

6186 [V]

tire *n.*

F pneu(s) *m. (pl.)*
E neumático *m.*
D Bereifung *f.;* Reife *f.*
R шина *ж.*
A إطار *m*

6187 [N]

TNT equivalent

F équivalence TNT *f.*
E equivalencia TNT *f.*
D TNT-Entsprechung *f.*
R тротиловый
  эквивалент *м.*
A مُكَافِـئ ثَالِث نَتْرَيْت
  التُّولُوين

6188

(color) tone

F ton *m.*
E tono *m.*
D Ton *m.*
R тон *м.;* цвет *м.*
A عُمْق اللَّوْن

6189

tonnage

F tonnage *m.*
E tonelaje *m.*
D Tonnage *f.*
R тоннаж *м.*
A زِنَة ـ حمولة السفينة بالطن *f*

6190

tool

F outil *m.*
E herramienta *f.*
D Werkzeug *neut.*
R прибор *м.;*
  инструмент *м.*
A أَدَاة *f*

6191

topographic

F topographique
E topográfico
D topographisch
R топографический;
  топо-
A طُبُغْرَافِي

6192

topographical crest

F crête topographique *f.*
E cresta topográfica *f.*
D topographischer Kamm
R топографический
  гребень *м.*
A ذروة طبغرافية

6193

topography

F topographie *f.*
E topografia *f.*
D Topographie *f.*
R топография *ж.*
A طُبُغْرَافِيَّة

6194

top secret

F très secret
E máximo secreto
D streng geheim
R совершенно секретный
A سِرِّي لِلْغَايَة

6195

top up

F rajouter; remettre
E llenar a tope
D nachfüllen *tr.*
R пополнять (запасами)
A مَلَأ

6196

torch

F torche électrique *f.*
E linterna eléctrica *f.;*
  lámpara de bolsillo *f.*
D Fackel *f.*
R фонарь *м.*
A مصباح

6197

tornado

F tornade *f.*
E tornado *m.*
D Wirbelsturm *m.*
R торнадо *ср.*
A ريح زَعْزَع ـ إعصار *m*

6198

torpedo *v.*

F torpiller
E torpedear
D torpedieren
R торпедировать
A نَسَف سَفِينَة بطوربيد

6199

torpedo *n.*

F torpille *f.*
E torpedo *m.*
D Torpedo *m.*
R торпеда *ж.*
A نَسِيفَة *f* ـ طوربيد *m*

6200

(homing) torpedo

F torpille à tête
  chercheuse *f.*
E torpedo teleguiado *m.*
D zielsuchender Torpedo
  m.
R самонаводящаяся
  торпеда *ж.*
A نسيفة موجهة

396

6201

torpedo defense net
F filet pare-torpilles *m.*
E red de defensa contra
 torpedos *m.*
D Torpedonetz *neut.*
R противоторпедная
 сеть *ж.;*
 противоминная сеть
 ж.
A شَبَكَة وَاقِيَة مِن الطوربيد

6202

torpedo tube
F tube lance-torpille *m.*
E tubo lanzatorpedos *m.*
D Torpedorohr *neut.*
R торпедный аппарат *м.*
A أنبُوب قَذف الطوربيد

6203

torque
F couple de torsion *m.*
E par de torsión *m.*
D Drehmoment *neut.*
R момент вращения *м.;*
 скручивание *ср.*
A مُزْدَوَجَة *f* عزم *m*

6204

torsion bar
F barre de torsion *f.*
E barra de torsión *f.*
D Drehstab *m.*
R штанга кручения *ж.;*
 противокрутительная
 штанга *ж.*
A قَضِيب التَّوَائي

6205

torture *v.*
F mettre à la torture
E torturar
D foltern
R пытать
A عذّب

6206

total *adj.*
F global; total
E total
D Gesamt- *komp.;* total;
 vollständig
R полный; тотальный
A كُلِّي

6207

total *v.*
F additionner; (se)
 monter (à)
E sumar
D (sich) belaufen (auf);
 betragen
R подсчитывать;
 подводить итог
A جَمَع

6208

total *n.*
F montant *m.;* somme *f.;*
 total *m.*
E suma *f.;* total *m.*
D Betrag *m.;* Summe *f.*
R сумма *ж.;* итог *м.;*
 целое *ср.*
A مَجْمُوع *m*

6209

total eclipse
F éclipse totale *f.*
E eclipse total *m.*
D totale Finsternis *f.*
R полное затмение *ср.*
A خسوف كامل *m* (الشمس)

6210

touch and go
F atterrissage et décollage
 sans interruption *m.*
 m.
E maniobra de aterrizar y
 despegar *f.*
D Lande- und
 Abflugmanöver *neut.*
R бреющая посадка *ж.*
A تمرين الهبوط والاقلاع

6211 [A]

touch down *v.*
F atterrir
E amerizar; aterrizar;
 tomar tierra
D aufsetzen
R посадиться;
 приземляться
A لَمَس الأرض

6212 [A]

touchdown *n.*
F posé *m.;* atterrissage
 m.; amerrissage *m.*
E amerizaje *m.;* aterrizaje
 m.; toma de tierra *f.*
D Aufsetzen *neut.*
R посадка *ж.;*
 приземление *ср.*
A لَمْس الأرْض

6213

(overseas) tour
F séjour outre-mer *m.*
E destino en ultramar *m.*
D Auslandsdienst *m.*
R объязанность за
 рубежом (NATO);
 интернациональный
 долг (Sov.) *ж. м.*
A دورة

6214

tow *v.*
F remorquer
E remolcar
D (ab)schleppen *tr.*
R буксировать; тянуть
A قَطَر

6215

tow *n.*
F remorque *f.*
E remolque *m.*
D Schleppen *neut.*
R буксир *м.*
A مَقْطُورَة *f* سَحْب *m*

397

6216 [A]
(control) tower
F tour de contrôle f.
E torre de control f.
D Kontrollturm
R командно-
диспетчерский пункт
м.
A برج المراقبة

6217
towrope
F (câble de) remorque
(m.) f.
E (cable de) remolque m.
D Schleppseil neut.
R буксирный трос м.
A حَبْل القَطْر

6218
toxic
F toxique m.
E tóxico
D ABC- komp.; giftig adj.
R ядовитый;
отравляющий
A سَامّ

6219 [C]
toxic agent
F agent toxique m.
E agresivo tóxico m.
D Kampfstoff m.
R отравляющее
вещество (ОВ) ср.
A عَامِل كِيمَاوِي سَامّ

6220
toxicity
F toxicité f.
E toxicidad f.
D Giftigkeit f.
R токсичность ж.
A سُمِّيَّة

6221 [S]
trace n.
F trace f.
E estela trazadora f.
D Leuchtspur
R трасса ж.
A أَثَر m

6222 [T]
trace n.
F calque m.
E trazo m.
D (Plan-)Pause
R чертёж м.
A تَخْطِيط m

6223 [T]
trace v.
F calquer
E calcar; trazar
D durchpausen
R чертить; сводить
чертёж;
калькировать
A إقْتَفَى الأَثَر

6224 [P]
trace v.
F suivre la piste de
E encontrar; seguir la
pista (de)
D (jemandem) nachspüren
tr.
R следить (за);
прослеживать
A تَتَبَّع

6225 [R]
trace n.
F trace f.
E señal f.
D Spur f.
R след м.
A دَرْب m

6226
trace elements pl.
F oligo-éléments m. pl.
E elementos vestijiales m.
D Spurenelemente neut.
pl.
R следы м. мн. ч.
A عناصر إسْتِشْفَافِيَّة f

6227 [S]
tracer (ammunition)
F munition traçante f.
E munición trazadora f.
D Leuchtspurmunition f.
R трассирующие
боеприпасы м. мн. ч.
A ذَخِيرَة مُذَنَّبَة f

6228 [E,N]
tracer (element)
F (élément) traceur m.
E (elemento) trazador m.
D Spurenelement neut.
R течёный атом м.
A عُنْصُر إسْتِشْفَافِي m

6229 [R]
track v.
F poursuivre
E perseguir; rastrear
D verfolgen
R сопровождать;
следить
A عَقَّب أَثَر

6230
track n.
F piste f.; voie f.; sentier
m.
E pista f.; senda f.;
camino m.
D Bahn f.; Spur f.; Weg
m.
R след м.; трасса ж.;
дорога ж.; тропа ж.
A دَرْب m

6231 [V]
track n.
F écartement m.
E longitud del eje
D Spurweite f.
R колея ж.
A المسافة بين المركبات

6232
(animal's) track
F trace *f.*
E huella *f.;* rastro *m.*
D (Tier)spuren *f. pl.*
R след *м.*
A أثر حيوان

6233
(computer or tape) track
F piste *f.*
E traza (en la cinta) *f.*
D (Komputer) Bandspur *f.*
R полоса *ж.*
A وجه الشريط

6234
(missile) track
F trajectoire *f.*
E trayectoria *f.*
D Flugkörperkurs *m.*
R направление *ср.;* сопровождение *ср.*
A مسار مقذوف

6235 [V]
track (of tank) *n.*
F chenille (de char) *f.*
E cadena *f.;* oruga (ambos de carro de combate) *f.*
D (Raupen-)Kette *f.*
R гусеница *ж.*
A جِنزِير *m*

6236
(person's) track
F trace *f.*
E pista *f.;* huella *f.*
D (Menschen)spuren *f. pl.*
R след *м.*
A أثر انسان

6237
(radar screen) track
F trajectoire *f.;* piste *f.*
E huella (en el radar) *f.*
D Radaraufzeichnung *f.*
R сопровождение *ср.*
A أثر راداري

6238
(railroad) track
F voie ferrée *f.*
E vía *f.*
D Eisenbahngleis *neut.*
R (железнодорожное) полотно *ср.*
A سكة حديد

6239
(ship's) track
F sillage *m.*
E estela *f.*
D Fahrwasser *neut.*
R кильватер *м.*
A مسار سفينة

6240
(significant) track
F piste significative *f.*
E huella identificada *f.*
D (positive) Radaraufzeichnung *f.*
R проводка цели *ж.*
A أثر راداري مهم

6241
(torpedo) track
F sillage *m.*
E surco *m.;* huella del torpedo en el agua *f.*
D Torpedokurs *m.*
R след *м.*
A مسار نسيفة

6242
(wheel) track
F trace *f.*
E rodada *f.*
D (Rad)spuren *f. pl.*
R колея
A أثر عجلة

6243 [P]
track down
F dépister
E encontrar
D aufspüren *tr.*
R выслеживать
A تَعَقَّب ـ إقتَفى أثر

6244
tracker
F traqueur *m.*
E rastreador *m.*
D Verfolger *m.*
R следопыт *м.;* следоискатель *м.*
A كشاف *m*

6245 [G,R]
tracker
F dispositif de poursuite *m.*
E seguidor (del mísil) *m.*
D Verfolgungsgerät *neut.*
R ориентатор *м.*
A مُتَعَقِّب *m*

6246
tracker dog
F chien policier *m.*
E perro rastreador *m.*
D Spürhund *m.*
R собака-искатель *м.*
A كلب كشاف *m*

6247 [R]
tracking *adj. n.*
F (de) poursuite *(adj.) n. f.*
E (de) persecución *(adj.) s. f.*
D (Zielweg-) Verfolgung(s-) *s. f. (komp.)*
R сопровождение *ср.;* проводка *ж.*
A تَعَقُّب ـ إقتفاء أثر

6248

track-link

F patin de chenille *m.*

E eslabón *m.*

D Panzerkettenglied *neut.*

R звено гусеницы *ср.*

A وصلة جنزير

6249 [A]

track mode

F contrôle automatique de la route *m.*

E control automático de la ruta *m.*

D gerichtete Flugbahn *f.*

R режим управления самонаведением *м.*

A طَرِيقَة المُلاَحَقَة

6250 [A,R]

track telling

F transfert de pistes

E traspaso de pistas *m.*

D Flugbahnbeschreibung *f.*

R поиск и сопровождение цели *м. ср.*

A طَرِيقَة اتِّصَال لِلدِّفَاع الجَوِّي

6251

(assault) trackway

F tapis articulé *m.*

E plancha (para permitir el desembarco de vehículos) *f.*

D Stahlmatten *f. pl.*

R колейный мост *м.*; колейный переход *м.*

A طريق حديد للاقتحام

6252

track width

F écartement *m.*

E distancia entre puntos medios de las ruedas *f.*

D Spurbreite *f.*

R колея *ж.*

A العرض بين العجلات اليمنى واليسرى

6253

(wheel) traction

F adhérence *f.;* traction *f.*

E tracción de ruedas *f.*

D (Reifen-)Griffigkeit *f.*

R тяга *ж.*

A عجلة الجَر

6254

(combat engineer) tractor

F véhicule de combat du génie *m.*

E tractor de ingenieros de combate *m.*

D Pionierzugmaschine *f.*

R инженерный тягач *м.*

A جرارة (هندسة قتالية)

6255

(soldier's) trade

F spécialité *f.*

E empleo (de soldado) *m.*

D Soldatenhandwerk *neut.*

R занятие *ср.*

A مِهنة

6256

traffic

F circulation *f.;* trafic [M] *m.*

E circulación *f.;* tráfico *m.*

D Verkehr *m.*

R движение *ср.*

A حَرَكَة مُرُور

6257

trafficability

F aptitude à la circulation *f.;* viabilité *f.*

E aptitud para la circulación *f.;* viabilidad *f.*

D Befahrbarkeit *f.;* Gangbarkeit *f.*

R проходимость *ж.*; грузоподъёмность (моста) *ж.*

A الصَّلاَحِيَة لِحَرَكَة المُرُور

6258

traffic capacity

F capacité de circulation *f.*

E capacidad de tráfico *f.*

D Verkehrskapazität *f.*

R пропускная способность *ж.*

A سَعَة المُرُور

6259

traffic circle

F rond-point *m.*

E circular *m.;* cruce giratorio *m.*

D Kreisverkehr *m.*

R круговой перекрёсток *м.*

A دَوْرَة المرور

6260

traffic control

F régulation *f.*

E regulación de tráfico *f.*

D Verkehrsregelung *f.*

R регулирование движения *ср.*; управление движением *ср.*

A مُرَاقَبَة حَرَكَة المُرُور

6261

traffic density

F densité du trafic *m.*

E densidad del tráfico *f.*

D Verkehrsdichte *f.*

R плотность движения *ж.*

A حَجَم السَّيْر

6262

traffic flow

F débit d'itinéraire *m.*

E intensidad de tráfico *f.;* arriente de tráfico *m.*

D Verkehrsfluss *m.*

R поток движения *м.*

A سَرَيان المرور

6263
traffic jam
F encombrement *m.*
E atasco de tráfico *m.*
D Verkehrsstau *m.*
R задержка *ж.*; пробка
　(движения) *ж.*
A إِزْدِحَام المرور

6264 [T]
trail *n.*
F voie *f.*
E senda *f.*
D (Feld-, Wald- usw.)
　Weg *m.*
R тропинка *ж.*; тропа
　ж.
A دَرْب *m*

6265 [P]
trail *v.*
F traquer
E vigilar
D verfolgen
R следить за;
　выслеживать
A لَاحَق

6266 [S]
trail *n.*
F flèche *f.*
E mástil *m.*
D Holm *m.*
R хобот *м.*
A مَسْنَد *m*

6267
(condensation) trail
F traînée de condensation
　f.
E estela de condensación
　f.
D Kondensstreifen *m.*
R конденсационный след
　м.
A آثار البُخَار

6268 [A]
(vapor) trail
F traînée de condensation
　f.
E estela de humo *m.*
D Kondensstreifen *m.*
R след *м.*; трасса *ж.*
A أثر بخار

6269
trailer
F remorque *f.*
E remolque *m.*
D Anhänger *m.*
R прицеп *м.*
A مَقْطورة *f*

6270 [A]
trail formation
F formation en colonne *f.*
E formación en columna
　f.
D (in)Reihe *f.*
R строй в колонне *м.*
A تَشْكِيل التَوَالي

6271 [V]
train *n.*
F train *m.*
E tren *m.*
D Zug *m.*
R поезд *м.*
A قِطَار *m*

6272
train *v.*
F former
E adiestrar; entrenar;
　instruir
D ausbilden *tr.*
R обучать; тренировать
A دَرَّب

6273 [B,S]
train *n.*
F traînée *f.*
E organización de la
　carga *f.*
D Leitfeuer *neut.*
R наводка *ж.*
A تسديد ـ تصويب

6274
trainee
F stagiaire *m.*
E aprendiz *m.*
D Anlernling *m.*
R обучаемый
A جندي تحت التدريب

6275
training
F entraînement *m.;*
　formation *f.;*
　instruction *f.*
E adiestramiento *m.;*
　instrucción *f.*
D Ausbildung *f.*
R обучение *ср.*; боевая
　подготовка *ж.*
A تَدْريب *m*

6276
(advanced) training
F perfectionnement *m.*
E instrucción superior *f.*
D (forgeschrittene)
　Ausbildung *f.*
R дальнейшее обучение
　ср.
A تدريب متقدم

6277
(adventure) training
F raid aventure *m.*
E entrenamiento para
　ejercicios arriesgados
　m.
D Wagnisübungen *f. pl.*
R дополнительная
　подготовка *ж.*
A تدريب المغامرة

6278
(basic) training
F instruction de base
　élémentaire *f.*
E formación básica *f.*
D Grundausbildung *f.*
R основная подготовка
　ж.
A تدريب اساسي

6279

(continuation) training
F   formation poussée *f.*
E   instrucción continuada *f.*
D   Weiterbildung *f.*
R   продолжение военного обучения *cp.*
A   تَدْرِيب إِسْتِمْراري *m*

6280

(flight) training
F   entraînement en vol *m.*
E   instrucción de piloto *f.*
D   Flugausbildung *f.*
R   лётное обучение *cp.*
A   تدريب الطيران *m*

6281

(ground) training
F   entraînement au sol *m.*
E   instrucción terrestre *f.*
D   Bodenausbildung (Fallschirmtruppen) *f.*
R   наземная подготовка *ж.*
A   التدريب على الارض

6282

(individual) training
F   formation individuelle *f.*
E   instrucción individual *f.*
D   Spezialausbildung *f.*
R   (одиночная) подготовка *ж.*
A   تدريب فردي

6283

(intensive) training
F   formation intensive *f.*
E   instrucción intensiva *f.*
D   Intensivausbildung *f.*
R   интенсивная подготовка *ж.*
A   تدريب شديد

6284

(joint) training
F   formation interarmées *f.;* instruction interarmées *f.*
E   instrucción conjunta *f.*
D   gemeinsame Ausbildung *f.*
R   совместная боевая подготовка *ж.*
A   تَدْرِيب مُشْتَرَك *m*

6285

(language) training
F   enseignement des langues *m.*
E   instrucción en idiomas *f.*
D   Fremdsprachenausbildung *f.*
R   учение языкам *cp.*
A   تعليم اللغات

6286

(military academic) training
F   enseignement militaire *m.;* formation *f.*
E   instrucción teórica militar *f.*
D   (militärakademische) Ausbildung *f.*
R   обучение *cp.*
A   تدريب/دِراسَات عَسْكَرِية

6287

(physical) training
F   éducation physique *f.*
E   gimnasia *f.*
D   Leibesausbildung *f.*
R   физкультура *ж.*
A   تربية بدنية

6288

training (skills)
F   entraînement *m.*
E   instrucción especializada *f.*
D   Ausbildungskenntnisse *f. pl.*
R   техническая подготовка *ж.*
A   التدريب المهني

6289

(survival) training
F   entraînement survie *m.*
E   ejercicios de supervivencia *m. pl.*
D   Überlebensausbildung *f.*
R   обучение при условиях, близко от боевых *cp.*
A   تدريب البقاء على قيد الحياة

6290

(universal military) training
F   instruction militaire obligatoire *f.*
E   servicio militar obligatorio *m.*
D   allgemeiner Militärdienst *m.*
R   всеобщая действительная служба *ж.*
A   تدريب عسكري عام

6291

(weapon) training
F   maniement d'armes *m.*
E   (arma de) instrucción *f.*
D   Waffenausbildung *f.*
R   огневая подготовка *ж.*
A   تدريب الاسلحة

6292

train on a target *v.*
F   pointer sur un objectif
E   apuntar a un blanco
D   (auf das Ziel) richten
R   наводить оружие
A   سدد مدفعا الى الهدف

6293

(combat) trains
F   trains de combat *m. pl.*
E   trenes de combate *m. pl.*
D   (vorgeschobene) Versorgungsdienste *m. pl.*
R   обоз *м.*
A   قدمة السند (في القتال)

6294

(field) trains

F trains de campagne *m. pl.*

E trenes de campaña *m. pl.*

D (rückwärtige) Versorgungsdienste *m. pl.*

R обоз *м.*

A قافلة تَمْوِين

6295

(unit) trains

F trains de combat *m. pl.*

E tren de transportes de una unidad *m.*

D Versorgungsdienste (der Einheit) *m. pl.*

R ремонтный обоз *м.;* замыкание *ср.*

A قدمات السند الداخلية

6296

trajectory

F trajectoire *f.*

E trayectoria *f.*

D Flugbahn

R траектория *ж.*

A مَسِير المقذوف *m*

6297

(curved) trajectory

F trajectoire courbe *f.*

E tiro curvo *m.*

D (gebogene) Flugbahn *f.*

R (кривая) траектория *ж.*

A محرك مُنْحني

6298

(flat) trajectory

F trajectoire tendue *f.*

E tiro rasante *m.*

D (flache) Flugbahn *f.*

R (настильная) траектория *ж.*

A محرك مسطَح

6299 [H]

tranquillizer

F tranquillisant *m.*

E tranquilizante *m.*

D Beruhigungsmittel *neut. pl.*

R успокаивающие средства *ср. мн. ч.*

A مهدّىء

6300 [R]

transceiver

F émetteur-récepteur *m.*

E transceptor *m.*

D Senderempfänger *m.*

R приёмо-передатчик *м.*

A جِهَاز مُرْسِل مُسْتَقْبِل

6301

transducer

F transducteur *m.*

E transductor *m.*

D Energieumwandler

R преобразователь *м.;* переприёмник *м.*

A مُحَوِّل طَاقة

6302 [H,L,S,U]

transfer *n.*

F transfert *m.*

E translado [H] *m.;* transporte [LS] *m.;* traspaso[U] *m.*

D Verlegung [LS] *f.;* Versetzung [H] *f.*

R перевод *м.;* перенос *м.;* передача *ж.;* перемещение *ср.*

A إِنْتِقَال *m*

6303 [H,L,S,U]

transfer *v.*

F muter [H]; transférer

E transladar [H]; transportar [LS]; traspasar[U]

D verlegen [LS]; versetzen [H]

R переводить; переносить; передавать; перемещать

A إِنْتَقَل

6304

transfer loader

F chariot de transbordement *m.*

E transportadora *f.*

D Verlader *m.*

R перевалочное судно *ср.;* перегрузочное судно *ср.;* лихтер *м.*

A حَامِل تَحْوِيل

6305

(blood) transfusion

F transfusion de sang *f.*

E transfusión de sangre *f.*

D Blutübertragung *f.*

R переливание крови *ср.*

A نقل الدم

6306

transient *n.*

F personne (personnel) de passage *f. (pl.)*

E transeúnte *m. f.*

D Durchgänger *m.*

R проезжий *м.*

A عَابِر *m*

6307

transient *adj.*
F transitoire
E fugaz; transitorio
D Augenblick- *komp.;*
 Durchgangs- [H]
 komp.; vorübergehend
 adj.
R переходный;
 временный
A عَابِر

6308

transistor
F transistor *m.*
E transistor *m.*
D Transistor *m.*
R транзистор *м.;*
 кристаллический
 триод *м.*
A تَرَانْزِسْتُور *m*

6309

transistorised *adj.*
F transistorisé
E transistorizado
D transistorisiert
R переведённый на
 транзисторы
A مزود بترانزستور

6310

transit area
F zone de transit *f.*
E zona de tránsito *f.*
D Durchgangzone *f.*
R участок переправы *м.*
A مِنْطَقة تَرْحِيل

6311 [A,M]

transit bearing
F alignement *m.;*
 relèvement en transit
 m.
E enfilación (en tránsito)
 f.
D Richtlinie *f.*
R угол пролёта *м.;*
 пеленг створа *м.*
A إتّجَاه تَرْحِيل

6312

transition *adj. n.*
F (de) transition (*adj.*) *n.*
 f.
E (de) transición (*adj.*) *s.*
 f.
D Übergang(s-) *s. m.*
 (*komp.*)
R переходный *м.;*
 переход
A تَحَوُّل *m*

6313

transition altitude
F altitude de transition *f.*
E altura de transición *f.*
D Übergangshöhe *f.*
R высота перехода *ж.*
A إِرْتِفَاع التَحَوُّل

6314 [A]

transition layer/level
F couche/niveau de
 transition *f. m.*
E capa/nivel de transición
 f. m.
D Übergangsraum *m.;*
 Übergangsniveau
 neut.
R слой перехода *м.;*
 уровень перехода *м.*
A مستوى التحول

6315

translation
F traduction *f.;* thème (à
 la langue étrangère)
 m.; version (à la
 langue maternelle) *f.*
E traducción *f.*
D Übersetzung *f.*
R перевод *м.*
A ترجمة

6316 [R,V]

transmission
F transmission *f.*
E transmisión *f.*
D Sendung [R] *f.;*
 Übersetzung z[V] *f.*
R передача *ж.*
A إِرْسَال *m*

6317 [E,R,U]

transmit
F émettre[R];
 transmettre[EU]
E transmitir
D senden [R]; übertragen
 [RE]
R передавать
A أَرْسَل

6318

transmitter
F émetteur *m.*
E transmisor *m.*
D Sender *m.*
R (радио)передатчик *м.*
A مُرْسِل *m*

6319

transmitter-receiver
F poste émetteur-
 récepteur *f.*
E transreceptor *m.*
D Sende-Empfangsgerät
 neut.
R приёмопередатчик *м.*
A جهاز مرسل ومستقبل

6320 [M]

transom
F arcasse *f.*
E travesaño *m.*
D Heckwerk *neut.*
R транец *м.*
A عَارِضَة *f*

6321

transonic
F transonique
E transónico
D schallnahe
R околозвуковой
A حَوْل سُرْعَة الصَّوْت

6322

transparency

F diapositive *f.*

E diapositiva *f.;*
   transparencia

D Dia(positiv) *neut.*

R диафильм *м.;*
   диапозитив *м.*

A الشَّفَافِيَّة *f*

6323

transparent

F transparent

E transparente

D durchsichtig

R прозрачный;
   просвечивающий

A شَفَّاف

6324

transponder

F transpondeur *m.*

E transmisor respondedor
   m.

D Antwortsender *m.*

R импульсный
   повторитель *м.;*
   транспондер *м.*

A جِهَاز سَائِل وَمُجِيب

6325

transport *v.*

F transporter

E transportar

D befördern

R перевозить;
   транспортировать

A نَقَل

6326

(air) transport

F transports aériens *m. pl.*

E transportes por avión
   m. pl.

D Lufttransport *m.*

R авиатранспорт *м.*

A نَقْل جَوِّي

6327

transport(ation) *adj.*
   n.

F (de) transport (*adj.*) *n.*
   m.

E (de) transporte (*adj.*) *s.*
   m.

D Transport(-) *s. m.*
   (*komp.*)

R транспорт(ный) *м.*
   (*прилаг.*);
   транспортные
   средства *ср. мн. ч.*

A نَقْل *m*

6328 [M]

(attack) transport

F navire de transport et
   de débarquement *m.*

E buque transporte de
   asalto *m.*

D Transportschiff *neut.*

R (поддерживающее)
   транспортное судно
   ср.

A سفينة انزال قوة التدخل

6329

(pack) transport

F transport à dos de
   mulet *m.*

E transporte a lomo *m.*

D Maultiertransport *m.*

R (вьючный) транспорт
   м.

A نقل على الحيوانات

6330

(strategic air)
   transport

F transport aérien
   stratégique *m.*

E transporte aéreo
   estratégico *m.*

D strategischer
   Lufttransport *m.*

R стратегическая
   транспортная
   авиация *ж.*

A النَّقْل الجَوِّي الإِسْتراتيجي

6331

(tactical air) transport

F transport aérien
   tactique *m.*

E transporte aéreo táctico
   m.

D taktischer Lufttransport
   m.

R тактический
   авиатранспорт *м.;*
   тактическая
   транспортная
   авиация *ж.*

A نَقْل جَوِّي تَعْبَوِي

6332

transportable

F transportable

E transportable

D transportierbar

R передвижной;
   перевозимый

A قَابِل لِلنَّقْل

6333 [A,V]

transport capacity

F capacité de transport *f.*

E capacidad de transporte
   f.

D Transportkapazität *f.*

R имеющиеся
   транспортные
   средства *ср. мн. ч.*

A سَعَة النَّقْل

6334 [A]

transport control
   center

F centre de contrôle de
   transport *m.*

E centro de control de
   transporte *m.*

D Transportleitstelle *f.*

R центр управления
   транспортом
   (транспортными
   средствами) *м.*

A مَرْكَز إِدَارَة النَّقْل

6335

(tank) transporter

F plate-forme porte-char
   f.
E portatanques *m. inv.*
D Panzertransportfahrzeug
   neut.
R (танковый)
   транспортёр *м.*
A ناقلة دبابات

6336 [M]

transport group

F groupe de transport
   amphibie *m.*
E agrupación de
   transporte anfibio *f.*
D Landungsschiffsgruppe
   f.
R группа транспортов
   ж.
A جَمَاعَة نَقْل

6337

transshipment

F transbordement *m.*
E transbordo *m.*
D Umladung *f.*
R перевалка грузов *ж.*
A تناقل (من سفينة إلى اخرى)

6338

transverse

F transversal
E transversal
D quer *adj.;* Quer- *komp.*
R поперечный *м.;*
   поперечный разрез
A مُسْتَعْرِض

6339

(booby) trap

F piège *m.*
E trampa explosiva *f.*
D (versteckte) Ladung *f.*
R мина-ловушка *ж.*
A فَخّ *m* . شَرَك *m*

6340

travelling

F déplacement normal
   d'une unité *m.*
E movimiento simultáneo
   m.
D (einheitliche)
   Truppenverlegung *f.*
R движение *ср.*
A سير الوحدة والاشتباك مع
   العدو غير المتوقع

6341 [B]

traverse *n.*

F traverse *f.*
E traviesa *f.*
D Schwenkbereich *m.*
R поперечина *ж.;*
   перекладина *ж.*
A مُعْتَرِضَة *f*

6342 [S]

traverse *v.*

F pointer en direction
E apuntar en dirección
D schwenken
R производить
   горизонтальную
   наводку
A سَدَّد بالاتِّجَاه

6343 [S]

traverse *n.*

F pointage en direction
   m.
E puntería en dirección *f.*
D Traverse *f.*
R горизонтальная
   наводка *ж.*
A التَّسْدِيد

6344

(all-round) traverse

F pointage tous azimuts
   m.
E movimiento circular de
   360 grados *m.*
D (allseitiger)
   Schwenkbereich *m.*
R (круговой) поворот
A التسديد في جميع الجهات

6345

trawl *v.*

F chaluter
E rastrear (con red
   barredera)
D (mit einem) Schleppnetz
   fischen
R тралить; ловить рыбу
   траловыми сетями
A صَاد بِشَبَكَة

6346

trawler

F chalutier *m.*
E barco rastreador *m.*
D Schleppnetzfischerboot
   neut.
R траулер *м.;* тральщик
   м.
A صَيَّاد بالشَّبَكَة *m*

6347 [B,U,V]

tread *n.*

F chape [V] *f.;* giron [B]
   m.; pas[U] *m.*
E huella [BV] *f.;* paso [U]
   m.
D Profil [V] *neut.;* Sprosse
   [B] *f.;* Tritt [U] *m.*
R ширина хода *ж.;*
   колея *ж.;* звено
   (гусеницы) *ср.*
A خُطْوَة *f*

6348

treaty

F convention *f.;* traité *m.*
E convenio *m.;* tratado
   m.
D Vereinbarung *f.;*
   Vertrag *m.*
R договор *м.*
A مُعَاهَدَة *f*

6349

tree

F arbre *m.*
E árbol *m.*
D Baum *m.*
R дерево *ср.*
A شجرة

6350

trench

F   tranchée *f.*
E   trinchera *f.*
D   Graben *m.*
R   траншея *ж.*; ров *м.*;
     окоп *м.*
A   خَنْدَق *m*

6351

**(communications)
     trench**

F   boyau *m.*
E   trinchera de
     comunicación *f.*
D   Verbindungsgraben *m.*
R   ход сообщения *м.*
A   خندق المواصلات

6352

(fire) trench

F   tranchée de tir *f.*
E   trinchera de combate *f.*
D   Schützengraben *m.*
R   (стрелковый) окоп *м.*
A   خندق الرمي

6353

(slit) trench

F   trancheé de tir *f.*
E   trinchera abrigo *f.*
D   Graben *m.*
R   щелевое убежище *ср.*
A   حفرة خندق

6354 [H]

trench foot

F   infection des pieds
     contractée dans les
     tranchées *f.*
E   pie de trinchera *m.*
D   Fussgraben *m.*
R   траншейная стопа *ж.*
A   قدم الخنادق

6355

trench knife

F   couteau à double
     tranchant *m.*
E   cuchillo de monte *m.*
D   Nahkampfmesser *neut.*
R   траншейный нож *м.*
A   سكين قتال

6356 [H]

triage

F   triage *m.*
E   clasificación (de bajas)
     f.
D   Sichtung (von
     Verwundeten) *f.*
R   сортировка раненых
     ж.
A   مُسْتَشْفَى إخْلَاء

6357

trials

F   évaluation opérationelle
     f.
E   pruebas *f. pl.*
D   Proben *f. pl.*
R   испытание *ср.*
A   تجربة

6358

(acceptance) trials

F   essai d' homologation
     m.
E   pruebas de recepción *f.*
     pl.
D   Abnahmeproben *f. pl.*
R   (приёмное) испытание
     ср.
A   إختبار القبول

6359

(troop) trials

F   essai opérationnel *m.*
E   maniobras reales *f. pl.*
D   Truppenversuch *m.*
R   полевое испытание *ср.*
A   التجربة مع القوات

6360

triangle

F   triangle *m.*
E   triángulo *m.*
D   Dreieck *neut.*
R   треугольник *м*
A   مُثَلَّث *m*

6361

triangular

F   triangulaire
E   triangular
D   Dreiecks- *komp.*;
     dreieckig *adj.*
R   треугольный
A   مُثَلَّث

6362

triangulation

F   triangulation *f.*
E   triangulación *f.*
D   Dreiecksaufnahme *f.*;
     Triangulation *f.*
R   триангуляция *ж.*
A   تَثْلِيث *m*

6363

trigger *n.*

F   détente *f.*; gâchette *f.*
E   disparador *m.*; gatillo
     m.
D   Abzug *m.*; Auslöser *m.*
R   крючок *м.*
A   زِنَاد *m*

6364

trigger guard

F   pontet *m.*
E   guardamonte *m.*
D   Abzugsbügel *m.*
R   спусковая скоба *ж.*
A   واقية الزناد

6365 [A,M]

trim

F   équilibre *m.*
E   equilibrio *m.*
D   Schiffslage *f.*
R   уравновешенность *ж.*
A   موازنة

6366 [A,M]

trim v.

F compenser [A];
équilibrer [M]

E compensar [A];
equilibrar [M]

D trimmen

R балансировать;
уравновешивать;
удифферентовывать

A وَازَن ـ عَدَّل

6367

(automatic) trim

F compensation
automatique

E equilibrio automático
m.

D automatisches
Trimmgerät neut.

R (автоматическая)
уравновешенность
ж.

A جهاز ضبط الموازنة

6368

triple

F triple

E triple

D dreifach adj.; Drillings-
komp.

R тройной; утроенный

A ثُلَاثِي

6369

tripod

F trépied m.

E trípode m.

D Dreifuss

R тренога ж.;
треножник м.

A مَسْنَد ثُلَاثِي m

6370

(chalk) troop

F troupe numérotée f.

E tropas en área de
embarque numeradas
y equipadas f. pl.

D (abgezählte)
Landungstruppen f.
pl.

R (десантные) войска ср.
мн. ч.

A قوات معينة للنقل في طائرة

6371

(air) trooping

F aérotransport de
personnel m.

E transporte aéreo de
tropas m.

D Luftverlegung f.

R (воздушная)
переброска ж.

A تَنَقُّل القوات جَوًّا m

6372

troops

F troupes f. pl.

E tropas f. pl.

D Truppen f. pl.

R войска ср. мн. ч.

A قطع / قوات

6373 [N]

troop safety

F mesures de securité
pour les troupes dans
la zone d'alerte
immédiate f. pl.

E distancia de seguridad
f.

D Truppensicherheit f.

R безопасность
неукрытых войск ж.

A امان القوات الصديقة

6374

troopship

F transport de troupes m.

E buque transporte de
tropas m.

D Truppentransportschiff
neut.

R транспорт для
перевозки войск м.

A ناقلة القوات

6375

tropical

F tropical

E tropical

D Tropen- komp.;
tropisch adj.

R тропический

A مَدَارِي . اسْتِوَائِي

6376

**Tropic of
Cancer/Capricorn**

F tropique du
Cancer/Capricorne m.

E trópico de
Cáncer/Capricornio
m.

D Wendekreis des
Krebses; Wendekreis
des Steinbocks

R тропик Рака; тропик
Козерога

A مَدَار الشَّرَطَان / الجَدْي

6377

tropopause

F tropopause f.

E tropopausa f.

D Tropopause f.

R тропопауза ж.

A التُّرْبُوبُوز

6378

troposphere

F troposphère f.

E troposfera f.

D Troposphäre

R тропосфера ж.

A التُّرْبُوسْفِير

6379 [R]
tropospheric scatter
F diffusion
 troposphérique *f.*
E dispersión troposférica
D troposphärische
 Streuung *f.*
R тропосферное
 распространение
 радиоволн *м.;*
 тропосферная связь
 ж.
A التَشْتِيت التَّروبُوسْفِيري *m*

6380
trouble spot
F point sensible *m.*
E centro de fricción *m.*
D Unruhengebiet *neut.*
R очаг беспорядков *м.*
A موضع اضطراب

6381
truce
F trêve *f.*
E tregua *f.*
D Waffenruhe *f.*
R перемирие *ср.*
A مهادنة

6382
truck *n.*
F camion *m.;* wagon
 (ferroviaire) *m.*
E camión *m.;* vagón
 (ferroviario) *m.*
D Lastkraftwagen (LKW)
 m.; Waggon
 (Eisenbahn) *m.*
R грузовик *м.;* вагон *м.*
A شَاحِنَة *f*

6383
truckload
F plein camion *m.*
E carga útil *f.;*
 cargamento *m.*
D Lastwagenladung *f.*
R полный груз
 грузовика *м.*
A حمولة عربة

6384
true airspeed
F vitesse propre *f.*
E velocidad propia
 verdadera *f.*
D wahre
 Fluggeschwindigkeit *f.*
R истинная воздушная
 скорость (ИВС) *ж.*
A سُرْعَة هَوَائِيَّة صَحِيحَة
 السرعة الحقيقية للهواء

6385
true altitude
F altitude vraie *f.*
E altura verdadera *f.*
D wahre Höhe (über NN)
 f.
R истинная высота *ж.*
A الإرْتِفَاع الحَقِيقِي

6386
true bearing
F relèvement vrai *m.*
E marcación verdadera *f.*
D wahre Peilung *f.*
R истинный азимут *м.*
A الاتِّجَاه الحَقِيقِي

6387 [B]
true level
F niveau de maçon *m.*
E nivel verdadero *m.*
D (wahre) Ebene *f.*
R истинный уровень *м.*
A المستوى الحقيقي

6388 [V]
trunk
F coffre *m.*
E portaequipaje *m.*
D Koffer *m.*
R багажник *м.*
A صُنْدُوق *m*

6389 [S]
trunnion
F tourillon *m.*
E muñón *m.*
D Geschützrohrdrehzapfen
 m.
R цапфа *ж.*
A مرتكز دوران الحامل

6390 [S]
tube
F canon *m.*
E cañón *m.*
D Rohr *neut.*
R ствол *м.*
A سَبَطَانَة

6391 [V]
tube
F chambre *f.*
E cámara (de aire) *f.*
D (Luft-)Schlauch *m.*
R камера *ж.*
A أنبوب *m*

6392 [R]
tube
F lampe *f.*
E lámpara *f.*
D Röhre *f.*
R лампа *ж.;* лампочка
 ж.
A صِمَام *m*

6393 [U]
tube
F tube *m.*
E tubo *m.*
D Rohr *neut.*
R труба *ж.;* трубка *ж.*
A أنْبُوب *m*

6394 [M]
tug(boat)
F remorqueur *m.*
E remolcador *m.*
D Schlepper *m.*
R буксир *м.;* буксирное
 судно *ср.*
A قَاطِرَة *f*

6395 [R]
tunable
F réglable
E capaz de sintonizarse
 m.
D einstellbar
R настраиваемый
A يمكن مؤالفته

409

6396

tundra
F toundra *f.*
E tundra *f.*
D Tundra *f.*
R тундра *ж.*
A إقْليم التّندرا بالمِنْطَقَة القُطبِيّة الشِّمَالِيّة

6397

tunnel *v.*
F percer un tunnel
E construir un túnel
D untertunneln *untr.*
R прокладывать туннель
A حَفَرَ نَفَقًا

6398

tunnel *n.*
F tunnel *m.*
E túnel *m.*
D Tunnel *m.*
R туннель *м.*
A نَفَق *m*

6399

turbine
F turbine *f.*
E turbina *f.*
D Turbine *f.*
R турбина *ж.*
A عَنَفَة ـ تُربِينة *f*

6400

turbocharger
F turbo-compresseur *m.*
E turbocompresor *m.*
D Turbolader *m.*
R турбонагнетатель *м.*
A شَاحِن تربِينِي *m*

6401

turbo-jet *adj.*
F (à) turbo-réacteur
E turboreactor *m.*
D Strahlturbinen- *komp.*
R турбореактивный
A نفّاث عَنَفِي

6402 [A,E]

turbulence
F turbulence *f.*
E remolino *m.;*
D Turbulenz [AE] *f.;*
  Wirbelbewegung [E] *f.*
R турбулентность *ж.*
A إضْطِرَاب *m*

6403 [P]

turbulence
F turbulence *f.*
E turbulencia *f.*
D Aufruhr *m.*
R беспорядки *м. мн. ч.;*
  беспокойство *ср.*
A إضْطِرَاب

6404 [P]

turbulent
F turbulent
E turbulento
D aufrührerisch
R беспорядочный;
  беспокойный
A مُضطَرِب

6405 [A,E]

turbulent
F turbulent
E turbulento
D verwirbelt
R турбулентный
A مُضْطَرِب

6406 [A,M]

turnaround *n.*
F rotation *f.*
E apartadero *m.*
D Bodenzeit [A] *f.;*
  Hafenliegezeit [M] *f.*
R время оборота *ср.;*
  оборачиваемость *ж.*
A دَوْرَة *f*

6407 [L]

turning movement
F mouvement tournant
  m.
E movimiento de
  envolvimiento
D Umfassungsbewegung *f.*
R обход *м.;* обходный
  манёвр *м.*
A حَرَكَة إلْتِفَاف

6408

turnout
F tenue *f.*
E apariencia personal *f.*
D Aussehen *neut.*
R выезд *м.*
A منظر عسكري

6409 [A,L,M,V]

turret
F tourelle *f.*
E torre [V] *f.;* torreta
  [ALM] *f.*
D Schützenstand [A] *m.;*
  Gefechtsturm [M] *m.;*
  (Panzer-)Turm [V] *m.*
R башня [V] *ж.;* турель
  [A] *м.*
A بُرْج *m*

6410

twilight
F crépuscule *m.*
E crepúsculo *m.*
D Dämmerung *f.*
R сумерки *ж. мн. ч.*
A شَفَق ـ غَسَق *m*

6411

twin-barrelled *adj.*
F bitube
E (de) dos cañones
D zweirohrig
R двухствольный
A مزدوج السبطانة

6412
twin-engined *adj.*
F bimoteur
E bimotor
D zweimotorig
R двухмоторный
A ذات محركين

6413
twin-gun *adj.*
F bitube
E (de) cañones gemelos
D Zwillings- *komp.*
R спаренный
A ذُو مِدْفَعَيْن

6414 [M]
twin-rudder *adj.*
F (à) deux gouvernails
E (de) dos timones
D Doppelruder- *komp.*
R двухрулевой
A ذُو دَفَّتين

6415 [M]
twin-screw *adj.*
F (à) hélices jumelles
E (de) hélices gemelas
D Doppelschrauben-
    komp.
R двухвинтовой
A ذُو مِرْوَحَتيْن

6416
two-seater *adj.*
F (à) deux places
E (de) dos plazas
D Zweisitzer- *komp.*
R двухместный
A ذات مقعدين

6417 [L]
two up
F par deux
E (por) dos
D Keilformation *f.*
R с двумя
    (подразделениями) в
    первом эшелоне; по
    двум; угол назад *м.*
A قِطْعَتَان إلى الأمام

6418
two-way traffic
F circulation dans les
    deux sens *f.*
E circulación en ambas
    direcciones *f.*
D Zweiwegverkehr *m.*
R движение в двух
    направлениях *ср.*
A المرور في اتجاه مزدوج

6419
type *v.*
F taper
E mecanografiar
D tippen
R писать на машинке
A كتب بآلة كاتِبَة

6420 [E,U]
type *n.*
F type *m.*
E tipo *m.*
D Typ *m.*
R тип *м.*; вид *м.*
A طِـرَاز *m*

6421
(blood) type
F groupe sanguin *f. m.*
E grupo sanguíneo *m.*
D Blutgruppe *f.*
R группа крови *ж.*
A تصنيف الدم

6422
typewriter
F machine à écrire *f.*
E máquina de escribir *f.*
D Schreibmaschine *f.*
R пишущая машинка *ж.*
A آلة كاتِبَة *f*

6423
typhoid fever
F (fièvre) typhoïde *f.*
E (fiebre) tifoidea *f.*
D Typhus *m.*
R сыпной тиф *м.*
A حُمَّى التيفُود

6424
typhoon
F typhon *m.*
E tifón *m.*
D Taifun *m.*
R тайфун *м.*
A إعْصَار مَدَاري *m*

6425
typical
F typique
E típico
D typisch
R типический; типичный
A نَمُوذَجِي

U

6426
ullage
F vide de citerne *m.*
E vacío de cisterna *m.*
D Schwund *m.*
R незаполненное
    пространство *ср.*
A تَفْرِيغ الكَمِّيَّة *m* ـ
    الناقصة *f*

6427
ultimatum
F ultimatum *m.*
E ultimátum *m.*
D Ultimatum *neut.*
R ультиматум *м.*
A بلاغ نهائي

6428
**ultra high frequency
(UHF)**
F (très haute) fréquence
E frecuencia ultraalta *f.*
D Ultrahochfrequenz *f.*
R ультравысокая
    частота (УВЧ)
A تردد ما بعد العالي

6429

ultrasonic

F ultrasonique;
ultrasonore

E ultrasónico

D Ultraschall- *komp.*

R сверхзвуковой

A فَوْق السَّمْعِي

6430

ultraviolet (UV) *adj.*

F ultra-violet

E ultravioleta *inv.*

D ultraviolett *adj.;*
Ultraviolett- *komp.*

R ультрафиолетовый

A فَوْق البَنْفْسَجِي

6431

(air) umbrella

F écran de protection
aérienne *m.*

E cobertura aérea *f.*

D Luftabschirmung *f.*

R (авиационное)
прикрытие *ср.*

A حماية جوية

6432

umpire

F arbitre *m.*

E árbitro *m.*

D Schiedsrichter *m.*

R посредник *м.*

A حَكَم

6433 [B,S]

unarmed

F non-armé

E (en) seguro; sin montar

D gesichert (Minen) [B];
ohne (eingeführte)
Sprengkapsel [S] *f.*

R безопасный

A غير محمي

412

6434 [A,H,M]

unarmed

F sans armes

E desarmado;
desmontado

D unbewaffnet

R невооружённый

A غَيْر مُسَلَّح

6435

unarmored

F non-blindé

E no blindado

D ungepanzert

R небронированный

A غير مدرع

6436

unauthorized

F non-autorisé

E desautorizado

D unbefügt

R неразрешённый;
неправомочный

A بِلا رُخْصَة

6437

**unauthorized
belligerent**

F franc tireur *m.*

E beligerante no
autorizado (ilegal) *m.*

D Freischärler *m.*

R непризнанная
воюющая сторона
ж.

A 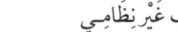 مُحَارِب غَيْر نِظامِي

6438

unavailable

F épuisé; non-disponible

E no disponible

D nicht vorhanden

R не имеющийся (в
распоряжении)

A غَيْر مُتَوَفِر

6439

unbreakable

F incassable

E irrompible

D unzerbrechbar

R неломкий

A غير قابل للكسر

6440

unbridgeable

F infranchissable

E donde no se puede
construir un puente

D unüberbrückbar

R непреодолимый

A غير قابل للتجسير

6441

unburied

F non-enseveli

E no enterrado

D unbegraben

R незарытый

A غير دفين

6442

unchallenged

F (sans être) interpellé

E inidentificado por el
centinela

D unangerufen

R неоспоренный

A بلا ايقاف من قبل الحارس

6443

uncharted

F inexploré; (qui n'est pas
sur la) carte

E inexplorado

D kartographisch
unerfasst

R неотмеченный

A غير مرسوم على الخريطة

6444

unclassified

F non-clasifié

E no clasificado

D offen

R несекретный

A غَيْر مُصَنَّف

6445

uncock *v.*

F désarmer
E desarmar
D entspannen
R разряжать
A ابطل النصب

6446

uncommitted force

F force qui n'est pas
prévue pour emploi
ou qui n'est pas en
contact avec
l'ennemi *f.*
E fuerza no empeñada *f.*
D (nichtzugeteilter)
Truppenteil *m.*
R резервы *м. мн. ч.*
A قوة غير معينة لمهمة

6447

unconditional
surrender

F reddition
inconditionnelle *f.*
E rendición incondicional
f.
D bedingungslose Aufgabe
f.
R безоговорочная
капитуляция *ж.*
A إسْتِسْلَام بِلا شُرُوط

6448

uncontaminated

F non-contaminé
E no contaminado
D unverseucht
R незаражённый
A غير ملّوث

6449

uncontrolled

F non gardé; non-
contrôlé
E libre
D unkontrolliert
R неконтрольный
A غير متحكم فيه

6450

uncontrolled mosaic

F mosaique sommaire *f.*
E mosaico sin puntos de
referencia *f.*
D unabgestimmtes
Reihenluftbild *neut.*
R накидной
аэрофотомонтаж *м.*
A فُسَيْفِسَاء غَيْر مُتَحَكَّم فِيه

6451 [A]

uncontrolled spin

F tonneau involontaire *m.*
E barrera accidental
(casual)
D unfreiwilliges Trudeln
neut.
R неуправляемый
штопор *м.*
A إنْهِيَار حَلْزُوني غَيْر مُتَحَكَّم فِيه

6452 [I,P]

uncorroborated

F non-corroboré
E no confirmado
D unbestätigt
R неподтверждённый
A غير مؤيد

6453

uncover

F découvrir
E descubrir; destapar
D aufdecken *tr.*
R открывать;
раскрывать
A كَشَف

6454

undecipherable

F indéchiffrable
E indescifrable
D nicht entzifferbar
R неразборчивый
A غير قابل لحل الرموز

6455

undefended

F non-défendu
E indefenso
D unverteidigt
R незащищённый
A بلا دفاع

6456

underarmed

F sous-armé
E sin las armas precisas
D (ungemäss) bewaffnet
R плохо вооружённый
A تسليح غير كافٍ

6457

undercarriage

F train d'atterrissage *m.*
E tren de aterrizaje *m.*
D Fahrwerk *neut.*
R ходовая часть *ж.*;
шасси *ср.*
A عَجَلَات الهُبُوط

6458 [P]

undercover

F clandestin
E clandestino; secreto
D geheim; heimlich
R тайный; секретный
A سِرِّي

6459

undercurrent

F courant de fond *m.*
E contracorriente *f.*;
corriente de fondo *f.*
D Unterströmung *f.*
R подводное течение *ср.*
A تَيَّار سُفْلي

413

6460

(to be) under fire

F essuyer le feu (de l'ennemi); (être) sous le feu

E (ser) batido por el fuego

D unter Feuer (liegen)

R (находиться) под огнём

A تحت النَّار

6461 [P]

underground

F clandestin

E clandestino

D Untergrund- *komp.*

R подпольный

A سِرِّي

6462 [T]

underground

F souterrain

E subterráneo

D unterirdisch

R подземный

A تَحْت الأَرْض

6463 [N]

underground burst

F explosion souterraine *f.*

E explosión subterránea *f.*

D Untererddetonation *f.*

R подземный (ядерный) взрыв *м.*

A تَفْجِير تحت الارض

6464 [P]

underground movement

F mouvement clandestin *m.;* résistance *f.*

E movimiento clandestino *m.;* resistencia *f.*

D Untergrundbewegung *f.*

R подпольное движение *ср.;* тайное движение

A مُنَظَّمَة سِرِّيَّة *f*

414

6465

undermanned

F (en) sous effectifs

E escaso de personal

D ungenügend bemannt

R неукомплектованный; имеющий некомплект

A بِه نَقْص في الأَيْدِي العَامِّلة

6466

undermine morale *v.*

F démoraliser

E desmoralizar

D (den Kampfgeist) unterminieren

R вносить разложение

A قوِّض المعنويات

6467

underpass

F passage inférieur *m.*

E paso inferior *m.*

D Unterführung *f.*

R подуличный переезд *м.;* тоннель *м.*

A مَمَرّ تَحْت الأَرْض

6468

under power

F sous pression

E bajo fuerza (presión)

D unter Kraft

R движущийся; на ходу

A تَحْت قِيَادَة

6469

undersea

F sous-marin *adj.*

E submarino

D Untersee- *komp.*

R подводный

A تحت سطح البحر

6470 [A,S]

undershoot

F atterrir trop court [A]; descendre trop court [S]

E planear corto [A]; tirar corto [S]

D (zu kurz) landen

R не долетать

A نبا عن المطار عند الهبوط

6471 [M]

under tow

F (en) remorque

E remolcado

D (im) Schlepp *m.*

R (на) буксире

A يُقْطَر/مقطور

6472 [M]

undertow

F courant de fond *m.;* ressac *m.*

E resaca *f.*

D Sog *m.*

R отлив прибоя *м.*

A التيار السفلي

6473

undertrained

F mal entrainé

E insuficientemente entrenado; bajo de forma

D (ungenügend) ausgebildet

R необученный

A تدريب غير كافي

6474

underwater

F sous-marin

E submarino

D Unterwasser- *komp.*

R подводный

A تَحْت المَاء

6475

underwater demolition
F destruction sous-marine *f.*
E destrucción submarina *f.*
D Unterwassersprengung *f.*
R подводное подрывание *ср*
A نَسْف تَحْت الماء

6476

under way
F (en) marche
E (en) marcha
D (in) Fahrt
R (на) ходу; движущийся
A قَيْد السَّفْر

6477

underway replenishment
F ravitaillement à la mer *m.*
E reabastecimiento en el mar *m.*
D Versorgung während der Fahrt *f.*
R пополнение (запасов) на ходу *ср.*
A التَّمْوِين في الطريق إلى

6478

undisciplined
F indiscipliné
E indisciplinado
D undiszipliniert
R недисциплинированный
A قليل الضبط

6479

undrinkable
F non-potable
E no potable
D untrinkbar
R непитевой
A غير صالح للشرب

6480 [R]

undulation
F ondulation *f.*
E ondulación *f.*
D Welligkeit *f.*
R неровность *ж.*
A تموج

6481

unexploded
F non-explosé
E sin explotar
D Blindgänger- *komp.*
R неразорвавшийся
A غَيْر مُتَفَجِّر

6482

unfit for service
F inapte au service
E inapto para servicio
D dienstuntauglich
R негодный при службе
A غير صالح للخدمة

6483

unguarded
F non-gardé
E no defendido
D unbewacht
R незащитный
A بلا حارس

6484

unidentifiable
F (qui ne peut pas être) identifié
E no identificado
D unerkennbar
R неопознаваемый
A غير معين

6485

unified command
F commandement unifié *m.*
E mando unificado *m.*
D einheitliche Führung *f.*
R единое командование *ср.*
A قِيَادَة مُوَحَّدَة *f*

6486

uniform *n.*
F uniforme *m.;* tenue (de...) *f.*
E uniforme *m.*
D Uniform *f.*
R форма *ж.*
A بِزَّة *f*

6487

unify
F unifier
E unificar
D vereinheitlichen
R объединять
A وَحَّد

6488 [P]

union
F union *f.*
E unión *f.*
D Bund *m.*
R профсоюз *м.;* тредюнион *м.*
A إتِّحَاد *m*

6489 [E,L]

unit *n.*
F groupe [E] *m.;* unité [L] *f.*
E grupo [E] *m.;* unidad [L] *f.*
D Baugruppe [E] *f.;* Einheit [L] *f.*
R часть *ж.*
A وَحْدَة *f*

6490

unit distribution
F distribution des ravitaillements sur place *f.*
E distribución *f.*
D Einheitsversorgung *f.*
R распределение по частям *ср.*
A طريقة التوزيع بواسطة الوحدات

6491

unite
F (s')unir
E unir(se)
D (sich) vereinigen
R соединять(ся)
A وَحَّد

6492

**United Nations
Organisation
(UNO)**
F ONU *f.*
E Organización de
Naciones Unidas
(ONU) *f.*
D Vereinten Nationen (die
UNO)
R Организация
Объединённых
Нации (ООН) *ж.*
A منظمة الأمم المتحدة

6493

unit in contact
F unité en contact avec
l'ennemi *f.*
E unidad en contacto *f.*
D Einheit in
Feindberührung *f.*
R часть в контакте *ж.*
A وحدة مشتبكة مع العدو

6494

unit of issue
F unité de dotation *f.*
E unidad de distribución
f.
D Ausgabeeinheit *f.*
R выдающая часть *ж.*
A الوحدة الموزعة

6495

universal
F universel
E universal
D Universal- *adj.;*
universal *komp.*
R всеобщий; всемирный;
универсальный
A عالمي

416

6496

universal joint
F joint universel *m.*
E junta cardán *f.*
D Kardangelenk *neut.*
R карданный шарнир *м.*;
универсальный
шарнир *м.*; кардан
м.
A وُصْلَة عالمية

6497 [T]

**universal transverse
mercator grid
(UTMG)**
F grille de mercator
transverse universelle
f.
E coordenadas mercator
universal transversas
f. pl.
D UTM-Gitter *neut.*
R универсальная
поперечно-
цилиндрическая
меркаторская
проекция *ж;* 'YTM-
сетка' *ж.*
A الطَرِيقَة العَامَّة لتَرْبِيع
مَرْكَاتُور

6498

Unknown Soldier
F Soldat Inconnu *m.*
E soldado desconocido *m.*
D unbekannter Soldat *m.*
R Неизвестный Солдат
м.
A الجندي المجهول

6499

unlash *v.*
F désarrimer; débréler
E desatar
D losbinden *tr.*
R развязывать
A فك

6500 [S]

unlimber *v.*
F décrocher l'avant-train
E desenganchar
D abprotzen *tr.*
R снимать с передка
A فك عربة المدفع

6501 [S]

unload
F décharger
E descargar
D entladen
R разряжать
A فَرَّغ

6502 [M,V]

unload
F décharger
E descargar
D ausladen; löschen
R разгружать
A فَرَّغ

6503

"Unload!"
F "Désapprovisionnez!"
E "¡Descarguen armas!"
D „Entladen!"
R "Разряжай!"; "Оружие
разрядить!"
A فَرَّغ

6504

unlock
F ouvrir (à clef)
E abrir (con llave)
D aufschliessen
R отпирать; открывать
A نَزَع القُفْل

6505

unmanned
F sans équipage
E no pilotado [A]; no
tripulado [M]
D unbemannt
R непилотируемый
A غير مزود برجال

6506

unmetalled
F non-goudronné
E sin asfaltar
D ungepflastert (Strasse)
R нешоссейный
A غير مرصوف

6507

unnavigable
F non-navigable
E innavegable
D (nicht) befahrbar
R несудоходный
A غير قابل للملاحة

6508

unoccupied
F (zone) libre; (maison) inhabitée; (position) inoccupée
E deshabitado; despoblado; vacante
D unbesetzt
R незанятый
A غير محتل

6509

unofficial
F non-officiel
E no oficial
D inoffiziel
R неофициальный
A غير رسمي

6510

unopposed
F sans rencontrer de résistance
E sin encontrar resistencia
D unbehindert
R несопротивлённый
A بلا اعتراض

6511 [S]

unregistered
F non-registré
E no registrado
D (nicht) aufgezeichnet
R нерегистрированный
A غير مسجّل

6512

unsafe
F dangereux
E peligroso
D unsicher
R опасный
A خطير

6513 [I]

unscramble *v.*
F déchiffrer
E descifrar
D entschlüsseln
R расшифровывать
A حل الرموز

6514

unscrew
F dévisser
E desatornillar
D (auf-)abschrauben *tr.*
R отвинчивать
A فَكَ الصامولة

6515

unseaworthy
F impropre à la navigation
E no apto para la navegación
D nicht seetüchtig
R негодный для плавания
A غَيْر صَالِح لِلإبْحَار

6516

unserviceability period
F délai d'immobilisation *m.*
E tiempo de reparación *m.*
D Instandsetzungsdauer *f.*
R срок неисправности *м.*
A مدة التصليح

6517

unserviceable
F hors d'état de servir; inutilisable
E inservible; inutilizable
D nicht einsatzbereit; unverwendbar
R негодный; неисправный
A غَيْر قَابِل لِلإسْتِعْمَال

6518

unsinkable
F insubmersible
E insumergible
D unsinkbar
R незатопляемый
A عديم القابلية للغطس

6519

unspent
F non-explosé; non-utilisé
E bala no cansada *f.*
D unverbraucht
R нерасходный
A غير مصروف

6520

unstable
F instable
E inestable
D instabil; unsicher
R неустойчивый; нестойкий
A غَيْر مُسْتَقِرّ

6521

unsterilized
F non-stérilisé
E no esterilizado
D unsterilisiert
R нестерилизованный
A غير عقيم

6522

unsubstantiated
F non-confirmé
E no confirmado
D unbestätigt
R неподтверждённый
A غير محسّد

6523

unsurveyed

F non-levé

E desnivelado

D unvermessen

R не привязанный топографически

A غَيْر مَمْسُوح

6524

untenable

F (position) insoutenable

E insostenible

D unhaltbar

R необороняемый

A لا يمكن الدفاع عنه

6525

unused

F neuf

E sin usar

D ungebraucht

R неиспользованный

A غَيْر مُسْتَعْمَل

6526 [C,N]

unwarned exposed

F exposé et non-alerté

E expuesto sin previo aviso

D ungewarnt und ohne Deckung

R непредупреждённый (личный состав) вне укрытия м.

A تَعَرُّض . بِدُون إنْذَار

6527 [S]

"Up…!"

F "Plus haut!"

E "¡Elevar!"

D „Höher!"

R "На …выше!"

A أزِدْ ـ إرفع

418

6528

uphill

F (en) montant

E cuesta arriba *adv.;* de arriba *prep.*

D bergauf

R вверх; вверху в гору

A إلى أعلى التل

6529

upper

F supérieur

E superior

D ober; Ober-

R высший; верхний

A عُلْوِي

6530

upper deck

F pont supérieur *m.*

E cubierta superior *f.*

D Oberdeck *neut.*

R верхняя палуба ж.

A السَّطْح العُلْوِي *m*

6531 [H,U]

upright *adj.*

F droit; vertical

E derecho; vertical

D aufrecht; senkrecht

R прямой; вертикальный

A قَائِم

6532

uprising

F insurrection *f.*

E insurrección *f.*

D Aufstand *m.*

R восстание *ср.*

A ثورة

6533

upstream *adj.*

F (en; vers l') amont (de)

E río arriba

D stromauf(wärts) (von)

R вверх по течению

A ضِدّ التَّيَار

6534

upwind

F au vent

E barlovento

D windwärts

R вверх по ветру; против ветра

A ضِدّ الرِّيح

6535

uranium

F uranium *m.*

E uranio *m.*

D Uran *m.*

R уран *м.*

A يُورَانِيُوم *m*

6536

urban complex

F agglomération *f.*

E conjunto de centros urbanos *m.*

D Ballungsgebiet *neut.*

R городская агломерация

A مِنْطَقَة حَضَرِيَّة *f*

6537

urban terrorism

F terrorisme urbain *m.*

E terrorismo urbano *m.*

D Stadtguerillakampf *m.*

R городской терроризм *м.*

A الإرْهَاب في المُدُن

6538

urgency

F urgence *f.*

E urgencia *f.*

D Dringlichkeit *f.*

R срочность ж.

A إسْتِعْجَال *m*

6539

urgent

F urgent

E urgente

D dringend

R срочный

A مُسْتَعْجِل

6540

utility

F service public *m.*

E servicio público *m.*

D Versorgungsbetrieb *m.;*
Stadtwerke *neut. pl.*

R коммунальное
предприятие *ср.;*
коммунальные
услуги *ж. мн. ч.*

A فَائِدَة

6541

utility helicopter

F hélicoptère de
manœuvre *m.*

E helicóptero de utilidad
(general) *m.*

D Mehrzweckhubschrauber
m.

R вертолёт общего
назначения *м.*

A طَائِرَة عَمُودِيَّة عَامَّة الفَائِدَة

V

6542

vaccination

F vaccination *f.*

E vacunación *f.*

D Impfung *f.*

R вакцинация *ж.*

A تلقيح

6543

vacuum *adj. n.*

F (à) vide (*adj.*) *n. m.*

E (de, al) vacío
(*adj.*) *s. m.*

D Vakuum(-) *neut.*
(*komp.*)

R вакуумный *прилаг.;*
вакуум *сущ. м.*

A فَرَاغ *m*

6544

vacuum bottle

F bouteille isolante *f.;*
bouteille thermos *f.*

E termo *m.*

D Thermosflasche *f.;*
Vakuumkolben [E] *m.*

R термос *м.;* вакуумная
фляжка *ж.*

A قِنِينَة مُفَرَّغَة *f*

6545

V agent

F agent V *m.*

E agente V *m.*

D Nervengas *neut.*

R OB нервно-
паралитического
действия

A عَامِل كِيمَاوِي مُثِير
لِلْأَعْصَاب

6546 [T]

valley

F vallon *m.;* vallée *f.*

E valle *m.*

D Tal *neut.*

R долина *ж.*

A وَادٍ

6547 [R]

valve

F lampe *f.*

E lámpara *f.*

D (Elektronen-) Röhre *f.*

R лампа *ж.*

A صِمَام *m*

6548 [B,E,U,V]

valve

F clapet *m.;* soupape *f.*

E válvula *f.*

D Ventil *neut.*

R клапан *м.;* вентиль *м.;*
золотник *м.*

A صِمَام *m*

6549

(propellor) vane

F pale *f.*

E paleta (de hélice) *f.*

D Propellerflügel *m.*

R крыло *ср.;* вертушка
ж.

A رِيشَة المِرْوَحة

6550 [S]

(steering) vane

F ailette *f.*

E ala de control de
dirección *f.*

D Steuerflügel *m.*

R руль *м.*

A جنيحة توجيه

6551

(turbine) vane

F aube *f.*

E paleta (de turbina) *f.*

D Turbinenschaufel *f.*

R лопатка *ж.*

A ريش العنفة

6552

(wind) vane

F aile *f.*

E veleta *f.*

D Windhahn *m.*

R (ветровое) крыло *ср.*

A دوارة الريح

6553

vanguard

F avant-garde *f.*

E vanguardia *f.*

D Vorhut *m.*

R авангард *м.*

A المُقَدَّمَة *f*

6554

vapor

F vapeur *f.*

E vapor *m.*

D Dampf *m.*

R пар(ы) *м.* (*мн. ч.*)

A بُخَار *m*

419

6555

variability
F variabilité *f.*
E variabilidad *f.*
D Variabilität *f.*
R изменчивость *ж.*
A قابلية التغيير

6556

variable *adj. n.*
F variable *adj. n. f.*
E variable *adj. s. f.*
D variabel *adj.;* Variable
   s. f.; veränderlich *s.*
R изменчивый;
   переменный
A مُتَغَيِّر

6557

variation
F déclinaison
   (magnétique) *f.;*
   variation *f.*
E declinación (magnética)
   f.; variación *f.*
D Abweichung *f.;*
   (magnetische)
   Deklination *f.*
R изменение *ср.;*
   склонение *ср.*
A تَبَدُّل *m*

6558

vector *n.*
F vecteur *m.*
E vector *m.*
D Vektor *m.*
R вектор *м.;* курс *м.*
A مُوَجَّه

6559

vector *v.*
F diriger
E dirigir; orientar
D leiten
R направлять; наводить
A وَجَّه

420

6560

vectored attack
F attaque téléguidée *f.*
E ataque teledirigido *m.*
D ferngeleiteter Angriff *m.*
R наведённый перехват
A هُجُوم مُوَجَّه

6561 [A]

vectored thrust
F poussée dirigée *f.*
E empuje dirigido *m.*
D gelenkter Schub *m.*
R направленная тяга *ж.;*
   дирекционная тяга
A دَفْع مُوَجَّه

6562

veer
F tourner (vent); virer
   [M]
E girar (viento); virar [M]
D ausscheren [M] *tr.;*
   rechtsdrehen (Wind)
   tr.
R менять направление;
   травить
A دار ـ إنحَرَف

6563

vehicle
F engin *m.;* véhicule *m.*
E vehículo *m.*
D Fahrzeug *neut.*
R машина *ж.;*
   автомобиль *м.;*
   корабль *м.*
A مَرْكَبَة *f*

6564

**(armored fighting)
   vehicle (AFV)**
F véhicule blindé de
   combat *m.*
E carro blindado de
   combate *m.*
D Panzerkampfwagen *m.*
R бронетанковая
   машина *ж.*
A مركبة قتال مدرعة

6565

(air cushion) vehicle
F véhicule à coussin d'air
   m.
E aerodeslizador *m.*
D Luftkissenfahrzeug
   neut.
R машина на воздушной
   подушке *ж.*
A حوامة

6566

**(armored
   reconnaissance
   scout) vehicle**
F véhicule blindé de
   reconnaissance *m.*
E carro blindado de
   reconocimiento *m.*
D Panzerspähwagen *m.*
R боевая
   разведывательная
   дозорная машина
   (БРДМ) *ж.*
A مركبة استكشاف مدرعة

6567

**(armored recovery)
   vehicle**
F véhicule blindé de
   dépannage *m.*
E vehículo blindado de
   recuperación *m.*
D Bergepanzer *m.*
R танковый тягач *м.*
A مَرْكَبَة إصْلاَح مُدَرَّعَة *f*

6568

(articulated) vehicle
F véhicule articulé *m.*
E vehículo articulado *m.*
D Gelenkfahrzeug *neut.*
R прицепная машина
   ж.; сочленённый
   автомобиль *м.*
A مَرْكَبَة مِفَصَّلية *f*

6569

(command) vehicle

F véhicule de
   commandement *m.*
E vehículo de mando *m.*
D Befehlsfahrzeug *neut.*
R командирская машина
   ж.
A عَرَبَة قِيَادة *f*

6570

**(command and
   control) vehicle**

F véhicule de
   commandement *m.*
E vehículo de comando y
   control *m.*
D Befehlsfahrzeug *neut.*
R командирская машина
   ж.
A عربة القيادة والمراقبة

6571

**(command post)
   vehicle**

F véhicule de
   commandement *m.*
E vehículo de puesto de
   mando *m.*
D Kommandofahrzeug
   neut.
R командирская машина
   ж.
A عربة قيادة

6572

(end) vehicle

F véhicule serre-file *m.*
E último vehículo *m.*
D Endfahrzeug *neut.*
R замыкание *ср.*
A سيارة المؤخرة

6573

**(infantry fighting)
   vehicle**

F véhicule de combat
   d'infanterie *m.*
E vehículo de combate de
   infantería *m.*
D Schützenpanzer *m.*
R боевая машина
   пехоты (БМП) *ж.*
A عَرَبَة قِتَال لِلمُشَاة *f*

6574

**(maneuverable
   reentry) vehicle**

F véhicule de rentrée
   guidé *m.*
E vehículo de reentrada
   dirigido *m.*
D lenkbare;
   wiedereintrittsfähige-
   Weltraumfähre *f.*
   (meist eng. Begriff)
R маневрирующий
   возвращаемый отсек
   м.
A مَرْكَبَة رَجْعَة قَابِلة لِلمَنَاوَرَة

6575

**(mechanised combat)
   vehicle**

F véhicule de combat
   mécanisé *m.*
E vehículo mecanizado de
   combate
D Schützenpanzer *m.*
R боевая машина *ж.;*
   боевая машина
   пехоты (БМП)
   (Сов.) *ж.*
A مركبة قتال آلية

6576

**(multiple
   independently
   targeted reentry)
   vehicle (MIRV)**

F missile à ogive à
   charges multiples et
   indépendantes *m.*
E arma con multiples
   cabezas dirigidas a
   diferentes objetivos *f.*
D MIRV *m.;* Flugkörper
   m. (meist eng Begriff)
R система
   разделяющихся
   самонаводящихся
   головных
   частей(MIRV) *ж.*
A مركبة رجعة متعددة ومستقلة
   التوجيه

6577 [G,N]

**(multiple independent
   reentry) vehicle
   (MIRV)**

F vecteur nucléaire à tête
   multiple et à objectifs
   indépendants; ICBM
   mirvé *m.*
E vehículo de reentradas
   múltiplas independente
   (MIRV) *m.*
D mehrere unabhängige
   Gefechtsköpfe (meist
   eng. Abkürzung) *f. pl.*
R система
   разделяющихся
   головных частей
   (боеголовок)
   индивидуального
   наведения *ж.*
A مَرْكَبَة مُسْتَقِلَّة
   مُتَعَدِّدَة الرَّجْعَة *f*

6578 [G,N]

**(multiple reentry)
vehicle (MRV)**

F véhicule de rentrée
    multiple

E vehículo de reentradas
    múltiplas (MRV) *m.*

D mehrere Gefechtskopfe *f.*

R комплект
    разделяющихся
    боеголовок; *м.*
    возвращаемый
    отсек с
    комплектомразделя-
    ющихся боеголовок
    м.

A 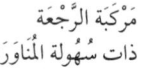
    ذات سُهُولة المُناوَرة

6579

**(reconnaissance)
vehicle**

F véhicule de
    reconnaissance *m.*

E vehículo de
    reconocimiento *m.*

D Aufklärungsfahrzeug
    neut.

R боевая
    разведывательная
    дозорная машина
    (БРДМ) *ж.*

A مَرْكَبَة إِسْتِطْلَاع *f*

6580 [A]

(reentry) vehicle

F véhicule de rentrée *m.*

E cuerpo (vehículo) de
    reentrada *m.*

D Wiedereintrittsflugkörper
    m.

R возвращаемый отсек
    м.; возвращаемая
    боеголовка *ж.*

A مَرْكَبَة رَجْعَة *f*

6581

**(remotely piloted)
vehicle (RPV)**

F engin téléguidé *m.*

E vehículo dirigido por
    control remoto *m.*

D (ferngelenktes)
    Aufklärungsflugzeug
    neut.

R (дистанционно-
    управляемый
    непилотируемый)
    самолёт *м.*

A مركبة موجهة عن بُعد

6582

(soft-skinned) vehicle

F véhicule non-blindé *m.*

E vehículo no blindado *m.*

D (ungeschütztes)
    Fahrzeug *neut.*

R небронированная
    машина *ж.*

A مركبة غير مدرعة

6583

(tanker) vehicle

F camion citerne *m.*

E petrolero *m.*

D Tankerfahrzeug *neut.*

R бензозаправщик *м.*

A ناقلة وقود

6584

**(tank transporter)
vehicle**

F plate-forme porte-char
    f.

E portatanques *m.*

D Panzertransportfahrzeug
    neut.

R танковый транспортёр
    м.

A ناقلة الدبابات

6585

(tracked) vehicle

F véhicule chenillé *m.*

E vehículo de oruga *m.*

D Kettenfahrzeug *neut.*

R (гусеничная) машина
    ж.

A مجنزرة

6586

**(tracked landing)
vehicle**

F véhicule amphibie de
    débarquement

E vehículo oruga de
    desembarco *m.*

D Amphibienketten-
    fahrzeug *neut.*

R гусеничная
    высадочная машина
    ж.

A مركبة انزال مجنزرة

6587

vehicle-borne

F motorisé

E motorizado

D motorisiert

R моторизованный;
    перевозимый
    машиной

A مَحْمُول في مَرْكَبَة

6588

velocity

F vitesse *f.*

E velocidad *f.*

D Geschwindigkeit *f.*

R скорость *ж.*

A سُرْعَة *f*

6589

(burnout) velocity

F vitesse en fin de
    combustion *f.*

E velocidad remanente *f.*

D Brennschluss-
    geschwindigkeit *f.*

R скорость в конце
    активного
    траектории *ж.*

A سرعة الصاروخ عند نهاية
    الوقود

6590
(cut-off) velocity
F vitesse à l'arrêt de
   propulsion *f.*
E velocidad de crucero *f.*
D Brennschluss-
   geschwindigkeit *f.*
R скорость в конце
   активного действия
   ж.
A سُرْعَة القَطْع *f*

6591
(escape) velocity
F vitesse de libération *f.*
E velocidad de escape *f.*
D Ausströmungs-
   geschwindigkeit *f.*
R вторая космическая
   скорость *ж.*;
   скорость убегания
   ж.
A سُرعة الإفلات (من جاذبية
   الأرض)

6592 [A]
(high) velocity
F (à) haute vitesse
E (de) alta velocidad
D hohe Geschwindigkeit
   f.
R высокая скорость *ж.*
A عالي السرعة

6593
**(muzzle) velocity
(MV)**
F vitesse initiale *f.*
E velocidad inicial *f.*
D Mündungs-
   geschwindigkeit *f.*
R начальная скорость
   ж.
A السُّرْعة الإبتِدَائِيَّه *f*

6594
(relative) velocity
F vitesse relative *f.*
E velocidad relativa *f.*
D Relativgeschwindigkeit
   f.
R относительная
   скорость *ж.*
A سُرْعَة نِسْبِيَّة *f*

6595 [S]
(terminal) velocity
F vitesse en fin de
   parcours *f.*
E velocidad emanente *f.*
D Endgeschwindigkeit *f.*
R конечная скорость *ж.*
A سُرْعَة نِهَائِيَّة *f*

6596
vent *n.*
F orifice *m.*
E orificio *m.;* válvula de
   purga *f.*
D Entlüfter *m.;* öffnung *f.*
R вентиляционное
   отверстие *ср.*
A ثُقْب ـ مَخْرَج *m*

6597
(gas) vent
F trou des gaz *m.*
E ventanas de ventilación
   f. pl.; boquetes de
   ventilación *m. pl.*
D Gasloch *neut.*
R газовод *м.*
A منفذ غاز

6598
ventilate
F ventiler
E ventilar
D lüften
R вентилировать;
   проветривать
A هَوَّى

6599
ventilator
F ventilateur *m.*
E ventilador *m.*
D Lüftungsklappe *f.*
R вентилятор *м.*
A مِرْوَحَة *f*

6600
verifiable
F vérifiable
E verificable
D kontrollierbar
R поддающийся
   проверке
A قَابِل للإثبات

6601
verify
F vérifier
E verificar
D prüfen
R проверять
A حَقَّق ـ أَثْبَت

6602
versatile
F polyvalent
E adaptable
D vielseitig
R многосторонний
A متعدد الاستعمال

6603
versatility
F souplesse d'emploi *f.*
E adaptabilidad
D Vielseitigkeit *f.*
R многосторонность *ж.*
A تعدد الاستعمال

6604
vertical *adj. n.*
F vertical(e) *adj. (n. f.)*
E vertical *adj. s. f.*
D senkrecht(e) *adj. (s. f.)*
   vertikal(e) *adj. (s. f.)*
R вертикаль(ный) *ж.*
   (прилаг.)
A عَمُودي

423

6605

vertical envelopment

F enveloppement vertical *m.*

E envolvimiento vertical *m.*

D Umfassung (aus der Luft) *f.*

R вертикальный охват *м.*

A تغطية الهدف بقوات منقولة جوا

6606 [M]

vertical replenishment

F hélitransfert *m.*

E suministro por helicóptero *m.*

D Hubschrauberversorgung *f.*

R пополнение с воздуха *ср.*; пополнение с вертолёта *ср.*

A تَمْوين عَمُودي

6607 [A]

vertical separation

F étagement en altitude *m.;* séparation verticale *f.*

E separación vertical *f.*

D Höhenabstand *m.;* Höhenstaffelung *f.*

R эшелонирование по высоте *ср.*

A الفَاصِلَة الرَّأسِيَّة

6608

very high frequency (VHF)

F (très haute) fréquence *f.*

E frecuencia muy alta (VHF) *f.*

D Ultrakurzwellen *f. pl.*

R очень высокая частота (ОВЧ) *ж.*

A تردد عال جدا

6609 [N]

(escort) vessel

F escorteur *m.*

E buque de escolta *m.*

D Begleitschiff *neut.*

R охранение конвоя *ср.*

A سفينة حراسة

6610

(fishery protection) vessel

F garde-pêche *m. inv.*

E buque guarda-pesca *m.*

D Fischereischutzfahrzeug *neut.*

R судно рыболовственного охранения *ср.*

A حَارِسة صَيْد البَحْر *f*

6611 [M]

(intelligence collection) vessel

F bâtiment espion *m.*

E barco de información e inteligencia *m.;* barco espía *m.*

D Nachrichtenerfassungs-schiff *neut.*

R разведывательное судно *ср.*

A سفينة جمع الاستخبارات

6612

(naval) vessel

F bâtiment de guerre *m.*

E buque de marina *m.*

D Marineschiff *neut.*

R военно-морское судно *ср.*; корабль *м.*

A سفينة حربية

6613

(sailing) vessel

F voilier *m.*

E velero *m.*

D Segelboot *neut.*

R парусное судно *ср.*; парусник *м.*

A مَرْكَب شِرَاعِي

6614

(salvage) vessel

F navire de relevage *m.*

E buque de recuperación *m.*

D Bergungsschiff *neut.*

R спасательное судно *ср.*

A مَرْكَب اسْتَعَادة

6615

veteran

F ancien combattant *m.*

E veterano *m.*

D Veteran *m.*

R ветеран *м.*

A عسكري متقاعد

6616

veterinarian

F vétérinaire *m. f.*

E veterinario *m.*

D Tierarzt *m.*

R ветеринар *м.*

A بَيْطَري *m*

6617

veterinary *adj.*

F vétérinaire

E veterinario

D tierärztlich

R ветеринарный

A بَيْطَري

6618

viaduct

F viaduc *m.*

E viaducto *m.*

D Viadukt *m.*

R виадук *м.*; путепровод *м.*

A قنطرة

6619

victory

F victoire *f.*

E victoria *m.*

D Sieg *m.*

R победа *ж.*

A انتصار

6620 [N]

victualling

F ravitaillement *m.*
E aprovisionamiento *m.*
D Verproviantierung *f.*
R снабжение
продовольствием *ср.*
A التموين بالطعام

6621

videotape

F bande magnétique
vidéo *f.*
E cinta de video *f.;* video
cassette *m.*
D Videoband *neut.*
R видеолента *ж.*
A شريط صوري

6622

**videotape recorder
(VTR)**

F magnétoscope *m.*
E grabador de video *m.*
D Videorekorder *m.*
R видеомагнитофон *м.*
A مُسَجِّل شَرِيط صُوري

6623

view

F vue *f.*
E vista *f.*
D Aussicht *f.*
R вид *м.*
A منظر *m*

6624

vigilance

F vigilance *f.*
E vigilancia *f.*
D Wachsamkeit *f.*
R бдительность *ж.*
A إنتِباه

6625

vigilant

F vigilant
E vigilante
D wachsam
R бдительный
A مُنتَبِه

6626

village

F village *m.*
E pueblo *m.;* aldea *f.*
D Dorf *neut.*
R село *ср.;* деревня *ж.*
A قرية *f*

6627

VIP

F personnage très
important *m.*
E persona importante *f.*
D Ehren- *komp.*
R высокопоставленное
лицо *ср.*
A شخص عظيم الشأن

6628

visibility

F visibilité *f.*
E visibilidad *f.*
D Sichtbarkeit *f.*
R видимость *ж.*
A رُؤْيَة *f*

6629

visible

F visible
E visible; visual
D sichtbar
R видимый
A مَرْئي

6630

visible spectrum

F spectre visible *m.*
E espectro visible (visual)
m.
D sichtbares Spektrum
neut.
R спектр видимости *м.*
A طَيْف مَرْئي *m*

6631

visit *n.*

F visite *f.*
E visita *f.*
D Besuch *m.*
R визит *м.;* посещение
ср.
A زيارة

6632

visit *v.*

F visiter
E visitar
D besuchen
R наносить визит;
посещать
A زار

6633

visor

F visière *f.*
E visor *m.*
D Schirm *m.*
R броневая ставня *ж.;*
щель *ж.*
A وَاقِيَة *f*

6634

visual

F visuel
E visual
D optisch *adj.;* Sicht-
komp.; visuell
R визуальный;
зрительный
A بَصْري

6635

visual aid

F aide visuel *m.*
E ayuda visual *f.*
D Anschauungsmaterial
neut.
R визуальное (учебное)
пособие *ср.*
A وَسِيلَة بَصْرِيَّة

6636

visual communication

F télégraphie optique *f.*
E transmisiones ópticas *f.
pl.*
D optischer
Fernmeldeverkehr *m.*
R визуальная связь *ж.;*
зрительная связь *ж.*
A إتّصَال بَصْري

6637

visual flight rules *pl.*

F réglements de vol à vue
m. pl.
E reglas para el vuelo
visual f. pl.
D Sichtflugregeln f. pl.
R правила визуального
полёта ср. мн. ч.;
устав визуального
полёта м.
A القَوانِين لِلطَّيْران بالنَظَر

6638 [A]

**visual identification
(mode)**

F (contrôle automatique
pour) identification à
vue m.
E identificación visual f.
D Augenaufklärung f.
R зрительное опознание
ср.; режим
зрительного
опознания м.
A طَريقة التَعَرُّف البَصَري

6639

visual inspection

F contrôle visuel m.
E inspección visual f.
D Sichtkontrolle f.
R наружный осмотр м.
A فَحْص بالعَيْن

6640

vital

F capital; vital
E esencial; vital
D lebenswichtig;
wesentlich
R жизненный;
важнейший
A حَبَّوي

6641

vocal

F vocal
E vocal
D Stimm- *komp.*
R голосовой
A صَوْتي

6642

voice

F voix f.
E voz f.
D Stimme f.
R голос м.
A صَوْت m

6643

voice *adj.*

F (en) phonie
E oral; a la voz
D Sprech- *komp.*
R голосовой
A صوتي

6644

volley

F volée f.
E descarga f.; salva f.
D Salve f.
R залп м.
A وَابِل m

6645

(to fire) volleys

F tirer par rafales; tirer
une salve
E (lanzar una) salva de
fusiles f.
D Salven (schiessen) f. pl.
R залпами водить огонь
A رمى وابلا من الرصاصات

6646

volt (V)

F volt m.
E voltio m.
D Volt neut.
R вольт м.
A فُولت m

6647

voltage

F tension f.; voltage m.
E voltaje m.
D Spannung f.
R вольтаж м.;
напряжение ср.
A جُهْد مُقاس بالفُولت m

6648

voluntary *adj.*

F volontaire
E voluntario
D freiwillig
R добровольный
A إخْتِياري

6649

volunteer *v.*

F (s')engager; (se) porter
volontaire
E alistarse
D freiwillig eintreten tr.;
(sich) freiwillig
melden
R предлагать (помощь);
поступать
добровольцем
A تَطَوَّع

6650

volunteer *adj. n.*

F volontaire adj. n. m. f.
E voluntario adj. s. m.
D freiwillig (-e, -er) adj.
(s. m.)
R добровольный м.;
доброволец
A مُتَطَوِّع m

6651

vortex

F tourbillon m.
E vórtice m.
D Wirbel m.
R водоворот м.; вихрь
м.
A زَوْبَعَة ـ دَوامه f

6652

voyage *n.*

F voyage m.
E viaje m.
D (See-)Reise f.
R (морское) путешествие
ср.
A رِحْلَة

6653

vulnerability

F vulnérabilité *f.*
E vulnerabilidad *f.*
D Verwundbarkeit *f.*
R уязвимость *ж.*
A تَعَرُّض

6654

vulnerable

F vulnérable
E vulnerable
D empfindlich;
  verwundbar
R уязвимый
A مُعَرَّض لِلهُجُوم

6655

vulnerable area

F zone vulnérable *f.;* zone
  sensible *f.*
E zona vulnerable *f.*
D empfindlicher Raum *m.*
R уязвимый район *м.*
A مِنْطَقَة مُعَرَّضَة لِلخَطَر

W

6656

wade

F passer à gué
E vadear
D waten
R переправляться;
  переезжать вброд
A عَبَر خَائِضاً

6657

wading crossing

F passage à gué *m.*
E paso vadeable *m.*
D Waten *neut.*
R переправа *ж.*
A عُبُور خَائِضاً

6658

waggon

F wagon *m.*
E vagón *m.*
D Waggon *m.*
R телега *ж.;* вагон *м.*
A عربة

6659

wait *v.*

F attendre
E esperar
D warten
R ждать; ожидать
A إنْتَظَر

6660

"Wait out"

F "Attendez-terminé!"
E "¡Espera!" (cuando
  haya una pausa larga)
D „Warten Ende!"
R "Подождите!"
A انتظر ـ انتهاء

6661 [A,M]

wake

F sillage *m.*
E estela *f.*
D Kielwasser [M] *neut.;*
  Nachstrom [A] *m.*
R кильватер *м.;*
  попутный поток *м.*
A أَثَر *m*

6662

wake up

F (se) réveiller
E despertar(se)
D aufwachen *tr.;* wecken
R просыпаться
A يَقِظ

6663

walk *v.*

F marcher
E andar; caminar
D gehen
R ходить; идти
A مَشَى

6664

walkie-talkie

F talkie-walkie *m.*
E transceptor portátil *m.*
D (tragbares)
  Sprechfunkgerät *neut.*
R портативный
  приёмопередатчик
  м.
A جهاز راديو صغير

6665

walking patient

F malade assis *m.;*
  malade non couché
  m.
E herido a pie *m.;*
  paciente a pie *m.*
D Leichtverletzte(-r) *m.;*
  Leichtkranke(-r) *m.*
R лёгкораненый *м.;*
  ходячий больной *м.*
A مَرِيض قَادِر على السَّير

6666

wall

F mur *m.;* paroi *f.*
E muro *m.;* pared *f.*
D Mauer *f.;* Wand *f.*
R стена *ж.*
A حَائِط *m*

6667

wander

F errer
E vagar
D wandern
R бродить
A تاه

6668

war

F guerre *f.*
E guerra *f.*
D Krieg *m.*
R война *ж.*
A حَرْب *f*

6669

(civil) war

F guerre civile f.
E guerra civil f.
D Bürgerkrieg m.
R гражданская война ж.
A حرب أهلية f

6670

(limited) war

F guerre limitée f.
E guerra limitada f.
D begrenzter Krieg m.
R ограниченная война ж.
A حَرْب مَحْدُودَة f

6671

war correspondent

F correspondant de guerre m.
E corresponsal de guerra m.
D Kriegskorrespondent m.
R военный корреспондент м.
A مراسل حربي

6672

war crime

F crime de guerre m.
E crimen de guerra m.
D Kriegsverbrechen neut.
R военное преступление ср.
A جريمة حرب

6673 [H]

ward

F salle f.
E sala f.
D Station f.
R палата ж.
A قِسْم في مُسْتَشْفى

428

6674

(air raid) warden

F chef d'ilôt (défense passive) m.
E vigilante de alarma antiaérea m.
D Luftschutzwart m.
R дружинник (противовоздушной обороны) м.
A مراقب الغارات الجوية

6675

(fire) warden

F responsable de la lutte anti-incendie m.
E oficial de bomberos m.
D Feuerwächter m.
R дружинник (противопожарной) м.
A خفير الاطفاء

6676 [M]

wardroom

F carré des officiers m.
E sala de oficiales m.
D Offiziersmesse f.
R офицерская кают-компания ж.
A مَطْعَم الضُّبَّاط (في سَفِينَة)

6677

warehouse

F entrepôt m.
E almacén m.
D Lagerhaus neut.
R склад м.
A مُسْتَوْدَع m

6678

warfare

F guerre f.
E guerra f.
D Kriegführung
R война ж.; ведение войны
A حَرْب f

6679

(antisubmarine) warfare

F lutte ASM f.
E operaciones antisubmarinas f. pl.
D Kriegsführung gegen U-Boote f.
R (противолодочная) война ж.
A الحرب المضادة للغواصات

6680

(biological) warfare

F guerre biologique f.
E guerra biológica f.
D biologische Kriegführung f.
R биологическая война ж.
A حَرْب أَحْيائية f

6681

(chemical) warfare

F guerre chimique f.
E guerra química f.
D chemische Kriegführung f.; Gaskrieg m.
R химическая война ж.
A حَرْب كِيمِيائِية f

6682

(conventional) warfare

F guerre conventionnelle f.
E guerra convencional f.
D konventioneller Krieg m.
R обычная война ж.
A حَرْب تَقْلِيدية f

6683

(counter terrorist) warfare

F contre-guérilla *f.*
E operaciones contra-
terroristas *f. pl.*
D (Antiterrorist)-
kriegführung *f.*
R (антитеррористическая)
война *ж.*
A الحرب المضادة للعصابات

6684

(electronic) warfare

F guerre électronique
E guerra electrónica *f.*
D elektronische
Kriegführung *f.*
R радиопротиводействие
ср.; радиовойна *ж.*
A الحَرْب الإلكتْرُونية *f*

6685

(germ) warfare

F guerre bactériologique
f.
E guerra biológica *f.*
D biologische
Kriegsführung *f.*
R биологическая война
ж.
A حرب الجراثيم *f*

6686

(guerrilla) warfare

F guérilla *f.*
E operaciones guerrilleras
f. pl.
D (Guerilla)kriegführung
f.
R (партизанская) война
ж.
A حرب العصابات

6687

(land) warfare

F guerre sur terre *f.*
E guerra terrestre *f.*
D Landkriegsführung *f.*
R война на суше *ж.*
A الحَرْب الأرْضِية / الحرب
البرّيّة *f*

6688

(mine) warfare

F guerre des mines *f.*
E guerra de minas *f.*
D Minenkrieg *m.*
R минная война *ж.*
A حَرْب الألْغام *f*

6689

(psychological) warfare (psywar)

F guerre psychologique *f.*
E operaciones
psicológicas *f. pl.*
D (psychologische)
Kriegführung *f.*
R (психологическая)
война *ж.*
A الحرب النفسية

6690

(subversive) warfare

F guerre subversive *f.*
E operaciones subversivas
f. pl.
D (subversive)
Kriegführung *f.*
R (подрывная)
деятельность *ж.*
A الحرب الهدامة

6691

(unconventional) warfare

F guerre non-classique *f.*
E guerra no-convencional *f.*
D Partisanenkrieg *m.*
R необычная война *ж.*
A حَرْب غَيْر تَقْليدِيَّة

6692

(urban guerrilla) warfare

F guerrilla urbaine *f.*
E guerra de guerrillas
urbana *f.*
D Stadtguerillabekämpfung
f.
R партизанская война в
городах *ж.*
A حَرْب العِصَابات في المُدُن

6693

(on a) war footing

F (sur le) pied de guerre
E (en) pie de guerra
D Kriegsfuss *m.*
R (боевое) положение
ср.
A حالة الحرب *f*

6694

war game

F kriegspiel *m.*
E simulacro de guerra *m.*
D Planübung *f.*
R военная игра *ж.*
A لعبة حرب

6695

(chemical) warhead

F tête chimique *f.*
E cabeza química *f.*
D Gassprengkopf *m.*
R химическая
боеголовка *ж.*
A رأس حَرْبي كِيمِيائي *m*

6696

(cluster) warhead

F charge explosive à
projectiles secondaires
f.
E multicabeza *f.*
D Bündelgefechtskopf *m.*
R (кассетная)
боеголовка *ж.*
A رأس حربي ذو مجموعة من
المتفجرات

6697 [G]

(missile) warhead

F ogive *f.*
E cabeza de guerra (de
misil) *f.*
D Flugkörpergefechtskopf
m.
R головная часть *ж.*
A رأس الصاروخ المتفجر

429

6698

(nuclear) warhead

F ogive nucléaire *f.*
E cabeza atómica *f.*
D Atomgefechtskopf *m.*
R (ядерная) головная
часть *ж.*
A رأس متفجر نووي

6699 [S]

**(shell or projectile)
warhead**

F tête militaire *f.*
E cabeza de guerra (de
proyectil) *f.*
D Gefechtskopf *m.*
R боеголовка *ж.*
A رأس القذيفة المتفجر

6700

warlike

F guerrier
E guerrero
D kriegerisch
R воинственный
A حَرْبي

6701

warm front

F front chaud *m.*
E frente cálido *m.*
D Warmluftfront *f.*
R тёплый фронт
(воздуха) *м.*
A جَبْهَة دَافِئَة *f*

6702

warn

F avertir
E advertir; alertar; avisar
D warnen (vor)
R предупреждать
A أَنْذَر

430

6703 [C,N]

warned exposed

F exposé et alerté
E expuesto y alertado
D gewarnt ohne Deckung
R предупреждённый
(личный состав) вне
укрытия *м.*
A مُعَرَّض مُنْذَر

6704 [C,N]

warned protected

F protégé et alerté
E protegido y alertado
D gewarnt in Deckung
R предупреждённый
(личный состав) в
укрытии *м.*
A مَحْمِي مُنْذَر

6705

warning *adj.*

F avertisseur
E (de) alarma; (de) aviso
D Warn- *komp.*
R предупредительный
A إنْذَاري

6706

warning *n.*

F alerte *f.;* (pré)avis *m.;*
mise en garde *f.*
E alerta *f.;* aviso *m.*
D Warnung *f.*
R предупреждение *ср.*
A إنْذار *m*

6707

(advance) warning

F préavis *m.*
E aviso previo *m.*
D Vorankündigung *f.*
R заблаговременное
предупреждение *ср.*
A إنْذَار مُتَقَدِّم

6708 [R]

(early) warning

F détection précoce *f.*
E alerta temprano *m.*
D Frühwarnung *f.*
R (благовременное)
оповещение *ср.*
A انذار مبكر راداري

6709

(early) warning

F alerte lointaine *f.*
E alarma temprana *f.*
D Frühwarnsignal *neut.*
R (благовременное)
предупреждение *ср.*
A انذار مبكر

6710

(local) warning

F alerte rapprochée *f.*
E alarma local *f.*
D (hiesige) Frühwarnung
R (местное)
предупреждение *ср.*
A انذار محلي

6711

**(nuclear strike)
warning**

F préavis d'attaque
nucléaire *m.*
E alarma de ataque
nuclear *f.*
D Atomwarnung *f.*
R воздушная тревога
ж.; "ядерная
тревога" *ж.*
A انذار هجوم نووي

6712

(strategic) warning

F alerte stratégique *f.*
E alerta estratégico *m.*
D strategische Warnung *f.*
R стратегическое
оповещение *ср.*
A انذار استراتيجي *m*

6713
warning device
F dispositif avertisseur *m.*
E dispositivo de alarma (alerta) *m.*
D Warnanlage *f.*
R тревожно-сигнальный аппарат *м.*
A جِهَاز إِنْذَار

6714 [A,E,M,U,V]
warning light
F lampe avertisseuse *f.*
E luz de alerta (alarma) *f.*
D Warnfeuer [AM] *neut.;* Warnlicht [EUV] *neut.*
R сигнальный огонь *м.;* предупредительный огонёк *м.*
A مِصْبَاح تَنْبِيه ـ ضَوء إنذار

6715 [N]
(minimum) warning time
F délai minimum d'alerte *m.*
E tiempo mínimo de aviso (de alerta) *m.*
D Mindestwarnzeit *f.*
R минимальное время предупреждения *ср.*
A أَدْنَى زَمَن لِلإِنْذار

6716 [M]
warp *n.*
F amarre *f.;* (h)aussière *f.*
E amarra *f.*
D Warp (-trosse) *m.* (*f.*)
R перлинь *м.;* трос *м.*
A حَبْل *m*

6717 [M]
warp *v.*
F touer
E atar
D warpen
R искажать(ся); коробить(ся)
A سَحَب

6718
warplane
F avion de guerre *m.;* avion militaire
E avión militar *m.*
D Kampfflugzeug *neut.*
R военный самолёт *м.*
A طائرة حربية

6719
(death) warrant
F ordre d'exécution *m.*
E sentencia de muerte *f.*
D Todesurteil *neut.*
R распоряжение смертного приговора *ср.*
A أمر تنفيذ

6720 [P]
warrant (of arrest)
F mandat d'arrêt *m.*
E orden de arresto *m.*
D Haftbefehl *m.*
R ордер (ареста) *м.*
A مُذَكَّرَة تَوْقِيف

6721
(search) warrant
F mandat de perquisition *m.*
E auto de registro *m.*
D Haussuchungsbefehl *m.*
R право на обыск *ср.*
A أمر تفتيش

6722
(travel) warrant
F bon de transport *m.*
E autorización para viajar *f.*
D Reiseberechtigungsschein *m.*
R путёвка *ж.*
A إجازة سير

6723
war reserves *pl.*
F réserves de guerre *f. pl.*
E reservas de guerra *f. pl.*
D Kriegsreserven *f. pl.*
R военные запасы *м. мн. ч.*
A إِحْتِيَاطَات الحَرْب

6724
warship
F navire de guerre *m.*
E buque de guerra *m.*
D Kriegsschiff *neut.*
R военный корабль *м.*
A سَفِينَة حَرْبِيَّة

6725
wash *v.*
F laver
E lavar
D waschen
R мыть(ся)
A غَسَل

6726 [A,M]
wash *n.*
F sillage [M] *m.;* souffle [A] *m.*
E estela (turbulenta) [A] *f.;* turbulencia [M] *f.*
D Kielwasser [M] *neut.;* Luftstrudel [A] *m.*
R кильватер *м.;* прибой *м.;* скос потока [A] *м.*
A أَثَر السَّفِينَة أَوْ الطَّائِرَة

6727
(brain) washing
F lavage de cerveau *m.*
E lavado de cerebro *m.*
D Gehirnwäsche *f.*
R мозгопромывание *ср.*
A غَسل الدماغ

431

6728

(nuclear) waste

F déchets radioactifs *m. pl.*

E desperdicios nucleares *m. pl.*

D Atomabfälle *m. pl.*

R радиоактивные отбросы *м. мн. ч.*

A بقايا نووية

6729 [M]

watch *n.*

F quart *m.*

E cuarto *m.*

D Wache *f.*

R вахта *ж.*

A نَوْبَة حِرَاسَة

6730 [U]

watch

F montre *f.*

E reloj *m.*

D (Armband-) Uhr *f.*

R часы *м. мн. ч.*

A سَاعَة *f*

6731

watch *v.*

F observer; veiller

E observar; vigilar

D achten; beobachten

R наблюдать; следить

A تَرَقَّب

6732

watchkeeper

F homme de quart *m.*

E hombre de cuarto [M] *m.;* imaginaria [L] *f.*

D Wachmann *m.*

R вахтенный офицер *м.;* сторож *м.*

A ضَابِط نَوْبة الحِرَاسَة

6733

watchtower

F tour de guet *f.*

E atalaya *f.*

D Aussichtsturm *m.*

R сторожевая вышка *ж.*

A برج مراقبة

6734

watchword

F mot de passe *m.*

E santo y seña *m.*

D Kennwort *neut.;* Zeichen

R пароль *м.*

A كلمة مرور

6735

water

F eau *f.*

E agua *f.*

D Wasser *neut.*

R вода *ж.*

A ماء *m*

6736

waterborne

F transporté par eau

E (por) transporte marítimo; transportado por mar

D (auf dem) Wasserweg befördert

R перевозимый по воде

A مَحْمُول بِطَرِيق الماء

6737

water bottle

F bidon *m.*

E cantimplora *f.*

D Wasserflasche *f.*

R фляга *ж.*

A مطرة

6738

water cooled

F (à) refroidissement par eau

E refrigerado por agua

D wassergekühlt

R водоохлаждённый; ...водяного охлаждения

A مُبَرَّد بِالماء

6739

watercourse

F cours d'eau

E curso de agua *m.*

D Wasserlauf *m.*

R ручей *м.;* река *ж.*

A مَجْرَى مائي

6740

water hole

F mare *f.*

E charco *m.*

D Wasserloch *neut.*

R оазис *м.;* водяная дыра *ж.*

A عين

6741

water jacket

F chemise d'eau *f.*

E camisa de agua *f.*

D Wasserkühlmantel *m.*

R водяная рубашка *ж.*

A كُمّ تَبْرِيد

6742

water level

F niveau de l'eau *m.*

E nivel de agua *m.*

D Wasserstand *m.*

R уровень воды *м.*

A منسوب الماء

6743

waterline

F ligne de flottaison *f.*

E línea de flotación *f.*

D Wasserlinie *f.*

R ватерлиния *м.;* уровень воды

A خَطّ الماء

6744

waterlogged
F imbibé; alourdi d'eau
E anegado; empapado; inundado
D sumpfig; vollgesogen
R полузатопленный
A مَغْمُور بِالْمَاء

6745

water main
F conduite principale d'eau f.
E cañería maestra de agua f.
D Hauptwasserrohr neut.
R главный водопровод м.
A قناة الماء الرئيسية

6746

waterproof adj.
F imperméable; (montre) étanche
E impermeable
D wasserdicht
R водонепроницаемый
A كتيم

6747

water purification unit
F groupe de purification d'eau m.
E purificador de agua m.
D Wasserreinigungsanlage f.
R водоочистительная установка ж.
A وَحْدَة تَنْقِيَة الماء

6748

(coastal) waters
F eaux côtières f. pl.
E aguas costeras f. pl.
D Küstengewässer f.
R побережные воды ж. мн. ч.
A مياه ساحلية

6749

(international) waters pl.
F eaux internationales f. pl.
E aguas internacionales f. pl.
D internationale Gewässer neut. pl.
R международные водные пути м. мн. ч.
A الْمِيَاه الدَّوْلِيَّة f

6750

(narrow) waters pl.
F passe navigable f.
E paso navegable m.
D Enge f.
R узкость ж.
A مَضِيق m

6751

(territorial) waters pl.
F eaux territoriales f. pl.
E aguas jurisdiccionales f. pl.
D Hoheitsgewässer neut. pl.
R территориальные воды ж. мн. ч.
A مِيَاه إِقْلِيمِيَّة f

6752

(upper) waters pl.
F amont (d'une rivière) m.
E aguas arriba f. pl.; vaguada f.
D obere Gewässer neut. pl.
R верхнее течение ср.
A الْمِيَاه الْقَلْوِيَة f

6753

watershed
F (ligne de) partage des eaux m.
E divisoria (de aguas) f.
D Wasserscheide f.
R водораздел м.
A خَطّ تَقْسِيم الْمِيَاه

6754

waterspout
F trombe f.
E tromba (marina) f.
D Wasserhose f.
R водяной смерч м.
A مِيْزَاب ـ عامود الماء
الاعصاري في المحيط m

6755 [M]

water terminal
F terminus maritime m.
E terminal marítima f.
D Umschlaghafen m.
R морской конечный пункт м.; порт пересадки м.
A مَأْخَذ الماء

6756

watertight
F étanche à l'eau
E estanco
D wasserdicht
R водонепроницаемый
A كتيم ـ مقاوم للماء

6757

water tower
F château d'eau m.
E arca de agua f.
D Wasserturm m.
R водонапорная башня ж.
A خزان للماء

6758

water truck
F camion citerne de ravitaillement en eau m.
E camión cisterna m.
D Wasserwagen m.
R водовоз м.
A سيارة ماء

433

6759

waterway
F voie d'eau f.
E vía acuática f.; vía navegable f.
D Wasserweg m.
R водный путь м.
A مَجْرى ماء

6760

wave
F onde f.
E ola [M] f.; onda [ELR] f.
D Welle f.
R волна ж.
A مَوْجَة f

6761

(assault) wave
F vague d'assaut f.
E ola de asalto f.
D Angriffswelle f.
R штурмовая волна ж.
A موجة الاقتحام

6762

(blast) wave
F onde de choc f.
E onda explosiva f.
D Druckwelle f.
R ударная волна ж.
A موجة انفجار

6763

(followup) wave
F vague de renforcement f.
E oleada de refuerzo f.
D nachdrängende Welle f.
R силы второго эшелона ж. мн. ч.
A المَوْجَة الثانية للهُجوم f

6764

(ground) wave
F onde troposphérique f.
E onda terrestre f.
D Bodenwelle f.
R поверхностная волна ж.
A مَوْجة سطحية

6765

(shock) wave
F onde de choc f.
E onda de choque f.
D Schockwelle f.
R ударная волна ж.
A مَوْجَة الصَّدْم

6766

(tidal) wave
F raz de marée m.
E maremoto m.
D Flutwelle f.
R стоячая волна ж.
A مَوْجَه مَدّ f

6767

wavelength
F longeur d'ondes f.
E longitud de onda f.
D Wellenlänge f.
R длина волны ж.; частота ж.
A طول الموجة

6768

wax skis v.
F farter
E encerar los esquíes
D einwachsen tr.
R вощить лыжи
A تشميع الزحافات

6769 [S]

"(On the) way!"
F "Coup parti!"
E "¡En el aire!"
D „Schuss!"
R "Выстрел!"
A في الطريق

6770

weak
F faible
E débil; flojo
D schwach
R слабый
A ضَعيف

6771

weaken
F (s')affaiblir
E debilitar(se)
D nachlassen; schwächen
R ослаблять; слабеть
A أضْعَف

6772

weapon
F arme f.
E arma f.
D Waffe f.
R оружие ср.
A سِلاح m

6773

(anti-tank) weapon
F arme anti-char f.
E arma antitanque f.; arma contracarros f.
D Panzerabwehrwaffe f.
R противотанковое оружие ср.
A سلاح مضاد للدبابات

6774

(automatic) weapon
F arme automatique f.
E arma automática f.
D automatische Waffe f.
R автомат м.; автоматическое оружие ср.
A سِلاح آلي m

6775

(biological) weapon
F arme biologique f.
E arma biológica f.
D biologische Waffe f.
R биологическое оружие ср.
A سِلاح أحْيائي m

6776

weapon (CBR)

F armes NBC *f. pl.*

E NBC arma *f.*

D NBC-Waffe *f.*

R оружие массового
поражения (Сов.) *ср.*

A سلاح نووي جرثومي
كيماوي

6777

(crew-served) weapon

F arme collective *f.*

E arma colectiva *f.*

D (mannschaftsbediente)
Waffe *f.*

R оружие с расчётом *ср.*

A سلاح ذو طاقم

6778

(enhanced radiation)
weapon

F arme à rayonnement
renforcée *f.*

E arma de radiación
intensificada *f.*

D Neutronenwaffe *f.*

R нейтронная бомба *ж.*;
нейтронное оружие
ср.

A سِلاَح مزود بالإشعاع *m*

6779

(fractional yield)
weapon

F arme inférieure à la
kilotonne *f.*

E arma inferior a la
kilotonelada (término
inglés) *f.*

D Sub-KT-Wert (meist
eng. Begriff) *m.*

R (ядерное) оружие
малой мощности *ср.*

A سِلاَح ذو إستجابة جُزْئِية

6780

(gas-operated) weapon

F arme automatique à
emprunt de gaz *f.*

E arma actuada por gases
f.

D (Gaslade)waffe *f.*

R оружие газового
действия *ср.*

A سلاح يُشغَّل بالغاز

6781

(guided) weapon

F arme téléguidée *f.*

E arma guiada *f.*

D (Lenk)waffe *f.*

R (управляемый) снаряд
м.

A سلاح موجه

6782

(individual) weapon

F arme individuelle *f.*

E arma individual *f.*

D (persönliche) Waffe *f.*

R (личное) оружие *ср.*

A سلاح فردي

6783 [N]

(kiloton) weapon

F arme kilotonnique *f.*

E arma kilotónica *f.*

D Kilotonnen-Waffe *f.*

R килотонное оружие
ср.

A سِلاَح كيلوطُنّ *m*

6784

(light) weapon

F arme légère *f.*

E arma ligera *f.*

D (leichte) Waffe *f.*

R (лёгкое) оружие *ср.*

A سِلاَح خَفِيف

6785

(megaton) weapon

F arme mégatonnique *f.*

E arma megatónica *f.*

D Megatonnenwaffe *f.*

R мегатонное (ядерное)
средство *ср.*;
мегатонное
(ядерное) оружие *ср.*

A سِلاح ميكاطَنّ *m*

6786 [N]

(nominal) weapon

F arme de puissance
nominale *f.*

E arma de potencia
nominal *f.*

D zustehende Waffe *f.*

R номинальное
(ядерное) оружие *ср.*

A سلاح (إسْمِي / رمزي) *m*

6787

(nuclear) weapon

F arme nucléaire *f.*

E arma nuclear *f.*

D Atomwaffe *f.*

R ядерное оружие *ср.*

A سِلاَح نَووي *m*

6788

(personal) weapon

F arme individuelle *f.*

E arma individual *f.*

D Handfeuerwaffe *f.*;
Stan-Waffe *f.*

R личное оружие *ср.*

A سِلاَح فُرْدي *m*

6789

(single action) weapon

F arme à répétition *f.*

E arma no automática *f.*

D Einzelschusswaffe *f.*

R однозарядный *м.*;
оружие
одностороннего
действия *ср.*

A سلاح ذو عمل بسيط

435

6790

(single shot) weapon
F arme à un coup
E arma no automática *f.*
D Einzelschusswaffe *f.*
R однозарядный
A سلاح ذو طلقة مفردة

6791

(slide-action) weapon
F (arme) semi-
   automatique *f.*
E arma de carga manual *f.*
D Schlittenwaffe *f.*
R полуавтоматическое
   оружие *cp.*;
   самозарядное
   оружие *cp.*
A سلاح ذو مزلقة

6792

(stand-off) weapon
F arme tirée à distance *f.*
E arma de largo alcance
   f.
D (Abstands) Waffe *f.*
R (авиационное
   самонаводящееся)
   оружие *cp.*
A سلاح مباعد

6793

(strategic) weapon
F arme stratégique *f.*
E arma estratégica *f.*
D strategische Waffe *f.*
R стратегическое оружие
   cp.; оружие
   стратегического
   назначения *cp.*
A سِلاح استراتيجي

6794

**(sustained fire)
  weapon**
F arme à grande cadence
   de tir *f.*
E arma de tiro sostenido
   m.
D (Dauerschuss)waffe *f.*
R (станковый) пулемёт
   м.
A سلاح رمى غزير

6795

"Weapons free!"
F "Tir libre!"
E armas autorizadas para
   el disparo *f. pl.*
D Feuererlaubnis *f.*
R разрешается стрелять
   по неопознованным
   самолётам
A مسموح اطلاق النار على أية
   طائرة غير صديقة

6796

"Weapons hold!"
F "Tir prescrit!"
E armas controlando el
   objetivo *f. pl.*
D Feuerverbot *m.*
R разрешается стрелять
   по входящим в зону
   самолётам только в
   интересах самозащиты
A ممنوع اطلاق النار على أية
   طائرة إلا للدفاع عن النفس

6797

(infrared) weaponsight
F viseur infrarouge *m.*
E visor (infrarojo) de
   arma *m.*
D Infrarotsichtgerät *neut.*
R инфракрасный прицел
   м.
A مسددة تحت الحمراء

6798

"Weapons tight!"
F "Tir interdit!"
E armas en posición *f. pl.*
D (bedingte)
   Feuererlaubnis *f.*
R разрешается стрелять
   только по точному
   опознанию самолёта
A مسموح اطلاق النار على
   طائرات العدو التي
   تمييزها أكيد

6799

weapon system
F système d'armes *m.*
E sistema de arma *m.*
D Waffensystem *neut.*
R боевой комплект *м.*
A نِظام أسْلِحَة

6800

(fair) wear and tear
F usure normale *f.*
E desgaste natural *m.*
D (annehmbarer)
   Verschleiss *m.*
R износ *м.*;
   действительный
   износ *м.*
A تدهور

6801

weather *n.*
F temps *m.*
E tiempo (atmosférico) *m.*
D Wetter *neut.*
R погода *ж.*
A جَوّ *m*

6802 [M]

weather *adj.*
F (du) vent
E (de) barlovento
D Luv- *komp.*
R наветренный
A جَوّي

6803

weather forecast
F prévisions
   météorologiques *f. pl.*
E pronóstico del tiempo
   m.
D Wettervorhersage *f.*
R прогноз погоды *м.*
A تَنبُّؤ جَوّي *m*

6804

webbing
F brêlage *m.*;
   équipements *m. pl.*
E cinturón *m.*; equipo *m.*
D Gurtband *neut.*
R тканая лента *ж.*
A أحْزِمَة كَتانِيَّة *f*

6805 [E,L,U]
wedge *n.*
F cale *f.;* coin *m.*
E cuña *f.*
D Keil *m.*
R клин *м.*
A إسْفِين *m*

6806
wedge *v.*
F caler; coincer
E acuñar
D (ein)keilen *tr.*
R вклинивать(ся);
  вбивать клин
A ثَبَّت

6807 [L]
wedge formation
F formation en triangle *f.*
E formación en cuña *f.*
D Keilform *f.*
R строй-клин *м.;* клин
  м.
A تَشْكِيل إسْفِينِي *m*

6808
weigh
F peser
E pesar
D wiegen
R взвешивать(ся)
A وَزَن

6809
weigh anchor
F lever l'ancre
E levar el ancla
D Anker lichten
R поднимать якорь
A رَفَع المِرْسَاة

6810
weight *v.*
F alourdir; pondérer [E]
E añadir peso; ponderar
D belasten; wiegen [E]
R нагружать;
  обременять
A وَزَن . رَجَّح

6811
weight *n.*
F poids *m.*
E peso *m.;* pesa *f.*
D Gewicht *neut.*
R вес *м.*
A ثِقْل *m* وَزْن *m*

6812
(gross) weight
F poids brut *m.*
E peso bruto *m.*
D Bruttogewicht *neut.*
R общий вес *м.*
A الوَزْن الإجْمَالِي

6813
(gross vehicle) weight
F poids total en charge
  (PTC) *m.*
E peso total en carga *m.*
D zulässiges
  Gesamtgewicht *neut.*
R общий вес машины *м.*
A وَزْن المَرْكَبَة الإجْمَالِي

6814 [V]
(maximum gross)
  weight
F poids total en charge
  m.
E peso bruto máximo *m.*
D Höchstbruttogewicht
  neut.
R (максимальный
  полный) вес *м.*
A الوزن الاجمالي

6815
(maximum landing)
  weight
F poids maximal à
  l'atterrissage *m.*
E peso máximo de
  aterrizaje *m.*
D Höchstlandegewicht
  neut.
R максимальный
  посадочный вес *м.*
A الوَزْن الأقْصَى عِند
  الإبْرَار *m*

6816
(maximum take-off)
  weight
F poids maximum au
  décollage
E peso máximo de
  despegue *m.*
D höchste Startgewicht
  neut.
R максимальный
  взлётный вес *м.*
A الوَزْن الأقْصَى فِي الإقْلَاع *m*

6817
(net) weight
F poids net à vide *m.*
E peso neto *m.*
D Nettogewicht *neut.*
R (чистый) вес *м.*
A الوزن الصافي

6818
weld *v.*
F souder
E soldar
D schweissen
R сваривать(ся)
A لَحَم

6819
weld *n.*
F soudure *f.*
E soldadura *f.*
D Schweissen *neut.*
R сварка *ж.*
A لَحْم ـ لِحَام *m*

6820
well *n.*
F puits *m.*
E pozo *m.*
D Brunnen *m.*
R колодец *м.;* водоём *м.*
A بِئْر *m*

6821
well-armed
F (bien) armé
E (bien) armado
D (gut)bewaffnet
R хорошо вооружённый
A مسلح جيداً

6822
well-defended
F (bien) défendu
E bien defendido
D (gut)verteidigt
R хорошо защищённый
A حسن الدفاع

6823
well-disciplined
F (bien) discipliné
E disciplinado
D (gut)diszipliniert
R дисциплинированный
A حسن القيادة

6824
well-equipped
F (bien) équipé; (bien)
　outillé
E bien equipado
D (gut)ausgerüstet
R хорошо
　оборудованный
A جيد التجهيز

6825
well-guarded
F (bien) gardé
E bien vigilado
D (gut)bewacht
R хорошо сохранённый
A حسن الحراسة

6826
west *adj.*
F (de l')ouest
E (del) oeste; occidental
D westlich
R западный
A غَرْبي

6827
wet
F mouillé
E húmedo; mojado
D nass
R мокрый
A رَطْب

6828
whale
F baleine *f.*
E ballena *f.*
D Wal *m.*
R кит *м.*
A بَال ـ حوت *m*

6829
whaler
F baleinier *m.*
E ballenero *m.*
D Walfangboot
R китобойное судно *ср.*
A مَرْكب صَيْد الحوت

6830
wharf
F quai *m.*
E muelle *m.*
D Kai *m.*
R пристань *ж.*; причал
　м.
A رَصِيف *m*

6831
wheel *n.*
F roue *f.*
E rueda *f.*
D Rad *neut.*
R колесо *ср.*
A عَجَلَة *f*

6832 [L,U]
wheel *v.*
F pousser [U]; tourner [L]
E empujar [U]; girar [L]
D schieben [U];
　schwenken [L]
R катить;
　поворачивать(ся);
　заходить флангом
A دار

6833 [M,V]
(steering) wheel
F roue de gouvernail [M]
　f.; volant [V] *m.*
E rueda de timón [M] *f.;*
　volante [V] *m.*
D Lenkrad; Steuerrad [M]
R рулевое колесо *ср.*
A عَجلة القِياده

6834
wheelbase
F empattement *m.*
E batalla *f.;*
　enfrentamiento *m.*
D Radstand *m.*
R колёсная база *ж.*
A قَاعِدَة العَجَلة

6835
wheeled
F (à) roues
E rodado
D Rad- *komp.*
R колёсный
A مُدَوْلَب ـ على عَجَل

6836
wheelhouse
F timonerie *f.*
E timonera *f.*
D Ruderhaus *neut.*
R рулевая рубка *ж.*
A مَبِيت عَجَلَة التّوْجِيه

6837
whirlpool
F tourbillon *m.*
E torbellino *m.*
D Strudel *m.*
R водоворот *м.*
A دوامة *f*

6838
whirlwind
F tourbillon (de vent) *m.*
E tornado *m.;* turbulencia
　f.
D Windhose *f.*
R вихрь *м.;* смерч *м.*
A زَوْبَعَة *f*

6839
whistle *v.*
F siffler
E silbar
D pfeifen
R свистеть
A صَفَر

6840
whistle *n.*
F sifflet *m.*
E silbato *m.*
D Pfeife *f.*
R свист *м.*; гудок *м.*
A صِفَارَة *f*

6841 [M]
white caps *pl.*
F moutons *m. pl.*
E cabrillas (mar) *f. pl.*
D (Wellen mit)
  Schaumkronen *f. pl.*
R барашки *м. мн. ч.*;
  белые гребни *м. мн.*
  ч.
A أَمْوَاج بِيض القِمَم

6842
white flag
F drapeau blanc *m.*
E bandera blanca *f.*
D weisse Fahne *f.*
R белый флаг *м.*;
  флажок *м.*
A رَايَة بَيْضَاء *f*

6843
whiten
F blanchir
E blanquear
D tünchen
R белить; белеть
A بَيَّض

6844
whiteout
F voile blanc *m.*
E nevisca *f.;* ventisca *f.*
D Sichtverlust *m.*
R белая тьма *ж.*; мгла
  ж.
A إبيضاض قُطبي *m*

6845
wide
F large
E ancho
D breit
R широкий
A عَرِيض

6846
wide angle
F grand angulaire
E gran angular
D weitwinklig
R широкоугольный
A زَاوِيَة (عريضة) *f*

6847
width
F largeur *f.*
E anchura *f.*
D Breite *f.*
R ширина *ж.*
A عَرْض *m*

6848 [R]
(beam) width
F ouverture du faisceau *f.*
E amplitud de haz *f.*
D Bandbreite *f.*
R угол раствора *м.*
A عَرْض الحُزْمة

6849
"Wilco!"
F "Aperçu!"
E "¡OK!"
D „Verstanden!"
R "Понял!"
A 'فهمت '

6850
win *v.*
F gagner
E ganar
D gewinnen
R выигрывать;
  побеждать
A رَبِح

6851
winch *n.*
F treuil *m.*
E torno *m.*
D Winde *f.*
R лебёдка *ж.*
A مِلْفَاف *m*

6852
winch *v.*
F (lever etc... à) treuil
E (elevar; levantar etc.
  con) torno
D (hoch usw.) winden
R поднимать лебёдкой
A ادار المِلْفَاف

6853
wind *n.*
F vent *m.*
E viento *m.*
D Wind *m.*
R ветер *м.*
A رِيح *m*

6854 [E,U]
wind *v.*
F enrouler [E]; tourner
E enrollar [E]; girar
D drehen [U]; (auf)spulen
  [E]; winden [E]
R наматывать; крутить
A لَفّ ـ لَوَى

6855 [G]
windage
F effets de vent *m.*
E corrección-viento *f.*
D Windeffekt *m.*
R снос ветром *м.*
A انحراف ناجم عن الريح

6856
**wind cone (sleeve,
  sock)**
F manche à air *f.*
E manga *f.*
D Windsack *m.*
R ветровой конус *м.*
A كُمّ الرِّيح

6857

wind dummy
F mannequin largué
   avant le stick (troupes
   aéroportées) *m.*
E muñeco (empleado en
   paracaidismo) *m.*
D Windanzeiger *m.*
R макет *м.*
A دُمْية مظلية لتقدير تأثير الريح

6858

winding *n.*
F bobine *f.*
E bobinado *f.*
D Spule *f.;* Wicklung *f.*
R катушка *ж.*
A مَلَفّ *m*

6859 [M]

windlass
F guindeau *m.*
E molinete *m.*
D Winde *f.*
R врашпиль *м.;* лебёдка
   ж.
A دَوَّارَة *f*

6860

windmill (windpump)
F moulin à vent *m.*
E molino (de viento) *m.*
D Windmühle *f.*
R ветровой
   (водонапорный)
   насос *м.*
A طَاحُونَة هَوَاء

6861

window
F ruban accordé *m.*
E cinta perturbadora *f.*
D Störfolie *f.;* Düppel *m.*
R дипольные
   отражатели *м. мн.*
   ч.; пассивная помеха
   ж.
A شرائح معدنية عاكسة

440

6862

window
F fenêtre *f.*
E ventana *f.*
D Fenster *neut.*
R окно *ср.*
A شباك

6863

windproof *adj.*
F (à l')épreuve de vent
E (a) prueba de viento
D winddicht
R ветронепроницаемый
A مانع للريح

6864

wind pump
F éolienne *f.*
E bomba eólica *f.*
D Windpumpe *f.*
R крыльчатый насос *м.*
A مضخة تُدار بطاحونة هواء

6865

windshield
F pare-brise *m. inv.*
E parabrisas *m. inv.*
D Windschutzscheibe *f.*
R переднее стекло *ср.*
A واقِيَة الرِّيح

6866

windshield wiper
F essuie-glace *m.*
E limpiaparabrisas *m. inv.*
D Scheibenwischer *m.*
R стеклоочиститель *м.;*
   дворник *м.*
A مَسَّاحَة الزُّجَاج

6867

windward
F (au) vent
E (a) barlovento
D windwärts
R наветренный
A في الرِّيح

6868

wing
F aile *f.*
E ala *f.*
D Flügel *m.*
R крыло *ср.;* звено *ср.*
A جَنَاح *m*

6869

(stub) wing
F aile courte *f.*
E muñón de ala *m.*
D Stumpfflügel *m.*
R стабилизатор *м.*
A جناح ابتر *m*

6870

(swept) wing *adj.*
F (à) aile en flèche
E (de) ala replegada
D pfeilförmig
R стреловидный
A جناح ممتد إلى الخلف

6871 [A]

wingman
F ailier *m.*
E piloto de flanco *m.*
D Rottenflieger *m.*
R ведомый (лётчик) *м.*
A مُسَاعِد *m*

6872

wing nut
F écrou papillon *m.*
E palomilla *f.;* tuerca de
   alas *f.*
D Flügelmutter(n) *f. pl.*
R барашек *м.*
A صَمُولَة مُجَنَّحَة *f*

6873

wings
F insigne de pilote *m.*
E divisa de piloto *f.*
D Schwinge *f.;*
   Pilotenabzeichen *neut.*
R крылья *ср. мн. ч.*
A أجنحة (شارة طيار)

6874
winter
F hiver *m.*
E invierno *m.*
D Winter *m.*
R зима *ж.*
A شِتَاء *m*

6875
wintry
F hivernal
E glacial
D winterlich
R зимний
A شِتْوِي

6876 [E]
wire *v.*
F poser des fils
E instalar el alambrado
D Leitungen verlegen
R монтировать провода
A رَبَط بِسِلْك

6877
wire *n.*
F fil *m.*
E alambre *m.*
D Draht *m.*
R проволока *ж.*; провод *м.*
A سِلْك *m*

6878 [L]
wire *v.*
F grillager
E alambrar
D verdrahten
R устраивать проволочные заграждения
A مَدَّ أَسْلَاكًا

6879
(barbed) wire
F fil de fer barbelé *m.*
E alambre de púas *m.*
D Stacheldraht *m.*
R (колючая) проволока *ж.*
A أَسْلَاك شائكة *f*

6880
(concertina) wire
F barbelé à boudin *m.*
E alambrada plegable *f.*
D Windedraht *m.*
R проволочное заграждение *ср.*
A شبكة منفاخية

6881
(trip) wire
F fil de déclenchement *m.*
E alambre de disparo *m.*
D Auslösedraht *m.*
R (натяжная) проволока *ж.*
A سلك التعثر

6882
wire cutters
F cisailles *f. pl.*
E cortaalambres *m. inv.*
D Drahtschere *f.*
R проволочные ножницы *ж. мн. ч.*
A مقراض أَسلاك

6883
wire entanglement
F réseau de barbelés *m.*
E alambrada *f.*
D Drahtverhau *m.*
R проволочное заграждение *ср.*
A شبكة أَسْلَاك شَائكة

6884 [G]
wire guided
F filoguidé
E filodirigido
D drahtgelenkt
R управляемый по проводам; проводоуправляемый
A مُوَجَّه بالسِّلك

6885
wire mesh
F treillage métallique *m.*
E tela metálica *f.*
D Maschendraht *m.*
R проволочная решётка *ж.*
A تداخل السلك

6886
wire rope
F câble d'acier *m.*
E cable de alambre *m.*
D Drahtseil *neut.*
R стальной трос *м.*
A حَبْل مَعْدَنِي *m*

6887 [P]
wire tapping
F captage *m.;* écoute clandestine *f.*
E captación por derivación *f.*
D Abhören *neut.*
R подслушивание телефонных разговоров *ср.*; перехват телефонных разговоров *м.*
A اسْتِرَاق السَمْع لِلمُحادثات التليفونية

6888
withdraw *v.*
F (se) replier
E retirar(se)
D zurückziehen *tr.*
R отступать
A سحب/انسحب

6889
withdrawal
F décrochage *m.;* repli *m.*
E retirada *f.;* repliegue *m.*
D Rückzug *m.*
R отступление *ср.*
A انسحاب

441

6890

withdrawal action
F décrochage *m.*
E retirada *f.*
D Ausweichbewegung *f.*
R отступление *ср.*
A عملية انسحاب

6891 [E,T,U]

wood
F bois *m.*
E bosque [T] *m.;* madera [EU] *f.*
D Holz [EU] *neut.;* Wald [T] *m.*
R лесоматериалы *м. мн. ч.;* лес [T] *м.*
A خَشَب *m*

6892

wooden
F (de, en) bois
E (de) madera
D Holz- *komp.*
R деревянный
A خَشَبِي

6893

wool
F laine *f.*
E lana *f.*
D Wolle
R шерсть *ж.*
A صُوف *m*

6894

woollen
F (de; en) laine
E (de) lana
D Woll- *komp.*
R шерстяной
A صُوفِي

6895

(code) word
F mot code *m.*
E palabra (clave) *f.*
D Kodewort *neut.*
R кодовое слово *ср.*
A كَلِمَة رَمْزِية *f*

6896

(pass) word
F mot de passe *m.*
E consigna *f.;* seña *f.*
D Kennwort *neut.*
R пароль *м.*
A كلمة مرور،

6897

work *n.*
F travail *m.*
E trabajo *m.*
D Arbeit *f.*
R работа *ж.;* труд *м.*
A شُغْل *m*

6898

work *v.*
F fonctionner; travailler
E funcionar; trabajar
D arbeiten; funktionieren
R работать
A إشْتَغَل

6899

workshop
F atelier *m.*
E taller *m.*
D Werkstatt *f.*
R мастерская *ж.;* цех *м.*
A مَعْمَل إصْلاح ـ ورشه *f*

6900

work up
F préparer
E preparar
D ausbilden *tr.*
R разрабатывать; осваивать (новый корабль)
A دَبَّر

6901

wound *n.*
F blessure *f.*
E herida *f.*
D Verletzung *f.;* Wunde *f.*
R рана *ж.;* ранение *ср.*
A جُرْح *m*

6902

wound *v.*
F blesser
E herir
D verletzen; verwunden
R ранить
A جَرَح

6903

wounded *adj.*
F blessé
E herido
D verwundet
R раненый
A جريح

6904

wrap *v.*
F envelopper
E envolver
D einwickeln *tr.*
R завёртывать; складывать
A غَلَّف

6905

wreck *n.*
F naufrage *m.*
E naufragio *m.*
D Wrack
R крушение *ср.;* остов *м.;* обломки *м. мн. ч.*
A حُطَام *m*

6906

wreck *v.*
F détruire [EU]; faire échouer [MU]
E estropear [EU]; hundir [M]
D zerstören; vernichten; (zum) Scheitern bringen [M]
R вызывать крушение; потоплять [M]; разрушать
A حَطَّم

6907

wrecker

F dépanneuse *f.*
E camión de recuperación *m.*
D Bergungsfahrzeug *neut.*
R эвакуационный тягач *м.*
A عَرَبَة جَرّ

6908

write-off *n.*

F perte totale *f.*
E baja *f.;* perdida total *f.*
D Totalverlust *m.*
R списание *ср.*
A إسْتِبْعَاد *m*

6909

write off *v.*

F amortir (entièrement)
E (dar de) baja; suprimir
D (vollständig) abschreiben *tr.*
R списывать
A إسْتَبْعَد

X

6910

X-axis

F axe des abscisses (des X) *m.*
E eje de abscisas (de las X) *m.*
D Abszisse *f.;* X-Achse *f.*
R ось абсцисс *ж.;* ось-икс *ж.*
A مِحْوَر السِّينَات

6911

X-ray *v.*

F radiographier
E radiografiar
D röntgen
R просвечивать рентгеновыми лучами
A صَوَّر بالأَشِعَّة السِّينِيَّة

6912

X-rays *pl.*

F rayons X *m. pl.*
E rayos X *m. pl.*
D Röntgenstrahlen *m. pl.*
R рентгеновые лучи *м. мн. ч.;* рентгеновские лучи *м. мн. ч.*
A الأَشِعَّة السِّينِيَّة *f*

6913

X-scale

F échelle en X *f.*
E escala de las X *f.*
D X-Masstab *m.*
R шкала-икс *ж.;* масштаб-икс *м.*
A مِقْيَاس السِّينَات

Y

6914

(marshalling) yard

F gare de triage *f.*
E playa (de formación de trenes) *f.*
D Rangierbahnhof *m.*
R (сортировочная) станция
A ساحة مناورة القطر

6915

(ship building) yard

F chantier de construction navale *m.*
E astillero *m.*
D Schiffswerft *f.*
R (судостроительная) верфь *ж.*
A مشغل البناء البحري

6916

yardarm

F bout de vergue *m.*
E verga *f.*
D Rahnock *m.*
R нок-рея *м.*
A ذِرَاع العَارِضَة

6917

yaw *n.*

F embardée [M] *f.;* lacet [AGS] *m.*
E guiñada [AGM] *f.;* desviación *f.;* inclinación de la tangente [S] *f.*
D Geschosspendelung [S] *f.;* Gierbewegung [M] *f.;* Scherung[AG] *f.*
R отклонение *ср.;* рыскание *ср.*
A إنعِرَاج

6918 [A,G,M,S]

yaw *v.*

F (faire un) mouvement de lacet[AGS]; embarder[M]; dévier de la route[M]
E zozobrar
D gieren
R раскачиваться [A]; подвергаться рысканием [GS]; отклоняться от курса [M]
A انعرج تمعج/زاغ

6919

Y-axis

F axe des ordonnées (des Y) *m.*
E eje de ordenadas (de las Y) *m.*
D Ordinate *f.;* Y-Achse *f.*
R ось-игрек *ж.*
A مِحْوَر العَيْنَات

6920

year

F an *m.;* année *f.*
E año *m.*
D Jahr *neut.*
R год *м.*
A سَنَة *f*

6921 [N]
yield
F puissance *f.*
E rendimiento *m.*
D Detonationswert *m.*
R мощность [S] *ж.*;
   тротиловый
   эквивалент *м.*
A حاصِل *m*

6922
yoke
F joug *m.*
E balancín *m.;* horquilla
   f; yugo *m.*
D Joch *neut.*
R иго *ср.*; скоба *ж.*
A نِير ـ ظُلم *m*

6923
Y-scale
F échelle en y *f.*
E escala de las y *f.*
D Y-Masstab *m.*
R шкала-игрек *ж.*;
   масштаб-игрек *м.*
A مِقْياس العَيْنات

Z

6924
Z-axis
F axe-Z *m.*
E eje Z *m.*
D Z-Achse *f.*
R ось-3 *ж.*
A المحور العيني

6925 [M]
Z-bar
F fer en Z *m.*
E hierro en Z *m.*
D Z-Eisen *neut.*
R зетовая балка *ж.*
A قَضِيب مَقْطَعُه على شَكْل Z

444

6926
zenith
F zénith *m.*
E cénit (zénit) *m.*
D Zenit *m.*
R зенит *м.*
A ذِرْوَه

6927
zero *n.*
F zéro *m.*
E cero *m.*
D Null *f.*
R нуль *м.*; ноль *м.*
A صِفْر *m*

6928 [E,S]
zero *v.*
F remettre au zéro [E];
   ajuster [S]
E poner en cero [E];
   ajustar [S]
D (auf) Null einstellen
   [E]; justieren [S]
R устанавливать на нуль
   [E]; прицеливаться
   [S]
A صَفَّر

6929 [N]
(desired ground) zero
F point zéro souhaité *m.*
E punto cero deseado *m.*
D geplanter
   Bodennullpunkt *m.*
R намеченный эпицентр
   взрыва *м.*
A نُقْطة الصِفْر الأَرْضِية
   المُعَيَّنة *f*

6930 [N]
(ground) zero
F point zéro
E tierra cero *f.*
D Nullpunkt *m.*
R эпицентр взрыва *м.*
A نُقْطة الصِفْر *f*

6931
zero in on *v.*
F pointer le tir vers le
   centre de la cible/ de
   l'objectif
E reglar el tiro sobre
D (auf Nullpunkt)
   einstellen *tr.*
R пристреливаться
A وجّه النار إلى مركز الهدف

6932 [G]
zero-length launching
F départ ponctuel *m.*
E lanzamiento puntual *m.*
D Punktstarten *neut.*
R запуск с места, с
   установки (ПУ) без
   направляющих *м.*
A إطْلاق صِفْري الإمْتِداد

6933
zigzag *v.*
F zigzaguer
E zigzaguear
D (im) Zickzack gehen
   (fahren usw.)
R делать зигзаги
A تَعَرَّج

6934
zonal
F zonal
E zonal
D Zonen- *komp.*
R зональный; районный
A مُقَسَّم الى مَناطِق

6935
zone *n.*
F zone *f.*
E zona *f.*
D Raum *m.;* Zone *f.*
R зона *ж.*
A مِنْطَقة

6936

zone *v.*

F répartir en zones
E dividir en zonas
D (in) Zonen einteilen *tr.*
R разделять на зоны
A قَسَّم إلى مَناطِق

6937

(beaten) zone

F zone battue par le feu *f.*
E zona batida *f.*
D Wirkungsbereich *m.*
R (поражаемая) площадь *ж.*
A منطقة مضروبة

6938

(blind bombing) zone

F zone de bombardement sans restriction *f.*
E zona (de bombardeo libre) *f.*
D (unbeschränkter) Bombenbereich *m.*
R (открытая) зона бомбометания *ж.*
A مِنْطَقَة قَصْف أَعْمَى *f*

6939

(buffer) zone

F zone tampon *f.*
E zona (tapón) *f.*
D Pufferzone *f.*
R (буферная) зона *ж.*
A دَوْلَة / مِنْطَقَة حاجِزة *f*

6940

(combat) zone

F zone de l'avant *f.*
E zona (de combate) *f.*
D Kampfgebiet *neut.*
R зона боевых действий *ж.*
A مِنْطَقَة القِتال *f*

6941

(communications) zone

F zone des communications (des arrières) *f.*
E zona de etapas *f.*
D Verbindungszone *f.*
R зона коммуникации *ж.*
A مِنْطَقة المُواصَلات *f*

6942

(demilitarized) zone

F zone démilitarisée *f.*
E zona desmilitarizada *f.*
D entmilitarisierte Zone *f.*
R демилитаризированная зона *ж.*
A مِنْطَقَة مُجَرَّدَة من السِّلاح *f*

6943

(drop) zone

F zone de largage *f.*
E zona (de lanzamiento) *f.*
D Abwurfgebiet *neut.*
R район сбрасывания *м.*
A مِنْطَقَة الهُبوط *f*

6944

(dropping) zone

F zone de parachutage *f.*
E zona (de salto) *f.*
D Absprunggebiet *neut.*
R район выброски десанта *м.*
A منطقة انزال المظليين

6945

(exclusion) zone

F zone d'interdiction *f.*
E zona de exclusión *f.*
D Ausschlussgebiet *neut.*
R зона исключения *ж.*; запретная зона *ж.*
A منطقة محظورة

6946

(free fire) zone

F zone d'ouverture de feu *f.*
E área de tiro libre *f.*
D (unbeschränkter) Wirkungsbereich *m.*
R зона непредельного огня *ж.*
A منطقة الرمي بلا تقييد

6947

(killing) zone

F zone dangereuse *f.*; zone d'embuscade battue par le feu *f.*
E zona donde debe ser conducido el enemigo para ejercecer fuego efectivo sobre él *f.*
D Feuerfeld *neut.*
R зона поражения *ж.*; огневой мешок *м.*
A منطقة قتل

6948 [A]

(landing) zone

F zone d'atterrissage *f.*
E zona de aterrizaje *f.*
D Landeraum *m.*
R площадка приземления *ж.*
A منطقة الهبوط *f*

6949

(safety) zone

F zone de sécurité *f.*
E zona de seguridad *f.*
D Sicherheitszone *f.*
R зона безопасности *ж.*
A مِنْطَقَة الامان

6950

(time) zone

F fuseau horaire *m.*
E huso horario *m.*
D Zeitzone *f.*
R часовой пояс *м.*
A توقيت المنطقة

6951 [A]
(touchdown) zone
F zone de poser *f.*
E area de aterrizaje *f.*
D Aufsetzzone *f.*
R район посадки *м.*
A مِنْطَقَة لَـمْـس الأَرْض

6952
(war) zone
F zone de guerre *f.*
E zona de guerra *f.*
D Kriegszone *f.*
R зона военных
  действий *ж.*
A منطقة حربية

6953
zone of attack
F zone d'attaque *f.*
E zona de responsibilidad
  f.
D Angriffsbereich *m.*
R полоса наступления
  ж.
A منطقة الهجوم

6954
zone of fire
F zone d'appui immédiat
  (artillerie) *f.*
E zona de tiro (de fuego
  de apoyo) *f.*
D Wirkungsbereich *m.*
R полоса обстрела *ж.*
A منطقة الرمي

6955
zone suffix
F suffixe de zone *m.*
E sufijo de zona *m.*
D Zonen-Suffix *neut.*
R суффикс зоны *м.*;
  указатель часового
  пояса *м.*
A لاَحِقَة المِنْطَقَة

6956
zone time
F heure du fuseau horaire
  f.
E hora de huso horario *f.*
D Zonenzeit *f.*
R поясное время *ср.*
A منطقة التوقيت

6957
zone time system
F système d'heures
  zonales *m.*
E sistema de husos
  horarios *m.*
D Zonenzeitsystem *neut.*
R система поясных
  времён (поясное
  время) *ж.(ср.)*
A نِظَام المَنَاطِق الوقتية

6958
zoom lens
F zoom *m.*
E objetivo telefotográfico
  m.
D Telelense *f.*
R зумлинза *ж.*
A عدسة تقريب الصورة

6959
Z-scale
F échelle en Z *f.*
E escala de las Z *f.*
D Z-Masstab *m.*
R шкала-зет *ж.*;
  масштаб-зет *м.*
A مِقْياس عَيْني

6960
zulu time
F heure zulu *f.*
E hora zeta *f.*
D mittlere Greenwich-Zeit
  f.; Z-Zeit *f.*
R время по Гринвичу *ср.*
A تَوْقِيت جِرِينِتْش المُتَوَسِّط

INDEX OF DEFINITIONS AND EQUIVALENTS
BRITISH-ENGLISH

USA USAGE	BRITISH EQUIVALENT OR DEFINITION
(to) Annex	(to) Attach/Append
(to) Battle	(to) Fight
Battalion Task Force	Battle Group
Body armor	Flak jacket
Buttoned up	Closed down (of armoured vehicles)
Combat maneuver forces	Forces which use fire and movement to engage the enemy with direct fire weapons.
Combat service support	The logistic and administrative support provided to sustain combat troops.
Combat trains	Elements in a unit which provide immediate combat service support to forward troops of the unit.
Combined Arms Team	Combat Team
Committed force	A force which has contact with the enemy or is engaged in or chosen for a specific task which prevents its employment elsewhere.
Control measures	The boundaries, objectives and other measures which help commanders allot responsibility, control operations and coordinate fire and movement.
Corrective maintenance	Light repair
Cross-attachment	Detached duty with another arm by a teeth-arm sub-unit.
Cross-servicing	Mutual logistic support
Dead space	Dead ground
Debarcation	Disembarkation
Divided highway	Dual carriageway
Dogtag	Identity disc
Draft	Selection for National Service
Draftee	A conscript
Dressing Station	RAP
Earmuffs	Ear defenders
Echelonment	The deployment of troops and equipment into assault, combat support and combat service support groupings.
Enlisted personnel	WOs, NCOs and men.
Fall	Autumn
Field shop	Field workshop
Field rations	Combat rations

USA USAGE	BRITISH EQUIVALENT OR DEFINITION
Field trains	The elements of a unit which provide the combat service support which is not immediately required for operations.
Flight strip	Temporary airstrip
Flight training	Flying training
Flightworthy	Airworthy
Follow and support	A force which follows an exploiting force and secures communications or eliminates bypassed forces.
Forward Arming and Refueling Point	A forward temporary facility which allows a commander to refuel and re-arm helicopters.
Freight car	Goods wagon
Fresh	Inexperienced
Furlough	Leave
General Orders	Standing Orders
Ground (trafficable)	Passable ground
Guard house	Guard room
Gym suit	Track suit
Habitual association	The close relationship fostered between combat and support units by frequent cross-attachments (q.v.).
Intrenching tool	Entrenching tool
Litter	Stretcher
Litter bearer	Stretcher bearer
Litter patient	A sick or wounded man who is not "walking wounded"
Mail	Post
Machine pistol	SMG
Manipulative electronic deception	The manipulation of own forces electronic emissions to deceive the enemy and mask friendly forces' intentions.
Mask clearance	The absence of any obstacle in the path of a weapon's trajectory.
Mask clearance	The amount of clearance with which a projectile passes above an object between the weapon and the target.
Meaconing	Enemy transmissions which confuse the navigation of friendly aircraft or ships.
Mess hall	Cookhouse
Military crest	The area on a forward slope at which the slope can be observed to the base without any dead ground.
Movement to contact	Advance to contact

USA USAGE	BRITISH EQUIVALENT OR DEFINITION
Multiple employment	The concept by which a unit with high mobility can be given more than one task during a single operation.
Naval Academy	Naval College
Off limits	Out of bounds
Objective area	The area around the object which will be beaten by fire.
Obscuration	The effects of mist, fog or other weather, dust and smoke on the battlefield.
Obstacle (cultural)	A man-made feature
Obstacle (existing)	A natural or cultural obstacle which exists when planning begins.
Obstacle (standard)	An obstacle on a standard list for preparation by an Engineer support unit.
Obstacle (tactical)	An active obstacle which channels the enemy and destroys him.
Officer of the guard	An officer acting under the officer of the day who is responsible for the instruction, discipline and performance of the guard.
Olive drab(s)	The colour and material used for US Army uniforms and the uniforms made from the material.
On limits	In bounds
Overwatch	Normal tactical movement or the readiness of a unit or sub-unit to support another unit with fire if required.
Overwatch (bounding)	A leap-frogging technique used when enemy contact is expected. The unit moves forward by bounds. Each sub-unit which moves is protected by another ready to give fire support.
Overwatch (travelling)	A technique used when enemy contact is expected. Following units or sub-units keep distance behind the leading elements and pauses to overwatch.
Program	Programme
Passage of lines (forward and rear)	The act of passing one unit through the position of another.
Passage point	The point at which passage of lines will occur.
Preplanned mission request	A request for air support which is received in time to allow coordination and preplanning.
Prescribed load	The quantity of essential combat supplies and spare parts held by a unit to make it self-supporting for a set number of days.

USA USAGE	BRITISH EQUIVALENT OR DEFINITION
Priority of fires	Directions given to a fire support planner which order the priority of fire support according to the relative importance of the task of the unit being supported.
Priorities of support	Priorities set for the provision of combat support and combat service support in accordance with the importance of the missions of the units being supported.
Quartermaster corps	Nearest equivalent may be Royal Army Ordnance Corps
Quartering party	Advance party
Railroad	Railway
Radiation status	The factors which permit a commander to calculate and measure a unit's exposure to radiation.
Raider	Commando or SAS-type soldier.
Ranger	US Special Forces soldier.
Enlisted men	Ranks (other)
Reconnaissance by fire	Fire on a suspected enemy position which may draw fire or reduce movement.
Reinforcing fire(s)	A mission in which the fire of one artillery unit is reinforced by that of another.
Resupply (vertical)	Resupply by helicopter
Retain mission	A mission in which a unit prevents occupation of a feature or position by the enemy.
Room (situation)	Operations room
Running light	Navigation light
Rupture force	A force which penetrates enemy lines in order to make a gap through which other forces can penetrate and exploit.
Sharpshooter	Sniper; crack shot
Signal corps	Royal Signals
Signal pistol	Verey pistol
Single envelopment	An operation against one flank of the enemy.
Situation room	Operations room
Sky-diving	Free-fall parachuting
Squad leader	Section commander
Station (gasoline)	Petrol station
Status (readiness)	The preparedness of a unit for an operation.
Subsequent operation phase	The post-assault phase of an airborne, airmobile or amphibious operation which may lead on to a further attack, a defence or withdrawal.

USA USAGE	BRITISH EQUIVALENT OR DEFINITION
Supply point distribution	Resupply from a central point
Support (general)	A total plan of orchestrated artillery support for an operation.
Support echelons	Units which give fire support to attacking or defending units.
Support (service)	Combat service support
Support (tactical)	Combat support
Supporting distance	The spacing between units which allows them to provide mutual fire support or to come to one another's aid during an operation.
Survivability operations	The development of physical and electronic protective measures to reduce the effectiveness of enemy weapon systems.
Task force	Battle group
Tailgate	Tailboard
Tail pipe	Exhaust pipe
Target (impromptu)	Target of opportunity
Target dossier	Target list
Task force (company)	Combat team
Team (company)	Combat team
Templet	Template
Terrain (key)	Vital ground
Terrain reinforcement	Mobility, counter-mobility and trenching operations.
Topographical crest	The real crest or highest point of a piece of high ground.
Trains (combat and field)	Unit elements which provide combat service support.
Travelling	Unit movement when enemy contact is unlikely and all unit elements can move simultaneously.
Utility helicopter	A general purpose battlefield helicopter.
Unit distribution	A method of supplying units in unit location using the transport of the supplying agency.
Unit in contact	A unit in contact with the enemy and thus not available for another mission.
Universal military training	A programme which requires all young male citizens eligible for duty, to serve a specific period of active and reserve duty.
Vacuum bottle	Vacuum flask
Vertical envelopment	An attack by airdropped or airlanded troops which envelops the enemy flanks or rear and encircles or cuts him off.
Veteran	A retired serviceman

USA USAGE	BRITISH EQUIVALENT OR DEFINITION
Walking patient	Out-patient; walking wounded
Windshield	Windscreen
Windshield wiper	Windscreen wiper
Zone of attack	The area of responsibility beyond the line of contact which is allotted to an attacking unit.
Zone of fire	The area within which an artillery or other fire support unit is ready to provide fire support.

INDEX DE TERMES FRANÇAIS

belligérant 441
benne preneuse 2044
berceau 1059
bernache 375
(en) berne 2151
besoin 4757
besoin opérationnel 3676
bête de somme 3811
béton 943
béton armé 4661
béton précontraint 4275
bidon 690, 1927, 6737
bief (canal) 4541
bifurcation 2642, 4901
bifurquer 563
(se) bifurquer 1326
bilan logistique 2946
bilatéral 462
bimoteur 6412
binoculaire 471
biologique 473
bit 477
bitube 6411, 6413
bivouac 479, 5279
bivouaquer 478
blanchir 6843
blessé 6903
blessé 722
blesser 2474, 6902
blessure 2475, 6901
blessure à l'oeil 1568
blessure superficielle 1778
blindage 256, 4086, 4346
blindage composite 921
blindage glacis 4081
blindé 531
(non)blindé 6435
blinder 5089, 5283
blip 494
bloc 495
bloc à clé nonréutilisable 3815
blocage 2928
blockhaus 639, 4006
blocus 497
(faire le) blocus 498
bloqué (port) 2382
bloquer 498, 808
bluff 509
bluffer 508
bobine 6858
bois 6173, 6891
(de, en) bois 6892
boîte 549
boîte de vitesses 2009
bombardement 523
bombardement à basse altitude 2974
bombardement aérien 4480

bombardement d'artillerie 5274
bombardement en déport 3614
bombardement en vol rasant 2744
bombardement par obus 5274
bombardement par ricochet 5438
bombardement par vagues 710, 3922
bombardement sur zone 243
bombarder 525, 5264, 5753
bombarder au mortiers 3329
bombardier 522, 528
bombe 521
bombe-H 2206
bombe à hydrogène 2367
bombe à retardement 6176
bombe photo-éclair 3982
bond 544
bon de transport 6722
bondon 637
boqueteau 5583
(à) bord 5, 3627
(de) bord 87, 3349, 5302
bord de la mer 5328
borne 6108
bouche 3385
bouché 3761
bouche d'incendie 2363
boucher 3554, 4094, 5113
bouchon 4093
bouchon-allumeur 1972, 5220
boucle 614
boucle mosaïque photo 5796
boucler 612, 1005
bouclier 5281
bouclier (char) 3056
bouclier de protection contre les vagues
  (véhicule amphibie) 3833
bouée 640
bouée-culotte 591
bouée d'ancre 193
bouée de corps-mort 3322
bouée lumineuse 2810
bougie 4096
boule de feu 1707
boulon 519
bourre 3816
boussole 915
bout de crosse 4790
bout de vergue 6916
bouteille 543
bouteille (de gaz) 1096
bouteille d'oxygène 3800
bouteille isolante 6544
bouteille thermos 6544
bouton 654
boutonner 653
(à) bout portant 4160, 4510

cellule photo-électrique   3981
censeur   740
centigrade   744
central   745, 3183
centraliser   746
central téléphonique   6084
centre   741, 3182, 5228
centre combiné d'opérations   5631
centre de contrôle aérien tactique   5972
centre de contrôle de transport   6334
centre de gravité   743
centre de la cible   626
centre de regroupement de prisonniers de
    guerre   4141
centre de résistance   4147
centre de transmission   5388
centre de transmissions   906
centre d'information de combat   870
céramique   749, 750
cercle   796
cérémonial   751
cerner   550, 1460, 5917
certificat   752
certificat de sécurité   5160
certifier   753
cerveau   554
cessation des hostilités   5937
cesser le feu   730
cessez-le-feu   729
chaîne   755, 765
chaland   2811
chaleur   2220
chaloupe   4020
chalumeau   505
chalumeau oxyacétylénique   3798
chaluter   6345
chalutier   6346
chambre   764, 6391
chambre de décompression   1142
chambre des cartes   3065, 4935
chambre des machines   4938
champ de bataille   405
champ de glace   2384
champ de mines de cloisonnement   3230
champ de mines   3229
champ de mines de harcèlement   3231
champ de tir   239
champ de vision   1619
champignon atomique   3377
chandelier   5649
(levier de) changement de vitesse   2011
changer   5950
changer de vitesse   5289
changer la classification d'un document
    4576
changer la direction   5948
changer le cap   164

chantier   1336, 5416
chantier (naval)   5311
chantier de construction navale   6915
chape   6347
chapelet de bombes   5725
char   6001
char amphibie   6002
charbon   840
char de bataille   6006
char embossé   6007
charge   768, 2891, 2893, 2895, 2907,
    5782
(en) charge   2890
chargé (de)   2690
(à) charge additionnelle   5846
charge creuse   2301
(à) charge creuse   5248
charge d'impulsion   2422
charge explosive à projectiles secondaires
    6696
charge maximum autorisée   2897
chargement   2908, 5303
chargement par convoi   2910
(fusil à) chargement par levier   4377
chargement par unité constituée   2912
chargement standard   2904
chargement tactique   2911
chargement vertical   2913
charge normalisée   2906
charge nucléaire   3507
chargé par la culasse   2909
charge platte   5452
charge pratique limite   2901
charge propulsive   4341
charger   769, 770, 2492, 2894, 2896,
    5779
charge réduite   4613
charger quelqu'un (de)   6054
charge sous élingue   2903, 2905
charge standard sur palette   2899
chargeur   2997
charge utile   3925
chariot de transbordement   6304
chariot élévateur   1870
char poseur de pont   597, 6003
chasse   2357
chasse-neige   5514
chasse aux mines   2359
chasser   2356
chasser sur l'ancre   1355
chasseur(s)-intercepteur(s) de nuit
    3458
chasseur de chars   6012
chasseur de pénétration   2569
châssis   775, 1892
château d'eau   6757
chaudière   516

chauffer 2221
chaussures de tennis 2139
(faire) chavirer 699, 2652, 3792
chef 2208, 2759
chef de corps 3579
chef de groupe 5617
chef de pièce 5617
chef d'état-major 785
chef de transport 759
chef d'ilôt (défense passive) 6674
chef largueur 1385, 2641
chemin à fortes pentes 4893
chemin de fer 4483
chemin de fer à voie étroite 3393
cheminée 1937
cheminement 2712
cheminement de repli 2867
chemise 2602
chemise d'eau 6741
chenal 765, 1581
chenal de sécurité 5016
chenal dragué 5946
chenal exploré 5125
chenille (de char) 6235
chercher l'autorisation de 822
cheval 2326
cheval (-vapeur) (CV) 2327
cheval de frise 4793
chevalet de pointage 4789
chevron 5795
chien de recherche 5501
chien policier 6246
chiffre 794, 1264, 3519
chimique 780
chiotte 2215
chirurgical 5909
chirurgie 5908
chirurgien 5907
choc 632, 2413, 5313
choisir (parmi) 5170
choix 5171
chronique 790
chronomètre 792, 5738
cible 6022
cible à éclipse 6035
cible aérienne remorquée 6041
cible facile 6039
cible réelle 6029
cicatrice 5060
ciel 5448
cime 3933
ciment 737
cimetière 738
circonférence 800
circuit 797
circuit de mise à feu 1724
circuit intégré 2503

circulation 6256
circulation aérienne 140
circulation dans les deux sens 6418
ciré 3625
cisailles 6882
citadelle 801
civil 804
(en) civil 805, 3362, 4045
clandestin 810, 6458, 6461
clapet 6548
clapotis 609
clarification 811
classe de chargement militaire 3198
classe de recrutement 69
classé hors service 459
classer 2045
classification 814, 815, 2052
classification de sécurité 5162
classification d'un itinéraire 4972
classifié 816
classique 5227
clavette 2656
clavier de touches 2660
clé 2655
(-)clé 2658
clef 2654, 2655
(-)clef (suffixe) 2658
cloche 440
cloison 625
clôture (en fil métallique) 1601
clôturer 1600
clouer au sol 4019
club militaire 5214
(se) coaguler 833
coaxial 846
cocktail Molotov 3314
(mettre en) cocon 849
code 850
code condensé 593
code de panneaux 3834
coder 1463
coefficient d'arrimage 4795
coffre 783, 6388
cohue 3288
coin 1006, 6805
coincer 5168, 6806
col 3894
collationner 854
collecteur d'admission et d'échappement 3050
colline 2275
collision 858
colonne 862, 2859
colonne (de fumée) 4005
colonne de direction 863
coma 864
combat 28, 865

congère 5510
congés de fin de service 6112
conique 956
conjugaison 2317
connaissance approfondie 3110
connaissement 467
conquérir 958
conscription 1350
conscrit 1353
conseil 51
conseiller 52, 53
consigne 2493
consigne de tir 5701, 5706
consigne permanente 3714
consignes générales 2016
console 960
consolider 5778
consommation 1545
consommer 1543
conspirer 4088
constante 962
constater 753
constituer des stocks de réserve 5731
construire 618
consultatif 54
contact 965
contacter 966
contamination 970
contamination résiduelle 4773
contaminer 969
conteneur 968, 1908
contenir 967
continent 971, 3017
contingent 975
contingent national 3396
contourner 658, 5446
contrainte logistique 2945
contraire 3690
contrat 976
contre-attaque 1022
contre-contremesures électroniques 1440
contre-espionage 1024, 2517
contre-guérilla 6683
contre-mesure 1027
contre-mesures électroniques 1441
contre-mesures radar 4433
contre-pente 5480
contre-plaqué 4101
contrebandier 5496
contrecarrer 1021
contre l'incendie 1716
contrôle 978, 4118
(point de) contrôle 778
contrôle automatique de la route 6249
contrôle automatique de navigation 3419

contrôle de la circulation aérienne 141
contrôle de la valeur physique individuel 6130
contrôle de manoeuvre d'identification 5497
contrôle d'émission 1457
contrôle de qualité 4396
contrôle de repérage 4641
contrôle de sécurité 5161
contrôle de stocks 5728
contrôle d'inventaire 2577
contrôle du sol 2097
contrôle intermittent 5598
contrôle opérationnel 3673
contrôler 776, 979, 1041, 3316
contrôle tactique 5976
contrôleur 986
contrôleur aérien avancé 1878
contrôleur aérien tactique 5973
contrôleur de secteur 5147
contrôle visuel 6639
convention 6348
convention d'échange 5070
Convention de Genève 2021
conventions de manoeuvres 4800
convergence 991
convergence de tir 938
(faire) converger 935
(faire) converger (sur) 990
convoi 992
convoi océanique 3562
coopération 999
coopération civilo-militaire 806
coordonnés 1001
coordonnées cartésiennes 713
coordonnées hectométriques 4621
coordonnées polaires 4172
coordonner 1000
coque 2352
coque du réacteur 4549
coque noyée 2353
corbeille 392
corde 4943
cordeau détonant 1953
cordite 1003
cordon 1004
corps 513
corps (d'armée) 1009
corps de garde 2115
corps expéditionnaire 1542
correct 4860
corrélation 1014
correspondant de guerre 6671
corridor aérien 91
(se) corroder 1015
corrosif 1017
corrosion 1016

destruction sous-marine   6475
(en) désuétude   3542
détaché   2956, 5194
(être) détaché   5136
détaché au près de   294
détachement   295, 1234, 1235, 3878
détachement de corvée   3885
détachement de débarquement   3887
détachement de protection d'un
   dispositif de destruction   2113
détachement postcurseur   3891
détachement précurseur   3880, 3889
détachement temporaire   1077
détacher   293, 1233
détail   579, 1236
détailler   578, 2600
détail topographique   3404
détecter   1239
détecteur   1242, 5189
détecteur au sol télécommandé   5190
détecteur de mensonges   2791
détecteur d'interception radioélectrique
   2526
détecteur radiologique   4440
détecteurs au sol télécommandés   4696
détection   1240
détection précoce   6708
détection radar   4436
(se) détendre   5454
detenir   1238
détente   6363
déterminé par travail topographique
   5926
détonateur   1247
détonation   364, 1246, 4729
détonation (sonique)   534
détonation par sympathie   5958
(faire) détoner   1245
détour   1248
détournement   2274
détourner (par la force)   2273
détroit   5538, 5756
détruire   1230, 2663, 5995, 6906
(à) deux gouvernails   6414
(à) deux places   6416
développement   1251
(se) développer   1250
devers   5457
déviant (du cap fixé)   1033
déviation   1253, 1323
dévier de la route   6918
dévisser   6514
diagonal   1255
dialogue   2533
diamètre   1257
diaphragme   1258
diapositive   5466, 6322

diesel-électrique   1259
différentiel   1260
diffuser   1309
diffusion   1310
diffusion troposphérique   6379
digital   1265
digue   5130, 5586
dimension   1266
diminuer   2977
direct   1268
directeur-général   5141
direction   982, 1270, 2008, 2760, 2827
direction assistée   5715
(en) direction du nord   3483
directionnel   1271
direction repère   2846
directive   1272
directives pour la mise-en-place   987
diriger   981, 1269, 2756, 3039, 6559
(se) diriger vers   3030
discipline   1283
(bien)-discipliné   6823
discrétion   5138
discrimination d'objectifs   6047
disculper   820
disparu au combat   3276
(se) disperser   1296, 5062
dispersion   1190, 1297, 5061, 5607
disponibilité   322, 4550, 5213
disponibilité instantanée   4552
disponibilité opérationnelle   4551, 4554
disponible   323
disposer (de)   1305
dispositif   1254, 1307
dispositif anti-lueur   5895
dispositif avertisseur   6713
dispositif d'affichage   1304
dispositif d'alerte anti-intrusion   2570
dispositif de destruction   5219
dispositif de post-combustion   4658
dispositif de poursuite   6245
dispositif de réchauffe   67
dispositif d'image thermique   6142
dispositif d'ouverture de parachute
   4679
dispositif tête haute   2217
dissident   1311
dissoudre (une unité)   1276
dissuader   1243
distance   1312, 4514
distance de décentrement   3615
distance de sécurité   5014
distance de sécurité et d'appui
   réciproque   5893
distance d'observation   1313
distance focale   1830
distance minimum de sécurité   3242

E

état-major de planification   5635
état-major intégré   5629
état-major interallié   5627
état-major interarmées   5630
état-major parallèle   5632
état constant   5708
état d'alerte   5662
état de guerre   5675
état de préparation   5705
état des effectifs   5670
état des pertes   2883, 4733
état insurrectionnel   2499
état major particulier   5634
état néant   4737
état nominatif   4925
éteindre   4409
étincelle   5553
étiqueter   2688
étiquette   2687
étoile   5664
étoile polaire   5665
être à l'ancre   4844
être défectueux   3033
étroit   3392
étude de facteurs   5638
étui (de revolver)   2302
évacuation   1521
évacuation (matériels)   340
évacuation par air   110
(d')évacuation sanitaire par air (évasan)
   3140
évacué   1522
évacuer   1520
évadé   1505
(s')évader (de)   1504
évaluation   282, 1525
évaluation de dommages nucléaires
   3505
évaluation de l'ennemi   2521
évaluation des dommages   1100
évaluation opérationelle   6357
évaluer   1524
évasion   1503
évasion et récupération   1527
éventualité   973
évitement   1526
éviter   1523
évolution   1534
examen médical   3994
excavateur   1530
excavatrice (pour tranchées)   1262
(d')exception   1451
exécuter   1532
(s')exercer (à)   4246
exercice   1372, 1534
(d')exercice   4245

exercice anti-incendie   1711
exercice de combat   1615
exfiltration   1535
exigences   1539
expédier   3009
expéditeur   3735
expédition   1541, 5304
expédition de secours   3892
expérience   1547
expérimental   1548
expérimenté   1546
expert   1549
expertise des dégâts   5923
exploitation   1552
exploiter   1551, 2189
exploratoire   1553
explorer   1554
(non) explosé   6519
explosible   1557
explosif   1558, 2262
explosif anti-chars   2264
explosif puissant   2263
explosifs et munitions   3723
explosion   1555
explosion à basse altitude   2973
explosion aérienne   90
explosion au sol   2095, 5901
explosion nucléaire aérienne   3502
explosion nucléaire en surface   3513
explosion nucléaire sous-marine   3515
explosion nucléaire souterraine   3514
explosion prématurée   4255
explosion souterraine   6463
exposé et alerté   6703
exposé et non alerté   6526
(s')exposer   1559
exposition   1560, 4264
extérieur   1564, 3743, 3745
externe   1565
extincteur   1714
extinction des feux   2726, 2818, 6060
extracteur   1566

F

fabriquer   4313
fac-similé   1573
facteur   1574
facteur de planification logistique   4070
facteur de securité   5015
(être de) faction   2114
factionnaire   5191
facultatif   3697
fading   1575
faible   6770

gite 2880
givrage 2388
(se) givrer 2387
glace 2379
glacier 2028
glacis 5478
glissant 5474
glissement 5472
glisser 5471
(faire) glisser 5465
glissière (fixe) 5464
global 6206
golfe 2126
gonflable 2456
gonfler 4378
goniomètre 1640
goniomètre boussole 1273
gorge 2040
goudron 6021, 6051
(non-)goudronné 6506
goulet 2127
goupille de sécurité 4017
(appareil à) gouvernail 2008
gouvernail (de direction) 4980
gouvernement 2042
gouvernemental 2041
gouvernement militaire 3195
gouverner 5713
gouverneur militaire 2043
grade 2047, 4524
grade à titre définitif 5823
grade d'officier 895
grade d'officier supérieure de marine 1748
graduel 2053
graduer 2054
grain 5618
grain de plomb 3937
graissage 2987
graisse 2065
graisser 2066, 2986, 3618
graisseurs 4170
grand angulaire 6846
grand cercle 2068
grande route 2272, 4894
grandes lignes 3750
grande tente 3089
(à) grande vitesse 2269
grange 374
graphique 2056
grappe 836
grappin 2057
gravier 2060
gravir un degré dans l'escalade 1501
gréement 4858
grenade (à main) 2072
grenade à manche 2074

grenade défensive 2071
grenade d'exercice 2070
grenade fumigène 2073
grenade neutralisante 2075
grenade sous-marine 1216
grenadier-voltigeur (GV) 4854
grève 5787
grièvement blessé 5201
grillager 6878
grille 2058, 3111
grille de mercator transverse universelle 6497
grille de navigation 3416
(uniforme de couleur) gris-vert (olive) 3626
gros 3015
gros de l'avant-garde 3016
grosse mer 2231
grossir 3007
grossissement 3005
gros temps 2233
groupe 2106, 5616, 6489
groupe (date/heure) 1111
groupe aérien embarqué 711
groupe allégé de commandement 891
groupe central de planification 6066
groupe chasseur de sous-marin 2358
groupe d'abordage 3881
groupe de destruction 3883
groupe de plage 414
groupe de purification d'eau 6747
groupe de support logistique 3291
groupe de transport amphibie 6336
groupement 2106, 2108, 6055
groupement blindé 2227
groupement d'emploi du génie 6067
groupement interarmées 6057
groupement logistique 2944
groupement mécanisé 2230
groupement naval d'assaut 3410
groupement naval de plage 3411
groupement tactique 396
groupe moteur nucléaire 3510
groupe motopropulseur 3810
groupe porte-avions 712
(se) grouper 2107
groupe sanguin 501, 6421
grue 1061
gué 1859
guéable 1860
guérilla 6686
guérite 5193
guerre 6668, 6678
guerre bactériologique 6685
guerre biologique 6680
guerre chimique 6681
guerre civile 6669

guerre conventionnelle 6682
guerre de mouvement 3292
guerre des mines 6688
guerre électronique 6684
guerre limitée 6670
guerre non-classique 6691
guerre psychologique 6689
guerre subversive 6690
guerre sur terre 6687
guerrier 6700
guerrilla urbaine 6692
guêtre 1975, 5554
guetteur 2953
guidage 2120, 2121
guidage de fin de trajectoire 6111
guidage en vol 3181
guidage par inertie 2449
guidage par laser 2722
guidage par télécommande manuelle
 3058
guidage passif 3906
guide 2124, 3806
guider 2122, 2123
guidé sur faisceau 430
guidon 1867, 1921, 2125, 2169
guindeau 6859
gymnase 2138
gyro-compas 2141
gyro-stabilisateur 2146
gyroscope 2144
gyroscopique 2145

H

habilitation de sécurité 5163
habitable 2147
habitacle 470, 848
hache 331
hachures 2149
haie 2234
haler 2199
Halte! 2157
hangar 2179
harcèlement 5594
harceler 2181
harnais 2191
harpon 2192
(au) hasard 4494
hâter 1540
hausse 342
(angle de) hausse 1446
hausse à charnière 5363
hausse à oeilleton 5365
hausse de combat 5362
haussière 2203, 3824
haut 2259

haut-fond 5247, 5261
haut-parleur 2970
haut commandement 2261
(de) haute mer 2268, 5109
(à) haute résistance 2229
(à) haute tension 2270
hauteur 2236
hauteur barométrique 4555
hauteur de largage 2238
hauteur de sécurité de retombée 2240
hauteur d'explosion convertie 2243
hauteur d'ouverture de parachute 2239
hauteur libre 3768
hauteur minimum d'explosion 3239
hauteur optimum d'explosion 2242
(à) haute vitesse 6592
havresac 2680
hayon 5983, 5984
hélice 4340, 5099
(à) hélices jumelles 6415
hélicoptère 97, 2245
hélicoptère armé 2136
hélicoptère de manœuvre 6541
héliporté 2244
hélitransfert 6606
hélitransport 2246, 2248
hélium 2249
(en) hérisson (défense) 158
hermétiquement 2254
heure-P 3991
heure de départ 5023
heure de Greenwich 2069
heure de réception 6181
heure de sortie d'un document 6180
heure du bord 5292
heure du fuseau horaire 6956
heure H 2255
heure limite 1122
heure locale 2922
heure N 3453
heure prévue d'arrivée 1516
heure prévue de départ 1517
heures hors service 5403
heure sur l'objectif 6182, 6183
heure zulu 6960
(se) heurter (contre) 633, 857
hinterland 2277
hisser 2284
hiver 6874
hivernal 6875
homme(s)-grenouille(s) 1915
homme de barre 2253
homme de base 3081
homme de quart 6732
homme de veille 2954
homme politique 4179
hommes de troupe 4525

homogène 2306
homologue 3691
hôpital de campagne 2332
hôpital de l'arrière 2330
hôpital de l'avant 2333
hôpital militaire médico chirurgical 2334
horaire 5064, 6184
horizon 2319, 5450
horizon apparent 2320
horizontal 2323
horizon visible 2322
horizon vrai 2321
hors bord 3743
hors de portée 4509
hors d'état de servir 6517
hors d'usage 5211
Hors limites! 3609
hospitaliser 2335
hostile 2337
hostilités 2340
houle 5905, 5945
houleux 4952
hublot 5103
huile 3619
humide 2354
humidité 2355
humidité relative 4669
hydraulique 2364
hydravion 5119
hydrofoil 2365
hydrogène 2366
hydroglisseur 2371
hydrographie 2369
hydrographique 2368
hydrophone 2370
hydrostatique 2373
hypersonique 2376
hypsométrie 2378

I

ICBM mirvé 6577
iceberg 2381
"Ici.." 6150
idée de manoeuvre 3751
identification 2390
identification amie ou ennemie 2391
(contrôle automatique pour) identification à vue 6638
identification radioélectrique 4460
(qui ne peut pas être) identifié 6484
(s')identifier 2392
identité 2393
ignifugé 1722, 1753
île 2589

illégal 2397
illimité 2434
illuminateur laser 2723
illumination du champ de bataille 406
image 2401
imbibé 6744
immédiat 2404
immersion 2410
immunisation 2412
impact 2413
impact normal 3480
impasse 5646
imperméable 6746
imploser 2416
implosion 2417
impracticable (route) 2415
impropre à la navigation 6515
improvisé 2421
improviser 2420
impulsion 4373
impulsion électromagnétique 1437
(à) impulsions 4372
inapte 3473
inapte au service 6482
incassable 6439
incendiaire 2425
incendie 1700
incident 2426
incident nucléaire 3508
inclinaison 1267, 5457
incliné 2429, 5458
incliner 2427
(s')incliner 5280, 5456, 6171
incursion 2432
indéchiffrable 6454
indéfendable 2433
indéfini 2434
indentation 4619
indépendant 2435, 5194
indicateur 2438
indicateur de position-sol 2439
indicatif (d'appel) de réseau 5376
indicatif (d'appel) international 5375
indicatif (d'appel) visuel 5378
indicatif d'appel 674, 5373
indication 4556
indice d'octane 3566
indiquer 4168
indiquer sur la carte 3084
indiquer exactement 4021
indirect 2442
indiscipliné 6478
indispensable 1511
induction 2446
induire 2444
inégal 4953
inerte 2447

lanterne 4992
lapider 5733
largable 2628
largage 109, 1382
largage à haute altitude ouverture à
  basse altitude 3847
largage de la charge 4678
largage en chute libre 1900
largage lourd 4085
large 603, 6845
(au) large 3608, 3616
largeur 570, 6847
larguer 1383, 2626, 4674
larguer les amarres 720
laser 2719
latéral 2728, 5351
latitude 2729
latrine 2730
latte 397
lavage de cerveau 6727
laver 6725
laver (à grande eau) 2329
légal 2771
légende 2657, 2772
législation militaire 3197
légitime défense 5174
lent 5482
lentille 2774
(dans) les forces armeés 5208
lest 352
(en) lest 2805
létal 2776
létalité 2777
leurre 1147
leurrer 1146
levé à la planchette 5924
levé d'itinéraire 5925
levée 5921
lever 2285, 2800
lever du soleil 5844
lever l'ancre 6809
lever le blocus 4486
levier 2784
levier d'armement 2165
levier de commande 984, 3650
levier de pointage 2175
(à) l'horaire 5068
liaison 2787, 2875, 2877
libération 4677
libérer 1280, 2788, 4676
liberté 1899, 2789
libre 1896
(zone) libre 6508
libre (itinéraire etc.) 824
libre de glace 2386
lier 468, 2724, 6165
lieu de franchissement 5417

Lieutenant de tir 3586
(en) ligne 2845
(hors) ligne 2844
ligne (de front) 2855
ligne avant 2869
ligne avant des forces amies 1881
ligne d'arrêt 2834
ligne de bataille 2830
ligne de compte-rendu 2850
ligne de conduite 4178
ligne de coordination des feux 2833
ligne de débouche 2860
ligne de flottaison 6743
ligne de foi 2984
ligne d'égale intensité (radioactive)
  2839
ligne de mire 2864
ligne de Plimsoll 4087
ligne de position 2848, 2863
ligne de projection 2861
ligne de référence 2849
ligne de sauvetage 2799
ligne de sécurité 2829, 2841
ligne de sécurité nucléaire 2842
ligne de sûreté 2837
ligne de tir 2835, 2862
ligne de visée 2864
ligne discontinue 2838
ligne d'observation 2843
ligne internationale de changement de
  date 2543
ligne isogone 2840
ligne médiane 742
ligne principale de résistance 4775
lignes de communication 2870
ligne tireur-but 2836
limite 545, 2821, 2822, 3117
limite avancée de la zone de bataille
  1880
limite d'action 2824, 2825
limite de bond 2847
limite de résistance 4112
limite des eaux territoriales 6154
limite de sécurité 5017
limite des neiges éternelles 5512
limite de tir 2826
limite forestière 6174
limiter (à) 2823
linéaire 2856
liquide 2878
liquide pour freins 559
liste 2881, 4922
liste de contrôle 777
liste d'objectifs 2884
lit 437
lit de marée 6164
littoral 2889

livrer 1189, 1606
local 2918
localisation radiogoniométrique 1741
localiser 2920, 2923
loch 2939
locomotive 2936
logarithme 2942
logement 14, 465, 4404
loger 13
logiciel 5518
logistique 2943, 2947, 5209
lointain 4693
lois de la guerre 2739
long 2948
(le) long du (bord, etc...) 162
longer 5446
longer la côte 842
longeur d'ondes 6767
long feu 2180
longitude 2949
longitudinal 2950
longueur 2773
longueur de colonne 4903
lot 2968
loupe 3006
lourd 2226
loxodromie 2851
lubrifiant 2985
lubrification 2987
lubrifier 2986
lueur de bouche 3387, 1766
lumière 2803
lune 3318
lunette à infrarouge 5504
lunettes (protectrices) 2039
lunettes fumées 5842
lutte ASM 6679
lutter 403

M

machine 2991
machine à écrire 6422
madrier 6173
magasin 2995, 5741
magasin à munitions 4415
magasin d'habillement 5743
magnétique 2999
magnétophone 6019
magnétoscope 6622
mailler 5235
maillon d'attache 5236
main 2159
(à) main 2161
main d'œuvre 3053, 3054

main sur main 2173
maintenance (de matériel) 3020
maintenir 3018
maintenir en attente 2288
maintenir l'ordre 4176
maintien de l'ordre 3023, 4873
-maître (suffixe) 3105
maître d'équipage 542
maître timonier 4402
maîtrise 884, 983, 3109
maîtriser 980, 981, 3103
majeur 3028
major de cantonnement 3576
malade 3910, 5348
(se faire porter) malade 4747
malade assis 6665
malade couché 2888
malade non couché 6665
maladie 1285
maladie des rayons 4445
(ayant le) mal de mer 5128
mal des caissons 455, 1143
mâle 3032
mal entrainé 6473
mal fonctionner 3033
mamelon 2684
manœuvrer 2162
manche à air 6856
manche à balai 2634, 5723
manchon 5462
mandat d'arrêt 6720
mandat de perquisition 6721
manette des gaz 6155
maniable 3045
maniement d'armes 2170, 6291
(habilité au) maniement des armes 5434
manier 2163
manifestant 1200
manifestation 1198
manifeste 3049
manifeste maritime 3564
manifester 1197
manipulateur 2659
manipulations de trafic (déception) 3051
manipuler 3647
mannequin largué avant le stick (troupes aéroportées) 6857
manoeuvre 3042, 3043, 3047
manoeuvre d'abordage 3655
manoeuvre de dépassement 3902
manoeuvre d'enveloppement 1489
manoeuvre de recueil 3902
manoeuvre d'évasion 3044
manoeuvre d'évitement 1528
manoeuvre en perroquet 3796

manoeuvre en tenailles   4018
manoeuvrer   2167, 2168
manoeuvre retardatrice   1185
manquant   8
manque   1166, 5331
manuel   3057
manutentionner   2163, 3048
maquette   3296
marais   3090, 5940
maraudeur   3066
marchand   3158
marche   3067
(en) marche   6476
marche (en avant)   2219
marche à l'ennemi   3360
marche arrière   2007
marche d'approche   235
marche forcée   1856
marche lente   2981
marcher   3068, 6663
marcher en pointe   5555
mare   4182, 6740
marée   6163
(à) marée   6161
marée basse   2982
marée descendante   1410
marée de vive eau   5611
marée haute   2271
marge   3070
marginal   3071
marin   3073
marine de guerre   3421
"marines"   3074
maritime   3073, 3075
marquage de sécurité   2852
marque de mer   5118
marque de Plimsoll   4087
marquer   3076
marquer à la craie   758
marqueur   3077
marqueur de distance   3082
marqueur d'extrémité de trouée de mine
   3078
marqueur laser   2721, 3080
masque   1070, 3056
masque à gaz   1997, 4781
masque à oxygène   3801
masquer   3095, 5092
massacre   3099
masse   3098
masse critique   1075
mat   3112
mât   3101
mât de charge   1219
mât de drapeau   1749
matelot   5024
matelotage   5117

matelots et petits grades   4535
matériau   3114
matériel   1498, 2188, 3116, 3722
matériel cryptographique   1087
matériel de guerre   252
matériel de pontage   599
matériel non-consommable   5861
matériel roulant   4931
matière   3114
matraque   394
mauvais fonctionnement   3034
maximum   3117
mazout   3622
mécanicien   3129
mécanicien-ajusteur   1737
mécanicien navigant   1790
mécaniser   3138
mécanisme   3130
mécanisme d'autodestruction   3134
mécanisme de culasse   3131
mécanisme de mise à feu   3133
mèche   1945
mèche lente   1966
médecin   1337
médecine de l'air   59
médecine préventive   4278
médecin militaire   3592
médecin militaire aéromédical   1794
médecin parachutiste   3860
médian   3141
médical   3144
médicament   3147
mégahertz   3153
mégatonne   3154
mégawatt   3155
mélange   3287
membrane   1258
membre   2819
"même portée!"   4714
Mêmes éléments!   4710
mémoire   3157
mémorandum   3156
mémoriser   5050
menace   6151
menacer (quelqu'un de)   6153
mener   2755, 2756
meneur   4870
(passer les) menottes (à)   2160
menton   786
(à la) mer (homme)   3760
mer   5104
mer-mer   5903
mercure   3160
mer houleuse   2232
méridien   3161
message   3163
message météorologique   3172

mess des officiers   3162
mesure   3128
mesure d'interdiction   1202
mesurer   2001, 3127
mesures de déception radio   4452
mesures de défense contre les mines
   3226
mesures de securité pour les troupes
   dans la zone d'alerte immédiate   6373
métal   3169
métallique   3171
météorologie   3175
météorologique   3173
mètre (linéaire)   3176
métrique   3177
(se) mettre à couvert   1050
mettre à la ration   4537
mettre à la torture   6205
mettre à l'eau   2734
mettre au point   1832
mettre au rebut   4664
mettre en cocon   3334
mettre en fusion   1946
mettre en ligne   2871
mettre en marche   5667
(se) mettre en panne   2225
mettre en place (préalablement)   4260
mettre en pool (en commun)   4184
mettre en réseau   3437
mettre en sûreté   5153
mettre hors d'état de fonctionner
   4698
mettre le cap (sur)   2210
mettre le cap vers   5225
meurtre   3375
meurtrier   2775, 3376
meurtrière   2955, 5477
mi-marée   2154
microfilm   3178
microphone   3179
microprocesseur   3180
midi   3478
milice   3200
milieu   3182
(au) milieu   180
milieu de chenal   3184
militaire   3191, 5202, 5208, 5215
(de) militaire   5526
militaires de tous grades   4526
militant   3190
mille marin   3188
mille terrestre   3189
millième   3187
mine   3201
mine (acoustique)   3203
mine (terrestre)   3219
mine à dépression   3223

mine à détonateur auxiliaire   3204
mine à effet dirigé   3209
mine à effet horizontal   3217
mine anti-personnel   3206
mine antipersonnel   3207
mine à retard   3210, 3212
mine armée   3205
mine à tête chercheuse   3216
mine autopropulsée   3222
mine de fond   3208
mine dérivante   3214
mine inerte   3218
mine magnétique   3221
mine neutralisée   3213
miner   3202
mine rampante   3211
mineur   3244
mine ventouse   3220
minimiser   3236
minimum   3238
Ministère de la Défense Nationale   3243
minute   3245
minuterie   3136
mire aérienne   5360
miroir   3247
mise au point   1251
mise en batterie   4216
mise en garde   6706
mise en place   4216
missile   3252, 3263
missile anti-missile   3258
missile anti-sous-marin lancé d'un
   aéronef   3256
missile antiaérien   3257
missile à ogive à charges multiples et
   indépendantes   6576
missile auto-guidé   3271
missile balistique   3260
missile balistique à moyenne portée
   3268
missile balistique de portée intermédiaire
   3266
missile balistique intercontinental   3265
missile de croisière   3261
missile de guidage automatique   1705
missile filoguidé   3274
missile 'lance et n'y pense plus'   3267
missile libre   3262
missile sous-marin-air   3275
missile surface-surface   3273
missile volant au ras de l'eau   3269
mission   3277, 3279
mission clandestine   3656
mission de conservation de terrain
   3280
mission de couverture éloignée   3795
mission de recherche   3281

(de) mission réservée   1149
mission sur demande urgente   3278
mitrailler   5753
mitrailleur   2132
mitrailleuse   2992
mitrailleuse légère   2813
mixte   3286
mobile   3289, 3353
mobilisation   3294
mobiliser   3295
mobilité   3293
mobilité opérationnelle   3675
mobilité tous terrains   407
mode de transport   3301
modèle   3297
modèle de série   3300
modeler (sur)   3299
modéré   3303
(se) modérer   3302
moderne   3304
modernisation   3305
moderniser   3306
modification   179, 3307
modification retrospective   4819
modifier   163, 3308
modulaire   3309
modulation   3310
modulation d'amplitude   3311
modulation de fréquence   3312
modulation par impulsions et codage
   3313
moniteur   3315
mono-   5411
montagne   3345
montagneux   3346
montant   5835, 6208
(en) montant   6528
monté   3347
monter   278, 1734, 1802, 3341, 3342
(se) monter (à)   6207
monter la garde (sur)   2653
monticule   2684
montre   6730
moral   3324
moratorium   3325
morse   3327
mort   722, 1121
mortes eaux   3422
mortier   3328
mort naturelle   3403
mosaique   3331
mosaique sommaire   6450
mot code   6895
mot de passe   3908, 6734, 6896
moteur   1472, 3336
(à) moteur   4243
moteur à réaction   2623

moteur auxiliaire   320
moteur en marche   4991
moteur hors-bord   3744
motif de camouflage   1308
motocyclette   3338
motorisé   6587
mou   5453
moufle   4367
mouillage   192
mouillage auxiliaire   1454
mouillage d'attente   2293
mouillé   6827
mouiller   456
mouiller (corps-mort; mine)   5413
mouilleur de mines   3233
moulin à vent   6860
mousse   1828
mousson   3317
moustiquaire   3333
moustique   3332
moutons   6841
mouvement   3356
mouvement clandestin   6464
(faire un) mouvement de lacet   6918
mouvement en navette   5347
mouvement navire-rivage   5309
mouvement par voie aérienne   125
mouvement rétrograde   4820
mouvement tournant   6407
moyen   326, 3122, 3148
moyen d'aide au combat de nuit   3460
moyenne   325
moyens   1571, 3124
moyens de lancement   1191
moyens de transport   908
moyeu   2350
mulet   3363
muletier   3364
multi-usages   3373
multilatéral   3368
multinational   3369
multiple   3370
multiplex   3372
munition   182
munition à douille séparée   5180
munition encartouchée   1743
munitions   3374
munition spéciale   5558
munition traçante   6227
mur   6666
mur du son   5540
musette   2679
muter   6303
mutilé   3381
(se) mutiner   3383
mutinerie   3382
mutuel   3384

N

nacelle 4104
nadir 3389
(au) napalm 3390
nation 3394
national 3395
nationalité 3398
naturel 3401
naufrage 5310, 6905
naufragé 719
nautique 3405
naval 3407
navigabilité, certificat de 143
navigable 144, 3413
(non) navigable 6507
(officier) navigateur 3420
navigation 3415
navigation à l'estime 4575
navigation astronomique 733
navigation hyperbolique 2374
naviguer au schnorkel 5505
(en état de) naviguer 3414
navire 5290
navire(s) marchand(s) 5300
navire-piège 5294
navire auxiliaire 321
navire de guerre 6724
navire de relevage 6614
navire détaché d'un convoi 2766
navire de transport et de débarquement 6328
navire gros porteur 5297
navire ralliant un convoi 995
navire ravitailleur 3335
navire se détachant d'un convoi 996
nécessaire de modification 2676
nécessiter 4756
négatif 3425
négociateur 3432
négociations 3431
négocier 3429
neige 5507
neiger 5508
nerf 3434
net 5249
netteté 1172
nettoyage 3323
nettoyer 819
neuf 6525
neutralisation 3448
neutraliser 3449
neutre 3447
neutron 3450
nid de mitrailleuses 3436, 4035
niveau 2778
(de) niveau 2779

niveau à bulle 2781
niveau de la mer 5115
niveau de l'eau 6742
niveau de maçon 6387
niveau des approvisionnements 2783
niveau de stocks 5729
niveau moyen de la mer 3125
niveau opérationnel 3674
niveler 2048, 2780
niveleuse 2051
nœud 2283, 2685
nœud coulant 5473
nœud de chaise 548
no man's land 3468
nombre 3520
nombre de mach 2993
nombre impair 3569
nominal 3469
nommer 3470
nommer officier 896
non-belligérant 3472
non-contaminé 6448
non-contrôlé 6449
non-éclaté 492
non-protégé 2480
non armé 6433
non autorisé 6436
non classifié 6444
non corroboré 6452
non défendu 6455
non disponible 6438
non explosé 6481
non gardé 6449
non levé 6523
non vu 2967
(du) nord 3482
normal 3479
normaliser 3481
norme 5650
notation 3489
notes annuelles 4734
notification 3492, 3498
nouer 2686
noyau du réacteur 4548
noyautage 2454
noyauter 2453
noyé 2409
(se) noyer 1386
nuage 835
nuage radioactif 3503
nucléaire 3500
nucléaire, biologique et chimique (NBC) 3501
nuit 3455
numération 3489
numérique 1265
numéro 3521

numéro-repère 760
numéro de fabrication 5199
numéro de référence 4623
numéro de série 5199
numéro de téléphone 6085
numéro d'ordre 5198
numéroter 3518
(en) nylon 3525

O

obéir (à) 3528
objecteur de conscience 3535
objectif 2774, 3532, 6022
objectif à battre à l'horaire 6037
objectif à la demande 6031
objectif de priorité absolue 4289
objectif immergé 6040
objectif initial 2038
objectif inopiné 6027, 6032, 6050
objectif linéaire 6028
objectif mouvant 6030
objectif non ponctuel 6024
objectif peu résistant 6039
objectif ponctuel 6034
objectif prévu 6033
objectif résistant 6026
objectif secondaire 6038
objection 3531
objet 3530
objet d'un document 5805
obligatoire 3041
oblique 3536
obscur 1105
obscurcissement 3537
obscurité 1107
observateur 3541
observateur avancé 1883
observation 3538
observation du tir 5602
observatoire 4230
observer 5596, 6731
obstacle 3544
obstacle actif 3553
obstacle artificiel 3545
obstacle dans la zone arrière 3551
obstacle de série 3552
obstacle existant 3546
obstacle infranchissable 3548
obstacle naturel 3549
obstacle passif 3550
obstacle passif de flanc 3547
obturateur 5345
obus 5263, 5268
obus à éjection par le culot 5267
obus anti-char 5265

obus chimique 5269
obus de mortier 3330
obus éclairant 5271
obus explosif à tête d'écrasement 5270
obus fumigène 5272
obusier 2348
obusier compact léger 2349
obus marqueur 5273
obus perforant 5266
obus perforant à sabot détachable 5004
occasion 3689
(à) occultations 3556
occupation 3557
occuper 3559
(s')occuper de 2166
occuper le terrain 2292
occuper une position 3560
océan 3561
océanographie 3565
(d')octane élevé 2267
octant 3567
oculaire 1569, 3568
oeil (yeux) 1567
oeilleton de visée 5368
offensif 3570
offensive 3571
officiel 3606
(non) officiel 6509
officier 3573
officier-adjoint 43
officier accompagnateur 3582
officier adjoint 3583
officier auto 3594
officier commandant 890
officier d'approvisionnements 3602
officier de carrière 5525
officier de jour 3603
officier de la garde 3604
officier de liaison 3591
officier de liaison (air) 3574
officier de liaison de l'armée de terre 3587
officier de marine 3595
officier de permanence 3581, 3598
officier de quart 3605
officier de renseignement 3589
officier d'état-major 3600
officier d'ordinaire 3578, 3593
officier du 3e Bureau 3597
officier du matériel 3599
officier instructeur 3588
officier observateur d'artillerie 3585
officier régulateur 3575
officier régulateur de plage 416
officier subalterne 3590, 3601
officier supérieure navale 3584
ogive 3487, 6697

ouverture automatique   3643
ouverture de route   4899
ouverture du faisceau   6848
ouvrage à destruction réservée   6036
ouvrir   3636
ouvrir (à clef)   6504
ouvrir la marche   2754
ouvrir le feu   3640
ouvrir une brèche   569
oxygène   3799
oxygène liquide   2983

P

pagaie   3818
pagayer   3817
paillet   3111
paix   3929
palan   499
palans   5970
pale   481, 6549
pale de rotor   4951
palette   3825
palettiser   3826
palmes   1795
panier   392
panne   576, 1580
(en) panne   3713
(être en) panne   577
panneau   1047, 3828, 3829, 5396
panneau de marquage de cheminement
   3079
panneau de sauvetage   1506
panneau de signalisation   3832
pannier de bât   3835
panoramique   3836
panser   468
papiers de bord   5291
paquebot   2868
paquetage   3807
paquet de pansements   1614
par-dessus le bord (jeter qch.)   3760
parachutage   3845, 3854
parachutage à ouverture retardée   3846
parachute   3837
parachute à matériel   3840
parachute à ouverture automatique
   3844
parachute à ouverture commandée
   3843
parachute de freinage   1380, 3839
parachute extracteur   3842
parachuter   3838
parachute ventral   3841
parachutiste   3848
parader   3849

parados   3853
paraffine   3855
parafoudre   2815
parallaxe   3856
parallèle   3857
paramilitaire   3861
parapet   3862
parasites   290, 5677
parasites électriques   2534
paratonnerre   2816
paravane   3866
parc automobile   1774
par deux   6417
pare-...   5282
pare-brise   6865
pare-chocs   634, 1604
parement   1572
parer   1602
parité   1499
parka   3868
Parlez!   3759
paroi   6666
parole   3871
part   3873
(ligne de) partage des eaux   6753
parti   3879
particularité du terrain   1597
particule   3876
partiel   3875
partie platte   2782
partir   1207
(faire) partir   1292
partisan   2118, 3877
pas   3802, 6347
(au) pas   5716
pas (d'une vis etc...)   4038
pas accéléré   4412
passage   3895, 3899
passage à gué   6657
passage à niveau   2050
passage en plongée   5811
passage en submersion   5811
passage inférieur   6467
passager   3903
passe navigable   6750
passer   1076
passer à gué   6656
(faire) passer à gué   1858
passerelle   1826
passerelle (de commandement)   596
passerelle (d'embarquement)   1979
passer par le contrôle   739
Passez!   3904
patente de santé   466
patient   3910
patin   5430
patin de chenille   6248

pied 1849
(mettre) pied (à terre) 1290
(sur le) pied de guerre 6693
piège 6339
pierre 5732
pieu 517
pile 401, 4279
pile (de pont) 4002
piller 2961, 4099, 5008
pilotage 4011
(en) pilotage sans visibilité (en PSV) 493
pilote 4008, 4010
pilote automatique 314
pilote d'essais 6137
piloter 4009
pilule 4004
pinceau lumineux 3939
pionnier 4024
piqué 3488
piquer 1320
piquet 3995
piquet d'honneurs funèbres 3882
piquet d'incendie 3996
piquet radar 3997
piste 4843, 6230, 6233, 6237
piste (d'envoi) 4993
piste balisée 1764
piste d'atterrissage 5790, 5793
piste de roulement (de circulation) 6063
piste hostile 2339
piste significative 6240
piste temporaire d'envol et d'atterrissage 5792
pistolet 4029
pistolet de signalisation 4032
pistolet graisseur 2067
pistolet lance-fusées 4030, 4033
pistolet mitrailleur 4031, 5807
pivot 4040, 4041
pivotant 4043
pivoter 5954
(faire) pivoter (sur) 4042
place forte 5799
placer une garnison (ville) 1986
plafond 731
plafond de vol stationnaire 2346
plage 413, 1866
plage arrière 4400
plaignant 4344
plaine 4044
plan 1365, 2778, 4047, 4063, 4064
plan-maître 4058
planche 4067
plancton 4068
plan d'arrimage 5751

plan d'atterrissage (troupes aéromobiles) 4056
plan de chargement 4057
plan de comparaison 1113
plan de contingence 4049
plan de convoi 5065
plan de déception 4051
plan de destruction 5069
plan de dissimulation 4050
plan d'emploi des feux 4053
plan d'enlèvement 4052
plan de pose de mines 3232
plan de référence 4622
plan des tirs observés 3540
plan de stockage 1395
plan de tir restrictif 4061
plan de vol 4054
plan directeur 3062, 3686, 5423
plan d'opération 4059
planer 2034, 4062
planer (moteur arrêté) 843
planète 4065
planeur 2037
plan fixe horizontal 5987
planifié 4069
planifier 4048
planning 5064
plan tactique au sol 4055
planton 3167, 3718
plaque de base 4078
plaque de blindage 4077
plaque de couche de crosse 4079
plaque d'identité 5981, 5982
plaquettes de brouillage 754, 4944
plasma 4073
plastic 2265
(explosif) plastic 4075
plastifiant 4076
(matière) plastique 4074
plastique renforcé de fibre de verre 2032
plat 1772, 5245
plate-forme 4084
plate-forme de lancement 3814
plate-forme porte-char 6335, 6584
plateau 4083
plateau continental 972
plateau de pression 4082
plateforme de poser (d'hélicoptères) 2247
plateforme petrolière 3624
platine 2931
plein bateau (personnel) 512
plein camion 6383
(en) pleine forme 1735
pleine lune 1930
plein jour 1117

programmeur 931, 4319
progrès 4323
progressif 4325
progressiste 4324
projecteur 1808, 5127, 5600
projectile 3251, 4329
projectile à fusée 4913
projectile à mitraille (d'habitude terme
  anglais) 837
projectile laser-guidé ou vidéoguidé
  5487
projection de mercator 3159
projet 4328
projeter 4048
prolongation 1563
prolongation (temps) 4333
prolongement (espace) 4333
prolonger 4332
promontoire 507, 2212, 4335
promotion 4336
propagande 4337
propergol 4339
propergol liquide 2879
propre 818
propulser 4338
propulseur 3337
propulseur auxiliaire 537
propulsion 4342
propulsion à fusée 4915
protecteur 4347
protection 4346, 5087, 5096
protection civile 802
protection contre l'observation 1046
protège-embruns 5588
protégé et alerté 6704
protéger 5090
protéger (contre) 4345, 5283
protéger les flancs 1758
protocole 4349
proton 4350
prototype 3298, 4351
provisoire 2535, 6097
prudence 727
Prudence! (international) 3237
psychiatrique 4360
psychologique 4361
public 4362
public-relations 4364
publication enregistrée 4643
publier 4676
puisard 5839
puissance 3517, 4239, 4244, 6921
puissance au frein 4241
puissance de combat 872
puissance de feu 1721
puissance nucléaire 3509
puits 5241, 6820

pulsoréacteur 4374
pulvérisation 5604
punir 1282, 4380
punissable 4381
punitif 4383
punition 4382
purger 489
pylône 3102
pylône pour roquettes 4417

Q

quadrant 4392
quadrillage 2076
quai 4001, 4407, 6830
quai flottant 1797
qualification 4394
qualité de transmission 5776
qualité d'un carburant 2049
quantum 4397
(en) quarantaine 4398
quart 6729
quart de 4h à 8h 3326
quart de minuit 3185
quart de nuit 3464
quartier général 2214
quille 2651
quote-part 4413
quotient 4414

R

raccord d'élingue 5938
rad 4418
radar 4419
radar à balayage électronique 4427
radar aéroporté à antenne latérale 4429
radar à fréquence variable 4424
radar anti-mortier 4426
radar d'acquisition 4420
radar d'acquisition à ondes entretenues
  4422
radar d'alerte avancée 4421
radar de recherche 4428
radar de veille 4430
radar d'observation verticale 4425
radar Doppler 4423
radar oscilloscope 2441
rade 4890, 4904
radeau 4476
radeau de sauvetage 4477
radiale 4442
radiateur 4446

T

tenue de cérémonie   1368
tenue de combat   866, 1613, 2673
tenue de combat complète   3706
tenue de corvée   1595
terme de procédure   4359
terminal   6106, 6109
terminal éloigné   4697
Terminé!   3742
(se) terminer   6113
terminus   6105
terminus maritime   6755
terrain   6115
terrain (d'aviation) opérationnel
   3670
terrain accidenté   2091, 2276
terrain d'atterrissage   134
terrain d'aviation   85, 111
terrain de dégagement   167
terrain de déroutement   170
terrain de manoeuvres   2089
terrain de redéploiement   4610
terrain de secours   1453
terrain d'essai   2090
terrain en surplomb   6116
terrain practicable   2093, 6118
terrain principal   3011
terrassement   1407
terre   1405, 2694, 5519
(à) terre   273
(de) terre   2082
(par voie de) terre   1406, 3771
terre-plein(s)   6114
terre molle   2092
territoire   6123
territoire occupé   3558
terroriser   6126
terrorisme   6124
terrorisme urbain   6537
terroriste   6125
tertre   3340
test   6129
test d'aptitude à l'engagement   6131
test d'aptitude opérationnelle   6133
test de cooper   6130
tester   6128
tête   2207, 2209
tête chimique   6695
(obus explosif à) tête d'écrasement
   5620
tête de ligne   4482
tête de pont   415, 598
tête de pont aérien   115
tête militaire   6699
texte en clair   4046
théâtre (d'opérations)   247, 6138
thème (à la langue étrangère)   6315
théodolite   6139

thermique   6140
thermoguidé   2223
thermomètre   6145
thermonucléaire   6146
timonerie   6836
tir   1645
tir(s) de renforcement   1693
tir à balles réelles   2892
tir à bout portant   1685
tirage   4284
tirailleur   5444
tir à la fourchette   1651
tirant d'air   3768
tirant d'eau   1352
(à égal) tirant d'eau   1529
tir anti-aérien   1751
tir au canon   2133
tir au radar   1688
tir à vue   1663
tir balayant   1689
tir continu   1655
tir contre la terre   5326
tir d'accompagnement   1647
tir d'arrêt   1666
tir de balisage   1677
tir de barrage   380, 1650
tir d'écharpe   1680
tir de concentration   1654
tir de contre-batterie   1657, 1658
tir de contrepréparation   1659
tir de destruction   1662
tir de destruction (de neutralisation)
   1025
tir défensif   1692
tir d'efficacité   1717
tir de flanquement   1667
tir de fouchage   1678
tir de harcèlement   1670
tir de neutralisation   1679, 1696
tir d'enfilade   1665
tir de préparation   4257
tir de protection   1660
tir de ratissage   1653
tir de régimage   1652
tir de réglage   1648, 1690, 1691
tir d'essai   1699
tir d'interdiction   1674
tire-feu   2717
tir échelonné   1694, 4522
tir éclairant   1672
tirer   1644, 4366, 5321
tirer à plein fouet   1720
tirer au flanc   3035
tirer à vue   1682
tirer long   3789
tirer par rafales   6645
tirer plus près   1384

U

veste   2603
vêtement de pressurisation   4273
vêtement protecteur   4348
vêtements   834
vétérinaire   6616, 6617
viabilité   6257
viaduc   6618
victoire   6619
(à) vide   6543
vide de citerne   6426
vieillissant   3542
(en) vigueur   1855
vigilance   6624
vigilant   6625
village   5228, 6626
ville ouverte   3639
virage   454
virage conventionnel   4302
virer   367, 453, 6562
virgule   4121
vis   5099
visée   5359
viser   81, 82
viseur   532
viseur à lunette   5503
viseur infrarouge   6797
viseur reflex   5367
viseur tachymétrique   5369
viseur télescopique   5370
visibilité   6628
visible   6629
visière   6633
vision nocturne   3462
vision stéréoscopique   5718
visite   5123, 6631
visite médicale   3852
visiter   5121, 6632
visser   5098
visuel   6634
vital   6640
vitesse   2004, 5566, 6588
vitesse à l'arrêt de propulsion   6590
vitesse ascensionnelle   4532, 5567
vitesse de convoi   5568
vitesse de croisière économique   5569
vitesse de largage   5571
vitesse de libération   6591
vitesse de marche   3803
vitesse de progression   4531
vitesse du son   5575
vitesse du vent   5573
vitesse en fin de combustion   6589
vitesse en fin de parcours   6595
vitesse indiquée   2436
vitesse initiale   6593
vitesse maximale permise   5574
vitesse maximum de croisière   5572

vitesse minimale de manoeuvre   5714
vitesse propre   6384
vitesse relative   132, 6594
vitesse sol   5570
vivres   4355
vocal   6641
voie   2715, 6230, 6264
voie (aérienne)   2713
voie aérienne principale   4970
voie d'eau   6759
voie de garage   5355
voie d'évacuation d'urgence   4965
voie ferrée   6238
voie hiérarchique   756
voile blanc   6844
voiler (lumière)   5237
voilier   6613
voilure (de parachute)   689
voiture de liaison   5636
voix   6642
vol   1781, 4907
(en) vol   86
vol à basse altitude   1785, 2975
volant   1827, 6833
volant de pointage en direction   2178
volant de pointage en hauteur   2177
vol aux instruments   1784
vol à vue   1788
vol d'essai   6132
volée   6644
vol en rase-mottes   3391
voler   1823
voler en rase-mottes   2235
volet   1761
(aux) volets fermés   655
vol inaugural   1786
volontaire   6648, 6650
vol plané   2033
vol rasant   1783
vol stationnaire   2345
volt   6646
vol tactique   1787
vol tactique en suivi de terrain   1782
voltage   6647
voyage   6652
voyager   3903
(en) vrac   624, 2958
vrille   5581
vue   6623
vue optique   2865
vulnérabilité   6653
vulnérable   6654

W

wagon   6658

INDICE ALFABETICO ESPAÑOL

adelantar 3791
¡Adelante! 3904
adiestramiento 6275
adiestrar 6272
adjuntar 208
adjunto 1217
administración 44
administración (de personal) 3040
administrativo 45
adoctrinamiento 2443
adquirir 4311
adquisición 4312
adquisición del blanco por laser 3080
adquisición de objetivo 6042
adversario 3688
advertir 6702
aéreo 55, 84, 3767
aerodeslizador 2101, 2347, 2365, 6565
aerodinámica 57
aerodinámico 56, 5770
aerofaro 420
aeromédico 58
aeromóvil 123
aeronáutica 61
aeronáutico 60
aerotransportable 128
aerotransportado 87, 88
afilar 5250
afirmar 63
afirmativo 64
aflojar 2960
aflojar(se) 5454
agarre 4384
agarrotarse 5168
agente 71, 1176, 1974
agente antisubversivo 4874
agente biológico 474
agente nervioso 3435
agente químico 781
agente V 6545
agitado 635
agotamiento 1538
agotar 1536
agregación 295, 1077, 5136
agregación habitual 2148
agregado 72, 294
agregado militar 3193
agregar 293
agresivo de hostigamiento 2182
agresivo fugaz 3474
agresivo persistente 3964
agresivo tóxico 6219
agrupación aérea embarcada 711
agrupación conjunta 6057
agrupación de transporte anfibio 6336
agrupación mixta de combate 6056,
6069

agrupación táctica 6055
agrupamiento táctico 6068
agrupar(se) 2107
agua 6735
agua contaminada 4180
agua destilada 1314
agua firma 5455
agua potable 1373
(de) agua salada 5027
aguas arriba 6752
aguas costeras 6748
aguas internacionales 6749
aguas jurisdiccionales 6751
agudo 38
aguja del seguro 4017
agujero 2298
¡ah del barco! 76
aherrumbrarse 4999
aire 84
aire-aire 138
aire-tierra 114, 139
aislado 3088
aislamiento térmico 6143
aislar 3087
ajustador 1737
ajustar 41, 42, 279, 6928
ajuste del giro 2143
ala 6868
ala de control de dirección 6550
al agua (hombre) 3760
alambrada 372, 6883
alambrada plegable 6880
alambrar 6878
alambre 6877
alambre de disparo 6881
alambre de púas 6879
(de) ala replegada 6870
alarma 145
(de) alarma 6705
alarma de ataque nuclear 6711
alarma de incendios 1704
alarma local 6710
alarma temprana 6709
a la voz 6643
alba 1115
álbum de fotos de personas buscadas
4918
alcance 2137, 4496, 4506
alcance de día 4499
alcance máximo eficaz 4507
alcance mínimo 4508
alcantarilla 5231
alcanzar 2278, 3763, 3765, 3790
alcázar 4400
aldea 6626
aleación 157
(de) aleación ligera 2809

alerón 80
alerta 146, 147, 5662, 6706
(estado de) alerta 4554
alerta en tierra 2094
alerta estratégico 6712
alerta general 2015, 4406
(estado de) alerta inmediato 4553
alertar 148, 6702
alerta rojo 4606
alerta temprano 6708
alerta y control lejano por medios
    aerotransportados 89
aleta 1637
aletas 1795
aletazo 1761
alfabeto fonético 3976
alfabeto Morse 3327
alférez 3601
alforja 3835, 5011
aliado 161
alianza 154
aliar(se) 160
alijo 664
alimentación 1598
alineación 152
alinear 151, 2872
alinearse 1582
alistamiento 1481
alistar 5398
alistar(se) 1479, 6649
aliviar 4690
alma 541
almacén 5741, 6677
almacenaje 5739
almacenar 1394
almacén de bombas 524
almirante 3584
alojamiento 14, 465, 4404
alojar 13, 464
alquilar 773
alquitrán 6021
alquitranado 6051
(de) alta graduación 1748
(de) alta mar 2268, 3563, 5109
(de) alta tensión 2270
(de) alta velocidad 2269, 6592
altavoz 2970
alternar 166
alternativa 168
alternativo 169
altímetro 171
altitud 172
altitud de crucero 1084
alto 2259
(dar el) alto 762
¡Alto! 2157, 5735
alto el fuego 729, 2290

alto explosivo 2263
alto mando 2261
(de) alto nivel 2266
(de) alto octanaje 2267
altos 2158
altura 2236
(de) altura 3616
altura barométrica standard 4272
altura de abertura de paracaídas 2239
altura de explosión 2243
altura de explosión libre 2240
altura de la marea 4881
altura de lanzamiento 2238, 2639
altura de transición 6313
altura máxima de nube nuclear 2237
altura mínima de explosión 3239
altura mínima segura 3241
altura óptima de explosión 2242
altura sobre el suelo 2096
altura verdadera 6385
alumbrar 2804, 2806
aluminio 173
alza 342, 1446, 5366
alza abatible 5363
alza abatida 5362
alza aérea 5360
alza de anillo 5368
alzamiento 4882
alza óptica de tirador 5503
alza óptica infrarroja 5504
alza posterior 4565, 5361
alzar 2285, 2800
alzar (con palanca) 2785
amarra 3824, 6716
amarradero 457, 3321
amarrado 1591
amarrar 439, 1592, 3319, 6167
amarre 2725
(materiales de) amarre 6166
amartillar 847
ambiente 174
ambulancia 175, 2332
(de) ambulancia 176
ambulante 1611
amenaza 6151, 6152
amenazar 1052, 6153
amerizaje 5589, 6212
amerizar 149, 1318, 2695, 6211
ametrallador 2132
ametralladora 2992
ametralladora ligera 2813
amianto 272
amordazar 1973
amortiguador 5316
amortiguador de fogonazo 5895
amotinado 4875
amotinarse 3383, 4872

apoyo militar 3192
apoyo mutuo 5887
apreciación de información 2521
apreciación de la situación 1518
apreciar 1514
aprender 2764
aprendiz 6274
apresar 4293
apretado 829
apretar 4366
apretar el paso 4269
aprobar 820, 822
aproche 233
aproches protegidos 1053
aprovisionamiento 6620
aproximación controlada desde tierra
 2098
aproximarse (a) 830
aptitud para la circulación 6257
apuñalar 5621
apuntador 2132
apuntar 82, 2741, 5357
apuntar (a, con) 4109
apuntar a un blanco 6292
apuntar en dirección 6342
apunte 1486, 3156
"aquí. ." 6150
árbitro 6432
árbol 5242, 6349
árbol de levas 682
arbusto 650
arcada (puente) 5547
arca de agua 6757
arco 238
arco frontal 1918
archivar 1629
archivos 1630
ardid 4996
área 241
área de aterrizaje 6951
área de fuego prohibido 3465
área de tiro libre 6946
área surcada 2149
arena 5039
arena movediza 4411
arenoso 5044
arma 250, 562, 6772
arma actuada por gases 6780
arma antitanque 6773
arma automática 4713, 6774
arma biológica 6775
arma colectiva 6777
arma con multiples cabezas dirigidas a
 diferentes objetivos 6576
arma contracarros 6773
arma de carga manual 6791
arma de disuasión 1244

arma de largo alcance 6792
arma de potencia nominal 6786
arma de radiación intensificada 6778
arma de tiro sostenido 6794
armado 253
(bien) armado 6821
arma estratégica 6793
arma guiada 6781
arma individual 6782, 6788
arma inferior a la kilotonelada (término
 inglés) 6779
arma kilotónica 6783
arma ligera 6784
arma megatónica 6785
armamento 251
arma no automática 6789, 6790
arma nuclear 6787
armar 249, 1942
armar (hormigón) 4660
armas 262
armas autorizadas para el disparo 6795
armas controlando el objetivo 6796
armas convencionales 989
armas de cinturón 5353
armas en posición 6798
armas nucleares tácticas 5977
armas portátiles 5485
armazón 1892, 1894
armazón (del fuselaje) 113
armería 261
armero 4416
armisticio 255
armón (de artillería) 2820
aro de seguridad 4369
aro de seguro 4869
aro guía 1378
arpón 2192
arrancar 5667
arranque 5668
arreglar por adelantado 4265
arremeter 6160
arremeter contra una posición 4997
arremetida 6159
arriente de tráfico 6262
arriesgar(se) 4884
arrollar 4933
arroyo 1067, 5768
arrumbar (a) 2210
arsenal 267, 1336
ártico 240
artículo 268, 2598, 2599
artículos de uso común 903
artificial 270
artificios pirotécnicos 4391
artillería 271, 3721
artillería pesada 2228
asa 2164

asaltar 275, 5747
asalto 274
asalto de fuego y maniobra 1625
ascenso 4336
as de guía 548
asegurado 5150
asegurar 5152
asegurar con listones 398
asesinato 3375
asesino 3376
asiento de atrás 4007
asiento proyectable 1433
asignación 156, 284
asignar 155, 283
aspillera 2955
aspirar 81
asta de bandera 1749
astillero 1336, 5311, 6915
asumir el mando 887
atacar 297, 1470
atadura 2725
atalaje 2190, 2191
atalaya 6733
ataque 296, 5786, 6166
ataque aéreo 133
ataque aéreo pedido a corto plazo
 3278
ataque con un disparo nuclear
 5788
ataque de detención 2295
ataque de sorpresa 5499, 5913
ataque diseminado 5605
ataque frontal 1919
ataque precipitado 2193
ataque principal 3012
ataque químico a dosis desprevenida
 5914
ataque teledirigido 6560
atar 468, 2724, 6165, 6717
atar con correa 5759
atasco de tráfico 6263
atenuación 299
atenuar 298
aterrizaje 2698, 6212
aterrizaje con instrumentos 2703
aterrizaje de asalto 2699, 2701
aterrizaje de panza 2702
aterrizar 2695, 6211
aterrorizar 6126
atmósfera 289
atómico 292
átomo 291
atornillar 5098
atracar 456, 6167
atracar al muelle 1333
atracar de costado 3320
atraer con señuelo 1146

atrás 287, 343, 4561
¡Atrás toda! 1933
atraversar a snorkel 5505
atrincherarse 1263
auditor de guerra 2635
aumentador 2522
aumentar 3007, 5717
aumentar (el alcance) 2431
aumento 3005
auriculares 1404, 2216
aurora 1115, 5844
ausencia sin permiso 7
ausente 8
autenticación 307
autenticador 308
autenticar 306
autentificación de red 3442
auto de registro 6721
autodirigido 5176
autoenderezante 5176
automático 313, 5173
automatización 315
automatizar 312
automotor 317
automóvil 316, 4243
automóvil del comando 5636
automóvil del Estado Mayor 5636
autonomía 4497
(de) autonomía limitada 2200
(de) autonomía reducida 4512
autónomo 1705, 2435
autopropulsado 5175
autoridad 309, 4984
autoridad de coordinación 1002
autorización 310, 3960, 3962
autorización para viajar 6722
autorización previa 4290
autorizar 311
autosuficiencia en combate 5177
auxiliar 5134, 5819
(de) auxilio 1451
avance 48, 4387
avance hacia el enemigo 3360
avante 75
¡Avante toda! 1932
avanzado 49, 1877
avanzar 47, 4321, 4389
avanzar separados 1588
ave 475
avería 576, 1099, 1596, 3034
avería causada por una ave dentro del
 reactor de un avión 476
averiado 607
averiarse 577, 3033
aviación 328
aviación militar 112
aviación táctica 5974

barra 370, 371
barraca 2362
barra de torsión 6204
barra de tracción 1363
barranco 2127, 4539
barrear 384
barrena 5581
barrenar 5103
barrer 823, 4488, 5943, 5944
barrera 385, 533
barrera accidental (casual) 6451
barrera natural 3402
barricada 383, 4895
barriga 442
basar (en) 387
base 386, 5679
base aérea 85
base de datos 1110
base hueca 2300
base principal 3013
básico 389
bastón de esquiar 5440
batalla 28, 402, 6834
batalla de ruptura 581
batallón 395
batería 400
batería de adquisición de blancos 2925
(ser) batido por el fuego 6460
batir (con) 1622
batirse en retirada 4815
bayoneta 411
beligerante 441
beligerante no autorizado (ilegal) 6437
bengala 1763
bidón 1927, 1928, 2617
bien defendido 6822
bien equipado 6824
bien vigilado 6825
bifurcación 1869
bifurcarse 1326
bilateral 462
bimotor 6412
binocular 471
biológico 473
bit 477
bitácora 470
bitio 477
bitoque 637
blanco 626, 6022
¡Blanco! 6023
blanco auxiliar 4144
blanco remolcado 6041
blanquear 6843
blindaje 256, 4077, 4086, 4346, 5282, 5285
blindaje compuesto 921
blindar 5089

blinder 5283
blof 509
(hacer un) blof 508
bloque 495
(puerto) bloqueado 2382
bloquear 498, 4486
bloqueo 497
bobinado 6858
boca 3385
boca de incendias 2363
(a) boca de jarro 4510
boca de riego 2363
bocina 2325
bodega 2287
bola de fuego 1707
bolardo 517
boletín 629
boletín meteorológico 4743
bolsa 4103
bolsillo 4103
bomba 521, 4375
bomba-H 2206
bomba beluga 837
bomba de hidrógeno 2367
bomba de iluminación 3982
bomba de incendios 1712
bomba de retardo 6176
bomba de succión 5833
bomba eólica 6864
bomba pequeña 530
bombardear 525, 5264
bombardear con morteros 3329
bombardeo 523
bombardeo a baja cota 2974
bombardeo aéreo 4480
bombardeo a ras de tierra 2744
bombardeo de la costa 5326
bombardeo de rebote 5438
bombardeo de zona 243
bombardeo en oleadas 710
bombardeo por visado indirecto 3614
bombardeo sistemático 3922
bombardero 522, 528, 4481
bomba rompadora de metralla 837
bombear 4376
bombero 1719
bombilla 623
boquetes de ventilación 6597
boquilla (de manga) 3499
borde anterior de la zona de combate 1880
bordear 5446
(a) bordo 5, 3627, 5302
borne 6108
borrar de la plantilla 5777
bosque 6891
bosquecillo 5583

caja de engranajes 2009
cajón 667
cala 1068
calado 1352
calar... 1362
calcar 6223
calcular 929
calcular el término medio 324
cálculo logístico -apreciación
  (estimación)logística 2946
caldera 516
¡calen bayoneta! 1742
calentar 2221
calibrador 671
calibrar 669, 670
calibre 668
(de) calibre reducido 5486, 5803
calidad 2049
calina 2205
calmante 3821
calmarse 3302
calor 2220
calorifugaje 6143
calzo 788
cama 437
cámara (cinematográfica) 793
cámara (de aire) 6391
cámara de combate 1624
cámara de descompresión 1142
cámara de mapas 3065
cámara de máquinas 4938
cámara de poscombustión 67
cámara de televisión 6092
cámara fotográfica 677
cámara polaroid 4173
camareta (de la tripulación) 4405
camarote 662
cambiar 163, 5950
cambiar de frente 4609
cambiar de marcha 5289
cambiar de sitio 1298
cambiar el rumbo 164
¡Cambio! 3759
camilla 2886, 5784
camillero 2887, 5785
caminar 6663
camino 4891, 6230
camino cerrado 4900
camino de acceso 4892
camino de fuertes pendientes 4893
camino de retiro 2867
camino libre 4897
camión 2962, 6382
camión cisterna 6016, 6758
camión de recuperación 6907
camión de suministros 5744
camisa de agua 6741

campamento de prisioneros de guerra
  4295
campaña 440, 681
(de) campaña 1611
campana (de paracaídas) 689
campo 680
(de) campo 1611
campo (de aviación) operativo 3670
campo (de aviación) principal 3011
campo de aterrizaje de tránsito 4226
campo de aviación 111
campo de batalla 405
campo de emergencia 1453
campo de hielo 2384
campo de minas 3229
campo de minas de hostigamiento 3231
campo de redespliegue 4610
campo de tiro 1618
campo de urgencia 170
campo eventual 167
(a) campo través (traviesa) 1078
campo visual 1619
camuflaje 679
camuflaje radar 4431
camuflar 678
caña del timón 6170
canal 683, 765
canal barrido 5946
canal de acceso 234
canal de marea 6164
canalete 3818
canal explorado 5125
canalizar 684
canalizo 1581
canal radio 4450
cancelar 685
candalero 5649
cañería maestra de agua 6745
canjear (por, con) 1531
cañón 687, 2040, 2128, 6390
cañón autónomo 4974
cañón de campaña 1616
cañón de carro 6015
cañón de colisa 5956
cañonear 399, 5753
cañoneo 2131
cañonero 2129, 2132
(de) cañones gemelos 6413
cantidad de radiación 5704
cantimplora 6737
cantina 690, 3168, 3388
capa 2745
capa/nivel de transición 6314
capacidad 694, 696
capacidad (de destrucción) 4237
capacidad (de puente) 814
capacidad (de tránsito) de camino 4896

capacidad anfibia 5947
capacidad de almacenaje 5740
capacidad de cruzar (zanjas) 5550
capacidad de llevar carga bajo eslinga 2902
capacidad de puerto 4192
capacidad de segundo ataque 5137
capacidad de supervivencia 5927
capacidad de tráfico 6258
capacidad de transporte 6333
capacidad de vadeo (sin preparación) 1862
capa protectora 5282
capaz (de) 695
capaz de sintonizarse 6395
capellán 766
capellán militar 3819
capitán 3104, 5441
capitán de puerto 2185
capitán preboste 4357
capitulación 5915
capó 2309
capoc 2648
capota 2310
capotillo 698
cápsula 701
cápsula (fulminante) 692
cápsula fulminante 3946
captación por derivación 6887
captafaros 4628
captar 3999, 5186
captarse (a) 2934
captura 5169
capucha 2308
característica naturel 3404
característica operativa 3671
carbón 840
carburador 706
carburante 1926
carburante hipergólico 2375
carburantes y lubricantes 3970
cardán 2025
carga 707, 768, 771, 2893, 2895, 2908, 3807, 5782
(de) carga 708
carga aérea peligrosa 4797
carga de impulsión 2422
carga de profundidad 1216
carga de proyección 4341
cargado 2891, 2907
cargado (con) 2690
cargador 718, 2995, 2997
carga explosiva 5452
carga hueca 2301
(a) carga hueca 2264
carga llevada bajo eslinga 2903
carga máxima 2897

cargamento 6383
cargamento (entero) 5303
cargamento vertical 2913
carga no esencial 3477
carga nuclear 3507
carga para un convoy 2910
(a) carga perforante 5248
carga por unidad 2906
carga práctica de seguridad 2901
cargar 769, 1632, 2894, 2896, 5779
carga reducida 4613
carga reglementaria 2904
cargar sobre plataformas 3826
carga standard en plataforma 2899
carga táctica 2911
carga transportada debajo de helicóptero 2905
carga útil 3925, 6383
cargo 1935
carlinga 848
carnicería 3099
carpeta 1628
carretera 2272, 4894
carretera cortada 4900
carretera de dos calles 1327
carro 5078
carro (enterrado y emplazado) 6007
carro-puente 597
carro blindado de combate 6564
carro blindado de reconocimiento 6566
carro de combate 6001, 6006
carro de combate anfibio 6002
carro tiendepuentes 6003
carta marítima 772
cartera 601
cartera portamapas 3063
cartográfico 714
cartuchera 4238
cartucho 715
cartucho de fogueo 483
casco 2251, 2351, 2352, 5591
castigable 4381
castigar 4380
castigo 4382
castillo 1864
catalisar 724
catalizador 723
catapulta 725
catapultar 726
categoría 5700
cauce 437
caucho 4978
caudal 1817
cautivo 702
cautivo 703
(dar) caz 4385
caza 2357, 4386

caza(minas) 2359
caza de ataque a tierra 1623
caza de penetración 2569
caza nocturno 3458
cazar 2356
cebar 4280, 4281
cebo 4283
ceder el paso (a) 2027
cédula personal 2394
cegadura por destello 1767
cegar 485
ceguera causada por los reflejos de la
  nieve 5509
celda 734
celestial 732
célula 734
celular 736
cementario 738
cemento armado 4661
cénit (zénit) 6926
censor 740
censurar 739
centígrado 744
centinela 5191
centrador de boca 3386
central 745, 3183, 4043
central (telefónica) 6084
central de tiro 1710
centralizar(se) 746
centro 741, 1513, 3182
centro común (de vehículos) 4185
centro de canal 3184
centro de control aéreo táctico 5972
centro de control de transporte 6334
centro de fricción 6380
centro de gravedad 743
centro de informaciones de combate
  870
centro de mensajes 3165
centro de operaciones conjuntas 2633
centro de transmisiones 906, 5388
cerámica 750
cerámico 749
cerca 3608
cerca (de alambre) 1601
cercado de tierra 2710
cercano 829
cercar 1007, 5917
cerco 5356
cerebro 554
ceremonial 751
cernerse 2344
cero 6927
cerradura 2927
cerrar 496, 828, 3554, 5110, 5113,
  5153, 5343
cerrar (con fuego) 1206

cerrar con llave 2929
cerrojo 520
certero 23
certificado 752
certificado de seguridad 5160
certificar 753
cesar el fuego 730
cesta 392
cicatriz 5060
ciclo de información 2520
ciclón 1095
(a) ciegas 493
ciego 491
cielo 5448
ciencia 5073
cieno 3634
científico 5074, 5075
cierre 590, 2931
cifra 794, 850, 3519
cifrar 1462, 1463, 5081
cilíndrico 1097
cilíndrico de caucho 4956
cilindro 1096
cilindro de oxígeno 3800
cinta 444
cinta de video 6621
cinta magnetofónica 6018
cinta metálica 447
cinta perturbadora 6861
cinturón 450, 451, 6804
cinturón de seguridad 449
cinturón salvavidas 2797
circuito 797
circuito de disparo 1724
circuito integrado 2503
circulación 6256
circulación aérea 140
circulación en ambas direcciones 6418
circular 158, 798, 6259
círculo 796
círculo máximo 2068
circunferencia 800
cirujano 5907
cirugía 5908
ciudad abierta 3639
ciudadela 801
civil 804
clandestino 810, 6458, 6461
claridad 1172
claro 818
clase 69
clase de abastecimientos 817
clases de tropa 1480
clases y soldados 1480
clásico 5227
clasificación 815
clasificación (de bajas) 6356

condensador 947
condensar(se) 946
(en) condiciones de navegar 5132
(en) condiciones de vuelo 144
condominio 949
conducción del tiro 1708
conducir 950, 1376, 2162, 2755, 3039, 5713
conducir en cañerías 4027
conducir un ataque 5555
conductor 1377
conducto regular 756
conectado 2890
conectar 4098
conectar a tierra 1406
conexión 2875, 4130
conexión de eslinga 5938
conferencia 953
confiable 4684
confidencial 954
configuración del terreno 2752, 2793
confirmación 955
¡Conforme! 4917
confraternizar 1895
congelación 1924
congelar 1904
cónico 956
conjunto 280, 2632
(de) conjunto 2632
conjunto de centros urbanos 6536
conjunto de modificación 2676
con motor 4243
conmutador 5951
cono 952
conocimiento 467
conquistar 958
consecutivo 5197
consejero 53
consejo 51
consejo de guerra 1038
conservación 3020, 3025
conservación (del orden público) 3023
conservar 5050
conservar (una posición) 2292
consigna 763, 3908, 5387, 6896
consola 960
consolidación 961
conspirar 4088
constante 962
construir 618
construir un túnel 6397
consultivo 54
consumir 1543
consumo 1545
contacto 965
contador Geiger 2012
container 968

contaminación 970
contaminación residual 4773
contaminar 969
contar 1019
contenedor 1908
contener 967, 2296
contestación del fuego 1025
contestar 1021, 4722
continente 971, 3017
contingencia 973
contingente 975
contingente nacional 3396
continuación 1847
contorno 3750
contra-contramedidas electrónicas 1440
contra-espionaje 1024
contra-subversivo 1026
contraalmirante y superior 1748
contraataque 1022
contrabandista 5496
contrabandista de armas 2135
contracarro 228
contracorriente 6459
contrainteligencia 2511
contrainteligencia táctica 2517
contramaestre 542
contramedida 1027
contramedidas electrónicas 1441
contramedidas radar 4433
contrapendiente 5480
contra personal 223
contraseña 1028, 4723, 4724
contraseña de servicio 5377
contrato 976
control 778, 978, 983
controlador 986
controlador aéreo táctico 5973
controlador avanzado de aviones 1878
controlar 776, 979, 3316
controlar el aterrizaje de un avión desde tierra 5997
control automático de la ruta 6249
control automático de navegación 3419
control de calidad 4396
control de emisión 1457
control de existencias 5728
control de la organización del movimiento 3357
control de maniobra de identificación 5497
control desde tierra 2097
control de seguridad 5161
control operativo 3673
control remoto 4694
(a) control remoto 4695
control táctico 5976
control tráfico aéreo 141

cuello 3423, 3424
cuello boyante 1811
cuenca 390
cuenta 16
(dar) cuenta de 18
cuenta hacia atrás 1020
cuerda 4943
cuerda salvavidas 2799
cuerdas de suspensión 5936
cuerdas de sustentación 4859
cuero 2256
cuerpo 513
cuerpo (de ejército) 1009
cuerpo (vehículo) de reentrada 6580
cuerpo expedicionario 1542
cuesta 2428, 4880
cuesta arriba 6528
cueva 5511
cuidar (a) 3522
cuivarse 367
culata 652, 657, 4079
cumbre 3933
cuña 1059, 6805
(en) cuña 3632
cuneta 1319
cuota 4413
cúpula 1088
cureña 2130
curso 1036
curso de agua 6739
curva 454
curva de nivel 2832
cúter 1094
chaleco blindado (antibalas) 259
chaleco salvavidas 2604
chapa 5260
chapa de identidad 5982
chaqueta 2602
charca 4182
charco 6740
charretera 1491
chasis 775
chatarra 5085
chaveta 2656
chimenea 1937
chispa 5553
chocar (con) 857
chocar (contra) 633
choque 632, 858, 5313
chorro 2621
chorro nuclear 5606
chubasco 5618

D

dado 2596

daño 1099
daño nuclear 3504
daños secundarios 855
dar 2595
dardo 1773
dar honores a (la bandera) 2307
dársena 391, 1334
dato 1112
datos 1109
datos de salida 4285
datos marginales 3072
datos meteorológicos 3174
de arriba 6528
débil 6770
debilitar 5517
debilitar(se) 6771
decimal 1132
decisión (por) 1136
declarar 753
declinación 1139
declinación (magnética) 6557
declinación magnética 3004
declinación magnética de cuadriculado
  2079
(datos) de entrada 2478
defecto 1154
defender 1158
defensa 1159, 1603
defensa antiaérea 108
defensa antiaérea pasiva 3905
defensa antigás 782
defensa a toda costa 6025
defensa avanzada 1879
defensa contra misiles 219
defensa de contrapendiente 1160
defensa de radar 3997
defensa en posición (organizada) 4213
defensa en profundidad 1161
defensa estática 1162
defensa improvisada 2196
defensa lineal (frontal) 2857
defensa litoral 3915
defensa móvil 3290
defensa nuclear 3506
defensa pasiva 802
defensa perimétrica 3952
defensa propia 5174
(a la) defensiva 1165
defensivo 1164
defile 3850
definición 1173
degradar 1177, 1346
delantero 1877
delatar (a) 2460
delegar 1186
deletreo 5576
deletreo fonético 3977

delicado 5187
de longitud 2856
demanda 1193
de mano 2161
de marineria 2790
demolición reservada 4768
demorar(se) 1182
densidad 1203
densidad del tráfico 6261
dentista 1205
dentro de los límites 3633
denunciante 2462
dependiente (de) 1209
depositar 1394
depósito 1212, 1393, 4771, 5727, 5739,
 6000
depósito cilíndrico 1387
depósito de combustible 6004
depósito de combustible deobturación
 automática 6009
depósito de combustible lanzable 6010
depósito de material (de guerra) 3724
depósito de munición 2996
depósito de municiones 183, 5742
depósito de obturación automática
 6008
depósito de vestuario 5743
depósito lanzable 6005
depresión 1214
deprimir 1213
(de) derecha 4863
derecho 4861, 6531
derecho (a la derecha) 4862
derecho de detención 4865
derecho de paso 4866
derecho de visita 4864
derecho internacional 2544
derecho militar 3197
deriva 1371, 2758, 2768
(a la) deriva 46
(ir a la) deriva 1369
derivación 1370
derivar 1369
derramar 5580
derrapar 5428
derrape 5429
derrelicto 1218
derribar 2663
derribo 2664
derrota 1153
derrotar 1152, 4961
derrumbar 3793
desactivador (de explosivos) 3577
desalación 1222
desalinar 1221
desamarrar 720
desarmado 6434

desarmar 1274, 4698, 6445
desarme 1275
desarrollar(se) 1250
desarrollar en fases 3971
desarrollo 1251
desatar 2959, 6499
desatornillar 6514
desautorizado 6436
desbordar 1802, 3746
desbordarse 1806, 5580
¡descansen armas! 3716
¡Descanso! 4788
descarga 6644
descarga de fuego 1723
descargar 1279, 1281, 3610, 4698, 6501,
 6502
descargo 4678
¡ descarguen armas! 6503
descenso automático (término inglés)
 2035
descentralizar 1130
descercar (ciudad etc.) 4689
descifrar 1135, 1140, 1148, 6513
desclasificar 1138
descomprimir 1141
descontaminación 1145
descontaminar 1144
descripción detallada 4316
descubrir 6453
desechado 2624
desembalar 574
desembarcadero 5642
desembarcar 1126, 2695
desembarco 1125, 2698
desembarco de asalto 2700
desembarque 1125
(en) desenfilada 1169
(a) desenfilada de casco 1168
(a) desenfilada de torre 1170
desenfilada de torreta 4212
desenfilar 1167
desenganchar 6500
deserción 1156
desertar 1155, 1223
desertor 1225
desertor 1157
desfiguración 1308
desfiladero 1171
desfile militar 3069
desgaste 302
desgaste natural 6800
deshabitado 6508
deshacerse (de) 1306
deshelador 1180
desherrumbrar 1220
deshidratación 1179
desierto 1224

desigual (camino)   4953
desincrustar   5054
desinfectante   1287
desintegración   1128
deslizamiento   5472
deslizar(se)   5465
deslizarse   5471
deslumbramiento   2029
deslumbrar   1118
desmantelar   1288
desmembración   586
desmilitarizar   1194
desmodulador   5083
desmodular   5081
desmontado   6434
desmontarse   1290
desmoralizado   1201
desmoralizar   6466
desmovilización   1195
desnacionalizado   5671
desnivelado   6523
despachar   1292
despacho   1293
despacho de oficial   895
desparecido en combate   3276
despedir   1289, 4675
despegar   2801, 5990
despegar inmediatamente   5082
despegue   2802, 5989
despegue de salto   5991
despegue en serie   5993
despegue inmediato   5080
despegue normal   5992
(de) despegue y aterrizaje cortos   5994
despejado (camino etc.)   824
despejar   821
despejo de caminos   4899
desperdicios nucleares   6728
despertar(se)   6662
despiezar   5791
despiojar   1192
desplazamiento   1301
desplazar   1300
desplazarse   3355
desplegar   1302
desplegar(se)   1210
despliegue   1211, 1303
despliegue de fuerzas   2753
despoblado   6508
desquite   4808
destacamento   1234, 1235, 3878
destacamento (pequeño)   5616
destacamento de abordaje   3881
destacamento de busquada (de
   exploración)   3892
destacamento de contraincendios   3996
destacamento de demolición   3883, 3884

destacamento de desembarco   3887
destacamento de entierro   3882
destacamento de mecánicos   5145
destacamento de playa   414, 3893
destacamento de retaguardia   3891
destacamento de trabajo   3885
destacamento retrasado de información
   3888
destacar   1233
destacar (para)   1237
destacarse   5405
destapar   6453
destello   1766
destinado en   5696
destinar   4222
destino   1229, 4234
destino en ultramar   6213
destrucción   1196, 1231, 2664
destrucción improvisada   2197
destrucción submarina   6475
destructor   1232
destructor de carros   6012
destructor de minas   3225
destruir   1230, 5995
(en) desuso   3542, 3543
desvanecimiento   1575
desviación   1175, 1252, 1253, 1323, 6917
desviado   1033
desviar   1325
desviar(se)   1174
desvío   3611
desvío de carretera   659
desvío de posición   4214
desvío de sextante   5234
desvío normal   5653
desvío probable   4297
desvío probable circular   799
detalle   1236
detaller   2600
detección   1240
detectar   1239
detective   1241
detector   1242, 5189
detector de explosivos   5500
detector de mentiras   2791
detectores terrestres teledirigidos   4696
detector radiológico   4440
detención   265
detener   266, 1238, 1602, 1746, 2288
detenido   4292
detonación   364, 1246
detonacíon por simpatía   5958
detonador   1247
(hacer) detonar   1245
detrás   4561
detruir   2663
de vigilancia   3888

escopeta   5338
escopeta (de cañón acortado)   5052
escopeta recortada   4876
escora   2880
escorar   2882
escotilla   2198
escotilla de escape   1506
(con) escotillas cerradas   655
escrutar   5058
escuadra   1774, 5616
escuadra portaaviones   712
escuchar   2885, 3316
escudo   5281
escudo térmico   5284
escuela   5072
escuela (superior)   9
escuela naval militar   3408
esencial   6640
esfera   5577
esférico   5578
eslabón   5955, 6248
eslinga   5470, 5760
eslinga de carga   709
eslingar   5468
espacial   62
espacio aéreo   131
espacio aéreo controlado   985
espacio muerto   1123
espacio para hacer maniobra   4942
espalda   335
espaldón   656
especialista   1549, 5560
especializarse   5561
especificación   5562
especificación técnica   6075
especificar   5564
espectro   5565
espectro visible (visual)   6630
espejo   3247
¡Espera! (cuando haya una pausa larga)
   6660
esperar   6659
espía   5614
espiar   5615
espionaje   1510
espiral   5585
espoleta   1945, 1948
espoleta a percusión   1954
espoleta autodestructora   1967
espoleta de alivio de presión   1961
espoleta de cápsula fulminante   1953
espoleta de contacto   1955
espoleta de culote   1950
espoleta de impacto   1959
espoleta de influencia   1969
espoleta de percusión   1958

espoleta de presión   1960
espoleta de proximidad   1952, 1962,
   1963
espoleta de radioproximidad   1971
espoleta de relojería   1951
espoleta de retardo variable   1970
espoleta de suelta   1965
espoleta de tiempo   1968
espoleta de tracción   1964
espoleta iniciadora   2395
espoleta mecánica   1957
espoleta por influencia   1956
espoleta secundaria   1949
espolón   5613
espuma   1828
espuma de caucho   1829
esqueleto   1893
esquema de superficie fotografiada
   3107
esquí   5427
esquí alpino   5433
esquiar   5426
esquina   1006
esquí nórdico   5432
estabilidad   5622
estabilizado por aletas   1643
estabilizador   5624
estabilizar   5623
estable   5625
establecer   1512
establecer enlace (con)   2786
establecer una cortina de fuego   2743
establecimiento   1513
establilizado por rotación   5584
establizador giroscópico   2146
estación   5681
estacionar   5678
estacionario   5695
estación de captación   2526
estación de detección   4231
estación de retransmisión   4712, 5693
estación de servicio   5688
estación directora   5691
estacíon radiogoniométrica   5686
estación satélite   3757
estadillo de formación   5670
estadillo del tiempo de paz   3932
estadística(s)   5699
estadístico   5698
estado   948, 5669
estado constante   5708
estado de efectivos   5773
estado de guerra   5675
estado de preparación   5672
estado de preparación armada   5673
estado de preparación de seguridad
   5674

estado mayor 5628
(de) estado mayor 5626
estado mayor conjunto 5627
estado mayor de conjunto (EMACON) 5630
estado mayor de operaciones conjunto 5631
estado mayor de planificación 5635
estado mayor integrado 5629
estado mayor paralelo 5632
(de) estado sólido 5528
estallar 582
(hacer) estallar 5226
estallido 1765
estampido 4729
estampido (sónico) 534
estanco 6756
estandardización 5654
estandardizar 5655
estanque 4182
estante de munición 4415
estar a la espera 5656
estar al ancla 4844
estar de guardia 2653
estar en declive 5280
estar investigando 3031
estar listo (para) 5656
estar listo para el movimiento (dentro de ....horas) 3497
estar sobre las armas 5663
estático 5676
estela 6239, 6661
estela (turbulenta) 6726
estela de condensación 6267
estela de humo 6268
estela trazadora 6221
estellar 1550
estibador 5722
estibar 5750
estimación 1515
estimación aproximada 2119
estimación de oficial del Estado Mayor 5638
estrangulador 789, 6155
estrategia 5766
estratégica 5764
estratégico 5761
estratosfera 5767
estrecho 3392, 5538, 5756
estrella 5664
estrella polar 5665
estrellarse 1062
estribo 4002
(a) estribor 286, 5666
estropeado 3713
estropear 6906
estuario 1519

estudiante militar 5801
estudio 282
etapa 2770, 5643
etapa de impulsión 5641
etiqueta 2687
etiquetar 2688
evacuación 1521
evacuación aérea 110
evacuación de bombas fallidas 526
evacuación de heridos 3140
evacuado 1522
evacuar 1520
evadido 1505
evadir 1523
evadirse 582
evaluación 1525, 4534
evaluación de averías 1100, 5923
evaluación de daños nucleares 3505
evaluar 1524
evasión 1503, 1526
evasión y escape 1527
evitar 658
excavadora 1530
excavadora con pala 5340
excavar 1261
excedente 5910
exceder al enemigo en artillería 3748
exceso de poder destructor 3770
exfiltración 1535
exhibición 1199
exigencias 1539
existencias 5727
existencias operativas 3678
expedición 1541
expedidor 3735
expediente 1343
expediente de objetivos 6048
expendable 1705
experimental 1548
experimento 1547
experto 1546
exploración 5059, 5122
exploración de sector 5149
exploración y rescate 5124
explorador 5076
explorador 5126
explorar 1554, 5058, 5077, 5121
exploratorio 1553
explosión 1555
explosión (granada) 647
explosión a baja cota 2973
explosión en el aire 90
explosión en superficie 2095
explosión nuclear de superficie 3513
explosión nuclear en el aire 3502
explosión nuclear submarina 3515
explosión nuclear subterránea 3514

fuente   5609
fuera-bordo   3743
fuera de acción   29
fuera de alcance   4509
¡Fuera de límites!   3609
fuera de los límites   546
fuera de ruta   1033
fuerte   5712, 5798
fuerza   1853, 5772
fuerza(s) de pacificación   3930
fuerza centrífuga   748
fuerza costera   844
fuerza de cobertura   1054
fuerza de desembarco   2707
fuerza de disuasión   5764
fuerza de intervención inmediata   4529
fuerza de maniobra   3046
fuerza de relevo   1843
fuerza de ruptura   4995
fuerza de tracción (al gancho)   1364
fuerza eventual   974
fuerza hipotecada   899
fuerza mínima   3240
fuerza no empeñada   6446
fuerzas   1857
fuerzas acorazadas   257
fuerzas armadas   254
fuerzas de maniobra   871
fuerzas de reserva   4769
fuerzas de seguridad local   5157
fuerzas especiales   5559
fugaz   6307
fumígeno   5488
función   1935
funcional   1936
funcionar   1934, 3647, 6898
funcionario   3607
fundir   1946
fusible   1943, 1966
fusil   4846
fusilar   5321
fusil automático   4847
fusil de repetición   4848
fusil de tirador apostado   4852
fusilero   4854
fusil lanzagranadas   4849
fusil provisto de visor   4851
fusil sin retroceso   4850
fusión   1947

G

gabarra   2811
gafas (submarinas)   2039
gafas de sol   5842

gafas nocturnas   3461
gálibo (de carga)   2003
galón   4840, 5795
galón de servicio militar   5216
gallardete   3943
gama de frecuencias   4500
ganar   6850
ganar una posición ventajosa   2937
gancho   2312
gancho de cola   2314
gancho de frenaje   2313
gancho de remolque   2315
garaje   1982
(al) garete   46
garita de centinela   5193
garrar   1355
gas   1988
gas-oil   3621
gaseoducto   4028
gas inerte   1992
gas irritante   1993
gas lacrimógeno   1995, 2990
gas mostaza   1994
gas nervioso   1991
gasolina   1998
gasolinera   5690
gas sofocante   1990
gastar   1543
gasto   1545
gas vesicante   1989
gatillo   6363
gato   2601
gaza   461
gemelos prismáticos   472
generador   2019
general   2014
generar   2018
¡gente de paz!   1913
geográfico   2022
gimnasia   6287
gimnasio   2138
girar   795, 5582, 5948, 5954, 6832, 6854
(hacer) girar   4948
girar (sobre)   4042
girar (viento)   6562
girar el motor   4837
girar sobre sí mismo   5603
girocompás   2141
giroscópico   2145
giróscopo   2144
glacial   6875
glaciar   2028
glacis   5478
globo   356
globo de barrera   381
gobernable   3413

H

habilidad   4315
habilitado   3926
habitable   2147
hacer flotar   1798
hacer fuego   1644
hacer fuego con puntería directa   1720
hacer rumbo   5225
hacerse a la mar   5661
hacer señales   5379
hacer un dibujo   5424
hacer un informe   4725
hacer un nudo   6165
hacia el este   1408
hacia el mar   5131
hacia el norte   3483, 3484, 3485
hacha   331
hangar   2179
(a) harnés del casco   2252
haz   428, 5253
haz paralelo   3859
hebilla   614
helada   1922
helado   1925, 2389
helar(se)   1904
helarse   2387
hélice   4340, 5099
(de) hélices gemelas   6415
helicóptero   97, 2245
helicóptero armado (de ametralladoras y
   misiles)   2136
helicóptero de utilidad (general)   6541
helio   2249
helipuerto   2247
herida   2475, 6901
herida cerebral   555
herida ocular   1568
herida superficial   1778
herido   6903
herido a pie   6665
herido transportado en camilla   2888
herir   2474, 6902
herir con la bayoneta   410
herméticamente   2254
herramienta   6190
herrería   1868
herrumbro   4998
herrumbroso   5002
hervir   5906
hidráulico   2364
hidroavión   5119
hidrofoil   2365
hidrófono   2370
hidrógeno   2366
hidrografía   2369
hidrográfico   2368

hidroplano   2371
hidrostático   2373
hielo   2379
hierro   2586
hierro en Z   6925
hilera   4976
hilera de minas   3234
hilera doble de mines   5794
hileros   609
hipersónico   2376
hipsometría   2378
hogar   1941
hoja   482, 1839, 4317, 5257
hoja (de premios y castigos)   5258
hoja (estanada)   5260
hombre(s)-rana   1915
hombre de cuarto   6732
hombrera   1491, 5339
homogéneo   2306
hora de a bordo   5292
hora de apagar las luces   2818
hora de huso horario   6956
hora de llegada (de recibo)   6181
hora de origen   1111, 6180
hora de salida   5023
hora estimada de llegada   1516
hora estimada de salida   1517
hora H   2255
hora local   2922
hora N   3453
hora P   3991
horario   5064, 6184
hora tope   1122
hora zeta   6960
horizontal   2323
horizonte   2319, 5450
horizonte aparante   2320
horizonte auténtico   2321
horizonte visual   2322
hormigón   737, 943
hormigón armado   4661
hormigón pretensado   4275
horno   1940
horquilla   553, 6922
horquillar   551
hospital (militar) general   2334
hospital de apoyo de campaña   2331
hospital de campaña   2333
hospital de campaña avanzado   5683
hospitalizar   2335
hospital principal   2330
hostigar   5459
hostil   2337
hostilidades   2340
hostilizar   2181
hoyo   2299
hoyo de protección   1889

huelga 5787
huella 1852, 6232, 6236, 6347
huella (en el radar) 6237
huella del torpedo en el agua 6241
huella identificada 6240
humedad 2355
humedad relativa 4669
húmedo 2354, 6827
humo 5489
humo de ocultación 5492
humo de oscurecimiento 5491
humo indicador 5490
hundido 5843
hundimiento 5818
hundir 6906
hundir(se) 5413
hundirse 1888, 5817
huracán 2360
huso horario 6950

I

¡(en) revista armas! 2489
iceberg 2381
iceberg que amenaza los buques 2109
idea de maniobra 941
idea general 3750
identidad 2393
identifación amigo o enemigo 2391
identificación 2390
identificación radioeléctrica 4460
identificación visual 6638
identificar(se) 2392
idioma 2716
ignifugo 1753
(en) iguales colados 1529
ilegal 2397
ilimitado 2434
iluminación 2400
(de) iluminación 1762, 2399
iluminación del campo de batalla 406
iluminador laser 2723
iluminar 2398, 2806
imagen 2401
imaginaria 6732
imán 2998
imbornal 5101
immovilizado por la niebla 1835
impacto 2413
impacto corto 2281
impacto directo 2279
impacto largo 2280
impacto normal 3480
impedir 1206, 4276
impermeable 6746
implosión 2417

implosionar 2416
improvisado 2421
improvisar 2420
impulsar 1375, 4338
impulsor (del disparador) 37
(a, de) impulsos 4372
imunización 2412
inactivar 2423
inapto para servicio 6482
inastillable 5252
incapacitado(-a) 3473
incendiario 2425
incendio 1700
incidente 2426
incidente nuclear 3508
inclinación 1267, 5457
inclinación de la tangente 6917
inclinado 2429, 5458
inclinar(se) 2427, 6171
inclinarse 5456
incombustible 1722
(estar) incomunicado 5529
incursión 2432, 4479
indefendible 2433
indefenso 6455
indefinido 2434
independiente 2435, 5194
indescifrable 6454
indicación 4556
indicación visual 494
indicador 2002, 2438, 4161
indicador (de dirección) 4232
indicador de dirrección 3092
indicador de posición en el suelo 2439
indicador de posición en planta 2441
indicar 4168
indicar con precisión 4021
indicar en el mapa 3083
indicativo de llamada 5373
indicativo de red 5376
indicativo internacional 5375
indicativo visual 5378
indirecto 2442
indisciplinado 6478
indispensable 1511
inducción 2446
inducir 2444
inercia 2448
inerte 2447
inestable 6520
inexplorado 6443
infalible 1848
infante 2452, 5524
infantería 2451
infantería de marina 3074
infantería mecanizada 3139
inferior 2976

L

línea de proyección 2861
línea de referencia 2849
línea de rumbo 2851
línea de seguridad 2829, 2841
línea de seguridad nuclear 2842
línea de situación 2863
línea de suministro 2866
línea de tiro 2862
línea frontal 2855
línea internacional de cambio de fecha 2543
línea isogónica 2840
lineal 2856
línea límite de fuegos 2837
línea loxodrómica 2851
línea mediana 742
línea observador-objetivo 2843
línea pieza-objectivo 2836
línea Plimsoll 4087
línea principal de resistencia 4775
línea quebrada 2838
líneas de comunicación 2870
linterna eléctrica 1770, 6196
líquido 2878
líquido derramado 5579
líquido para frenos 559
líquido propulsor 2879
liso 676
lista 2881, 4922, 4927, 4946
lista de control 777
lista de objetivos 2884
lista de personal 4925
listo 4557
¡listo! 4746
listón 397
listo para marchar .4906
litera 437, 638, 2886
litoral 2484, 2889
local 2918
(a) localización 2933
(exacta) localización de un emplazamiento 4149
localización óptica 1771
localización radar 4436
localizar 2920, 2923, 2924, 2938
localizar por radiogoniometría 1741
loción para ahuyentar los insectos 4716
locomotora 2936
logaritmo 2942
logística 2947
logístico 2943, 5209
lona 691
longitud 2949
longitud del eje 6231
longitud de onda 6767
longitudinal 2950
lote 2968

lubricación 2987
lubricante 2985
lubricar 2986, 3618
luces de navegación 3418
luces de situación 5354
luces laterales 5354
luchar 403
lugar 5416
lugar de desembarco 5419
lugar de lanzamiento 5420
lugar protegido 5418
luna 3318
luna llena 1930
lupa 3006
luz 2803, 4992, 5547
luz (del día) 1117
luz centelleante 4410
luz de alerta (alarma) 6714
luz de cola 5985
luz de destellos 1769
luz de fondeadero 194
luz de popa 5721
luz de situación 4217
luz fija 1744
luz libre 3768
luz piloto 4015
llama 1752
llamada 4568
llamada general 3443
llamada telefónica 6083
llama piloto 4014
llamar 672, 4567
llamarada 1765
llamar al servicio militar 675
llano 1772, 2782, 4044, 5245
llave 2655
llave de cierre 5736
llegar a 4542
llenar (de) 1631
llenar (una brecha) 594
llenar a tope 6195
llevar 1606, 1825
llevar (armas) 432
llevar ... grados 431
llevar la delantera 2754
lluvia radiactiva 4485

M

macarrón 631
machila 2680
macho 3032
madera 6173, 6891
(de) madera 6892
madera contraplacada 4101
maestre 3113

movimiento de pinza   4018
movimiento envolvente   1489
movimiento retrogrado   4820
movimiento simultáneo   6340
muchedumbre   1082
muelle   2629, 4407, 6830
muelle rígido   2186
muelles   1334
muerte por causas naturales   3403
muerto   1121
muerto en combate   2665
muesca   3490
mulatero   3364
mulo   3363
multicabeza   6696
(hacer a) multicopista   1397
multilátero   3368
multinacional   3369
múltiplex   3372
múltiplo   3370
multitubo   3365
muñeco (empleado en paracaidismo)   6857
munición APDS   5004
munición con vaina   4957
municiones   182, 3374
munición especial   5558
munición fija   1743
munición semifija   5180
munición trazadora   6227
muñón   6389
muñón de ala   6869
muro   6666
mutilado   3381
mutuo   3384

N

nación   3394
nacional   3395
nacionalidad   3398
nadir   3389
(de) napalm   3390
natural   3401
naufragio   5310, 6905
náufrago   719
náutica   5117
náutico   3405
navaja   813, 2606
naval   3407
navegable   3414
navegación   3415
navegación astronómica   733
navegación de estima   4575
navegación radiogoniométrica   4457

navegar a snorkel   5505
navigabilidad, certificado de   143
navegación hiperbólica   2374
navigante   3420
navio de ronda   2117
NBC arma   6776
neblinoso   1836
necesitar   4756
negativo   3425
negociación(es)   3431
negociador   3432
negociar   3429
nervio   3434
neumático   4102, 6186
neumático de nieve   5516
(que padece) neurosis de guerra   5275
neutral   3447
neutralización   3448
neutralizar   485, 3449
neutrón   3450
nevar   5508
nevisca   6844
nido de ametralladoras   3436
nido de heridos   4113
niebla   1834, 3285
nieve   5507
(de) nilón   3525
ningún informe   4737
nítido (foto etc)   5249
nivel   2778, 2781, 4064
(a) nivel   2779
nivelado   2779
niveladora   2051
nivelar   2048, 2780
nivel de abastacimiento   2783
nivel de agua   6742
nivel de comparación   1113
nivel de existencias   5729
nivel del mar   5115
nivel de puntería   4392
nivel de seguridad   5163
nivel medio del mar   3125
nivel operativo   3674
nivel verdadero   6387
no apto para la navegación   6515
no blindado   6435
no clasificado   6444
no combatiente   3472
no confirmado   6452, 6522
no contaminado   6448
noche   3455
no defendido   6483
no disponible   6438
no enterrado   6441
no esterilizado   6521
no funcionar   5211
no identificado   6484

nombrar 1228, 3470
nombrar oficial 896
nombre en clave 851
nómina 4926
nominal 3469
no oficial 6509
no pilotado 6505
no potable 6479
no registrado 6511
norma 5650
normal 3479
normalizar 3481
normas de conducta 4178
(del) norte 3482
nota 215
notación 3489
no tripulado 6505
(sin) novedad 3491
nube 835
nube atómica 3377
nube nuclear 3503
nuclear 3500
nuclear, bacteriológica e química (NBQ) 3501
núcleo de reactor 4548
núcleo de resistencia 4776
nudo 2685
nudo corredizo 5473
numerar 3518
número 3520, 3521
número de bajas 2883
número de mach 2993
número de orden (en una marcha) 5198
número de referencia 4623
número de serie 760, 5199
número de teléfono 6085
número impar 3569

O

obedecer 3528
objeción 3531
objetivo 83, 2038, 2774, 3532, 6022
objetivo acorazado 6026
objetivo aislado 6034
¡Objetivo batido! 6023
objetivo constituído por una demolición reservada 6036
objetivo de oportunidad 6050
¡Objetivo destruído! 5587
objetivo extenso 6024
objetivo imprevisto 6027, 6032
objetivo inmediato 6031
objetivo lineal 6028
objetivo móvil 6030

objetivo previsto 6033
objetivo prioritario 4289
objetivo programado 6037
objetivo real 6029
objetivo secundario 6038
objetivo sorpresa 6035
objetivo submarino 6040
objetivo telefotográfico 6958
objetivo vulnerable 6039
objeto 3530
objeto arrojadizo 3251
oblicuo 3536
obligacíon 898
obligatorio 3041
observación 3538, 5602
observación (continua) del campo de batalla 408
observador 3541
observador avanzado 1883, 3585
observar 3539, 5596, 6731
observar clandestinamente 5645
observatorio 4230
obsoleto 3543
obstaculizar con cráteres 1065
obstáculo 3544
obstáculo artificial 3545
obstáculo de profundidad 3551
obstáculo en un flanco 3547
obstáculo estandar 3552
obstáculo infranqueable 3548
obstáculo natural 3546, 3549
obstáculo protector 3550
obstáculo táctico 3553
obstrucción 3555
obstruir 369, 3554
obstruir(se) con sedimentos 5408
obturador 5345
obturar 5113
obús 2348
obús desmontable 2349
ocaso 5845
occidental 6826
océano 3561
oceanografía 3565
octanaje 3566
octante 3567
ocular 1569, 3568
(de) ocultaciones 3556
ocultado en niebla 1835
ocultar 932, 5092
ocupación 3557
ocupar 3559, 5167
ocupar una posición 3560
odontologio 1204
(del) oeste 6826
ofensiva 3571
ofensivo 3570

orden de batalla   404
orden de ejecución   3707
orden de fuego   3708
orden del día   3719
orden de movimiento   3710
orden de operaciones   3712
orden de régimen interior   3714
orden logística   3705
orden que prohibe los vuelos   3709
oreja   1402
orejeras   1403
orejeta   2988
orgánico   3725
organismo   70, 1513
organización   3726
organización de la carga   6273
organización de la defensa   3728
organización del escalón   1420
organización del terreno   3729
Organización de Naciones Unidas
   (ONU)   6492
organizar   3730
órgano   70
orientación   3733
orientar   3731, 3732, 6559
orientar un mapa   5223
orificio   6596
origen   3734
orilla   1866, 5324
(en la) orilla   273
oruga (ambos de carro de combate)   6235
oscilación   3738
oscilador   3739
oscilante   3737
(hacer) oscilar   3736
oscilógrafo   3740
osciloscopio   3741
oscurecer   1106
oscurecimiento   3537
oscuridad   1107
oscuro   1105
OTAN   3400
otero   2684
otoño   318
oxidación   4998
oxidado   5002
oxidarse   4999
oxígeno   3799
oxígeno líquido   2983

P

paciente   3910
paciente a pie   6665
pacifista (que se niega a tomar) las
   armas   3535

Padre   3819
pagar   3923
país   1029
país de origen   1030
pala   481
palabra (clave)   6895
palabra de honor   3871
palabra de servicio   4359
pala de rotor   4951
palanca   2164, 2784
palanca (de montar)   2165
palanca (para mover un obús)   2175
palanca de comando   984
palanca de control   2634
palanca de gases   6155
palanca de gobierno   5723
palanca de mando   3650
palanca de velocidades   2011
paleta   480
paleta (de hélice)   6549
paleta (de turbina)   6551
palete   3111
palo   3101, 5551
palomilla   6872
panel   3829
panel de navegación   3833
panorámico   3836
pantalla   5091
pantalla de flotación   5095
pantalla de radar   4438
pantalla polyédrica   1008
pantano   3090, 5940
pantoque   463
panza   442
papel   4919
parabrisas   6865
paracaída acrobática   5449
paracaídas   3837
paracaídas-bolsa neumática   3863
paracaídas de apertura automática
   3844
paracaídas de caída libre   3843
paracaídas de carga   3840
paracaídas de frenado   1380
paracaídas de freno   3839
paracaídas de pecho   3841
paracaídas extractor   3842
paracaídas pilotillo   3842
paracaídas ventral   3841
paracaidista   3848, 3864
parachoques   634, 1604, 5316
parada   3850
parados   3853
paralaje   3856
paralelo   3857
paramentos   1572
paramilitar   3861

parapeto 3862
parar 5344
parar(se) 2156, 5647, 5734
pararrayos 2815, 2816
parásitos 290, 5677
paraván 3866
parcial 3875
parche de color 860
par de torsión 6203
pared 6666
parientes más próximas 3452
parka 3868
parlamentar 3870
parlamento 5998
parque de bomberos 5689
parte 1293, 3873, 4728
parte (logística) 5463
parte al vuelo 4735
parte de ausentes 4730
parte de bajas 4733
parte del estado de fuerzas 4741
parte meteorológico 3172
partícula 3876
partida 1208, 2598
partida de alojamiento 3889
partida de incursión 3890
partido 3879
partir 1207
partisano 3877
pasaje 3895, 3899
pasajero 3903
pasar de contrabando 5495
pasar de escalón 2763
pasarela 1826
pasar por encima 3788
(hacer) pasar por una esclusa 2935
pasar rasando 5435
pasar un obstáculo 3430
pase 3897, 3962
¡Pase! 3904
paseo entablado 1390
paso 3802, 4038, 6347
(al) paso 5716
paso a nivel 2050
paso de escalón 3902
paso deliberado 1188
paso de línea 3902
paso de río 4887
paso improvisado 2195
paso inferior 6467
paso navegable 6750
paso vadeable 6657
pastilla 5967
patente de sanidad 466
patín 5430
patio de maniobra 3094
patrón 1056, 5441, 6096

patrulla 3911
patrulla de aguas limítrofes 3915
patrulla de captura 3918
patrulla de cazacarros 3920
patrulla de combate 3914
patrulla de reconocimiento 3916
patrulla de vigilancia 3919
patrulla en tierra 3917
patrulla fronteriza 3913
patrullar 3912
patrulla retardadora 5707
pausado 1187
paz 3929
pecho 784
pedal 3936
pedal de freno 561
pedido 4759
pedir 673, 4755, 4758
pedir el 'santo y seña' 762
película original 3106
peligro 1102
peligroso 1104, 6512
pelotón de ejecución 1727, 3886
pendiente 2046, 4880
pendiente gradual 5479
penetración 3941
penetrar 2568, 3940, 4003
península 3942
percebe 375
percutor 4016
percha de reabastecimiento 4298
perder 2963
perder contacto 2964
perder terreno 2965
perdidas 721, 2966
perdidas masivas 3100
perdida total 6908
perdido 2967
perdigón 3937
perecer ahogado 1386
perfil de vuelo 1793
perforación 3948
perforante 260
perforar 3947
perigeo 3950
perímetro 3951
periódico 3954
período 3953
período de radioactividad 2794
periscópico 3957
periscopio 3955
perito 1546
permanecer en tierra 2084
permanecer inmóvil 1905
permanente 3958, 5676
permiso 2765, 3896
permiso de paseo en tierra 5327

permiso final   6112
permitir   3961
perno   519
perpendicular   3963
perro detector de explosivos   5501
perro rastreador   6246
persecución   1844
(de) persecución   6247
perseguidor   5240
perseguir   774, 1845, 1846, 4322, 4385, 5239, 6229
persona importante   6627
personal   3054, 3966
(número de) personal   5639
personal de tierra   2100
personal en plantilla   5774
personal permanente   5633
personal prestado   2917
(en) perspectiva   3968
perturbador de barreamiento   2613
perturbador de búsqueda automática   2609
perturbador de exploración   2611
perturbador de repetición   2610
pesa   6811
pesado   2226
pesar   6808
pescantes   1114
pescar   1730
peso   6811
peso bruto   6812
peso bruto máximo   6814
peso específico   5563
peso máximo de aterrizaje   6815
peso máximo de despegue   6816
peso neto   6817
peso total en carga   6813
pesquera   1777
petardo   3086
(a) petición   3629
petición prevista   3283
petróleo   3623, 3855, 3969
petrolero   3620, 6014, 6583
picado   3488, 4952
picar   1320
pico   3933
(a) pico   5712
pico de lanzamiento   2738
pie   1849, 1851
(en) pie de guerra   6693
pie de imprenta   2418
pie de trinchera   6354
piedra   5732
piel   5436
pieza   920, 3872
pieza de recambio   4718
piezas de recambio de a bordo   3628

pila   401, 4279
píldora   4004
píldora anti-fatiga   3945
pilotar   4009
piloto   4008, 4010
piloto automático   314
piloto de flanco   6871
piloto de pruebas   6137
pinaza   4020
pinchazo   4379
pintar   3823
pintura   3822
piojo   2971
piquete   3995
piso   5749
pista   237, 6230, 6236
pista (de despegue)   4993
pista de aterrizaje   134
pista de rodaje   6063
pistola   4029
pistola ametralladora   5807
pistola automática   4031
pistola de señales   4030, 4032
pistola engrasadora   2067
pistola Verey de señales   4033
pistolera   2302
pivote   4040
placa base   4078
placa de presión   4082
plancha (de atraque)   1979
plancha (para permitir el desembarco de vehículos)   6251
plancha de cubierta   4080
plan de almacenamiento   1395
plan de aterrizaje (de tropas aerotransportadas)   4056
plan de carga   4057
plan de decepción   4051
plan de demolición   5069
plan de embarque   4052
plan de emergencia   4049
plan de encubrimiento   4050
plan de fuegos   4053
plan de maniobra   5071
plan de vuelo   4054
planeado   4069
planeador   2037
planear   2034, 4062
planear corto   6470
planeo   2033
planeo largo   3786
planeta   4065
planificar   4048
plankton   4068
plano   4047, 4063, 4064
plano (mapa) en relieve   4688
plano (topográfico)   3060

plano a grandes rasgos 4060
plano artillero de tiro observado 3540
plano de cola 5987
plano de estiba 5751
plano de la situación 5422
plano de operación 4059
plano de referencia 4622
plano de tiro restrictivo 4061
plano maestro 4058
plano planimétrico 4066
plan táctico de operaciones 4055
plantilla 6096
plantilla al completo 5775
plasma 4073
plástico 4074
plástico explosivo (término inglés) 2265
plástico reforzado con fibra de vidrio
  2032
plastificador 4076
plataforma 3825, 4084, 5261
plataforma continental 972
plataforma de aterrizaje 3813
plataforma de lanzamiento 3814
plato blindado inclinado 4081
playa 413, 1866
playa (de formación de trenes) 6914
plaza de armas 2089
plaza fuerte 5799
plazo 3953
(a) plazo corto 5334
plazo de anticipación 2761
pleamar 1803, 2271
(a) pleno rendimiento 1931
pliegue 1840
plomo 1944
pluma de carga 1219
plutónio 4100
población 4186
poco profundo 5246
poder 4239
poder separador en alcance 4521
polaina 1975, 4390
polaina corta 5554
polar 4171
polea 4367
policía 4177
policía militar 3199
poliestireno 4181
política de defensa 1163
político 4179
polivalente en combustible 3367
polo 4174
polo magnético 3002
polvillo radioactivo 1583
pompa 4375
ponderar 6810
poner al corriente (de) 600

poner(se) en contacto (con) 966
poner en capullo 849
poner en cero 6928
poner en libertad 1280, 2788
poner en marcha 5667
poner en servicio activo 897
poner pantalla a (luz) 5237
ponerse al pairo 2225
pontón 1797, 4183
pop-up point 4140
popa 5719
(a) popa 1, 66
(de) popa a proa 1863
por la borda (echar algo) 3760
porra 394
porta(a)viones 105
portaequipaje 6388
portafusil 5469
portamáquinas 2979
portatanques 6335, 6584
portátil 3052, 4189
portavoz 2969, 5595
porteo 4190
por tierra 3771
poscombustión 68
posición 1740
posición (de vuelo) 301
posición 4193
posición actual 4209
posición camuflada 2258
posición clave (estratégica, vital) 4204
posición de apresto 4211
posición de ataque 4196
posición de bloqueo 4198
posición de fuego 4202
posición de lanzamiento 2640
posición de la placa base 4197
posición de predicción 4207
posición de rodillas 4205
posición desenfilada 4199
posición de tendido 4210
posición de vigilancia 4206
posición dominante 4200
posición en desenfilada de casco 4203
posiciones sucesivas 5832
posición establecida 1850
posición previamente reconocida 4208
posición simulada 4201
posición ventajosa 4156
posición 2926
positiva 4284
positivo 4218
poste 4175
potencia 3517, 4239
potencia al freno 4241
potencia de combate 872
potencia de fuego 1721

potencia de penetración   4240
potencia de señal   5776
potencia específica   4244
potencia nuclear   3509
pozo   5241, 6820
pozo de ametralladoras   4035
practicaje   4011
practicar   4009
practico   4010
prebostal   4356
precendencia   4248
precesión   4249
precisión   21, 22, 4250
(con) precisión   4022
preciso   23
predictor   4251
preestablecido   4266
prefigurar   3749
prefijo de llamada   674
preliminar   4254
prender   704
(de) prensa   4268
preparación   4256
preparado   4557
preparar   4258, 6900
preparar una emboscada   5230
preparar una paracaída   3812
preparativos   3350
preparatorio de vuelo   4253
preponderante (en carros)   2227
preponderante (en infantería)   2230
prescindible   1544, 2628
presentarse   4727
presentarse en el puesto   4744
presente   4263
presión   4271
presionar (sobre, por)   4270
presión de aire   129
presión en tierra   2102
presión sanguínea   502
(buque) preso (entre)   2382
préstamo   2915
prestar   2916
presupuesto   615
pretexto   1049
prevención   4277
preventivo   4252
previsto   4069, 4247, 4259, 5068
primera curación   1729
primera línea   2869
primer plano   1865
principal   3010, 3028
principal (sufijo)   3105
(orden de) prioridad(es)   4286
prioridad de apoyo   4288
prioridad de fuegos   4287
prioridad de movimiento   3358

prisión   4291
prisionero de guerra   4294
proa   547
probabilidad   4296
probabilidad de destrucción   2667
probar   6128
procedimiento   4299
procedimiento cuadriculado de objetivo
   6049
procedimiento de identificación   763
procedimientos operativos normales
   4301
procesamiento   4309
procesamiento automático de datos
   4308
procesar   4343
procesar información   4310
proceso   4306
producción   3754, 4314
producto   900, 3300
productor de humo   2020
productos de fisión   1733
profundidad   1215
(en) profundidad de periscopio   3956
profundidad de vadeo   1861
profundo   1150
programa   4318, 5064
programa de conservación   5066
programa de convoy   5065
programador   4319
programador   931
progresista   4324
progresivo   4325
progresos   4323
prohibición   359
prohibición del uso de armas cont-ι
   aviones   2289
prohibir   358, 4326
prolongación   1563
prolongación (tiempo)   4333
prolongar (tiempo)   4332
promedio   325
promontorio   2212, 4335
pronóstico del tiempo   6803
pronóstico de zona   245
propaganda   4337
proporcionar   4353
propulsar   4338
propulsión   4342
propulsión por reacción   4915
propulsor   4339
protección   1039, 1045, 4346
protección corporal personal   514
protector   4347
proteger   5090
proteger (contra)   4345, 5283
proteger con sacos de arena   5041

proteger contra (sol) 5237
protegido y alertado 6704
protocolo 4349
protón 4350
prototipo 3298, 4351
proveedor 5858
proveer 4353
provisional 2535, 4354, 6097
próximo 829
proyección Mercator 3159
proyectar 4048, 5063
proyectil 4329
proyectil (contracarros) 5265
proyectil (de bote de metralla) 5268
proyectil (de expulsión de culote) 5267
proyectil (perforante) 5266
proyectil (teledirigido) contracarro 3259
proyectil ayudado por cohete 4913
proyectil balístico 3260
proyectil balístico intercontinental 3265
proyectil de iluminación 5271
proyectil fumígena 5272
proyectil guiado 3263
proyectil HESH 5270
proyectil indicadora de objetivos 5273
proyectil inteligente 5487
proyectil químico 5269
proyecto 4328
proyector 5127
prudencia 727
prueba 6129
(de) prueba(s) 6127
prueba de aptitud física 6130
(a) prueba de bombas 531
(a) prueba de choques 5319
(a) prueba de la inmersión 2411
(a) prueba de mal trato 1848
(a) prueba de patinazos 5431
(a) prueba de viento 6863
prueba en tierra 6135
prueba operacional de recepción 6133
pruebas 6357
pruebas de recepción 6358
(hacer) pruebas de tierra 2103
pruebas preparatorias de vuelo 6134
psicológico 4361
psiquiátrico 4360
publicar 4676
público 4362
pueblo 5228, 6626
puente 595
puente aéreo 117
puente aéreo táctico 5975
puente colgante 5935
puente de mando 596
puente giratorio 5949

puerta 1339
puerto 2184, 3894, 4187
puerto de matrícula 2303
puerto de refugio 4638
puesta del sol 5845
puesto 4221, 5680, 5682
puesto avanzado 3752
puesto de carga 4115
puesto de control 4116
puesto de enlace 6103
puesto de mando 4227, 5685
puesto de mando aéreo 4224
puesto de observación 4229
puesto de observación aérea 4225
puesto de socorro 5684, 5687
puesto en tierra (por avión/helicóptero) 2697
puesto escucha 4228
puesto eventual 4194
puesto fronterizo 4111
puestos de abandano (del buque) 3
puestos de combate 30, 5697
puestos de engrase 4170
pulmón 2989
pulsador 4388
pulso 4373
pulso electromagnético 1437
pulsoreactor 4374
pulverización 5604
pullup point 4142
puñal 1098
punitivo 4383
punta 4105
(a) punta de pistola 2134
puntal 2630
puntales 5329
puntería en dirección 6343
puntería indirecta 2748
puntiagudo 5249
punto 1344, 4106
punto/campo de recogida 4000
punto avanzado de municionamiento y repostaje 4126
punto cero deseado 6929
punto clave 4131
punto crítico 4119
punto de abastecimiento 4148
punto de aguada 4159
punto de ajuste 4133
punto de amarre 1738, 4152
punto de anclaje 1738
punto de atadura 4133
punto de aterrizaje 4132
punto decisivo 4155
punto de concentración de prisioneros de guerra 4141
punto de contacto (de unión) 4117

reconocimiento radioeléctrico 4460
reconquistar 4569
recreo educativo del soldado (RES) 4495
rectificación 1012, 4602
rectificador 4603
recuperable 4596
recuperación 4598, 5034
recuperación (de material) 340
recuperar 4594, 4595, 5031
recursos 4780
rechazar 2296, 3783, 4664
rechazar un ataque 4715, 4753, 6157
red 3438, 3444
red (de mando) 3439
red de alarma 3441
red de defensa contra torpedos 6201
red de fijación 3445
red de retransmisión 4461
red de transporte 908, 3446
redesplegar 4609
redondo 4955
reducción 4615, 6149
reducción de grado 4616
reducir 1384, 4612
reducir a chatarra 5084
¡Reducir al minimo! 3237
reemplazar 4717
reemplazos 4720
reentrada 4620
reentrante 4619
reentrar 4618
reexpedir 4611
referencia cartográfica 3064
referencia terrestre 2711
reflector 4629
reflejar 4626
reflejo (imagen) 4627
reflexión 4627
reflexión acústica 5394
reflujo 1410
reforzamiento 4662
reforzar 4659
refracción 4631
refractar 4630
refrigerado por agua 6738
refuerzo 6147
refuerzo del terreno 6122
refuerzos 4662
refugiado 4637
refugio 639, 1045, 1046, 5277
(buscar) refugio 4636
(lugar de) refugio 4635
refugio subterráneo 1391
regar (con manga) 2329
régimen ascensional 4532
régimen necesario de abastecimiento 5870

regimiento 4639
registrar 1914, 2941, 4305, 4640, 5121
registrar a fondo 4988
registrar un blanco 4593
registro 4641, 5123
registro de minas 3232
registro de navegación 2939
registro de objetivos 4644
regla 4985
regla de cálculo 5467
reglamento(s) 4650
reglamentos 4986
reglamentos de la circulación 4652
reglamentos del cuartel general 2016
reglamentos de seguridad 4651
reglamentos por identificar buque o avión enemigo 2338
reglar el tiro (sobre blanco auxiliar) 4642
reglar el tiro sobre 6931
reglas de tiro antiaéreo 4987
reglas para el vuelo visual 6637
reglas permanentes de servicio 5659
regulación 4649
regulación de tráfico 6260
regulador 4653
regular 4646, 4648
regular el tiro 4498
rehabilitación 4655
rehabilitar 4654
rehacerse 4489
rehén 2336
reja 2058
relación(es) 4665, 4666
relación de compresión 926
relaciones civilo-militares 807
relaciones públicas 4364
relámpago 2814
relevar 4689
relevo 4685
relieve 4686
reloj 6730
relojería 3136
relleno 3816
remache 4889
remacher 4888
remar 4366, 4975
remar con canalete 3817
remo 3526, 3818
remolcado 6471
remolcador 6394
remolcador de salvamento 5036
remolcar 6214
remolino 6402
remolque 6215, 6269
(cable de) remolque 6217
remoto 4693

rezagarse 5754
rezón 2057
riachuelo 1067
ribera 366
ribera enemiga 5325
riesgo 2204, 4883
riesgo excepcional 1455
riesgo insignificante 3428
rígido 4867
rincón 1006
río 4886
río arriba 6533
riostra 5800
risco 507
ritmo 3805
ritmo de consumo 964
ritmo de paso 4412
rizo escollo 4617
roblón 4889
robo 4907
roca 4909
rocoso 4916
rodada 6242
rodado 6835
rodar 6062
rodear 5446
rodeo 1248
rodillo 4928
rodillo (para alambre de púas) 4929
¡Roger! 4917
rompehielos 2383
rompeolas 587, 5130
romper 572
romper el contacto 1286
romper filas 1584
rompientes 580, 5898
ropa de cama 438
ropa protectora 4348
rotación 4949
rotativo 4947
roto 607
rotor 4950
rotular 2688
rótulo 2687
rotura 583
rozar 2064
rueda 6831
rueda de repuesta 5552
rueda de timón 6833
rueda motriz 1379
ruído 3467
ruinas 4983
rumbo 1032, 2211
rumbo de caza 1035
ruptura 573, 585, 4994
ruta 4962
(en) ruta 1034

ruta (de navegación) 2714
ruta aérea 4963
ruta aérea principal 4970
ruta de desvío 659
ruta de emergencia 4964
ruta del convoy 4966
ruta de navegación 5307
ruta de retiro de emergencia 4965
ruta de seguridad 5016
ruta ortodrómica 4967
ruta principal 4894
rutina 4973

S

saber 2764
sabot (término francés) 5003
sabotaje 5005
saboteador 5007
sabotear 5006
sacar 1361
sacar (etc.) con una bomba 4376
sacar (información) 4817
sacar con sifón 5414
sacar de servicio 3334
saco 346, 2679
saco-manta 5461
saco de dormir 5461
saco terrero 5040
sacudida 486
sala 6673
sala de control 4936
sala de máquinas 4937
sala de oficiales 6676
sala de operaciones 4939, 4941
salida 1208, 5536, 6109
salida (del ordenador) 3755
salida del sol 5844
saliente 4330, 4331, 5025
salino 5026
salir 1207, 1450, 5020
salir a la superficie 5900
saltar 2638
saltar en paracaídas 3845
salto 544, 2637
salto a gran altura y apertura baja
    3847
salto de apertura retardada 3846
saludar 1649, 5028
salva 6644
salva de ....cañonazos 5030
(lanzar una) salva de fusiles 6645
salvamento 5035
(de) salvamento 4763
salvamento aeronaval 130
salvar 4762, 5032

salvoconducto   5012
sancionar   5037
sanciones   5038
sangrar   490
sangre   500
santabárbara   2994
santo y seña   6734
'santo y seña'   763
saqueador   3066
saquear   2961, 4099, 5008
satélite   5047
satélite de caza   2666
satélite en órbita   3702
saturación   5049
saturar   5048
sección   5142
sección de pontoneros   5143
sección transversal   5144
(en) seco   2260
secretario   5140
secretario general   5141
secreto   816, 1055, 5139, 6458
sector   5146
sector de tiro   5148
secuestrar   2273
secuestro   2274
secundario   5134
sedimento   5407
seguidor (del mísil)   6245
seguimiento   1844, 1847
seguimiento inmediato   2342
seguir   1842
seguir la pista (de)   6224
segundo   3113, 5133
segundo oficial   3583
según horario   5067
seguridad   5154
seguridad de funcionamiento   4682
seguridad de transmisiones   5156, 5159
seguridad física   5158
seguro   4684, 5013, 5151
(en) seguro   1577, 1578, 6433
selección   5171
seleccionar (entre)   5170
selectivo   5172
selva   2643
(de) selva   2644
sellar   5110
sello   5112
sello (de plomo)   5111
sembrador de minas   2747
sembrar   2740, 2742
sembrar minas   2750
sembrar minas (de forma dispersa)   2751
sembrar minas según un plan   2749
semiautomática   5179

semicargada   2150
semigrosor   2153
semioruga   2155
semiremolque   5181
seña   6896
señal   3077, 5380, 5381, 6225
señalamiento del camino   5397
señal convencional   5374
señal de fin   5383
señal de identificación   5392
señal del hueco en un campo minado   3078
señal de niebla   5384
señal de reconocimiento   5387
señal de socorro   5382
señalero   5391
señal que jalona el camino por un campo minado   3079
señal terrestre   5385
senda   2715, 6230, 6264
sendero   3909
sendero seguro   2852
sensibilidad   5188
sensor terrestre autónomo   5190
sentencia de muerte   6719
sentido   5185
sentir   5186
señuelo   1147
separación vertical   6607
separado   5194
separar   5195
separarse   563, 5592
sepultura   2059
séquito   4811
ser de grado superior   3756
serie   5198, 5200
serrar   5051
(de) servicio   1400
servicio de asistencia   5212
servicio de estado mayor   5637
servicio de información   5207
servicio de intendencia   4403
servicio de material (de guerra)   5210
servicio de personal   3967
servicio de transmisiones   5389
servicio militar   5203
(en el) servicio militar   5208
servicio militar obligatorio   6290
(de) servicio pesado   2229
servicio público   6540
servicios especiales   5557
servicios públicos   4365
servidumbre de paso   4866
servir   3038
servodirección   5715
servomecanismo   3135
seto vivo   2234

sextante 5232
shock eléctrico 5314
silbar 6839
silbato 6840
silenciador 3361, 5400, 5401
silencio 5399, 5403
silencio de radar 4439
silencio de radio 4462
silencioso 5402
silo 5406
silueta 5404
silla 5009
sillín 5009
símbolo 5957
simulacro de guerra 6694
simulado 3978
simulador 5410
simular 5409
sin asfaltar 6506
sin cargo 1903
sincrónico 5960
sincronizar 5959
sin encontrar resistencia 6510
sin explotar 6481
sin las armas precisas 6456
sin montar 6433
sintonizar (a) 3437
sin usar 6525
sirena 5415
sirena (de niebla) 1837
sísmico 5165
sismógrafo 5166
sistema 5961
sistema amplificador 606
sistema articulado 2877
sistema de adaptación al terreno 6121
sistema de alerta de intrusos 2570
sistema de arma 6799
sistema de aterrizaje por instrumentos 2498
sistema de bombardeo de órbita parcial (término inglés) 1890
sistema de esquivar el terreno 6119
sistema de husos horarios 6957
sistema de intercomunicación 2530
sistema de misiles antibalísticos 219
sistema de navegación a inercia 2450
sistema de regaderas 5612
sistema de visión nocturna 3463
sistema de vuelo a altura constante 6120
sistema georef 2023
sistema nervioso central 747
sistema no repetible 3631
sitiar 2580
sitio 5356, 5416

situación 2926, 4193, 5416, 5421
situación (de la unidad) 5705
situación de combustible 5703
situación de equipos 5702
situar de antemano 4260
sobre (cerrado) 1488
sobrecarga 3782
(de) sobrecarga 5846
sobrecargar 3781
sobreponer 5847
sobrepresión 3780
sobrepresión de aire libre 1897
sobrepresión máxima 3935
sobreviviente 5930
sobrevivir 5929
sobrevuelo 3778
socavón 5818
socorro 4687
software 5518
sol 5840
solar 5416, 5521
soldadesca 5527
soldado 5523
soldado (profesional) 5525
soldado de fuerzas especiales estadounidenses 4520
soldado de infantería 2452
soldado desconocido 6498
soldado estadounidense 2024
soldados 4527
soldados rasos 4525
soldadura 5116, 6819
soldar 5522, 6818
solicitud 4754
soltar 4674, 4676
solución anticongelante 221
solvente 5530
sollado de marinería 3166
(a la) sombra 5238
someter(se) 5814
sonar 5531
sonar pasivo 5534
sonar submergido 5532
sonar suspendido de helicóptero 5533
sonda acústica 1423
sondaje acústico 1424
sondar 540, 1554, 5539
sondeo 5541
sonido 5537
(barrera del) sonido 5540
sonoboya 5535
soplar 503
soplete 505
soplete oxiacetilénico 3798
soportar 432
soporte 518, 552
soporte de culatín 4790

T

técnicas de combate 1612
técnico 6074, 6076
techo 731
techo de vuelo estacionario 2346
tela impermeable 5256
tela metálica 6885
telecomunicaciónes 6077
telecontrol 4694
teledirección 4451
teledirección manual 3058
teledirigido 4695
telefonear (a) 6080
teléfono 6081
teléfono del carro 6082
telegrafía 6078
teleimpresor 6086
telemetria 4516
telemetría 6079
telemetria radar 4437
telémetro 1642
telescopar(se) 6088
telescópico 6089
telescopio 6087
teletipo 6090
televisar 6091
televisor 6093
tema 5805
témpano de hielo 2385
témpanos flotantes 2380
temperatura 6095
tempestad de arena 5043
templado 6094
templar 4409
temporal 5748, 6097
tendedor (sembrador) de minas 3233
tender 2740, 2742
tender una emboscada 178
tender un puente sobre 594
teneduría de libros 20
tener correlación (con) 1013
tener un alcance de 4523
tensión 5783, 6101
tenso 6061, 6099
tensor 6100
teodolito 6139
térmico 6140
terminal 6105, 6106, 6107, 6109
terminal marítima 6755
terminar(se) 6113
término 6105
término medio 325
termo 6544
termómetro 6145
termonuclear 6146
terral 3617
terraplén 365, 1407, 4492, 6114
terreno 6115

terreno accidentado 2276
terreno blando 2092
terreno dominante 6116
terreno escabroso 2091
terreno medio 2088
terreno muerto 2086
terreno transitable por vehículo 6118
terreno transitable por vehículos 2093
territorio 6123
territorio ocupado 3558
terrorismo 6124
terrorismo urbano 6537
terrorista 6125
test de selección de reclutas 6131
texto claro 4046
tiempo 6175
tiempo (atmosférico) 6801
tiempo concedido (para una acción)
  6178
tiempo de exposición 1562
tiempo de paz 3931
tiempo de reacción 4545
tiempo de reparación 6516
tiempo de vida 2795, 5262
tiempo medio de Greenwich (TMG)
  2069
tiempo medio entre fallos 3126
tiempo mínimo de aviso (de alerta)
  6715
tiempo muerto 1124
tiempo pesado 2233
(en) tiempo real 4560
tiempo sobre el objetivo 6183
tienda 6102
tiendas de campaña 6104
tiendepuentes 2746
tierra 1405, 2694
(de) tierra 273, 2082
tierra-aire 2104, 5902
tierra-tierra 2105
tierra cero 6930
tierra de nadie 2696, 3468
tierra excavada 5593
(fiebre) tifoidea 6423
tifón 6424
timón 1636, 2250
timón (de dirección) 4980
timón de picado 4981
timonel 2253
timonera 6836
timonero 4402
timones de profundidad 2372
típico 6425
tipo 6420
tipo de posición de las armas 5701,
  5706
tira de guarnición 1984

transferir 4609
transfusión de sangre 6305
(de) transición 6312
transistor 6308
transistorizado 6309
transitorio 6307
transladar 6303
transladar una unidad (a otro sitio)
  4692
translado 6302
transmisión 1374, 3164, 5380, 6316
transmisión de datos 2876
transmisión en circuito 2317
transmisiones 907, 6077
transmisiones ópticas 6636
transmisor 6318
transmisor respondedor 6324
transmitir 5182, 6317
transónico 6321
transparencia 6322
transparente 6323
transportable 6332
transportable por avión 142
transportado en avión 116
transportado en helicóptero 2244, 2246
transportado por mar 5107, 6736
transportador 3858, 4352
transportadora 6304
transportador de rodillos 4930
transportar 1606, 6303, 6325
transportar el tiro 5288
transportar en avión 1824
transportar en viajes de retorno 5346
transporte 2201, 6302
(de) transporte 6327
transporte aéreo 117
transporte aéreo de tropas 6371
transporte aéreo estratégico 6330
transporte aéreo táctico 6331
transporte a lomo 6329
transporte de ida y vuelta 5347
transporte del radar 5286
transporte en helicóptero 2248
(por) transporte marítimo 6736
transporte oruga acorazado (TOA) 258
transportes por avión 6326
transreceptor 6319
transversal 6338
trasera 4561
traslapo 3772
traspaís 2277
traspasar 6303
traspaso 6302
traspaso de pistas 6250
tratado 6348
tratamiento 4307
tratamiento médico 3145

tratar 4304
(por el) través 4
travesaño 518, 6320
travesía a snorkel 5811
traviesa 6341
trayectoria 6234, 6296
trayectoria de acercamiento 236
trayectoria de planeo 2036
trayectoria de vuelo 1792
trayectoria hostil 2339
traza (en la cinta) 6233
trazado 4089
trazado general 3750
trazado general de la operación 3751
(elemento) trazador 6228
trazar 4091, 6223
trazo 6222
tregua 6381
tren 6271
tren de aterrizaje 2005, 6457
tren de combate 874
tren de transportes de una unidad 6295
tren de víveres y bagajes 874
trenes de campaña 6294
trenes de combate 6293
trepar 826
treta 1599
triangulación 6362
triangular 6361
triángulo 6360
trinchera 6350
trinchera abrigo 6353
trinchera de combate 6352
trinchera de comunicación 6351
trineo 5460
triple 6368
trípode 6369
tripulación 912, 1072
tripulación de avión 106
tripulación de carro de combate 6011
tripular 3036
trocha 2712
tromba (marina) 6754
tronera 5477
tropa numerada 761
tropas 6372
tropas auxiliares 319
tropas de choque 5320
tropas divisionarias 1332
tropas en área de embarque numeradas
  y equipadas 6370
tropas esquiadoras 5447
tropas inexpertas 4540
tropas paracaidistas 3865
tropel 1082
tropical 6375
trópico de Cáncer/Capricornio 6376

tropopausa   6377
troposfera   6378
tubería   4028
tubo   4026, 6393
tubo de aire   119
tubo de desogüe   1358
tubo de escape   5986
tubo lanzatorpedos   6202
tubo snorkel (de respiración)   5506
tuerca   3524
tuerca de alas   6872
tundra   6396
túnel   6398
turba   3288
turbina   6399
turbina a gas   2000
turbocompresor   6400
turboreactor   6401
turbulencia   6402, 6403, 6726, 6838
turbulento   6404, 6405
(trabajo por) turno   5287

U

ultimátum   6427
último vehículo   6572
(de) ultramar   3785
ultrasónico   6429
ultravioleta   6430
único   5411
unidad   6489
unidad autónoma   19
unidad contigua   40
unidad de apoyo   5894
unidad de distribución   6494
unidad de duchas   393
unidad de elite   1058
unidad de pertenencia   3867
unidad de salida   3747
unidad de sanidad   3146
unidad en contacto   6493
unidad en reserva   3475
unidad entrante   2430
unidad especial   1058
unidad inmediata   3433
unidad motorizada   3339
unidad residente   4772
unidad subordinada   5827
unificar   6487
uniforme   6486
uniforme(s) verde oliva   3626
uniforme de combate   1613
uniforme de etiqueta   1368
unión   6488
unir(se)   6491
unir(se) (a)   2631

universal   6495
uranio   6535
uranio enriquecido   1482
urgencia   6538
(de) urgencia   1451
urgente   2405, 6539
usar   1543
útil de mango corto   2567
utilidad   5213
utilizar   2189

V

vacante   6508
(de, al) vacío   6543
vacío de cisterna   6426
vacunación   6542
vadeable   1860, 3898
vadear   1858, 6656
vado   1859
vagar   6667
vagón   6658
vagón (ferroviario)   6382
vagón de mercancias   1907
vaguada   6752
vaina   716, 5254
vaina (de bayoneta)   5053
valor   1977, 4683
válvula   6548
válvula de purga   6596
válvula de seguridad   5018
valle   6546
vanguardia   50, 6553
vano (techo)   5546
vapor   6554
(de) vapor   5709
(buque de) vapor   5710
varadera   5430
varar   412
variabilidad   6555
variable   6556
variación   6557
vector   6558
vehículo   6563
vehículo anfibio   184
vehículo articulado   6568
vehículo blindado de recuperación   6567
vehículo de comando y control   6570
vehículo de combate de infantería   6573
vehículo de mando   6569
vehículo de oruga   6585
vehículo de puesto de mando   6571
vehículo de reconocimiento   6579
vehículo de reentrada dirigido   6574
vehículo de reentradas múltiplas (MRV)
   6578

vehículo de reentradas múltiplas
  independente (MIRV) 6577
vehículo dirigido por control remoto
  6581
vehículo mecanizado de combate 6575
vehículo no blindado 6582
vehículo oruga de desembarco 6586
vehículo para la nieve 5513
velar (por) 2653
velero 6613
veleta 6552
velocidad 3804, 5566, 6588
velocidad absoluta 5570
velocidad baja 2981
velocidad de ascenso 5567
velocidad de convoy 5568
velocidad de crucero 6590
velocidad de escape 6591
velocidad del sonido 5575
velocidad de progresión 4531
velocidad de salto 5571
velocidad de viento 5573
velocidad económica de crucero 5569
velocidad emanente 6595
velocidad indicada 2436
velocidad inicial 6593
velocidad máxima de crucero 5572
velocidad máxima permitida 5574
velocidad propia verdadera 6384
velocidad relativa 6594
velocidad relativa al aire 132
velocidad remanente 6589
vencer 958, 1152
vencer (al enemigo) 436
vendaje de campaña 1614
vendar 468
ventana 6862
ventana de expulsión 5481
ventanas de ventilación 6597
ventilador 1587, 6599
ventilar 6598
ventisca 6844
ver 5358
verde-oliva 3626
vereda 4843
verga 6916
verificable 6600
verificar 6128, 6601
vertical 6531, 6604
vértice geodésico 4153
vestido de paisano 3362
vestidos 834
veterano 6615
veterinario 6616, 6617
vía 6238
via (aérea) 2713
vía acuática 6759

viabilidad 6257
viaducto 6618
viaje 6652
viajero 3903
vía navegable 6759
vía principal de abastacimiento 4968
vías de paso 3900
victoria 6619
vida 2796
video cassette 6621
vidrio 2030
vidrio cilindrado 2031
viento 6853
viento de cola 5988
viento de tierra 3617
viento duro 1976
viga 427, 2026
vigente 1855
vígia 2953, 2954
vigilancia 3795, 6624
(de) vigilancia 5918
vigilancia aérea 137
vigilancia móvil 3797
vigilancia por saltos 3796
vigilante 6625
vigilante de alarma antiaérea 6674
vigilante de muelle 195
vigilar 3316, 4176, 5239, 6265, 6731
vigor 5648
viraje convencional 4302
virar 6562
virutas 754
(ristra de) virutas 4944
visibilidad 6628
visibilidad mutua 2565
visible 6629
visión nocturna 3462
visita 6631
visitar 6632
visor 5359, 6633
visor (infrarojo) de arma 6797
visor de bombardeo 532
visor nocturno 5364
vista 6623
vista estereoscópica 5718
vista óptica 2865
visual 6629, 6634
vital 6640
vivaque 479, 5279
vivaquear 478
víveres 4355
víveres de campaña 1620
vocal 6641
volante 1379, 1827, 6833
volante de puntería (en dirección) 2178
volante de puntería (en elevación) 2177
volar 487, 506, 1823

DEUTSCHES WÖRTERVERZEICHNIS

absprengen 1093
abspringen 2638, 3838
Absprung 1382, 2637, 3854
Absprunggebiet 6944
Absprunggruppe 5724
Absprunghöhe 2238
Absprungraum 2705
Absprungsgeschwindigkeit 5571
(Zeit-)Abstand 2559
Abstandflugkörper 3271
Abstandzünder 1963
abstellen (zu) 1237
Abstellung 5136
Absturz 1063
abstürzen 1062
Absuchfeuer 1653
Abszisse 6910
Abtasten 5058, 5059, 5942
Abtastung 5941
Abteilung 914
abtreten 1584
Abtreten! 1584
Abtriebswinkel 201
Abtrift 2768
Abwärtssichtradar 4425
Abwasserkanal 5231
abwechseln (lassen) 166
abwechselnd 165
(Spionage-)Abwehr 1024
abwehren 1602, 2296
Abwehrfeuer 1692
(U-Boot-) Abwehrrakete
    4911
abweichen 1174
(vom Kurs) abweichend 1033
abweichender Zielpunkt 4136
Abweichung 1175, 1284, 6557
abwerfbar 2628
abwerfbarer Zusatztank 6005
abwerfen 1383, 2626, 3838, 4674
Abwesende(r) 8
Abwesendenmeldung 4730
abwickeln 3974
Abwind 1348
Abwurf 1382, 3854
Abwurf (aus der Luft) 109
Abwurf (ohne Fallschirm) 1900
Abwurfgebiet 6943
Abwurfhöhe 2238
Abwurfpunkt 4145
(Motor) abwürgen 5647
Abzeichen 344
(Dienstgrad-)Abzeichen
    5795
Abzug 4284, 6363
Abzugsbügel 6364
Abzugsleine 2717

abzweigen 563
Abzweigung 1869
Achse 332, 333, 742
achten 6731
achter(n) 66
Achterdeck 4400
achtern 1
Adjutant 43, 79
Aerodynamik 57
aerodynamisch 56
aeronautisch 60
Agent 71
Aggregat 72, 4071
ahoi! 76
Akademie 9
Akten 1343, 1630
Aktenbündel 1343
Aktenmappe 601
Aktionsradius 4471
aktiv 34
aktivieren 31
Aktivierung 32
Aktivierungsmittel 33
Akustik 27
akustisch 26
Alarm 145, 146, 5662
alarmbereit 147
alarmieren 148
(in) Alarmstand sein 5663
Alarmstart 5080
(einen) Alarmstart durchführen 5082
Alarmstufe rot 4606
alleinfahrendes Schiff (z.B. aus den
    Geleitzug) 2766
alle Mann 153
allgemein 2014
Allgemeinbefehle 2016
allgemeine Befehlsgewalt (meist eng.
    Begriff) 3673
allgemeiner Alarm (meist eng. Begriff)
    2015
allgemeiner Militärdienst 6290
Alliierte(r) 161
allmählich 2053
Allwetter- 159
Alternative 168
Aluminium 173
amerikanischer Soldat 2024
Amperemeter 181
Amphibienkettenfahrzeug 6586
amphibisch 185
amphibische Operation 3653
Amplitudenmodulation 3311
amtlich 3606
Amts- 3606
Analogrechner 186
Analyse 188

analysieren 187
Analytiker 189
an bestimmten Auftrag
    gebundene Befehlsgewalt (meist
    eng.Begriff) 3672
anbohren 540, 5102
anbrechen 575
(Zünder) anbringen an 1942
ändern 163
andersdenkend(-er -e) 1311
Anfahrtfunkfeuer 419
Anflugbeginn 2466
Anflugfunkfeuer 419
Anflugschneise 236
anfordern 673, 4755, 4758
Anforderung 1193, 4759
(Luft-)Anforderung 4754
angegliedert 294
Angehörige 3452
angepasste Reaktion (meist
    eng.Begriff) 4784
angerreichertes Uran 1482
angezeigte Eigengeschwindigkeit 2436
angliedern 293
Angliederung 295
angreifen 275, 297
(im Tiefflug) angreifen 5753
Angriff 274, 296, 5786
Angriffsachse 2854
Angriffsbeginn 2255
Angriffsbereich 6953
Angriffsgrenze 2824
Angriffslandung (amphibisch) 2700
Angriffslandung (aus der Luft) 2701
Angriffsplan 5071
Angriffsspitze 5556
Angriffsstellung 4196
Angriffstrupp 6065
Angriffstruppen 5320
Angriffsvorbereitung durch
    überraschenden C-Waffeneinstaz
    5914
Angriffswelle 1412, 6761
Angriff zurückschlagen 4715
anhaken 2316
anhalten 1746
Anhaltepunkt 4142
Anhalterecht 4865
anhängefähig 2933
anhängen 2311, 2316, 2934
Anhänger 6269
(auf) Anhieb schiessen 1682
Anker 190
Ankerboje 193, 3322
Ankerlicht 194
Anker lichten 6809
(vor) Anker liegen 4844

ankern 191
Ankerplatz 192
Ankerwache 195
Anklage 767
Anklagevertreter 4344
Ankündigung 3492
Anlage 209
(Triebwerk-)anlage 3810
Anlage über Logistik in Befehlen 210
Anlanden 5309
anlassen (eine Pumpe) 4281
anlassen (Motor) 5667
Anlasser 5668
Anlaufzeit 2761
anlegen 1333, 1335, 3319
anlegen (an) 3320
Anleihe 2915
Anlernling 6274
Anmarsch 235
Anmerkung 215
Annäherung 233, 991
Annäherungszünder 1962, 1971
Annahmeprobe 6133
annullieren 685
anpassen 41
Anreger 2469
Anruf 763, 6083
anrufen 672, 762, 6080
ansammeln 5731, 5746
Ansatz 2988
anschäkeln 5235
Anschauungsmaterial 6635
Anschlag liegend 4210
Anschlagtafel 630
anschliessen 4098
(sich einem Funknetz) anschliessen| 4745
(sich) anschliessen (an) 2631
Anschlussversorgung 4802
ansetzen 2733
(etwas) ansteuern 5225
ansteuern (nach) 2210
anstreichen 3823
answesend 4263
Antarktis- 216
antarktisch 216
Anteil 4413, 5463
Antenne 217
Antennendom 4475
Anti-G-Anzug 2110
Antibiotikum 220
antiseptisch 225
antreten 1582
Antrieb 1374, 4342
Antwort 4723
Antwortbake 424
(be-)antworten 4722

Antwortsender 6324
Anwassern 5589
Anweisungsschild 4232
(sich) anwerben (lassen) 1479
Anwerbung 1481
Anzahl 3520
anzeigen 2460
Anzeiger 1242
anzünden 2804
Apogäum 230
Appell 3851, 4927
Äquator 1493
äquatorial 1494
Äquinoctium 1496
Arbeit 6897
arbeiten 6898
Arbeitsanzug 1595
arbeitsaufwendig 3055
Arbeitsdienst 1594
Arbeitskommando 3885
Arbeitskräfte 3053
Archiv 1630
arktisch 240
Arm (Gewässer) 565
Armaturenbrett 3831
Armband 567
Armbinde 567
Armee 263
armieren (Beton) 4660
Armierung 5254
Arrest 265
Arrestant 4292
Arsenal 267
Artikel 2599
Artikel für gemeinsamen Gebrauch 903
Artillerie 271
Artilleriebekämpfung 1657
Arzneimittel 3147
Arzt 1337
Arzt bei der Fallschirmtruppe 3860
ärztlich 3144
ärztliche Behandlung 3145
ärztliche Untersuchung 3994
Asbest 272
Asphalt 6051
Astrokompass 288
Astronavigation 733
Atemgerät 589
atmen 588
Atmosphäre 289
Atom 291
Atom- 292, 3500
Atom-Sicherheitsgrenze 2842
Atomabfälle 6728
Atomantrieb 3510
atomar, biologisch und chemisch (ABC) 3501

atomarer Zwischenfall 3508
atomare Verteidigung 3506
Atomgefechtskopf 6698
Atommacht 3509
Atomschäden 3504
Atomschädenbewertung 3505
Atomsprengkörper 3507
Atomwaffe 6787
Atomwaffenunfall 3516
Atomwarnung 6711
Aufbau 620, 1872
Aufbau(ten) 5852
aufbauen 621
aufblasbar 2456
Aufbrausen 1765
aufdecken 6453
auf ein Minimum herabsetzen 3236
auffassen 1239, 5186
Auffassungsgerät (meist eng. Begriff) 5189
auffrischen 4654
(einzeln) aufführen 2600
Aufgabe 1935, 6053
Aufgeber 3735
aufgelöst 3973
aufgesessen 3347
(nicht) aufgezeichnet 6511
aufgreifen 3999
Aufhängung 5934
aufklären 5077
aufklärende Batterie 2925
Aufklärung 4583
Aufklärung (durch Beschuss) 4590
Aufklärung (durch Gewalt) 4591
Aufklärungsfahrzeug 6579
(ferngelenktes) Aufklärungsflugzeug 6581
aufladen 2894, 2896
Auflegekarte 3774
Auflieger 5181
auflockern 1296
Auflockerung 1297
Auflockerungsraum 1295
auflösen 1276, 3974
Auflösungsvermögen 1172, 4779
aufmachen 3636
Aufmarschgebiet 939
Aufnahmepunkt/-platz 4000
aufnehmen 3998, 6017
aufpassen (auf) 2653
aufpumpen 4378
Aufputschungsmittel 3945
aufrecht 6531
aufrechterhalten 3018
Aufrechterhaltung 3023
aufrollen 4933
aufrufen 675

Aufruhr 2499, 4871, 4877, 6403
Aufruhrbekämpfung 4873
Aufruhrbekämpfungsmittel 4874
Aufrührer 2500, 4875
aufrührerisch 6404
Aufsatzwinkel 4393
Aufschlag 2413
Aufschlag! 5587
Aufschläge 1572
Aufschlagzünder 1954, 1955, 1958
aufschliessen 6504
Aufseher 5854
Aufsetzen 6211, 6212
Aufsetzzone 6951
aufspüren 6243
Aufstand 2499, 4833, 4882, 6532
Aufsteckmunition 5180
aufstellen 1512, 2871, 3091, 3730
(sich) aufstellen 1874
(als Posten) aufstellen 4220
Aufstellung 1307
Aufstieggeschwindigkeit 4532
Auftanken (bei laufenden Motoren) 4634
auftauchen 1450, 5900
aufteilen 686
Auftrag 285, 2493, 3277
Auftreffpunkt 4164
aufwachen 6662
Aufziehkabel 4879
Auge 1567
Augen- 3568
Augenabstand 2547
Augenaufklärung 6638
Augenblick- 6307
augenblicklich 1091
Augenverletzung 1568
ausbaggern 1366
ausbauen 3730
ausbilden 2492, 6272, 6900
Ausbilder 2495
Ausbildung 2493, 6275
(forgeschrittene) Ausbildung 6276
(militärakademische) Ausbildung 6286
Ausbildungs- 2494
Ausbildungskenntnisse 6288
Ausbildungsoffizier 3588
ausbrechen 582
Ausbruch 583
Ausdauer 1464
Ausdruck 4285
ausfahren 2977
Ausfälle ohne Feindeinwirkung 3471
ausfliegen 1824
ausfragen 1127
ausführen 1532
ausführender Truppenführer (meist eng.
  Begriff) 1533

Ausführungsbefehl (meist eng. Begriff)
  3707
Ausgabe 2594
Ausgabeeinheit 6494
Ausgabestelle 4123
Ausgang 6109
Ausgangsbautyp 4351
Ausgangspunkt 4122, 4163
Ausgangswerkstoff 2898
Ausgangsziel 2038
ausgeben 2595
(ungenügend) ausgebildet 6473
Ausgehverbot 1089
(gut) ausgerüstet 6824
Ausgleich 3612
ausgleichen 349, 3613
Ausgleichsstück 5938
Ausguck 2954
aushalten 1465
aushalten (gegen) 2297
Aushang 3493
Aushängeschild 5396
Ausklinkung 4678
ausladen 1279, 6502
Auslagerung 1393
Auslagerungsplan 1395
Auslandsdienst 6213
Auslaufbefehl 5022
auslaufen 5020
auslaufen (nach) 5019
Ausleger 2630
Auslösedraht 6881
Auslösemechanismus 3137
auslösen 4673
Auslöser 6363
Auslösevorrichtung 4680
Auslösezünder 1965
Auslösungsvermögen 4521
ausmessen 2001
ausmustern 1278, 4664
ausnutzen 1551
Ausnutzung 1552
ausplündern 5008
Auspuff- 1537
Auspufftopf 3361, 5401
Ausrichten 152, 2872
ausrichten (nach) 151
ausrüsten (mit) 1497
Ausrüstung 252, 1498
Ausrüstungsgegenstände stillegen
  3334
Ausrüstungszustand 5702
ausschalten 3449, 5995
ausscheidendes Schiff 996
ausscheren 6562
ausschiffen 1126
Ausschiffung 1125

ausschlachten 686
Ausschlussgebiet 6945
Ausschuss 4663
ausschwärmen 1588
Aussehen 6408
Aussen- 1564, 1565, 3745
aussenbord(s) 3743
Aussenbordmotor 3744
Aussenstation 3757
ausser Betrieb 29
ausser Dienst 5211
äusserer 3745
äusserlich 1564, 1565
ausser Schussweite 4509
aussetzen 1559, 3087
aussetzend 2539
Aussicht 6623
(in) Aussicht 3608
Aussicht gewähren auf 3776
Aussichtspunkt 4156
Aussichtsturm 6733
aussonderungsreif 459
aussteigen 1249, 1290
ausstrahlen 1458
Ausstrahlung 1456
ausstreichen 5777
Ausströmungsgeschwindigkeit 6591
austauschbar 2529
austauschbar (meist eng. Begriff) 2549
Austauschbarkeit 2528
Austauschbarkeit (meist eng. Begriff)
    2548
austauschen 2527
austauschen (mit, gegen) 1531
Austauschplan 5070
ausüben 4246
Auswahl 5171
Auswahl- 5172
Ausweich- 5826, 5857
Ausweichbewegung 6890
ausweichen 1523, 1526, 2027
Ausweichflugplatz 167
Ausweichmanöver 1528
Ausweichplan 4049
Ausweichstellung 4194
Ausweichstrecke 4964, 4965
Ausweichziel 6038
Ausweis 2394, 3897
(sich) ausweisen 2393
auswerfen 1432
Auswerfer 1434
auswerten 1524
Auswertung 1525
Auszieher 1566
Auszieherluke 5481
Ausziehschirm; (ein kleiner) Schirm der
    den Hauptschirm herauszieht 3842

authentisieren 306
Authentisierung 307
Automation 315
automatisch 313, 5173
automatische
    Datenübertragungsverbindung 2876
automatische Datenverarbeitung 4308
automatischer Landeanflug (meist eng.
    Begriff) 2035
automatisches Trimmgerät 6367
automatische Waffe 4713, 6774
automatisieren 312
Autorität 309
Axt 331
Azimut 334

B

Bach 1067, 5768
Back 1864
backbord 4188
Backbord- 4188
(nach) Backbord halten 231
Backenauflage 4791
Bade- und Wascheinheit 393
Bagger 1367, 1530, 5340
Bahn 6230
(Umlauf-)Bahn 3699
Bahn- 3701
bahnen 821
Bahnhof 5681
Bahnübergang 2050
Bajonett 411
(mit dem) Bajonett erstechen 410
Bajonettscheide 5053
Bake 417
Balken 427
Ballast 352
Ballistik 355
ballistisch 354
ballistische Bordrakete 3253
Ballon 356
Ballungsgebiet 6536
Balsaholz 357
Bandbreite 6848
Bandit 362
(Komputer) Bandspur 6233
Bank (Bänke) 368
Baracke 2362
Barkasse 2736
Barometer 377
Barometerstand 4555
barometrisch- 378
Barre 370
Barrikade 383
Base 5679

bewaffnet 253
(gut) bewaffnet 6821
(ungemäss) bewaffnet 6456
bewaffnete Luftaufklärung 4585
bewaffneter Hubschrauber 2136
bewaffneter Zusammenstoss 812
Bewaffnung 251
(sich) bewegen 3354
beweglich 3289, 3353
bewegliche Einsatzkräfte 3046
bewegliche Überwachung 3797
bewegliche Versorgungsgruppe 3291
bewegliche Verteidigung 3290
Beweglichkeit 3293
Beweglichkeitbegrenzungoperation
   3661
(stark) bewegt 4952
Bewegungsfreiheit 4942
Bewegungskrieg 3292
bewehren 5089
Beweis 1199
Bewerter 4787
bewohnbar 2147
bewölkt 3761
Bezeichnung 3489
beziehen 1360
Beziehungen 4666
Beziehungen zur Öffentlichkeit 4364
Bezugsebene 1113, 4622
Bezugsnummer 4623
Bezugspunkt 1112
(sich) biegen 453
Biegung 454
bilateral 462
Bild 2401
(Luft-)Bildaufklärung 3990
(Luft-)Bildkarte 3989
bildliches Nachrichtenmaterial 2513
Bildreihe 5796
Bildübertragung(-sgerät) 1573
billigen 5037
binden 468, 6165
Bindung 469
binokular (für beide Augen) 471
biographische Informationen 2508
biologisch 473
biologische Kriegführung 6680
biologische Kriegsführung 6685
biologischer Kampfstoff 474
biologische Waffe 6775
Birne 623
Bit 477
Biwak 479
biwakieren 478
(Luft-)Blase 610
blasen 503
Blasengas 1989

Blatt 482
Blech 5260
Blende 1258, 3056
Blenden 1118, 1119
Blendung 2029
Blickfeld 207
blind 491, 493
blinder Alarm 1586
Blindflug 1784
Blindgänger 492
Blindgänger- 6481
Blindheit durch Blendung 1767
Blindlandung 2703
Blinkfeuer 1769
Blitz 2814
Blitzableiter 2816
Blitzlichtbombe 3982
Blitzschutz 2815
Block 495
Blockade 497
(eine) Blockade aufheben 4486
blockieren 498
blosstellen 928
Blut 500
Blutdruck 502
bluten 490
Blutgruppe 6421
Blutgruppen 501
Blutübertragung 6305
Bö 5618
Boden 2083, 5519
(weicher) Boden 2092
Boden- 2082
Boden-Boden- 2105, 5904
Boden-Bodenrakete 3273
Boden-Luft- 2104, 5902
Boden-Signal 5385
Bodenausbildung (Fallschirmtruppen)
   6281
Bodenaushub 5593
Bodendetonation 3513, 5901
Bodendruck 2102
Bodenerfassung 4644
Bodenfalte 1840
Bodenfreiheit 2096
bodengeleitetes Abfangen 2099
(Flugzeug-)Bodenkontrolleur 3092
Bodenpersonal 2100
Bodenplatte 4078
Bodenplattenstellung 4197
Bodenprobe machen 2103
bodenseitig geleiteter Anflug 2098
(unabhängiger) Bodensensor 5190
Bodensprengpunkt 2095
Bodenstoss 5315
Bodenventil 5108
(an) Boden verlieren 2965

Bodenwelle 6764
Bodenwiederempfang 4823
Bodenzeit 6406
Bogen 238
Bohrturm 3624
böig 635
Boje 640
Bolzen 519
Bombardement 523
bombardieren 399, 525
Bombe 521
Bombenabwurf mit Verzögerungszünder
im Tiefflug (meist eng. Begriff)
2744
Bombenabwurfpunkt 4110
Bombenabwurfwinkel 529
(unbeschränkter) Bombenbereich
6938
Bombenentschärfungsoffizier 3577
Bombenflächenwurf 243
Bombengruppe 5725
Bombenpunkt 4145
Bombenräumung 526
Bombenschacht 524
Bombenschädenbericht 4732
Bombenschütze 522
bombensicher 531
Bombensprengkommando 527
Bombenzielgerät 532
Bomber 528
Bomblet 530
Boot 511, 1060
Bootsführer 1056
Bootsladung 512
(Ober-)Bootsmann 542
(an) Bord 5, 3627, 5302, 5305
Bord- 87, 3349
Bord-Boden 114
Bord-Boden- 139
Bord-Bodenrakete 3255
Bord-Bordrakete 3254
Bord-Unterwasserrakete 3256
Bordbuch 2939
Bordersatzteile 3628
(Panzer-) Bordkanone 6015
Bordmechaniker 1790
Bordschütze 2132
Bordsprechanlage 2530
Böschung 365, 1507
Bote 3167
Botschaft 1449
Brand 1700
Brand- 2425
Brandbekämpfungs- 1716
Brandung 580, 5898
Brandwache 3996
Brandwunde 645

brechen 572, 4630
Brechung 4631
breit 603, 6845
Breite 570, 2729, 6847
Breitenstreuung 2324
Breitenziel 6028
Bremsbelag 560
Bremse 557
bremsen 556
Bremsflüssigkeit 559
Bremsklotz 788
Bremspedal 561
Bremsrakete 4821
Bremsschirm 1380, 3839
Bremsvorrichtung (für Flugzeuge) 104
brennbar 877, 1755, 2455
(ver)brennen 644
Brennpunkt 1831
Brennschluss 646
Brennschlussgeschwindigkeit 6589,
6590
Brennstab 1929
Brennweite 1830
Bresche 568
Breschenmarkierung 3078
(eine) Bresche schlagen 569
Brise 592
Bruch 573, 4994
bruchfest 5252
Bruchgrenze 4112
Brücke 595
Brückenabschnitt 5143
Brückenbogen 5547
Brückenkopf 598
Brückenlegepanzer 597, 2746, 6003
Brückenpfeiler 4002
Brunnen 5609, 6820
Brust 784
Brustfallschirm 3841
Bruttogewicht 6812
Bücherrevision 303
Bücherrevisor 305
Buchhaltung 20
Buchse 651
(Ich) buchstabiere 5576
Bucht 409, 1068
Bug 547
Bund 6488
Bündel 836
Bündelgefechtskopf 6696
bündeln 1832
bündig (mit) 1822
Bündnis 154
Bunker 639, 4006, 5277
(U-Boot-)Bunker 3938
Bürgerkrieg 6669
Büro 643, 3572

durchwatbar 1860
durchwaten 1858
Düse 2620, 2622, 3499
Düsen- 2619
Düsenmotor 2623
Dwarslinie 2855
dynamisch 1401

E

Ebbe 1410, 2982
ebben 1409
eben 2779
Ebene 2778, 2782, 4044
(wahre) Ebene 6387
Echo 1422
Echolot 1423
Echolotung 1424
Echtzeit- 4560
Ecke 1006
(in die) Ecke treiben 1007
Edelgas 1992
Effekt 1428
Ehren- 6627
Ehren erweisen 2307
Ehrensalutkommando 3886
Ehrenwache 2116
eichen 670
Eilbote 1031
Eilmarsch 1856
Eimer 611
Ein- 5411
einberufen 675, 1351
Einberufung 1350, 4601
einbrechen 3940
Einbruch 3941
Einbruchskampf 581
eindringen 2568
Eindringling 2569
einer(-e,-es) voraus 3632
einfach 5411
Einfach- 5411
Einfahrt 1485
Einfall 2432
einfallen 2571
Einfallswinkel 203
Einfangsgebäude 6105
einfliegen 3940
Einflug 3941
einfügen 2481
Eingabe (-daten) 2478
eingabeln 551
Eingang 1485
eingefroren 2382
eingestufte Explosionshöhe 2243

Eingezogener 959
eingliedern (in) 2502
(sich) eingraben 1263
Eingreifen 2563, 2564
(den Kurs) einhaltend 1034
Einheit 6489
(einziehende) Einheit 40, 2430, 3747
Einheit in Feindberührung 6493
einheitliche Führung 6485
Einheits- 3309
einheitseigen 3725
Einheitsführer 3579
Einheitsversorgung 6490
(eine) Einheit verlegen 4692
einholen 3763, 3765, 3790
einkesseln 1460, 4933, 5917
Einkesselung 1461
einlaufen 1484
Einmal(schlüssel)system 3631
Einmal- Schlüsselblock 3815
einnebeln 5092
ein Programm aufstellen 4320
einquartieren 464, 4399
einrichten 3731
einrichten (nach) 3299
Einrichtung 2491
Einrichtungen 1571
Einsatz 5536
Einsatz-kampfverband (auf Bataillons
    Ebene) 396
Einsatzalarm 4406
Einsatzbefehl (mit voller
    Kampfausrüstung) 3706
Einsatzbefehl geben 3717
Einsatzbereitschaft 4551
einsatzfähig 3669
Einsatzflugplatz 3670
Einsatzführer 759
Einsatzgliederung 404
Einsatzplan 1037
Einsatzverband 6055
einschalten 5953
Einschiesschuss 5372
Einschiessen 1690, 4498
(sich) einschiessen 4642
Einschiessgeschoss 4958
Einschiesspunkt 4144
(sich) einschiffen 1447
Einschiffung 1448
Einschiffungsraum 3352
einschliessen 2580
einschränken 4796
einschränken (auf) 2823
einsetzen 2482
einsickern 2453
einsperren 2419

einspritzen 2472
(Benzin) einspritzen 4281
einsteigen 1484, 3341
einstellbar 6395
einstellen 42, 4648, 5933
(einen Zünder) einstellen 5218
(auf Nullpunkt) einstellen 6931
einstellen (auf) 3437
Einstellung 300, 4649, 5934
Einstellung der Feindseligkeiten 5937
Einstellungsmechanismus 3136
Einstellungsprüfung 6131
Einstiegklappe 1047
einstufen 2045
Einstufung 815
(sich) eintauchen 2408
einteilen 578
einteilig (mit) 2501
Einteilung 579
eintragen 2941, 4091, 4640
einwachsen 6768
Einwand 3531
Einwand erheben 3529
einweisen 600
Einweisung 602
einwickeln 849, 6904
einzaunen 1600
Einzelhaft 5529
Einzelheit 1236
Einzelkanone 4974
einzeln 5411
Einzelreihe 5412
Einzelschlagwaffe 5788
Einzelschusswaffe 6789, 6790
einziehbar 4814
Eis 2379
Eisberg 2381
Eisbrecher 2383
Eisen 2586
Eisenbahn 4483
Eisenbahn- 4484
Eisenbahngleis 6238
Eisfeld 2384
eisfrei 2386
elektrisch 1435
Elektrizität 1436
elektromagnetischer Impuls 1437
Elektron 1438
Elektronen- 1439
Elektronenrechner 930
Elektronik 1442
elektronisch 1439
elektronische Gegenmassnahmen 1441
elektronische Kriegführung 6684
elektronischer Suchradar 4427
elektronische Schutzmassnahmen 1440
Elektroschock 5314

Element 1443
Emission 1456
Empfang 4573
Empfang bestätigen 24
empfangen 4571
Empfänger 4572
Empfänger bzw. Sender 6106
Empfangsbestätigung 25, 4570
Empfangsstörung 2534
Empfangzeit 6181
empfehlen 4581
Empfehlung 4582
empfinden 5186
empfindlich 5187, 6654
empfindlicher Punkt 4158
empfindlicher Raum 6655
Empfindlichkeit 5188
End- 1638, 6107
Endbahnhof 6105
Ende! 3742
Endfahrzeug 6572
Endgeschwindigkeit 6595
Endkoordinationslinie 2833
Endphasenlenkung 6111
Endphasenwirkung 6110
Endstation 4482
Energie 1467
Energieumwandler 6301
eng 3392
Enge 6750
(Meer-)Enge 5756
Engpass 1171
entbehrlich 1544
Entdeckung 1240
Enteiser 1180
Entenmuschel 375
Enterkommando 3881
entern 510
Enteroperation 3655
Entfernung 1312
(Ziel-)Entfernung 4514
(Beobachter-Ziel-)Entfernung 1313
Entfernungsmesser 1642
Entfernungsmessmarkierung 3082
Entfernungsmessung 4516, 6079
entführen 2273
Entführung 2274
entgegengesetzt 3690
entgegnen 1021
enthalten 967
entladen 1281, 6501, 6503
entlassen 1278, 1280, 1289
Entlassung 1195
Entlassungsurlaub 6112
Entlaubungsmittel 1176
entläusen 1192
entlüften 489

Entlüfter 6596
entmilitarisieren 1194
entmilitarisierte Stadt (einem Angriff
 nicht ausgesetzt) 3639
entmilitarisierte Zone 6942
entmutigt 1201
entrosten 1220
Entsalzen 1221, 1222
Entschädigung 4708, 4794
entschärfen 1274, 4698
entschärfte Mine 3213
Entscheidung (über) 1136
Entschluss 942
entschlüsseln 1148, 6513
entsetzen (Stadt usw.) 4689
entseuchen 1144
Entseuchung 1145
entspannen 6445
entstrahlen 1144
Entstrahlung 1145
entwaffnen 1274
entwässern 1356
Entwässerung 1179
entwerfen 1227
(sich) entwickeln 1210, 1250
Entwicklung 1211, 1251
Entwurf 1226
entziffern 1135, 1140
Entzündungspunkt 4125
erblicken 5358
Erde 1405
erden 1406
Erdhügel 3340
Erdkampfflugzeug 94
Erdnähe 3950
Erdöl 3623, 3969
Erdöl- 3620
erfahren 1546, 2764
erfassen 2938
Erfassungsfeuer 1691
erfinden 2573
Erfinder 2575
Erfindung 2574
Erfordernisse 1539
erforschen 1554
Erfrieren 1560
Erfrierung 1924
ergänzen 5855
(auf Antrag) ergänzen 4807
Ergänzung 179
(automatische) Ergänzung 4804
Ergänzungsmeldung 4731
Ergebnis 942
Ergreifung 5169
Erhebungen anstellen 3031
erhitzen 2221
erhöhen 1445

(sich) erhöhen 2241
Erhöhungswinkel 202
(sich) erholen 4604
Erholung 4605
erkennen 4578
Erkennung 4577
Erkennungsmarke 5981, 5982
Erkennungsrauch 5490
Erkennungssignal 5387
Erklärung 811
erklettern 5055
erkunden 4592
erlahmen 5454
ermitteln 2581
Ermittlungen 2582
ermorden 3375
Ermüdung 1593
ernennen 1228, 3470
erneuern 4701
erobern 958
erproben 6128
erreichen 4542
Ersatz- 5826
Ersatzrad 5552
Ersatzstoff 5825
Ersatzteil 4718
Ersatzteil(e) 3874
erschiessen 5321
erschöpfen 1536
Erschöpfung 1538
ersetzen 4717
Erstattung 4822
-Erstattung 4822
erste Hilfe 1729
Ertrag 3754
ertränken 1386
ertrinken 1386
erwidern 4722
(Feuer) erwidern 4825
erzeugen 2018
(Rauch-) Erzeuger 2020
Essen 3121
Essgeschirr 2675
Etappe 2770, 5643
Etappen 2158
etikettieren 2688
Ettikett 2687
evakuieren 1520
Evakuierte(r) 1522
Evakuierung 1521
Evakuierung durch die Luft 110
Exekutionskommando 1727, 3886
Exerzierausbildung 1372
Exerzierplatz 2089
Expeditionskorps 1542
Experte 1549
explodieren 1550

Explosion 1246, 1555
explosionsgeschützt 1556
Explosionswirkung 488
explosiv 1557

F

Fächerfunkfeuer 418
Fachmann 1549
Fachmann(-leute) 5560
Fackel 6196
(Glüh)-Faden 1626
Fading 1575
fähig 1735
fähig (zu) 695
Fähigkeit 694
Fahnenflucht 1156
Fahnenflüchtige(r) 1157
Fahnenflüchtiger 1225
Fahnenmast 1749
Fahnenträger 5652
Fahrbahn 2715
Fahrbeleuchtung 4992
Fahrbereich 4471, 4497
(planmässiges) Fahrbereitmachen 5066
Fahrbereitschaft 4185
Fähre 1605
fahren 1376
Fahrer 1377
Fahrgast 3903
Fahrgestell 775, 2005
(das) Fahrgestell einfahren 4487
fahrplanmässig 5067
Fahrplanziel 6037
Fahrrinne 765, 2714
(in) Fahrt 6476
Fahrt (voraus) 2219
Fahrwasser 1581, 6239
Fahrwassermitte 3184
Fahrwerk 6457
Fahrzeug 6563
(ungeschütztes) Fahrzeug 6582
Fahrzeugheber 2601
Faktor 1574
Fallschirm 3837
(mit dem) Fallschirm abspringen 3845
(ein mit dem) Fallschirm abwerfbares
    Schlauchboot 3863
Fallschirmauslöser 4679
Fallschirmentfaltungshöhe 2239
Fallschirmjäger 3864, 3865
Fallschirmkappe 689
Fallschirmleinen 4859
Fallschirmpacker 4857
Fallschirmspringen 5449
fallschirmspringen (aus hoher Flughöhe

mitFallschirmeröffnung aus neidriger
    Höhe) 3847
Fallschirmspringen (verzögerter
    Absprung) 3846
Fallschirmspringer 3848
Fallschirmtragseile 5936
falsch 1585
Faltboot 2649
Fangleine 3824
Farbe 859, 3822
Farbfleck (Tarnung) 860
Farbmarkierung 5118
Faschine 1589
Faser 1608
(Betriebstoff-) Fass 1928
Fassungsvermögen 5740
Feder 5608
federführende Dienststelle 1002
(unter) Federspannung 5610
Fehlanzeige 4737
Fehlschlag 1579
fehlschlagen 6
Fehlzündung 3248
feierlich 751
Feind 1466, 6152
Feindlagebeurteilung 2521
feindlich 1466, 2337
feindliche Merkmale 2338
Feindseligkeiten 2340
Feindsignal 2339
Feld- 1611
Feldanzug 1613
Feldflugplatz 134
Feldgeschütz 1616
Feldjäger 3199
Feldjäger- 4356
Feldkampffertigkeit 1612
Feldkaplan 766
Feldkarte 3062
Feldküche 1617
Feldlazarett 2331, 2332, 2333
Felduniform 2673
Feldverpflegung 1620
Feldwache 3995
Feldwerkstatt 1621
Feldzeugmaterial 3723
Feldzug 681, 1541
Fels 4909
felsig 4916
Fender 1603
Fenster 6862
Fern- 4502, 4503
Fernaufklärung 4586
Fernbediengerät 4697
fernbedienter Bodensensor
    4696
ferngeleiteter Angriff 6560

Fischereiflotte 1777
Fischereischutzfahrzeug 6610
Fla- 5902
flach 1772, 4063, 5245
Fläche 241, 4064
Flächenbombardierung 710, 3922
Flächenziel 6024
Flachfeuer 1668
Flagge 1483, 1745
Flaggoffizier 3584
Flaggrang 1748
Flaggschiff 5296
Flak 1751
Flakrakete 3257
Flamme 1752
Flammenwerfer 1754, 6158
Flanke 1757
Flankeinheit 1759
flanken 1758
Flankendeckung 1760
Flankendeckungsfeuer 1667
Flankhindernis 3547
flankieren 1756
flankierende Einheiten 3433
(unter) flankierendes(-em) Feuer 1469
flankierendes Feuer 1680
flankierend schiessen 1468
Flasche 543
Flaschenzug 499, 4367
Flegel 1750
Fleischwunde 1778
fliegen 1823, 5435
Flieger 122
Fliegerabwehr- 218
Fliegerangriff 4480
Fliegerbodenangriff 1623
Fliegerdarstellung 126
Fliegereinsatz (auf Abruf) 3278
Fliegergegenangriff 3659
fliegerisch 55
fliehen 1504
fliessen 1814
Floss 4476
Flosse 1636
Flossen 1795
Flotille 1812
Flotte 1775
(Handels-)Flotte 5306
Flucht 1503
Fluchten und Ausweichen 1527
Flüchtige(r) 1505
Flüchtling 4637
Flug 1781
Flugabwehr- 218, 5902
Flugabwehr- (Fla-) 107
Flugarzt 1794
Flugauftankung 4298

Flugausbildung 6280
Flugbahn 6296
(flache) Flugbahn 6298
(gebogene) Flugbahn 6297
Flugbahnbeschreibung 6250
Flugbenzin 327
Flugbereich 4497
Flügel 481, 1637, 6868
Flügelmann 3081
Flügelmutter(n) 6872
flügelstabilisiert 1643
Fluggast 3903
Fluggeschwindigkeit 132
Fluginformationen 1791
Flugkörper (meist eng Begriff) 6576
Flugkörpergefechtskopf 6697
Flugkörperkurs 6234
Fluglage 301
Fluglinie 118
Flugplan 4054
Flugplatz 111
Flugprofil 1793
(Ab, An, Ein) Flugschneise 2713
Flugsicherung 141
Flugstrecke 4963
Flugüberwachung (vom Boden) 2097
Flugverbotsbefehl 3709
Flugweg 1792
Flugwesen 61, 328
Flugzeit 6179
Flugzeug 92
(im) Flugzeug befördern 1824
Flugzeugbesatzung 106
(mit) Flugzeugen gelandet 116
(mit) Flugzeugen transportfähig 142
Flugzeugführer 4010
Flugzeughalle 2179
Flugzeug mit pfeilförmigen
    Tragflächen 100
Flugzeugträger 105
Flugzeugträgergruppe 712
Flugzeugverladungsplan 4052
Flugzeugzelle 113
Fluoreszenz 1820
fluoreszierend 1821
Fluss 4886
Fluss- 4885
flüssig 1818, 2878
Flüssigkeit 1819, 2878
Flüssigkeitsrakete 4912
Flüssigkeitstreibstoff 2879
Flüssigsauerstoff 2983
Flussmündung 1519
Flusstrecke 4541
Flut 1803, 1807
fluten 1802
Fluthöhe 4881

Fussgraben 6354
Fusspunkt 3389
Fusssoldat 5524
Fussspur 1852

G

Gabel 1378
Gabelstapler 1870
Gamasche 1975, 5554
Gammastrahl 1978
Gang 2004
Gangbarkeit 6257
(in) Gang setzen 2467, 5667
Gangspill 882
Garage 1982
Garnison 1987, 4219
(mit einer) Garnison belegen 1986
Gas 1988
Gasabwehr 881
(von) Gas getrieben 3649
Gasgranate 5269
Gashebel 6155
Gaskrieg 6681
Gasloch 6597
Gasmaske 1997
Gaspedal 1999
Gasschleuse 782
(Minen) Gasse 2712
Gassenmarkierung 3079
Gassprengkopf 6695
Gasturbine 2000
Gebäude 700
Gebiet 619, 6123
(taktisches) Gebiet 2919
Gebiets-Sperroperation 3654
Gebietsaufklärung 4589
Gebietswettervorhersage 553
Gebirgs- 3344
Gebirgsvorsprung 5613
gedeckt 1051
(in) gedeckte(r) Stellung 1170
gedeckter Anmarschweg 1053
gedeckte Stellung 4199
geeignet 4395
Gefahr 1102, 2204
Gefahrenzone 1103
gefährlich 1104
(im Felde) gefallen 2665
Gefangene(r) 703
gefangennehmen 704, 4293
Gefängnis 4291
Gefecht 421, 425, 1471
Gefechtsfeld 405
Gefechtsfeldbeleuchtung 406
Gefechtsfeldüberwachung 408

Gefechtsgruppe 871
Gefechtskopf 6699
Gefechtsnachrichtenstelle 870
Gefechtsstand 4227
Gefechtsstationen 402
(gemeinsamer) Gefechtsverband 6057
Gefolge 4811
gefrieren 1904
(ein, zu)gefroren 1925
Gegenangriff 1022
Gegenfeuer 1658
Gegenmassnahme 1027
(Insekten-) Gegenmittel 4716
gegen Personal 245
gegenseitig 3384, 4574
Gegensprechanlage 2530
Gegenstand 3530
Gegenüber 3691
gegenüberliegend 3690
(dem Feind) gegenüberstehen 1570
gegenwärtig 1091, 4263
gegenwärtige Stellung 4209
gegisstes Besteck 4575
Gegner 3688
(an)gehalten 5695
Gehäuse 717, 2343
geheim 816, 6458
geheim(nis) 5139
Geheim- 810
Geheimhaltung 5138, 5154
Geheimhaltungsstufe 2052, 5162
(die) Geheimhaltungsstufe aufheben 1138
Geheimoperation 3656
Geheimschreibekunst 1086
Geheimschrift 794
Geheimschriftanalyse 1085
geheimschutzbedürftiges Material 1087
Geheimzeichen 1028
gehen 6663
Gehirn 554
Gehirnwäsche 6727
(jemanden) gehorchen 3528
Geigerzähler 2012
Geisel 2336
geladen 2907
Gelände 6115
(unebenes) Gelände 2091
(fahrbares) Gelände 6118
(befestigtes) Gelände 6122
Geländeausnutzung 3729
geländegängig 1078
Geländegängigkeit 407
Geländekampfstoff 3964
Geländeschilauf 5432
(per Luft) gelandet 2697
Geländewinkel 242

Gelegenheit 3689
Gelegenheitsziel (beobachtetes Schiessen)
  6050
Geleit 1509
Geleitschutz 994, 5094
Geleitzug 992
Geleitzugbeladung 2910
Geleitzugfahrplan 5065
Geleitzuggeschwindigkeit 5568
Geleitzugs- 993
Geleitzugweg 4966
Gelenkfahrzeug 6568
gelenkter Kreisverkehr 1457
gelenkter Schub 6561
gemässigt 6094
gemeinsam 901, 2632
gemeinsame Ausbildung 6284
gemeinsame logistische Unterstützung
  1081
gemeinsamer Gefechtsstand 2633
gemeinsamer Stab 5630
Gemeinsamkeit 902
Gemisch 3287
gemischt 3286
Gemischtbeladung 2906, 2911
genau 223
genauer Standort 4023
genau festlegen 4021
Genauigkeit 206
Gendarm 2013
genehmigen 820, 822
(etwas) genehmigen 311
Genehmigung 310, 3960
General- 2014
Generalplan 4058
Generalsekretär 5141
Generalstab 5628
Generator 2019
Genfer Konvention 2021
(in) geöffneter Ordnung 3644
geographisch 2022
Georef-System 2023
Gepäck 347, 3807
Gepäckkommandant (veraltet)
  348
geplant 4069
geplanter Bodennullpunkt 6929
Gerätedepot 3724
geräumter Kanal 5125
geräumtes Fahrwasser 5946
Geräusch 3467
gerichtete Flugbahn 6249
gerichtlich vorgehen 4343
geringfügig 3071, 3244
gerinnen (lassen) 833
Gerippe 1894
Geröll 5086

Gerüst 1894, 1980
Gesamt- 6206
geschätzte Abflugzeit 1517
geschätzte Ankunftszeit 1516
geschätzter Ausfall 4719
Geschirrlast 2902, 2903
(mit) geschlossenen Luken 655
geschlossene Ortschaft 622
geschmeidig 1780
Geschoss 4329, 5263
(panzerbrechendes) Geschoss 5004,
  5266
Geschossführungspring 5003
Geschosskanister 5268
Geschosspendelung 6917
Geschütze 3721
Geschützfeuer 2131
Geschützrohrdrehzapfen 6389
geschützt 1051, 5278
Geschwindigkeit 3804, 5566, 6588
Geschwindigkeitsgrenze 5574
Geschwindigkeit über Grund 5570
gesetzlich 2771
gesichert 2447
gesichert (Minen) 6433
Gesichtsfeld 1619
gespannt 6061, 6099
gestaffelt 1415
Gestell 1059, 1892
gestrandet 121, 2260, 5757
Gesträuch 650
Gestrüpp 5100
Gesundheitspass 466
Gesundheitsvorsorge 4278
getrennt 5194
Getriebe 2009
Gewalt 1853
Gewaltmarsch 1856
gewarnt in Deckung 6704
gewarnt ohne Deckung 6703
Gewässerkunde 2369
Gewehr 4846
(rückstossfreies) Gewehr 4850
Gewehr-Spannhahn 2165
Gewehr ab! 3716
Gewehrgranate 4849
Gewehr mit Rahmenzuführung 4848
Gewehr mit Zielfernrohr 4851
Gewicht 6811
gewinnen 6850
Gewohnheit 4973
geworfene Ladung 2624
Gezeit 6163
Gezeiten- 6161
Gezeitenmitte 2154
Gezeitentafel 5965
gezogener Lauf 4853

Halbinsel 3942
Halbketten- 2155
(auf) Halbmast 2151
halbmilitärisch 3861
Halbwertschicht (HWS) 2153
Halbwertzeit 2794
Hals 3423, 3424
Halt 1850, 2937
Halt! 2157, 5735
Haltbarkeitsdauer 2796
halten 2288
(an)halten 5734
(eine Stellung) halten 2292
(an)halten (lassen) 2156
Hand 2159
Hand- 2161, 3057
Handbedienung 3059
Handbuch 3057
Handels- 893, 3158
Handelsschiff(e) 5300
handelsüblich 893
Handfernlenkung 3058
Handfeuerwaffe 6788
Handfeuerwaffen 5485
Handgranate 2072
Handhabung einer Waffe 2170
Handschellen anlegen 2160
Hand über Hand 2173
Hängebrücke 5935
Hängeladung 2905
Harpune 2192
Haubitze 2348
Haupt- 2658, 3010, 3028, 3105
Hauptangriff (veraltet) 3012
Hauptbaugruppe 3029
Hauptflugplatz 3011
Hauptgefechtsfeld 3014
Hauptgruppe 3029
Hauptnachschubstrasse 4968
Hauptquartier 2214
Hauptrotorwelle 5243
hauptsächlich 3010
Hauptschalter 3108
Hauptsicherung 3016
Hauptsperrzone 448
Hauptstrasse 2272, 4894
Hauptstützpunkt 3013
Haupttrupp 3015
Hauptwasserrohr 6745
Hauptwiderstandslinie 4775
Hauptwiderstandspunkt 4776
Haushalt 615
Haussuchungsbefehl 6721
Haut 5436
Hebel 2784
Hebelkraft 4384
heben 2800

(mit einer Stange usw.) heben 2785
Hebestange 2175
Hebevorrichtung 4384
Hebewerk 2286
Hebewinde 2006
Heck 5719
Heckanker 5720
Hecke 2234
Heckenhüpfen 1783, 2235
Heckhaken 2314
Hecklampe 5721
Heckwerk 6320
Heer 264
Heeresverbindungsoffizier 3587
Heimatshafen 2303
heimlich 810, 6458
heimlich herausbringen 1535
heisser Draht 2341
heisslaufen 3769
Heizkörper 4446
Heizöl 3622
helfen 95
Helium 2249
Helm 2251
Helmeinsatzstück 2252
hemmen 3554
herabsetzen 1177, 1346, 2977
Herbst 318
herfallen über 3784
Herkunftsland 1030
hermetisch 2254
herrenlos 1218
Herrschaft 983, 3109, 4984
Herrschaft über 884
herstellen 4313
(ein Flugzeug) heruntersprechen 5997
„Hier ist…" 6150
hieven 2224
Hilfe 78, 4687
Hilfs- 5047, 5134, 5819
Hilfsmotor 320, 537
Hilfsschiff 321
Hilfstruppen 319
Himmel 5448
Himmels- 732
Himmelsrichtungen 4169
Hinausschiessen 3786
Hindernis 3544, 3555
Hindernis (im rückwärtigen
    Kampfgebiet) 3551
Hindernis (künstlich) 3545
Hindernis (natürlich) 3546, 3549
Hindernis (ständig) 3552
Hindernis (taktisch) 3553
Hindernis (unpassierbar) 3548
Hindernisanzeige für Tiefflug 6119
Hindernisräumung 3660

hinhalten 1183
hinrichten 1532
hinter 4561
hinter- 4561
Hintergrund 337
Hintergrundstrahlung 338
Hinterhalt 177
(aus dem) Hinterhalt überfallen 178
Hinterhang 5480
Hinterhangverteidigung 1160
Hinterland 2277
(auf etwas) hinweisen 4168
Hirnverletzung 555
hissen 2284
Hitzerschöpfung 2222
Hitzeschutz 5284
Hitzestrahlung 6144
Hitzewirkung 6141
hoch (hohe-) 2259
Hochebene 4083
hochexplosiver Sprengstoff 2263
Hochleistungs- 2229
Hochschuss 1683, 2280
Hochsee- 2268, 3563, 5109
Hochseegeleitzug 3562
Hochspannung 2270
Höchst- 3117
Höchst- Spitzen- 3934
Höchstbruttogewicht 6814
höchste Marschgeschwindigkeit 5572
höchste Startgewicht 6816
Höchstladelinie 4087
Höchstlademarke 4087
Höchstlandegewicht 6815
(bei) Höchstleistung 1931
höchstzulässige Dosis 3118
hochtransformieren 5717
Hochwasser 2271
hochziehen 2285
Höhe 172, 2236
hohe Geschwindigkeit 6592
Hoheitsgewässer 6751
(auf) hohem Niveau 2266
Höhenabstand 2562, 6607
Höhenflosse 5624, 5987
Höhenlinie 2832, 2862
Höhenlinienabstand 2560
Höhenmesser 171
Höhenmessung 2378
Höhenpunkt 5599
Höhenschreiber 376
Höhenstaffelung 6607
Höhenstellrad 2177
(automatische) Höhensteuerung für
  Tiefflug 6120
Höhenwinkel 1446
Höhepunkt 230

höher! 6527
höherer Dienstgrad 5184
höheres Dienstalter 5184
(auf) Höhe richten 4517
hohe See 2268
Hohlladung 2301
Hohlladung(s-) 5248
Hohlladungs- 2264
Hohllage 2300
(ein-, auf-, ab-) holen 2199
Holm 6266
Holz 6891
Holz- 6892
homogen 2306
horchen 2885
Horchposten 4228
Hörer 2174
Horizont 2319, 5450
(scheinbarer) Horizont 2320
Hörnachrichtenmaterial 2506
Hosenboje 591
Hubschrauber 97, 2245
Hubschrauberaufgabe 2246
(per) Hubschrauber befördert 2244
Hubschrauberlandeplatz 2247
Hubschraubertransport 2248
Hubschrauberversorgung 6606
Hügel 2275
hügeliges Gelände 2276
Hulk 2351
Hülse 4104
Hupe 2325
Hydrant 2363
hydraulisch 2364
Hydrographie 2369
hydrographisch 2368
hydrostatisch 2373
Hyperbelverfahren 2374
hypergolischer Treibstoff 2375

I

identifizieren 2392
Identität 2393
IL-Triebwerk (intermittierende
  Luftstrahl-) 4374
Immunisierung 2412
Impfung 6542
implodieren 2416
Implosion 2417
Impressum 2418
improvisieren 2420
Impuls 4373
Impuls- 4372
Impulsmodulation 3313
inaktivieren 2423

Kettenfahrzeug 6585
Kraft 1853
Kiel 2651
Kiellinie 2859
Kielwasser 6661, 6726
Kies 2060
Kilotonnen-Waffe 6783
Kilowatt 2670
Kimm 2322
(hinter der) Kimm 2353
Kimme 342, 4565, 5366
kinetische Energie 2671
Kinn 786
Kinnriemen 787
kippen 4036
Klammerfeuer 1651
Klampe 809
Klappe 1761
Klappmesser 813, 2606
Klappvisier 5363
Klartext 4046
Kleider 834
Kleidersack 2679
Kleidungslager 5743
Klein-U-Boot 3186
kleiner 3244
kleinkaliber 5486
Kleinkaliber- (z.B.
   Ausbildungsvorrichtung) 5803
Kleinspurbahn 3393
Kleinstbombe 530
Klemme 6108
Klippe 4909
Klo 2215
Klosett 2215
Knall 364, 4729
(Überschall-) Knall 534
knapp an Arbeitskräften 5333
knebeln 1973
Kniestellung 4205
Knopf 654
(zu-)knöpfen 653
Knoten 2685
Knoten machen 6165
Knotenpunkt 2642
Knüppel 394
koaxial 846
Koch 998
kochen 997
Kode 850, 2654
Kodewort 6895
Koffer 6388
Kohle(n) 840
Kohlefaser 705
Koje 638
(Gewehr-) Kolben 652

Kolbenkappe 4079
Kolbenplatte 4790
Kolbenschlag 5797
Kolonne 862, 992
Koma 864
Komma 4121
Kommandant 888
Kommandeur 890, 3579
Kommandeur (der Militärpolizei) 4357
kommandieren 879, 3704
Kommando 881, 1234, 1235, 3878
(kleines) Kommando 5616
Kommando- 882
Kommandobrücke 596
Kommandofahrzeug 6571
(das) Kommando haben 880
Kommandoraum 4936
Kommandostab 891
Kommandoturm 957, 5021, 5685
Kommen! 3759
kommentieren 214
Kommissar 894
Kompanie 911
(verstärkte) Kompanie 6069
Kompaniechef 913
Kompass 915
(Magnet-)Kompass 3001
Kompassfehler 917
Kompasshaus 470
Kompasslinie 2851
Kompresse 924
Kompressor 927
Kondensation 945
Kondensator 947
kondensieren 946
Kondensor (Linse) 947
Kondensstreifen 6267, 6268
Kondominium 949
Konferenz 953
Konossement 467
konsekutiv-Dolmetscher 2553
Konstante 962
konstruieren 1227
Konstruktion 1226
Kontakt 965
Kontakt verlieren 2964
Kontingent 975
Konto 74
Kontroll- 3105
Kontrollampe 4015
Kontrolle 978
kontrollierbar 6600
kontrollieren 776, 979, 2486, 6128
Kontrolliste 777
Kontrollmassnahmen 987
Kontrollpunkt 4116, 4118
Kontrollstelle 778

599

Kurzbiographie 4316
kurzfristig 5334
kurzläufige Flinte 4876
Kurzschluss 5332
Kurzschuss 5330
Kurzstreckenflugzeug 2200
Kurzurlaub (für die Nacht) 3777
Kurzwelle- 5335
Küste 841, 5324
(an der) Küste entlang fahren 842
(die) Küste meiden 5660
Küsten- 2484
Küstenbeschuss 5326
Küstengewässer 6748
Küstenlinie 5328
(in) Küstennähe 3616
Küstennavigation 4011
Küstenschutz 3915
Küstenstreifen 2889
Küstenverband 844
Küstenwacht(-mann) 845
Kutter 1094

L

Ladebaum 1219
Ladehemmung 5737
Ladehemmung haben 2607
Ladeklappe 5983, 5984
Ladeliste 3049
Lademass 2003
laden 2894, 2896, 4280
(auf)laden 769
Ladeplan 4057
Ladepunkt 4134
Laderaum 764, 2287
Ladestock 4493
Ladung 2893, 2895
(mit verstärkter) Ladung 5846, 6339
Lafette 2130, 3351
Lage 2926, 5421
Lagebericht 4740
Lagebesprechung 3720
Lagebeurteilung 1518
Lagekarte 5422, 5423
Lagemeldung aus der Luft (meist eng.
   Begriff) 4735
Lager 664, 680, 1393
(motorisiertes) Lager 5744
Lagerbestand 2576
Lagerbeständigkeit 5262
Lagerhaus 6677
lagern 1394
Lagerraum 5741
Lagerung 5739
Lagezimmer 4941

Lagune 2691
Lähmung 3448
Lähmungsgas 1991
Lamellen- 2693
lamelliert 2693
Land 1029, 2694
(an) Land 273
Lande- und Abflugmanöver 6210
Landefeuer 2708
Landehilfsmittel 2704
Landekopf 415
Landematte 3111
Landematte (veraltet) 2709
landen 1126, 2695
(zu kurz) landen 6470
Landeplatz 5790
Landepunkt 4132, 4165
Landerampe 3813
Landeraum 2705, 6948
landern 149
Landestag 5642
Landesteg 4001
Landestelle 2186, 5419
Landestreifen 5792
Landgängerboot 2790
Landkriegsführung 6687
Landmine 3219
Landung 1125, 2698
Landungsboot 2706
Landungsschiff 5298
Landungskommando 3887
Landungsplan 4056
Landungsschiffsgruppe 6336
Landungstruppe 2707
(abgezählte) Landungstruppen 6370
Landungstruppen (meist eng. Begriff)
   761
Landungsverband 3410
Landurlaub 5327
Landzunge 5586
lang 2948
Länge 2773, 2949
(auf) Länge richten 4519
Längs- 2856, 2950
Längsachse 2951
langsam 5482
langsame Fahrt 2981
längschiff 1863
Längsfeuer 1665
Längsneigung 4039
längsseits 162
Langstrecken- 4502
Langstrecken-Luftweg 4970
Langwellen- 2952
Lasche 2725
Laser 2719
Laser-(Zielflug-)Lenkung 2722

601

nass 6827
Nation 3394
national 3395
nationale Freiheitsfront 3399
nationale Infrastruktur 3397
nationaler Anteil 3396
NATO 3400
Natur- 3398, 3401
natürlich 3398, 3401
natürliche Deckung 1048
natürlicher Tod 3403
natürliches Merkmal 3404
natürliche Sperre 3402
nautisches Zwielicht 3406
Navigation 3415
Navigationsgitter 3416
Navigationsmarke 3417
Navigationsoffizier 3420
Navigator 3420
NBC-Waffe 6776
Nebel 1834, 5489
Nebel- 5488
(durch) Nebel behindert 1835
(im) Nebel gehüllt 1835
Nebelgeschoss 5272
Nebelhorn 1837
Nebellampe 1838
Nebelsignal 5384
Nebelwand 5088, 5097
Neben- 5134
Nebenkarte 2483
Nebenschaden 855
neblig 1836
Negativ 3425, 3425
negativer Geländewinkel 200
(Deckung) nehmen 1050
(Stellung) nehmen 5694
(sich) neigen 2427, 5280
Neigung 2428, 5457
Nenn- 3469
Nerv 3434
Nervengas 6545
Nervengas (z.B. GB, GX) 1974
Nervenkampfstoff 3435
Nettogewicht 6817
Netz 3438
Netzauthentisierung 3442
Netzrufzeichen 5376
neugliedern 4704
Neugliederung 4703
neu klassifizieren 4576
(Radar) neu richten 5286
neutral 3447
Neutralisierung 3448
Neutron 3450
Neutronenwaffe 6778
neutroninduzierte Aktivität 3451

neu zuteilen 4558
Neuzuteilung 4559
nicht-kriegführend 3472
nicht einsatzbereit 6517
nicht entzifferbar 6454
nicht in Ordnung 3713
nicht lebenswichtige Ladung 3477
nicht seetüchtig 6515
Nichtverbrauchsgüter 5861
nicht vorhanden 6438
nicht zu verteidigen 2433
nieder 2972
niederdrücken 1213
Niederhalten 1025, 4019
(Truppen) niederhalten 6168
Niederhaltungsfeuer 1679
Niederlage 1153
niederlegen (Waffen) 2085
(radioaktiver) Niederschlag 1583
niederwerfen 3793
niedriger 2973
Niemandsland 2696, 3468
Niete 4889
(ver)nieten 4888
Nippflut 3422
Niveau 4064
Nivelliergerät 2781
Nivellierungszeichen 452
Nockenwelle 682
nominal 3469
nominal- 3469
Nord(en) 3482
(nach) Norden 3485
(nach) Norden unterwegs 3483
nördlich 3482, 3485
Nordwert 3484
Norm 5650
normal 3479
normal- 3479
Normalabweichung 5653
Normalaufschlag 3480
normalisieren 3481
Normalnull (NN) 3125
Not- 1451
Notanker 5259
Notankerplatz 1454
Notausstieg 1506
Notfall 1452
Notflugplatz 1453
Notlandeplatz 170
Notrisiko 1455
Notsignal 5382
Nottauchen 1064
Nottreppe 1713
Notverband 1614
notwassern 1318
Null 6927

Ortszeit 2922
Ortungsfehler 4214
Ortungsgerät 2439
Ortungsstandort 5686
Ortungsstelle 4231
Ostwert 1408
Oszillator 3739
Oszillograph 3740
Oszilloskop 3741
Ozean 3561
Ozeanographie 3565

P

Packeis 2380
packen falten (ein Fallshirm) 3812
Packhaubitze 2349
Paddel 3818
paddeln 3817
Palette 3825
palettieren 3826
palettierte Sammelpackung 2899
(einfacher) Palstek 548
Panne 576
(eine) Panne haben 577
Panzer (in Deckungsstellung) 6007
Panzerabwehr- 228
Panzerabwehrgeschoss 5265
Panzerabwehrrakete 3259
Panzerabwehrtrupp 6073
Panzerabwehrwaffe 6773
Panzerbesatzung 6011
panzerbrechend 260
Panzergrenadiere 3139
Panzerkampfwagen 6564
Panzerkettenglied 6248
Panzerspähwagen 6566
Panzersprechgerät 6082
panzerstark 2227
Panzersuchpatrouille 3920
Panzertransportfahrzeug 6335, 6584
Panzertruppe(n) 257
Panzerung 256, 4086, 4346
Panzerweste 259
Parade 3850, 4830
(die) Parade abnehmen 5029
Parademarsch 3069
Paradeplatz 2089
Paradestand 5670
Paradeuniform 1368
Parados 3853
Paraffinöl 3855
Parallaxe 3856
parallel 3857
Parallellineal 3858
Parallelstab 5632

Parka(s) 3868
Parkbremse 3869
Parlamentärflagge 1747
Parole 3871
Partei 3879
Partisan 2118, 3877
Partisanenbekämpfungs- 1026
Partisanenkrieg 6691
Paspelierung 5795
Pass 3894
Passage 3899
passierbar 3898
Passierschneise 3900
passive Luftverteidigung 3905
passives Sonar 5534
passive Zielsuchlenkung 3906
Patient 3910
Patrone 715
Patronengurt 363
Patronenhülse 716
Patronenmunition 1743
Patronentasche 4238
patrouillieren 3912
Patt 5646
(Plan-)Pause 6222
Pedal 3936
peilen 2923
(den Standort) peilen 1739
Peiler 1640
Peilnetz 3445
Peilstandort 1740
Peilstrich 2984
Peilung 434
Peilwinkel 434, 4667
pendeln 5346
Pendelverkehr 5347
pensionieren 4813
Perforierung 3948
Perigäum 3950
periodisch 3954
Periskop 3955
Periskop- 3957
(an)Periskoptiefe 3956
perpendikular(-) 3963
Personal 3966
Personalabteilung 3967
Personalersatz 4720
Personalführung 3040
persönliche Habe 3965
persönlicher Stab 5634
persönliches Platzfunkfeuer 423
Perspektive 3968
perspektivisch 3968
Pfad 3909
Pfarrer 3819
Pfeife 6840
pfeifen 6839

Pfeilchen 1773
pfeilförmig 6870
Pferd 2326
Pferdestärke (PS) 2327, 4241
Pflanzendecke 1046
Pflege 5217
pflegen 3522, 5206
Phänomen 3975
Phase 3972
(in) Phasen einteilen 3971
phonetisches Alphabet 3976
phonetische Schrift 3977
Phosgen 3979
Phosphor 3980
Photogrammetrie 3983
Photozelle 3981
physikalisch 3992
physisch 3992
Pier 2629
Pilotenabzeichen 6873
Pilotendruckanzug 2110
Pilzwolke 3377
Pinasse 4020
Pionier 4024
Pionier(e) 1474
Pioniersicherungsoperationen 3683
Pionierwesen 868, 1476
Pionierzugmaschine 6254
Pistole 4029
Pistolengriff 2081
Pistolentasche 2302
Plan 940, 4047, 5064
Plan- 5068
Plane 688, 6052
planen 4048
(zeitlich) planen 5063
Planet 4065
planieren 2048, 2780
Planiermaschine 2051
Planierraupe 627
planimetrische Karte 4066
Planke 4067
plänkeln 5442
Plänkler 5444
Plankton 4068
planmässig 1187
Planpause 3773, 3775
Planschiessen 1686
Planübung 6694
Planungsfaktor 4070
Planungsstab 5635
Plasma 4073
plastischer Sprengstoff 4075
Plattenladung 5452
Plattenpanzerung 4077
Plattform 4084
Plattform-Abwurf 4085

Platz 2926
Platzpatrone 483
Plünderer 3066
plündern 2961, 4099
Plutonium 4100
pneumatisch 4102
Pöbel 3288
Pol 4174
polar 4171
Polar- 240, 4171
Polarkoordinaten 4172
Polaroidkamera 4173
Polarstern 5665
Politik 4178
Politiker 4179
Polizei 4177
Polizeiwache 5692
Poller 517
Polsterung 3816
Ponton 4183
Portage 4190
Position 4193, 5680
Positionslampe 4217
Positionslampen 3418
positiv 4218
Post 3008
Postbote (des Bataillons) 4236
Postdienst 4235
Posten 2598, 5191
Posten(bereich) 4221
Präzession 4249
Präzision 4250
Presse(-) 4268
Primärzelle 4279
Probe 4657, 6129
Probebank 4856, 6136
Probeflug 6132
Proben 6357
Probeschiessen 1699
Produktion 4314
Profil 6347
Programm 4318
programmieren 4320
Programmierer 931, 4319
Progressist 4324
Projekt 4328
Propaganda 4337
Propeller 4340
Propellerbö 5475
Propellerflügel 6549
(Sitzungs)Protokoll 3246
Protokoll (Verhandlungsprotokoll) 4349
Proton 4350
Protze 2820
Proviant 4355
Proviantoffizier 3578, 3593

Provianttasche 2202
Prüf- 6127
prüfen 739, 776, 6128, 6601
Prüfer 2490
Prüfung 6129
psychiatrisch 4360
psychologisch 4361
Puffer 617
Pufferzone 6939
Puls 4373
Pumpe 4375
pumpen 4376
Punkt 268, 1344, 4106
Punkt- 4022
Punktaufschlagzünder 1959
Punktdrehung 5603
Punktstarten 6932
Punktstörung 2614
pyrotechnische Gegenstände 4391

Q

Quadrant 4392
Quadrat 5619
Qualifikation 4394
qualifiziert 4395
Qualitätskontrolle 4396
Quant(en) 4397
(unter) Quarantäne 4398
Quartier 465, 4404, 4405
Quartiermeister 4401
Quartieroffizier 3576
Quecksilber 3160
quer 6338
Quer- 6338
querab 35
Querfeuer 1698
Querruder 80
Querschnitt 5144
Querstrebe 518
Quetschkopf- 2265
Quotient 4414

R

Rad 4418, 6831
Rad- 6835
Radar(-) 4419
Radar-Gegenmassnahmen 4433
Radarauffassung 4420
Radaraufzeichnung 6237
(positive) Radaraufzeichnung 6240
Radarbake 424
Radarecho 4435
Radarerfassungsbereich 4434

Radarpeilung 4436
Radarposten 3997
Radarschiessen 1688
Radarschirm 4438
Radarstille 4439
Radarstörung 4432
Radarwiederempfang 4824
Rädelsführer 4870
radial 4442
Radio(-) 4447
radioaktiv 4448
radioaktiver Niederschlag 4485
Radioisotop 4454
Radiokanal 4450
radiologisch 4456
Radius 4467
Radstand 6834
Rahmen 1893
Rahnock 6916
Rakete 3260, 4910
(ferngesteuerte) Rakete 3252
Rakete (TOW) 3274
Rakete (über das Meer streichend) 3269
Rakete (ungelenkt nach dem Abschuss) 3267
-Rakete 6012
Raketenabwehrflugkörper 3258
Raketenabwehrsystem 219
Raketenantrieb 4915
Raketen mit erdnaher Umlaufbahn (meist eng. Begriff) 1890
Raketenmotor 3337
(mit) Raketen nachbeschleunigtes Geschoss 4913
Raketenrüstsatz 4417
Raketenwerfer 4914
Rampe 4491
Rand, Spielraum 3070
randalieren 4872
ranghöher 5183
(im) Rang höherstehen 3756
Rangierbahnhof 6914
Rangieren 5342
Rangiergleis 5355
rangniedrig 2645
Rangniedriger(e) 2645
Rastraum 5644
Rat(schlag) 51
Ration 4538
rationieren (d.h. beschränken) 4537
Raub 4907
Rauch 5489
Rauchgranate 2073
Rauchtopf 5493
Rauchwolke 4005
Raum 6935

richtig   1011, 4860
Richtigstellung   179
Richtkreis   1273
Richtlinie   1272, 2865, 6311
Richtpunkt   4138
Richtschütze   2132
(mit) Richtstrahler senden   429
Richtung   1270
Richtungslinie   2846
(Gitter-)Richtungswinkel   2077
Riemen   445, 5469, 5758
Riff   4617
Ring   361
Ringvisier   5368
Rinne   1389, 2127
Risiko   4883
Risikoradius   4474
Riss   1057
Rohr   4026, 6390, 6393
Röhre   6392
(Elektronen-) Röhre   6547
(durch) Rohre leiten   4027
Rohrleitung   4028
Rollbahn   6063
Rolle   4367, 4919, 4923, 4928
rollen   4920, 6062
Rollenbahn   4930
rollendes Material   4931
röntgen   6911
Röntgenstrahlen   6912
Rost   4998
(ver)rosten   4999
rostfrei   5001
rostig   5002
Rostschutz-   224
Rostschutzmittel   5000
Rotation   4949
Roter Halbmond   4607
Rotes Kreuz   4608
rotierend   4947
Rotor   4950
Rotorblatt   4951
Rotte   1627
Rottenflieger   6871
Route   4962
Routine   4973
Rücken   335, 4845
Rückenwind   5988
rückgängig machen   685
Rücklauf   4579
Rückmeldung   4724
Rucksack   4979
Rückschlag   5224
Rückstoss   2661
Rückstossdämpfer   616
Rückstrahler   4628
Rückstreuung   341

Rückvisier   5361
rückwärtiges Gebiet   4562
rückwärts   287, 343
Rückwärtseinschnitt   4766
rückwärts fahren   4827
Rückwärtsgang   2007
Rückzug   4816, 6889
Rückzugbewegung   4820
Rückzugslinie   2867
Ruder   3526
Ruderhaus   6836
rudern   4366, 4975
Ruderpinne   6170
Rufzeichen   5373
(im) Ruhestand   3944
Ruhe und Erholung   4495
Ruhezeit   5403
Ruinen   4983
Rumpf   2352
rund   798, 4955
Rundfunksendung   605
Rundholz   5551
Rundsicht   3836
Rundumsichtradarschirm   2441
Rundumverteidigung   3952
Runzeln   3527
Rutschbahn (fest)   5464
rutschen   5428, 5429

S

Sabotage   5005
Sabotage-Alarm-System   2570
Saboteur   5007
sabotieren   5006
Sachverständige(r)   5560
sägen   5051
salutieren   5028
(ein) Salut schiessen   1649
Salut von .... Schuss   5030
Salve   6644
Salven (schiessen)   6645
Salz-   5026
salzig   5026
Salzwasser   5027
sammeln   5031, 5034
(sich) sammeln   856
Sammelpackung   2906, 2911
(Kriegsgefangenen-)Sammelstelle   4141
Sand   5039
Sandbank   5042
sandig   5044
Sandsack   5040
(mit) Sandsäcken befestigen   5041
Sandsturm   5043
Sanitäts-   176, 3144

schilaufen 5426
Schild 5281
Schilderhaus 5193
Schirm 6633
(Bild-)Schirm 5091
Schistock 5440
Schitruppen 5447
Schlacht 402
Schlachtflugzeug 99
Schlafsack 438, 5461
Schlagbolzen 4016
schlagen 436
Schlagkräfte 5789
Schlagseite 2880
Schlamm 5407
Schlauch 2328
(Luft-)Schlauch 6391
Schlauchboot 2457
Schlauchendstück 3499
(im) Schlepp 6471
schleppen 1354, 6215
(ab)schleppen 6214
Schlepper 6394
Schlepphaken 2315
(mit einem) Schleppnetz fischen 6345
Schleppnetzfischerboot 6346
Schleppseil 6217
schleuderfrei 3476
schleudern 5428, 5429
schleudersicher 5431
Schleudersitz 1433
Schleuderstarthilfe 725
Schleuse 2932, 5484
Schlick 3634
(ver)schliessen 828 , 5343
Schlingern 4920, 4924
Schlitten 5430, 5460
Schlittenwaffe 6791
schlitzen 5459
Schloss 520, 2927
Schlucht 1389, 2040, 4539
Schlupf 339, 5472
schlüpfen 5471
schlüpfrig 5474
Schlussdeckungsfeuer 1666
Schlüssel 2654, 2655
Schlüssel- 2658, 4043
Schlüsselgelände 6117
Schlüsselpunkt 4131, 4157
Schlüsselstellung 4204
Schlusslicht 5985
Schlussrohr 5986
Schmerz 3820
schmerzstillendes Mittel 3821
Schmiede 1868
schmieren 2066, 3618
(ab)schmieren 2986

Schmierpresse 2067
Schmierpunkt 4170
Schmierstoff 2985
Schmierung 2987
schmuggeln 5495
Schmuggler 5496
Schnabel 2738
Schnalle 614
(an, um, zu)schnallen 612
Schnappatrouille 3918
Schnee 5507
Schneeblindheit 5509
Schneegrenze 5512
Schneeloch 5511
Schneemobil 5513
Schneepflug 5514
Schneereifen 5516
Schneeschuhe 5515
Schneewehe 5510
Schneidbrenner 3798
schneiden 1092
schneien 5508
Schneise 4843
schnell 1590
Schnell- 2269
Schnellangriff 2193
schnelle Einsatztruppe 4529
schneller schiessen an Schuss 3748
schnelles Marschtempo 4412
Schnellfeuer 1687
Schnellschiessen 5323
Schnellschuss 5498
Schnellsprengung (veraltet) 2197
Schnelltauchen 1064
(Quer-)Schnitt 5142
Schnittpunkt 4130
Schnorchel 5506
schnorcheln 5505
Schnüffeler 5500
Schnüffelhund 5501
Schnur 1779, 2718, 2725
Schock 5313
schocksicher 5319
Schockwelle 5318, 6765
Schockwirkung durch Feuer und/oder
    Bewegung 5317
Schornstein 1937
Schott 625
Schottpanzerung 921
Schraffierung 2149
schräg 1255, 2429, 3536, 5458
Schrägentfernung 4513
Schrägluftaufnahme 3985
schräg sein 6171
schräg stellen (sein) 5456
Schrapnell 5341
Schraube 5099

schrauben 5098
Schreiber 825
Schreibmaschine 6422
Schritt 3802
(im Gleich-) Schritt 5716
Schrittmacher 3806
Schrotflinte 5338
(kurzgesägte) Schrotflinte 5052
Schrott 5085
Schule 5072
Schultergurt 5339
Schulterstück 1491
Schulung 2443
Schuss 4954, 5336
Schuss! 6769
Schussangabe 2437
Schussbereich 5607
Schusseigenschaft 5395
(gleiche) Schussentfernung 4714
Schussfeld 239
Schusslinie 2827
Schuss pro Waffe 4960
Schussrichtung 2836
Schussweite 2137
(Höchst-)Schussweite 4506
Schusswerte 1725
Schütt- 2958
Schutz 4346, 5282
Schutz- 4347
Schutzblech usw. 5282
Schutzbrille 2039
Schütze 4854
schützen 4176
schützen (vor) 4345
Schützendeckungsloch 1889
Schützengraben 6352
Schützenmine 3206, 3207
Schützenpanzer 6573, 6575
Schützenstand 6409
Schutzhafen 4638
Schutzhindernis 3550
Schutzidentität 1049
Schutzkleidung 4348
Schutz suchen 5276
Schutzwall 4492
schwach 6770
schwächen 6771
Schwallbrett 5588
Schwarm 5312
schwarzes Brett 630, 3494
Schwebeflug 2344, 2345
Schwebgipfelhöhe 2346
Schweigen 5399
schweigend 5402
Schweissen 6818, 6819
Schwenkbereich 6341
(allseitiger) Schwenkbereich 6344

schwenken 5954, 6342, 6832
(ein)schwenken 5948
Schwenkflügelflugzeug 102
Schwenkstellrad 2178
schwer 2226
schwer beschiessen 5048
Schwere 2063
schwere Artillerie 2228
schwerer Seegang 2231
schwere See 2232
schweres Wetter 2233
Schwerguttransporter 5297
Schwerkraftsanzug 222
Schwerpunkt 743
Schwerpunktbildung 936
schwerverwundet 5201
Schwimm- 185
Schwimmbalg 5095
Schwimmdock 1801
schwimmen 1796, 1809, 1810
schwimmen (aufwärts) 1800
schwimmen (stromabwärts) 1799
schwimmend 65, 330
Schwimmer 1797
schwimmfähig 642
Schwimmfähigkeit 5947
schwimmfähig machen 1798
Schwimmkasten 667
Schwimmkraft 641
Schwimmkragen 1811
Schwimmpanzer 6002
Schwimmsteg 1797
Schwimmtaucher 5437
Schwimmweste 2604
Schwinge 6873
schwingen (lassen) 3736
schwingend 3737
Schwingung 3738
Schwund 1575, 6426
Schwungrad 1827
Schwimmfahrzeug 184
See 2692, 5104
See- 3073, 3075, 3405
Seeaufklärung 4587
(in) See gehen 5661
Seekampf 3409
Seekarte 772
seekrank(heit) 5128
Seeladeverzeichnis 3564
Seele 541
seemännisch 3405
Seemanstum 5117
Seemeile 3188
seetüchtig 5132
Seetüchtigkeit 5132
Seeversorgungsraum 3412
seewärts 5131

(auf dem) Seeweg befördert   5107
Seezeichen   5118
Segelboot   6613
Segelflugzeug   2037
Segeltuch   691
Sehachse   3693
Sehstreifen   2843
seicht   5246
Seil   4943
seismisch   5165
Seismograph   5166
Seite   5352
Seiten   2728
Seiten-   5351
Seitenabweichung   1370
Seitengewehr aufpflanzen!   1742
Seitenlampen   5354
Seitenruder   4980
Seitenwaffen   5353
seitlich   2728, 5351
seitlich gerichteter Bordradar   4429
Sekretär   5140
Sektor   5146
Sektorenabtastung   5149
Sektorkontrolleur   5147
sekundar   5134
Sekundarstrahlung   5135
Sekunde   5133
selbstabdichtender Kraftstoffbehälter
   6008
selbständig   2435
selbstaufrichtend   5176
selbstbeweglich   317
Selbstfahr-   5175
Selbsthilfesatz   2674
selbstöffnender Fallschirm   3844
selbstspannend   5173
selbstständig   5194
Selbststeuergerät   314
Selbststeuerung durch automatische
   Peilung (meist eng. Begriff)   3419
selbsttätige Aufziehleine   2853
Selbstverteidigung   5174
Selbstzerlegungs-   1578
Selbstzerstörungsmechanismus   3134
Selbstzerstörungszünder   1967
Selbsuchmine   3222
Sende-Empfangsgerät   6319
senden   3009, 5182, 6317
Sender   4466, 5682, 6318
Senderanflug   2304
Senderempfänger   4465, 6300
Sendung   6316
Senfgas   1994
Senk-   667
senken   1213
senkrecht   6531

senkrecht(e)   6604
Senkrechte   3963
Senkrechtstarter (meist eng. Abkürzung)
   103
Senkung   1214, 1840
(Boden-) Senkung   5818
Senkungswinkel   200
Serie   5200
Servomechanismus   3135
Sextant   5232
Sextantabweichung   5234
Sextantwinkel   5233
sicher   5151
sichere Schiffahrtsroute   5016
Sicherheit   5154
(enge) Sicherheit   5155
(personelle) Sicherheit   5158
Sicherheits-   1577
Sicherheitsabstand   3242, 5014
Sicherheitsbescheid   5163
Sicherheitsbescheinigung   5160
Sicherheitsgrenze   2826, 2841, 5017
Sicherheitsgurt   449, 450
Sicherheitshöhe   2240
Sicherheitskoeffizient   5015
Sicherheitslinie   2852
Sicherheitsradius   4473
Sicherheitsüberprüfung   5161
Sicherheitsventil   5018
Sicherheitsvorschriften   4651
Sicherheitswinkel   205
Sicherheitszone   6949
Sicherheitszünder   1966
sichern   5090
sicherstellen   5153
Sicherung   1943, 1944, 5013
Sicherung gegen-U-Boote   5094
Sicherungsanlage   1972
Sicherungsauftrag   3280
Sicherungsoperationen   3681
Sicherungsschleier   5087, 5096
Sicherungssplint   4017
Sicherungssplintring   4869
Sicherungstruppen   1054
Sicht-   6634
sichtbar   6629
sichtbares Spektrum   6630
Sichtbarkeit   6628
Sichtflug   1788
Sichtflugregeln   6637
Sichtgerät   960, 1304
Sichtkontrolle   6639
Sichtung (von Verwundeten)   6356
Sichtverlust   6844
Siedlung   5228
Sie dürfen passieren   3904
Sieg   6619

Steigerung 1502
Steiggeschwindigkeit 5567
Steigpunkt (eines Flugzeuges) 4140
Steigung 2046, 4038, 4880
steil 5255, 5712
Steilabhang 1507
Steilfeuer 1671
Steilluftbild 2980
Steilufer 507
Stein 5732
steinigen 5733
stellen 2923, 6171
(befestigter) Stellplatz 2187
Stellung 2926, 4193, 5416
(teilgedeckte) Stellung 4203
(eine) Stellung beziehen 3560
Stellungen 1407
(folgende) Stellungen 5832
stellvertretend 1217
Stellvertreter 5824
stereoskopische Sicht 5718
Stern 5664
Steuer 2250
Steuerbord 5666
(nach) Steuerbord halten 286
Steuerbremse 558
Steuereinrichtung 2008
Steuerflügel 6550
Steuerknüppel 2634, 5723
Steuerkompass 916
Steuerkurs 2211
Steuermann 2253, 4402
steuern 5713
Steuerorgan 977
Steuerrad 6833
Steuerruder 4980
Steuersäule 984
Steuerungen 988
Steuervermögen 5714
Stich 2283
Stichprobe 5598
Stickgas 1990
Stiefel 538
Stielgranate 2074
still 676
stillegen (Flugwerkehr, Flugzeug) 2084
stillstehen 1905
Stimm- 6641
Stimme 6642
Stock (-werk) 5749
stopfen 2290
Stoppuhr 5738
Stöpsel 4093
Stör- 2539
Störangriff 5594
stören 2181
(absichtlich) stören 2608

Störfeuer 1670
Störflugzeug 2569
Störfolie 6861
Störminenfeld 3231
(automatischer) Störsender 2609
Störung 1580, 1596, 2183, 2612, 3467
Störungen 290, 5677
Störzeichen 839
Stoss 632, 2413, 4387, 5313
Stossdämpfer 5316
stossen (gegen) 633
stossendes Schiff 995
Stossfuge 657
Stossstange 634, 1604
Stosstrupp 4995
Stosswirkung 4377
strafbar 4381
Strafe 4382
strafend 4383
straff 6061
Strahl 428, 1766, 2621
Strahlenbündel 3939
Strahlendosimeter 1633
Strahlendosis 1561, 4443
Strahlendosisleistung 4444
strahlenförmig 4442
Strahlturbinen- 6401
Strahlungsgrad 5704
Strahlungskrankheit 4445
Strahlungsmesser 4441
Strahlungsmessgerät 4440
Strand 413
stranden 412
Strandgut 2624
Strandkommando 414, 3887, 3893
Strandmeistergruppe 3411
Strasse 4891
Strasse frei 4897
Strasse gesperrt 4900
Strassenbelastbarkeitsbewertung 4972
Strassenbeschilderung 5397
Strassenfreigabezeit 4898
Strassenkampf 5771
Strassenkapazität 4896
Strassenkreuzung 1080
Strassenmarkierung 4902
Strassenräumung 4899
Strassensperre 4895
Strassenvermessung 5925
Strategie 5766
strategisch 5761
strategische Abschreckungswaffe 5764
strategische Konzeption 5763
strategische Reserve 2017, 5765
strategischer Lufttransport 6330
strategische Waffe 6793
strategische Warnung 6712

(durch Fernsehsender) übertragen   6091
übertragen (mit Relais)   4671
übertreffen   3748
überwachen   2112, 2653, 3316, 5239,
   5645
Überwacher   5240
überwachter Luftraum   985
Überwachung   3795
(chemische) Überwachung   5922
Überwachung(s-)   5918
Überwachungsmassnahmen   987
Überwachungsradar   4430
Überwachungsstellung   4206
überwältigen   3779
überwinden (Minenfeld)   569
Überwinden eines Minenfelds ohne
   grössere Vorbereitungen   2194
überwinden ein Hindermiss   3430
(Truppen-)Übung   1534
(Feld-)Übungen   3047
Übungs-   4245
Übungsbeschränkungen   4800
Ufer   366, 5324
(feindliches) Ufer   5325
Uferkommandant   416
Uferkommando   414, 3893
Uferland   1866
Uhr   792
(Armband-) Uhr   6730
Uhrmine   3210
Uhrwerkzünder   1951
Uhrzeitseinstellung   6177
Ultimatum   6427
Ultrahochfrequenz   6428
Ultrakurzwellen   6608
Ultraschall-   6429
ultraviolett   6430
Ultraviolett-   6430
Umdrehungen pro Minute (U/Min.)
   4835
Umfang   800
umfassen   1487, 3746
umfassend   158, 922
Umfassung   1490
(einseitige) Umfassung   3668
Umfassung (aus der Luft)   6605
Umfassungsbewegung   1489, 6407
umgebend   174
Umgebungs-   174
umgehen   658, 5446
umgehen mit   2167
Umgehungsstrasse   659
umgekehrt   2579
umgruppieren   4645, 4704
Umgruppierung   4703
Umhang   698
umkehren   2578

Umkehrflug   4302
Umkehrgrenzpunkt   4167
umkippen (lassen)   3794
Umkreis   3951
umkreisen (Satellit)   3700
umladen   4691
Umladung   6337
Umladungspunkt   3417
Umleitung   1248, 1323
umreissen   3749
(grobe) Umriss(e)   3750
(versiegelter) Umschlag   1488
Umschlaghafen   6755
umschlingen   5468
umstellen   5950
umsteuern   3308
Umsturz   5828
umstürzlerisch(-e-r)   5829
unabgestimmtes Reihenluftbild   6450
unabhängig   2435
unangerufen   6442
unbefügt   6436
unbegraben   6441
unbegrenzt   2434
unbehindert   6510
unbekannter Soldat   6498
unbeladen   2805
unbemannt   6505
unbeobachtes Schiessen   1702
unbesetzt   6508
unbestätigt   6452, 6522
unbestimmt   2434
unbewacht   6483
unbewaffnet   6434
undiszipliniert   6478
uneben   4953, 4982
unentbehrlich   1511
unerfahrene Truppen   4540
unerkennbar   6484
unerlaubte Abwesenheit   7
Unfall   11
Unfall-   12
unfreiwilliges Trudeln   6451
ungebraucht   6525
ungelenkte Rakete   3262
ungenügend bemannt   6465
ungepanzert   6435
ungepflastert (Strasse)   6506
ungerade Zahl   3569
ungesichert   2480
ungewarnt und ohne Deckung   6526
unhaltbar   2433, 6524
Uniform   6486
universal   6495
Universal-   6495
unklarer Anker   1887
unkontrolliert   6449

Ventilator 1587
veraltend 3542
veraltet 3543
veränderlich 6556
verankern 191
verantwortlich 17
Verantwortung 4786
Verantwortungsbereich 248
Verantwortungswechsel 3901
verarbeiten 4304
Verarbeitung 4307
Verband 2108
(vorgeschobener) Verbandsposten 5683
Verbandstelle 5687
Verbergen 932, 933
(sich) verbergen 2257
Verbesserung 1012
verbieten 358, 4326
verbinden 2874
verbindlich 3041
Verbindung 2787, 2875
Verbindung aufnehmen (mit) 2786
(sich in) Verbindung setzen (mit) 966
Verbindungsflugzeug 93
Verbindungsgraben 6351
Verbindungslinien 2870
Verbindungsnetz 3444
Verbindungsoffizier 3591
Verbindungssicherheit 5156
(in) Verbindung stehen (mit) 905
Verbindungstrupp 6103
Verbindungszone 6941
verborgen 1055
verborgene Strahlenbelastungsdosis
    2727
Verbot 359
Verbrauch 1545
verbrauchen 1543
Verbrauchsgeschwindigkeit 964
(Mengen- u. Einzel-)
    Verbrauchsgüter 5862
Verbrecheralbum 4918
verbreiten 1309
Verbreitung 1310
Verbrennung 878
Verbrennungsvorrichtung 876
(sich) verbünden (mit) 160
verbündete Kampfgruppe 6068
verbündete Operation 3657, 3663
Verdächtige(r) 5932
verdächtigen 5931
Verdeck 2310
verdichten 923
Verdichtung 925
Verdichtungsverhältnis 926
(die) Verdichtung vermindern 1141
verdoppeln 1398

verdrahten 6878
verdrängen 1300
verdunkeln 1106
Verdunkelung 3537
Verdunkelungsrauch 5491
Vereinbarkeit 918
vereinbart 4247
Vereinbarung 73, 6348
vereinheitlichen 6487
(sich) vereinigen 6491
Vereinten Nationen (die UNO) 6492
vereisen 2387
vereist 2389
Vereisung 2388
Verfahren 4299
Verfolgekurs 1035
verfolgen 774, 1845, 4385, 6229, 6265
Verfolger 6244
Verfolgung 1844, 4386
(Zielweg-)Verfolgung(s-) 6247
Verfolgungs- und
    Vernichtungsgeschwader 2358
Verfolgungsgerät 6245
Verfolgungsgrenze 2825
verfügbar 323
verfügbare Kampfkraft 3054
Verfügbarkeit 322
(über....) verfügen 1305
(zur) Verfügung stehen 883
Vergaser 706
Vergeltung 4808
Vergeltungsmassnahmen 4752
vergleichen 854
vergraben 649
vergrössern 1477, 2431, 3007
Vergrösserung 1478, 3005
verhaften 266
Verhältnis 4536, 4665
verhandeln 3429, 3870
Verhändler 3432
Verhandlungen 3431, 5998
verhindern 4276
Verhör 2556
verhören 2555
Verkantungswinkel 6172
Verkehr 6256
Verkehrsbeschränkung 3359
Verkehrsdichte 6261
Verkehrsfluss 6262
Verkehrskapazität 6258
Verkehrsleitpunkt 4233
Verkehrsnetz 908, 3446
Verkehrsregelung 6260
Verkehrsstau 6263
Verkehrsvorschriften 4652
Verkettung 2877
verkleiden 4828

Versorgungsraum 3027
Versorgungsschiff 5295
Versorgungssystem 2866
Versorgungsvorrang 4288
Versorgung während der Fahrt 6477
Versprengter 5755
Verstanden! 4917, 6849
verstärken 4659, 5778, 6147
Verstärker 2522
Verstärkung(en) 4662
Verstärkungsauftrag (Artilleriefeuer) 1693
verstauen 5750
Versteck 664, 2256
verstreutes Minenlegen 2751
verstümmelt 3381
Verstümmelung 1983
Versuch 1547
Versuchs- 1548, 4008
Versuchsgelände 2090
versunken 5843
verteidigen 1158
(gut)verteidigt 6822
Verteidigung 1159
(in der) Verteidigung 1165
Verteidigung aus Stellungen 4213
Verteidigung ohne grössere
    Vorbereitungen (veraltet) 2196
Verteidigungs- 1164
Verteidigungsgürtel 3951
Verteidigungsministerium 3243
Verteidigungspolitik 1163
verteilen 1315, 2595
Verteiler 1317
Verteilerrohr 3050
Verteilung 1316, 2594
Vertiefung 2299
vertikal(e) 6604
Vertikalladung 2913
Vertikalluftaufnahme 3986
Vertrag 976, 6348
Verträglichkeit 918
vertraulich 954
vertrauliche Information 4799
vertreiben 4961
vertreten 4748
Vertreter 4749
vervielfältigen 1397
Verwaltung 44
Verwaltungs- 45
Verwaltungsbefehl 3705
verwaltungsmässig 45
Verwaltungsoffizier 3583
Verweis erteilen 4751
Verwendbarkeit 5213
verwerfen 4664
verwirbelt 6405

verwundbar 6654
Verwundbarkeit 6653
Verwundbarkeitskreis 4474
verwunden 6902
verwundet 6903
Verwundete(r) 722
Verwundetenabtransport per
    Hubschrauber 3140
Verwundetenmeldung 4733
Verwundetenrolle 2883
Verwundetensammelpunkt 4113
Verwundetentransportpunkt 4114
Verwürf(e)lungsvorrichtung 5083
verwürfeln 5081
Verzeichnis 4641
(sich) verziehen 613
verzögern 4809
(sich) verzögern 1182
verzögerte Eröffnung (eines Fallschirms) 3642
Verzögerung 1181
Verzögerungs- 1184
Verzögerungstaktik 1185
(regelbarer) Verzögerungszünder 1970
Verzugszeit 1124
Vessorgungsschiff 5301
Veteran 6615
Viadukt 6618
Videoband 6621
Videorekorder 6622
vielseitig 6602
Vielseitigkeit 6603
Vielstoff- (z.B. Motor) 3367
Viereck 5619
viereckig 5619
Visier (-einrichtung) 5359
Visierantenne 5360
Visierkimme 3490
Visierlinie 2864
(ohne) Visier schiessen 1720
Visierschuss 5371
Visierskala 2055
visuell 6634
Vogel 475
Vogelschlag 476
Völkerrecht 2544
volle Einsatzbereitschaft 4552
Volle Kraft voraus! 1932
Volle Kraft zurück! 1933
vollgesogen 6744
vollkommene Kampfkraft 872
Vollmond 1930
vollständig 6206
Volltreffer 2279
Volt 6646
Volumen 696
vom Lande abhalten 2792

629

W

(Gaslade)waffe   6780
(leichte) Waffe   6784
(Abstands) Waffe   6792
(Dauerschuss)waffe   6794
(persönliche) Waffe   6782
(mannschaftsbediente) Waffe   6777
Waffe mit Hinterladung   2909
Waffen   262
Waffen (mit bedingter Feuererlaubnis)
   6169
Waffenausbildung   6291
Waffenbereitschaftszustand   5706
Waffengattung   250, 562
Waffengeschicklichkeit   5434
Waffenkammer   261
Waffenruhe   6381
Waffenschmuggler   2135
Waffenständer   4416
Waffenstillstand   255
Waffensystem   6799
(mehr) Waffenwirkung als nötig   3770
wagen   4884
Wagenpark   1774
Waggon   6658
Waggon (Eisenbahn)   6382
Wagnisübungen   6277
(aus)wählen (aus)   5170
wahre Fluggeschwindigkeit   6384
wahre Höhe (über NN)   6385
wahre Peilung   6386
wahrer Horizont   2321
wahrscheinliche Abweichung   4297
wahrscheinliche kreisförmige Streuung
   799
Wahrscheinlichkeit   4296
Wahrzeichen   2711
Wal   6828
Wald   6891
Walfangboot   6829
Wall   3862
Walze   382, 4928
walzen   4921
Wand   6666
wandern   6667
(Öl) Wanne   5839
(Panzer-)Wanne   2352
Waren   900
Wärme   2220
Wärme-   6140
Wärmebildgerät   6142
Wärmeschutz   6143
wärmesuchend   2223
Wärmezerstreuer   5896
Warmluftfront   6701
Warmluftströmung   6140
Warn-   6705
Warnanlage   6713

warnen (vor)   6702
Warnfeuer   6714
Warnlicht   6714
Warnnetz   3441
Warnung   6706
Warp (-trosse)   6716
warpen   6717
Wartelinie   2837
warten   3019, 5206, 6659
„Warten Ende!"   6660
Wartepunkt   4127
Warteraum   4137
Wartung   3021, 5217
(planmässige) Wartung   3026
(routinemässige) Wartung   3025
waschen   6725
Wasser   6735
Wasser-   2373
wasserdicht   2411, 6746, 6756
Wasserdruck-   2364
Wasserflasche   6737
Wasserflugzeug   5119
Wasserfontaine   5606
wassergekühlt   6738
Wasserhose   6754
Wasserkühlmantel   6741
Wasserlauf   6739
Wasserlinie   6743
Wasserloch   6740
wassern   149
Wasserreinigungsanlage   6747
Wasserscheide   6753
Wasserstand   6742
Wasserstelle   4159
Wasserstoff   2366
Wasserstoffbombe   2367
Wasserstoffbombe (H-Bombe)   2206
Wasserturm   6757
Wasserungszone   150
Wasserverdrängung   1301
Wasserwagen   6758
Wasserweg   6759
(auf dem) Wasserweg befördert   6736
waten   6656, 6657
Watfähigkeit   1861
Watfähigkeit (ohne Vorbereitung)   1862
Watt   1866
Wechselangliederung   1077
Wechselbeziehung (zu)   1014
(in) Wechselbeziehung bringen   1013
Wecken   4826, 6662
Weg   6230
(Feld-, Wald- usw.) Weg   6264
Wegerecht   4866
weglaufen   1155, 1223
Wehrdienstpflichtiger   959
Wehrmachtsangehörige(r)   5208

Wehrpflichtige(r) 1353
Wehrpflichtsgruppe 69
Weichkopf 5620
Weichkopfgeschoss 5270
Weichmacher 4076
weisse Fahne 6842
Weiterbildung 6279
weiterdrängen 4269
Weitschuss 3758
weitwinklig 6846
Well- 1018
Welle 5242, 6760
Wellen-Schutzblech 3833
Wellenbrecher 587
Wellenlänge 6767
Welligkeit 6480
Wendekreis des Krebses 6376
Wendekreis des Steinbocks 6376
Wendepunkt 4155
Wenderadius 4470
werfen 6156
Werfer 2737
Werft 1334, 1336, 5311
Werkstatt 6899
Werkstoff 2188
Werkzeug 6190
Werkzeugsatz 2678
wesentlich 6640
Weste 5445
westlich 6826
Wetter 6801
Wetterbericht 4743
Wetterkunde 3175
Wettermeldung 3172
Wettervorhersage 6803
Wickelgamasche 4390
Wicklung 6858
widerhallen 1421
Widerruf 4568
widerrufen 4567
Widerrufvermögen 3157
widerspiegeln 4626
Widerstand 3692
Widerstandsnest 4103
(die Depots) wiederauffüllen 4803
wiederaufrüsten 4564
wiederbeleben 4831
wiederbewaffnen 4564
wieder eintreten 4618
Wiedereintritt 4620
wiedereintrittsfähigeWeltraumfähre
   (meist eng. Begriff) 6574
Wiedereintrittsflugkörper 6580
wieder ergreifen 4569
wiedererlangbar 4596
Wiedergabe 1303
wiedergeben 1302

wiedergewinnen 4594
Wiedergewinnung 4597
Wiederherrstellung 4656
wiederholen 4709
Wiederholungsimpfung 536
wieder laden 4691
wiegen 6808, 6810
Wimpel 3943
Wind 6853
Windanzeiger 6857
winddicht 6863
Winde 6851, 6859
Windedraht 6880
Windeffekt 6855
winden 6854
(hoch usw.) winden 6852
Windgeschwindigkeit 5573
Windhahn 6552
Windhose 6838
Windmühle 6860
Windpumpe 6864
Windsack 6856
Windschutzscheibe 6865
Windstärkemesser 196
windwärts 6534, 6867
Windwinkel 201
Winkel 198
Winkelreflektor 1008
Winter 6874
winterlich 6875
Wirbel 6651
Wirbelbewegung 6402
wirbeln 5582
Wirbelsturm 6197
wirklich 35
wirksam 1429
wirksame Höchstschussentfernung 4507
wirksamer und tödlicher Radius 4468
Wirkung 1428
Wirkung im Ziel 4785
Wirkungsbereich 6937, 6954
(unbeschränkter) Wirkungsbereich 6946
Wirkungsfeuer 1717
Wirkungsradius 4472
wirtschaftlich 1425
Wirtschaftstruppenteil 19
(Natur-) Wissenschaft 5073
Wissenschaftler 5075
wissenschaftlich 5074
Woge 5905
wogen 5906
Wolke 835
Wolkendeckehöhe 2237
Woll- 6894
Wolle 6893
Wrack 6905

УКАЗАТЕЛЬ РУССКИХ ТЕРМИНОВ

А

абонент 3757
авангард 50, 6553
аванпост 3752
аварийное прекращение 6
аварийный 5928
аварийный люк 1506
авария 11, 1063
авиабаза 85
авиалиния 118
авиамедицина 59
авиамедицинский 58
авианосец 105
авианосная авиационная группа 711
авианосная группа 712
авиаприцел 532
авиатранспорт 125, 6326
авиатранспортёр 101
авиационная поддержка войск 5873
авиационная спасательная служба
 130
авиационное горючее 327
авиационное прикрытие 1039
авиационный 60, 88
авиация 328
авионика 329
астронавигация 733
авто- 5347
автобаза 4185
автовоз РОРО 4932
автоколонна 992
автокомпас 1639
автомат 4847, 5807, 6774
автоматизация 315
автоматизировать 312
автоматическая передача данных
 2876
автоматическая пушка 687
автоматическая система приземления
 2035
автоматический 313
автоматическое оружие 6774
автомобиль 316, 6563
автомобильный 317
автономная мина 2359
автономный 2844
автопилот 314
автострада 2272

агенство 70
агент 71
агент-сыщик 1241
агрегат 72, 280, 3029, 5221
агрегат приборов 2678
адаптор 4097
административный 45
администрация 44, 3967
адмиральское звание 1748
адъютант 43, 79
азбука Морзе 3327
азимут 334, 434
айсберг 2381
академия 9
Академия Военно-Морского Флота
 3408
аккумулятор 401
акселератор 1999
активатор 33
активировать 31
активная оборона 3290
активные помехи 2612
активный 34
активный аэродром 3670
акустика 27
акустический 26
аллея 4843
альбом фотоснимков преступников
 4918
альтернатива 168
альтернативный 169
альтиметр 171
альянс 154
алюминий 173
амбар 374
амбулаторный приём 3852
американский солдат 2024
амортизатор 616, 5316
амперметр 181
амфибия 184
анализ 188, 579
анализировать 187, 578
анализ цели 6043
аналитик 189
аналогичная электронная
 вычислительная машина 186
аналогичный компьютер 186
ангар 2179
анемометр 196

аннексировать 211
аннотация 215
аннотировать 214
аннулировать 685
антарктический 216
антенна 217
антибиотик 220
антикоррозийное средство 5000
антиобледенитель 1180
антиракета 3258
антисептический 225
антисептическое средство 225
антифриз 221
апогей 230
арест 265
арестовывать 266, 1238
арктический 240
армия 263
арсенал 267
артиллерийский огонь 5274
артиллерийско-техническое снабжение 3723
артиллерийское дело 2133
артиллерист 2132
артиллерия 271, 3721
артподдержка (морского) десанта 5326
архив 1630
арьергард 4563
асбест 272
астрокомпас 288
асфальт 6051
атака 274, 296, 771, 5786
атаковать 275, 297
атмосфера 289
атмосферные помехи 290, 5677
атом 291
атомный 292
атомный котёл 4547
аттестация 4734
аэровокзал 6105
аэродинамика 57
аэродинамический 56
аэродром 111
аэродромный пункт управления 4107
аэромобильный 88, 123
аэронавигационные огни 3418
аэронавтика 61
аэроперевозка 5347
аэростат 356
аэростат заграждения 381
аэротранспортабельный 142
аэрофотоснимок 3984
аэрофотосъёмка 3987

Б

багаж 347

багажник 6388
база 386, 5679
база данных 1110
бак 1864, 6000
(сбрасываемый топливный) бак 6010
(самозатягивающийся топливный) бак 6009
(к) бакпорту 231
баланс 350
балансировать 6366
балисаж 351
балка 427, 2026
балласт 352
баллистика 355
баллистическая ракета 3260
баллистическая ракета подводного базирования 3272
баллистическая ракета средней дальности 3268
баллистический 354
баллон с кислородом 3800
бальза 357
банда 4868
бандит 362
банка 368
банно-дезинфекционный взвод 393
барабан 1387
барак 2362
барашек 6872
барашки 6841
баржа 373
барограф 376
барометр 377
барометрическая высота 4272
барометрический 378
барометрическое показание 4555
баррикада 383
баррикадировать 384
барьер 385
бассейн 390
батальон 395
батальонная группа специального назначения 396
батарея 400, 401, 4446
батарея, отыскивающая целей 2925
башней укрытой 4212
башня 6409
бдительность 6624
бдительный 147, 6625
беглый огонь 1687
бегство 1503, 1526
бежавший 1505
бежать 1504
беженец 4637
безмолвный 5402
безоговорочная капитуляция 6447
безопасная высота взрыва 2240

быть в режиме висения 2344
быть вырисованный (на фоне
    горизонта) 5405
быть готовым 5656
быть наготове 5656
быть старше 3756
бюджет 615
бюллетень 629, 3493
бюро 643, 3572

В

вагон 6382, 6658
важнейший 6640
важнейший пункт 4157
вакуум 6543
вакуумная фляжка 6544
вакуумный 6543
вакцинация 6542
вал 4492, 4928, 4929
вал винта 5244
валик 1096
вариант 3696
вариант обстановки 973
варить 997
ватерлиния 6743
вахта 6729
вахта на якорной стоянке 195
вахтенный журнал 2939
вахтенный офицер 3605, 6732
вбивать клин 6806
введение 2469, 2482
введённая в бой часть 899
вверх 6528
вверх по ветру 6534
вверх по течению 6533
вверху в гору 6528
ввод 2478, 6109
(сидеть) в воде на ... 1362
вводить 2481
вводить в строй 897
вводить по этапам 3971
вводная 3751
ведение войны 6678
ведение огня при быстром
    развёртывании и перемещении
    5322
ведения в бой 4987
ведомый (лётчик) 6871
ведро 611
ведущее колесо 1379
ведущий переговоры 3432
ведущий поясок 1378
ведущий самолёт 95
веер 938, 5253
веерный (маркерный) радиомаяк 418

вездеходный 1078
вектор 6558
величина мёртвой зоны 5439
вентилировать 6598
вентиль 6548
вентилятор 1587, 6599
вентиляционное отверстие 6596
вербовка 4601
верёвка 4943
вероятная радиальная ошибка 799
вероятное отклонение 4297
вероятность 4296
вероятность поражения 2667
вероятный противник 6152
вертеться 4042, 5954
вертикаль(ный) 6604
вертикальная погрузка 2913
вертикальный 6531
вертикальный охват 6605
вертлюг 5955
вертолёт 97, 2245
вертолётная площадка 2247
вертолётно-десантный 2244
вертолётный транспорт 2248
вертолёт общего назначения 6541
вертушка 6549
верфь 1334, 5311
(судостроительная) верфь 6915
верхнее течение 6752
верхний 3767, 6529
верхний горизонт воды 2218
верхний клиренс 3768
верхняя граница леса 6174
верхняя палуба 6530
верховное командование 2261
вес 6811
(чистый) вес 6817
(максимальный полный) вес 6814
весло 3526
вести 2122, 2162, 4009
вести бой 403, 1622
вести огонь 1644, 1682, 5321
вести огонь прямой наводкой 1720
вести переговоры 3429, 3870
вести себя 2168
вести судебное преследование 4343
вестовой 3718
ветер 6853
(под) ветер 2767
ветеран 6615
ветеринар 6616
ветеринарный 6617
ветерок 592
ветер с берега 3617
ветка 564
ветровой (водонапорный) насос 6860
ветровой конус 6856

внезапная поливка   5914
внезапная тактика   2282
внезапное нападение   5913
внезапность   5912
внезапный   5912
внезапный химический налёт   5914
внешний   1564, 1565, 3745
внешняя граница обороны   3951
вниз   1347
внизу   443, 1347
вносить в журнал   2941
вносить разложение   6466
внутренний   2477, 2537, 2541
внутривенный шприт   2566
вода   6735
(на) воде   65
водитель   1377
водительство   2760
водить   2754, 2755, 2756, 5713
водить доворот   5288
водный путь   6759
водовоз   6758
водоворот   6651, 6837
водоём   6820
водоизмещение   1301, 1352
(иметь) водоизмещение (в)   1300
водолаз   1321
водонапорная башня   6757
водонапорный кран   2363
водонепроницаемый   2411, 6746, 6756
водоохлаждённый   6738
водоочистительная установка   6747
водопроток   5101
водораздел   6753
водораспределительный пункт   4159
водород   2366
водородная бомба   2206, 2367
водосточная канава   1357
водохранилище   4771
водяная дыра   6740
водяная рубашка   6741
...водяного охлаждения   6738
водяной   2364
водяной смерч   6754
водяной столб   5606
воевать   1622
военная администрация   3195
военная игра   6694
военная полиция   3199
военная помощь   3192
военная разведка   2514
военно-гражданские отношения   807
военно-морское судно   6612
военно-морское училище   3408
военно-морской   3407
военно-морской флот (ВМФ)   3421
военно-полевой суд   1038

военновоздушные силы (ВВС)   112
военное время   5675
военное право   3197
военное преступление   6672
военнопленный   4294
военнослужащие   1480
военнослужащий   5215, 5523
военнослужащий американских войск
  специального назначения   4520
военные действия   2340
военные запасы   3374, 6723
военный   3191, 5202
военный атташе   3193
военный аэромедицинский врач   1794
военный врач   3592
военный зубной врач   3580
военный корабль   6724
военный корреспондент   6671
военный полицейский   4356
военный прокурор   2635
военный самолёт   6718
военный священник   766
возбуждать   36
возбуждающее средство   3945
возврат   4620
возвращаемая боеголовка   6580
возвращаемый отсёк   6578, 6580
возвращаться (в атмосферу)   4618
возвращение   4620, 4794
возвращение данных   4817
(угол) возвышения   1446
возвышенность   4880
воздействие   1428
воздействие на цель   4785
воздействовать   4543
воздерживаться от огня   2290
воздух   84
воздух-воздух   138
воздух-земля   114, 139
(в) воздухе   86
воздухопровод   119
воздушная перевозка   125
воздушная подвижность   124
воздушная поддержка   5871
воздушная пробка   121
воздушная разведка   4584
воздушная РЛС бокового действия
  4429
воздушная скорость   132
воздушная трасса   4963
воздушная тревога   6711
воздушная яма   127
воздушно-космический   62
воздушнодесантный   87
воздушное господство   136
воздушное движение   140
воздушное наблюдение   137

время запаздывания 1124
время облучения 1562
время оборота 6406
время отплытия 5023
время передачи 6180
время по Гринвичу 6960
время полёта 6179
время принятия 6181
время прохождения 3907
время прохождения по дороге 4905
вровень (с) 1822
вровень с поверхностью воды 330
всасывающий насос 5833
всемирный 6495
все наверх! 153
всеобщая действительная служба
  6290
всеобщий 6495
всепогодный 159
"Всплеск!" 5587
всплывать 5900
вспомогательная точка 4135
вспомогательные войска 319
вспомогательный 5819
вспомогательный (корабль) 321
вспомогательный (мотор) 320
вспышка 1765, 1766
вставлять 2481
вставлять (штепсель) 4098
встреча 3151, 4699
встречать(ся) 3150
встречный бой 3152
вступление в связь 3443
вся команда 153
вторая космическая скорость 6591
вторгаться 2568, 2571
вторжение 2432, 2572
вторичное излучение 5135
вторичный 5134
вторичный план 4049
второй эшелон 1413, 1416
второстепенный 5134
второстепенный груз 3477
втулка 637, 651, 5462
вход 1485, 2476
вход военнослужащим воспрещён 3609
входить 1484
входить в док 1333
входной порт 1979
входные данные 2478
вход с трапа 1979
входящая долина 4619
входящий в состав 2501
въезд 1485
въезжать 1484
выбирать (из) 5170
выбор 3696, 5171

выборочная проверка 5598
выбрасыватель 1566
выбрасывательный паз 5481
выбрасывать 1432, 2626
выброс 494
выброска 1382, 2637, 3854
выброска с воздуха 109
выброшенный груз 2624
выброшенный на берег 5757
выведение 4597
выведенный 3973
выведенный из строя 607
вывеска 5396
вывод 3755, 6109
выводить 3974
выводить (из фигуры) 4594
выводить среднее число 324
вывоз 1535
вывозить 5031
(давать) выговор 4751
выгружать 3610
выгружать(ся) 1126, 1249
выгрузка 1125
выдавать 2595
выдавать паёк 4537
выдача 2594
выдача из пунктов снабжения 5869
выдающая часть 6494
выдающийся 4330
выделение целей 6047
выделенный 1149
выделять 4675
выдерживать 432, 2297
выдержка 5648
выезд 6408
выживать 5929
выжидательная позиция 4195
выжидательная стоянка 2293
выжидательное положение 4211
выжидательный пункт 4137
выздоровление 4605
вызов (по телефону) 6083
вызов больных (к врачу) 5350
(по) вызову 3629
вызывать 672
вызывать крушение 6906
выигрывать 6850
выключатель 5951
выключать 5344, 5952
выкуп 4528
вылазка 5536
выламывать 582
вылет на поиск 3281
вымпел 3943
выносить 1465
(делать) вынужденную посадку на
  воду 1318

Г

гидролокатор (на) буксире   5533
гидроплан   2371
гидросамолёт   5119
гидростатический   2373
гидрофон   2370
гильза   5462
(патронная) гильза   716
гильзовый снаряд   4957
гиперболическая система навигации
   2374
гиперголическое топливо   2375
гиперзвуковой   2376
гирогоризонт   2320
гирокомпас   2141
гиромагнитный компас   2142
гироскоп   2144
гироскопический   2145
гиростабилизатор   2146
глава   2208
главарь   4870
главная база   3013
главная нервная система   747
главная полоса обороны   4775
главная станция сети   5691
главная точка сопротивления   4776
главная цель   4289
главное поле боя   3014
главные румбы   4169
главные силы   3015
главные силы авангарда   3016
главный   2658, 3010, 3028, 3105
главный аэродром   3011
главный водопровод   6745
главный выключатель   3108
главный командный пункт   4941
главный планшет   3107
главный путь подвоза   4968
главный путь снабжения   4968
главный сбор   3029
главный удар   3012
главный штаб   5627
гладкоствольный   5494
глаз   1567
глазной   3568
глазок   2988, 4152
гласис   4081
глетчер   2028
глиссер   2371
глубина   1215
глубина колонны   4903
глубинная (подлодочная) бомба   1216
глубокий   1150
глубокий тыл   2277
глубоководный водолаз   1151
глушитель   3361, 5400, 5401
(пулемётное) гнездо   3436
год   6920

годный   1735
годный к полёту   144
голова   2209
головная часть   3487, 6697
(ядерная) головная часть   6698
головной дозор   4105
головной телефон   2216
головной фонарь   2213
голос   6642
голосовой   6641, 6643
гончарный   749
гора   3345
горение   878
горизонт   2319, 2321, 5450
горизонталь   2832
горизонтальная наводка   6343
горизонтальное отклонение   2324
горизонтальный   2323
гористый   3346
горн   1940
горная дорога   4893
горно-вьючная гаубица   2349
горнолыжный спорт   5433
горный   3344
городская агломерация   6536
городской терроризм   6537
горчичный газ   1994
горючее   1926, 3622
горючее и смазочные материалы
   (ГСМ)   3970
горючий   877
горячая линия   2341
горячее пятно   5597
госпитализировать   2335
госпиталь   2330
(эвакуационный) госпиталь   2333
господство   884, 4984
господствовать   3103, 3776
господствовать (над)   886
господствующая местность   6116
государство   5669
готовность   4550, 4552, 5705, 5706
готовность 1 (один)   5674
готовность 2 (два)   5673
готовность к оказанию тактического
   защитного обеспечения с ходу
   3797
готовность объекта к подрыву
   (безопасная)   5674
готовность объекта к подрыву
   (боевая)   5673
готовые материальные средства   5860
готовый   4557
готовый (к боевым действиям)   147
гофрированный   1018
грабёж   4907

грабить 2961, 4099, 5008
гравий 2060
градуировать 2054
градус 1178
(№) градусов мороза 1923
гражданин 804, 3395
гражданская война 6669
гражданская оборона 802
гражданские беспорядки 803
гражданско-военное сотрудничество 806
(ручная) граната 2072
(осколочная) граната 2071
граната-болванка 2070
гранатомёт 2737
граница 539, 545, 1920
граница безопасной зоны 5017
граница обстрела 2826
график 772, 1256, 1816, 2056, 4089, 5064
график технического обслуживания 5066
гребень 1071, 4845
(боевой) гребень 3194
гребок 3818
грести 4975
грести гребками 3817
(ядерная) грибовидная туча 3503
грибовидное облако 3377
гриф секретности 5162
громить 4961
громкоговоритель 2970
грудь 784, 2489
груз 707, 1906, 2895
грузить 2896
грузить(ся) 1447
груз на поддонах 2899
грузовая марка 4087
грузовая сетка 709
грузовик 2962, 6382
грузовик со складом готовых имуществ 5744
грузовое судно 1910
грузовой 708
грузовой вагон 1907
грузовой самолёт 1909
грузоподъёмность (моста) 6257
грузоподъёмность вертолёта 2902
грунт 2083
(слабый) грунт 2092
группа 2106, 3878
(штурмовая) группа 6065
группа (парашютистов) 5724
группа, действующая в тылу противника 3888
группа крови 501, 6421
группа транспортов 6336

группа управления 891
группировать 2107
групповое бомбометание 3922
(военный) губернатор 2043
губительный огонь 1703
гудок 2325, 5415, 6840
гудрон 6051
гусеница 6235
гусеничная высадочная машина 6586
густота 1203
гуськом 5412

Д

давать взаимы 2916
давать задний ход 4827
давать знак 5379
давать обратный ход 4827
давать осечку 3249
давать сигнал 5379
давить 4267
давление 4271
давление воздуха 129
давление крови 502
дальнейшее обучение 6276
дальний 4503, 4693
дальнобойный 4501, 4502, 4503
дального действия 4502
дальномер 1642
дальность 1312, 4496, 4514
(максимальная) дальность 4506
дальность безопасного удаления 4473
дальность выстрела 2137
дальность действительного огня днём 4499
дальность действия 4497
дальность наблюдатель-цель 1313
дальность по прежнему! 4714
дальняя разведка 4586, 5079
дальняя связь 6077
дамба 5130
данные 1109, 2461
данные для стрельбы (огня) 1725
данные о цели 6045
дата начала действий 6046
дата удара 6046
датчик 5189
"даю по буквам" 5576
дверь 1339
двигатель 1472, 3336, 3337
двигатель (танка) 3810
двигатель) 320
двигать(ся) 3354
двигаться перекатами 2763
двигаться по инерции 843
двигаться по направлению ... градусов 431

двигаться походным порядком 3068
движение 3356, 3710, 6256, 6340
движение в двух направлениях 6418
движитель 4340
движущая сила 4342
движущаяся цель 6030
движущийся 6468, 6476
двоичная цифра 477
двойная автострада 1327
двойное шоссе 1327
двойной 1345
двойной охват 4018
дворник 6866
(с) двумя подразделениями в первом
    эшелоне 6417
двусторонний 462
двухвинтовой 6415
двухместный 6416
двухмоторный 6412
двухрулевой 6414
двухствольный 6411
деблокировать 4486
девиация (компаса) 1252
девиация компаса 917
дегазация 1145
дегазировать 1144
дежурный 1400
дежурный офицер 3581, 3598, 3603
дежурный офицер по военным
    действиям 3597
дезактивизация 1145
дезактивировать [HN] 1144
дезертир 1225
дезертировать 1223
дезертирство 1156
дезинсектировать 1192
дезинфицирующее средство 1287
дезинформация 4051
дезориентирование 1131
действие 28, 3651
(воздушнодесантное) действие 3652
действие взрывной волны 488
(в) действии 1855
действительное воинское звание 5823
действительный 35, 1429
действительный износ 6800
действительный чин 5823
(тайные) действия 3656
(подрывные) действия 3656
действия обнаружения и уничтожения
    противника 3667
(воздушные) действия по
    воспрещению местности 3654
действия по подавлению беспорядков
    4873
действия по поддержанию мира 3664
действия по поддержке порядка 3665

действия сохранения и безопасности
    3681
действия специального назначения
    3682
действовать 1934, 2168, 3645, 3647
действовать в первом эшелоне 5555
действовать клином атаки 5555
действуемый газом 3649
действующий 1936, 2891, 3669
декомпрессионная камера 1142
делать донесение 4725
делать зигзаги 6933
делать короткое замыкание 504
деление 2055
делительный циркуль 1328
делить(ся) 1326
дело 1343, 1630
дело целей 6048
дельта 1519
демилитаризированная зона 6942
демилитаризировать 1194
демобилизация 1195
демонстрант 1200
демонстрация 1198
демонстрировать 1197
демонтировать 1288, 5791
деморализованный 1201
день 1116
день-Д 1120
день-К 2650
день выдвижения конвоя 2650
день достижения полной мощности
    3928
день начала мобилизации 3119
день начала операций 1120
депеша 1293
депо 1212
деревня 6626
дерево 6349
деревянный 6892
держать 2288
держать в надёжном месте 5153
держать в руках 2163
держать курс (на) 2210
держать под надзором 5645
держать под обстрелом 1040
держать под прицелом 1052
держаться от берега 5660
деривация 1370
деррик-кран 1219
десант 2698
десантная группа 761
десантная партия 3887
десантник 3848, 3864
десантно-диверсионный отряд 892
десантное средство 2706
(морско-)десантные войска 2707

десантный катер   276
десантный настил   2709
десятичная дробь   1133
десятичный   1132
деталь   920, 3872
детектирование   4602
детектор   4603
детонация   1246
детонировать   1245
дефиле   1171
дефолиант   1176
децентрализировать   1130
дешифрование   1085
(подрывная) деятельность   6690
джунгли   2643
диагональный   1255
диалоговый режим   2533
диаметр   1257
диапазон   4500, 5565
диапозитив   5466, 6322
диафильм   6322
диафрагма   1258
диверсант   5007
диверсионная группа   3890
диверсионные войска   5559
диверсионные группы (НАТО)   5707
диверсионный   5829
диверсия   5005, 5828
дивизионное довольствие   1331
дивизионные части   1332
дивизионный   1330
дивизия   1329
дизельно-электрический   1259
дизельное топливо   3621
динамический   1401
дипольные отражатели   6861
директива   1272
дирекционная тяга   6561
дирекционный угол   2077
дислокация   1307
дислоцирующий   5696
дисплей   1303, 1304
диссидент(ный)   1311
дистанционное управление (ДУ)   4694
дистанционные наземные детекторы   4696
дистанционный   4693
дистиллированная вода   1314
дисциплина   1283
дисциплинированный   6823
дисциплинировать   1282
дифференциал   1260
дифференциальный   1260
дифференцированные ответные действия   4784
диффузор   1317
длина   2773, 2949

длина волны   6767
длинноволновой   2952
длинный   2948
дневальный почтовой конторы   4236
дневной свет   1117
днище   463
дно   437
дно моря   5106
добавлять   5855
доброволец   6650
добровольный   6648, 6650
доводить до минимума   3236
довольствие   5820
догадка   2119
договор   73, 6348
догонять   3765, 3790
дождевое платье   3625
дождливый сезон   3317
дожигание топлива   68
доза   1341
доза (радиоактивного) облучения   4443
доза облучения   1561
(воздушная) дозаправка   4633
дозаправка в воздухе   2458
дозаправлять (топливом)   4632
дозволенная скорость   5574
дозвуковой   5821
дозиметр   1342, 4441
дозор   3911
(боевой) дозор   3914
(разведывательный) дозор   3916
дозорный катер   3921
док   1334
доклад   4728
доклад (донесение) о применении ОВ   4742
доклад неизменной обстановки   4737
доклад об обнаружении   4739
доклад об обстановке   4740
доклад о результатах бомбометания   4732
доклад при выходе на цель   4735
докладывать   4726
(нечего) докладывать   3491
докование   1335
докритический   5804
доктрина   1338
долгий   2948
долгота   2949
должностное лицо   3607
должность   4221
долина   6546
доложите о готовности!   4746
доля   4413
доля (части)   5463

З

замещающий 5826
замёрзший 1925
заминированный участок 3227
замирание 1575
замок 2927
замораживать 1904
замороженный 1925
заморский 3785
замыкание 6295, 6572
(короткое) замыкание 5332
замыкать рот 1973
замысел 940
замысел боевых действий 941
замышлять 4088
занесение 1486
занимать 3037, 3559
занимать место 2292, 5694
занимать позицию 2292, 3560, 5694
заниматься контрабандой 5495
занимающий соответствующий пост 3691
занос 5429
заносить 5428
заносить в список 4640
занятие 1534, 3557, 6255
западный 6826
запаковать парашют 3812
запал 1944, 5220
запальная свеча 4096
запас 3070, 5730
запас(ный) 4767
запас(ы) 5727
запасать 5746
запасная позиция 4194
запасная стоянка 1454
запасная цель 6038
запасное колесо 5552
запасные войска 319
запасный 1451, 5657
запасный аэродром 170, 1453
запасный аэродром для перебазирования 4610
запасный путь 5355
запасный становой якорь 5259
запасы 4780
запекаться 833
запечатанные приказы 5114
запирать 2930
запирать на замок 2929
запирающий механизм 2931
записывать 2938
записывать на плёнке 6017
записывать целью 4593
заплечный ранец 2202
запорный кран 5736
заправлять 5206
заправляться 1632

заправочная станция 5688, 5690
заправочный пункт 4124
запрашивать (цену, плату) 770
запрет 359, 2289
запрет (полёта) 3709
запретная зона 832, 4327, 4798, 6945
запретная зона (для стрельбы) 3465
запретный район 3466
запрещать 358, 2531, 4326
запрещать полёты 2084
запрещение 359
запрещён(ный) 546
запрограммированный 4266
запрос 1193
запугивание 509
запугивать 508
запуск 2802, 5336
запускать 2735, 4674, 5667
запускаться 2801
запуск с места, с установки (ПУ) без направляющих 6932
запутываться 1886
запчасти 3874
запятая 4121
зарамочное оформление карты 3072
зарамочный 3071
заранее наведённый 4266
заранее наводить 4265
заранее подготовленный 4247
заранее устанавливать 4260
зарплата 3924
(утренняя) заря 1115
заряд 768, 2893
заряжать 769, 2894, 4281
заряженный 2907
заряженный с казённой части 2909
засада 177
засекречивать по высшей степени 4576
(радио)засечка 1741
засечка цели 4089
заслонка 789, 5345
заслонять (от света) 5237
засорять(ся) илом 5408
(сторожевая) застава 3995
заставать врасплох 5911
заставлять 1854
застёгивать(ся) на пуговицы 653
застёгивать пряжку 612
затвор 520, 590, 2931, 4093, 5345, 5484
затворный механизм 3131
затемнять 1106
затенять 5237
(корабль) затертый льдами 2382
затмевающий (огонь, маяк) 3556
затопление 1803

затоплять 1806, 5102
затоплять(ся) 5810
затор 495
затухание 299
затухать 298
затыкать 4094
затылок 4079
затычка 637
затягивание 3772
затяжной выстрел 2180
затяжной прыжок 1901
захват 5169
захватный 2933
захватывать 2934, 5167
захватывать в вилку 551
захватывающий 2933
заходить флангом 6832
заход на цель 3895
заход солнца 5845
зацепление 2010
зацеплять 2311, 2316
зацечка 1740
зашифровывать 1463, 5081
защёлка 838, 4679
защита 1159, 4346, 5282, 5285
защитная одежда 4348
защитная толща 5285
защитные очки 2039
защитный 1164, 4347
защищать 1041
защищать (от) 4345, 5283
защищать(ся) 1158
защищать мешками с песком 5041
защищать фланг 1758
защищённый 5151
заявка 4754, 4759
заявляться 4727
заявляться на службу 4744
звать 672
звезда 5664
звездочка (форма) 5664
звенитная артиллерия 1751
звено 6868
звено (гусеницы) 6347
звено гусеницы 6248
звено системы снабжения 1419
звенчатая (патронная) лента 447
звонить 6080
звонок 440
звук 3467, 5537
звук взрыва 4729
звуковая волна 5544
звуковое отражение 5394
звуковой 26
звуковой барьер 5540
звуковой буй 5535
звуковой толчок 534

звукоизоляция 5542
звукометрия 5543
звукоусилитель 2969
здание 619
здоровый 1735
'зелёная улица' (разг.) 824
земленосный мешок 5040
землеройная машина 1262
землечерпалка 1367
земля 1405, 2083, 2694, 5519,
    6123
земля-воздух 2104, 5902
земля-земля 2105, 5904
землянка 1391
земляные укрепления 1407
зенит 6926
зенитная ракета 3257
зенитный 218
зеркало 3247
зеркальное стекло 2031
зеркальный прицел 5367
зетовая балка 6925
зима 6874
зимний 6875
знак 5957
знак различия 345
знаменосец 5652
знамя 861, 1745
знаток 1549
значение 5780
значок 344
золотник 5464, 6548
зона 242, 6935
(буферная) зона 6939
зона безопасности 6949
зона боевых действий 6940
(открытая) зона бомбометания
    6938
зона военных действий 6952
зона исключения 6945
зона коммуникации 6941
зональный 6934
зона молчания 1123
зона непредельного огня 6946
зона опасности 1103
зона первичных разрушений 448
зона поражения 6947
зона разрушений 446
зонд 356
ЗОС 4006
зрительная связь 6636
зрительное опознание 6638
зрительный 6634
зрительный позивной (сигнал) 5378
зубной врач 1205
зубоврачебный 1204
зумлинза 6958

зыбучие пески   4411
зыбь   5905, 5945

И

иго   6922
идёт снег   5508
идти   6663
идти вдоль края   5446
идти вкось   5456
идти вразброд   5754
идти на лыжах   5426
идти ко дну   1888
идти первым   2754
избегать   1504, 1523
избежание   1526
избежание и побег   1527
извещение   3492
извещения лётчикам   3495
извещения мореплавателям   3496
изводить   2181
изгородь   2234
(проволочная) изгородь   1601
изготовка   4211
изготовление   4314
издавать   1458
изделие   2599
излишек   5910
излишний   5910
излучать   429, 1458
излучение   1456
изматывающие помехи   2183
изменение   6557
изменник   4700
изменчивость   6555
изменчивый   6556
изменять   3308
изменять(ся)   163
измерение   1266, 3128
измерение глубины   5541
измерение эхолотом   1424
измерительный прибор   2002
измерять   2001, 3127
измерять глубину   5539
изнеможение   1538
износ   6800
изнурение   1538
изнурительно-защищающие действия   3666
изобара   2590
изображать в определённом масштабе   5056
изображение   2401
изобретатель   2575
изобретать   2573
изобретение   2574

изогоническая линия   2840
изолировать   2531
изоляция   2532
изометрический   2591
изотерма   2592
изотермический   2592
изотоп   2593
из строя   29, 5211
ил   3634, 5407
иметь дальнобойность (…) метров   4523
имеющиеся транспортные средства   6333
имеющий некомплект   6465
имеющий перевес брони   2227
имеющий перевес пехоты   2230
имитировать   5409
имитирующее устройство   5410
иммунизация   2412
имплозия   2417
импровизированная поддержка   5881
импровизированный   2421
импровизировать   2420
импульс   4025, 4373
импульсный   4372
импульсный повторитель   6324
имущества   3116, 3374
инакомыслящий   1311
инвентарь   2576
индивидуальный   5194
индикатор   1242, 2438, 6142
(лазерный) индикатор (цели)   3080
индикатор движущейся цели   2440
индикатор кругового обзора (ИКО)   2441
индикаторная мощность   4241
индикаторный огонь   4015
индикаторный экран РЛС   4438
индоктринация   2443
индуктировать   2444
индукция   2446
инертный   2447
инертный газ   1992
инерциальная система навигации   2450
инерциальное наведение   2449
инерция   2448
инженер   1473
инженерно-сапёрное искусство (дело)   1476
инженерно-сапёрные войска   1474
инженерное защитное сооружение   3683
инженерное искусство   1475
инженерный тягач   6254
инициатива   2470
инициатор   2471

К

канонерская лодка 2129
канцелярия 4940
капеллан (устар.) 766
капитан 3104
капитанские документы 5291
капитуляция 5915
капок 2648
капор 2308
(складной) капот 2310
капот (двигателя) 2309
капрал артиллерии 522
капсула 701
капсюль 692, 701, 1247, 4283
капюшон 2308
карантинное свидетельство 466
карательный 4383
караул 2111
караульная служба 2114
караульное помещение 2115
карбюратор 706
кардан 6496
карданный шарнир 6496
карданов подвес 2025
карман 4103
карта 772, 3060, 5257
карта боевой обстановки 5423
карта обстановки 5422
картечный снаряд 5268
картографический 714
карточка взысканий поощрений 5258
кассета 718, 5466
кассетная бомба 837
кассир 3926
катализатор 723
катализировать 724
катапульта 725
катапультировать 726, 2732
катапультируемое сиденье 1433
категория снабжения 817
катер 1094, 2736
катить 6832
катить(ся) 4920
катушка 6858
каучук 4978
качание 3738
качать(ся) 3736, 4908
качать насосом 4376
качающийся 3737
качество 2049
качка 5905
(килевая) качка 4924
каюта 662
каяк 2649
"к бою!" 5697
квадрант 4392
квадрат 5619
квадратный 5619

квадратурный прилив 3422
квалификация 4394
квалифицированный 4395
квант 4397
квантовый 4397
карантин 4398
(в) карантине 4398
квартирмейстер 4401
квартиры 4404
квота 4413
кепка 693
керамика 750
керамический 749
кессон 667
кессонная болезнь 455, 1143
киловатт 2670
(...) километров в час 2669
килотонное оружие 6783
киль 1636, 2651
кильватер 6239, 6661, 6726
кильватерный огонь 4217
кингстон 5108
кинетическая энергия 2671
кинжал 1098
киноаппарат 793
кислород 3799
кислородная маска 3801
кисть 836
кит 6828
китобойное судно 6829
кладбище 738
клапан 6548
класс 2047
классификация 814
классификация боевой части 3198
классификация дорог 4972
классификация моста 3198
классифицировать 2600
клеёная фанера 4101
клемма 6106, 6108
клетка 734, 4279
клеточный 736
клещи 4018
клика 4868
клин 788, 2656, 5025, 6805, 6807
  6417
клинок 481
клиренс 2096
кличка 3454
клуб военнослужащих 5214
ключ 2654, 2655, 2659
ключевая доска 2660
ключевая позиция 4204
ключевой 2658
ключевой пункт 4131
(направленный) к морю 5131
кнопка 654, 2659, 4388

линия 'орудие-цель' 2836
линия прицеливания 2864
линия равных уровней радиации 2839
линия цели 2863
лист 482
листва 1841
лист настилки палубы 4080
листовой 2693
листовой металл 5260
литник 1938
лихорадка 1607
лихтер 2811, 6304
личина 1049
личное оружие 5353, 6788
личность 2393
личные вещи 3965
личный (опознавательный) знак 5982
личный (радио-)маяк 423
личный знак 5981
личный номер 5199
личный состав 1480, 3966, 4525, 4526, 4527
личный состав в кадрах 5633
личный штаб 5634
личные столовые принадлежности (амер.) 2675
лишать 1206
лобовая часть 1918
ловить рыбу 1730
ловить рыбу траловыми сетями 6345
ловить (цель) 2934
логарифм 2942
логарифмическая линейка 5467
лодка 511
лодка на подводных крыльях 2365
ложа 5726
ложиться в дрейф 2225
ложная атака 1199
ложная тревога 1586
ложная цель 1147
ложный 1324, 1585, 3978
ложный манёвр 1599
локальный 2918
локомотив 2936
локсодромия 2851
ломать 572
ломкий 1891
лопасть 480
лопасть винта 4951
лопата 2567
(механическая) лопата 5340
лопатка 480, 6551
лоцман 4010
лоцманский флаг 4013
лоцманское дело 4011

лоцманское судно 4012
лощадиная сила 2327
лощадь 2326
лощина 2127, 2299, 4539
лубок 5590
луна 3318
лупа 3006
луч 428
(инфракрасный) луч 2464
лучевая болезнь 4445
луч точного наведения 3939
лыжа 5427
лыжная палка 5440
лыжные войска 5447
люверс 4133
люк 2198, 5103
люлька 1059

М

магазин 2995, 2997, 5741
магазинная коробка 2995
магазинное оружие 4713
магистраль 4894
магистральная трасса 4970
магнит 2998
магнитная буря 3003
магнитная мина 3221
магнитное склонение 3004
магнитное склонение сетки 2079
магнитный 2999
магнитный азимут 3000
магнитный полюс 3002
магнитофон 6019
магнитофонная запись 6020
макет 1147, 3296, 3297, 6857
максимальная высота 731
максимальная высота облачности 2237
максимальная дальность 4504
максимальная дальность действительного огня 4507
максимальная дальность действия 4505
максимальная дальность полёта 4505
максимально-допустимая нагрузка 2897
максимальный 3117, 3934
максимальный взлётный вес 6816
максимальный посадочный вес 6815
максимум 3117, 3933
малая вода 2982
малая скорость 2981
малокалиберный 5486
малый 2972

маневрирование   5342
маневрировать   3042
маневрирующий возвращаемый отсек
   6574
манёвр   3043
манёвренная война   3292
манёвренная группа   3046
манёвренность   3293
манёвренный   3045
манёвр "змейка"   5497
манёвроспособная местность   4942
манёвр уклонения   1528
манёвры   3047
манифест   3049
манифестация   1198
мантелет   3056
маркировать   2688
маркировка дороги   4902
мародёр   3066, 4481
марш   3067
маршевый эшелон   5198
маршировать   3068
маршрут   4962
маршрут конвоя   4966
маршрутная аэрофотосъёмка   3988
маршрутная карта   4971
маршрутная сеть   3446
маршрутная съёмка   5925
маска   1049, 5088
маскировать   1118, 3095
маскировать(ся)   678
маскировка   679
маскировка краской   1119
маскировочная вплетённая материя
   1984
маскировочный узор   860, 1308
маскирующий дым   5491, 5492
масло   3619
масса   3097
массирование   3097
массировать   3098
массовые потери   3100
мастерская   6899
масштаб   5057
масштаб-зет   6959
масштаб-игрек   6923
масштаб-икс   6913
(плетёный) мат   3111
материал   3114
материальная часть   3116
материально-техническое обеспечение
   5876, 5886, 5890
материальные средства   5745, 5859
материк   971, 3017
матка   3335
матовый   3112
матрос   5024

матросы   4535
маховик   1827
маховик поворотного механизма
   2178
маховик подъёмного механизма   2177
мачта   3101, 3102
машина   316, 2991, 6563
(гусеничная) машина   6585
машина на воздушной подушке
   2101, 2347, 6565
машинное отделение   4937, 4938
маяк   417, 2812
мгла   2205, 6844
мегаватт   3155
мегагерц   3153
мегатон   3154
мегатонна   3154
мегатонное (ядерное) оружие   6785
мегатонное (ядерное) средство   6785
медиана   3141
медицинская помощь   3145
медицинская эвакуация   3140
медицинский осмотр   3994
медицинский пункт   5684
медленный   5482
медсестра   3523
международная линия суточного
   времени   2543
международное право   2544
международные водные пути   6749
международный   2542
международный позывной (сигнал)
   5375
международный свод   5375
междупалубное пространство   458
межень   3125
межконтинентальная баллистическая
   ракета   3265
мел   757
(на) мели   74
мелкая бухта   391
мелкий   3244, 5245, 5246
мелководье   5247, 5312
мембрана   1258
мензульная съёмка   5924
меньший   3244
менять курс   164
менять направление   6562
меняться   1531
мера   2002
мера по воспрещению   1202
мера противодействия   1027
меридиан   3161
мерить   3127
меркаторская проекция   3159
меры борьбы с
   радиопротиводействием   1440

мобильный 3289
могила 2059
модель 3297, 3300
модернизация 3305
модернизировать 3306
модификационный комплект 2676
модификация 3307, 4819
модифицировать 3308
модулирующий сигнал 5386
модульный 3309
модуляция 3310
(частотная) модуляция 3312, 3313
(амплитудная) модуляция 3311
мозг 554
мозгопромывание 6727
мокрый 6827
мол 587, 2629, 4001, 4407
молниеотвод 2815, 2816
молниеуводитель 2816
молния 2814
молчание 5399
"Молчать!" 3237
момент вращения 6203
(телевизионный) монитор 6093
монтаж аэроснимков 5796
монтировать 3342
монтировать провода 6876
моральное состояние 3324
мораторий 3325
море 5104
мореходность 5132
мореходный 3405, 5109, 5132
мороз 1922
морская болезнь 5128
морская высадочная группа 3411
морская десантная группа 3410
морская десантная операция 3653
морская миля 3188
морская пехота 3074
морская практика 5117
морская уточка 375
морские сумерки 3406
морские части прибрежного действия
844
морской 3073, 3075, 3405, 5107, 5109
морской бой 3409
морской конечный пункт 6755
морской путь 5307
морской флот 1775
морской эшелон 1417
моряк 5024
москит 3332
мост 595
мостик 596
мостоукладчик 2746
мотопехота 3139
мотопехотный 3139

мотор 1472, 3336
моторизованная часть 3339
моторизованный 3349, 6587
моторный 4243
моторный катер 4242
мотоцикл 3338
мотоциклист связи 1294
мощность 4239, 6921
мощность дозы облучения 4444
мощность ядерного оружия 3517
мужской 3032
мул 3363
муссон 3317
муфта 5462
мушка 1867, 1921
мыс 697, 2212, 4335
мыть(ся) 6725
'мякина' 754
мятеж 2499, 3382, 4833, 4871
мятежник 2500, 4875

Н

набережная 365, 4407
набивка 3816
набирать 3379
наблюдатель 2953, 2954, 3541
наблюдательный пост 4229
наблюдательный пункт (НП) 4230
наблюдать 1041, 3539, 5058, 6731
наблюдать за 2653
наблюдение 3538, 5918
наблюдение за полем боя 408
навалом 624
наведение 2121
наведение на маршевом участке
полёта 3181
наведение по лучу 430
наведённая радиоактивность 2445
наведённый перехват 6560
наветренный 6802, 6867
навигатор 3420
навигационная (координатная) сетка
3416
навигационный знак 5118
навигационный огонь 4992
навигация 3415
наводить 82, 151, 1269, 2122, 2123,
2741, 3731, 4109, 5357, 6559
наводить оружие 6292
наводить справки 3031
наводка 152, 3733, 6273
наводнение 1804
наводный борт 1898
наводнять 1802, 1805
наводчик 2745

"на... выше" 6527
наготове к (тактическому защитному обеспечению) 4206
наготове к перевозке 3497
награда 4838
нагревать(ся) 2221
нагружать 2896, 6810
нагруженный 2690
нагрузка 2895
надевать наручники 2160
надёжность 4682
надёжный 1848, 4684
надзиратель 5854
надзирать 5853
надзор 5918
надир 3389
надстройка 5852
надувать 4378
надувная лодка 2457
надувной 2456, 4102
(парашютный) надувной плотик 3863
нажим 4271, 5782
нажимная плита 4082
назад 343
наземная подготовка 6281
наземная тревога 2094
(номинальное) наземное давление 2102
наземное наведение истребителей-перехватчиков 2099
наземное управление 2097
наземное управление посадкой 2098
наземный 2082
наземный взрыв 2095, 5901
наземный сигнал 5385
наземный скачок уплотнения 5315
наземный ядерный взрыв 3513
назначать 155, 283, 770, 1228, 3470
назначать (в,на) 1237
назначать на должность 4220
назначение 156, 284, 285, 1935
назначенный 1149
наибольшая скорость хода 5572
найтов 2725
наказание 4382
наказуемый 4381
наказывать 1282, 4380
накапливать (запасы) 5731
накачивать 4376, 4378
накидка 698
накидной аэрофотомонтаж 6450
накладная 467
наклон 2428, 4039
наклонение 1267
наклонная дальность 4513
наклонная установка 4491

наклонный 2429, 3536, 5458
наклонять(ся) 2427, 5456, 6171
наклоняться 4036
накрывать 5752
нактоуз 470
налево 4188
налёт 4479, 5786
наличие 322
наличие топлива 5703
наличные боеприпасы 2898
наличный 323
наличный (боевой) состав 1430
наматывать 6854
наметить в общих чертах 3749
намеченный эпицентр взрыва 6929
намёк 6185
нанимать 773
наносить 2731, 4091, 5847
наносить (удар) 1189
наносить визит 6632
наносить на карту 3061, 3083
наносить поражение 1152
наносить удар 5621
наносы 5407
нападать 275, 297
нападение 296, 771, 4479
напалм 3390
напалмовый 3390
написанный на полях 3071
наплаву 65
наполнять(ся) 1631
напорный 2062
направление 1270, 2211, 2827, 6234
направление течения прилива (отлива) 6164
направленная тяга 6561
направленный 1271
направленный на курсе 1034
направлять 2122, 4109, 6559
направлять в бой 3717
направлять(ся) (к) 2210
направлять вкось 5456
направляться (на) 3030
направляющийся на север 3483
направо 4862
напряжение 5782, 6101, 6647
(под) напряжением 2890
напряжённый 6099
напряжённый бетон 4275
наращивание 620
наращивать 621
нарезка 4855
нарезной канал ствола 4853
народ 3394
народный 3395
нарты 5460
наружный 3745

нейтронно-индуктированная радиоактивность 3451
нейтронное оружие 6778
неконтактный взрыватель 1962
неконтрольный 6449
нелегальный 2397
неломкий 6439
немедленная временная доза 2407
немедленная постоянная доза 2406
немедленное снабжение по воздуху 5865
немедленный 2404
ненадёжный 2480
ненатянутый 5453
необороняемый 6524
необученный 6473
необходимый 1511
необычная война 6691
необязательный 3697
неопознаваемый 6484
неопределённый 2434
неоспоренный 6442
неотмеченный 6443
неотступное преследование 2342
неофициальный 6509
непилотируемый 6505
непитевой 6479
непланированное минирование 2750
неплановая цель 6050
неподвижный 5676, 5695
неподвижный дозор 3919
неподтверждённый 6452, 6522
непосредственная поддержка 5872, 5877
непосредственное охранение 5158
неправильно срабатывать 3033
неправомочный 6436
непредупреждённый (личный состав) вне укрытия 6526
непреодолимый 6440
не привязанный топографически 6523
непризнанная воюющая сторона 6437
неприкосновенный 1451
неприятельский 2337
непробиваемый бомбами 531
непроходимый 2415
непрямая наводка 2748
непрямой 2442
неразборчивый 6454
неразорвавшийся 492, 6481
неразрешённый 6436
нерасходный 6519
нерасходуемые предметы снабжения 5861
нерв 3434

нерегистрированный 6511
неремонтируемый на месте 460
нержавеющий 5001
неровность 6480
несгораемый 1722
несекретный 6444
нескользящий 3476, 5431
неснаряжённый 2447
несопротивлённый 6510
несражающийся 3472
нестерилизованный 6521
нести (флаг) 1825
нести полевой караул 5192
нести тяжесть 432
нестойкий 6520
нестойкое отравляющее вещество (ОВ) 3474
несудоходный 6507
несущий винт 5243
несчастный случай 11
неудача 1579, 5224
неукомплектованный 6465
неуправляемая (баллистическая) ракета 3262
неуправляемый штопор 6451
неустойчивый 6520
нефтеналивное судно 6014
нефть 3623, 3969
(плавучая) нефтяная буровая платформа 3624
нефтяной 3620
нечётное число 3569
нечистый якорь 1887
нешоссейный 6506
нивелировать 2048, 2780
нивелировщик 2051
нивелирующая машина 2051
нижеенулевой 5831
нижний 2976
низкие частоты 2978
низкий 2972
низкий перспективный аэроснимок 2980
низкорамочный (полу)прицеп 2979
низший 2976
ниспровергать 5830
нить 1626
ничейная местность 3468
ничья земля 2696, 3468
новобранец 4600
новобранцы 4540
нога 1849
(в) ногу 5716
"(К) ноге!" 3716
нож 2681
ножны 5053, 5254
нок-рея 6916

ноль 6927
номер 3521
номинальное (ядерное) оружие 6786
номинальный 3469
норма 4413, 5650
нормализовать 3481
нормальное отклонение 5653
нормальный 3479
нормальный уровень воды 3125
норма снабжения 2783
норма содержания 5729
нос 547, 2212, 3486
носилки 2886, 5784
носилочный раненый 2888
носильщик 2887
носовой 3486
носовой конус 3487
нотация 3489
ночная адаптация глаз 3456
ночная вахта 3185, 3464
ночное видение 3462
ночное воздействие 3457
ночной (инфракрасный) прицел 5364
ночной бинокль 3461
ночной бой 3459
ночной истребитель 3458
ночной эффект 3457
ночь 3455
нужда 4757
нуждаться 4756
нуждающийся 5333
нужды 1539
нулевая линия 2846, 2849
нулевой уровень 1113
нуль 6927
нумеровать 3518
нутромер 671
ныряльщик с аквалангом 1915
"нюхатель" 5500

О

оазис 6740
обвинительный акт 767
обгонять 3763, 3791
обезболивающее средство 197
обезвоживание 1179
обезвреживать (бомбу) 4698
обезвреживание мин 3228
обезвреживание неразорвавшихся
    боеприпасов 526
обеспечение (высокого) качества
    4396
обеспечивать 4311, 4353, 5863
обзорный радиолокатор 4430
обивка 2873

обитаемый 2147
обкладывать 2580
облако 835
облачность 1042
облегчать 4690
обледенение 2388
обледенеть 2387
облицовка 4829
облицовывать 4828
обломки 6905
облучать 2587, 5942
облучение 1560, 2400, 5941
обман 509, 1131
обманывать 508, 1129
обменивать 1531
обмениваться 2527
обмен приданными подразделениями
    1077
обморожение 1924
обмотка (обмотки) 4390
обнажить оружие 1361
обнаружение 1240
обнаруживание 2926
обнаруживать 1239, 2923, 2924, 3998,
    5358, 5596
обновлять 4701
обогащенный уран 1482
обоз 6293, 6294
обозначать 3076, 3084
оболочка 717
оборачиваемость 6406
оборона 1159
оборона на обратном скате 1160
оборона переднего края 1879
(на) обороне 1165
оборонительная подготовка местности
    6122
оборонительная политика 1163
оборонительный 1164
оборонять(ся) 1158
обороняться 2291
(№) оборотов в минуту 4835
оборудование 1498
оборудование компьютера 2188
оборудование местности 3729
оборудование обороны 3728
оборудовать 1497, 3730
обострять 5250
обоюдный 3384, 4574
обрабатывать 4304
обрабатывать оборону 5517
обработка 4307
обработка данных 4308
обработка сведений 4309
обработка слов 4310
образ действий 1037
обратная засечка 4766

обратно 343
обратное пламя 336
обратное рассеяние 341
обратный скат 5480
обратный ход 2007
обращение 2170
обременять 6810
обрыв 507
обследованный фарватер 5125
обслуживание 3020
(техническое) обслуживание 5217
обслуживать 3019
обслуживающий 5209
(наземный) обслуживающий экипаж 2100
обстановка 5421
обстреливание берега 5326
обстреливать 5264, 5943
обстреливать (продольным огнём) 1468, 4488
обстреливать из миномёта 3329
обстреливать с воздуха 5753
обтекаемый 5770
(носовой) обтекатель антенны 4475
обтюратор 5112
обтюрировать 5113
обучаемый 6274
обучать 2492, 6272
обучение 2493, 6275, 6286
обучение при условиях, близко от боевых 6289
обход 659, 1248, 1323, 6407
обходить 658, 5446
обходить фланг 1756, 3746
обходный манёвр 6407
обход по одному флангу 3668
обшивка 717, 4829
общая тревога 2015
общественная информация 4363, 4364
общественный 4362
общий 901, 2014
общий вес 6812
общий вес машины 6813
общий план 4058
общий резерв 2017
общий стационарный госпиталь 2334
общность 902
объежать 658
объединение 2504
объединённая поддержка 5885
объединённая сеть 2503
объединённый 2632
объединённый штаб 5630, 5631
объединять 2502, 4184, 6487
объединять(ся) 875
объезд 1248
объект 2491, 3530, 3532

объём 696
объявление 3493
объявлять невиновным 820
объявлять отбой 4567
обязанность за рубежом (NATO)
обыкновенный 3479
обыск 5123
обыскивать 1914, 4988, 5121
обыскивающий 5126
обычная близость 2148
обычная война 6682
обычные виды оружия 989
обязательный 3041
обязательство 898
ОВ нервно-паралитического действия 6545
овраг 2127, 4539
"оглушительная" граната 2075
огневая мощь 1721
огневая подготовка 6291
огневой бой 1715
огневой вал 380
огневой мешок 2087, 6947
огневой стержень 1929
огневой шар 1707
огнемёт 1754, 6158
огнеопасный 1755, 2455
огнестойкий 1753
огнетушитель 1714
огнеупорный 1722
(под) огнём 1701
(обеспечение) огнём 1660
огнённый шторм 1723
огни 2708
огонёк 2803
огонь 1645, 1647, 1656, 2803
(круговой) огонь 1698
(учебный) огонь 1684, 1699
(продольный) огонь 1665, 1668, 1669, 1674, 1689
(губительный) огонь 1678
(управляемый) огонь 1655, 1675, 1677, 1688
(расходящий) огонь 1650, 1664, 1672, 1690
(подавляющий) огонь 1658, 1661, 1670, 1696
(сосредоточенный) огонь 1654
(прочёсывающий) огонь 1653, 1695, 1697
Огонь! 1646
огонь в упор 1685
огонь и манёвр 1706
огонь на подавление 1679
огонь на поражение 1717
огонь на разрушение 1662
огонь непрямой наводкой 1673

осциллятор 3739
(каменистая) осыпь 5086
ось 332, 333, 742, 4040, 5242
ось-З 6924
ось-игрек 6919
ось-икс 6910
ось абсцисс 6910
ось наступления 2854
отбивать 1021
отбирать 5170
отблеск 4627
отбой 4568
отбор 5171
отборный 5172
отбрасывать 4664
отброска 4663
отбросы 4663
отбывать 4569
отвергать 3783
отверстие 229, 568, 2298, 3641,
    3948
отвесный 5255
отвесный берег 507
ответ 4723, 4724
ответная детонация 5958
ответный удар 4808
ответственность 4786
ответственный начальник 1533
ответчик 4782, 4787
отвечать 4722
отвечать на огонь 4825
отвинчивать 6514
отвлекающий 1324
отвлекающий манёвр 1322
отвод 1323, 3612
отводить 1325
отголосок 1422
отгонять 1602
отдавать на слом 5084
отдаваться эхом 1421
отдавать честь 5028
отдавать швартовы 720
отдача 2661, 3753, 4579
отделение 914, 5616
(противотанковое) отделение 6073
отделение обезвреживания 527
отделение учёта и отчётности 20
отделка (мундира) 1572
отдельный 5194
отделять 1233
отделять(ся) 5195
отделяющийся от конвоя 996
отдых! 4788
отдых и выздоровление 4495
отзыв 1028
отказ 1580
отказывать 4664

отказывающийся (от военной службы)
    3535
откат 2661, 4579
откатываться 4580
откидной борт 5983, 5984
откладывать 5746
отклонение 1175, 1252, 1253, 1371,
    6917
отклонение по секстанту 5234
отклонение 4214
отклонение эпицентра (от цели) 3615
отклонённый от курса 1033
отклонять 1174
отклоняться откурса 6918
откос 1507
открывать 821, 3636, 6453, 6504
открывать огонь 3640
открытие вытяжным фалом 3643
открытое море 2268
открытый 824, 2480, 3635, 3637
открытый город 3639
открытый текст 4046
отлив 1410, 6163
отливать 1409
отлив прибоя 6472
отличительный 1260
отличный стрелок 3085
отлого спускаться 5280
отмель 370, 5586
отменять 685
отметка 494, 4138, 4143, 4151, 4556
отметка (высоты) 4623
отметка высоты 452, 2378, 5599
отметка по решётке 4621
отметка цели 2339
отмечать 3076
относительная биологическая
    эффективность 4668
относительная влажность 4669
относительная скорость 6594
относительный номер 4623
отношение 300, 4536
отношение (пропорция) сжатия 926
отношения 4665, 4666
отозвание 4568
отпечаток 2418
отпирать 6504
отплывать 5020
отправитель 3735
отправка 1293
отправление 1208
отправлять 1292
отправлять(ся) 5019
отпуск 1939, 2594, 2765, 4495
отпускать 1289, 2595, 2959, 2960, 4674
отпуск на берег 5327
отпускной билет (на (…) часов) 3777

оцеплять 1005
осечка 2180
очаг 4103
очаг беспорядков 6380
очень высокая частота (ОВЧ) 6608
очередной вираж 4302
очередность 4286
очередность движения 3358
очередность перевозки 3358
очередь 4960
очередь (пулемёта) 647
очередь срочности 4286
очерёдность обеспечения огнём 4288
очертание 3750
очистка 3323, 5033
очищать 819
очищать от накипи 5054
очищать от ржавчины 1220
ошибка доставки 1190
ошибка по секстанту 5234
ощущать 5186

П

павший (в бою) 2665
паёк 4538
паз 5116
паковый лёд 2380
пал 517
палата 6673
палатка 6102
палаточное имущество 6104
палуба 1137
(на) палубе 3630
память 3157
панель 960, 3829
панорамный 3836
пар 5709
пар(ы) 6554
параван 3866
парад 3850, 4830
парадная обеденная форма (англ.) 2675
парадная форма одежды 1368
парадное положение 5670
парадос 3853
параллакс 3856
параллель 3857
параллельные линейки 3858
параллельные потери и ущерб 855
параллельный 3857
параллельный веер 3859
параллельный штаб 5632
парафин 3855
парашют 3837

(грузовой) парашют 3840
(вытяжной) парашют 3842
(управляемый) парашют 3843
(дополнительный) парашют 3841
(саморазвётывающий) парашют 3844
парашютизм 5449
парашютировать 3838
парашютист 3848
парашютнодесантные войска 3865
парашютное снаряжение 4859
парить 2344
парк 1774
парка 3868
парламентёрский флаг 1747
паровой 5709
пароль 3908, 6734, 6896
паром 1605, 1797, 4476
пароход 5710
партизан 2118, 3877
партизанская война в городах 6692
партия 2968, 3879
партия груза для части 2906
парусник 6613
парусное судно 6613
пасмурный 3761
пассажир 3903
пассажирский пароход 2868
пассивная гидролокация 5534
пассивная помеха 6861
пассивная противовоздушная оборона (ПВО) 3905
пассивное самонаведение 3906
патент на офицерский чин 895
патрон 715
патронная сумка 4238
патронник 764
патронташ 363
патрулировать 1041, 3912
патрулирующий полёт 3919
патрулирующий часовой 4358
пациент 3910
паяльная лампа 505
паять 5522
педаль 3936
пеленг 3000
пеленгатор 1640, 1641
пеленгаторная станция 5686
пеленг створа 6311
пена 1828
(быть на) пенсии 3944
пенсионный 3944
первая помощь 1729
первичный элемент 4279
первое ориентирование гироскопа 2143
первоначальное пополнение 4805
первоначальное сближение 2466

планирование 2033
планированная артиллерийская
поддержка 5879
планированное минирование 2749
планированный 4069, 5227
планировать 2034, 4048, 4062, 5063
план конвоя 5065
планктон 4068
план обеспечения 4050
план обмена 5070
плановая цель 6037
планово-перспективный
аэрофотоснимок 3985
плановое техническое обслуживание
3026
плановые донесения 4317
плановые поиски 3658
плановый 4069, 4247, 5067, 5068
плановый аэроснимок 3986
плановый вызов огня 3283
плановый коэффициент 4070
план огня 4053
планомерное форсирование (реки)
1188
планомерный 1187
план перевозки 5964
план погрузки 4052, 4057
план подрыва 5069
план полёта 4054
план по складыванию 1395
планшет 4092, 5963
планшет воздушной обстановки 126
план штивки 5751
пластификатор 4076
пластический 4074
пластичное бризантное ВВ 2265
пластичное взрывчатое вещество
4075
пластичный 4074
пластмасса 4074
платёжная ведомость 4926
платить 3923
плато 4083
платформа 4084
плац-парад 2089
плацдарм 598
плацдарм высадки 415
плацдарм десантирования 115
плащ-палатка 5256
пленник 703
пленочный дозиметр 1633
плечевой ремень 5339
плечо приклада 4791
плёнка 6018, 6020
плёс 4541
плоский 1772, 4063, 5245
плоскогорье 4083

плоскость 4064
плот 1797, 4476
(спасательный) плотик 4477
плотность 1203
плотность движения 6261
плохо вооружённый 6456
(стартовая) площадка 3814
(посадочная) площадка 3813
площадка десанта 5419
площадка приземления 6948
площадь 241
(поражаемая) площадь 6937
плутоний 4100
плыть по инерции 843
пляж 413
пневматический 4102
пневмокостюм 222
побег 1503
победа 6619
побеждать 1152, 6850
побережные воды 6748
побережье 841, 2889, 5324, 5328
побивать камнями 5733
побочный (вопрос) 5351
побуждать 36
повар 998
поведение 951
поверхностная волна 6764
поверхность 5899
повиноваться 3528
повод 5805
поворачивать(ся) 6832
поворот 4302
(круговой) поворот 6344
поворот на месте 5603
поворотный мост 5949
поворотный пункт 4155
повреждение 1099, 2475
повреждение мозга 555
повседневная форма 3626
повстанец 2500, 4566
повторение 4657
Повторите! 4710
повторитель 4711, 4712
повторять 4709
повышать 1445
повышать(ся) 2241, 5717
повышение давления 3780
повышенное давление (на открытом
пространстве) 1897
погибший 1121
погода 6801
погон 5339
погонщик мулов 3364
пограничный дозор 3913
пограничный контрольный пункт
4111

порт приписки   2303
портфель   601, 3063
поручать   283, 1186
поручение   284
порыв   486
порядковый номер   760, 5199
порядок   5196
порядок ведения огня   4987
посадиться   6211
посадка   1448, 2698, 6212
посадка на брюхо   2702
посадка с убранным шасси   2702
посадочная площадка   150
посадочно-десантный   116, 2697
посадочные фары   2708
посадочный маяк   421
посадочный передатчик   421
посаженный   3347
посаженный на машине   3349
поселение   5228
посередине судна   180
посещать   6632
посещение   6631
последний заградительный огонь
    1666
последний рубеж огневого вала   2833
последовательность   5196
последовательный   5197
последствия   1847
последующий этап действий   5816
посольство   1449
поспешная оборона   2196
поспешнопроведённое разрушение
    2197
поспешный проход   2194
посредник   6432
посредничать   3143
пост   4219, 4221, 5679
поставлять (оружие) на
    предохранитель   4698
поставлять бомбу   4072
постановка на якорь   3321
постель   437
постепенный   2053
постепенный скат   5479
постоянная величина   962
постоянная часть   4772
постоянно действующаяся инструкция
    5659
постоянное отражение   3959
постоянное положение   5708
постоянный   3958
постоянный огонь   1744
постоянный прицел   5362
пост подслушивания   4228
пост регулирования движения   4233
пост регулировщиков   4233

построение   1872, 3851
постройка   619
поступательное движение   2219
поступать (в армию)   5398
поступать добровольцем   6649
поступать на военную службу   1479
поступление на военную службу
    1481
посылать   1291
посылать по почте   3009
посыльный   1031, 4990
потери   721, 2966
потерпевший кораблекрушение   719
потерпеть аварию   1062
потерпеть неудачу   1576
потерять контакт   2964
поток высотного ветра   2625
поток движения   6262
потолок   731
потолок висения   2346
потопленный   2409
потоплять   6906
потоплять(ся)   2408
потребности   1539
потребность   4757
поход   681
походное движение   3067
походный мешок   4979
походный шаг   4412
похоронная команда   3882
почва   5519
(оказывать) почести   2307
почётный караул   2116
починка   4625, 4705
починять   4702, 4706
почта   3008
появляться   4727
появляющаяся мишень   6035
пояс   444
пояснение   811
поясное время   6956
поясной ремень   451
пояс плавучести   1811
правая нарезка   4863
правила   4986
правила визуального полёта   6637
правило   2175
правильный   1011
правительственный   2041
правительство   2042
право   4861
правого борта   5666
(на) правом борту   286
право на обыск   4864, 6721
право на остановку   4865
право остановки   4865
право проезда   4866

(радио)приём   4573
приёмная контрольная проверка
   новобранцев   6131
(радио)приёмник   4572
приёмо-передатчик   6300
(радио)приёмо-передатчик   4465
приёмопередатчик   6319
приземление   6212
приземляться   149, 2695, 6211
признавать   4578
призовая команда   3881
призыв   1350
призывать   675, 1351
призывник   959, 1353
призывной возраст   69
приказ   885, 3703
(постоянный) приказ   3714
приказ выполнения   3707
приказ на перевозку   3710
приказ об открытии огня   3708
приказ о выходе в море   5022
приказ по части   3719
(по) приказы   3711
приказывать   879, 3704
приклад   657
приклад (винтовки)   652
приковываться   1905
прикомандирование   295, 5136
прикомандировывать   293
прикреплять   293
прикреплять ярлык   2688
прикрывать   1044, 5090
прикрывать завесой   5092
прикрывать огнём   5092
прикрытие   688, 5087, 5096
(авиационное) прикрытие   6431
прикрытие с воздуха   1039
прикрытый   5278
прилагать   208
прилив   6163
приливный бассейн   6162
приливо-отливный   6161
приложение   209, 5856
приложение по тылу   210
приманивать   1146
примечание   215, 3489
"Примкнуть штык!"   1742
приморье   2889
принимать   3999, 4571
принимать (на палубу)   4594
принимать командование   887
принимать по этапам   3971
принимать честь   5029
принудительный   3041
принятие   4597
природная (окружающая) радиация
   338

природный   3401
присасывающийся (подрывной) заряд
   3220
присваивать офицерское звание   896
присвоение звания   4336
присоединять   211
присоединять(ся)   2872
присоединяться к сети   4745
присоединяющийся к конвою   995
приспособленный для воздушной
   перевозки   128
приспособлять   1734
приспущённый   2151
пристань   1334, 2629, 5642, 6830
пристань снабжения   3417
пристреливать   4642
пристреливать на угол возвышения
   4517
пристреливать по трубочной
   дальности   4518
пристреливаться   6931
пристреливаться по дальности   4498
пристрелка   1648, 1691, 5336
пристрелка по реперу   4644
пристрелочный выстрел   5372
пристрелочный репер   4144
приступать (к)   2467
присутствующий   4263
притворяться больным   3035
приток   1067
прихватка   1738, 6166
приходить в соприкосновение   966
приходить на помощь   4690
прицел   342, 4565, 5359, 5361, 5366
(складной) прицел   5363
(угол) прицеливания   1446
прицеливать   2741
прицеливаться   82, 5357, 6928
прицельная веха   4223
прицельная помеха   2614
прицеп   3065, 6269
прицеплять   2311
прицепная машина   6568
причал   457, 2186, 3321, 4407, 6830
причаливать   3319
пришвартовать(ся)   3320
пробивать   3947, 4003
пробивать брешь   569
пробивать огнём   399
пробка   637, 1943, 4093
пробка (движения)   6263
проблесковый огонь   1769
проблесковый свет   1769
пробный   6127
пробовать   6128
пробой   4379
провал   1579

проверка   303, 2488, 6129
проверка безопасности   5161
проверка времени   6177
проверка секретности   5161
проверять   304, 776, 2486, 6128, 6601
проверять на земле   2103
проветривать   6598
провиант   5859
провод   2757, 6877
(газо-, нефте-, водо-)провод
   4028
проводить   950
проводить бреющий полёт   2235
проводить манёвры   3042
проводить поиск   5942
проводка   6247
проводка судов   4011
проводка цели   6240
проводник   2124
проводоуправляемый   6884
провозить контрабандой   5495
проволока   6877
(колючая) проволока   6879
(натяжная) проволока   6881
проволочная решётка   6885
проволочное заграждение   372, 6880,
   6883
проволочные ножницы   6882
прогноз погоды   6803
прогноз погоды по округам   245
програмист   4319
программа   4318
программировать   4265
программист(ка)   931
программное обеспечение   5518
програмный язык   4320
прогресс   4323
прогрессивный   4324, 4325
прогрузка части на одно средство
   транспорта   2912
продвигать (вперёд)   4338
продвигаться   4269, 4321
(стремительно) продвигаться   6160
продвигаться вперёд   47
продвигаться на местонахождение
   4303
продвижение   4342, 6159
продвижение вперёд   48
продлевать   4332
продление   1563, 4333
продолжать   4332
продолжение   4333
продолжение военного обучения   6279
продолжительность   1464
продолжительность полёта   6179
продольная ось   2951
продольный   1863, 2950

продольный мостик   1826
(под) продольным огнём   1469
продувать   489
продукты питания   963, 4355
продукты расщепления   1733
проезжий   6306
проект   1226, 4328
проектирование   4331
прожектор   1808, 5127
прозрачный   6323
произведения расщепления   1733
производить   2018, 4313
производить горизонтальную наводку
   6342
производить декомпрессию   1141
производить налёт   4478
производить осмотр   2485
производить перегруппировку   4609
производить поиск   5058
производить разведку   4592
производить смену   4689
производить топосъёмку   5919
производство   3754, 4314
происшествие с ядерным оружием
   3508, 3516
прокладка   1996
прокладывать   2740
прокладывать (курс)   4091
прокладывать кабель   2742
прокладывать курс (на)   5225
прокладывать туннель   6397
прокол   4379
прокурор   4344
проламывать брешь   569
пролёт   5547
пролив   5538, 5756
пролом   573, 4994
промах   3787
промахиваться   3284, 3789
промежуток   2559
промежуточный   2535, 2538
промежуточный аэродром   167
промежуточный уравнительный рубеж
   2847
промеренный   5926
проникание   3941
проникать   2453, 3940, 4003
проноситься   1800
пропавший без вести   3276
пропаганда   4337
пропитание   5820
пропорция   4536
пропуск   3897, 3962
пропускная способность   6258
пропускная способность дорог   4896
прореживание   6149
прорез   5477

прорыв   571, 573, 585, 3941
прорывать   584, 3940
прорывная сила   4995
просачивание   2454
просачиваться   2453
просверливание   3948
просверливать   3947
просвет   2096, 3096
просвечивать рентгеновыми лучами
   6911
просвечивающий   6323
прослеживать   6224
просматривать   739
проспект   3827
простой   5411
простреливать   5943
просыпаться   6662
протест   3531
протестовать   3529
против ветра   6534
противник   1466, 3688
противовоздушная оборона (ПВО)
   108
противовоздушные действия   3659
противовоздушный   107
противогаз   1997, 4781
противозаносный   226
противокрутительная штанга   6204
противолодочное прикрытие   5094
противолодочный   227, 5903
противоминная сеть   6201
противоминные действия   3226
противопартизанский   1026
противоперегрузочный комбинезон
   2110
противопехотная мина   3207
противопехотный   223
противоповстанческий полицейский
   отряд   4878
противопожарная система   5612
противопожарные действия   1711
противопожарный   1716
противоположный   3690
противорадиолокационная маскировка
   4431
противорадиолокационные меры
   4433
противоракетная система   219
противоржавый   224
противотанковая самоходка   6012
противотанковое оружие   6773
противотанковый   228
противотанковый разведдозор   3920
противотанковый снаряд   5265
противотанковый управляемый
   реактивный снаряд (ПТУРС)   3259
противоторпедная сеть   6201

противохимическая оборона   782
противоядерная защита   3506
противоядерная оборона   3506
протирка   4370
протокол   3246, 4349
протон   4350
прототип   3298, 4351
протрактор   4352
протраленный фарватер   5946
профилактическая медицина   4278
профилактический ремонт   3024
профиль   4316
профиль полёта   1793
профсоюз   6488
проход   2712, 2714, 3899
проход (в минном поле)   1981
проходимость   6257
проходимый   3898
"Проходите"   3904
проход с ходу   2194
(тыловое) прохождение   3902
(передовое) прохождение   3902
процедура   4299
процесс   4306
процесс разведывательных данных
   2520
прочёсывание   3323
прочёсывать   5093
прочность   4682
прочность почвы на сдвиг   5520
прочность почвы на срезывание   5520
прочный   4684, 5798
проявление   1303
проявлять   1302
пруд   4182
пружина   5608
пружинный   5610
прыгание с парашютом   3845
прыгать (с парашютом)   2638
прыжок   2637
прыжок с больших высот с
   замедленным развёртыванием
   парашюта   3847
прыжок с замедленным
   развёртиванием парашюта   3846
пряжка   614
прямая наводка   4510
прямое визуальное изображение
   2217
прямое попадание   2279
прямое снабжение   5866
прямой   1268, 6531
прямой наводкой   4160
прямоточный воздушнореактивный
   двигатель (пврд)   4490
прямоугольная (координатная) сетка
   3196

прямоугольная система координат
713
прятать(ся) 2257
психиатрический 4360
психическая травма 869
психологический 4361
птица 475
ПТУРС (противотанковый
управляемый реактивный снаряд)
типа TOW 3274
публичный 4362
пуговица (одежды) 654
пузырь 610
пулемёт 2992
(станковый) пулемёт 6794
пулемётчик 2132
пульсирующий воздушно-реактивный
двигатель (ПВРД) 4374
пульт 960, 3828, 3831
пульт управления 3830
пуля 628, 4954
пуля 'дум-дум' 1392
пункт 268, 4106, 4219
пункт водоснабжения 4159
пункт высадки 4132
пункт зарядки 4115
пунктирная линия 2838
"пункт невозврата" 4167
пункт ожидания 4127
пункт оповещения 4231
пункт переправы 5417
пункт перехвата 4129
пункт перехода к атаке 4140, 4154
пункт перехода к набору высоты
4142
пункт погрузки 4134
пункт прохождения 4139
пункт распределения 4123
пункт регулирования 4118
пункт сбора 4195
пункт сбора военнопленных 4141
пункт сбора донесений 3165
пункт сбора раненых 4113
пункт сбрасывание 4145
пункт снабжения 4148
пункт соприкосновения 4117
пункт управления 4936
пункты смазывания 4170
пункт эвакуации раненых 4114
Пуск! 5337
пускать(ся) в ход 5667
пусковая установка 4914
пусковой бугель 2738
пустая порода 5593
пустотелое дно 2300
пустыня 1224
путевая карта 4971

путевая скорость 5570
путевой компас 916
путепровод 6618
(морское) путешествие 6652
путёвка 3896, 6722
пути продвижения 3900
пути прохождения 3900
пути сообщения 908, 2870
путь 2714, 4962
путь обхода 4964
путь отхода 2867, 4965
путь подвоза 2866
путь подхода 234
путь снабжения 2866
пучок 836
пытать 6205

Р

работа 6897
работать 3646, 3647, 6898
работающий двигатель 4991
рабочая карта 3686
рабочая команда 3885
рабочая одежда 1595
рабочий 3669
равнина 2782, 4044
равновесие 350
равновесие (в равновесии) 1495
равноденствие 1496
равноценность 1499
рад 4418
радиальный 4442
радиатор 4446
радио 4447
радио- 4447
радиоактивные отбросы 6728
радиоактивный 4448, 4456
радиоактивный дождь 4485
радиовещательная сеть 606
радиовзрыватель 1971
радиовойна 6684
радиограмма 3164, 5380
радиодезинформация 3051
радиодорожка 4450
радиоизотоп 4454
радиологический 4456
радиолокатор обнаружения цели
4420
радиолокатор с непрерывным
излучением 4422
радиолокационная станция (РЛС)
4419
радиолокационная станция (РЛС)
дальнего обнаружения 4421
радиолокационное наблюдение 4434

радиолокационное опознавание 2391
радиолокационное определение
　дальности 4437, 4459
радиолокационный 4419
радиолокационный дозор 3997
радиолокационный маяк 424
радиолокация 4436, 4455
радиомаяк 425
радиомаяк-3 426
радиометр 4440
радиомолчание 4462
радионавигационные средства 4449
радионавигация 4457
радиообман 4452
радиоопознавание 4460
радиопеленгатор 1640
радиопеленгация 4453
радиопередатчик 4466
радиопередача 605
радиоперехватчик 3315
радиоприёмник 5222
радиопротиводействие 6684
радиоразведка 2510, 2512, 2516
радиоразведка и
　радиопротиводействие 1441
радиорелейная линия 2875
радиорелейная станция 5693
радиосвязная работа 4300
радиосвязь 4464
радиосеть 3440
(командная) радиосеть 3439
радиостанция 5682
радиотелеграфия 4463
радиотелепринтер 6090
радиоуправление 4451
радиоуправляемый самолёт 1381
радист 3687, 4458
радиус 4467
радиус действий 4471
радиус поражения 4472
радиус разворота 4470
радиус сплошного поражения 4468
радиус уязвимости 4474
разбивать 572
разбивать (на категории) 578
разбирать 1288
разбитый 607
разблокировать 4673
разбойник 362
разборка 579
развал 586
развалина 4983
разведка 2505, 4583
(боевая) разведка 2509
(акустическая) разведка 2506
(фотографическая) разведка 2515
разведка боем 4591

разведка маршрута 4588
разведка огнём 4590
разведка района 4589
разведка цели 2518
разведчик 5076
разведывательная машина 5078
разведывательная служба 5207
разведывательное донесение 4736
разведывательное судно 6611
разведывательный радиоприёмник
　2526
разведывать 1554, 5077
разветвляться 563
развёртка 5059, 5941
развёртывание 1211, 4216
развёртывать(ся) 1210
развёртывать войска 4215
развёртывать поиск 5942
развёртывать силы 4215
развивать(ся) 1250
развивать наступление 1845
развиваться 4322
развилка 1869, 2988
развитие 1251, 4323
развитие наступления 1844
развязывать 575
развязывать(ся) 2959, 2960
разгар 3933
разгонная ступень 5641
разгонять 1296
разгружать 1279, 1394, 6502
раздавать 1315
разделять 575
разделять(ся) 1326, 5195, 5592
разделять на зоны 6936
раздражающее ОВ 2182
раздражающее отравляющее вещество
　1993
разжаловать 1177
разлёт 1297
разлёт (осколков) 5604
разливаться 5580
разложение 4779
размах 1266, 5546, 5549
размер (груза) 2003
размер запаса 5729
размеры 3128
размещать 13, 464, 2924, 4399,
　5678
размещение 14, 465, 4404
разминирование 3224
разминированный 3225
разминировать 823
разногласие 1284
разогревать 2221
"Разойдись!" 1289, 1584
разоружать(ся) 1274

ремонтировать 3764, 4624, 4706
ремонтно-восстановительные работы 1101
(сухой) ремонтный док 2061
ремонтный комплект 2676
ремонтный обоз 6295
рентгеновские лучи 6912
рентгеновые лучи 6912
реорганизация 4703
реорганизовывать 4704
реостат 4839
репер 452, 1112, 4138
репетиция 4657
репрессалия 4752
ресурсы 4780
ретранслировать 4671
ретрансляционная станция 5693
рефлектор 4628, 4629
рефракция 4631
рецепт 4262
речка 5768
речной 4885
речной порог 4530
решение (по,о) 1136
решётка 2058, 2076, 2078
решётчатый настил 1390
ржаветь 4999
ржавчина 1016, 4998
ржавый 5002
рикошет 4842
(бить) рикошетом 4841
риск 2204, 4883
рисковать 4884
риск первой степени 1455
рисовать 1359
рисовать кроки 5424
рисовать мелом 758
рисунок 1365
риф 4617
РЛС вертикального обзора 4425
РЛС определения местности (при маловысотном полёте) 6120
РЛС с фазово-антенной системой 4427
ров 1319, 6350
ровный 2779
(на) ровный киль 1529
рогатка 4793
рогожка 3111
род войск 250, 562, 5204
родная страна 1030
родные 3452
ролик 4367, 4928, 4929
роликовый конвейер 4930
роль 4919
рота 911
ротационный 4947

ротация 4949
ротор 4950, 4951
рощица 5583
ртуть 3160
рубеж 539, 545, 2869, 4146
рубеж безопасной зоны 5017
рубеж безопасности 2852
рубеж безопасности ведения огня 2841
рубеж отправки докладов 2850
рубеж сопротивления 2837
рубеж ядерной безопасности 2842
рубец 5060
рубить 5459
ружьё 5338
ружьё с пропиленными стволами 5052
рука 2159
рукав (реки) 565
рука за рукой 2173
руководитель 2759
руководить 981, 2167, 2756, 3039
руководство 982, 2760
рукопашный бой 831, 2176
рулевая колонка 863
рулевая рубка 6836
рулевое колесо 6833
рулевое управление с усилителем 5715
рулевое устройство 2008
рулевой механизм 2008
рулевой тормоз 558
рулежная полоса 6063
рулежно-подходная полоса 237
рулить 6062
руль 481, 1636, 2169, 4980, 4981, 6550
румпель 2250, 6170
ручей 1067, 5768, 6739
ручка 2164, 2784
ручка управления 3650, 5723
ручки 988
ручная граната 2074
ручное командное наведение 3058
ручное командное управление 3058
ручное оружие 5485
ручное управление 3059
ручной 2161, 3057
ручной пулемёт 4031
ручной телефон 2174
ручной тормоз 3869
рыболовственная флотилия (рыбфлот) 1777
рыскание 6917
рыть 1261
рычаг 978, 2784
рычаги 988

рычаг переключения передач   2011
рычаг управления   984, 2634, 3650, 5723
рычаг управления газом   6155
"рэнджер"   4520
рюкзак   4979
ряд   1627, 2715, 4976, 5200
рядом   162

С

саботаж   5005
саботировать   5006
садить(ся) на корабль   1447
садиться   2695, 3341
сальник   1996
салют   5030
салютная команда   3886
салютовать   1649
самовзводный   5173
самовольная отлучка   7
самовоспламеняющееся топливо   2375
самовыпрямляющийся   5176
самодвижущийся   317
самозарядное оружие   6791
самозатягивающаяся бочка   6008
(дистанционно-управляемый непилотируемый) самолёт   6581
самолёт(ы)   92
самолёт-заправщик   6013
самолёт-корректировщик   5601
самолёт вторжения   2569
самолёт наведения   96
самолётовылет   5536
самолёт с вертикальным взлётом и посадкой   103
самолёт связи   93
самолёт с изменяемой геометрией крыла   102
самолёт со стреловидными крыльями   100
самоликвидатор   1967, 3134
самоликвидация   3134
самонаведение   2304, 2305
самонаводящаяся ракета   3270
самонаводящаяся торпеда   6200
самооборона   5174
самостоятельное орудие   4974
самостоятельность   5177
самостоятельный   2435
самостоятельный наземный элемент   5190
самотёком   2062
самоуправляемый снаряд   1705
самоходная мина   3222

самоходное противотанковое орудие   6012
самоходный   4243, 5175
самум   5043
сани   5460
санитар(ка)   3523
санитар-носильщик   5785
санитарная машина   175
санитарная сумка   2674
санитарная часть   3146
санитарный   176
(медицинско-)санитарный   3144
санкции   5038
санкционировать   311
сапёр   4024
сапёрные действия   868
сапёры   1474
сапог   538
сарай   374
сбивание (пеленга)   3120
сближаться   830
сближение с противником   48
сбор   3151, 3380, 3720, 4699, 5034
сборище   3288
сборка   280
сбор конвоя   993
сборщик парашютов   4857
сбрасываемое оружие   1705
сбрасываемый   2628
сбрасываемый топливный бак   6005
сбрасывание   1382, 3854, 4678
сбрасывать   1383
сбрасывать(ся) на парашюте   3838
сбрасывать за борт   2627
сваривать(ся)   6818
сварка   6819
сведения   2461
сведения для служебного пользования   4799
сведения из изображений   2513
сведения о полёте   1791
(на) свежем воздухе   3638
свежий   1912
свержение   5828
сверлить   540
свёртывать   4933
сверхдавление   3780
сверхзвуковой   5851, 6429
сверхкомплектный   5850
сверхмалая подлодка   3186
сверять (часы)   5959
свет   2803
светить   2804
светомерие   1771
свечение   1820
свидетельство   752
свист   6840

свистеть 6839
свита 4811
свобода 1899, 2789
свободное падение 1901
свободный 824, 1896, 2956
свободный от льда 2386
сводить чертёж 6223
сводка 4822, 5838
сводка больных 4747
сводный штаб 5629
"Свой!" 1913
связанный с джунглями 2644
связист 3687, 4458, 5391
связной 3167
связывать 468
связывать(ся) 2874
связь 469, 2875
связь взаимодействия 2787
сгиб 454
сгибаться 453, 613
сгорание 878
сгущать 946
сгущение 945
сдаваться 5916
сдвиг 5286
сдерживать 967, 4796
сдерживающая атака 2295
сдерживающие действия 1185
север 3482
(на) север 3485
северное отшествие 3484
северный 3482
Североатлантический союз 3400
(к) северу 3485
седло 5009
седловина 5010
сейсмический 5165
сейсмограф 5166
секрет 5139
секретарь 5140
секретность 5138
секретные сведения 4799
секретный 816, 5139, 6458
секстан 5232
секстант 5232
сектор 5146
секторный поиск 5149
сектор обстрела 1618, 5148
сектор стрельбы 239, 2414
секунда 5133
секундомер 5738
селективный 5172
селекторная связь 2530
село 6626
сервомеханизм 3135
середина 741, 3182
сержант 3596

серия 5200
серия бомб 5725
сесть на корабль 510
сесть на мель 412
сетка 3438
(координатная) сетка 2076
сетка (оптического прибора) 4810
сетка от москитов 3333
(радио-)сеть 3438
сеть (сообщений) 3444
сеть оповещения 3441
сеть радиопеленгаторных постов 3445
сечение 5144
сжатие 925, 4271
сжигать 644
сжимать 923
сигнал 5380
сигнал-команда по самоликвидации 5381
сигнал бедствия 5382
сигнализировать 5379
сигнал общей тревоги 2015
сигнал о конце связи 5383
сигнал опознания 5387
сигнал тревоги 146
сигнальная ракета 4033
сигнальный (ракетный) пистолет 4032
сигнальный огонь 420, 6714
сигнальный флажок 5390
сиденье 5129
сидящий на мели 5757
сизигийный прилив 5611
сила 1467, 1853, 4239, 5772
(пробивная) сила 4240
сила сигнала 5776
(в) силе 1855
силой оружия 2134
силуэт 5404
силы боевого манёврирования 871
силы второго эшелона 6763
силы по поддержанию мира 3930
сильное волнение 2232
сильно теснить 4267
сильный 5798
символ 5957
симпатичная детонация 5958
симулировать 5409
синхронизировать 5959
синхронный 5960
сирена 5415
система 5961
система разделяющихся
    головных частей (боеголовок)
    индивидуального наведения 6577
система ликвидации 1232

система наведения по земным
 параметрам 2023
система обнаружения проникновения
 2570
система обстрела с частичной орбиты
 1890
система подрыва 1232
система посадки по приборам 2498
система поясных времён (поясное
 время) 6957
система радиорелейной связи 4461
система разделяющихся
 самонаводящихся головных
 частей 6576
система раннего воздушного
 оповещения и наведени.` 89
система самонаведения 3419
система управления огнём 1709
системный анализ 5962
СК 3742, 5383
скала 4909
скалистый 4916
сканирование 5059
скат 4491
скатанная постель 438
скафандр 222
скача 5915
скачок 2636
склад 1212, 1393, 5741, 6677
склад артиллерийско-технического
 имущества 3724
склад боеприпасов 2996, 5742
(полевой) склад боеприпасов 183
складка местности 1840
складной 3370
складной нож 813, 2606
склад оружия 261
складывать 4184, 5050, 6904
складывать(ся) 6088
склон 2428, 5457
склонение 6557
(магнитное) склонение 1139
склонный 2429
склонять(ся) 2427
склянка 440
скоба 552, 5236, 6922
(быстродействующая) скоба 4680
(вода) скованная льдом 2382
сковывать огнём 4019
скольжение 5472
скольжение (винта) 339
скользить 4062, 5465, 5471
скользить (по) 5435
скользкий 5474
скользящий узел 5473
скопление 3380
скорая помощь 1729

скороподъёмность 4532
скорость 3804, 5566, 6588
скорость ветра 5573
скорость в конце активного действия
 6590
скорость в конце активного
 траектории 6589
скорость звука 5575
скорость конвоя 5568
скорость подъёма 5567
скорость расхода 964
скорость самолёта 132
скорость убегания 6591
скос потока 6726
скреплять 808
скручивание 6203
скрывать 932, 3095
скрывать(ся) 2257
скрытая смертельная доза 2727
скрытие 933
скрытность 5138
скрытность связи 5159
скрытный 810, 1055
"скрытый" самолёт 98
скрытым корпусом 2353, 4203
скучиваться 636
слабеть 6771
слабина 5453
слабый 5453, 6770
след 1852, 6225, 6230, 6232, 6236,
 6241, 6268
следить 6229, 6731
следить (за) 5615, 6224
следить за 6265
следовать 1846, 5239
следовать (за) 1842
следоискатель 6244
следопыт 6244
следопытная группа 6072
следы 6226
слезоточивое отравляющее вещество
 1995
слезоточивый газ 2990, 4874
слепое приземление 2703
слепой 491, 493
слепой полёт 1784
слепота от снега 5509
слесарно-механическое отделение
 5145
слесарь-монтажник 1737
слип 5476
сличать 854, 1013
слоистый 2693
слой 2745
слой перехода 6314
сломанный 608
служащий 3607

столовая 3168
стоп! 5735
стопор 2928, 4093
стормовой ветер 1976
сторож 6732
сторожевая вышка 6733
сторожевая застава 3752
сторожевой (корабль) 2117
сторожевой катер 3921
сторожить 2112
сторона 5352
сточная труба 5231
стояние отлива 5455
стояние прилива 5455
стоянка 5680
стоять биваком 478
стоять лицом (к противнику) 1570
стоять на месте ("стойте!") 5658
стоять на часах 2112
стоять на якоре 4844
стоячая волна 6766
страдающийся от контузии 5275
страна 1029
страны света 4169
стратегическая транспортная авиация 6330
стратегический 5761
стратегический замысел 5763
стратегический резерв 5765
стратегическое оповещение 6712
стратегическое оружие 6793
стратегическое сосредоточение 5762
стратегическое средство устрашения 5764
стратегия 5766
стратосфера 5767
стрела 2630
стрелка 4161
стрелковая ячейка 1889, 4035
стрелковое дело 3378
стрелковое мастерство 5434
стрелковое оружие 5485
стреловидный 6870
стрелок 4854, 5444
стрельба без наблюдения 1702
стрельба навскидку 5323
стрельба непрямой наводкой 1673
стрельба по исчисленным данным 1686
стрельба с наблюдением 1681
стрельба шкалой 1694
стреляный патрон 4959
стрелять 1277, 1644, 2735, 5321
стреляющая группа 1718
стремительно атаковать 5747
стремительное нападение 2193
стремительный бой 1625

стремительный марш 1856
стремиться 81, 4269
стремнина 4530
стресс 5783
строевая подготовка 1372
строить 618
строить(ся) 3849
строй 1872
строй-клин 6807
строй в колонне 6270
строй кильватера 2859
строй фронта 2855
строп 5470
стропка 2718
струя 5769
ступень 5640
ступенька 4989
ступица 2350
ступня 1849
стык 2533, 4130
стык дорог 2642, 4901
стычка 812, 1471, 5443
стягивать ремнем 5759
сугроб 5510
суда 5306
суда первого эшелона десанта 277
судно 5290
судно (суда) 1060
(противолодочное) судно-ловушка 5294
судно переснаряжения 5301
судно рыболовственного охранения 6610
судовое время 5292
судовой 3073
судовой груз 5303
судоремонтный завод 1336
судостроительная верфь 1336
судостроительный завод 5311
судоходный 3413, 3414
судоходство 5306
сумерки 1399, 6410
сумка 346, 3807
сумма 5835, 6208
суммировать 5837
суточная норма снабжения 5867
суффикс зоны 6955
суходол 1389
сухой док 1388
сухой овраг 1389
сухопутные войска 264
сухопутный 3771
суша 2694
(на) суше 2260
существенный 1511
сфера 5577
сферический 5578

схватка   812, 5443
схема   772, 797, 1256, 1816, 3062,
    4047, 5425
схема организации   3727
схема целей   3775
схема штивки   5751
сходиться (в, на)   990
сходной трап   910
схождение   991
сцепление   838, 2877
счёт   16
счёт времени готовности   1020
счётная линейка   5467
счётная часть   19
счётчик Гейгера   2012
(навигационное) счисление   4575
считать   929, 1019, 3518
сыпаться снегом   5508
сыпной тиф   6423
сырой   2354
сырость   2355

Т

табель времён полных вод   5965
таблетка   5967
таблица   5963
таблица поправок (огня)   3540
таблица приливов   5965
табличный   5968
тавотный шприц   2067
тайна   5139
тайное движение   6464
тайный   1055, 5139, 6458
тайный склад оружия   664
тайфун   6424
такелаж   499, 2190, 4858, 5970
тактик   5979
тактика   5980
тактическая воздушная поддержка
    5892
тактическая контрразведка   2517
тактическая мобильность   407
тактическая перевозка   5975
тактическая погрузка   2911
тактическая транспортная авиация
    6331
тактические боевые действия   3684
тактические военно-воздушные силы
    (ВВС)   5974
тактические воздушные действия   3685
тактический   5971
тактический авиатранспорт   6331
тактический резерв   5978
тактическое защитное наблюдение
    3795

тактическое обеспечение   3796
тактическое ядерное оружие   5977
тали   499
(тросовый) талреп   2718
тальк   5996
тамбур газоубежища   121
тангаж   4039
танк   6001
(основной) танк   6006
(окопанный) танк   6007
танкер   6014
танковая пушка   6015
танковый мостоукладчик (ТМУ)   6003
танковый транспортёр   6584
танковый тягач   6567
татуировка   6058
тахометр   5969
тахометрический визир   5369
тащить   2199, 3048, 4366
тащить(ся)   1354
(в) твёрдом состоянии   5528
твёрдопокрытая площадка   2187
твёрдый   1728, 5528
театр боевых действий   6138
текущее техническое обслуживание
    3025
текущий   1091
текущий ремонт   3021, 3766
телевизионная камера   6092
телега   6658
телеграфия   6078
телеметрия   6079
телескоп   6087
телескопический   6089
телескопический прицел   5370
телеуправление   4694
телеуправляемый   4695
телеурегулирование   4694
телефон   6081
телефонная станция   6084
телефонный номер   6085
телефон танка   6082
тело   513
телохранитель   515
тема   5805
темнеть   1106
тёмное время   5403
темнота   1107
тёмные очки   5842
тёмный   1105
темо   1817
темп   3805, 5566
температура   6095
темп марша   3803
темп наступления   4531
темп огня   4533
темп потребления   964

(в) тени 5238
теодолит 6139
тёплая струя 6140
тепловое излучение 6144
тепловое истощение 2222
тепловое облучение 6141
тепловой 6140
тепловой экран 5284
теплоизоляция 6143
теплосамоуправляемый 2223
тёплый фронт (воздуха) 6701
тереть(ся) 4977
терминальный 6107
термическая изоляция 6143
термический 6140
термометр 6145
термос 6544
термоядерный 6146
терпеть 1465
терраса 6114
территориальные воды 6751
территория 6123
терроризировать 6126
терроризм 6124
террорист(ка) 6125
террористический 6125
терять 2963
терять скорость 5647
теснина 1171
тесный 829
техник 269, 6076
техника 1475
техника штабной службы 2947
техническая подготовка 6288
техническая разведка 2519
техническая скорость самолёта 2436
технические спецификации 6075
технические требования 6075
технические условия 6075
технический 6074
техническое обеспечение 3020
течение 1090, 1815
(по) течению 46, 1347
течения 1817
течёный атом 6228
течь 1814, 2762
тина 3634, 5407
тип 6420
типический 6425
типичный 6425
типовой 5651
тире 1108
тихий 676, 5482
тканая лента 6804
товар 900
товарный вагон 1907
ток 1090

токсичность 6220
толкать 4338
толпа 1082
толчея 609
толчок 4387, 5313, 5314
толщина защитного слоя 2153
тон 6188
тоннаж 6189
тоннель 6467
тонуть 1386
(за)тонуть 5413
топить 1386
(за)топить 5413
топка 1941
топливная банка 1927
топливная бочка 1928
топливний бак 6004
топливо 1926
топо- 6191
топографическая съёмка 5921
топографический 5920, 6191
топографический гребень 6192
топография 6193
топор 331
топосъёмка 5921
топь 5940
торговое судоходство 5300
торговые суда 5300
торговый 893
(морско-)торговый 3158
торжественная заря 6059
торжественный 751
торжественный марш 3069
тормоз 557
(парашютный) тормоз 3839
тормозить 556
тормозная жидкость 559
тормозная колодка 788
тормозная педаль 561
тормозная подкладка 560
тормозная ракета 4821
(палубное) тормозное устройство 104
тормозной гак 2314
тормозной крюк 2313
тормозной парашют 1380
торнадо 6197
торпеда 6199
торпедировать 6198
торпедный аппарат 6202
тотальный 6206
точечный 4022
точечный ориентир 4023
точить 5250
точка 1344, 4025, 4106
точка возгорания 4125
точка вращения 4041
точка запуска 2831

точка захода на цель 4150
точка наводки 4108
точка опоры 4384
точка попадания 4128, 4164, 4165
точка приземления 4128
точка прицеливания 4108
точка прохождения 4139
точка разрыва 4162
точка сбрасывания 4110
точка сброса 2831
точно определять 4021, 5564
точность 21, 4250
точность стрельбы 22
точный 23
(на) траверзе 4
травить 6562
травить (трос) 3927
траектория 6296
(кривая) траектория 6297
(настильная) траектория 6298
траектория планирования 2036
траектория полёта 1792
трал 1750
траление 3235
тралить 5944, 6345
тральщик 6346
транец 6320
транзистор 6308
трансляция 4672
транспондер 6324
(вьючный) транспорт 6329
транспорт(ный) 6327
транспорт для перевозки войск 6374
(танковый) транспортёр 6335
транспортир 4352
транспортировать 6325
(поддерживающее) транспортное
    судно 6328
транспортные средства 6327
транспортный самолёт 101
траншейная стопа 6354
траншейный нож 6355
траншея 6350
трап 2689
трасса 2713, 2715, 4089, 6221, 6230,
    6268
трассирующие боеприпасы 6227
тратить 1543
траулер 6346
требование 4754
требование срочного воздушного
    налёта 3282
требовать 673, 4270, 4755, 4756
требуемая норма снабжения 5870
тревога 145
тревога 'красная' 4606
тревожно-сигнальный аппарат 6713

тредюнион 6488
тренажер 5410
тренировать 6272
тренировать(ся) 4246
тренировка 1534
тренога 6369
треножник 6369
треугольник 6360
треугольный 6361
трещина 1057
трёхмильный рубеж 6154
триангуляция 6362
тройной 6368
тропа 3909, 6230, 6264
тропик Козерога 6376
тропик Рака 6376
тропинка 6264
тропический 6375
тропопауза 6377
тропосфера 6378
тропосферная связь 6379
тропосферное распространение
    радиоволн 6379
трос 663, 2203, 4943, 6716
тротиловый эквивалент 6187, 6921
труба 4026, 6393
трубка 6393
(воспламенительная) трубка 4283
(ударная, снарядная) трубка 1948
трубопровод 4026
труд 6897
труп 1010
трюм 442, 2287
тряский 635
тугой 6061
туман 1834, 3285
туманная лампа 1838
туманная сирена 1837
туманный 1836
туманный сигнал 5384
тумба 517
тундра 6396
туннель 6398
тупик 5646
турбина 6399
турбонагнетатель 6400
турбореактивный 6401
турбулентность 6402
турбулентный 6405
турель 6409
туча 835
тушить 4409
тыл 388, 4562
тыл и снабжение 2947
тыловой 2943, 4561, 5209
тыловой гак 2314
тыловой район 388, 4562

У

узнавать 4578
указание 2437, 3705
указание пути 351
указанное время 6178
указанное время налёта 6183
указанное расстояние взаимного
  действия 5893
указатель 2438, 3077, 4161, 5396
(маршрутный) указатель 3079
указатель (минного поля) 3078
указатель (наземного)
  местоположения 2439
указатель курса 422
указатель маршрута 4232
указатель часового пояса 6955
указывание маршрутов 5397
указывать 4021, 4168
укладывать 5750
уклон 2046, 5457
уклонение 1526, 3044
уключины 3527
укомплектованность 5775
укомплектовывать 3038
укомплектовывать (личным составом)
  3036
укрепление, прикрывающее фланг
  1759
укреплённая площадка 5418
укреплять 1876, 4659, 4660, 5778
укрывать(ся) 5276
укрывать рельефом 1167
укрываться 1050
укрытие 1045, 2256, 4635
(естественное) укрытие 4199
укрытие (в укрытии) 1169
укрытие (от наблюдения) 1046
укрытие-нора 5511
укрытие для подлодок 3938
(в) укрытии 1051
укрытый 5278
укрытый подступ 1053
укрытым корпусом 1170
улетать 1207
уличные бои 5771
уловка 4996
ультиматум 6427
ультравысокая частота (УВЧ) 6428
ультрафиолетовый 6430
умение 4315
уменьшать 4612
уменьшать давление 1141
уменьшать напряжение 4670
уменьшение 4615
уменьшенный заряд 4613
умеренный 3303, 6094
умерять 3302

универсальная поперечно-
  цилиндрическая меркаторская
  проекция 6497
универсальный 6495
универсальный шарнир 6496
унитарные боеприпасы с переменными
  зарядами 5180
унитарные снаряды (патроны) 1743
уничтожать 212, 1134, 1230, 2663,
  5995
уничтожать паразитов 1192
уничтожение 213, 1231, 2664
упаковывать 3808
уполномочивать 311, 1186
(прикладный) упор 4790
управление 44, 882, 977, 982, 2042,
  2120, 2121, 3109
управление (личным составом) 3040
управление воздушным движением
  141
управление движением 3357, 6260
управление излучениями 1457
управление огнём 1708
управление операциями 987
управление учётом материальных
  средств 2577
управляемая ракета 3263
управляемый 2845
управляемый по проводам 6884
управлять 980, 981, 1269, 1376, 2123,
  2162, 2167, 3039, 4009, 5713
упредительное время 2761
упредительный механизм 4251
упреждающий 4252
упреждающий удар 5594
упреждение 2758
упреждённая точка 4207
уравновешенность 6365
(автоматическая) уравновешенность
  6367
уравновешивать 349, 6366
ураган 2360
уран 6535
уровень 2778, 2781, 4064
уровень воды 6742, 6743
уровень моря 5115
уровень перехода 6314
(на) уровне воды 330
усеивать воронками 1065
усиление 3005
усиление изображения 2402
усиленный заряд 5846
усиливать 4659
усиливать(ся) 2241
усилитель 3006, 4711, 4712
усилитель изображений 2522
усилитель изображения 2403

фокусировать 1832
фокусное расстояние 1830
фольга 1839
фон 337
фонарь 6196
фонарь молния 2361
фонетическое правописание 3977
форма 6486
формат 1871
формуляр минного поля 3232
форсажёр 4658
форсажная камера 67
форсирование 4887
форсирование с ходу 2195
форсунка 2620, 2621, 3499
фосген 3979
фосфор 3980
фотоаппарат 677
фотоаппарат Поларойд 4173
фотобомба 3982
фотограмметрия 3983
(воздушная) фотокарта 3989
фотоплан 3331
(воздушная)(аэро-)
    фоторазведка 3990
фотосхема 3331
фотоэлемент 3981
фрахт 1906
фрахтовать 773
фронт 1916
Фронт (для) Народного
    Освобождения 3399
фронтальное наступление 1919
фронтальный удар 1919
фронт ударной волны 5318
фугас 1885
функциональный 1936
функционировать 1934
функция 1935
фуражка 693
фуфайка 2618

Х

характеристика 3949, 5392
характеристика оружия 5395
хватать 5167
хвостовой костыль 5430
хвостовой фонарь 5985
химикалии 779
химическая боеголовка 6695
химическая война 6681
химическая разведка 5922
химический 780
химический снаряд 5269
(боевое) химическое вещество 781

химическое обследование 5922
хирург 5907
хирургический 5909
хирургия 5908
хитрость 4996
хлопушка 3086
хмурый 3761
хобот 6266
ход 3356, 5566
ход, достаточный для управления
    5714
ходить 6663
ходить на лыжах 5426
ходовая часть 6457
ходовые огни 3418
ход сообщения 6351
(на) ходу 6476
ходьба на лыжах 5432
ходячий больной 6665
хозяйственная работа 1594
холм 2275, 3340
холмистая дорога 4893
холмистая местность 2276
холодный 852
холодный пуск 853
холостой 2957
холостой патрон 483
холст 691
хоронить 649
хорошо вооружённый 6821
хорошо защищённый 6822
хорошо оборудованный 6824
хорошо сохранённый 6825
храбрость 1977
хранение 5739
хранить 1629, 3334
хранить на складе 1394
хребет 4845
хроническая доза 791
хронический 790
хронометр 792
хрупкий 1891
хунта 2646

Ц

цапфа 6389
цвет 859, 6188
целеуказание 2437, 6042
целеуказательный снаряд 5273
целик 3490, 5361, 5366
целиться 81
целое 6208
(по) Цельсию 744
цель 83, 3530, 3532, 6022
(живая) цель 6029

(боевая) цель 6029
(срочная) цель 6031, 6033
(твёрдая) цель 6026, 6040
(одиночная) цель 6034
(неплановая) цель 6027, 6032
(площадная) цель 6024
(незащитная) цель 6039
Цель! 6023
цель (воспрещения) 6025
(запасная) цель подрывания 6036
цемент 737
цензор 740
центр 741
централизировать 746
центральная 6084
центральная группа планирования
    6066
центральная линия 742
центральный 745, 4043
центр мишени 626
центробежная сила 748
центр совместных операций 2633
центр тяжести 743
центр управления огнём 1710
центр управления тактической
    авиацией 5972
центр управления транспортом
    (транспортными средствами) 6334
цепь 755, 797
церемониальный 751
цех 6899
циклон 1095
цилиндр 1096, 1387
цилиндрический 1097
циновка 3111
циркулярный 798
цистерна 6000, 6016
цифра 1264, 3519
цифровой 1265

Ч

час-Н 3453
час-П 3991
час-Ч 2255
часовой 5191
часовой отряд подрывных зарядов
    2113
часовой пояс 6950
(по) часовой стрелке 827
(против) часовой стрелки 1023
частица 3876
частичный 3875
частопроблесковый огонь 4410
частота 1911, 6767
частото-поворотливая РЛС 4424

часть 1443, 3396, 3872, 3873, 5142,
    6489
часть в контакте 6493
часть резерва 3475
часы 6730
чаща 6148
чека 2928
чередовать(ся) 166
чередующий 165
черпак 2044
чертёж 6222
чертить 6223
черты 3750
честное слово 3871
чёткость 1172
чехол 2602
чинить 4624
численность 5772
численность сил мирного времени
    3932
численный масштаб 4750
число 3520
число-М 2993
число Маха 2993
чистить 819
чистить драгой 1354
чистый 818
член 2819
чувствительность 5188
чувствительный 5187
чувствительный элемент 5189
чувство 5185
чувствовать 5186
чужой берег 5325

Ш

шаблон 3774
шаг 3802, 4038, 5716
шар 5577
шарикоподшипник 353
шарообразный 5578
шасси 775, 6457
шатающийся 2956
шатёр 3089
шахта 5241
шашка 5452
швартовка 1335
шекл 5236
шельф 5261
шерсть 6893
шерстяной 6894
шест 5551
шея 3423, 3424
шина 5590, 6186
ширина 6847

ширина фронта   1917
ширина хода   6347
широкий   603, 6845
широкоугольный   6846
широта   570, 2729
шифр   794
шифровальное устройство   5083
шифровальный материал   1087
шифрование   1086
шкала   5057
шкала-зет   6959
шкала-игрек   6923
шкала-икс   6913
шканцы   4400
шквал   5618
шкив   4367
школа   5072
шлаг   461
шлагбаум   385
шланг   2328
шлем   2251, 2252
шлюз   2932, 5484
шлюзовать   2935
шлюпбалки   1114
шлюпка с увольняемыми на берег
   2790
шноркель   5506
шнур   2717
(гибкий) шнур   1779
(огнепроводный) шнур   1966
шов   5116
шомпол   4493
шоссе   2272, 4894
шоссейная дорога   4894
шофёр   1377
шпигат   5101
шпиль   700
шпиндель   5242
шпион   5614
шпионаж   1510
шпионить   5615
шпонка   2656
шрапнель   5341
штаб   2214, 5626
штабная оценка   5638
штабное дежурство   5637
штабной   5626
штабной офицер   3600
штабный отдел   3065
штаб оперативного планирования
   5635
штанга   371
штанга кручения   6204
штат   1513
штатная численность   5774
штатно-организационное расписание и
   табель имущества   5966

штатный   3725
штатская одежда   4045
штатская одежда (в штатском)   3362
(в) штатской одежде   805
(в) штатских   4045
штепсель   4095
штивать   5750
штопор   5581
шторм   1976
(жестокий) шторм   5748
шторм-трап   4945
штриховка   2149
штроп   5760
штурман   2253, 3420
штурманская рубка   4935
штурмовать   5747
штурмовая волна   6761
штурмовой самолёт   94
штурмовой эшелон   1412
штык   411
шум   3467
шунтирование   5342

Щ

щелевое убежище   6353
щель   3641, 6633
щит   3056, 3833, 5281, 5282
щитка   3833
щиток   3829
(заправочный) щуп   4298

Э

эвакуационный тягач   6907
эвакуация   340, 1521, 5034
эвакуация по воздуху   110
эвакуированный   1522
эвакуировать   1520
эвакуируемый   4596
эда   3121
экватор   1493
экваториальный   1494
эквивалент   1500
эквивалентность   1499
экипаж   912, 919, 1072, 6064
экипаж самолёта   106
экипаж танка   6011
экономить   1426
экономический   1425
экономия сил и средств   1427
экономный   1425
экран   5091
экран индикатора   1304

وَسَائِل القَذْف 1191	وحده أساسيه 3867
وَسَط 3182	وحده أوليه في منظمه 734
وَسَّعَ 1477	وحدة تنقية الماء 6747
وسيط كيماوي 723	وحدة الجناح 1759
وسيله بصريه 6635	وحدة حَمَّام 393
وشم 6060	وحدة الحِمْل 2906
وشى 2460	الوحده الخلفيه تساند الوحده الأماميه في حالة
وَصَف 4261	الاشتباك بالعدو 3797
وَصَف فنى 6075	وحدة ذاكرة الكومبيوتر 477
وصفه 4262	وحدة صغيره 5827
وَصَل 950، 4098	وحده طبيه 3146
وَصَل إلى 4542	وِحْدة عَرْض 1304
وَصَل بالأرض 1406	وحده غير مُدَّرجه 3475
وصله 1738، 5112	وحدة قياس الجرعه الشعاعيه 4418
وصلة جنزير 6248	وحدة مجاوره 3433
وصله عالميه 6496	وحدة محاسبه 19
وصلة مُعطيات 2876	وحده محاربة 40
وَصَّى 4581	وحده مساندة 5894
وَضَع 2740، 4220	الوحده المسانده تتقدم بقفزات 3796
وضع 4216، 5700	وحده مستجيبه 4782
وضع الإستعداد 5705	وحده مشتبكه مع العدو 6493
وضع الإستعداد لإطلاق المقذوف 5706	وحدة مغادِرة 3747
وضع اشارات الطرق 5397	وحده مقيمه 4772
وضع الالغام بشكل منتشر 2751	الوحده الموزعه 6494
وضع الالغام بلا خطه 2750	الوحده الناريه 2898
وضع الانبطاح 4210	وحْدة النخبه 1058
وضع بطاقه على 2688	وحده وارده 2430
وَضَع تمهيدياً 4260	وراء 66
وضع الجثث 4205	وراء البحار 3785
وضع الجهاز 5702	ورشه 6899
وضع جهازاً محفوظاً في	ورق الشجر 1841
مكان واقي 3334	وَرَقه فلزيَّه 1839
وَضَع حَاميه 1986	وزارة الدَفاع 3243
وضع السلاح على الأرض 2085	وَزَعَ 1210
وضع السيطره على الاسلحه	وَزَّع 1315، 2595
المضادة للطائرات 5701	وزن 6808، 6810، 6811
وضع العلامات 3084	الوزن الاجمالي 6812، 6814
وضع علامه على الخريطة 3083	الوزن الأقصى عند الإبرار 6815
وضع على مِنَصّه 3826	الوزن الأقصى في الإقلاع 6816
وضع في الملفات 1629	الوزن الصافي 6817
وضع القفز 2640	وزن المركبه الإجمالي 6813
وضع قنبله 4072	وزن الملاحين والآلات 3925
وضع القوات 4215	وسائل 3124

هيدروليكي 2364		هَدف 3532, 6022, 6023

هيدروليكي 2364
هَيْكَل 775, 1893, 1894, 2352
هيكل سفينه 2351
هيكل الطائره 113
هيئة اركان مُوَحَّده 5629
هيئة التنسيق 1002

و

وابل 6644
واجب 6053
واجبات الأركان 5637
واجه العدو 1570
واحد إلى الأمام 3632
واد 4619, 6546
وازَن 349, 6366
واش 2462
واصَف 5564
واق 4347
واقِيه 6633
واقية الامواج 3833
واقية الريح 6865
واقية الزناد 6364
واقية صواعق 2815
واقيه من الصدم 634
واقية الوحل 5588
وبَّخ 4751
وَتَد 3995
وتد التسديد 4223
وثائق السفينه 5291
وثَّق 306
وثيقه مسجله 4643
وجه 3121
وَجَّه 1269, 2123, 2741, 3731, 3732, 5713, 6559
وجه خريطه 5223
وجه الشريط 6233
وجه النار إلى مركز الهدف 6931
وَحَّد 875, 5655, 6487, 6491
وَحْده 6489
وَحده آلِيَّه 3339

هَدف 3532, 6022, 6023
الهدف الأولى 2038
هدف تحت سطح الماء 6040
هدف تدريب لرمي الخطف 6035
هدف تقطره طائره للتدريب 1380
هدف ثانوي 6038
هدف حي 6029
هَدَف خطي 6028
الهدف الرئيسي 4289
هدف صُلب 6026
هدف ضعيف 6039
هدف عارض 6050
هدف مانع 6025
هدف مباغت 6032
هدف مُباغت 6027
هدف مُخطط 6033
هدف مطلوب 6031
هدف معيَّن 6037
هدف مقطور 6041
هَدَف منطقه 6024
هدف منطقه 6044
هدف مهدد للتدمير 6036
هَدَف نقطي 6034
هَدم 5830
هُدنه 255
هذا ... واحد 6150
هراوه 394
هَرَب 1504
هرَّب 5495
هُروب 1503
هَزَّ 4908
هزم 1152, 4961
هزيمه 1153
هَش 1891
هضبة 4083
الهلال الأحمر 4607
هِلْيُوم 2249
هندسه 1475, 1476
هندسة القتال 868
هوائي 55, 217
هويه 2393
هَوَى 6598
هياج 4871

نقطة إشعاع شديد

|---|---|
نقطة الشحن 4115	نقطة إشعاع شديد 5597
نُقطة الصدم 4164	نقطة الإصابه 4165, 4128
نقطة الصعود من إرتفاع	نقطة الإصابه الوسيطه 4166
منخفض لبدء الهجوم 4140	نقطة الإعتراض 4129
نُقطة الصفر 6930	نقطة الإقتراب من الهدف 4150
نُقطة الصفر الأرضيه المُعيّنه 6929	نُقطة (الالتهاب ـ الوميض) 4125
نقطة العبور 4120	نقطة الالقاء 4145
نُقطة عبور الحدود 4111	نقطة القاء القنابل 4110
نقطه (أو نبضه) على شاشة الرادار 494	نُقطه أماميه للتسليح
نقطة الفحص 4116, 778	والتموين بالوقود 4126
نُقطة الكسر 4112	نُقطة إنزال 4132
نقطة اللاعوده 4167	نقطة الإنطلاق 4163
نقطه للدلاله على الهدف 4151	نقطة إنعطاف 4154
نقطة مخططة 4149	نقطة الإنفجار 4162
نُقطة مدار 4137	نُقطة أو ميدان الالتقاط 4000
نُقطة المراقبه 4118	نُقطة الأوج 230
نقطة مرجع 4143	نقطة (البدء ـ الرحيل) 4122
نُقطه مُعرّضه للخطر 4158	نقطة التثليث 4153
نُقطه مُعْلمه 4023	نُقطة تجمُّع 4127
نُقطه مُعينه 1741	نقطة التحميل 4134
نُقطة المَلْء 4124	نُقطة تحَوُّل 4155
نقطة ممسوحة 4149	نقطة التزييت 4170
نُقطى 4022	نُقطة التسجيل 4144
نَقَل 6325, 3354	نقطة التسديد 4108
نَقَل 6327	نقطة تسديد غير مباشر 4136
نقل باليد 3048	نقطه تعبويه 2919
نقل بطائره 1824	نُقطة التماس 4117
النقل بطائرات عموديه 2248	نُقطة تمـوين 4148
النقل جواً 124	نُقطة التوجيه 4138
نَقل جَوّي 6326	نُقطة توزيع 4123
النقل الجوي الاستراتيجي 6330	نقطة توزيع الماء 4159
نقل جوي تعبوى 6331	نقطة توزيع التموينات 5869
نَقل الحركه 1374	نقطة جمع أسرى الحرب 4141
نقل الدم 6305	نقطه حائده 4135
نقل الرادار 5286	نُقطه حرجه 4119
نقل على الحيوانات 6329	نُقطه حصينه 4147
نَقل من سفينه إلى أخرى 5308	نُقطة الحضيض 3950
نقل من سفينه إلى الشاطىء 5309	نقطه حيويه 4157, 4131
نقل النيران 5288	نُقطه خارجيّه 3752
نموذج 3297	نقطة ربط 4152
نُموذج انتاج 3300	النقطه الرئيسيه للمقاومه 4776
نموذج بالحجم الحقيقي 3296	نُقطة سيطره جويه 4107
نموذجي 6425	نقطة الشبوب 4142

713

ن (ن ـ أ ـ ن ظ)

مَنيع على النَّار 1722	منطقة مُجرَّده من السلاح 6942
مهاتفه لاسلكيه 4464	منطقه محرمه 4327
مهادنه 6381	منطقه محظوره 6945, 4798
مهارات البحر 5117	منطقة المراحل 5644
(مهارة) إستعمال الأسلحه 2170	منطقة المسؤوليات 248
مهاره في إستعمال الاسلحه 5434	منطقة مضروبه 6937
المهاره في الميدان 1612	منطقه مُعرضه للخطر 6655
مُهايىء للقابس 4097	منطقه مُغلقه 832
مهبط طائرات عموديه 2247	منطقه مَلْغومه 3227
مَهجُور 1218, 3088	منطقة المواصلات 6941
مَهْد 1059	منطقه الميناء 4191
مُهدّىء 6299	منطقة نار حره 1902
مُهرَّب 5496	منطقة هُبُوط 150
مُهرب أسلحه 2135	منطقَة الهبوط 2705
مهْماد 617	منطقة الهبوط 6948, 6943
مهمه 285, 3277, 3278	منطقة الهجوم 6953
مهمة الاعاقه 3280	منطقة الهدف 6044
مُهمَّه تفتيش 3281	منطقة الهدف 3533, 3534
مُهنْدِس 1474	منْظار (ثنائي العينيه) 472
مهندس طيران 1790	منظار الرَّصْد الليلي 3461
مهندس عسكري 4024	منظر 6623
مهنه 6255	منظَر عسكري 6408
مواد إستهلاكيه 963	مُنظِّم 4653
مواد الكتابه الرمزيه 1087	مُنظِّم السُّرعه 3806
مواد للإستعمال المشترك 903	منظمة الامم المتحده 6492
موارد 4780	منظمة حلف شمال الأطلسي 3400
موازن 3611	مُنَظَّمه سريه 6464
موازنه 1492, 6365	منظوري منظور 3968
مواصفات 5562	مَنَع 1206, 4276
مواصلات 908	منع التجوُّل 1089
مواقع ترك السفينه 3	مَنَع طائره من الطيران 2084
مواقع القتال 30	منع من الإستراق التليفوني 5081
مؤامره 4090	مُنْعَطف 454
المـوت الطبيعي 3403	مُنْفَذ غاز 6597
مَوَتِّر 6061	مُنَفِّر 4716
مؤتمر 953	مُنْفَصِل 5194
مُوجَب 4218	مِنْقَل 3567
موجه 5905, 6760	مِنقَله 4352
مُوَجَّه 6558	منقول بحراً 5107
موجة الاقتحام 6761	منقول في طائره عموديه 2244
موجه إلى الشمال 3483	مُنَكَّس 2151
موجة إنفجار 6762	منيع على الإنفجار 1556
مُوَجَّه بالرادار 430	منيع على اللهب 1753

مقياس الحرارة

مسطره حاسبه 5467	مستودع الذخيره 5742
مسطرة رسم خطوط متوازيه 3858	مُسْتَودَع شَحْن 1908
مُسَكَّن 5164	مستودع العتاد والذخيره 3724
مُسَلَّح 253	مَسْتور 1051
مسلح جيداً 6821	مستويات الإمداد 817
مسلفه كاشطه 627	مُستوى 2778
مسمار الأمان 5013	مستوى البحر 5115
مُسمار برشام 4889	مستوى التَحَوُّل 6314
مِسْمَاع مائي 2370	مستوى التموين 2783
مسموح اطلاق النار	المستوى الحقيقي 6387
على أيه طائره	مستوى عال 2266
غير صديقه 6795	مستوى المُخزونات 5729
مسموح إطلاق النار على	مستوى المُعطى 1113
طائرات العدو التي	مُسَجِّل 6019
تمييزها اكيد 6798	مُسَجِّل الذبذبه 3740
مَسند 518، 6266	مُسَجِّل شريط صُورى 6622
مسند الاسلحه 4416	مَسْح 5059
مسند ثلاثي 6369	مَسَح 5058، 5919
مسند الذخيره 4415	مسح القطاع 5149
مسند الرمى 4792	مسح الطرق 5925
مُسَن 2004	مسحوق الطلق 5996
مسنه السير المعكوس 2007	مُسَدَّد 5359
مِسواة 2781	مسدّده أماميه 1867، 1921
مساواة التسديد 4392	مسده إنعكاسيه 5367
المسؤول عن الأمتعه 348	مُسَدَّده تاكومتريه 5369
مسؤوليه 4786	مسده تحت الحمراء 6797
مسير الإقتراب 235	مسدده التدقيق 5365
مَسير دائره عُظمى 4967	مسده تليسكوبيه 5370
مسير المقذوف 6296	مسده جويه 5360
مُسيطر جوي أمامي 1878	مسده خلفيه 342، 4565، 5361، 5366
مُسيطر جوى تعبوى 5973	مسده ذات حدقه 5368
	مُسَدَّده ذات صُفيحه 5363
	مسده قتاليه 5362
م (م ش - م ك)	مُسَدَّده ليليّه 5364
	مُسَدَّس 4029
	مُسَدَّس 4836
مشاغب 4875	مسدس إشارات ضوئيه 4033
مُشاغبه 4877	مُسَدَّس إشاره 4032
مُشاه 2451	مسدس تشحيم 2067
مُشاه آليه 3139	مسدس للإشاره 4030
مُشاة البحريه 3074	مسرح العمليات 6138
مِشبك 614	مُسَطَّح 1772
مشبوه 5932	مَسْطَره 4985

مسار مقذوف 6234	مركز قياده جويه 4224
مسار نسيفه 6241	مركز الكشاف 4229
مُساعد 43, 3113, 6871	مركز المرور 4233
مُساعد هُبُوط 2704	مركز معلومات القتال 870
مُساعده 78	مركز مقابل 3757
مساعده عسكريه 3192	مَركز الهدف 626
مَسَافه 1312	مَركزي 745
مسافة الإزاحه 3615	مَركم 401
مسافة الأمان 5014	مرمى 5337
المسافه بين المركبات 6231	مَرن 1780
مسافة التفويت 5439	مروحه 1587, 4340, 6599
مسافة السند 5893	مُرور 3899
مساكن 4404	المرور الجوِّي 140
مساندة القتال الجوى 5875	المرور في إتجاه مزدوج 6418
المسانده القتاليه 5874	مرؤوس 5815
مسبار للتموين بالوقود في الطيران 4298	مريض 3910, 5348
مُسبِّب 3735	مريض قادر على السير 6665
مُستتِر 1169	مريض النقاله 2888
مُستخرِج 1566	مريله 237
مُستدير 798, 4955	مَرثي 6629
مُستشار 53	مرآة 3247
مستشفى إخلاء 6356	مُزْدوج 1345
مستشفى أمامي 2333	مزدوج السبطانه 6411
مستشفى سند القتال 2331	مُزدوجه 6203
مُستشفى عام 2334	مُزَرَر 655
مستشفى القاعده 2330	مزرور 655
مُستشفى الميدان 2332	مَزْلقه 4491
مُستعجِل 6539	مزلقه 5464
مُستعد 4557	مُزمِن 790
مُستعرض 6338	مزود بـ 2596
مستعمره 5228	مزود بترانزستور 6309
مُستقبِل 4572	مِزْوده 2202
مُستقبِل رسائل مُلتقِطه 2526	مزِيج 3287
مُستقِر 5625	مزِيد الشده 2522
مُستقِل 2435	مزيد شدة الصوره 2403, 5504
مستنقع 3090, 5940	مساحه 5920, 5921
مستنكِف ضميرياً 3535	مساحه تصويريّه 3983
مستهلك 1544	مَسَاحة الزجاج 6866
مُستَوٍ 2779, 4063	مساحة العطب 5923
مستودع 1212, 1393, 2287, 6677	مساحة كيميائية 5922
مستودع إسطواني 5406	مساحه مشيده 622
مُستودَع أسلحه 261	مسار جوي 120
مستودع الألبسه 5743	مسار سفينه 6239

مدرج المطار 134		مدى (عـن كُتب) 4510	
مَدْرج مُنور 1764		المدى في النهار 4499	
مِدْرجه جانبيّه 6063		المدى المائل 4513	
مَدْرسه 5072		مدى المدفع 2137	
المُدرَّعات 256		مدى (الهدف) 4514	
مِدفع 2128		مذبحه 3099	
مدفع دبابه 6015		مُذبِذب 3739	
مِدفع سريع الطلقات 687		مُذكره 3156	
مدفع عديم الإرتداد 4850		مُذكرة توقيف 6720	
مِدفع قوس 2348		مُذيب 5530	
مدفع مُتحرك 4974			
مِدفع مُركب على قُطب 5956			
مِدفع المَيدان 1616		م (م ر - م س)	
مِدْفعي 522,2132			
المدفعيه 271			
المدفعيه الثقيله 2228		مَرّ بالسد 2935	
مدني 804		مُرى يا صَديق 3904	
مُدَّه 3953		مِرآه 3247	
مدة الإبحار 5023		مراسل حربي 6671	
مدة التصليح 6516		مراسم 4349	
مُدة الطيران 6179		مَرَافع 1114	
مُدَّة المرور 4905		مرافق 79	
مدوَّره يدويه للإرتفاع 2177		مُراقب 740, 2953, 2954, 3541	
مدوره يدويه للتسديد 2178		مراقب الاسْقاط 1385	
مُدَولب 6835		مراقب حريق 6675	
مُدير 986		مراقب الغارات الجويه 6674	
مدير البرنامج 4319		مُراقبه 977, 5918	
مدير برنامج الحاسبه 931		مُراقبة الإرساء 195	
مُدير السُّكان 2253		مُراقبه أرْضيَّه 2097	
مدير السكان في السفينة 4402		مراقبة الجرد 2577	
مُدير القطاع 5147		مراقبة الجوده 4396	
مدينه مفتوحه 3639		مراقبة حركة المرور 6260	
مُديه 813		المراقبه للعمليات 3673	
مديه الجيب 2606		مُراقبة المخزون 5728	
مَدى 4497		مراقبة المرور الجوّي 141	
المدى الأدنى 4508		مراكز القتال 5697	
المدى الأقصى 4504, 4505		مراكز النوتيه 4405	
المدى (الأقصى) 4506		مَرْبَط 3321	
المدى الأقصى المؤثر 4507		مربط 6106, 6108, 6109	
مدى الأمان 4473		مِرْبَط بعيد 4697	
مدى التدمير 4472		مُرَبَّع 5619	
مدى السلاح حتى . . . 4523		مُرتب الحموله 5722	
مدى العمل 4496		مُرتب الذخيره 2898	

محطة تحديد الإتجاه

مُحتل 3713	محطة تحديد الإتجاه 5686
مُخَلَّط 3286	محطه الترحيل 5693
مختلف 3286	محطة توليد طاقه نوويه 3510
مُخَدِّر 197	مَحَطَّة السيطره على الشبكه 5691
مُخَرَّب 5007	محطة قياده 5685
مُخْرَج 6596	مَحظور 816
مخرج من برنامج تدريجيا 3973	محظور على الجند 3609
مخروط 952	محفار ثلج 5514
المخروط الأمامي 3487	محفظة الخرائط 3063
مخروطي 956	مِحقَنه 2377
مخزن 2994، 2995، 2996	محكمه عسكريه 1038
2997، 5741	مُحَلِّل 189
مخزن الأسلحه 267	مَحَلِّي 2918
مخزن القنابل 524	مُحَمَّل 2907
مخزون 5727	مُحَمَّل (بـ) 2690
مخزونات 5727	مُحَمَّل بنابض 5610
مخزونات العمليات 3678	مُحَمَّل كُرَيَّات 353
مخطاف 2057	محمول بطريق الماء 6736
مُخَطَّط 4069، 4089	محمول جواً 86، 87، 88
مخفر الإطفاء 5689	محمول في مركبه 6587
مخفر الحرس 2115	محمي 5151، 5278
مخفر شرطه 5692	محمي مُنْذِر 6704
مُخفف الصدمه 616	مِحْوَر 332، 333، 2025، 4040، 4041
مُخَلِّي 1522	المحور البصري 3693
مُحَمَّد 616	محور التموين 4968
مِخنقه 789	محور الدفع 2854
مُخَيَّم 479، 680	محور السينات 6910
مَدّ 1807	محور طولي 2951
مَدّ أسلاكاً 6878	محور العَيِّنَات 6919
المَدّ الأصغَر 3422	المحور العيني 6924
المد الأكبر 5611	مِحْوَري 4043
مد إلى الأمام 5717	محول طاقه 6301
مَدّ حبلاً 2742	مُحيد 1253
المد والجزر 6163	مُحيط 174، 3561، 3951
مَدَار 3699	محط الدائره 800
مدار السرطان/الجدي 6376	مخاضه 1859
مداري 3701، 6375	مُخاطره إضطراريه 1455
مُدَبَّر 1187، 4247، 5227	مَخْبَأ 2256
مُدَبِّر سابقاً 4259	مخبأ 664، 2258
مَدْخَل 1485	مُخْبِر 1241، 2462
مدخنه 1937	مُخْتَرَع 2575
مُدَرَّب 2495	مختص بالهواء والغازات 4102
مَدْرَج 4993	مختص بعلم النفس 4361

727

لَوْحه 960
لوحه 5091، 3828، 3829
لوحة الأسماء الرمزيه 3834
لوحة إعلانات 3494
لوحه بيانيه لاهداف المدفعيه 4092
لوحة تشخيص 3832
لوحة خرائط 5963
لوحة الدلاله 5396
لوحة رادار 4438
لوحة عرض في مستوى رأس الطيار 2217
لوحة القياده 3831
لوحة مساحة 5924
لوحة مفاتيح 2660
لوحه ملونه 860
لوحة النار المرصوده 3540
لوحة النشرات 630
لُوغارِتْمَه 2942
لولب 5099
لولبي 5585
لون 859
لوّى 6854
ليزر 2719
ليف 1608
ليفه كربونيه 705
ليل 3455

م(م أ – م ذ)

ما يكفي من السرعه لجعل السفينه
تحت سيطرة الدّفه 5714
ماج 5906
مأخذ الماء 6755
ماده 900، 2599، 3114، 5822
ماده كيميائيه 779
مازوت 3622
مال 2427، 2882، 5456، 6171
مال/يميل 367
مانع 385، 625، 4277
مانع اصطناعي 3545
مانع تعبوي 3553
مانع الحراره 5896

لعبة حرب 6694
لُغْم 3201
لَغْم 3202
لُغْم أرضي 3219
لغم افقي الفعل 3217
لغم باطل 3218
لغم (تدريب) 3215
لغم (زاحف) 3211
لغم (ساعه) 3210
لُغْم صغير 1885
لغم صوتي 3203
لغم ضد الأشخاص 3206
لغم ضغطي 3223
لغم عائم 3214
لغم (قاع) 3208
لغم (كليمور) 3209
لغم مُبْطل 3213
لغم (مُتأخر الفعل) 3212
لُغْم مُتحرّك 3222
لغم مجهز بصاعق إضافي 3204
لغم مسلح 3205
لُغْم مُضادّ للإشخاص 3207
لغم مغنطيسي 3221
لغم موجه 3216
لغه 2716
لفَّ 6854
لَفّ الجبل 439
لفافة الساق 4390
لفحة الحر 5120
لقب 3454
للتمرين 4245
لَمَس الأرض 6211
لَمَس الأرض 6212
لَهَب 1752
لوازم التعديل 2676
لوازم التنظيف 2672
اللوازم للبقاء على قيد الحياه 2677
اللوازم لوضع برنامج الآله الحاسبه 5518
لَوَّث 969
لوح الإداره 3830
لوح خشب 4067
لَوْح سَطْحي 4080
لَوْح للسياره بالوُقُوف 1746

730

ل

قنطره 6618	قَلْعه 801
قنيله 530	قَلِل 3237
قنينه 543	قَلَّل 3236
قنينه مُفرَّغه 6544	قَلْنْسُوه 692
قُوَّات 1857	قليل الضبط 6478
قوات الإنتشار السريع 974	قليل العمق 5312، 5246
قوات التزحلق 5447	قماش صوفي 2618
قوات خاصه 5559	قَمَر 5047، 3318
قوات خاصه تبقى في منطقه بعد	القمر التام 1930
إستيلاء العدو عليها 5707	قمر صناعي قتّال 2666
قوات الصدم 5320	قمر صناعي مداري 3702
قوات الطوارىء 975	قَمَط 808
قوات الفرقه 1332	قُمْع 1938
قوات مُخصَّصه 899	قمع 4408
القوات المسانده للقتال 5876	قمله 2971
القوات المُسلَّحه 254	قنّاص 5502
قوات معينه للنقل في طائره 6370	قناع اكسجين 3801
قوات مكافحة الشغب 4878	قناع التنفس 4781
قوات المناورات القتاليه 871	قناع الغاز 1997
قواعد الإشتباك 4987	قناه 683
قوانين/قواعد 4986	قناة تصريف 1357
قوانين الحرب 2739	قناة لاسلكيه 4450
القوانين للطيران بالنظر 6637	قناة الماء الرئيسيه 6745
قوس 238	قناة ملاحيه 1581
قوس جَبْهى 1918	قنبله 521
قوس النيران 239	قنبلة بندقيه 4849
قوَّض المعنويات 6466	قنبلة بنزين 3314
قوقعه 2352	قنبلة دخان يدويه 2073
قوَّه 5772، 4384، 1853	قنبلة شظايا 5341
قوة الإختراق 4995	قُنبُلة صغيره 530
قوة الإشاره 5776	قنبله عنقوديه 837
قوة إنزال 2707	قُنبُله ملتصقه 3220
قوة تَحمُّل لِلتربه 5520	قنبله موجهه بالليزر 5487
قوة التغطيه 1054	قنبله موقوته 6176
قوه جويه تعبويه 5974	قُنبله ناريه للتصوير في الليل 3982
قوه حافظه على الامن 3930	قُنبله نوويه 2206
قوة الحملة 1542	قُنبُلة هاوُن 3330
قوه ساحليه 844	قُنبله هيدروجينيه 2367
القوه السريعة الانتشار 4529	قُنبله يدويَّه 2072
قوة سند لاحقه 1843	قنبله يدويه (شظايا) 2071
قوه ضاربه 5789	قُنبله يدويه (لاصقه) 2074
قوه طارده مركزيه 748	قُنبله يدويه (للتعليم) 2070
قوه غير معينه لمهمه 6446	قنبله يدويه مدوِّخه 2075

غير مُصَنف 6444

غير معين 6484

غير ملوث 6448

غير ممسوح 6523

غير مُؤثِّر 3473

غير مؤيد 6452

غير نظامي 2588

غير مِن تصنيف 1138

غيضه صغيرة الاشجار 5583

غيم 835

ف

فاحِص بالصَّدى 1423

فاحص حسابات 305

فأر 1225

فأس 331

فاسد 2447

الفاصل بين المركبه والأرض 2096

فاصل رأسي 2562

الفاصل الزمني لرَدّ الفِعل 4545

فاصله 2559

فاصلة التربيع 2561

الفاصله الرأسيه 6607

فاصله ما بين المنحنيات 2560

فاض 1802، 1805

فاعليه إشعاعيه مُستحثّه 2445

فاوض 3429، 3870

فائده 6540

فائض 5910

فائض عن العدد 5850

فتح 958، 3636، 5953

فتح الأنوار 2807

الفتح بالحبل القراري 3643

فتح ثغره 569

فتح ثغره بِسُرعه 2194

فتح الطريق 4899

فتح مُتأخر 3642

فتحه 229

فتحة رمي 2955

فتّش 1914، 2485

غَمَر 1806، 5810

غِمْد 5254

غمد الحربه 5053

غوّاصه 5808

غواصه صغيره 3186

غوغاء 3288

غياب بدون إذن 7

غيبوبة 864

غَيَّر 163

غير آمن 2480

غير دفين 6441

غير رسمي 6509

غير صالح للإبحار 6515

غير صالح للإستعمال 6517

غير صالح للخدمه 6482

غير صالح للشرب 6479

غير عقيم 6521

غير قابل لحل الرموز 6454

غير قابل للتجسير 6440

غير قابل للكسر 5252، 6439

غير قابل للملاحه 6507

غير قانوني 2397

غير لامع 3112

غَير مُباشِر 2442

غير مُتحكم فيه 6449

غير مُتفَجِّر 6481

غير مُتوَفِّر 6438

غير مجسَّد 6522

غير محارب 3472

غير محتل 6508

غير مُحدَّد 2434

غير محظور على الجند 3633

غير مُحْكم 2956

غير مَحْمي 6433

غير مدرع 6435

غير مرسوم على الخريطه 6443

غير مرهصوف 6506

غير مزوَّد برجال 6505

غير مُستعمل 6525

غير مُستَقِر 6520

غير مُسجَّل 6511

غير مُسَلَّح 6434

غير مصروف 6519

ضد التيار 6533	ضابط (أسنان) 3580
ضد الريح 6534	ضابط الاعاشه 3578
ضَرَب 399, 436	ضابط الإيواء 3576
ضَرَب 364	ضابط بحري 3595
ضربه 5786	ضابط التخلص من القنابل 3577
ضربة أخمص 5797	ضابط تموين 4401
ضربة بسلاح واحد 5788	ضابط التموين 3602
ضربه جويه 133	ضابط الحراسه 3604, 3605
ضَرَر 2080	ضابط الخفر 3598
ضرر الدماغ 555	ضابط الرصد الأمامي 3585
ضروره 1452	ضابط رُكن 3600
ضَروري 1511	الضابط السياسي 894
ضعيف 6770	ضابط (صغير) 3590
ضَغْط 4271, 5780, 5783	ضابط صَف 3596
ضغط 923, 4267, 5781	ضابط طبي 3592
الضغط الأرضي 2102	ضابط العتاد والذخيره 3599
ضغط الدم 502	ضابط العمليات 3597
الضغط الذروي 3935	ضابط عون 3601
ضغط زائد 3780	ضابط للحراسه 3582
ضغط زائد للمحيط 1897	ضابط مدرب 3588
ضغط عال 2270	ضابط المطعم 3593
ضغط الهواء 129	ضابط من مرتبة القاده 3584
ضُفدَع بَشَري 1915	الضابط المناوب 3581
ضَفه 365, 366	ضابط مُنفذ 3583
ضَمَّ 211	ضابط موقع الرمى 3586
ضماد الميدان 1614	ضابط النقل 3594
ضوضاء 3467	ضابط نقل جوي 3575
ضوء الإرساء 194	ضابط نوبة الحراسه 6732
ضوء إنذار 6714	ضابط اليوم 3603
ضوء غامر 1808	ضادّ 1021
ضوء فلورسنت 1821	ضاغط 927
ضوء فلورسنت 1820	ضباب 1834, 3285
ضوء المناره 420	ضباب خفيف 2205
ضوء الموضع 4217	ضبابي 1836
ضوء موضعي 5600	ضباط مرافقون 5634
ضَوْء النهار 1117	ضَبَط 41, 42, 1832
ضوء وَمّاض 1769	ضبط الشغب 4873
ضَيِّق 3392	ضبط الصمامه بحسب المدى 4518
	ضبط العيار 670
	ضبط متقدماً 4266
ط	ضَخّ 4376
	(ضخ) ورفع بمضخّه 4378
طابعه برقيه 6086	ضد الإنزلاق 3476

745

ض

صاد

صَحراء 1224	صاد 2356
صحفي 4268	صاد بشبكه 6345
صحيح 1011, 4860	صادر 889, 4758, 4760
صحيفة خريطه 5257	صار 3102, 3101
صَخر 4909	صاروخ 3252
صخري 4916	صاروخ 4910
صَدّ 2296	صاروخ أرض ـ أرض 3273
صد هجوماً 4715, 4753	صاروخ باحث عن الهدف 3270
صَدَأ 4998	صاروخ جو ـ أرض 3255
صُدْر 784	صاروخ (جو ـ إلى تحت الماء) 3256
صُدره مقاومه للرصاص 259	صاروخ جو ـ جو 3254
صِدريه 237	صاروخ ذو طيران حُر 3262
صَدْع 1057	صاروخ ذو وقود سائل 4912
صَدَفه 375	صاروخ ضد الطائرات 3257
صدَّق 822	صاروخ ضد المقذوفات القذافيه 3258
صدَّق على 306, 5037	صاروخ عابر للقارات 3261, 3265
صَدم 633	صاروخ قذافي متوسط المدى 3266
صَدْمه 632, 2413, 5313	صاروخ قذافي مطلق من الجو 3253
صدمة أرضيه 5315	صاروخ كابح 4821
صدمه كهربائيه 5314	صاروخ للاطلاق والترك 3267
صديق 1913	صاروخ مضاد للغواصات 4911
صَدىً 1422	صاروخ مطلق من أنبوب متعقب
صدى أرضي 4823	بصرياً موجه بالسلك 3274
صدىً دائم 3959	صاروخ مطلق من مدى بعيد 3271
صدى الرادار 4435	صاروخ مقذوف من غواصه 3272
صدى راداري 4824	صاروخ مقذوفي 3260
صَدِيء 4999, 5002	صاروخ مُوجّه 3263, 3264
صَرَف 489, 1289, 1356	صاروخ مُوجه مضاد للدبابات 3259
صَعَّد 1501	صاروخ يطير فوق سطح الماء 3269
صَعدَ 826	صالِب 2651
صَعَد 4371	صالح لسير السُفن أو الطائرات 3413
صعد إلى السطح 5900	صالح للسكن 2147
صَغُر 4612	صالح للطريق 4906
صغير 2645, 3244	صالح للمرور 3898
صغير العيار 5486	صالحه للطيران 144
صَفّ 152, 4976	صالحون للخدمه 1430
صَفّ 151, 2871, 2872	صامِد للقنابل 531
صَفّ 1627	صان 3019, 5206
صَفّ الغام 3234	صانع 269
صفاره 6840	الصِحافه 4268
صفارة الإنذار 5415	صحح الرمي بالحصر 4515
صَفد 5236	صحح الرمى بحسب الخط 4519
صِفر 6927	صحح الرمي بزاوية الارتفاع 4517

747

سهام نارية

السيطره 884	سِهام ناريه 4391
السيطره التعبويه 5976	سَهْل 2782,4044
السيطره على الرَّمى 1708	سهم جوّي 1773
سيطره من بُعد 4695	سُهُولة الحركه 3293
السَّيطره من بُعد 4694	سهولة الحركه في ساحة المعركه 407
سَيْل 1815	سُوُر 1600
	سور 1601
	سُونار 5531
ش	سُونار سَلْبي 5534
	سونار معلق بطائره عموديه 5532
شاحن تُربيني 6400	سونار مقطور خلف سفينه 5533
شاحنه 6382	سونكي 411
شاحنه للدبابات 6016	سَوّى 2045,2780
شارَك 4184	سُوء الأداء 3034
شاره 344,3493,5795	سياج 1601,2234
شارة الرُّتبه 345	سياده 3109
شارة فيلميَّه 1633	سياده جويه 136
شاسيه 775	سياده مشتركه 949
شاطىء 413,1866,5324	سَيّار 4065
شاطىء العدو 5325	سياره 316
شامل 158,922	سيارة إستكشاف 5078
شباك 6862	سيارة إسعاف 175
شبح 5404	سيارة إطفاء الحريق 1712
شبكة 3438	سيارة المؤخرة 6572
شبكة الإتصالات إلى الخلف 3440	سيارة ماء 6758
شبكة أُسلاك شائكه 372,6883	سياسه 4178
شبكة إنذار 3441	سِيَاسه دِفَاعيه 1163
شبكة خطوط 2076	سياسي 4179
شبكة راديو لتعيين النقطه 3445	سِياق 4299
شبكة القياده 3439	سَيّب 4675
شبكة منفاخيه 6880	سَيَج 1600
شبكة المواصلات 3444	سير 445
شبكة نقل 3446	سير 3067
شبكه واقيه من الطوربيد 6201	سير إضطراري 1856
شبه تلقائي 5179	سير تعليق 5938
شبه جزيره 3942	سَيْر الكتف 5339
شبه عسكري 3861	سير نقل الحركه 1378
شبيكه 4810	سير الوحده والاشتباك مع العدو
شتاء 6874	غير المتوقع 6340
شتوي 6875	سير الوحده والاشتباك مع العدو
شجاعه 1977	المتوقع 3797
شجره 6349	سَيْطَر 3103
شجيرات قصيره 5100	سيطره على 983

سقاطة الامان 4017
سَقط 1576
سقوط ذاتي 1900
سُكّان 4186
سُكّان 4980
سَكَن 14
سكه إضافيه 5355
سكه حديد 4483، 6238
سكة الحديد 4484
سكه حديد ضيقه 3393
سكوت 5399
سِكّين 2681
سكين قتال 6355
سلاح 250، 251، 562، 6772
سلاح آلي 6774
سلاح إحيائي 6775
سلاح إستراتيجي 6793
سلاح (إسمي/رمزي) 6786
سلاح الإشاره 5389
سلاح التموين 4403
سلاح الجو 112
سلاح خفيف 6784
سلاح ذو إستجابه جُزئيه 6779
سلاح ذو طاقم 6777
سلاح ذو طلقه مفرده 6790
سلاح ذو عمل بسيط 6789
سلاح ذو مزلقه 6791
سلاح رمى غزير 6794
سلاح فردى 6782، 6788
سلاح الفرسان (الدبابات) 728
سلاح كيلوطُنّ 6783
سلاح مُباعد 6792
سلاح مزود بالإشعاع 6778
سلاح مضاد للدبابات 6773
سلاح موجه 6781
سلاح ميكا طن 6785
سلاح نووي 6787
سلاح نووي جرثومي كيماوي 6776
سلاح يُشغّل بالغاز 6780
سلب 2961، 4099
سِلبي 3425
سَلَح 249
سلسله 755، 5200

سِلسِلَه القياده 756
سُلطَه 309
سلعه 900
سِلك 6877
سلك التعثر 6881
سلك كهربائي معزول 1779
سُلّم 910، 2689
سُلّم 3929
سُلّم الترددات 4500
سُلّم حِبال 4945
سُلّم النجاه 1713
سَلّه 392
سُلوك 300، 951
سَمّاعات 1404
سماعات الرأس 2216
سَماوي 732
سَماء 5448
السَمْت 334
سَمَح 3961
سَمعي 26
سمه مُميزه 5392
سُمّيّه 6220
سَنّ 5250
سنتيجراد 744
سَنَد 552
سَنَد إداري 5886
سَنَد ارتجالي 5881
سَنَد تعبوي 5891
سند جوي 5871
سند جوى تعبوى 5892
السند الجوى التعرضي 5889
سَنَد جوي غير مُباشر 5883
السَنَد الجَوّى الفَورى 5880
سَنَد الخدمة 5890
سَنَد عام 5879
سَنَد غير مُباشر 5882
سَنَد مُباشر 5872، 5877
سَنَد مُتَبادَل 5887
سَنَد متكامل 5885
سَنَد مدفعّية البَحرّيه 5888
سند جوي قريب 5873
سنه 6920
سِنّي 1204

رَاقب 739, 980, 3316	ر (ر أ - ر ش)
4176, 5645	
راكب 3903	
راكد 5453	راتب 3924
رام 522	رادار 4419
رام ماهِر 3085	رادار الانذار 4421
رام ماهر 5251	رادار باحث 4428
رامي 4854	رادار تحديد اماكن الهاونات 4426
رايه 3943	رادار التفتيش 4420
راية إستدعاء المرشد 4013	رادار تفتيش ذو موجات مُضمنه 4422
رايه بيضاء 6842	رادار جوي يبحث إلى الجانب 4429
رايه صغيره 2125	رادار دُوبْلر 4423
راية المُفاوضه 1747	رادار الرصد 4430
رُباط ضاغط 924	رادار طائره يراقب نحو الأرض 4425
رُبّان 3104, 5441	رادار قافز الترددات 4424
ربح 6850	رادار متطاور الصف 4427
رَبط 2724	رَادع 1244
رَبط 1591, 1592	رادع إستراتيجي 5764
رَبط 468, 6167	راديو 4447
رَبط 469, 2725, 6166	راديو تحديد الإتجاه 1641
رَبط الأحزمه 2189	رأس 697, 2207, 2208
رَبط بالرصيف 456	2209, 2212
رَبط بسلك 6876	رأس الإطلاق 2738
رَبط بسير 5759	رأس جِسْر 598
رَبط على طول 3320	رأس جسر جوّى 115
رَبوه 3340	رأس جِسْر الملاحه 3417
رتاج 2932	رأس حربي ذو مجموعه من المتفجرات 6696
رَتّب 3091, 5750	رأس حربي كيميائي 6695
الرتب الأُخرى 4527	رأس سكه حديد 4482
رتبه 4524	رأس الشاطىء 415
رتبة فعلية 5823	رأس الصاروخ المتفجر 6697
رتبه بحريه عاليه 1748	رأس القذيفه المتفجر 6699
رتل 862, 1627	رأس متفجر نووي 6698
رتل أُحادي 5412	رأس مهروس 5620
رَجّح 6810	راصِد 3541
رَجع 4618	راصِد أمامي 1883
رجعه 4620	رافده 427
رَجُل الحَريق 1719	رافدة القَصّ 2651
رَجَم 5733	رافدة قص سويّه 1529
رحّل 4671	رافِعْه 1061, 1219, 2784
رحله 6652	رافعة سياره 2601
رَحَويّه 700	رافعه شوكيه 1870
رُخصة 3960	زَافِق 15

خ

حرب أهلية

جزام	361, 451, 5760	حرب أهليه	6669
جزام الأمان	449	الحرب البرية	6687
حزام الخدمه	5046	حرب تقليديه	6682
جِزام العوم	1811	حرب الجراثيم	6685
حزام المقعد	450	حرب الحركه	3292
جزام النجاه	2797	حرب العصابات	6686
حزب	3879	حرب العصابات في المُدن	6692
حُزمه	428, 5253	حرب غير تقليديه	6691
حزمة أسلاك	2757	حرب كيميائيه	6681
حُزمة أغصان	1589	حرب محدوده	6670
حزمة ليزر	2720	الحرب المضاده للعصابات	6683
حُزمه متوازيه	3859	الحرب المضادة للغواصات	6679
حُزمة نور ضيقه	3939	الحرب النفسيه	6689
حِسَاب	16	الحرب الهدامه	6690
حَسَّاس	5187	حربه	411
حساسيه	5188	حربي	5205, 6700
حَسَب	929	خرج	1074
حُسن الإستخدام	3110	حرر	2788
حسن الحراسه	6825	حَرَس	1509, 2111
حسن الدفاع	6822	حَرَس	1508, 2112
حسن القياده	6823	حرس التدمير	2113
حَشَدَ	3091, 3295	حرس الشرف	2116
حَشْد	936, 1082, 3093, 3294	حَرَس القافله	994
حَشْد إستراتيجي	5762	حَرَس وطني	3200
حشو	3816	جِرش	6148
حُشوه	768	حَرَف	1174
حشوه مخفضه	4613	حرف الجبال	2683
حشوه مُشكله	5248	خَرق	645
حشوه مفرغه	2301	حرق الوميض	1768
حشوه ناقصه	4613	حُرقه	645
حَشيَّه	1996	حركه إرتجاعيه	339
حِصار	497, 5356	حركة التراجع	4820
حِصان	2326	حركة التفاف	1489, 6407
حَصباء	2060	حركه مُتناوبة النقل	5347
حصباء مُقيَّره لرصف الطُرُق	6051	حركة مرور	6256
حَصَدَ	5943	خَرَكي	1401
حَصَر	496	خَرَكي جَوّي	56
حَصَر	5752	خَرَّم	358, 1206, 2531
حَصَر العَطَب	1101	حريه	1899, 2789
حَصَر في زاويه	1007		
حصر في موضع	2920		
حَصَر (الهدف)	551	ح (ح ز - ح ي)	
حَصَل	4311	حَزْ	3490

763

جهاز الملاحه بالقصور الذاتي 2450

جهاز نووي 3507

جهاز يتيح تضاريس الأرض 6121

جهاز يخطى تضاريس الأرض 6120

جهاز يدوي 2174

جُهد 5782

جُهْد القَطْر 1364

جُهد مقاس بالفولت 6647

جَهَّز 1497

جهز بصمامه 1942

جَهَّز (سفينه) 897

جَهَّز (صمامه) 5218

جَوّ 84، 289، 6801

جو ـ أرض 114

جَوّ عاصف 2233

جواب 4723، 4724

جواب سلبي 3426

جواز المرور 5012

جوف 541، 2299

جوف أملس 5494

جوف محلزن 4853

جوي 6802

جَوّي 3173، 62

جيب 4103

جيب هوائي 127

جيد التجهيز 6824

جيروسكوب 2144

جيروسكوبي 2145

جيش 264

جيش الميدان 263

جيش نظامي 4647

ح (ح أ - ح ر)

حاجب 3718

حاجز 383، 385، 625

3230، 3544

حاجز الأمواج 587

الحاجز الصوتي 5540

حاجز ضد الغواصات 5094

حَاجز طبيعي 3549

حاجز عائم 533

حاجز لا يمكن عبوره 3548

حاجه 4757

حاد 38، 3613، 5712

حَاد 5249

حادث 11

حادث بسلاح نووي 3516

حادث نووي طارىء 3508

حادثه 2426

حاده 973

حَاذى 151

حارس 5191

حارس جائل 4358

حارس الجنب 1760

حارس شخصي 515

حَارسة صيد البحر 6610

حاسب 18

حاسبه 930

حاسبه بالتَشابُه 186

حاسه 5185

الحاشيه 4811

خَاصَر 498، 1005، 2580

الحاصِره 553

حاصل 6921

الحاصل النَّووى 3517

حاضِن 3351

حاضِن المدفع 2130

حَافظ على 3018

حافله شحن 2962

حافلة نقل المؤن 5744

حافه/طرف 5445

حاكم 4343

حاكِم الإصدار 1457

حاكِم عسكري 2043

خَالَف 160

حاله 5700

حالة الاستعداد 5672، 5673

حالة الإستعداد المأمونه 5674

حالة إستقرار 5708

حالة الإنذار الفورى 4552

حالة الحرب 5675، 6693

حاله صحّيّه 948

حالة الصلابه 5528

جهاز مرسل مستقبل

جهاز إختبار 4856	جماعة تنظيم الشاطىء 3893
جهاز الإرتداد 3132	جماعة تنفيذ الإعدام 3886
جهاز إرسال 4466	جماعة الدفن 3882
جهاز إرشاد الطائرات 422	جماعة سَنْد مُتنقله 3291
جهاز إشعال 5220	جماعة القياده 3720
جهاز إطلاق النار 1726	جماعة مُتقدِّمه 3880
جهاز إمتصاص الصدمات 5316	جماعه مُغيره 3890
جهاز إنذار 6713	جماعة المؤخره 3891
جهاز تثبيت 3350	جماعة نقل 6336
جهاز تدمير 5219	جُند 1904, 2387
جهاز تسييب 3137	جَمع 279, 856, 1874
جهاز تسييب سريع 4680	2107, 3379, 6207
جهاز تشويش باحث 2611	جمع القطع 4489
جهاز تشويش تلقائي البحث 2609	الجمهور 4362
جهاز تشويش مكرر 2610	الجميع 153
جهاز تصويب القنابل 532	جميع الرتب 4526
جهاز تَعْويم (دبابه) 5095	جَناح 1757, 6868
جهاز تَفَاضل 1260	جناح أبتر 6869
جهاز تلقائي أرضي للإحساس 5190	جناح المرضى 5349
جهاز تليفزيون 6093	جناح ممتد إلى الخلف 6870
جهاز التَّنبؤ 4251	جنا سلاح 3716
جهاز التَّنفُّس 589	الجند 5527
جهاز التَّوجيه 2008	جُند 1479, 4599
جهاز التوجيه 2305	جُندي 5523
جهاز توجيه سلبي 3906	جندي أمريكي 2024
جهاز الحفر 3624	جندي تحت التدريب 6274
جهاز راديك 4440	الجندي المجهول 6498
جهاز راديو 5222	جُندي مشاه 2452, 5524
جهاز راديو صغير 6664	جندي مظلي 3864
جهاز الرؤيه الليليه 3463	جندي من المغاوير 4520
جهاز سائل ومجيب 6324	جندي نظامي 5525
جهاز شخصي لتعيين المكان 423	جنزير 6235
جهاز ضبط الموازنه 6367	جنسيّة 3398
الجهاز العصبي المركزي 747	جنوب 5545
الجهاز الفاصل 838	جنوبي 5545
جهاز فتح المظله 4679	الجنود 4525
جهاز لإجتناب تضاريس الأرض 6119	جنود مستجدون 4540
جهاز مداري جُزئي للقصف 1890	جُنيح 80
جهاز مدمر ذاتيا 3134	جنيحة توجيه 6550
جِهَاز مُدِير 1273	الجهات الاربع 4169
جهاز مُراقبه 3315	جهاز 1498, 3896
جهاز مراقبة النيران 1709	جهاز إتلاف الصاروخ 1232
جهاز مُرسل مُسْتقبِل 4465 6300, 6319	جهاز إحساس 5189

جبهة الصدم 5318

جبيره 5590

جثّه 1010

جدار حجز الطلقات 656

جدّد 4624,4701

جدوّل 2881,5064

جدول أهداف 2884

جدول بياني 1816

جدول التنظيم والتجهيزات 5966

جدول التنقل 5964

جدول التوقيت 6184

جدول الخدمه 4946

جدول الرواتب 4926

جدول الصيانه 5066

جدول المد والجزر 5965

جدولى 5968

جديد 1912

جذع 5242

جذع الإداره 5244

جذع الدوار الرئيسي 5243

جذف 4975

جذَف 3817

جذُف 3817

جَرّ 2224

جَرّ 2199

جراب 2302

جراب (الذخيرة) 4238

جرّاح 5907

جرّاح في الطيران 1794

جراحه 5908

جراحي 5909

جراره (هندسه قتاليه) 6254

جرّافه 3866

جرّب 6128

جربنديّه 4979

جَرَح 2474,6902

جُرْح 6901

جرح بسطح الجلد 1778

جُرح العين 1568

جَرْد 2576

جرس 440

جُرْعه 1350

جرعه قاتله كامنه 2727

الجرعه القصوى المسموح بها 3118

جَرَف 4488

جرف شديد الإنحدار 507

جرموق 4390

جُرْن 374

جريح 6903

جريمة حرب 6672

جريه سريعه للنهر 4530

جَزْر 1410

جُزْء 3873

الجزء الطافي 1898

جُزْئي 3875

جزيره 2589

جسَّر 594,5548

جِسْر 595,5549

جسر جَوّى 117

جسر دَارج 1980

جسر مُعَلَّق 5935

جسر وَقْتي 1826

جسر ينزل من مركبه مدرعه 597

جِسْم 513

جِسر دوّار 5949

جسم كُرويّ 5577

جسم مَسْتور 1168

جعبه (الخرطوش) 4238

جعل تِلقائياً 312

جعله سويّاً 3481

جغرافي 2022

جغرافيه المحيطات 3565

جفاف 1179

جلا 1520

جلد 5436

جليد 2379

جليدي 2389

جماعه 2106,3878,3995,5616

جماعة إقتحام بحريه 3410

جماعة الايواء 3889

جماعه باقيه وراء خطوط العدو 3888

جماعة تامين القنابل 527

جماعة تدمير 3884

جماعة التدمير 3883

جماعة شغل 3885

جماعة تفتيش 3892

770

779

فهرس المصطلحات العربية

Appendix A

RANKS – GRADES – GRADOS – DIENSTGRADE – ВОИНСКИЕ ЗВАНИЯ – الرتب

Notes Rank structure and names differ between armies with a common language. Air force ranks are shown in italics where they differ. The ranks shown as matching in this table do not always precisely correspond in status and function; in particular the position of warrant officers differs between forces. United States Marine Corps and British Royal Marine ranks are not included; they resemble army ranks, but a given designation usually represents one actual rank higher.

Remarques La structure ainsi que les appellations de grade diffèrent entre les armées des pays de même langue. Les grades de l'armée de l'air sont indiqués en italiques lorsqu'ils sont différents. Les grades montrés comme équivalents dans le tableau suivant ne correspondent pas toujours exactement en ce qui concerne le statut et le rôle; en particulier la place qu'occupent les sous-officiers supérieurs varie d'une armée à l'autre. Les grades des personnels du United States Marine Corps et des Royal Marines britanniques ne sont pas mentionnés; ils sont similaires à ceux de l'armée de terre. Toutefois à appellation semblable, ils représentent généralement un grade au-dessus.

Notas La estructura y denominación de las graduaciones difieren en los diversos ejércitos con un lenguaje común. Las graduaciones que figuran aquí no siempre corresponden exactamente en rango y función, especialmente entre los suboficiales de los diferentes ejércitos. La Infantería de Marina de los Estados Unidos y la Real Infantería de Marina Británica no están incluidos. Sus graduaciones son similares a las del Ejército, pero una determinada denominación representa normalmente una graduación mas elevada.

Anmerkungen Selbst bei Streitkräften, die eine gemeinsame Sprache sprechen, unterscheiden sich die Dienstgradstruktur und die Bezeichnung. Luftwaffen-Dienstgrade sind im Fall von Abweichungen kursiv eingetragen. Die auf dieser Tabelle als einander entsprechend gezeigte Dienstgrade stimmen nicht immer genau hinsichtlich Status und Rolle überein; vor allem ist die Stellung der höchsten Feldwebeldienstgrade ("warrant officers") in den verschiedenen Streitkräften oft unterschiedlich. Die Dienstgrade des United States Marine Corps und der britischen Royal Marines sind nicht eingeschlossen; diese ähneln den Heeres-Dienstgraden, wobei jedoch die jeweilige Tätigkeit *de facto* meist dem nächsthöheren Heeresdienstgrad entspricht.

Структура воинских званий и их названия различаются между армиями, употребляющими общий язык. Воинские звания военно-воздушных сил, если различаются, написаны курсивом. Звания, представленные в этом списке сходными, не во всех случаях соответствуют друг другу по статусу и назначению; в частности, различается между вооруженными силами положение старшин. Не включаются здесь звания морских десантников Соединенных Штатов Америки и Великобритании; они похожи на армейские звания, но одно назначение обычно равняется следующему высшему званию.

المذكرات تختلف بنية الرتب واسماؤها بين جيوش تستعمل نفس اللغة . تظهر رتب سلاح الجو بحروف مائلة حيث تختلف من رتب الجيش . ان الرتب التي تجري معاً في هذا الجدول لا تتساوي دائماً بالضبط بمناسبة العمل والوضع وخصوصاً يختلف وضع نواب الضباط بين القوات . لا تظهر رتب المشاة البحرية الأمريكية والبريطانية . انها تشبه رتب الجيش ولكن اسم واحد عادة يمثل رتبة واحدة أعلى .

Appendix A *(cont)*

1. Army and *Air Force* – armée de terre et *de l'air* – ejército y *aviación* – الجيش وسلاح الجو
Heer und *Luftwaffe* – Сухопутные войска и *военно-воздушные силы* –

United States	Great Britain	France (tous *m*)	España (todos *m*)
private third class *airman*	private, gunner, trooper (etc.) *aircraftman*	soldat	soldado de 2ª, artillero
private first class *airman second class*	lance-bombardier/ - corporal *leading aircraftman*	soldat de 1ère cl.	soldado de primera
corporal *airman first class*	bombardier, corporal	caporal, brigadier	cabo
—	[lance-sergeant]	caporal-chef, brigadier-chef	cabo primero (1º)
sergeant *staff sergeant*	sergeant *senior technician*	sergent, maréchal de logis	sargento
staff sergeant *technical sergeant*	quartermaster sergeant, staff sergeant *flight sergeant* *chief technician*	sergeant chef	sargento primero
platoon sergeant, sergeant first class *master sergeant*	—	sergent-chef, maréchal des logis- chef	brigada
first sergeant, master sergeant *senior master* *sergeant*	warrant officer 2nd class, squadron/ battery/company sergeant major *master aircrew (pilot* *etc.), master technician*	adjudant	brigada
sergeant-major *or* warrant officer *chief master sergeant* or *warrant officer*	orderly room quarter master sergeant	adjudant	subteniente
command sergeant- major	regimental sergeant- major, warrant officer 1st class *warrant officer*	adjudant-chef	—
sergeant-major *or* chief warrant officer	—	adjudant-chef major	—
ensign; commissioned warrant officer (wo)	—	—	—
second lieutenant	second lieutenant *pilot officer*	sous-lieutenant	alférez
first lieutenant	lieutenant *flying officer*	lieutenant	teniente

Bundesrepublik Deutschland (BRD) (durchweg *m*)	Советский Союз	العربية
Jäger, Kanonier, Panzergrenadier, Schütze usw.	рядовой	جندي
Gefreiter, Obergefreiter, Hauptgefreiter	ефрейтор	جندي أول
Unteroffizier	младший сержант	نائب عريف
Unteroffizier	—	عريف
Stabsunteroffizier, Feldwebel	сержант	رقيب
Feldwebel, Oberfeldwebel	старший сержант	—
Oberfeldwebel	старший сержант	رقيب أول
Hauptfeldwebel	—	—
Hauptfeldwebel	старшина	وكيل ثاني
Hauptfeldwebel	—	—
Hauptfeldwebel	—	وكيل أول
—	Прапорщик	—
Leutnant	младший лейтенант, лейтенант	ملازم ثاني
Oberleutnant	старший лейтенант	ملازم أوّل

Appendix A (cont)

**1. Army and *Air Force* – armée de terre et *de l'air* – ejército y *aviación* –
Heer und *Luftwaffe* – Сухопутные войска и *военно-воздушные силы* – الجيش وسلاح الجو**

captain	captain *flight lieutenant*	capitaine	capitán
major	major *squadron leader*	commandant, chef de bataillon/ d'escadron(s)	comandante
lieutenant-colonel	lieutenant-colonel *wing commander*	lieutenant-colonel	teniente coronel
colonel	colonel *group captain*	colonel	coronel
brigadier general	brigadier *air commodore*	général de brigade/ *de brigade aérienne*	general de brigada
major general	major-general *air vice-marshal*	général de division/ *de division aérienne*	general de división
lieutenant general	lieutenant-general *air marshal*	général de corps d'armée/*de corps d'armée aérienne*	teniente general
general	general *air chief marshal*	général d'armée/ *d'armée aérienne*	general
general of the army *general of the air force*	field marshal *marshal of the Royal Air Force*	—	capitán general
—	—	—	—
—	—	—	—
—	—	—	—

**2. Navy – marine nationale – marina de guerra – Kriegsmarine – Военно-морской
флот (ВМФ) – البحرية**

United States	Great Britain	France (tous *m*)	España (todos *m*)
seaman recruit	ordinary rating	matelot non breveté	recluta
seaman apprentice	able rating	matelot breveté	aprendiz
seaman	leading rating	quartier maître de 2ème classe	marinero
—	—	quartier maître de 1ère classe	cabo 2° de marinería
petty officer third class	petty officer	second maître	cabo 2° especialista

Hauptmann	капитан	نقيب
Major	майор	رائد
Oberstleutnant	подполковник	مقدم
Oberst	полковник	عقيد
Brigadegeneral	—	عميد
Generalmajor	генерал-майор	لواء
Generalleutnant	генерал-лейтенант, генерал- полковник	فريق
General	—	فريق أول/قائد عام
[Feldmarschall]	генерал армии , *генерал авиации*	مشير
—	маршал рода войск, маршал авиации	
—	главный маршал рода войск, *главный маршал авиации*	
—	Маршал Советского Союза	

Bundesrepublik Deutschland (BRD) (durchweg *m*)	*Советский Союз*	العربية
Matrose	матрос	نوتي مبتدىء
Gefreiter	старший матрос	نوتي مدرب
Obergefreiter	старшина второй статьи	نوتي ماهر
Hauptgefreiter	—	—
Maat	старшина первой статьи	—

Appendix A *(cont)*

2. Navy – marine nationale – marina de guerra – Kriegsmarine – Военно-морской флот (ВМФ) – البحرية

petty officer second class	—	second maître	cabo 1° especialista
petty officer first class	—	—	sargento
chief petty officer	chief petty officer	maître	sargento 1°
senior chief petty officer	—	premier maître	brigada
master chief petty officer	—	maître principal major	subteniente
midshipman	midshipman	aspirant	guardia marina
(warrant officer grades)	—	—	—
ensign	—	—	—
lieutenant junior grade	sub-lieutenant	enseigne de vaisseau de 2ème classe enseigne de vaisseau de 1ère classe	alférez de fragata
lieutenant	lieutenant	lieutenant de vaisseau	alférez de navío
lieutenant-commander	lieutenant-commander	capitaine de corvette	teniente de navío
commander	commander	capitaine de frégate	capitán de corbeta
—	—	—	capitán de fragata
captain	captain	capitaine de vaisseau	capitán de navío
—	commodore	—	—
rear admiral	rear-admiral	contre-amiral	contraalmirante
vice admiral	vice-admiral	vice-amiral vice-amiral d'escadre	vicealmirante
admiral	admiral	amiral	almirante
fleet admiral	admiral of the fleet	—	Capitán General de la Armada
—	—	—	—

Obermaat	главный старшина	رقيب بحري
Bootsmann	главстаршина	—
Oberbootsmann	главный корабельный старшина	—
Hauptbootsmann	—	رقيب أول بحري
—	—	—
Fähnrich zur See	—	طالب بحرية
—	мичман, старший мичман	—
—	прапорщик	—
Leutnant zur See	младший лейтенант	—
Oberleutnant zur See	лейтенант, старший лейтенант	ملازم ملازم أول
Kapitänleutnant	капитан-лейтенант, капитан третьего ранга	رائد بحري
Korvettenkapitän	капитан второго ранга	مقدم بحري
Fregattenkapitän	—	—
Kapitän zur See	капитан первого ранга	عقيد بحري
Flotillenadmiral	—	—
Konteradmiral	контр-адмирал	لواء بحري
Vizeadmiral	вице-адмирал	فريق بحري
Admiral	адмирал	فريق أول بحري
—	адмирал флота	قائد القوات البحرية
—	Адмирал Флота Советского Союза	—

Appendix B

UNITS/FORMATIONS – UNITES/FORMATIONS – UNIDADES/FORMACIONES – EINHEITEN/VERBÄNDE – ЧАСТИ/СОЕДИНЕНИЯ – الوحدات والتشكيلات

Notes Organizations and unit/formation names differ between arms of an army and between armies with a common language; not all levels shown as matching in this table precisely correspond. Air force units are shown in italics where their names differ. The British Army uses cavalry designations almost throughout except in the artillery and the infantry; the United States, French and Spanish armies use them only in the cavalry and the Bundeswehr not at all (except *Regiment*).

Remarques L'organisation de même que la désignation des unités et des formations sont différentes suivant les armes et les armées et varient aussi entre pays de même langue. Les unités de l'armée de l'air sont indiquées en italique dans les cas où leur appellation diffère. L'armée britannique utilise les appellations de la cavalerie dans la plupart des armes à l'exception de l'artillerie et de l'infanterie; l'armée américaine, française et espagnole ne les emploie que dans l'arme blindée cavalerie et la Bundeswehr jamais excepté *Regiment*.

Notas Las organizaciones así como la designación de unidades y de formaciones difieren entre los diferentes ejércitos de un país e incluso entre ejércitos de paises con el mismo idioma; no todos los niveles expuestos en este cuadro se corresponden exactamente. Las unidades aéreas se ponen en bastardilla cuando sus nombres difieren. El Ejército de Tierra Británico usa las denominaciones de Caballería casi para todo excepto en Artillería e Infantería; los Estados Unidos, Francia y España solo los emplean en Caballería y los de Alemania en nada (excepto *"Regiment"*).

Anmerkungen Die Bezeichnungen von Verbänden und Einheiten der gleichen Grösse und Aufgabe unterscheiden sich schon innerhalb der gleichen Teilstreitkraft und natürlich auch bei verschiedenen, die gleiche Sprache benutzenden Streitkräften. Daher stimmen die auf dieser vergleichenden Tafel gezeigten Begriffe nicht in jedem Fall genau überein. Luftwaffenbezeichnungen sind im Fall von Abweichungen kursiv eingetragen. Das britische Heer gebraucht die ehemaligen Kavallerie-Bezeichnungen bei den meisten Waffengattungen ausser bei der Artillerie und der Infanterie, das amerikanische, französische und spanische Heer dagegen nur bei der Kavallerie, und die Bundeswehr gebraucht sie gar nicht mehr, wenn man von der Bezeichnung *Regiment* absieht.

Названия организаций, частей и соединений различаются как между родами войск армии, так и между армиями, употребляющими общий язык; вот почему не все степени, представляющиеся в этом списке одинаковыми, в действительности точно аналогичны. Названия частей военно-воздушных сил, если различаются, написаны курсивом. Британская армия использует кавалерийские названия по всей армии, за исключением лишь артиллерии и пехоты; армии Соединенных Штатов Америки, Франции и Испании используют их только в отношении кавалерии, а вооруженные силы ФРГ не используют их вообще (за исключением *Regiment*).

المذكرات تختلف أسماء المنظمات والوحدات والتشكيلات بين أسلحة الجيش وبين الجيوش التي تستعمل نفس اللغة فليس كل المراتب التي تظهر جنباً في هذا الجدول تتساوى بالضبط وتظهر وحدات سلاح الجو حيثما تختلف اسماؤها من وحدات الجيش ويستعمل الجيش البريطاني اسماء وحدات الفرسان في كل الأسلحة باستثناء المدفعية والمشاة فإن الجيوش في الولايات المتحدة وفرنسا واسبانيا تستعملها في الفرسان فقط فإن الجيش الالماني لا يستعملها مطلقاً باستثناء كلمة : 'Regiment' .

Appendix B *(cont)*

1. Army and *Air Force* – armées de terre et *de l'air* – ejército y *aviación* –
Heer und *Luftwaffe* – Сухопутные войска и *военно-воздушные силы* – الجيش وسلاح الجو

United States	*Great Britain*	*France*	*España*
squad	section	groupe *m*	pelotón *m*
platoon *section*	troop, platoon *section, pair*	peloton *m*, section *f*	sección *f*
company, battery, troop *flight*	squadron, battery, company *flight*	batterie *f*, escadron *m*, compagnie *f*, *escadrille f*, *escadron m*	compañía *f*, escuadrón *m*, *escuadrilla f*
battalion, squadron *squadron*	regiment, battalion *squadron*	régiment *m*, bataillon *m*, *escadron m*	regimiento *m* *batallón m* *escuadrón m*
brigade, regiment *group*	brigade [*wing*, now specialist subunit]	brigade *f* *escadre f*	brigada *f* *agrupación f*
division *wing*	division *group*	division *f*	división *f* *ala f*
corps	corps	corps d'armée *m*	cuerpo de ejército *m* *división aérea f*
army	army	armée *f*	ejército *m*
army group *air force*	army group *air force*	groupe d'armées *m* forces aériennes (*tactiques*) *fp*	grupo de ejércitos *m* *fuerza aérea f*
—	—	—	—
—	—	—	—

Bundesrepublik Deutschland (BRD)	Советский Союз	العربية
Gruppe *f*	отделение (с), отряд (м)	حظيرة
Zug *m* Rotte *f* (2 Flugzeuge)	взвод (м)	فصيلة فئة
Kompanie, Batterie *f* Schwarm *m*	рота (ж), батарея (ж)	سرب (جوي) سرية
Bataillon *n* (Abteilung *f*) Staffel *f*	батальон (м), дивизион (м) (арт.)	كتيبة
Brigade *f*, Regiment *n* Geschwader *n*	полк (м) бригада (ж) (арт.)	لواء
Division *f*	дивизия (ж)	فرقة
Korps *n*	[корпус (м)]	فيلق
[Armee *f*]	армия (ж)	جيش
[Armee-Gruppe *f*] [Luftflotte *f*]	[группа армии (ж)]	مجموعة جيوش
—	Группа Войск (ж)	—
—	Фронт (м)	—

Appendix B *(cont)*

2. Types of warship – classes de navire de guerre – tipos de buque de guerra –
Kriegsschiffklassen – Виды военных кораблей – السفن الحربية

Great Britain	*France*	*España*
fast patrol boat	patrouilleur *m*	patrullero *m*
fast attack boat	patrouilleur *m*	torpedero *m*
minelayer	poseur de mines *m*	buque minador *m*
minehunter	chasseur de mines *m*	—
minesweeper	dragueur de mines *m*	dragaminas *m*
fishery protection vessel	navire de surveillance de pêche *m*	guardacostas *m inv*
corvette	corvette *f*	corbeta *f*
destroyer	destroyer *m*	destructor *m*
frigate	frégate *f*	fragata *f*
cruiser	croiseur *m*	crucero *m*
light aircraft carrier/ helicopter carrier	porte-hélicoptères *m inv*	porta-helicopteros *m inv* portaviones ligero *m*
aircraft carrier	porte-avions *m inv*	portaviones *m inv*
(nuclear) hunter-killer submarine	sous-marin nucléaire d'attaque *m*	submarino (nuclear) de ataque *m*
(nuclear) missile-launching submarine	sous-marin nucléaire lance engins (SNLE) *m*	submarino (nuclear) lanzamisiles *m*
landing craft	engin de débarquement *m*	barcaza de desembarco *f*
landing ship	bâtiment de débarquement *m*	buque de desembarco *m*
headquarters ship	bâtiment amiral *m*	buque de mando *m*
support ship	ravitailleur *m*	buque de apoyo *m*
auxiliary	navire auxiliaire *m*	buque auxiliar *m*
tanker	pétrolier *m*	petrolero *m*

Bundesrepublik Deutschland (BRD)	Советский Союз	العربية
Schnellboot *n*	ракетный катер (м)	زورق دورية سريع
Schnellboot *n*	ракетный катер (м)	طراد
Minenleger *m*	минный заградитель (м) минзаг (м)	زارعة الغام
Minensuchboot *n*	тральщик- искатель мин (м)	سفينة صيد الغام
Minenräumboot *n*	морской тральщик (м) базовый тральщик (м) рейдовый тральщик (м)	كاسحة الغام
Fischereischutzboot *n*	рыболовный сторожевой корабль (м)	حارسة صيد
Korvette *f*	сторожевой корабль (м)	سفينة حارسة صغيرة
Zerstörer *m*	эсминец (м)	مدمرة
Fregatte *f*	противолодочный фрегат (м) многоцелевой фрегат (м)	فرغاط
Kreuzer *m*	крейсер	طرادة
Geleitflugzeugträger *m*	легкий авианосец (м) десантный, противолодочный вертолетоносец (м)	حاملة طائرات عمودية
Flugzeugträger *m*	авианосец (м)	حاملة طائرات
(Atom-) U-Boot-Jäger *m*	атомная подлодка (ж)	غواصة قانصة نووية
(Atom-) Raketen-U-Boot *n*	атомная подлодка (ж) с баллистическими ракетами	غواصة قاذفة صواريخ نووية
Landungsboot *n*	десантный катер (м)	زورق إنزال
Landungsschiff *n*	десантный корабль (м)	سفينة إنزال
[Kommandoschiff *n*]	плавучая база, плавбаза (ж)	سفينة قيادة
Versorgungsschiff *n* Tender *m*	плавучая мастерская (ж)	سفينة سند
Hilfsschiff *n*	военный транспорт (м)	سفينة مساندة
Tankschiff *n* Tanker *m*	военный танкер (м)	ناقلة نفط

Appendix C

NUMERALS/NOTATION – NOMBRES/SYMBOLES – NUMEROS/SIGNOS – ZAHLEN/ ZEICHEN – ЦИФРЫ/СИСТЕМА СЧИСЛЕНИЯ – الإعداد

Notes The significance of the full stop (decimal point) and the comma is the opposite in English and in most other languages. In English a full stop means a decimal; in other languages a comma represents the decimal. In English the comma when used represents thousands, millions etc. (i.e. a means of dividing into groups of 3 digits); in other languages these groups are still sometimes divided by a full stop.

It is best to use just a full stop or comma, according to language, for decimals and to group figures in accordance with the standards associated with the International SI System, i.e. with a space between each group of 3 figures working outwards from the decimal point or comma. If there is one figure over, it is grouped with the outermost 3-group; if 2, they form a separate outermost group.

In spoken German tens and units are reversed, e.g. *dreiundzwanzig*. When working between German and another language, it is therefore best either to write figures down or to speak each digit separately in their written order (as one does on the radio).

Remarques Le point (anglais: *decimal point*) et la virgule ont une signification opposée en anglais et dans la plupart des autres langues. En anglais un point indique une décimale; dans les autres langues c'est la virgule qui représente la décimale. En anglais la virgule lorsqu'elle est employée indique les milliers, les millions etc. (c'est un moyen de diviser un nombre en groupes de trois chiffres). Dans les autres langues les groupes sont séparés souvent par un point.

Il est donc judicieux d'utiliser un point ou une virgule, suivant la langue, pour indiquer les décimales et de se servir du système SI (Système International d'Unités) pour diviser les nombres en groupes de chiffres. Il suffit de regrouper les chiffres par trois à partir de l'inclination de la décimale en laissant un espace entre chacun de ces groupes. S'il reste un seul chiffre il se rattache au groupe extérieur. S'il y en a deux ils forment un groupe séparé.

En allemand parlé les dizaines et les unités sont interverties, p. ex. *dreiundzwanzig*. Pour passer de l'allemand à une autre langue il est bon soit d'écrire le nombre en chiffres soit d'énoncer chaque chiffre séparement dans l'ordre écrit, comme on le fait à la radio.

Notas El significado del punto y de la coma es opuesto en inglés a la mayor parte de otros idiomas. En inglés un punto significa un decimal; en otros idiomas una coma representa un decimal. En inglés la coma cuando se emplea representa miles, millones etc. (entre grupos de tres cifras); en otros idiomas estos grupos están también divididos por un punto.

Es mejor usar un punto o coma de acuerdo con las normas asociadas al sistema SI Internacional, es decir con un espacio entre cada grupo de 3 cifras a derecha e izquierda a partir del punto decimal o coma. Si resta sobra una cifra se agrupa con el grupo de 3 cifros mas cercano; si 2, forman un grupo separado.

En alemán hablado las decenas y unidades estan invertidas, por ejemplo *dreiundzwanzig*. Cuando se trabaja con el alemán y otro idioma es mejor por lo tanto escribir las cifras o decir cada cifra separadamente en su orden de escritura (como se hace en la radio).

Anmerkungen Der Punkt (engl. *decimal point*) und das Komma werden im Englischen völlig anders als in den meisten Sprachen benutzt. Im Englischen deutet der Punkt einen Dezimalbruch an, was bei den anderen Sprachen durch ein Komma getan wird. Im Englischen zeigt ein Komma dagegen die Tausender, Millionen usw. an: d.h. Dreier-Ziffern-Gruppen einer ganzen Zahl, die bei den anderen Sprachen manchmal durch einen Punkt getrennt werden, werden im Englischen durch das Komma getrennt.

Daher ist ratsam, nur einen Punkt bzw. ein Komma (je nach Sprache) für einen Dezimal-bruch zu verwenden und sonst die in dem "SI" Internationalen Einheitensystem standardis-ierten Gruppierungen zu gebrauchen. Dabei geht man vom Komma bzw. Punkt in beide Richtungen und trennt jede Dreier-Ziffern-Gruppe durch einen Zwischenraum. Ist eine Ziffer übrig, wird sie mit der äussersten Dreier-Gruppe verbunden; sind 2 Ziffern übrig, bilden sie eine eigene Gruppe.

Im Deutschen werden die Zehner und Einer nicht in der Reihenfolge gesprochen bzw. in Buchstaben geschrieben, in der die Zahlen hintereinander stehen. Beispiel: dreiundzwan-zig. Es empfiehlt sich deswegen, bei Übersetzungen die Zahlen als Zahl und nicht in Buchstaben zu schreiben oder sie einzeln nacheinander – wie in einem Funkspruch – zu sprechen. Beides ist gebraüchlich und wird verstanden.

Примечания Употребление точки (десятичной точки) и запятой отличается в английском языке от их употребления в других языках. В английском языке десятичная дробь указывается точкой, а в других языках – запятой. В английском языке запятая является средством разделения чисел на трехцифровые группы, т.е. на тысячи, миллионы и т.д.; в других языках числа иногда разделяются точкой.

Чтобы выразить десятичные дроби, лучше всего употреблять или точки или запятые, в соответствии с правилами языка; кроме того, писать числа по стандартам международной системы счисления (SI), т.е., с промежутком после каждой трехцифровой группы от точки (запятой). Одиночная остальная цифра присоединяется к дальнейшей трехцифровой группе; две остальные цифры образуют отдельную дальнейшую группу.

При разговоре на немецком языке десятки и единицы ставятся в обратном порядке, как, например, 23 – *dreiundzwanzig*. Поэтому при устном переводе с немецкого на другой язык лучше или написать цифры, или произнести каждую цифру отдельно по порядку написания, как принято в радиовещании.

المذكرات ينعكس معنى النقطة والفاصلة باللغة الانكليزية في لغات أخرى . ان النقطة بالانكليزية تعني الكسر العشري ولكن الفاصلة تمثل الكسر العشري في لغات أخرى . في حالة استعمال الفاصلة بالانكليزية هي تمثل الآلاف والملايين (وسيلة تفصيل المجموعات المتألفة من ثلاثة أرقام) ففي لغات أخرى تم تفصيل هذه المجموعات بعض الأوقات بواسطة النقطة .

من الأفضل استعمال النقطة أو الفاصلة على حسب اللغة لتفصيل الكسر العشري مع جمع الأرقام وفقاً لنظام (SI) الدولي ـ وضع فراغ بين كل مجموعة من ثلاثة أرقام من النقطة أو الفاصلة نحو الخارج في حالة وجود رقم زيادة يضع في المجموعة القصوى فإذا يبرز رقمان يشكلان مجموعة قصوى منفردة .

في اللغة الالمانية تتعاكس العشرات والأرقان تحت العشرة (dreiundzwanzig) لذلك عندما يعمل واحد بين الالمانية ولغة أخرى من الأفضل أن يكتب الأرقام أو يتكلم الأرقام فرداً فرداً كالطريقة اللاسلكية .

Appendix C *(cont)*

1. Cardinal numbers – nombres cardinaux – números cardinales – Kardinalzahlen – Количественные числительные (с.мн.) – الأعْدَاد الأصْلِيَة

0	nought, zero	zéro	cero
1	one	un(e)	uno(a)
2	two	deux	dos
3	three	trois	tres
4	four	quatre	cuatro
5	five	cinq	cinco
6	six	six	seis
7	seven	sept	siete
8	eight	huit	ocho
9	nine	neuf	nueve
10	ten	dix	diez
11	eleven	onze	once
12	twelve	douze	doce
13	thirteen	treize	trece
14	fourteen	quatorze	catorce
15	fifteen	quinze	quince
16	sixteen	seize	dieciséis
17	seventeen	dix-sept	diecisiete
18	eighteen	dix-huit	dieciocho
19	nineteen	dix-neuf	diecinueve
20	twenty	vingt	veinte
21	twenty-one	vingt et un	veintiuno
22	twenty-two (etc.)	vingt-deux (etc.)	veintidós (etc.)
30	thirty	trente	treinta
31		trent et un	treinta y uno
32		trente-deux (etc.)	treinta y dos (etc.)
40	forty	quarante	cuarenta
50	fifty	cinquante	cincuenta
60	sixty	soixante	sesenta
70	seventy	soixante-dix	setenta
71		soixante et onze	
72		soixante-douze (etc.)	
80	eighty	quatre-vingts	ochenta
81		quatre-vingt-un	
90	ninety	quatre-vingt-dix	noventa
91		quatre-vingt-onze (etc.)	
100	a/one hundred	cent	ciento
101	one hundred and one (etc.)	cent un (etc.)	ciento uno (etc.)
110	one hundred and ten	cent dix	ciento diez
140	one hundred and forty	cent quarante	ciento cuarenta

null	ноль (м), нуль (м)	صفر	0
eins (einer, eines, eine)	один (м), одна (ж), одно (с), одни (мн)	أحد	1
zwei (zwo)	два (м,с), две (ж)	اثنان	2
drei	три	ثلاثة	3
vier	четыре	أربعة	4
fünf	пять	خمسة	5
sechs	шесть	ستة	6
sieben	семь	سبعة	7
acht	восемь	ثمانية	8
neun	девять	تسعة	9
zehn	десять	عشرة	10
elf	одиннадцать	أحد عشر	11
zwölf	двенадцать	اثنا عشر	12
dreizehn	тринадцать	ثلاثة عشر	13
vierzehn	четырнадцать	أربعة عشر	14
fünfzehn	пятнадцать	خمسة عشر	15
sechzehn	шестнадцать	ستة عشر	16
siebzehn	семнадцать	سبعة عشر	17
achtzehn	восемнадцать	ثمانية عشر	18
neunzehn	девятнадцать	تسعة عشر	19
zwanzig	двадцать	عشرون	20
einundzwanzig	двадцать один	واحد وعشرون	21
zweiundzwanzig (usw.)		اثنان وعشرون والخ	22
dreissig	тридцать	ثلاثون	30
einunddreissig (usw.)	тридцать один	واحد وثلاثون	31
		اثنان وثلاثون والخ	32
vierzig	сорок	أربعون	40
fünfzig	пятьдесят	خمسون	50
sechzig	шестьдесят	ستون	60
sieb(en)zig	семьдесят	سبعون	70
		واحد وسبعون	71
		اثنان وسبعون والخ	72
achtzig	восемьдесят	ثمانون	80
		واحد وثمانون	81
neunzig	девяносто	تسعون	90
		واحد وتسعون	91
hundert (einhundert)	сто	مئة	100
hunderteins (usw.)	сто один	مئة وواحد	101
hundertzehn	сто десять	مئة وعشرة	110
hundertvierzig	сто сорок	مئة وأربعون	140

Appendix C *(cont)*

1. Cardinal numbers – nombres cardinaux – números cardinales – Kardinalzahlen –
Количественные числителвные (с.мн.) – الأَعْدَاد الأَصْلِيَة

141	one hundred and forty-one (etc.)	cent quarante et un	ciento cuarenta y uno
142		cent quarante-deux (etc.)	
200	two hundred (etc.)	deux cent(s) (etc.)	doscientos(-as)
300			trescientos
400			cuatrocientos
500			quinientos
600			seiscientos
700			setecientos
800			ochocientos
900			novecientos
1000	a/one thousand	mille	mil *inv*
1001	one thousand and one (etc.)	mille un (etc.)	mil uno (etc.)
2000	two thousand (etc.)	deux mille (etc.)	dos mil (etc.)
10 000	ten thousand	dix mille	diez mil
100 000	a/one hundred thousand	cent mille	cien mil
1000 000	a/one million	un million (de) *m*	un millón (de) *m*
1000 000 000	a/one billion (US)	un milliard (de) *m*	mil millones *mpl*
1000 000 000 000	(a/one billion (Br.) (a/one trillion (US)	un billion (de) *m*	un billón (de) *m*

2. Fractions – fractions – fracciones – Brüche – Дроби – الكُسُور

$\frac{1}{2}$	half	moitié *f*, demi-	medio(-a)
$1\frac{1}{2}$	one and a half	un(e) ... et demi(e)	uno(-a)... y medio (-a)
$2\frac{1}{2}$	two and a half	deux ...s et demi(e)	dos ...s y medio(-a)
$\frac{1}{3}$	a/one third	un tiers *m*	un tercio *m*
$\frac{1}{4}$	a/one quarter	un quart *m*	un cuarto *m*
$\frac{1}{5}$	a/one fifth	un cinquième (etc.)	un quinto *m* (etc.)
$\frac{1}{100}$	a/one hundredth	un centième *m*	un centésimo *m*
$\frac{1}{1000}$	a/one thousandth	un millième *m*	un milésimo *m*

hunderteinundvierzig (usw.)		مئة وواحد وأربعون	141
		مئة واثنان وأربعون والخ	142
zweihundert (usw.)	двести	مئتان	200
	триста	ثلاث مئة	300
	четыреста	أربع مئة	400
	пятьсот	خمس مئة	500
	шестьсот	ست مئة	600
	семьсот	سبع مئة	700
	восемьсот	ثمان مئة	800
	девятьсот	تسع مئة	900
tausend (eintausend)	тысяча (ж)	ألف	1000
tausendeins (usw.)	тысяча один	واحد وألف	1001
zweitausend (usw.)	две тысячи	ألفان	2000
zehn tausend	десять тысяч	عشرة آلاف	10 000
hunderttausend	сто тысяч	مئة ألف	100 000
eine Million f	один миллион (м)	مليون	1000 000
eine Milliarde f	один миллиард (м)	ألف مليون	1000 000 000
eine Billion f	один триллион (м)	بليون	1000 000 000 000
ein Halb n, eine Hälfte f, halb-	(одна) половина (ж)	نصف	$^1/_2$
eineinhalb, anderthalb	полтора (с)	واحد ونصف	$1^1/_2$
zweieinhalb	два с половиной	اثنان ونصف	$2^1/_2$
ein Drittel n (usw.)	одна третья (ж)	ثلث	$^1/_3$
	одна четвертая (ж)	ربع	$^1/_4$
	одна пятая (ж)	خمس	$^1/_5$
ein Hundertstel n	одна сотая (ж)	المئة	$^1/_{100}$
ein Tausendstel n	одна тысячная (ж)	الألف	$^1/_{1000}$

Appendix C *(cont)*

3. Ordinal numbers – nombres ordinaux – números ordinales – Ordnungszahlen – Порядковые числителвные – الأَعْدَاد التَّرْتِيبَّة

1	first 1st	premier (-ère) 1er, 1ère, 1°	primer (o) 1°, 1ª (etc.)
2	second 2nd	deuxième 2ème, 2° (etc.)	segundo
3	third 3rd	troisième	tercer(o)
4	fourth 4th (etc.)	quatrième (etc.)	cuarto
5	fifth		quinto
6	sixth (etc.)		sexto
7			séptimo
8			octavo
9	ninth		noveno, nono
10	tenth (etc.)		décimo
11			undécimo
12			duodécimo
13			decimotercero (etc.)
20	twentieth		vigésimo
21	twenty-first (etc.)	vingt et unième	vigésimo prim(ér)o (etc.)
22		vingt-deuxième (etc.)	

4. Mathematical notation – symboles mathématiques – signos matemáticos – mathematische Zeichen – Математическая система счисления – الرُّمُوز الرِّيَاضِيَّة

+	plus	plus	más
−	minus	moins	menos
×	times, by	fois	por
÷, :	divided by, over	divisé par, sur	dividido por, (sobre)
=	is/are, make(s), equals	est/sont, égale	es/son, es igual a, igual a
$\sqrt{}$	square root, surd	racine carrée *f*	raíz quadrada *f*
2^2	two squared	deux au carré	dos al cuadrado
2^3	two cubed	deux au cube	dos al cubo
2^4	two to the fourth	deux à la quatre (puissance)	dos a la cuarta
∞	infinite (-y) *adj (n)*	infini *adj n m*	infinito *adj s m*

Deutsch	Русский	العربية	
der Erster, erstens, 1.	первый, первая, первое, первые	أول	1
der Zweiter, zweitens, 2.	второй, вторая, второе, вторые	ثانٍ	2
der Dritter, drittens, 3.	третий, третья, третье, третии	ثالث	3
der Vierter, viertens, 4. (usw.)	четвертый (и т.д.)	رابع	4
	пятый	خامس	5
	шестой	سادس	6
	седьмой	سابع	7
	восьмой	ثامن	8
	девятый	تاسع	9
	десятый	عاشر	10
	одиннадцатый	حادي عشر	11
	двенадцатый	ثاني عشر	12
	тринадцатый	ثالث عشر	13
	двадцатый	العشرون	20
		الحادي والعشرون	21
		الثاني والعشرون	22

Deutsch	Русский	العربية	
plus, und	плюс	زائد (و)	+
minus, weniger	минус	ناقص	−
mal, multipliziert mit	умноженное на	يزداد (في)	×
(dividiert) durch	делённое на	يقسم	÷, :
ist/sind, macht, gleich	равняется	يساوي	=
Quadratwurzel f, die zweite Wurzel aus...	квадратный корень (м)	الجذر التربيعي	$\sqrt{}$
zwei (ins) Quadrat, zwei hoch zwei	два в квадрате	اثنان مربع	2^2
zwei Kubik, zwei hoch drei	два в кубе	اثنان مكعب	2^3
zwei hoch vier	два в четвертой степени	اثنان إلى الرابع	2^4
unendlich (Keit) adj (sf)	бесконечность (ж) бесконечный	لا نهائي	∞

Appendix D

COLORS – COULEURS – COLORES – FARBEN – ЦВЕТА (м.мн) – الأرقام

azure	d'azur	azul celeste
black	noir	negro
blue	bleu	azul
brown	brun	castaño
crimson	cramoisi	carmesí
gray	gris	gris
green	vert	verde
indigo	indigo *inv*	añil
mauve	mauve	malva
orange	orange	naranjado
pink	rosé	rosado
purple	pourpre	morado
red	rouge	rojo
violet	violet	violado
white	blanc	blanco
yellow	jaune	amarillo
-ish	-âtre	(diversas terminaciones)
dark-	foncé	oscuro
light-	clair	claro
medium	moyen	mediano
infrared	infra-rouge	infrarrojo
ultraviolet	ultra-violet	ultravioleta *inv*

himmelblau	лазурный, голубой	الازوردي
schwarz	черный	اسود
blau	синий	أزرق
braun	коричневый	أسمر
blutrot	малиновый	قرمزي اللون
grau	серый	رمادي
grün	зеленый	أخضر
indigoblau	индиго	اللون النيلي
malvenfarbig	розовато-лиловый	حباري
orange	оранжевый	برتقالي
rosa	розовый	قرنفلي
purpurn	пурпурный	ارجواني
rot	красный	أحمر
violett	фиолетовый	بنفسجي
weiss	белый	أبيض
gelb	желтый	أسفر
(Wurzel ggf. mit Umlaut) + -lich	-оватый (напр. зеленоватый) из-, ис- (напр. иззелена-)	مثل . . .
dunkel-	темно-	قائم
hell-	светло-	فاتح
mittel-	средний	متوسط
Infrarot-	инфракрасный	تحت الحمراء
Ultraviolett-	ультрафиолетовый	فو بنفسجي

Appendix E

POINTS OF THE COMPASS	**ROSE DES VENTS**	**PUNTOS CARDINALES**
Grid North	nord de la carte	norte geográfico
Magnetic North	nord magnétique	norte magnético
True North	nord vrai	norte verdadero

cardinal point	point cardinal *m*	punto cardinal *m*
magnetic variation	déclinaison magnétique *f*	variación magnética *f*
point (11°15′)	quart *m*	cuarta *f*
degree	degré *m*	grado *m*
radian	radian *m*	radián *m*
mil	millième *m*	milésima *f*

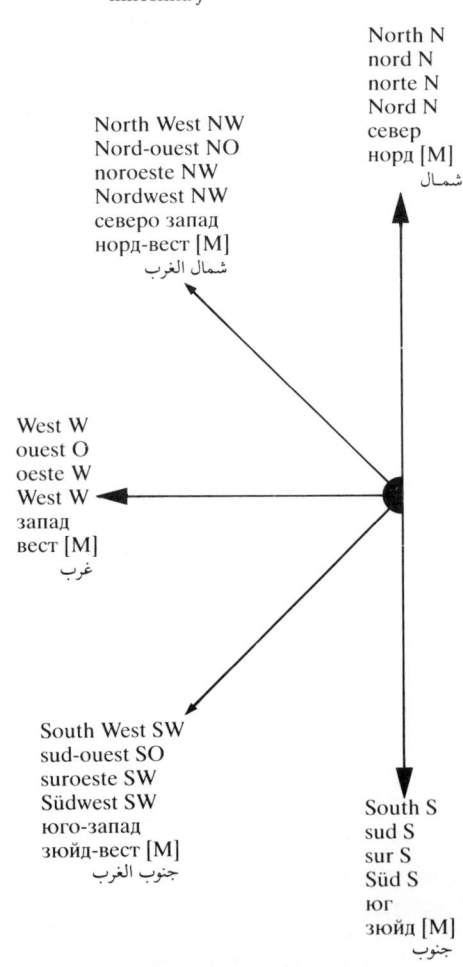

North N
nord N
norte N
Nord N
север
норд [M]
شمال

North West NW
Nord-ouest NO
noroeste NW
Nordwest NW
северо запад
норд-вест [M]
شمال الغرب

West W
ouest O
oeste W
West W
запад
вест [M]
غرب

South West SW
sud-ouest SO
suroeste SW
Südwest SW
юго-запад
зюйд-вест [M]
جنوب الغرب

South S
sud S
sur S
Süd S
юг
зюйд [M]
جنوب

KOMPASSTRICHE	ГЛАВНЫЕ РУМБЫ (м.мн.)	الجهات الاربع
Gitternord	север сетки (м)	شمال التربيع
missweisender Nord M	магнитный север (м)	الشمال المغنطيسي
magnetischer Nord L		الشمال الحقيقي
rechtweisender Nord M	истинный север (м)	
geographischer Nord L		
Himmelsrichtung *f*	главный румб (м), страна света (ж)	جهة من الجهات الأربع
Missweisung *f*	магнитное склонение (с)	انحراف مغنطيسي
Strich *m*	точка (ж)	نقطة
Grad *m*	градус (м)	درجة
Bogen *m*	радиан (м)	زاوية نصف قطري
Strich *m*	тысячная (ж)	مـل

North N
nord N
norte N
Nord N
север
норд [M]
شمال

North East NE
nord-est NE
nordeste NE
Nordost NO
северо-восток
норд-ост [M]
شمال الشرق

East E
est E
este E
Ost O
восток
ост [M]
شـرق

South East SE
sud-est SE
sudeste SE
Südost SO
юго-восток
зюйд-ост [M]
جنوب الشرق

South S
sud S
sur S
Süd S
юг
зюйд [M]
جنوب

Appendix F

TOOLS	OUTILS	HERRAMIENTAS
Earth	**Terre**	**Tierra**
auger	tarière *f*	tientaaguja *f*
pick	pioche *f*	zapapico *m*
shovel	pelle *f*	pala *f*
trenching tool	pelle-pioche individuelle *f*	herramienta de zapa *f*
Wood/metal	**Bois/métal**	**Madera/metal**
hammer	marteau *m*	martillo *m*
mallet	maillet *m*	mallete *m*
pliers *pl*	pince(s) *f* (*pl*), tenailles *fpl*	pinzas *fpl*
screwdriver	tournevis *m*	tornillador *m*
vise	étau *m*	tornillo de banco *m*
Wood	**Bois**	**Madera**
awl	poinçon *m*	lezna *f*
brace/bit	vilebrequin *m*/à main	berbiquí *m*/barrena *f*, mecha *f*
chisel	ciseau *m*	cincel *m*
gimlet	vrille *f*	barrenita *f*
rasp	râpe *f*	escofina *f*
saw	scie *f*	sierra *f*
square	équerre *f*	escuadra *f*
Metal	**Métal**	**Metal**
calipers *pl*	compas *m*	calibrador *m*
center punch	emporte-pièce *m*	granete *m*
cold chisel	burin *m*, ciseau à froid *m*	buril *m*
cutting/welding torch	chalumeau *m*	soplete *m*
drift	broche *f*	punzón *m*
drill	perceuse *f*	taladradora *f*/broca *f*
feeler gauge	calibre d'épaisseur *m*	calibre de laminas *m*
file	lime *f*	lima *f*
hacksaw	scie à métaux *f*	sierra de arco *f*
micrometer	palmer *m*	micrómetro *m*
soldering iron	fer à souder *m*	hierro de soldar *m*
Stilson wrench	serre-tube *mi*	llave Stilson *f*
tommy bar	broche *f*	broca *f*
wrench:	clef *f*:	llave *f*:
adjustable	anglaise/réglable	inglesa
box	à douille	tubular
open-ended	à fourche(s)	de tuercas
ring	fermée	dentada
socket	à douille	tubular

WERKZEUGE	ИНСТРУМЕНТЫ	الأدوات
Erde	**Земля**	أرض
Erdbohrer *m*	сверло (с)	مثقب
Pickel *m*	кирка (ж)	معول
Schaufel *f*	лопата (ж)	رفش
Klappspaten *m*	шанцевый инструмент (м)	رفش خفيف
Holz/Metall	**Дерево/металл**	خشب/معدن
Hammer *m*	молот (м), молоток (м)	مطرقة
Schlegel *m*	молоток (м), киянка (ж)	مطرقة خشبية
Zange *f*	клещи (мн), кусачки (мн)	ملقط
Schraubenzieher *m*	отвертка (ж)	مفك اللوالب
Schraubstock *m*	тиски (мн)	ملزمة
Holz	**Дерево**	خشب
Ahle *f*, Pfriem *m*	шило (с)	مخرز
Bohrleier *f*	коловорот (м), дрель (ж)	ملفاف/لقمة
Meissel *m*	долото (с), зубило (с)	أزميل
Holzbohrer *m*	буравчик (м), бурав (м)	برية
Raspel *f*	рашпиль (м)	مبرد خشب
Säge *f*	пила (ж)	منشار
Anschlagwinkel *m*	плотницкий угольник (м)	كوس رسم
Metall	**Металл**	معدن
Greifzirkel *m*	кронциркуль (м), нутромер (м)	فرجار
Körner *m*	кернер (м)	دنابة وضع علامة الركز
Kaltschrotmeissel *m*, Hartmeissel *m*	слесарное зубило (с)	أزميل قطع على البارد
Schneid-/ Schweissbrenner *m*	газовый резак (м)/ сварочная горелка (ж)	مشعل قطع/لحام
Dorn *m*	штрек (м), выколотка (ж)	أداة توسيع الثقوب
Bohrer *m*	сверло (с), дрель (ж)	مثقب
Lehre *f*	толщиномер (м)	مقياس الضبط
Feile *f*	напильник (м)	مبرد
Bügelsäge *f*	ножовка (ж)	منشار معادن
Mikrometerschraube *f*	микрометр (м)	ميكرومتر
Lötkolben *m*	паяльник (м)	كاوية لحام
Rohrzange *f*	разводной ключ (м)	مفتاح ستلسون
Drehstift *m*, Knebel *m*	установочный штифт (м)	قضيب ادارة يدوي
Schlüssel *m*:	ключ (м):	مفتاح :
Engländer *m*	разводной ключ (м)	ربط انضباطي
Steckschlüssel *m*	торцевой ключ (м)	صندوقي
Gabelschlüssel *m*	трубный ключ (м)	ربط مفتوح الفك
Ringschlüssel *m*	торцевой ключ (м)	على شكل حلقة
Steckschlüssel *m*	торцевой ключ (м)	ربط صندوقي

Appendix F *(cont)*

Electrical	Electrique	Eléctrico
ammeter	ampèremètre *m*	amperímetro *m*
circuit tester	vérificateur de circuit *m*	comprobador de circuitos *m*
ohmmeter	ohmmètre *m*	ohmiómetro *m*
voltmeter	voltmètre *m*	voltímetro *m*

Vehicles	**Véhicules**	**Vehículos**
jack	cric *m*, vérin *m*	gato *m*
slave lead	"fils de raccordement de batterie" *mpl (ou terme anglais)*	cables de forma de corriente *mpl*
towrope	remorque *f*	remolque *m*
wheelbrace	vilebrequin *m*	llave de ruedas *f*

Parts etc.	**Pièces etc.**	**Piezas etc.**
adhesive	adhésif *m*	adhesivo *m*
bolt	boulon *m*	perno *m*
bulb	lampe *f*	bombilla *f*
canvas	grosse toile *f*	lona *f*
clamp	bride de serrage *f*	abrazadera *f*
clip	attache *f*	grapa *f*
connector	raccord *m*	conectador *m*
fan belt	courroie de ventilateur *f*	correa de ventilador *f*
gasket	joint *m*	junta *f*
inner tube	chambre *f* à air	cámara *f*
insulating tape	chatterton *m*	cinta aislante *f*
lead (elec.)	câble *m*	cable *m*
nail	clou *m*	clavo *m*
nut	écrou *m*	tuerca *f*
repair kit	trousse de réparation *f*	herramientas de reparación *fp*
rivet	rivet *m*	roblón *m*
rubber patch	rustine *f*	parche *m*
rubber solution	dissolution *f*	disolución de goma *f*
screw	vis *f*	tornillo *m*
sealing compound	mastic *m*	mezcla obturadora *f*
shackle	maillon d'attache *m*	grillete de unión *m*
sparking plug	bougie *f*	bujía *f*
split pin	clavette fendue *f*	pasador *m*
stud	goujon *m*	remache *m*
terminal	borne *f*	borne *m*
tire	pneu *m*	neumático *m*
washer	rondelle *f*	arandela *f*
welding rod	baguette *f*	varilla *f* electrodo *m*
wire	fil *m*	alambre *m*

Elektrisch	Электрические инструменты	الكهرباء
Amperemeter n, Strommesser m	амперметр (м)	أمير متر
Leitungsprüfer m	испытательный прибор цепи (м)	جهاز اختبار الدوائر
Ohmmeter n	омметр (м)	أو متر
Voltmeter n	вольтметр (м)	فلطامتر

Fahrzeuge	Машины	سيارات
Wagenheber m	домкрат (м)	رافعة
Fremdstartkabel n	вспомогательный проводник (м)	أسلاك وصل
Abschleppseil n	буксирный канат (м)	حبل للقطر
Radmutterschlüssel m	коловорот колес (м)	مفتاح صمائل العجلة

Teile usw.	Детали и т.д.	القطع والخ
Klebstoff m	клей (м), клейкое вещество (с)	لصوق
Bolzen m	болт (м), стержень (м)	محزقة/ترباس
Birne f	колба (ж), лампочка (ж)	بصلة
Segeltuch n	брезент (м)	نسيج
Klammer f, Zwinge f	зажим (м)	قامطة
Klammer f, Rohrschelle f	зажим (м)	رباط/كلاب
Verbindungsstück n	соединитель (м)	وصلة
Ventilatorriemen m	приводной ремень (м)	نطاق المروحة
Dichtung f	прокладка (ж)	حشية
Reifenschlauch m	камера шины (ж)	إطار داخلي
Isolierband n	изоляционная лента (ж)	شريط عازل
Leitung f	проводник (м)	سلك
Nagel m	гвоздь (м)	مسمار
Mutter(-n) f	гайка (ж)	صمولة
Flickzeug n, Reparaturkasten m	набор инструментов (м)	عدة اصلاح
Niet n	заклепка (ж)	دسار
Gummiflicken m	резиновая заплата (ж)	قرصة اصلاح من المطاط
Gummilösung f	резиновый раствор (м)	محلول من المطاط
Schraube f	винт (м)	لولب
Dichtungsmasse f	заливочная масса (ж)	مركب ساد
Schäkel m	обойма (ж), скоба (ж)	حلقة
Zündkerze f	запальная свеча (ж)	شمعة اشعال
Splint m	шплинт (м)	شكة مشقوقة
Stiftschraube f	штифт (м), шпилька (ж)	مسمار/لسان
Endstück n, Klemme f	зажим (м), клемма (ж) полюс (м)	مربط
Reifen m	шина (ж), резина (ж) (разг.)	إطار
(Unterleg-) Scheibe f	шайба (ж), прокладка (ж)	فلكة
Schweisstab m	присадочный пруток (м)	قضيب لحام
Draht n	проволока (ж), провод (м) (элек.)	سلك